T0335469

THE MOLECULAR IMMUNOLOGY OF COMPLEX CARBOHYDRATES

ADVANCES IN EXPERIMENTAL MEDICINE AND BIOLOGY

Recent Volumes in this Series

Volume 222
OXYGEN TRANSPORT TO TISSUE X
Edited by Masaji Mochizuki, Carl R. Honig, Tomiyasu Koyama, Thomas K. Goldstick, and Duane F. Bruley

Volume 223
UREMIC TOXINS
Edited by Severin Ringoir, Raymond Vanholder, and Shaul G. Massry

Volume 224
UROGENITAL INFECTIONS
Edited by Amedeo Bondi, Donald D. Stieritz, Joseph M. Campos, and Linda Ann Miller

Volume 225
IMMUNOBIOLOGY OF PROTEINS AND PEPTIDES IV: T-CELL RECOGNITION AND ANTIGEN PRESENTATION
Edited by M. Zouhair Atassi

Volume 226
MOLECULAR MECHANISM OF MUSCLE CONTRACTION
Edited by Haruo Sugi and Gerald H. Pollack

Volume 227
OXYGEN TRANSFER FROM ATMOSPHERE TO TISSUES
Edited by Norberto C. Gonzalez and M. Roger Fedde

Volume 228
THE MOLECULAR IMMUNOLOGY OF COMPLEX CARBOHYDRATES
Edited by Albert M. Wu and L. Garry Adams

Volume 229
LIPOXINS: Biosynthesis, Chemistry, and Biological Activities
Edited by Patrick Y.-K. Wong and Charles N. Serhan

A Continuation Order Plan is available for this series. A continuation order will bring delivery of each new volume immediately upon publication. Volumes are billed only upon actual shipment. For further information please contact

THE MOLECULAR IMMUNOLOGY OF COMPLEX CARBOHYDRATES

Edited by

Albert M. Wu

Texas A&M University
and Texas Agricultural Experimental Station
College Station, Texas

Associate Editor

L. Garry Adams

Texas A&M University
and Texas Agricultural Experimental Station
College Station, Texas

PLENUM PRESS • NEW YORK AND LONDON

Library of Congress Cataloging in Publication Data

International Symposium on Molecular Immunology of Complex Carbohydrates (1985: College Station, Tex.)
The molecular immunology of complex carbohydrates.

(Advances in experimental medicine and biology; v. 228)
"Proceedings of an International Symposium on Molecular Immunology of Complex Carbohydrates, held September 13–15, 1985, in College Station, Texas"—T.p. verso.
"Satellite meeting of the 8th International Glycoconjugate Conference held on September 8–13, 1985, in Houston, Texas, U.S.A."—Pref.
Includes bibliographies and index.
1. Carbohydrates—Immunology—Congresses. 2. Glycoproteins—Immunology—Congresses. 3. Glycolipids—Immunology—Congresses. I. Wu, Albert M. II. Adams, L. Garry. III. International Symposium on Glycoconjugates (8th: 1985: Houston, Tex.) IV. Title. V. Series. [DNLM: 1. Glycoconjugates—immunology—congresses. W1 AD559 v.228 / QU 75 I614m 1985]
QR186.6.C37I57 1985 574.2′9 88-2532
ISBN 978-0-306-42818-0

Proceedings of an International Symposium on Molecular Immunology of Complex Carbohydrates, held September 13–15, 1985, in College Station, Texas

DEDICATION

This book is dedicated to my parents, Mr.& Mrs. Hong Wu (吳鴻, 吳張妙), as well as to unnamed colleagues and friends who have sincerely helped me in my scientific career. Their valuable assistance has made my research possible.

Albert M. Wu

吳 明 道 (臺灣、臺南)

Associate Professor

Department of Veterinary Pathology

Texas A & M University

College Station, Texas

U.S.A.

The carbohydrate moieties of ovine salivary mucus glycoprotein must be *O*-glycosidically linked to Ser or Thr of the protein core. Dr. W. Pigman (right) giving his structural concept of glycoproteins to Dr. Wu (editor) at New York Medical College and Flower and Fifth Avenue Hospitals, New York, N.Y. in 1972 (see Section II-6).

Pioneers of Quantitative Immunochemistry met at Michael Heidelberger's lecture at the College of Physicians and Surgeons of Columbia University, New York, N.Y. in 1978. Dr. Elvin.A. Kabat (right) born Sept. 1, 1914; Dr. M. Heidelberger (left) born April 29, 1888.

PREFACE

During the past three decades, the sugar moiety of complex carbohydrates has been found to be involved in important interactions of immunological specificity of antigens and to participate in a variety of cellular functions. The long polysaccharide side chains of the lipopolysaccharides on the outer membrane of Gram negative organisms provide surface antigens for differential serodiagnosis. Bacterial surface lectins are important in mediating the attachment of bacteria to host cells in the pathogenesis of infectious diseases. The carbohydrate moieties of cell surface glycoconjugates (glycoproteins and glycolipids) of mammals are the sites for intercellular recognition and for the regulatory molecular interactions such as interaction of complex carbohydrate with hormones or hepatic lectins. The carbohydrate side chains of many complex carbohydrates play essential roles as antigenic determinants of human blood group ABH, Le^a, Le^b, I, and i activities, as the Forssman specific determinant, and as tumor associated antigenic determinants. Prompted by these and other advances in the field, a Symposium on Molecular Immunology of Complex Carbohydrates was organized as a satellite meeting of the 8th International Glycoconjugate Conference held on September 8-13, 1985, in Houston, Texas, U.S.A. Many eminent scientists contributed their knowledge at this meeting. The lecture and poster materials of the symposium are contained in this proceeding book, which is divided into four Sections and one Appendix. Section I is entitled Antibody Specificity, Epitope, and Lectinology. Dr. Elvin A. Kabat presented a paper on "The Structural Basis of Antibody Specificity" in which three dimensional structures of monoclonal antibody combining sites from a variety of sources were demonstrated. Dr. Roland Schauer from FRG described how sialic acids can function as antigenic determinants and also as anti-antigens by masking the antigenic sites of antigens. Dr. Arne Lundblad from Sweden discussed ways in which blood group active haptens can be isolated from urine and faeces. These represent examples of the current concepts of antibody specificities and epitopes. Lectins used to separate immunocyte subpopulations, lectins of animal and plant origins, the function of bacterial lectins and classifications of GalNAc/Gal specific lectins, which provide pioneer information in lectinology, are also reported in this Section. Complex carbohydrates as antigens are described in Section II, which includes glycolipids, blood group glycoproteins and bacterial antigens. Ten articles and five abstracts reviewing current knowledge in this area are included in this Section. Glycoconjugates as cancer and/or tumor antigens, an important subject in current cancer research, comprises Section III. The articles in this Section provide fundamental information for cancer diagnosis. Methods for structural analysis of the hapten moieties of complex carbohydrate antigens are described in Section IV. Included are recent methods which clarify the structure of oligosaccharides by nuclear magnetic resonance spectroscopy and the direct chemical ionization (DCI), fast atom bombardment (FAB) and collisionally induced decomposition (CID) mass spectrometry. Finally, guidance for the carbohydrate specificity of lectins is illustrated in the Appendix. The symposium organizers are grateful to the Texas Agricultural Experiment Station and the Department of Veterinary Pathology, College of Veterinary Medicine, Texas A&M University for supporting this work. We thank Dr. Raymond W. Loan (Associate Dean for Research and Graduate Studies, College of Veterinary Medicine) for providing his editing experience, and Ms. Clara Scott (Departmental secretary, Veterinary Pathology) for her assistance in the arrangement of manuscript typing. We are indebted to the many editing and typing helpers, especially Ms. Doretha Jones and Ms. Lee Park. We appreciate the art work for this book served by the

Biomedical Learning Resources Center and the photographic and video expertise of Mr. Wayne Zander and Mr. William Blanton.

Albert M. Wu and Garry Adams
at Texas A&M University
College Station, Texas
U.S.A.
October 1987

CONTENTS

I. ANTIBODY SPECIFICITY, EPITOPE AND LECTINOLOGY

II. COMPLEX CARBOHYDRATES AS ANTIGENS

III. GLYCOCONJUGATES AS CANCER/OR TUMOR ANTIGENS

IV. STRUCTURAL ANALYSIS OF HAPTEN MOIETY OF COMPLEX CARBOHYDRATE ANTIGENS

APPENDIX

ANTIBODY COMBINING SITES:
HOW MUCH OF THE ANTIBODY REPERTOIRE ARE WE SEEING? HOW DOES IT INFLUENCE OUR UNDERSTANDING OF THE STRUCTURAL AND GENETIC BASIS OF ANTIBODY COMPLIMENTARITY?

Elvin A. Kabat

Deptartments of Microbiology, Human Genetics and Development
College of Physicians and Surgeons Columbia University
New York, N.Y. 10032
National Institute of Allergy and Infectious Diseases
Bethseda, Md. 20205 USA

Knowledge of monoclonal antibody combining sites and of their specificities and three dimensional structures has been obtained using a variety of materials and sources: These include (1) collections of monoclonal immunoglobulins of various classes from patients with multiple myeloma and Waldenström macroglobulinemia (Eisen et al., 1967; Seligmann and Brouet, 1973; Yoo and Franklin, 1971; Harboe et al., 1975; Freedman et al., 1976; Emmrich et al., 1985; Fink and Galanos, 1985), and cold idiopathic hemagglutinin disease (Williams, 1971; Feizi and Kabat, 1972; Roelcke, 1974), and from BALB/c (Potter, 1971,1977; Cohn, 1967)or NZB mice (Warner, 1975) injected with paraffin oil; (2) from infrequent mono- or pauciclonal antipolysaccharide responses in rabbits, most frequently to immunization with bacterial vaccines(Osterland et al., 1966; Krause, 1970; Braun et al., 1969; Haber et al., 1975); (3) in humans immunized with various polysaccharides, notably dextrans and levans(Yount et al., 1968); and most recently (4) to almost any antigen in mice by the hybridoma technique(Kohler and Milstein, 1975). The first of these sources provides no intrinsic information as to the specificity of the monoclonal immunoglobulin produced and only in a small proportion of the human and a somewhat larger proportion of the mouse myeloma proteins and macroglobulins have antibody activities been identified. These are almost always directed toward substances with repeating units such as polysaccharides and polynucleotides as well as to proteins on to which multiple haptenic groups have been attached; the sole myeloma found producing antiprotein reacted with flagellin, a polymeric molecule built of identical subunits (Smith and Potter, 1975). The fine specificities of many of the different combining sites have been mapped by quantitative immunochemical methods using quantitative precipitin and precipitin inhibition or radio- or enzyme immunoassays but unless data on a large number of active and inactive oligosaccharides are obtained, sufficient to provide assurance that one or more fill the combining site completely, one remains uncertain as to the true specificity (Kabat, 1983; Lemieux et al., 1984). Even when this is accomplished, one may find that several apparently structurally unrelated compounds cross react extensively; these most often involve charged moieties (Kabat et al., 1984; Naparstek et al., 1985).

On the other hand the inhibition studies frequently provide evidence of specificity differences among monoclonal antibodies (Kabat, 1976). Thus several mouse myeloma anti-$\alpha(1\to6)$dextrans were found to differ in combining site size (extent of the area complementary to the antigenic determinant) or in the shape of the combining site, whether specific for terminal nonreducing ends of chains (cavity-type sites) or for internal sequences of sugar chains (groove-type sites)(Cisar et al., 1975; Wu et al., 1978). Monoclonal human anti-I and anti-i sera have been found to show different specificities toward various internal sugar determinants of the blood group A, B, H, Le^a and Le^b substances (Feizi et al., 1971; Kabat et al., 1978; Wood and Feizi, 1979; Watanabe et al., 1979; Lemieux, et al., 1984). Myeloma antibodies reacting with chondroitin sulfates were inhibitable by disaccharide sulfates, by the desulfated disaccharides, by sugar phosphates and by sodium sulfate and sodium phosphate (Kabat et al., 1984). Rabbit antisera could distinguish the phospho- and dephospho-forms of a protein (Nairn et al., 1982). X-ray crystallographic studies with the phosphorylcholine-binding myeloma protein McPC603 showed that a sulfate ion in the crystal occupies the site in which the phosphate residue of phosphorylcholine would sit (Padlan et al., 1973).

The second source has yielded a substantial number of rabbit antibodies to pneumococcal and streptococcal polysaccharides in amounts sufficient for sequencing (see Kabat et al., 1983).

The third source has yielded most of the original insights (Kabat, 1985) into the sizes and shapes of antibody combining sites but until human-human hybridomas to the human antibodies such as the anti-$\alpha(1\rightarrow6)$ dextrans, are studied only limited data are obtainable.

It is to the hybridoma technique (Köhler and Milstein, 1975) that one now looks for the production of antibodies to any desired antigenic determinant; it has spawned a good-sized segment of modern biotechnology and has opened the way to the study of cell surfaces, cell differentiation, embryogenesis etc. It has led to increased estimates for the number of distinct antibody specificities by several orders of magnitude. It has also yielded a wide spectrum of monoclonal antibodies from non-immunized BALB/c mice (Dighiero et al., 1985), or following polyclonal activation with lipopolysaccharide or with mitogens but their specificities are often difficult to identify. Nevertheless one wonders whether these four sources of antibody immunoglobulins have made it possible to obtain a reliable estimate of the size of the repertoire to a single antigenic determinant and a reasonably comprehensive understanding at the gene level of how such a repertoire is maintained and expressed.

Among the uncertainties are the following: the malignant transformation leading to myeloma and macroglobulinemia is considered to be random, so that the nature of the antibody specificities seen in the mouse and in the human would most probably differ. In addition, prior contact with various antigens is almost certainly not the same for the two species. Their past history with expansion of clones due to environmental contact with various antigens may have increased the probability that a cell from an expanded clone was being triggered to induce antibody. The failure to detect antibodies to antigens not built up of repeating units and the limited technics and panels of non-polysaccharide antigens used in testing myeloma proteins for antibody activity may contribute to the differences. Although coupling of haptens to proteins provides multiple determinants, the size of the haptenic group is generally small so that it represents but a small portion of the determinant and since the structures on to which it has been attached differ, one has complicated the repertoire problem by the creation of a multiplicity of antigenic determinants each having the same introduced group. On immunization, these give rise to selected populations by triggering one or another clone present in high frequency. Hybridomas from such immunized mice generally do not reflect the sequences of the antibodies present in serum (Slaughter and Capra, 1983) even after enrichment to select for antibodies bearing the Ars cross reacting idiotype.

Another influence restricting elucidation of the repertoire has emerged from the study of antibodies formed after immunization with the stearylisomaltosyl oligosaccharides (Wood and Kabat, 1981a,b). These synthetic glycolipids from stearylisomaltose to stearylisomaltoheptaose were antigenic in rabbits and in C57BL/6 mice (Lai and Kabat, 1985) giving rise to antidextrans specific for $\alpha(1\to6)$dextran as evidenced by the similarity in quantitative precipitin assays with various dextrans to those with antibodies obtained after injection of dextran NRRL B512 with 96 per cent $\alpha(1\to6)$ linkages and 4 per cent $\alpha(1\to3)$ branches.

In studies of the antibody response in rabbits to the above series of stearylisomaltose oligosaccharides, we had found earlier (Wood and Kabat, 1981a,b) that only 3 of the 15 rabbits analyzed yielded antisera with antibody populations about 1/2 of which precipitated with linear dextrans D3 or LD7 (Ruckel and Schuerch, 1966) thus having some sites specific for internal chains of $\alpha(1\to6)$-linked glucoses (e.g. groove-type sites) similar to myeloma antidextran QUPC52 (Cisar et al., 1975); the remaining antibody in these three antisera and all of the antidextran in the other 12 antisera did not react with linear dextran and thus were specific for the terminal non-reducing ends of chains (e.g. cavity-type sites).

The data on the combining site sizes of these antibodies yielded an important additional insight. Another three of the 15 rabbit antibodies had combining sites complementary to more sugar rings than were present in the immunizing stearyl glycolipid (Wood and Kabat, 1981b). This had been noted earlier (Arakatsu et al., 1966; Outschoorn et al., 1974). In three rabbits immunized with isomaltotrionic acid-BSA; two formed antibody complementary to four glucoses and one formed antibody complementary to five glucoses although the immunizing antigen had only two intact sugar rings plus an α-linked open chain coupled to BSA. Evidently, the antigenic challege with the stearylisomaltosyl oligosaccharide and the isomaltotrionic acid-BSA had triggered proliferation of pre-existing clones with combining sites larger than the immunizing antigen; this provides proof with a structurally defined antigen of "original antigenic sin" (Fazekas de St. Groth and Webster, 1966).

When hybridoma antibodies to the stearylisomaltosyl oligosaccharides were produced in C57BL/6 mice a quite different pattern was seen (Lai and Kabat, 1985). Unlike the rabbit antibodies, the mouse hybridoma antibodies all reacted with linear dextran LD7 and thus were directed toward internal chains of α(1→6) linked glucose (groove-type sites). One of the 12 had a combining site as small as four glucose residues; the immunizing antigen was stearyl-IM5 which would have four glucose rings plus an open chain coupled to the stearyl moiety. Inhibition studies showed the specificity to be directed toward the three subterminal glucoses plus the open chain. The remaining hybridoma antibodies had combining sites larger than the sugar determinant in the immunizing glycolipid again indicating the non-randomness of the findings and making it difficult even to approximate the repertoire. Although the 12 hybridomas have groove-type sites, myeloma protein W3129 has been shown to have a cavity-type site in which two glucoses at the terminal non-reducing end contribute the bulk of the binding energy (Cisar et al., 1975; Bhattachargee et al., 1981).

The hybridoma technic both in the isolation of antibodies following injection of dextran and of the stearylisomaltosyl glycolipids yielded a surprising selection with respect to immunoglobulin classes of the (Sharon et al. 1981; Newman and Kabat, 1985; Lai and Kabat, 1985) hybridoma antibodies. All 32 hybridoma proteins were either IgM or IgA; recently an IgG3 was also found in a C58 mouse (Makover et al., in preparation). Five monoclonal anti-idiotypic antibodies to the anti-α-(1→6)dextran from a hybridoma were IgM and one was IgG1 (Borden and Kabat, in preparation).

Sera from humans immunized with dextrans also had populations of antibodies with groove- and cavity-type sites; in some instances linear dextran D3 precipitates all of the antidextran whereas in others it precipitates only a fraction of the antibody (Cisar et al., 1975). When only a portion of the poly- or pauciclonal antibodies in antisera are precipitated by linear dextran, the aggregates formed would have multiple non-reducing ends on to which antibodies specific for the non reducing termini could attach. Thus, in such mixtures the estimate of cavity-type sites would tend to be low. This problem would not occur with monoclonal antibodies.

Recently we have identified a human monoclonal, IgM^{NOV}, which reacts specifically with E. coli K1 and with meningococcal group B capsular polysaccharides composed entirely of poly-N-acetylneuraminic acid linked $\alpha(2\rightarrow8)$ (Kabat et al. in preparation). This 81 year old individual had a benign monoclonal gammapathy with about 23 mg of monoclonal IgM per ml and the monoclonal peak in cellulose acetate immunoelectrophoresis was completely removed by precipitation with the $\alpha(2\rightarrow8)$ poly-N-acetylneuraminic acid; no cross reactivity was seen with poly-N-acetylneuraminic acid linked $\alpha(2\rightarrow9)$ (meningococal group C polysaccharide). Although group B meningococcal meningitis is a major cause of meningitis in the world today, immunization with the group B specific polysaccharide has been unsuccessful (Wyle et al., 1972; Mandrell and Zollinger, 1982); thus one wonders about the basis for this relatively enormous antibody response. A report by Finne et al. (1983) indicates that there are antigenic similarities between brain components and bacteria causing meningitis.

A substantial proportion of hybridomas have been obtained from non-immunized mice and from mice previously injected with polyclonal activators. Thus 24 of 384 hybridomas from non-immunized 6 day old mice were found to be specific for a variety of antigens including actin, DNA, tubulin, myosin, TNP-BSA, neurofilaments etc. (Dighiero et al., 1985) Numerous hybridomas cross reacted extensively with one or more of these antigens especially actin, tubulin, myosin and TNP-BSA; others were specific and did not cross react. The cross reactions were seen with the isolated IgM in non- competitive and competitive enzyme immunoassay and in immunoblots. The structural basis for these cross reactions is unknown and thus it is difficult to evaluate the extent to which they affect the repertoire.

Idiotypes found on the $\alpha(1\rightarrow3)\alpha(1\rightarrow6)$ myeloma antidextrans have also been found by immunofluorescence in low frequency (0.2 to 1.8 percent) on the membranes of splenic B cells from non-immunized mice (Victor et al., 1983). A problem in evaluating the repertoire derives from the finding that in wild mice, the W3129 idiotypic specificity of anti-$\alpha(1\rightarrow6)$dextran produced by immunization with dextran was quite common whereas the QUPC52 idiotype has only been found on immunization of inbred mice (D'Hoostelaere and Potter, 1982). Unselected collections of IgM-secreting hybridomas from spleen cells of non-immunized newborn mice or from adult spleen cells polyclonally stimulated with LPS were screened for antibodies reactive to monoclonal anti-idiotypic antibodies and to haptens. Hapten specific antibodies were found with equal frequency in both collections but newborn antibodies reacting to the anti-idiotypic monoclonals occurred ten times more frequently (Holmberg et al., 1985).

Recently an estimate of the size of the repertoire of monoclonal antibodies to phosphorylcholine was made by a splenic focus assay (Owen et al., 1985). Spleen cells from non-immune adult BALB/c mice were injected intravenously into irradiated BALB/c recipients that had been immunized 1-2 months previously with hemocyanin. Splenic fragment cultures were stimulated _in vitro_ with a phosphorylcholine containing hapten coupled to hemocyanin to obtain foci secreting monoclonal antibodies. Their fine structure was probed, by inhibition of binding to phosphorylcholine coupled to bovine serum albumin, with phosphorylcholine, α-2-glycero- phosphorylcholine and choline. In addition they were tested for the TEPC 15 (T15) idiotype and their Ig isotype was determined. The frequency of clones having the T15 and non-T15 idiotypes was comparable; 109 non-T15 IgM monoclonals from 14 donors were examined and gave 23 distinct patterns which together with isoelectric points and partial cross reactivity with rabbit anti-T15, gave 60 distinct sets, and from the frequencies of single and recurrent clonotypes, the IgM primary B cell repertoire was estimated as $>10^2$ and the total primary IgM repertoire as $>10^7$. With a repertoire of this size it becomes impossible to make precise comparisons of the immune responses of different individuals.

It is clear that estimates of the repertoire of antibody combining sites to known antigenic determinants is constantly increasing and that much more must be done to obtain a thorough understanding of the three dimensional structures of antibody combining sites of known specificities and of how the repertoire is determined at the gene level.

I X-RAY CRYSTALLOGRAPHIC FINDINGS

Although enormous progress has been made during the past two decades in determining the amino acid and nucleotide sequences of antibodies and of their genes, their three dimensional structures from x-ray crystallographic studies at high resolution (Davies and Metzger, 1983), and by immunochemical mapping of the fine structure of antibody combining sites (Kabat, 1980,1982,1983), we do not yet have an intimate picture of antigen-antibody interaction with a determinant which fills the antibody combining site completely. I recently summarized the situation indicating that we still needed to know (Kabat, 1983) "the range of variability which antibody combining sites can manifest, for example, their limits of size and shape, whether they are obligatively female in character, the nature of the forces determining binding of different antigenic determinants in the site, the location and relation of idiotypic determinants to the site, how the synthesis of the amino acids which make up the variable region and more specifically the site itself is controlled at the DNA level and most important, how the enormous and seemingly limitless diversity of antibody combining sites is generated." Since this was written, a Bence Jones dimer has been found by x-ray crystallography to have a protruding convex binding site with grooves on two sides of the protrusion rather than a groove, cavity or pocket (Chang et al., 1985). One would certainly like to identify a ligand with which this would react and to learn how the amino acid sequence of the CDR2 differs from other Bence Jones dimers. If this type of site is obtained for an antibody, this would necessitate extensive revision of our thinking.

Bence Jones Proteins: All x-ray structures on light chain dimers and Fab fragments confirmed the prediction from varability plots that the combining sites were formed by the hypervariable regions of the light and heavy chains(Fig 1)(Wu and Kabat, 1970; Kabat and Wu, 1971); these are termed complementarity-determining regions (CDR) with the remainder of each chain, being the framework (FR). The light chain dimers appear to

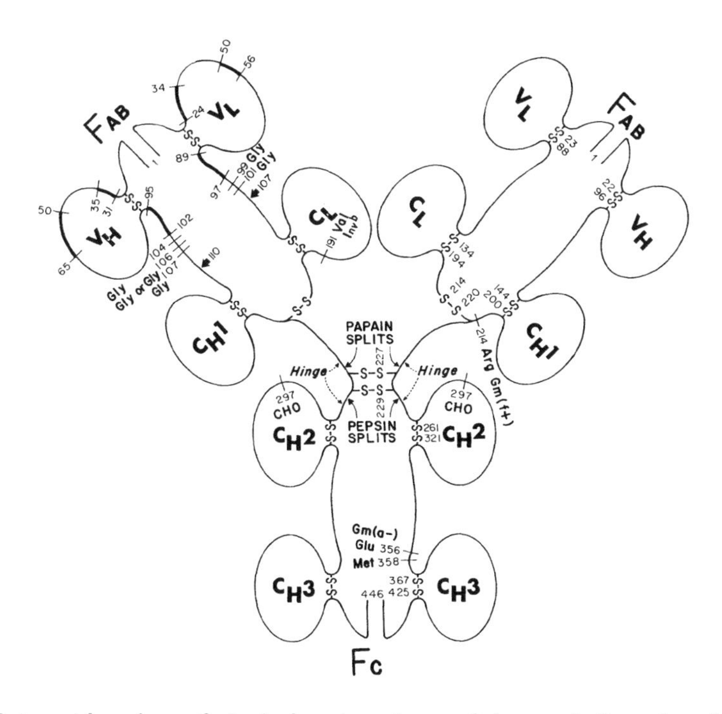

FIG. 1 Schematic view of 4-chain structure of human IgGκ molecule. Numbers on right side: actual residue numbers in protein Eu (Edelman et al., 1968, 1970); Numbers of Fab fragment on left side aligned for maximum homology; light chains numbered as in Wu and Kabat (1970) and Kabat and Wu (1971). Heavy chains of Eu have residue 52A, 3 residues 82A,B,C and lack residues termed 100 A,B,C,D,E,F,G,H and 35A, B. Thus residue 110 (end of variable region) is 114 in actual sequence. Hypervariable regions e.g. complementarity-determining segments of regions (CDR): heavier lines. V_L and V_H: light- and heavy-chain variable region; C_H1, C_H2 and C_H3: domains of constant region of heavy chain; C_L: constant region of light chain. Hinge region in which 2 heavy chains are linked by disulfide bonds is indicated approximately. Attachment of carbohydrate is at residue 297. Arrows at residues 107 and 110 denote transition from variable to constant regions. Sites of action of papain and pepsin and location of a number of genetic factors are given. From Kabat, E.A. Adv. in Protein Chemistry 32:1-75 (1978). Reproduced with permission of Academic Press. Modified from Kabat, E.A. Proc. 3rd Int. Convocation Immunology. S. Karger, Basel. 1972, pp4-30.

be primitive antibody combining sites since a given site binds a variety of ligands. The first Fab combining sites were from myeloma proteins so that the antigenic determinants were unknown. Although screening established that one bound phosphorylcholine (Padlan et al., 1985) and the other an hydroxy derivative of vitamin K (Saul et al., 1985); these ligands filled only a portion of the site and there is only a very low probability that a site filling ligand for them will be discovered. This may also prove a limitation to finding a site-filling ligand for Fab sites to most chemically modified proteins onto which a small haptenic group has been attached. The less specific light chain dimer site of the Mcg Bence Jones protein has also been examined for fine specificity largely with fluorescent ligands(Edmundson et al., 1984). Earlier studies had shown it to bind dinitrophenyl compounds, ε-dansyllysine, colchicine, methadone, morphine, menadione and others (Edmundson et al., 1974). The site had three subsites. The main binding cavity, a truncated cone was attached to a deep binding ellipsoidal pocket; the mouth of the truncated cone defines subsite A, the body of the cone subsite B and the ellipsoidal pocket subsite C. In searching for site filling ligands by diffusing various compounds into the crystal, it was observed that 1-anilinonaphthalene-8-sulfonate was initially bound in the truncated cone but with time entered the deep binding pocket (Edmundson et al., 1984) . This left the truncated cone unoccupied despite the presence of excess ligand in the mother liquor surrounding the crystals most likely indicating a conformational change in the orientation of the CDR. A variety of other ligands filled more of the site. The dimer of 6-carboxytetramethylrhodamine filled the site completely as did the heterodimer of 5- and 6- carboxytetramethylrhodamine but in different orientations; the heterodimer assumed more than one orientation in the site and diffused out of the site on dialysis whereas the 6-homodimer could not be removed from the site by 63 days of perfusion. Small conformational changes in the C-domain have also been observed when ligand was bound and appeared to vary with different types of ligand and predominantly involved monomer 1 which played the role of the heavy chain. It must be remembered that the Mcg IgG1 myeloma protein does not bind dinitrophenyllysine which binds to the Mcg Bence Jones light chain dimer (Firca et al., 1978). Novotny et al. (1983) have shown that V_H has five β-strands whereas V_L has but four making the L-chain dimer less analogous to the Fv than previously thought; Novotny and Haber (1985) have compared superimposed V_L-V_H and V_L-V_L domains; one V_L-V_L dimer (Rhe) did not superimpose closely with the V_H-V_L domains of the Fab fragments.

Another important development has been the preparation of heterodimers of Bence Jones proteins for study of their binding activity. This could be accomplished in good yield by reducing one dimer to the monomer and by S-sulfonation of the other dimer; when the disulfide bond was allowed to form, the yield of heterodimer was greater than 80 percent (Peabody et al., 1980). Equilibrium dialysis studies of a heterodimer of Mcg and Weir, the latter having no known binding activity; showed that the the heterodimer gave essentially the same binding.curve for di-DNP lysine as the Mcg dimer, which as a control had been reduced and reoxidized in the same manner.

More recent studies have yielded crystals of heterodimers of Mcg with other λ and κ Bence Jones proteins (Ely et al., 1985a). Detailed analysis of the reduction step showed that the intrachain disulfide bond, probably of the V-domain, was opened and this facilitated crystallization of the heterodimer. When suitable crystals of these heterodimers are obtained, substantially increased understanding of how each chain contributes to the binding of various ligands will become possible. One hopes that V_H-V_L or Fab heterodimers will also be crystallized.

The hybrid Weir-Mcg crystals bound di-DNP lysine in subsites A and B with the Mcg chain monomer 2, playing the role of the light chain. Weir has 36 amino acid differences from Mcg; the bilateral, but not the unilateral replacement of Val 48 and Ser 91 in Mcg by Leu and Met in Weir blocks access to subsite B (Ely et al., 1985b).

Fab fragments: Seven monoclonal Fab fragments from hybridoma proteins have been crystallized, five anti-egg white lysozyme (Silverton et al., 1984; Amit et al., 1984), one anti-azophenylarsonate and one anti-single-stranded DNA, IgG2bκ (Gibson et al., 1985). Three of the antilysozyme Fab's were also crystallized with lysozyme in the site (two were IgGκ). A 6Å electron density map of one (D1.3) of the Fab-lysozyme complexes (Amit et al., 1984) showed that the three CDRs of the light and three CDRs of the heavy chain were involved in the site unlike the previous findings with two myeloma proteins in one of which, McPC603(Davies et al., 1975), CDR 2 of V_L was blocked from the site region by the insertion of an additional six residues in CDR1 and in the other, Newm (Amzel et al., 1974) CDR2 was pulled away from the site region by a deletion in V_L of the seven residues following CDR2. Determination of the extent to which such insertions and deletions occur in monoclonal antibodies and myeloma proteins is of importance for understanding antibody diversity, site structure and the repertoire. Glu 121 in hen egg white lysozyme occupied a central position in the anti-lysozyme D1.3 binding area. The site involved non-contiguous areas of the antigen and was essentially topographical and conformational. Inhibition of the antilysozyme -egg white lysozyme reaction fails when Glu 121 is replaced by His as in partridge or turkey lysozyme or by Asn in Japanese quail lysozyme. The other anti-lysozyme is clearly not directed toward the region recognized by D1.3 (Silverton et al., 1984). Darsley and Rees (1985) have sequenced five additional monoclonal antilysozymes. When the high resolution data become available, we should learn toward which amino acids of lysozyme the other hybridoma antibodies are directed as well as the precise delineation of the site contours. Detailed data on the anti-single stranded DNA site are not available (Amit et al., 1985).

Energy minimizations of earlier structures: Energy minimization of the Fv portions of the three previously studied Fab fragments (McPC603, KOL and Newm) showed that very little adjustment of coordinates was needed (Novotny et al., 1983). A most important finding was that the V_L had four and the V_H chain had five antiparallel β-strands; the β sheets thus formed are curved giving rise to a β-barrel. The boundaries of the β-strands and hypervariable loops were considered to be part of the β-sheets and thus structural; this would shorten some of the CDR's by one to three residues at one or at both ends. An earlier statistical examination of each residue in the CDR (Kabat et al., 1977) to establish whether it was playing a structural or a contacting or conformational role had identified some of the same residues in accord with the x-ray crystallographic findings but also included residues playing a structural role within the CDR's themselves since the nature of the peptide bond makes it impossible for all side chains of three of four sequential residues to point into the site. Indeed some side chains in the CDRs which are external to those forming the site include idiotypic determinants.

An interesting exercise illustrating the difficulties of model building has been carried out (Novotny et al., 1984). The V_L of

McPC603 and the hemerythrin of Themiste dyscritum have 113 amino acids yet are known from crystallographic studies to be very different, the former being β-sheet and the latter an α-helical protein. It proved readily possible to model the former into the α helical form and the latter into the β-sheet. "Without violating the normal packing density and optimal van der Waals' contact distances, stereochemically correct structures are obtained after only small structural adjustments. Energy minimization is applied to both the correct and incorrectly folded structures and in all cases satisfactory energies are obtained."..."The incorrectly folded models however tended to have a larger solvent-accessible surface and a greater fraction of non-polar side chains exposed to solvent." Studies of this type provide substantial insight as to current limitations of modeling and energy minimization analyses.

Ohno et al. (1984, 1985) have attempted to model anti-phosphorylcholine combining sites on the assumption that CDR1 and CDR2 of V_H are of predominant importance to antigen binding and that FR2, residues 35 to 49, was crucial to their function. This had already been clear from both the x-ray crystallographic studies (Davies and Metzger, 1983) as well as the findings (Kabat et al., 1980,1983) and most recent data that the identical amino acid sequence of FR2 of V_L was present in one human, 25 mice and 15 rabbits, that FR2 of mouse germ-line V_H corresponded in nucleotide sequence with a rabbit cDNA (Bernstein et al., 1982) for nucleotides coding for amino acids, 36-47 with one and two nucleotide differences for residues 48 and 49 and that the amino acid sequence of FR2 of V_H was found in mouse and rabbit chains (Kabat, 1985). What is more controversial is the inference that V_L is of negligible or minimal importance to complementarity. Rudikoff(1983) has pointed out that the portions of the V_L of the three different subgroups $V_\kappa 8$, $V_\kappa 22$ and $V_\kappa 24$ involved in phosphorylcholine binding have the same sequence Tyr-Pro-Leu at residues 94,95,96 and all use J_5, the only J which provides Leu 96. The side chains of Leu 96 and of Tyr 94 are involved in contact with phosphorylcholine in the site; the Tyr-Pro-Leu sequence is not found in non-phosphorylcholine binding myelomas (Kabat et al., 1983). Thus it would appear that the Ohno hypothesis is an over-simplification, but significant in renewing attention to the important role of CDR1 and CDR2 which has largely been ignored since the J and D minigenes (Kabat et al., 1978; Sakano et al., 1979, 1980; Max et al.,

1979; Early et al. 1980) provided hypothetical multiplication factors which, if experimentally validated, would greatly augment the magnitude of the repertoire of antibody combining sites.

In an x-ray crystallographic study of β_2-microglobulin, the light chain of class 1 antigens of the major histocompatibility complex, was found to be a monomer with a striking similarity to an immunoglobulin C_H3 domain (Becker and Reeke, 1985).

The T cell receptor for antigen (Hedrick et al., 1984; Yanagi et al., 1984; Patten et al., 1984; for a review see Möller, 1984) as well as the T cell membrane, proteins, T8 (Litman et al., 1985) and T4 (Maddon et al., 1985) all have the domain structures of the immunoglobulin superfamily.

II FINE STRUCTURE AND TOPOGRAPHY OF ANTIBODY COMBINING SITES

In the absence of an x-ray crystallographic structure of an antibody combining site for which a ligand completely filling the site is available, studies of the fine structure of antibody combining sites have essentially been made by immunochemical mapping as described earlier. The compound active at the lowest molar concentration is considered to approximate most closely the structure of the antigenic determinant. The most definitive data have been obtained with antibodies to polysaccharide antigens since these are largely sequential determinants and the low molecular weight ligands are more easily obtainable by partial hydrolysis or synthetically than with conformational determinants formed by non-contiguous segments of a molecule which are brought into apposition by three dimensional folding. The prototypes of sequential antigenic determinants are found among polysaccharides. Conformational determinants are usually proteins although the $\alpha(1\rightarrow4)$-linked glucose polymers form helices and some antigenic determinants of proteins may be sequential (Atassi, 1975; Benjamin et al., 1984). Even sequential determinants may exist in one or more preferred conformations but little is known as to whether antibody combining sites may exist to more than one conformer and whether they tend to form most frequently to the conformer present in highest concentration in solution.

Among the antigen-antibody systems for which data are available on the fine structure of monoclonal antibody combining sites are the anti-$\alpha(1\to6)$dextrans (Sharon et al., 1982 a,b; Cisar et al., 1975; Wu et al., 1978; Newman and Kabat, 1985; Lai and Kabat, 1985), the anti-$\beta(1\to6)$galactans (Rudikoff, 1983; Hartman and Rudikoff, 1984; Glaudemans and Kovac, 1985; Pawlita et al., 1981), anti-blood group I and i (Lemieux et al., 1984; Wood and Feizi, 1979; Watanabe et al., 1979), anti-blood group A (Chen and Kabat, 1985; Gooi et al., 1985) and anti- B (Chen and Kabat, 1985); antibodies to the pyruvylated sugars of Klebsiella κ polysaccharides (Rao et al., 1984) and anti-chrondroitin sulfates (Kabat et al., 1984). Limited data have been obtained for anti-$\alpha(1\to3)\alpha(1\to6)$-dextrans (Newman et al., 1983, 1986); anti-inulins (Johnson et al., 1982); anti-Ars (Capra et al., 1982; Manser et al., 1984); anti-NP (Bothwell et al., 1981); anti-phosphorylcholine (Perlmutter et al., 1984; Rudikoff, 1983); anti-oxazolone(Kaartinen et al., 1983; Griffiths et al., 1984) and on anti-idiotopes on murine anti-streptococcal group A carbohydrates (Greenspan and Davie, 1985). Although as already noted, the last four suffer from the difficulty that the introduced group fills only a small portion of the site, they are all providing important sequence data and information about idiotypic determinants. The dansyl system (Fan and Karush, 1984) is providing a set of monoclonals which show a broad range of affinity constants and thus sequence data may prove especially useful, although the determinant may not be site filling.

The most studied anti-protein systems (for data see Reichlin, 1975; Arnon, 1977; Crumpton, 1974; Benjamin et al., 1984) are anti-myoglobins, anti-lysozyme, anti-cytochrome c, and anti-serum albumin (Doyen et al., 1985) and anti-influenza (Wiley et al., 1981; Coleman et al., 1983; Clarke et al., 1985).

The earlier studies on antibodies in whole antisera tended to emphasize sequential determinants and limited numbers of determinants (Atassi, 1975), but studies with monoclonal antibodies have established that almost all areas of the surface of a protein may be antigenic and that, even to a sequential determinant, individual monoclonals may see individual amino acid residues as immunodominant. The distribution of antigenic determinants on one surface of the lysozyme molecule as mapped

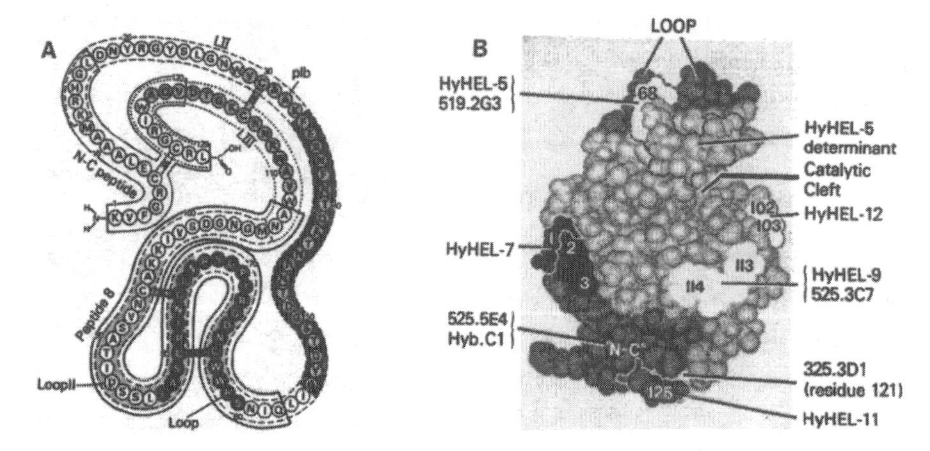

Fig. 2 A: Primary structure of chicken lysozyme (HEL) with eight peptides that have been shown to be antigenic when tested against anti-HEL: N-C peptide, solid outline; LII, dashed outline; plb, stippled; a continuous region (amino acids 34-54) within plb, black box; peptide 8, heavy black outline; the Loop, black with white lettering; Loop II, stippled box; and LIII, dotted outline.

B: Space-filling model of HEL, computer generated as described previously. The Loop and N-C peptide are dark grey, with residues 1-3 black. Specific residues recognized by monoclonal antibodies are colored or outlined in white; a hypothesized unit determinant for antibody HyHEL-5 is outlined in a dotted black line. HyHEL antibodies are from the Smith-Gill laboratory. Antibody Hyb.C1 is from Fujio's laboratory. All other antibodies are from the Sercarz laboratory. [From Benjamin et al., 1984; Courtesy Dr. Sandra Smith-Gill]. Reproduced with permission from the Annual Review of Immunology, 2:67-101. c 1984 by Annual Reviews Inc.

by monoclonal antibodies made in several laboratories is shown in Fig. 2, together with the primary sequence of hen egg white lysozyme (Benjamin et al., 1984; see also Moudallal et al., 1985 on tobacco mosaic virus protein; Berzofsky, 1985).

In addition to mapping antigenic determinants with monoclonal antibodies prepared by immunizing with the intact protein, many studies have been carried out by immunization using peptides from regions considered to be antigenic (Lerner, 1982; Tainer et al., 1984, 1985). In a fair number of instances such antisera and hybridoma antibodies reacted with the intact protein although the small peptides could exist in many conformations, whereas the mobility of the region in the protein occupied by the peptide was substantially restricted. Westhof et al. (1984) and Tainer et al. (1984) reported with polyclonal antisera that those regions of tobacco mosaic virus protein and of hemerythrin which were relatively more mobile tended to react with antisera to peptides from such regions in contrast to antisera to peptides from more well ordered regions. Such interactions might well contribute to the rapidity of antigen-antibody interaction. The large number of specificities in antisera to the peptides to antigenic regions which do not react with the intact protein further increases the size of the repertoire.

III CORRELATION OF SITE STRUCTURE WITH AMINO ACID SEQUENCES OF VARIABLE REGIONS

For an adequate understanding of the relation of amino acid sequence of the variable region of the light and heavy chain to the parameters determining binding of the antigenic determinant in the antibody combining site, it is important to have available a large number of monoclonal antibodies from various sources which will provide some degree of confidence that one has a random sample of the repertoire to the antigenic determinant involved as well as a site-filling ligand. It should be borne in mind that sequence differences do not necessarily have to be related to functional properties of the combining site and also that sequence differences in one portion of the V-region may induce changes in another area which may or may not affect binding or specificity. The anti-$\alpha(1\rightarrow6)$dextran system appears ideally suited to such an investigation.

In the anti-$\alpha(1\rightarrow6)$dextran system we have two IgAκ mouse myeloma proteins (Cisar et al., 1975), W3129 with a cavity-type site complementary to five glucoses, QUPC52 with a groove-type site complementary to six glucoses; the 32 IgMκ and IgAκ hybridoma antibodies made by immunization with dextran in BALB/c and C57BL/6 (Sharon et al., 1981; Newman and Kabat, 1985) and by immunization with stearyl isomaltose oligosaccharides (Lai and Kabat, 1985), all with groove-type sites as described above. Their site sizes, idiotypic specificities, binding constants, and the relative contributions of each sugar to the total binding energy have been measured. We are systematically making cDNAs and are cloning and sequencing them to derive the amino acid sequences. To date we have sequenced two BALB/c and one C57BL/6 hybridoma proteins obtained from spleens of mice immunized with α(1→6)dextran N279 with 96% α(1→6) and 4% α(1→3) linkages (Sikder et al., 1985).

The V-regions (residues 1 to 95) of the κ chains of all three are identical with the germ-line sequence of BALB/c monoclonal antibodies to 2-phenyloxazolone (Griffiths et al., 1984). Indeed, the C57BL/6 anti-α-(1→6)dextran, which uses the same J minigene as the BALB/c anti-2-phenyloxazolone, is identical throughout the entire V-J region including the nucleotides involved in V-J joining (Fig. 3).

This surprising identity led us to test all of our 32 hybridomas with oxazolone-BSA and four of the anti-oxazolone hybridomas with α(1→6)dextran; no cross reactivity was seen. This posed a dilemma with respect to the role of V_L in the antibody combining site. Since the two BALB/c anti-α(1→6)dextran hybridomas have combining sites complementary to an internal chain of six α-(1→6) glucoses and the C57BL/6 to an internal chain of seven α(1→6) glucoses, the complementary area of their sites is much larger than that of 2-phenyloxazolone. However all three anti-dextran sites are very similar except for their idiotypic specificity which is most frequently associated with the D minigene of V_H. It might thus be argued, since sequence similarity is generally found among antibodies of similar specificity (Rudikoff, 1983), that V_κ is important in determining the anti-dextran specificity. However the absence of cross reaction might not be consistent with this although V_H would obviously also be essential. It seems highly improbable that a combining site complementary to six or seven sugar rings could be formed without contacting residues from the CDRs of both chains.

Alternatively, one could ascribe specificity to V_H with V_L playing at most a modulatory role. This is essentially what has been done by default in the anti-NP (Bothwell et al., 1981) and the anti-α(1→3)α(1→6)dextran systems which have V_λ chains and there has been substantial reluctance to sequencing the light chains. Our earlier study of hybridomas to anti-α(1→3)α(1→6)dextrans and their five precipitin groups (Newman et al., 1983,1985) strongly suggested that sequence differences in V_L had to be involved in specificity to some degree. Moreover, those investigators interested in systems involving

```
                                                                                      <------------CDR1-
                            1         2         3         4         5         6         7         8         9
Sequence           1........0.........0.........0.........0.........0.........0.........0.........0.........0
14.6b.1κ   BALB/c  caaattgttctcacccagtctccagcaatcatgtctgcatctccaggggagaaggtcaccatgacctgcagtgccagctca---agtgta
                   GlnIleValLeuThrGlnSerProAlaIleMetSerAlaSerProGlyGluLysValThrMetThrCysSerAlaSerSer---SerVal
                                             10                            20                            30
26.4.1κ    BALB/c  .................................................................................---......
45.21.1κ   C57BL   .................................................................................---......
Vκ-Ox1     BALB/c  .................................................................................---......

                   ---------->                                          <--------CDR2------->
                            1         1         1         1         1         1         1         1         1
                   9        0         1         2         3         4         5         6         7         8
                   1........0.........0.........0.........0.........0.........0.........0.........0.........0
14.6b.1κ           agttacatgcactggtaccagcagaagtcaggcacctcccccaaaagatggatttatgacacatccaaactggcttctggagtccctgct
                   SerTyrMetHisTrpTyrGlnGlnLysSerGlyThrSerProLysArgTrpIleTyrAspThrSerLysLeuAlaSerGlyValProAla
                                             40                            50                            60
26.4.1κ            ..........................................................................................
45.21.1κ           ..........................................................................................
Vκ-Ox1             ..........................................................................................

                                                                                                   <-----
                   1        1         2         2         2         2         2         2         2         2
                   8        9         0         1         2         3         4         5         6         7
                   1........0.........0.........0.........0.........0.........0.........0.........0.........0
14.6b.1κ           cgcttcagtggcagtgggtctgggacctcttactctctcacaatcagcagcatggaggctgaagatgctgccacttattactgccagcag
                   ArgPheSerGlySerGlySerGlyThrSerTyrSerLeuThrIleSerSerMetGluAlaGluAspAlaAlaThrTyrTyrCysGlnGln
                                             70                            80                            90
26.4.1κ            ..........................................................................................
45.21.1κ           ..........................................................................................
Vκ-Ox1             ..........................................................................................

                                 <----------------J----------------->
                   ------CDR3---------->
                   2        2         2         3         3         3         3
                   7        8         9         0         1         2         3
                   1........0.........0.........0.........0.........0.........0
14.6b.1κ           tggagtagtaacccgtacacgttcggaggggggaccaagctggaaataaaacgg      J2
                   TrpSerSerAsnProTyrThrPheGlyGlyGlyThrLysLeuGluIleLysArg
                                             100                   108
26.4.1κ            ......................................................      J2
45.21.1κ           ..............act.........t.ct..............gc.g......      J5
                                 Leu         Ala               Leu
Vκ-Ox1   (a)       ..............act.........t.ct..............gc.g......      J5
                                 Leu         Ala               Leu
```

Fig. 3 Nucleotide sequences of V_K cDNA from anti-α-(1→6)dextrans and amino acids deduced from them. (From Sikder et al., 1985). Reproduced with permission of the Journal of Immunology.

V_λ have tended to rely on isoelectric focusing studies to support their contention that the chains were identical. Studies on a large number of myelomas (Gibson, 1984) have shown that "proteins differing only by neutral substitutions will exhibit electrophoretic identity". Of the mouse Vλ chains sequenced (Cohn et al., 1974, Cesari and Weigert, 1973; see Kabat et al., 1983) six of the seven somatic mutants did not differ in charged residues; thus one would have found only two bands for the seven mutant chains.

Studies of mouse V_λ chains (Eisen and Reilly, 1985) have completely changed our perspectives as to their role in the generation of diversity. Azuma et al. (1984) have one instance of a $>10^3$ fold increase in affinity per DNP-aminocaproate in chain recombination experiments involving Phe-Tyr replacement at the V-J junction of V_λ. It is not clear whether this difference results from a major conformational change in the site and hence on affinity due to alteration of V_L-V_H contacts or whether it is a direct change in the side chain contacting the antigen. Moreover, Tamoto, Reilly, Azuma and Eisen (cited as in preparation by Eisen and Reilly, 1985) have demonstrated substantial somatic mutation especially in the CDRs of V_λ chains and have calculated that between junctional diversity and somatic mutation in BALB/c mice, the number of V-region variants of just this one isotype (λ_1) is close to and probably exceeds the total number of B-cell clones in the individual mouse. It becomes clearly of major importance in the $\alpha(1\rightarrow3)\alpha(1\rightarrow6)$dextran system to sequence the λ chains as previously emphasized to account for the precipitin groups (Newman et al. 1983, 1986) and the same holds true for the anti-NP and other specificities. The chicken has but a single J_λ-C_λ sequence. A cDNA differed from the germ-line V_λ sequence in seven nucleotides giving six amino acid replacements of the 66 amino acids in the FRs and 21 nucleotides giving 11 amino acid differences of the 23 in the CDRs (Reynaud et al., 1985). Thus in chicken V_λ immunoglobulins somatic mutation also must play a predominant role.

In the oxazolone studies (Griffiths et al., 1984; Milstein, 1985; Berek et al., 1985) a large number of somatic mutants in both V_L and V_H have been obtained and on secondary immunization, a number of other germ-line genes are utilized. The problem of the role of V_κ in anti-$\alpha(1\rightarrow6)$dextran specificity could be resolved if chain recombination studies were carried out using defined V_κ mutants of anti-oxazolone with V_H of the anti-$\alpha(1\rightarrow6)$dextran and determining whether such hybrid chains reacted with dextrans. If in V_κ chains with small numbers of substitutions, the CDRs affected reactivity with dextran, this would indicate a complementarity- determining role for the germ-line V_κ. Site directed mutagenesis could also be used to introduce substitutions in the germ-line V_κOX-1 gene. These, together with the V_H genes of anti-$\alpha(1\rightarrow6)$dextran, would be cloned into an expression vector, transfected into a myeloma cell line and the specificity of the antibodies synthesized would be examined. An NMR method using deuterated L and H chains and chain recombinations can provide an evaluation of the role of each chain in binding if the hapten or antigen affects the NMR signal (Anglister et al., 1985).

TABLE 1

Amino Acids in CDRs of the V_H Chains. Physico-chemical and Immunological Properties of the Hybridoma Antibodies to $\alpha(1 \rightarrow 6)$ Dextran

	Residue	BALB/c 14.6b.1	BALB/c 26.4.1	C57BL 45.21.1
CDR1	31	Ser	Ser	Asp
	32	Tyr	Tyr	Tyr
	33	Trp	Trp	Tyr
	34	Ile	Ile	Met
	35	Glu	Glu	Asn
CDR2	50	Glu	Glu	Asp
	51	Ile	Ile	Ile
	52	Leu	Leu	Asn
	52A	Pro	Pro	Pro
	53	Gly	Gly	Asn
	54	Ser	Ser	Asn
	55	Gly	Gly	Gly
	56	SER	ASN	Gly
	57	Thr	Thr	Thr
	58	Asn	Asn	Ser
	59	Tyr	Tyr	Tyr
	60	THR	SER	Asn
	61	Glu	Glu	Gln
	62	LYS	ASN	Lys
	63	Phe	Phe	Phe
	64	Lys	Lys	Lys
	65	Gly	Gly	Gly
CDR3	95	HIS	TYR	Tyr
	96	Tyr	Tyr	Tyr
	97	Tyr	Tyr	Tyr
	98	Gly	Gly	Gly
	99	Ser	Ser	Ser
	100	Ser	Ser	Ser
	100A	SER	HIS	---[a]
	100B	Phe	Phe	---
	101	Ala	Ala	Leu
	102	Tyr	Tyr	Val
		J3	J3	J1
Reaction with Ox-BSA		-	-	-
Site Size [b]		6	6	7
Ratio IM4/IM3 [b]		7.5	14.9	6.7
K^a Dextran X 10^{-5} [b]		4.43	4.14	2.09
K^a IM7 x 10^{-4} [b]		5.76	7.01	1.78
Idiotypic Specificity[c]				
QUPC52 anti-QUPC52		0.06	0.26	0
14.6b.1 anti-QUPC52		0.72	0.54	0

[a] deletion; [b] from Sharon et al., 1982a,b;
[c] from Sharon et al., 1982.
(From Sikder et al., 1985)

```
                          1         2         3         4         5         6         7         8         9
Sequence       1........0.........0.........0.........0.........0.........0.........0.........0.........0
14.6b.1α       caggttcagctgcagcagtctggagctgagctgatgaagcctggggcctcagtgaagatatcctgcaaggctactggctacacattcagt
               GlnValGlnLeuGlnGlnSerGlyAlaGluLeuMetLysProGlyAlaSerValLysIleSerCysLysAlaThrGlyTyrThrPheSer
                                           10                            20                            30
26.4.1α        ..........................................................................................

45.21.1α       g....c........a.........c........g............................t......t....a.....g....c.
               Glu                     Pro      Val                                 Ser                Thr
pCH 105 (a)    .....c...........a......c........g.............t.......................t.......t..c....ca
                                       Pro      Val                                 Ser                Thr
B1-8    (b)    .....c..a.........c....g........tg.............t.........c.g............t..........c....cc
                                  Pro           Val                      Leu        Ser                Thr
VH-Ox1 (c)     .....g......a..g....a...c...gc...g..gc...ctcacagagcc..tcc..ca.t....ct.tct....g.ttt....a.cc
                           LysGlu      ProGly   ValAla   SerGlnSerLeuSer Thr  ThrValSer   PheSerLeuThr

               <---CDR-1---->                                             <--------------CDR-2-----------
                        1         1         1         1         1         1         1         1         1
               9        0         1         2         3         4         5         6         7         8
               1........0.........0.........0.........0.........0.........0.........0.........0.........0
14.6b.1α       agctactggatagagtgggttaagcagaggcctggacatggccttgagtggattggagagatttacctggaagtggtagtactaactac
               SerTyrTrpIleGluTrpValLysGlnArgProGlyHisGlyLeuGluTrpIleGlyGluIleLeuProGlySerGlySerThrAsnTyr
                                           40                            50     52A
26.4.1α        ....................a...................................................a..........
                                                                                   Asn
45.21.1α       ga.....ac..ga.c.....g........c.a....a.ga...................t...aat...aac.ac...g......g....
               Asp   TyrMetAsn             SerHis   LysSer                Asp   Asn   AsnAsn   Gly   Ser
pCH 105        .......at...c.c.....g.................g...................t.t....at...a..ga............t...
                     Tyr   His                      Gln                   Tyr   Tyr   ArgAsp
B1-8           ...........gc.c.....g................ga...................ag....gat...aat......g.......g...
                        MetHis                      Arg                   Arg   Asp   Asn      Gly
VH-Ox1         .....tggtg.ac.c......cgc...cct..a...a.g..t..g......c.g....ta..a.gg---.ctg....a..c..a..t..t
                     GlyValHis      Arg   Pro       Lys           Leu   Val   Trp   AlaGly

               ------------------>
               1        1         2         2         2         2         2         2         2         2
               8        9         0         1         2         3         4         5         6         7
               1........0.........0.........0.........0.........0.........0.........0.........0.........0
14.6b.1α       accgagaagttcaagggcaaggccacattcactgcagatacatcctccaacacagcctacatgcaactcagcagcctgacatctgaggac
               ThrGluLysPheLysGlyLysAlaThrPheThrAlaAspThrSerSerAsnThrAlaTyrMetGlnLeuSerSerLeuThrSerGluAsp
                60                        70                            80       82A82B82C
26.4.1α        .gt.....t.................................................................................
               Ser   Asn
45.21.1α       .a.c.............t...........g....t...c.ag.......g..............gg....c...................
               AsnGln                       Leu   Val   Lys       Ser            Gly   Arg
pCH 105        .at.......................c.g........c..........g................g.......................
               Asn                          Leu                   Ser
B1-8           .at............a...........c.g....t...c.a.c......g................g.......................
               Asn            Ser           Leu   Val   LysPro    Ser
VH-Ox1         .attc.gctc...t.tc..gactg.gca...gcaa...c.ac...aag.g.ca..tt.t.t.aa..a.g.a...t...ca.a....t...
               AsnSerAlaLeuMetSerArgLeuSerIleSerLys   Asn   LysSerThrValPheLeuLysMetAsn       GlnThrAsp

                                                         <-----------------J---------------------------
                                          <---------CDR-3------------------->
               2        2         2         3         3         3         3         3         3         3
               7        8         9         0         1         2         3         4         5         6
               1........0.........0.........0.........0.........0.........0.........0.........0.........0
14.6b.1α       tctgccgtctattactgtgcaagacattactacggtagtagctcctttt------gcttactggggccaagggactctggtcactgtctctgca    J3
               SerAlaValTyrTyrCysAlaArgHisTyrTyrGlySerSerSerPhe------AlaTyrTrpGlyGlnGlyThrLeuValThrValSerAla
                         90                         100 A  B                                110
26.4.1α        .......................t.................cat..------.......................................    J3
                                      Tyr                 His
45.21.1α       .....a..............tc.ct................------------ctagt.......ac......cac............ct..    J1
                                      Tyr                                 LeuVal       Thr      Thr             Ser
pCH 105        .....a.......t..........
                           Phe
B1-8           .....g........t.........t.cg.t...tacg....tag..acttt---.ac..............c..cactc.c..a.....ct..    J2
                                       TyrAsp   TyrGly    SerTyrPhe---Asp                ThrLeu           Ser
VH-Ox1         a.a...a.g..c........c...g..cggggg-------------------.......................................    J3
               Thr   Met               AspArgGly
```

Fig. 4 Nucleotide sequences of V_H cDNA from anti-α(1→6)dextrans and amino acids deduced from them. (From Sikder et al., 1985). Reproduced with permission of the Journal of Immunology.

The sequences of the three V_H chains (Fig. 4) lead to several interesting inferences. The two BALB/c V_H chains differ only at three residues in CDR2 and two in CDR3. Since their V_κ chains are identical throughout V and J, differences in size and specificity of their combining sites must be related to these differences. Table I summarizes these findings as well as the biological properties. The two BALB/c anti-α(1→6)dextran hybridomas have groove-type sites complementary to an internal chain of six glucoses, their binding constants are within experimental error assayed both with dextran and with isomaltoheptaose. They differ in the ratio of inhibitory power of IM4/IM3 but the significance of this difference is not clear. However, they show a significant difference in idiotypic specificity with hetero anti-idiotypic serum to QUPC52. Since most anti-idiotypic sera detect differences in the D-minigene, some of the differences in CDR3 are most likely to be responsible. It should be noted that there is no a priori necessity for every change in a CDR to affect site specificity. Indeed, Radbruch et al. (1985) have isolated a somatic mutant β1-881-V3 in which Arg replaces Gly at sequential amino acid 103 (residue 99 in Kabat et al., 1983) in CDR3(D) with no change in antigen binding, heterocliticity or fine structure of the binding site but with complete loss of idiotypic specificity as tested with an antibody to a hapten-inhibitable idiotype Ac 146.

The most striking finding is seen with V_H of the C57BL/6 mouse (Table I and Fig. 4). It is clear that a very different germ-line gene is used. There are substantial numbers of differences from the two BALB/c sequences in the CDRs including a deletion of two residues in CDR3. Yet the differences in site specificity of all three anti-α(1→6)dextrans are generally small relative to the number of contacting residues that one would expect for groove-type sites complementary to determinants of six and seven glucoses. Only one of five residues in CDR1, eight of 17 in CDR2, and five of seven in CDR3 excluding the deletion of two amino acids are identical with those of BALB/c. Thus the sites are built of very different amino acids. It is possible that the site complementarity is determined by these identical residues with the non-identical residues not influencing shape or complementarity. This would be analogous to the role of V_L residues 94,95 and 96 of the three different V_κ chains being involved in the complementarity of anti-phosphorylcholine antibodies (Rudikoff, 1983). However since anti-α(1→6)-dextran sites are much larger, it is worth considering that very similar sites may be formed by different amino acid side chains. Sequencing of the remaining anti-α(1→6)dextrans may resolve this question.

IV EFFECTS OF AMINO ACID AND CHAIN SUBSTITUTIONS ON SPECIFICITY

Despite the enormous amount of sequence variation in immunoglobulins, chance has permitted a series of simple correlations of structure and antibody or idiotypic specificity. There are several examples of an identifiable antibody or idiotypic change ascribable to a one or a few amino acid substitutions. The high degree of sequence identity among the phosphorylcholine-binding monoclonal antibodies has shown that as little as a single amino acid difference can make for an entirely different specificity. Thus, the U4 mutant of the phosphorylcholine-binding myeloma S107, which differs in CDR1 (residues 31-35) in that Glu 35 had been replaced by Ala, no longer bound phosphorylcholine but had acquired the capacity to bind double-stranded DNA, phosphorylated protamine and cardiolipin (Diamond and Scharff, 1984); the charged Glu, however, does not appear to be specificity-determining since another monoclonal DNA-binding protein has Asn at position 35 (Eilat et al., 1984). Glu 35 in McPC603 is a key residue in the crystal binding site for phosphorylcholine and 61 phosphorylcholine binding proteins have Glu 35. Yet five monoclonal proteins with Ser 35, one each with Ile and Gly replacing Met 34 and one with Ala 31 instead of Asp, all of the V_H III subgroup, were found to bind phosphorylcholine. In addition, two phosphorylcholine-binding V_HII mouse antibodies had Gly-Tyr-Ile-Thr-Asn in CDR1 again suggesting, unless CDR1 is not important to specificity, that similar combining sites may be formed by very different germ-line genes and with different amino acids in the CDRs. Unfortunately, most of the light chains have not been sequenced beyond residues 35 to 40 so that correlations of the sequence differences of V_H in CDR1 with V_L such as that noted above cannot be made. Many V_H chains have also not been sequenced beyond residue 40. The information derived from such partial sequences is so limited that one wonders why referees accept them and why journals publish them. Thus, there are no data on residues 94, 95 and 96 of V_L for any of the five phosphorylcholine binding hybridomas with Ser 35 replacing Glu 35.

Chain recombination studies between anti-arsonate (93G.7) and anti-2-phenyloxazolone (NQ5/89.4) hybridomas (Jeske et al., 1984) which share the same germ-line V_κ gene indicated that a single substitution at amino acid 96, the site of V→J joining could abolish the arsonate cross reacting idiotype as well as the anti-arsonate activity; although the anti-2-phenyloxazolone binding was abolished, the recombined control anti-oxazolone chains had 30 percent less activity perhaps ascribable to incomplete H-L assembly. NQ5/89.4 is a very different germ-line gene from V_{κ}-Ox1 in Fig. 3.

That a single amino acid substitution in the D-minigene can abolish idiotypic specificity without affecting antibody specificity was elegantly demonstrated (Radbruch et al., 1985); a somatic mutant was selected by cell sorting with anti-idiotypic sera from a hybridoma cell line B1-8.δ1 which produced a heteroclitic anti-4-hydroxy-3-nitro-5-iodo-phenylacetyl (NIP) of high affinity. One anti-idiotypic serum Ac38 defining a determinant outside the antibody combining site was selected against and the other Ac146 defining an idiotype in the binding site was selected for. The somatic mutant B1-8δ1-V3 had a single amino acid substitution of Arg for Gly at position 99 in the D-region (Kabat et al., 1983) (sequential position 103), and had lost its Ac146 idiotypic specificity but retained its antibody activity and its heterocliticity. Position 99 in the D-region is a surface residue (Padlan, 1977) and thus is consistent with the hypothesis that the side chains of idiotypic determinants are on the exterior of the antibody combining site (Kabat 1984).

In the anti-dextran B1355S with alternating $\alpha(1\to3)\alpha(1\to6)$ linkages two hetero anti-idiotypic rabbit sera have been used, anti-IdI MOPC104E and anti-IdI J558 which recognize different determinants. Both of these determinants correlate very well at positions 96 and 97 of the D minigene (Schilling et al., 1980; Clevinger et al., 1980.). Newman et al. (1983) have found that the various monoclonal anti-$\alpha(1\to3)\alpha(1\to6)$dextrans fall into five precipitin groups depending on the extent of cross reactivity with dextrans not having the alternating $\alpha(1\to3)\alpha(1\to6)$ structure. Of special interest is the finding that the IdI J558 found in all five precipitin groups at residues 96 and 97 is associated strongly with Arg-Tyr and less strongly with Asn-Tyr. The IdI MOPC104E, however, is restricted only to precipitin group 5 and is associated exclusively with Tyr-Asp (YD) at these positions also in Ac146 (Cumano and Rajewsky, 1985). A cross reacting idiotype associated with CDR2, residues 54 and 55 having Asn-Asn, is not found in one hybridoma with Lys-Lys. It would be of substantial interest to attempt to obtain anti-idiotypic sera to those monoclonals with other amino acids at positions 96 and 97.

Two distinct antibody specificities anti-GAT, (F17.170.200) with its cross reacting common (public) anti-idiotype (anti-pGAT) and anti-NP^b were found to use the same germ-line V_H gene (V186.2) the former with V_κ and the latter with V_λ chains (Rocca-Serra et al., 1983). The V_H regions (excluding D and J) were identical in amino acid sequence except for a Ser→Asn substitution at residue 60 in CDR2. The D and J sequences associated with V186.2 in this instance are not known; since the D minigene is closely identified with many idiotypic specificities the respective contributions of the κ and λ chains to the idiotype is difficult to assess.

Reth et al. (1981) had inferred by comparison of the NP sequences with those of MOPC 104E that His 35, Arg 50, Asp 52 and Tyr 99 were contacting residues for NP binding; these amino acids were also found in the translated nucleotides of an NP^a BALB/c expressed gene (Loh et al., 1983; Bothwell, 1984). Another comparison of B1-8, B1-48, and 17.2.25 with the three anti-α-(1→6)dextran V_H sequences (Fig 4) shows that in CDR2 six residues 51, 52A, 55, 57, 59 and 63 had identical amino acids in all instances. Since the determinants of the anti-NP^b and anti-NP^a are also very different structurally from anti-(α1→6)dextrans these residues are probably not contacting.

An important insight was gained by the study of eight monoclonal anti-NP antibodies from C57BL/6 mice, six of which had been suppressed by anti-idiotype administered at birth (Cumano and Rajewsky, 1985). All eight antibodies had the V_H 186.2 and seven had the DFl16.1 minigene (Kurosawa and Tonegawa, 1982); six used J_H2 and two J_H4. There are extensive variations at the V-D junction and residue 95 (sequential 99), the first residue of CDR3 appears to involve addition of N sequences (Alt and Baltimore, 1982; Desiderio et al., 1984) between V_H and D. Four of the six anti-idiotype suppressed mice have Tyr replacing Asp at position 96 (sequential 100), one has a deletion of residue 96 and one has Cys 96 followed by J with the remaining six residues of D missing; the sixth has Asn 96, a point mutation giving Ile at residue 101 and a deletion of the following two residues. The substitution at position 96 of Tyr, Cys and Asn for Asp would explain the ability of the idiotype suppressed mice to escape suppression, to make a functional antibody and provides further evidence for association of the anti-NP idiotype with the D minigene. An antibody produced by one, N1C12, of the two normal mice however, also had a deletion of residue 95, a Tyr at residue 96, a deletion of the last residue of D and the first residue of J; it, however, lacked the Ac146 inhibitable idiotope.

The sequences of the four of the $V_{\lambda}1$ chains from suppressed mice beginning at amino acid residues 45,48,50 and 57 to the end of CDR3 and of the J_{λ} minigenes were determined and were identical to B1-8V1 and $J_{\lambda}1$ corresponding to the germ-line sequences (Cumano and Rajewsky, 1985).

Additional sequence data on anti-NP^a (BALB/c) (Boersch-Supan, 1985; Bothwell, 1984) and anti-NP^b (C57B1) monoclonals have shown that V_H of the NP^b family can be divided into six subgroups defined serologically; groups V and VI differing from each other and from groups I to IV. The nucleotide sequences of the three BALB/c anti-NP^a, those of group VI and of group V differed among themselves by two to four amino acid residues from -19 to sequential residue 98, but differed from one another by six to nine and by 10 to 13 amino acids (see Dildrop, 1984; Brodeur and Riblet, 1984). Thus they are the products of distinct germ-line V_H genes as are the BALB/c and C57BL/6 anti-α-(1→6)dextran hybridomas (Fig. 4 and Table I).

Fougereau et al. (Roth et al., 1985; Fougereau et al., 1985) have examined nucleotide sequences in the idiotypic cascade Ab1→Ab2→AB3 (Jerne, 1974; Jerne et al., 1982; Urbain et al., 1981) in BALB/c mice employing the pGAT system (pGAT=public eg cross reacting antibody defined by hetero guinea pig or rabbit antisera), Ab1 being anti-GAT (polyGlu-^{60}Ala^{30}Tyr^{10}), Ab2 anti-idiotype and Ab3 anti-anti-idiotype. The population of Ab3 hybridomas obtained fell into two groups termed Ab1' and Ab3, the former reacting with GAT and having the pGAT idiotype and Ab3 the anti-anti-idiotype (anti-Ab2) which reacts with Ab2 which is considered to be the internal image and hypothesized (Jerne, 1974) to have similarities to the antigen. Two anti-idiotopes define two pGAT idiotopes termed Id20 and Id22.

Nucleotide sequences of six Ab1' V_H chains beginning at codon 29 were largely identical through the end of FR3, except for one chain having four amino acid substitutions in CDR2; all used J4 but the D-minigene segments of Ab1 and Ab1' differed both in sequence and in length. Four μ and one γ chains from two fusions had V sequences identical to two Ab1, but other Ab1s had more amino acid substitutions largely in CDR2 than did the Ab1's. The V_H sequence of one Ab3 from codon 29 to the end of FR3 was very different only 23 of 58 amino acids being identical; the Ab3 and the Ab1' were all members of the same V_H subgroup. The D minigene was very unusual, there being a substantial number of Gs indicating the presence of an N region, (Alt and Baltimore, 1982), the translated sequence being Gly-Gly-Tyr-Gly-Gly-Val. Segments of the various Ab1' D minigenes resemble the germ-line, DSP.2, DFl.16 and DQ.52. (Kurosara and Tonegawa, 1982)

Partial sequences of five Ab1' V_κ chains were identical from amino acid 35 through the end of FR3 and showed very few differences, they all used $J_\kappa 2$, whereas two Ab1's each used a different J_κ. The Ab3 differed from Ab1 and Ab1' in about 40 per cent of the amino acids from residues 35 through 95.

Fougereau et al. (1985) have called attention to the presence in the D region of two monoclonal Ab2s, 2-22 and 2-20, of Tyr-Tyr-Glu and Glu-Glu-Tyr which would be expected to occur in the GAT antigen and which they thought might be a internal image of an antigenic determinant. One wonders what the internal image might be when the antigenic determinant is a polysaccharide, nucleotide or introduced haptenic group. Unfortunately, with the interest on D and J minigenes in this system, investigators are again content with incomplete sequences, often not sequencing FR1 and CDR1.

V ANTIBODY DIVERSITY AND ANTIBODY COMPLEMENTARITY

Cloning and sequencing of V_L and V_H chains of immunoglobulins has provided many sequences of germ-line and expressed immunoglobulin chains and has established new mechanisms for the generation of the protein diversity of antibody combining sites. Immunoglobulin genes are unique in that the expressed genes are formed by the joining of V_L to one of the J_L minigenes and by V_H to one each of the D_H and J_H minigenes. The number of expressed clonotypes in the primary IgM repertoire of an inbred mouse is estimated to be $>10^7$ (Owen et al., 1985). To generate so large a number of specificities a considerable number of germ-line V_H genes each may combine with one of an undetermined but small number of D_H minigenes and one of four J_H genes to give an expressed V_H-gene and a considerable number of V_κ germ-line genes may join to one of four functional J_κ genes to give an expressed V_κ-gene. There are two germ-line V_λ genes, $V_\lambda 1$ and $V_\lambda 2$ and four J_λ minigenes associated as two separate clusters; $V_\lambda 1$ may combine with $J_\lambda 1$ or $J_\lambda 3$, but $V_\lambda 2$ thus far associates only with $J_\lambda 2$ (Eisen and Reilly, 1985). If there is completely random assortment of V_H,D_H and J_H, as well as of V_κ and J_κ and V_λ with J_λ, one could multiply the number of each to estimate the upper limits for the numbers of expressed genes. If any V_H and V_L can assemble one would increase the number of V-regions by the sum of the products of $V_H \cdot V_\kappa$ and $V_H \cdot V_\lambda$ (combinational diversity). Somatic mutation has clearly been recognized as playing an important part in further expanding the repertoire (Weigert et al., 1970,1976, Cohn et al., 1974; Baltimore, 1981a). Moreover, in V-D-J and V-J joining, the site of joining of the segments may vary (junctional diversity) and in many instances there are added nucleotides at the V-D and D-J junctions, termed N sequences which may affect the length of D and the translated amino acids; N sequences are characterized by high usage of Gs. Fig 5 shows the effects on the length of the D minigene of N sequences in anti-Ars binding antibodies using the V_H cross-reacting idiotype (Manser et al., 1985).

There is very little evidence on the functionality of these various chain combinations. Admittedly they generate diversity but how many of them generate functional antibody combining sites? The original minigene concept based on assortment of FR and CDR segments of light and heavy chains (Kabat et al., 1978,1979,1980) proposed that all FR and CDR segments were coded for by minigenes and that assortment of CDRs could generate much additional diversity. This was confirmed by the finding of the J segments in V_L and D and J segments in V_H; similar minigenes exist in the T-cell receptor. However, the finding that germ-line V_H genes coded only up to the end of FR3 and that V_L genes coded up to the last one or two residues in CDR3 indicated that the assortment data were recognizing a different type of phenomenon than was seen in D and J. Egel (1981), Baltimore (1981b) and Gough (1982) proposed that gene conversion could account for these data, the latter two emphasizing that the conversion occurred during evolution. However, Dildrop et al. (1982) and Krawinkel et al. (1982) sequenced a somatic mutant which was a double recombinant in which CDR1 and CDR2 of one germ-line gene were joined to

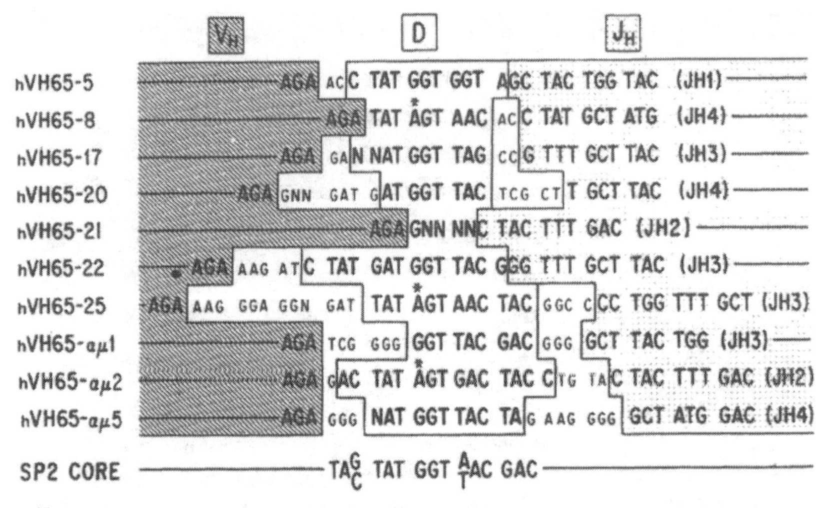

Fig. 5 Nucleotide sequences of the V_H-D-J_H region in the expressed V_H region genes of preimmune hybridomas that utilize the $V_H Id^{CR}$ gene segment. The nucleotide sequences are compared, beginning with the 3' terminal AGA codon derived from the $V_H Id^{CR}$ gene segment, to the "core" sequence of the germ-line D gene segment family of BALB/c termed SP2 and the consensus D sequence found in the expressed V_H genes of hybridomas that produce Id^{CR}-bearing, Ars-binding antibodies. Sequences have been aligned to produce the maximum homology with the SP2 core sequence. Nucleotides that appear to be derived from germ-line V gene segments are boxed. Nucleotides that cannot be accounted for by the sequences of the putative germ-line gene segment that were fused to form these functional V_H genes are shown in small lettering. Nucleotides that could not be unambiguously identified are indicated by "N." Three nucleotides that may represent a G to A polymorphism between some BALB/c and A/J SP2 D gene segments are indicated with an overlying asterisk. (From Manser et al.,). Science, 226:1283-1288, 1984. c by the AAAS.

CDR3 of another; the mechanism is not established; it could be by recombination or by gene conversion.

The controversy about minigenes other than D and J, including the data from blotting by Komoromy and Wall (1981) was nicely resolved by the Zachau group (Jaenischen et al., 1984) who sequenced a number of human $V_κ I$ genes and found a series of chains in which assortment of FR or CDR segments occurred. However, the boundaries of the assorting segments did not correspond to the FR or CDR boundaries; they could further demonstrate that the translated amino acid sequences would show the assortment which we ascribed to minigenes. The sequence differences were considered to be due to other mechanisms such as gene conversion which could have generated increased diversity of germ-line genes in evolution. Clarke and Rudikoff (1984) have also ascribed to gene conversion short sequences of amino acids in the phosphorylcholine system. Whether such events could occur somatically, as well, deserves further study. Baltimore and Loh (1983) in searching for mechanisms underlying gene conversion noted that the frequency of inverted repeats, eg. ATGCGA and TCGCAT, showed a complementary relationship to variablility (Wu and Kabat, 1970) in V_H (Kabat and Wu, 1971).

In so complex a system as the immunoglobulin V-regions, for which one can obtain only a minute portion of the sequence data, one must pay attention to unusual or unexpected relationships occurring at low frequencies since they may indicate mechanisms for the generation of diversity and for site complementarity, which may be occuring to a considerable extent.

One important finding needing further evaluation is that the human D-2 minigenes detected by probes with recognition and spacer sequences, (Siebenlist et al., 1981) were found to code for 14 nucleotides in CDR2 (Wu and Kabat, 1982) of a human V_HIII germ-line gene rather than in CDR3 as would have been expected; subsequently stretches of 13 (Rechavi et al., 1983) 12 (Loh et al., 1983) and 14 (K.E. Bernstein and R. Mage, personal communication, 1983) of the human D2 minigene matched a human V_HII genomic clone and mouse and rabbit cDNA. Eight nucleotides of a human D1 minigene matched a rabbit cDNA, also in CDR2 (Bernstein et al., 1982). Takahashi et al. (1984) have described a rearranged immunoglobulin pseudogene with all of CDR2 deleted. These findings were all made before the sequencing of the T-cell receptor whose $V_T\alpha$, $V_T\beta$, $D_T\alpha$, $D_T\beta$, $J_1\alpha$ and $J_T\beta$ chains have the same recognition and spacer sequences as the human V_H, D_H, J_H, V_L, J_L segments. (Hedrick et al., 1984; Yanagi et al., 1984; Patten et al., 1984). Thus, the use of probes containing these sequences may be detecting many other instances of this joining mechanism not only in the immunoglobulin superfamily but possibly throughout the genome.

Our finding (Chen et al., 1985) that the complementary strands of the D2 and D4 human minigenes coded for a portion of CDR1 in the V_κ chains provides support for the hypothesis of Ohno et al. (1982) that the heavy chain evolved from a 48 base pair segment of the complementary strand of a primordial light chain. These findings also raise the question of movement of small segments of DNA into regions other than CDR3 of V_H.

The extraordinary preservation of FR2 of V_κ at the amino acid level in one human, 25 mouse and 15 rabbit chains despite extensive variation in the rest of the V-regions (Kabat et al., 1979, 1980,1983 and additional data) and the finding in both mouse and rabbit of infrequently occurring V_κ chains in which FR2 may have from one to five substitutions requires more understanding. A similar preservation of FR2 of V_H in both mouse and rabbit was seen at the amino acid as well as at the nucleotide level in that identity of codons for amino acids 35 through 47 with one and two nucleotide substitutions of codons 48 and 49 were seen. This is the most preserved FR or CDR segment. Since all these FR2 segments in V_H and V_L were from expressed functional chains, one wonders why one chain is used with such high frequency and why a more uniform equilibration of these FR2s has not been arrived at after 80 million years. Yet another finding which could have much structural significance is that in two instances, a monoclonal cold agglutinin (Kobzik et al., 1976) and a mouse monoclonal autoantibody to thyroglobulin (Zanetti et al., 1985), the same idiotype was expressed on the separated heavy and light chains. There is clearly much more to be explained before we arrive at an understanding of the antibody diversity and antibody complementarity.

ACKNOWLEDGEMENTS

Work in the laboratories is suppported by Grants PCM 81-02321 from the National Science Foundation and 1R01 AI-19042 from the National Institute of Allergy and Infectious Diseases to E.A.K.; by Cancer Center Support Grant CA 13696 to Columbia

University, and Program Project Grant CA 21112 to Dr. Elliot F. Osserman. Work with the PROPHET computer system is supported by the National Cancer Institute, National Institute of Allergy and Infectious Diseases, National Institute of Arthritis, Diabetes and Digestive and Kidney Diseases, the National Institute of General Medical Sciences, and the Division of Research Resources (Contract NO1-RR-8-2118) of the National Institutes of Health.

We thank Drs. David R. Davies, Rose G. Mage and Richard Parker for helpful suggestions and Darryl J. Guinyard for his excellent typing of the manuscript.

REFERENCES

Alt, F. and Baltimore, D. (1982). Joining of immunoglobulin heavy chain gene segments: Implications from a chromosome with evidence of three D-J_H fusions. Proc. Natl. Acad. Sci. 79:4118-4122.

Amit, A.G., Boulot, G., Comarmond, M.B., Harper, M., Mariuzza, R.A., Phillips, S.E., Saludjian, P., Saul, F.A., Conger, J.D. and Nisonoff, A. (1985). X-Ray diffraction studies of an anti-azophenylarsonate antibody and of an antigen-antibody complex. Ann. Inst. Pasteur Immunol. 136:121-129.

Amit, A.G., Mariuzza, R.A., Phillips, S.E.V. and Poljak, R.J. (1984). Three-dimensional structure of an antigen-antibody complex at 6 Å resolution. Nature 313:156-158.

Amzel, L.M., Poljak, R.J., Saul, F., Varga, J.M. and Richards, F.F. (1974). The three-dimensional structure of a combining region-ligand complex of immunoglobulin New at 3.5 Å resolution. Proc. Nat. Acad. Sci. 71:1427-1430.

Anglister, J., Frey, T. and McConnell, H.M. (1985). NMR technique for assessing contributions of heavy and lightchains to an antibody combining site. Nature 315:65-67.

Arakatsu, Y., Ashwell, G. and Kabat, E.A. (1966). Immunochemical studies on Dextrans. V. Specificity and cross-reactivity with dextrans of the antibodies formed in rabbits to isomaltonic and isomaltotrionic acids coupled to bovine serum albumin. J. Immunol. 97:858-866.

Arnon, R. (1977). Immunochemistry of lysozyme. in: "Immunochemistry of enzymes and their antibodies". Salton, M.R.J., ed. New York: Wiley p. 1-28.

Atassi, M.Z. (1975). Antigenic struture of myoglobin: The complete immunochemical anatomy of a protein and conclusions relating to antigenic strutures of proteins. Immunochemistry 12:423-438.

Azuma, T., Igras, V., Reilly, E.B. and Eisen, H.N. (1984). Diversity at the variable-joining region boundary of λ light chains has a pronounced effect on immunoglobulin ligand-binding activity. Proc. Natl. Acad. Sci. 81:6139-6143.

Baltimore, D. (1981a.). Somatic mutation gains its place among the generators of diversity. Cell 26:295-296.

Baltimore, D. (1981b.). Gene conversion: Some implications for immunoglobulin genes. Cell 24:592-594.

Baltimore, D. and Loh, D.Y. (1983). Evolution of heavy-chain variable-gene segments. Prog. in Immunology 5:115-121.

Becker, J.W. and Reeke, G.N. (1985). Three-dimensional structure of β_2-microglobulin. Proc. Natl. Acad. Sci. 82:4225-4229.

Benjamin, D.C., Berzofsky, J.A., East, I.J., Gurd, F.R.N., Hannum, C., Leach, S.J., Margoliash, E., Michael, J.G., Miller, A., Prager, E.M., Reichlin, M., Sercarz, E.E., Smith-Gill, S.J., Todd, P.E. and Wilson, A.C. (1984). The antigenic structure of proteins. A reappraisal. Ann. Rev. Immunol. 2:67-101.

Berek, C., Griffiths, G.M. and Milstein, C. (1985). Molecular events during maturation of the immune response to oxazolone. Nature 316:412-418.

Bernstein, K.E., Reddy, E.P., Alexander, C.B. and Mage, R.G. (1982). A cDNA sequence encoding a rabbit heavy chain variable region of the V_Ha2 allotype showing homologies with human heavy chain sequences. Nature 300:74-76.

Berzofsky, J.A. (1985). Intrinsic and extrinsic factors in protein antigenic structure. Science 229:932-940.

Bhattachargee, A.K., Das, M.K., Roy, A. and Glaudemans, C.P.J. (1981). The binding sites of the two monoclonal immunoglobulins as J539 and W3129. Thermodynamic mapping of a groove- and a cavity-type immunoglobulin, both having antipolysaccharide specificity. Mol. Immunol. 18:230-277.

Boersch-Supan, M.E., Agarwal, S., White-Scharf, M.E. and Imanishi-Kari, T. (1985). Heavy chain variable region. Multiple gene segments encode anti-4-(hydroxy-3-nitrophenyl)acetyl idiotypic antibodies. J. Exp. Med. 161:1272-1292.

Bothwell, A.L.M., Paskind, M., Reth, M., Imanishi-Kari, T., Rajewsky, K. and Baltimore, D. (1981). Heavy chain variable region contribution to the NP^b family of antibodies: Somatic mutations evident in a λ2a variable region. Cell 24:625-637.

Bothwell, A.L.M. (1984). The genes encoding anti-NP antibodies in inbred strains of mice. in: "The Biology of Idiotypes". M.I. Greene and A. Nisonoff, eds. Plenum Publishing Corp. p. 19-34.

Braun, D.G., Eichmann, K. and Krause, R.M. (1969). Rabbit antibodies to streptococcal carbohydrates. Influence of primary and secondary immunization and possible genetic factors on the antibody response. J. Exp. Med. 129:809-830.

Brodeur and Riblet (1984). The immunoglobulin heavy chain variable region (Igh-V) locus in the mouse I. One hundred Igh-V genes comprise seven families of homologous genes. Eur. J. Immunol. 14:922-930.

Capra, J.D., Slaughter, C., Milner, E.C.B., Estess, P. and Tucker, P.W. (1982). The cross reactive idiotype of A-strain mice. Serological and structural analysis. Immunology Today 3:332-339.

Cesari, I.M. and Weigert, M. (1973). Mouse lambda-chain sequences. Proc. Natl. Acad. Sci. 70:2112-2116.

Chang, C., Short, M.T., Westholm, F.A., Stevens, F.J., Wang, B., Furey, W., Solomon, A. and Schiffer, M. (1985). A novel arrangement of immunoglobulin variable domains: X-ray crystallographic analysis of the λ chain dimer, Bence-Jones protein Loc. Biochemistry 24:4890-4897.

Chen, H. and Kabat, E.A. (1985). Immunochemical studies on blood groups LXXIII: The combining site specificities of mouse monoclonal hybridomas anti-A and anti-B. J. Biol. Chem. 260:13208-13217.

Chen, P.P., Kabat, E.A., Wu, T.T., Fong, S. and Carson, D.A. (1985). Possible involvement of human D minigenes in the first complementarity-determining region of κ light chains. Proc. Natl. Acad. Sci. 82:2125-2127.

Cisar, J., Kabat, E.A., Dorner, M. and Liao, J. (1975). Binding properties of immunoglobulin combining sites specific for terminal or non-terminal antigenic determinants in dextran. J. Exp. Med. 142:435-459.

Clarke, S.H., Hüppi, K., Ruezinsky, D., Staudt, L., Gerhard, W. and Weigert, M. (1985). Inter- and intraclonal diversity in the antibody response to influenza hemagglutinin. J. Exp. Med. 161:687-704.

Clarke, S.H. and Rudikoff, S. (1984). Evidence for gene conversion among immunoglobulin heavy chain variable region genes. J. Exp. Med. 159:773-782.

Clevinger, B., Schilling, J., Hood, L. and Davie, J.M. (1980). Structural correlates of cross-reactive and individual idiotypic determinants on murine antibodies to α-(1→3) dextran. J. Exp. Med. 151:1059-1070.

Cohn, M. (1967). Natural history of the myeloma. Cold Spring Harbor Symposia on Quantitative Biol. 32:211-222.

Cohn, M., Blomberg, B., Geckeler, W., Raschke, W., Riblet, R. and Weigert, M. (1974). First order considerations in analyzing the generator of diversity. in: "The Immune System: Genes, Receptors, Signals". in: ICN-UCLA Symposium on Molecular Biology. Acad. Press. New York pp.89-117.

Coleman, P.M., Varghese, J.N., Laver, W.G. (1983). Structure of the catalytic and antigenic sites in influenza virus neuraminidase. Nature 303:41-44.

Crumpton, M.J. (1974). Protein antigens: The molecular bases of antigenicity and immunogenicity. in: "The Antigens", Sela, M. ed. Academic New York: II:1-78.

Cumano, A. and Rajewsky, K. (1985). Structure of primary anti-(4-hydroxy-3-nitrophenyl)acetyl (NP) antibodies in normal and idiotypically suppressed C57BL/6 mice. Eur. J. Immunol. 15:512-520.

Darsley, M.J. and Rees, A.R. (1985). Nucleotide sequences of five anti-lysozyme monoclonal antibodies. EMBO J. 4:393-398.

Davies, D.R. and Metzger, H. (1983). Structural basis of antibody function. Ann. Rev. Immunol. 1:87-117.

Davies, D.R.,Padlan, E.A. and Segal, D.M. (1975). Three-dimensional structures of immunoglobulins. Ann. Rev. Biochem. 44:639-667.

Desiderio, S.V., Yancopoulos, G.D., Paskind, M., Thomas, E., Boss, M.A., Landau, N., Alt, F.W. and Baltimore, D. (1984). Insertion of N regions into heavy-chain genes iscorrelated with expression of terminal deoxytransferase in B cells. Nature 311:752-755.

D'Hoostelaere, L. and Potter, M. (1982). Genetics of the α1,6-dextran response: Expression of the QUPC idiotype in different inbred and congenic strains of mice. J. Immunol. 128:492-497.

Diamond, B. and Scharff, M.D. (1984). Somatic mutation of the T15 heavy chain gives rise to an antibody with autoantibody specificity. Proc. Natl. Acad. Sci. 81:5841-5844.

Dighiero, G., Lymberi, P., Holmberg, D., Lundquist, I., Coutinho, A. and Avrameas, S. (1985). High frequency of natural autoantibodies in normal newborn mice. J. Immunol. 134:765-771.

Dildrop, R. (1984). A new classification of mouse V_H sequences. Immunology Today 5:85-86.

Dildrop, R., Bovens, J., Siekevitz, M., Rajewsky, K. and Beyreuther, K. (1982). Immunoglobulin V-region variants in hybridoma cells II. Recombination between genes. EMBO J. 1:635-640.

Doyen, N., Lapresle, C., Lafaye, P. and Mazie, J.C. (1985). Study of the antigenic structure of human serum albumin with monoclonal antibodies. Mol. Immnol. 22:1-10.

Early, P., Huang, H., Davis, M., Calame, K. and Hood, L. (1980). An immunoglobulin heavy chain variable region gene is generated from three segments of DNA: V_H, D, and J_H. Cell 19:981-992.

Edelman, G.M. (1970). The covalent structure of a human γG-immunoglobulin XI. Functional implications. Biochemistry 9:3197-3205.

Edelman, G.M., Cunningham, B.A., Gall, W.E. Gottlieb, P.D., Rutishauser, V. and Waxdal, M.J. (1968). The covalent structure of an entire γG immunoglobulin molecule. Proc. Natl. Acad. Sci. 63:78-85.

Edmundson, A.B., Ely, K.R., Girling, R.L., Abola, E.E., Schiffer, M., Westholm, F.A., Fausch, M.D. and Deutsch, H.F. (1974). Binding of 2,4-dinitrophenyl compounds and other small molecules to a crystalline λ-type Bence-Jones dimer. Biochem. 13:3816-3827.

Edmundson, A.B., Ely, K.R. and Herron, J.N. (1984). A search for site-filling ligands in the Mcg Bence-Jones dimer: Crystal binding studies of fluorescent compounds. Mol. Immunol. 21:561-576.

Egel, R. (1981). Intergenic conversion and reiterated genes. Nature 290:191-192.

Eilat, D., Hochberg, M., Pumphrey, J. and Rudikoff, S. (1984). Monoclonal antibodies to DNA and RNA from NZB/NZW F_1 mice: Antigenic specificities and NH_2 terminal amino acid sequences. J. Immunol. 133:489-494.

Eisen, H.N., Little, J.R., Osterland, C.K. and Simms, E.S. (1967). Cold Spring Harbor Symposia Quant. Biol. 32:75-81.

Eisen, H.N. and Reilly, E.B. (1985). Lambda chains and genes in inbred mice. Ann. Rev. Immunol. 3:337-365.

Ely, K.R., Peabody, D.S., Holm, T.R., Cheson, B.D. and Edmundson, A.B. (1985a). Accessible intrachain disulfide bonds in hybrids of light chains. Mol. Immunol. 22:85-92.
Ely, K.R., Wood, M.K., Rajan, S.S., Hodsdon, J.M., Abola, E.E., Deutsch, H.F. and Edmundson, A.B. (1985b). Unexpected similarities in the crystal structures of the Mcg light-chain dimer and its hybrid with the Weir protein. Mol. Immunol. 22:93-100.
Emmrich, F., Bundle, D., Zee, J., Out, T., Zenke, G. and Eichmann, K. (1985). Two human IgM myeloma proteins with unusual specificities for streptococcal carbohydrate-associated epitopes. Scand. J. Immunol. 21:119-126.
Fan, S.T. and Karush, F. (1984). Restriction IgM expression--VI. Affinity analysis of monoclonal anti-dansyl antibodies Mol. Immunol. 21:1023-1029.
Fazekas de St. Groth, S. and Webster, R.G. (1966). Disquisitions on original antigenic sin. I. Evidence in Man; II. Proof in lower vertebrates. J. Exp. Med. 124:331-345; 347-361.
Feizi, T., Kabat, E.A., Vicari, G., Anderson, B. and Marsh, W.L. (1971). Immunochemical studies on blood groups XLIX. The I antigen complex: Specificity differences among anti-I sera revealed by quantitative precipitin studies; Partial structure of the I determinant specific for one anti-I serum. J. Immunol. 106:1578-1592.
Fink, P.C. and Galanos, C. (1985). Serum anti-lipid A antibodies in multiple myeloma and Waldenstrom's macroglobulinaemia. Immunol. 169:1-10.
Finne, J., Leinonen, M. and Mäkelä, P.H. (1983). Antigenic similarities between brain components and bacteria causing meningitis. The Lancet pp. 355-357.
Firca, J.R., Ely, K.R., Kremser, P., Westholm, F.A., Dorrington, K.J. and Edmundson, A.B. (1978). Interconversion of conformational isomers of light chains in the Mcg immunoglobulins. Biochemistry 17:148-158
Fougereau, M., Corbet, S., Ollier, P., Rocca-Serra, J., Roth, C., Schiff, C., Somme, G., Theze, J. and Tonnelle, C. (1985). The internal image and the structural idiotypic network (Ab1, Ab2, Ab3) in the GAT system. Ann. Inst. Pasteur/Immunol. 136:143-156.
Freedman, M., Merrett, R. and Pruzanski, W. (1976). Human monoclonal immunoglobulins with antibody-like activity. Immunochemistry 13:193-202.
Gibson, A.L., Herron, J.N., Ballard, D.W., Voss, E.W., He, X.M., Patrick, V.A. and Edmundson, A.B. (1985). Crystallographic characterization of the Fab fragment of a monoclonal anti-ss-DNA antibody. Mol. Immunol. 22:499-502.
Gibson, D.M. (1984). Evidence for 65 electrophoretically distinct groups of light chains in BALB/c and NZB myelomas. Mol. Immunol. 21:421-432.
Glaudemans, C.P.J. and Kovac, P. (1985). Probing thecombining site of monoclonal IgA J539 using deoxyfluoro- and other galactosides as ligands. Mol. Immunol. 22:651-653.
Gooi, H.C., Hounsell, E.F., Picard, J.K., Lowe, A., Voak, D., Lennox, E. and Feizi, T. (1985). J. Biol. Chem. 260:13218-13224.
Gough, N. (1982). Gene conversion and the generation of antibody diversity. TIBS. 7:307-308.

Greenspan, N.S. and Davie, J.M. (1985). Serologic and topographic characterization of idiotopes on murine monoclonal anti-streptococcal group A carbohydrate antibodies. J. Immunol. 134:1065-1072.

Griffiths, G.M., Berek, C., Kaartinen, M. and Milstein, C. (1984). Somatic mutation and the maturation of immune response to 2-phenyl oxazolone. Nature 312:271-275.

Haber, E., Margolies, M., Cannon, L.E. and Rosemblatt, M.S. (1975). Restricted clonal responses: A tool in understanding antibody specificity. Miami Winter Symposium 9:303-338.

Harboe, M., Deveril, J. and Eriksen, J. (1975). Capsular swelling and passive haemagglutination induced by monoclonal IgM reactiong with acid polysaccharides with *Klebsiella*. Acta Pathol. Microbiol. Immunol. Scand. Sect. C. Immunol. 83:97-105.

Hartman, A.B. and Rudikoff, S. (1984). VH genes encoding the immune response to ß-(1→6)-galactan: somatic mutation in IgM molecules. EMBO J. 12:3023-3030.

Hedrick, S.M., Nielsen, E.A., Kavaler, J., Cohen, D.I. and Davis, M.M. (1984). Sequence relationships between putative T-cell receptor polypeptides and immunoglobulins. Nature 308:153-157.

Holmberg, D., Wennerström, G., Andrade, L. and Coutinho, A. (1985). The high idiotypic connectivity of "natural" newborn antibodies is not found in adult mitogen-reactive B cell repertoires. Ann. Immunol. (Inst. Pasteur) (in press).

Jaenichen, H., Pech, M., Lindenmaier, W., Wildgruber, N. and Zachau, H.G. (1984). Composite human Vκ genes and a model of their evolution. Nucl. Acids Res. 12:5249-5263.

Jerne, N.K. (1974). Towards a network theory of the immune system. Ann. Immunol. (Inst. Pasteur) 125:373-389.

Jerne, N.K., Roland, J. and Czzaenave, P.A. (1982). Recurrent idiotopes and internal images. EMBO J. 1:243-247.

Jeske, D.J., Jarvis, J., Milstein, C. and Capra, D. (1984). Junctional diversity is essential to antibody activity. J. Immunol. 133:1090-1092.

Johnson, N., Slankard, J., Paul, L. and Hood, L. (1982). The complete V domain amino acid sequences of two myeloma inulin-binding proteins. J. Immunol. 128:302-307.

Kaartinen, M., Griffiths, G.M., Markham, A.F. and Milstein, C. (1983). mRNA sequences define an unusually restricted IgG response to 2-phenyloxazolone and its early diversification. Nature 304:320-324.

Kabat, E.A. (1976). Structural Concepts in Immunology and Immunochemistry. 2nd Edition. Holt Rinehart and Winston. New York.

Kabat, E.A. (1978). The structural basis of antibody complementarity. Adv. in Protein Chemistry, Academic Press, New York 32:1-75.

Kabat, E.A. (1980). Opinion. Antibodies, hypervariable regions and minigenes. J. Immunol. 125:951-969.

Kabat, E.A. (1982). Antibody diversity versus antibody complementarity. Pharmacological Rev. 34:23-38.

Kabat, E.A. (1983). The antibody combining site. Fifth International Congress. Immunology Academic Press. V pp. 67-85.

Kabat, E.A. (1984). Idiotypic determinants, minigenes and the antibody combining site. in: "The Biology of

Idiotypes," M.I. Greene and A. Nisonoff, eds. Plenum Publishing Corporation, New York, pp, 3-17.

Kabat, E.A. (1986). Antibody combining sites - Past, Present and Future. 9th International Subcellular Methodology Forum. Guildford, Surrey U.K. Sept. 3-6, 1984. in press.

Kabat, E.A., Liao, J., Bretting, H., Franklin, E.C., Geltner, D., Frangione, B., Koshland, M.E., Shyong, J. and Osserman, E.F. (1980). Human monoclonal macroglobulins with specificity for *Klebsiella* K polysaccharides that contain 3,4-pyruvylated-D-galactose and 4,6-pyruvylated-D-galactose. J. Exp. Med. 152:979-995.

Kabat, E.A., Liao, J.,Sherman, W.H. and Osserman, E.F. (1984). Immunochemical characterization of the specificities of two human monoclonal IgM's reacting with chondroitin sulfates. Carbohydrate Res. 130:289-297.

Kabat, E.A. and Wu, T.T. (1971). Attempts to locate complementarity-determining residues in the variable positions of light and heavy chains of immunoglobulins. in: "Immunoglobulins" S. Kochwa and H.G. Kunkel, eds. Ann. N.Y. Acad. Sci., 190:382-393.

Kabat, E.A., Wu, T.T. and Bilofsky, H. (1977). Unusual distributions of amino acids in complementarity-determining (hypervariable) segments of heavy and light chains of immunoglobulins and their possible roles in specificity of antibody-combining sites. J. Biol. Chem. 252:6609-6616.

Kabat, E.A., Wu, T.T. and Bilofsky, H. (1978). Variable region genes for the immunoglobulin framework are assembled from small segments of DNA - A hypothesis. Proc. Natl. Acad. Sci. 75:2429-2433.

Kabat, E.A., Wu, T.T. and Bilofsky, H. (1979). Evidence supporting somatic assembly of the DNA segments (minigenes), coding for the framework, and complementarity-determining segments of immunoglobulin variable regions. J. Exp. Med. 149:1299-1313.

Kabat, E.A., Wu, T.T. and Bilofsky, H. (1980). Evidence indicating independent assortment of framework and complementarity-determining segments of the variable regions of rabbit light chains. Delineation of possible J minigene. J. Exp. Med. 152:72-84.

Kabat, E.A., Wu, T.T., Bilofsky, H., Reid-Miller, M. and Perry, H. (1983). Sequences of proteins of immunological interest. Tabulation and analysis of amino acid andnucleic acid sequences of precursors, V-regions, C-regions, J-chain, β_2-microglobulins, major histocompatibility antigens, Thy-1, complement, C-reactive protein, thymopoietin, post-gamma globulin, and α_2-macroglobulin. U.S. Dept. Health and Human Services, Public Health Service, National Institutes of Health, Bethesda, MD.

Kobzik, L., Brown, M.C. and Cooper, A.G. (1976). Demonstration of an idiotypic antigen on a monoclonal cold agglutinin and on its isolated heavy and light chains. Proc. Natl. Acad. Sci. 73:1702-1706.

Köhler, G. and Milstein, C. (1975). Continuous cultures of fused cells secreting antibody of predefined specificity. Nature. 256:495-497.

Komaromy, M. and Wall, R. (1981). Organization of light chain variable region genes. in: "Immunoglobulin

idiotypes" ICN-UCLA Symposia on Molecular and Cellular Biology. 20:59-64.
Krause, R.M. (1970). The search for antibodies with molecular uniformity. Adv. Immunol. 12:1-56.
Kurosawa, Y. and Tonegawa, S. (1982). Organization, structure, and assembly of immunoglobulin heavy chain diversity DNA segments. J. Exp. Med. 155:201-218.
Lai, E. and Kabat, E.A. (1985). Immunochemical studies of conjugates of isomaltosyl oligosaccharides to lipid: Production and characterization of mouse hybridoma antibodies specific for stearyl-isomaltosyl oligosaccharides. Mol. Immunol. 22:1021-1037.
Lemieux, R.U., Wong, T.C., Liao, E. and Kabat, E.A. (1984). The combining site of anti-I Ma (Group 1). Mol. Immunol. 21:751-759.
Lerner, R.A. (1982). Tapping the immunological repertiore to produce antibodies of predetermined specificity. Nature 299:592-596.
Littman, D.R., Thomas, Y., Maddon, P.J., Chess, L. and Axel,R. (1985). The isolation and sequence of the gene encoding T8: A molecule defining functional classes of T lymphocytes. Cell 40:237-246.
Loh, D.Y., Bothwell, A.L.M., White-Scharf, M.E., Imanishi-Kari, T. and Baltimore, D. (1983). Molecular basis of a mouse strain-specific anti-Hapten response. Cell 33:85-93.
Maddon, P.J., Littman, D.R., Godfrey, M., Maddon, D.E., Chess, L. and Axel, R. (1985). The isolation and nucleotide sequence of a cDNA encoding the T cell surface protein T4: A new member of the immunoglobulin gene family. Cell 42:93-104.
Makover, S.D., Chen, H.C. and Kabat, E.A. in preparation.
Mandrell, R.E. and Zollinger, W.D. (1982). Measurement of antibodies to meningococcal group B polysaccharide: Low avidity binding and equilibrium binding constants. J. Immunol. 129:2172-2178.
Manser, T., Huang, S.Y. and Gefter, M. Influence of clonal selection on the expression of immunoglobulin variable region genes. Science 226:1283-1288.
Max, E.E., Seideman, J.G. and Leder, P. (1979). Sequences at the somatic recombination sites of immunoglobulin κ constant region gene. Proc. Natl. Acad. Sci. 76:3450-3454.
Milstein, C. (1985). From the structure of antibodies to the diversification of the immune response. EMBO J. 4:1083-1092.
Möller, G. (1984). T Cell Receptors and Genes. Immunol. Rev. 81:21-258.
Moudallal, Z., Briand, J.P. and Van Regenmortel, M.H.V. (1985). A major part of the polypeptide chain of tobacco mosaic virus protein is antigenic. EMBO J. 4:1231-1235.
Nairn, A.C., Detré, J.A., Casnellie, J.E. and Greengard, P. (1982). Serum antibodies that distinguish between the phospho- and dephospho-forms of a phosphoprotein. Nature 299:734-736.
Naparstek, Y., Duggan, D., Schattner, A., Madaio, M.P., Goni, F., Frangione, B., Stoller, B.D., Kabat, E.A. and Schwartz, R.S. (1985). Immunochemical similarities between monoclonal anti-bacterial Waldenstrom's macroglobulins and monoclonal anti-DNA lupus autoantibodies. J. Exp. Med. 161:1525-1538.

Newman, B.A. and Kabat, E.A. (1985). An immunochemical study of the combining site specificities of C57BL/6J monoclonal antibodies to $\alpha(1\rightarrow6)$-linked dextran B512. J. Immunol. 135:1220-1231.

Newman, B.A., Liao, J., Gruezo, F., Sugii, S., Kabat, E.A., Torii, M., Clevinger, B.L., Davie, J.M., Schilling, J., Bond, M. and Hood, L. (1986). Immunochemical studies of mouse monoclonal antibodies to dextran B1355S II. Combining site specificity, sequence, idiotype and affinity. Mol. Immunol. (in press).

Newman, B., Sugii, S., Kabat, E.A., Torii, M., Clevinger, B.L., Schilling, J., Davie, J.M. and Hood, L. (1983). Combining site specificities of mouse hybridoma antibodies to dextran B1355. J. Exp. Med. 157:130-140.

Novotny, J., Bruccoleri, R. and Karplus, M. (1984). An analysis of incorrectly folded protein models. J. Mol. Biol. 177:787-818.

Novotny, J., Bruccoleri, R., Newell, J., Murphy, D., Haber, E. and Karplus, M. (1983). Molecular anatomy of the antibody binding site. J. Biol. Chem. 23:14433-14437.

Novotny, J. and Haber, E. (1985). Structural invariants of antigen binding: Comparison of immunoglobulin V_L-V_H and V_L-V_L domain dimers. Proc. Natl. Acad. Sci. 82:4592-4596.

Ohno, S., Matsunaga, T. and Lee, A.D. (1984). The invariably present tryptophan loop as the core of all divergent antigen-binding pockets. Scand. J. Immunol. 20:377-388.

Ohno, S., Matsunaga, T. and Wallace, R.B. (1982). Identification of the 48-base-long primordial building block sequence of mouse immunoglobulin variable region genes. Proc. Natl. Acad. Sci. 79:1999-2002.

Ohno, S., Mori, N. and Matsunaga, T. (1985). Antigen-binding specificities of antibodies are primarily determined by seven residues of V_H. Proc. Natl. Acad. Sci. 82:2945-2949.

Osterland, C.K., Miller, E.J., Karakawa, W.W. and Krause, R.M. (1966). Characteristics of streptococcal group-specific antibody isolated from hyperimmune rabbits. J. Exp. Med. 123:599-614.

Outschoorn, I.M., Ashwell, G., Gruezo, F. and Kabat, E.A. (1974). Immunochemical studies on dextrans VIII. Specificity and cross-reactivity with dextrans of the antibodies formed in rabbits to isomaltohexaonic acid coupled to bovine serum albumin. J. Immunol. 113:896-903.

Owen, J.A., Sigal, N.H. and Klinman, N.R. (1982). Heterogeneity of the BALB/c IgM anti-phosphorylcholine antibody response. Nature 295:347-348.

Padlan, E.A. (1977). Structural basis for the specificity of antigen - antibody reactions and structural mechanisms for the diversification of antigen-binding specificities. Quart. Rev. Biophys. 10:35-65.

Padlan, E.A., Cohen, G.H. and Davies, D.R. (1985). On the specificity of antibody/antigen interactions: Phosphocholine binding to McPC603 and the correlation of t hree-dimensional structure and sequence data. Ann. Inst. Pasteur/Immunol. 136:271-276.

Padlan, E.A., Segal, D.M., Spande, T.F., Davies, D.R., Rudikoff, S. and Potter, M. (1973). Structure at 4.5 Å resolution of a phosphorylcholine-binding Fab. Nature (Lond.) New Biol. 145:165-167.

Patten, P., Yokota, T., Rothbard, J., Chien, Y., Arai, K. and Davis, M.M. (1984). Structure, expression and divergence of T-cell receptor ß-chain variable regions. Nature 312:40-46.
Pawlita, M., Mushinski, E., Feldmann, R.J. and Potter, M. (1981). A monoclonal antibody that defines an idiotope with two subsites in galactan-binding myeloma proteins. J. Exp. Med. 154:1946-1956.
Peabody, D.S., Ely, K.R. and Edmundson, A.B. (1980). Obligatory hybridization of heterologous immunoglobulin light chains into covalently linked dimers. Biochem. 19:2827-2834.
Perlmutter, R.M., Crews, S.T., Douglas, R., Sorensen, G., Johnson, N., Nivera, N., Gearhart, P.J. and Hood, L. (1984). The generation of diversity in phosphorylcholine-binding antibodies. Academic Press, New York. Advances in Immunology 35:1-37.
Potter, M. (1977). Antigen-binding myeloma proteins of mice. Adv. Immunol. Academic Press. 25:141-211.
Potter, M. (1971). Antigen-binding myeloma proteins in mice. Immunoglobulins Ann. N.Y. Acad. Sci. 190:306-321.
Radbruch, A., Zaiss, S., Kappen, C., Brüggemann, Beyreuther,K. and Rajewsky, K. (1985). Drastic change in idiotypic but not antigen-binding specificity of an antibody by a single amino-acid substitution. Nature 315:506-508.
Rao, A.S., Liao, J., Kabat, E.A., Osserman, E.F., Harboe, M. and Nimmich, W. (1984). Immunochemical studies on human monoclonal macroglobulins with specificities for 3,4-pyruvylated D-galactose and 4,6-pyruvalated D-glucose. J. Biol. Chem. 259:1018-1026.
Rechavi, G., Ram, D., Glazer, L., Zakut, R. and Givol, D. (1983). Evolutionary aspects of immunoglobulin V_H gene subgroups. Proc. Natl. Acad. Sci. 80:855-859.
Reichlin, M. and Eng, J. (1977). Identifying antigenic determinants on cytochrome c for B and T cells. in: "Immunobiology of proteins and peptides I." M.Z. Atassi and A.B. Stavitsky, eds. Plenum, N.Y. pp. 5-18.
Reth, M., Bothwell, A.L.M. and Rajewsky, K. (1981). Structural properties of the hapten binding site and of idiotypes in the NP^b antibody family. In Immunoglobulin Idiotypes and Their Expression. C. Janeway, H. Wigzell and C.F. Fox, eds. (New York: Academic Press), pp.169-178.
Reynaud, C., Anquez, V., Dahan, A. and Weill, J. (1985). A single rearrangement event generates most of the chicken immunoglobulin light chain diversity. Cell 40:283-291.
Rocca-Serra, J., Tonnelle, C. and Fougereau, M. (1983). Two monoclonal antibodies against different antigens using the same V_H germ-line gene. Nature 304:353-355.
Roelcke, D. (1974). Cold agglutination antibodies and antigens. Clin. Immunol. Immunopathol. 2:266-280.
Roth, C., Rocca-Serra, J., Sommé, G., Fougereau, M. and Theze, J. (1985). Gene repertoire of the anti-poly(Glu^{60} Ala^{30} Tyr^{10}) (GAT) immune response: Comparison of V_H, V_K, and D regions used by anti-GAT antibodies and monoclonal antibodies produced after anti-idiotypic immunization. Proc. Natl. Acad. 82:4788-4792.
Ruckel, E.R. and Schuerch, C. (1967). Chemical synthesis of a dextran model, poly-α-(1→6)-anhydro-D-glucopyranose. Biopolymers 5:515-523.

Rudikoff, S. (1983). Immunoglobulin structure-function correlates: Antigen binding and idiotypes. Contemporary Topics in Molecular Immunology 9:169-209.

Sakano, H., Hüppi, K., Heinrich, G. and Tonegawa, S. (1979). Sequences at the somatic recombination sites of immunoglobulin light-chain genes. Nature 280:288-294.

Sakano, H., Maki, R., Kurosawa, Y., Roeder, W. and Tonegawa, S. (1980). Two types of somatic recombination are necessary for the generation of complete immunoglobulin heavy-chain genes. Nature (Lond.) 286:676-683.

Saul, F.A. and Poljak, R.J. (1985). Three-dimensional structure and function of immunoglobulins. Ann. Inst. Pasteur/Immunol. 136:259-294.

Schilling, J.B., Clevinger, B., Davie, J.M. and Hood, L. (1980). Amino acid sequence of homogeneous antibodies to dextran and DNA rearrangements in heavy chain V-region gene segments. Nature 283:35-40.

Seligmann, M.. and Brouet, J.C. (1973). Antibody activity of human myeloma globulins. Semin. Hematol. 10:163-177.

Sharon, J., D'Hoostelaere, L., Potter, M., Kabat, E.A. and Morrison, S.L. (1982). A cross-reactive idiotype, QUPC52 Idx, present on most but not all anti-$\alpha(1\rightarrow6)$ dextran-specific IgM and IgA hybridoma antibodies with combining sites of different sizes. J. Immunol. 128:498-500.

Sharon, J., Kabat, E.A. and Morrison, S.L. (1982a). Immunochemical characterization of binding sites of hybridoma antibodies specific for $\alpha(1\rightarrow6)$-linked dextran. Mol. Immunol. 19:388-389.

Sharon, J., Kabat, E.A. and Morrison, S.L. (1982b). Association constants of hybridoma antibodies specific for $\alpha(1\rightarrow6)$-linked dextran determined by affinity electrophoresis. Mol. Immunol. 19:389-397.

Sharon, J., Kabat, E.A. and Morrison, S.L. (1981a). Studies on mouse hybridomas secreting IgM or IgA antibodies to $\alpha(1\rightarrow6)$-linked dextran. Mol. Immunol. 18:831-846.

Siebenlist, U., Ravetch, J.V., Korsmeyer, R.S., Waldmann, T. and Leder, P. (1981). Human immunoglobulin D segments encoded in tandem multigenic families. Nature 294:631-635.

Sikder, S.K., Akolker, P.N., Kaladas, P.M., Morrison, S.L. and Kabat, E.A. (1985). Sequences of variable regions ofhybridoma antibodies to $\alpha(1\rightarrow6)$ dextran in BALB/c and C57BL/6 mice. J. Immunol. 135:4215-4221.

Silverton, E.W., Padlan, E.A., Davies, D.R., Smith-Gill, S. and Potter, M. (1984). Crystalline monoclonal antibody Fabs complexed to hen egg white lysozyme. J. Mol. Biol. 761-765.

Slaughter, C.A. and Capra, J.D. (1983). Amino acid sequence diversity within the family of antibodies bearing the major anti-arsonate cross-reactive idiotype of the A strain mouse. J. Exp. Med. 158:1615-1634.

Smith, A.M. and Potter, M. (1975). A BALB/c mouse IgA myeloma protein that binds *Salmonella* flagellar protein. J. Immunol. 114:1847-1850.

Tainer, J.A., Getzoff, E.D., Alexander, H., Houghten, R.A., Olson, A.J. and Lerner, R.A. (1984). The reactivity of anti-peptide antibodies is a function of the atomic mobility of sites in a protein. Nature 312:127-133.

Tainer, J.A., Getzoff, E.D., Paterson, Y., Olson, A.J. and Lerner, R.A. (1985). The atomic mobility component of protein antigenicity. Ann. Rev. Immunol. 3:501-535.

Takahashi, N., Noma, T. and Honjo, T. (1984). Rearranged immunoglobulin V_H pseudogene that deletes the second complementarity determining region. Proc. Natl. Acad. Sci. 81:5194-5198.

Urbain, J., Wuilmart, C. and Cazenave, P.A. (1981). Idiotypic regulation in immune networks. Contempor. Topics Mol. Immunol. 8:113-148.

Victor, C., Bona, C. and Pernis, B. (1983). Idiotypes on B lymphocytes: Association with immunoglobulins. J. Immunol. 130:1819-1825.

Warner, N.W. (1975). Autoimmunity and the pathogeneis of plasma cell tumor induction in NZB inbred and hybrid mice. Immunogenetics 2:1-20.

Watanabe, K, Hakomori, S., Childs, R.A. and Feizi, T. (1979). Characterization of a blood group I-active ganglioside. J. Biol. Chem. 254:3221-3228.

Weigert, M., Cesari, I.M., Yonkovich, S.J. and Cohn, M. (1970). Variability in the lambda light chain sequences of mouse antibody. Nature 228:1045-1047.

Weigert, M. and Riblet, R. (1976). Genetic control of antibody variable regions. Cold Spring Harbor Symp. Quant. Biol. 41:837-846.

Westhof, E., Altschuh, D. Moras, D., Bloomer, A.C., Mondragon, A., Klug, A. and Van Regenmortel, M.H.V. (1984). Correlation between segmental mobility and the location of antigenic determinants in proteins. Nature 311:123-126.

Wiley, D.C., Wilson, I.A., Skehel, J.J. (1981). Structural identification of the antibody-binding sites of Hong Kong influenza haemagglutinin and their involvement in antigenic variation. Nature 289:373-378.

Williams, R.C. (1971). Cold agglutinins: Studies of primary structure, serologic activity and antigenic uniqueness. in: "Immunoglobulins." S. Kochwa and H.G. Kunkel eds. Ann. N.Y. Acad. Sci. 190:330-341.

Wood, C. and Kabat, E.A. (1981a). Immunochemical studies of conjugates of isomaltosyl oligosaccharides to lipid: Fractionation of rabbit antibodies to stearylisomaltosyl oligosaccharides and a study of their combining sites by a competitive binding assay. Arch. Biochem. 212:277-289.

Wood, C. and Kabat, E.A. (1981b). Immunochemical studies on conjugates of isomaltosyl oligosaccharides to lipid. I. Antigenicity of the glycolipids and the production of specific antibodies in rabbits. J. Exp. Med. 154:432-449.

Wood, E. and Feizi, T. (1979). Blood group I and i activities of straight chain and branched synthetic oligos accharides related to the precursors of the major blood group antigens. FEBS Letters 104:135-140.

Wu, A.M., Kabat, E.A. and Weigert, M.G. (1978). Immunochemical studies on dextran-specific and levan-specific myeloma proteins from NZB mice. Carbohydrate Res. 66:113-124.

Wu, T.T. and Kabat, E.A. (1970). An analysis of the sequences of the variable regions of Bence-Jones proteins and myeloma light chains and their implication for antibody complementarity. J. Exp. Med. 132:211-250.

Wu, T.T. and Kabat, E.A. (1982). Fourteen nucleotides in the second complementarity-determining region of a human heavy-chain variable region gene are identical with a sequence in a human D minigene. Proc. Natl. Acad. Sci. 79:5031-5032.

Wyle, F.A., Artenstein, M.S., Brandt, B.L., Tramont, E.C., Kasper, D.L., Altieri, P.L., Berman, S.L. and Lowenthal, J.P. (1972). Immunologic response of man to group B meningococcal polysaccharide vaccines. J. Infectious Diseases 126:514-522.

Yanagi, Y., Yoshikai, Y., Leggett, K., Clark, S.P., Aleksander, I. and Mak, T.W. (1984). A human T cell-specific cDNA clone encodes a protein having extensive homology to immunoglobulin chains. Nature 308:145-149.

Yoo, T.J. and Franklin, E.C. (1971). Lack of antibody activity in human myeloma globulins. J. Immunol. 107:365-367.

Yount, W.J., Dorner, M.M., Kunkel, H.G. and Kabat, E.A. (1968). Studies on human antibodies. VI. Selective variations in subgroup composition and genetic markers. J. Exp. Med. 127:633-646.

Zanetti, M., Liu, F.T., Rogers, J. and Katz, D.H. (1985). Heavy and light chain of a mouse monoclonal autoantibody express the same idiotype. J. Immunol. 135:1245-1251.

SIALIC ACIDS AS ANTIGENIC DETERMINANTS OF COMPLEX CARBOHYDRATES

Roland Schauer

Biochemisches Institut, Christian-Albrechts-Universität
Olshausenstraße 40, D-2300 Kiel
Federal Republic of Germany

Carbohydrates have been known for long as potent antigens, described in numerous publications and extensively being discussed in this book. It became evident in the last years that sialic acids play an important role in this field, too, although their exact function is not yet understood in every case (Reutter et al., 1982; Schauer, 1982; Schauer, 1983; Schauer, 1985). Scheme 1 gives a survey of the influence of carbohydrates in general and of sialic acids in special on immunological reactions. On the one hand carbohydrates act as antigens, e.g. as differentiation and onco-developmental antigens (Feizi, 1985), and on the other hand they can mask antigenic sites on proteins, lipids and carbohydrate oligo- and polymers and thus can represent "anti-antigens". It should be mentioned already here that the latter role is often due to, or at least strengthened by, the presence of sialic acids in glycan chains. Such a masking effect of carbohydrate chains may be illustrated by Figs. 1 and 2, showing, from different views, branched, N-glycosidically linked oligosaccharide chains, which cover the protein parts of the glycoprotein molecules like clouds and in this way may mask antigenic sites of the protein part. The bulky and hydrophilic glycan chains have also been found to protect proteins from other attack, e.g. proteases (Faillard & Schauer, 1972; Schauer, 1982, Semino et al., 1985). Many examples are known where sialic acids enhance this masking effect of the carbohydrate moiety (Schauer, 1982; Schauer,

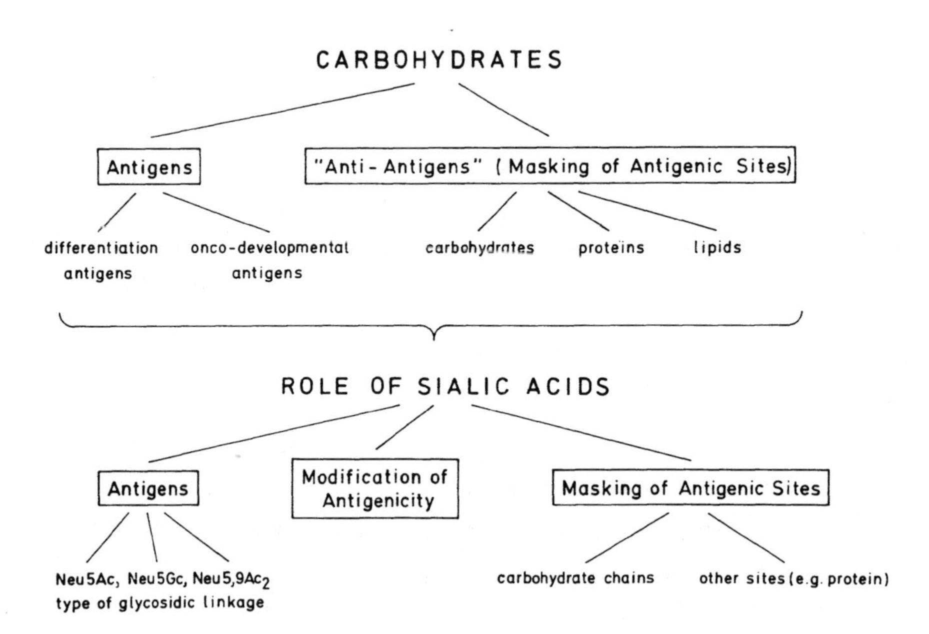

Scheme 1. Role of oligosaccharide and sialic acid residues in immunological recognition and masking.

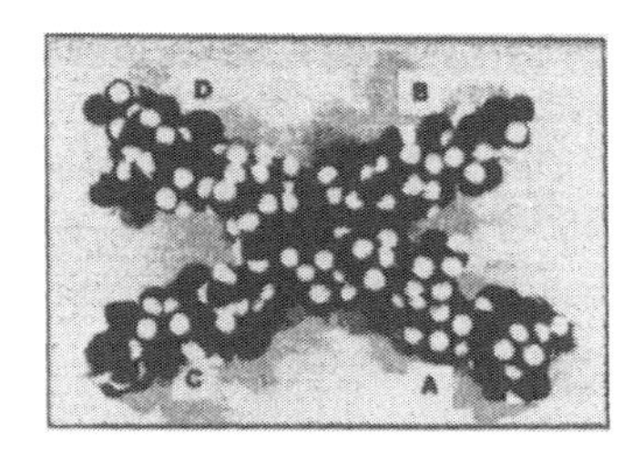

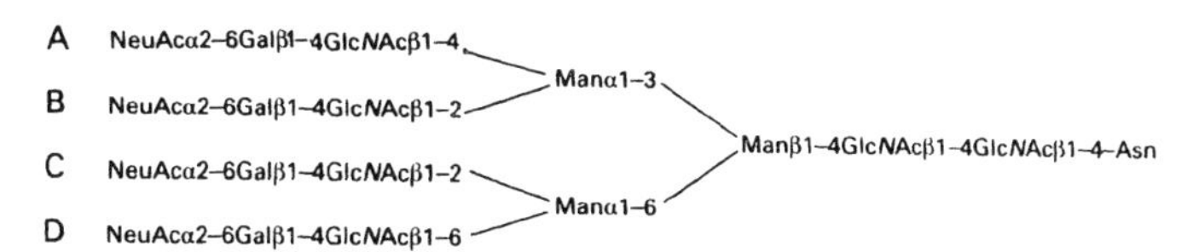

Fig. 1. "Umbrella-conformation" of a tetraantennary glycan of the N-acetyllactosamine type (from Montreuil et al., 1983)

1985). This may be due to the negative charge of this sugar residue or to an increase of the shielding effect of the carbohydrate chain by linking it more closely to the protein part mediated by the interaction of the negatively charged sialic acid with positively charged amino acid residues.

Like the total oligosaccharide chains sialic acids themselves play a dual role in immunology, too. They either represent antigens or they mask antigenic sites (Scheme 1). Both effects may be interwoven. I shall therefore focus on immunological and other recognition phenomena which are exclusively or mainly due to sialic acids and begin with an introduction to the biochemistry of these substances.

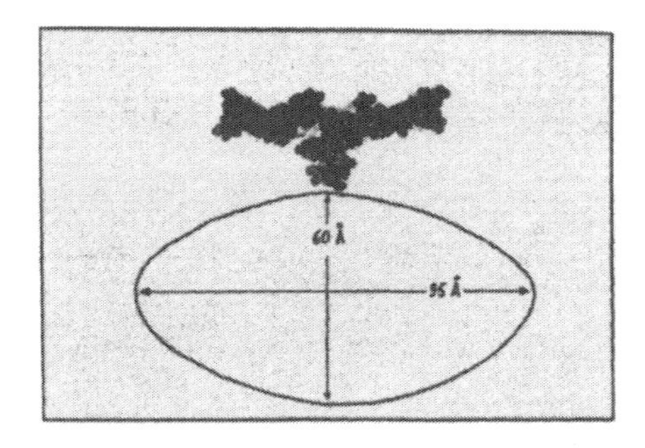

Fig. 2. Relative size of biantennary glycan and protein in human serum transferrin (from Montreuil et al., 1983)

STRUCTURE AND OCCURRENCE OF SIALIC ACIDS

Sialic acids comprise a family of more than 20 derivatives of neuraminic acid with acetyl or glycolyl groups at the amino function and acetyl, L-lactyl, methyl, sulfate or phosphate residues at the hydroxyl groups (Fig. 3) (Schauer, 1982; Corfield & Schauer, 1982). These substituents may be combined or several *O*-acetyl groups occur in one sialic acid molecule (maximum 3 in the case of *N*-acetyl-7,8,9-tri-*O*-acetylneuraminic acid of bovine submandibular gland glycoprotein), which leads to this high number of sialic acids. The largest number (fourteen) of sialic acids was found so far in bovine submandibular gland glycoprotein. The three main sialic acids are, arranged according to their frequency in nature, *N*-acetylneuraminic acid (Neu5Ac)>*N*-glycolylneuraminic acid (Neu5Gc)>*N*-acetyl-9-*O*-acetylneuraminic acid ($Neu5,9Ac_2$).

The distribution of the sialic acid species is species- and tissue-specific. For instance, in horse the *O*-acetyl group is mainly at *O*-4 of neuraminic acid, while the amino

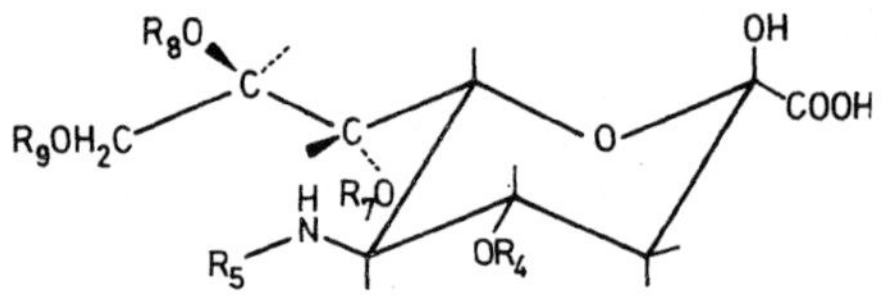

R_5	$R_{4,7,8,9}$	
$-C(=O)-CH_3$	$-H$	(4,7,8,9)
$-C(=O)-CH_2OH$	$-C(=O)-CH_3$	(4,7,8,9)
	$-C(=O)-CH(OH)-CH_3$	(9)
	$-CH_3$	(8)
	$-SO_3H$	(8)
	$-PO_3H_2$	(9)

Fig. 3. *N*- and *O*-substituents of naturally occurring sialic acids. The numbers in brackets indicate the positions of these residues on the neuraminic acid molecule.

group bears acetyl or glycolyl groups. In cow, in a variety of other animal species and in man, sialic acids are frequently O-acetylated in their side chain. From an immunological point of view, the sialic acids of man are most important. The main human sialic acid is Neu5Ac, followed by Neu5,9Ac_2 and small amounts of N-acetyl-7,9-di-O-acetylneuraminic acid, N-acetyl-9-O-lactylneuraminic acid and N-acetyl-2-deoxy-2,3-didehydroneuraminic acid (Schauer, 1982; Corfield & Schauer, 1982). Amounts less than 1% of the total sialic acid fraction of N-glycolylneuraminic acid were unequivocally detected in glycoproteins of healthy human individuals by a special procedure for the isolation and by combination of gas-liquid chromatography and mass spectrometry for the identification of this sialic acid (Nöhle & Schauer, 1984).

All sialic acids shown in Fig. 3, with the exception of the 9-phosphate derivative and unsaturated sialic acids found in body fluids (Corfield & Schauer, 1982; Schauer et al., 1984), not shown in this figure, are glycosidic components of oligosaccharides, polysaccharides, glycoproteins and gangliosides (Corfield & Schauer, 1982). As is shown in Fig. 4 the glycosidic linkage is always α, but variable with regard to the position and nature of the sugar to which sialic acid is bound (Montreuil, 1980; Schauer, 1982; Corfield & Schauer, 1982). However, α(2->3)linkages and galactose as binding partner prevail. Sialic acids occur most frequently in terminal position of oligosaccharide chains, sometimes in side position, e.g. in gangliosides, or internal, i.e. within a

position in oligosaccharide chains	: terminal, side-positioned, internal
linked to monosaccharides	: Gal, GalNAc, GlcNAc, Sia
nature of linkages	: α 2-3, 2-6, 2-8, 2-9

Fig. 4. The linkages of sialic acids found in complex carbohydrates.

carbohydrate chain, as in colominic acid, many gangliosides, glycoproteins with oligosialyl chains and some glycoconjugates of echinoderms (Schauer, 1982; Corfield & Schauer, 1982; Schauer, 1985; Finne, 1982; Sugita, 1979; van der Meer et al., 1983).

Thus, sialic acids by their multiple forms and manyfold linkages to oligosaccharide chains enormously contribute to the structural diversity of glycoconjugates. In this way sialic acids may exert manyfold and delicate influences on the numerous physico-chemical, biological and especially immunological functions of glycoconjugates.

SIALIC ACIDS AS ANTIGENS

Although free sialic acids are not antigens, in glycosidic linkage different sialic acids exert specific immunological functions. This is summarized in the Table. For example, a Waldenström macroglobulin specific for Neu5Ac has been reported (Tsai et al., 1977; Picard et al., 1985). Antibodies raised against various oligosaccharides bound to proteins were shown to distinguish between positional isomers (α(2->3) and α(2->6)linkages) of sialic acids (Smith & Ginsburg, 1980; Hakomori et al., 1983). Neu5Ac plays a well-known role in the specificity of MN blood group substances and participates in the cold-agglutination of erythrocytes mediated by different antibodies. The specificity of these many antibodies detected so far, which have been designated with 2 letters, e.g. Pr or Gd cold agglutinins, is directed against the terminal part of the N- or O-glycosidic oligosaccharide chains of erythrocyte membrane glycoproteins (Anstee, 1981; Roelcke, 1984; 1985; Roelcke et al., 1980). The optimum structure for cold agglutinins was found to be Neu5Ac α-glycosidically linked to galactose. Correspondingly, the antigen-antibody reaction can be inhibited by sialyllactose. However, Staub (1985) reported a "Gd-like" cold agglutinin which is not inhibited by this trisaccharide. Detailed investigations by Roelcke (1984) and Anstee (1981) have shown that the specificity of the great variety of cold agglutinins is influenced by the structure of the whole

Table. Sialic Acids as Antigens

Neu5Ac	Waldenström macroglobulin cold agglutinins (e.g. Pr, Gd, Sa) antibodies for 2->3 or 2->6-linked Neu5Ac M,N blood group substances
Neu5Gc	serum sickness antibodies Hanganutziu-Deicher antibodies blood group substances of East-Asian dogs
Neu5,9Ac$_2$	bacterial polysaccharides human colon mucin

glycan chain, whether it is short or long, linear or branched or of the lactosamine type. The antigenicity of these structures is dependent on the presence of Neu5Ac, as it is destroyed by sialidase treatment or by reduction of the carboxylic group of sialic acid to an alcohol residue. Another example of a strong influence of Neu5Ac on the immunogenicity of complex carbohydrates are capsular polysaccharides from various Streptococcus and Neisseria meningitidis species containing sialic acid in different molecular environments (Jennings et al., 1984). Sialic acid has been revealed as an important factor in the virulence of these microorganisms and in the human antibody response to their polysaccharide antigens. It controls the antigenic determinants on the polysaccharides by interaction with remote glycosyl units of the polysaccharides, as was studied using NMR spectroscopy.

N-Glycolylneuraminic acid, differing from Neu5Ac only in one oxygen atom, has been recognized as exerting immunological effects on dogs and man. This sialic acid as component of gangliosides, especially hematoside, was found in East-Asian including Japanese dog erythrocytes, and represents a determinant of blood group specificity (Yasue et al., 1978). In these animals, the appearance of Neu5Gc follows an autosomal, dominant inheritance, while the gene for Neu5Ac of hematoside is recessive. European dogs contain only Neu5Ac; the genotype is homozygous in these animals with regard to the nature of sialic acids in erythrocytes.

Antibodies against Neu5Gc have also been detected in man, so-called Hanganutziu-Deicher (HD) antibodies. A heterophilic HD antibody was first observed by Hanganutziu (1924) and Deicher (1926) in sera of patients who had received therapeutic injections of horse serum antitoxoid. The antigen was found to consist of gangliosides and glycoproteins containing Neu5Gc bound to C-3 of galactose. Human HD antibodies were also found in sera of patients with various diseases, including cancer, who had not received sera from animals (Kasukawa _et al._, 1976; Nishimaki _et al._, 1979, Higashi _et al._, 1984). Four species of antigenic gangliosides containing Neu5Gc (for the structure of one of them, the GM2 II^3Neu-5Gc-$GgOse_3$Cer, see Fig. 5) were isolated from human colon cancer (Higashi _et al_, 1985). Corresponding tissue samples from individuals without colorectal cancer did not contain these antigenic compounds. These authors identified in the tumor material 1% or less of the total lipid-bound sialic acid as Neu5Gc. However, although it had been believed for long that glycoconjugates from normal human individuals do not contain Neu5Gc, we have unequivocally detected minute amounts of Neu5Gc (about 0.01% of the total sialic acid fraction) in glycoproteins from serum of healthy men, by application of an improved isolation procedure of sialic acids and analysis by capillary gas-liquid chromatography/mass spectrometry (Schröder _et al._, 1983).

The origin of Neu5Gc both in glycoproteins of healthy individuals and gangliosides of colon tumor is unknown. The production of the small amounts of this sugar in normal tissues may be due to residual activities of Neu5Ac monooxygenase (EC 1.14.99.18). According to the wide occurrence of

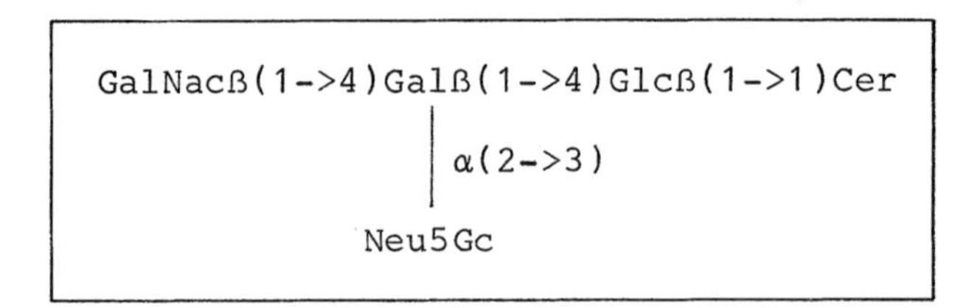

Fig. 5 Structure of ganglioside Neu5Gc-GM2, one of the "Hanganutziu-Deicher"-antigens isolated from human colon tumor (Higashi _et al._, 1985)

Neu5Gc in the animal kingdom, this enzyme is also frequent. Some of its properties have been studied in porcine submandibular glands (Schauer, 1978). The monooxygenase hydroxylates the N-acetyl group of Neu5Ac to N-glycolyl with the aid of NADPH and/or ascorbate and molecular oxygen (Fig. 6). It is probably a non-haem hydroxylase requiring ferrous ions for activity. It is conceivable that the gene of this enzyme is repressed in healthy human tissues, but may be activated in malignant cells, leading to gangliosides containing Neu5Gc. There is, however, experimental evidence from studies with mice and rats that Neu5Gc in glycoconjugates might be of nutritional origin (Nöhle & Schauer, 1984). A small proportion of both free and glycosidically (to submandibular gland mucin) bound Neu5Gc applied orally to the animals was resorbed from the intestine and incorporated into liver glycoproteins. These experiments show that bound sialic acids of the food stuffs can be liberated in intestine by sialidase and resorbed as intact sialic acid. An appreciable and variable proportion of the sialic acid liberated is cleaved by N-acylneuraminate-pyruvate lyase (EC 4.1.3.3) occurring in the intestinal lumen, too. N-Acetyl- or N-glycolylmannosamine derived from Neu5Ac and Neu5Gc, respectively, are taken up into the blood stream. In contrast to free sialic acids,

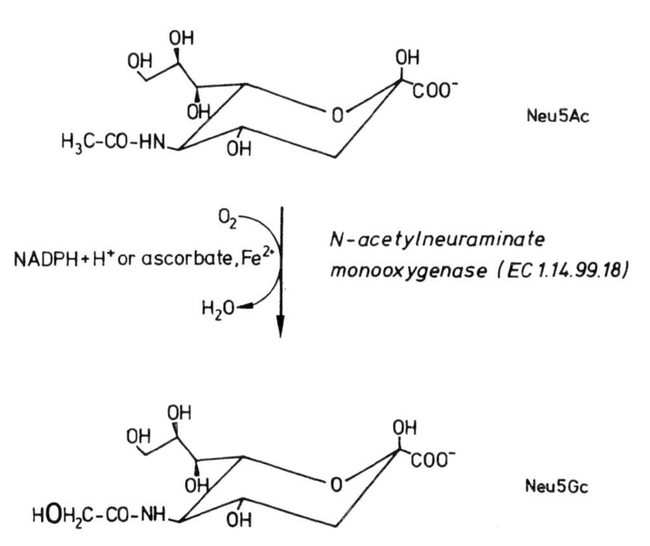

Fig. 6 Oxidation of Neu5Ac to Neu5Gc by a specific monooxygenase

which, as far as has been investigated, are not permeable through cell membranes and are rapidly excreted by the urine, acylmannosamines can be taken up by cells and used for sialic acid synthesis. In this way, N-glycolylmannosamine from mucin-bound Neu5Gc fed to the animals was shown to function as precursor of Neu5Gc of liver glycoconjugates in mice (Nöhle & Schauer, 1984). This metabolic fate of orally administered sialic acids has been elucidated with radioactive sialic acids. The pathway for glycosidicaly bound sialic acids is shown in Fig. 7. It is imaginable that a similar mechanism exists in man, leading to the small amount of Neu5Gc found in glycoconjugates. There is sufficient Neu5Gc in food, as milk (sialyllactose) and meat from e.g. cow and pig contain much of this sialic acid (Corfield & Schauer, 1982). The higher relative amount of Neu5Gc in cancer cells may be due to a more rapid uptake of acylmannosamines circulating in the blood stream, or to increased permeability of plasma membranes of these cells for sialic acids.

O-Acetyl groups in sialic acids have also been recognized to specifically influence the immunogenicity of glyco-

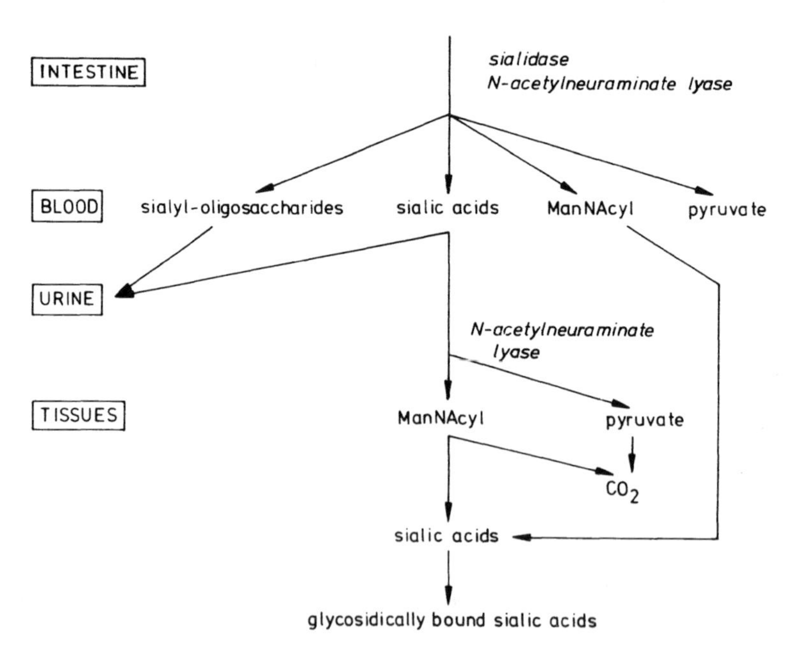

Fig. 7 Metabolism of orally administered, glycosidically bound sialic acids, e.g. Neu5Gc, in the mouse (from Nöhle & Schauer, 1984). ManNAc = N-acylmannosamines. For further details see the text.

conjugates. First evidence for an immunological role of O-acetylated sialic acids came from Ørskov et al. (1979) who reported that colominic acid O-acetylated at O-9 of Neu5Ac is more immunogenic than the non-O-acetylated polysaccharide. The best example found so far for an immunological role of O-acetylated sialic acids is the ganglioside GD3 containing $Neu5,9Ac_2$ in terminal position (Cheresh et al.,1984) (Fig.8). This is a tumor-specific antigen of human melanoma cells. Monoclonal antibodies were raised against this ganglioside demonstrating that the presence of the 9-O-acetyl group is essential for its immunological reaction. The antibody does not react with the saponified ganglioside. 1 mg of pure ganglioside could be isolated from 23 g of packed melanoma cells, and analyzed by immunostaining, NMR- and fast-atom bombardment mass spectrometry (Thurin et al., 1985).

O-Acetylated sialic acids can also be considered as differentiation antigens. Evidence for such a role came from studies of sialic acids of erythrocytes from chicken and hen, respectively, showing that only the cells from the adult bird contain $Neu5,9Ac_2$ (Reuter et al., unpublished results). Similar observations were made earlier with Neu5Gc, the relative amount of which increases during sexual maturation in serum glycoproteins of cow (Faillard & Cabezas, 1963).

The involvement of O-acetylated sialic acids in immunological reactions has stimulated interest in sialic acid modification by O-acetylation. This is mediated by the action of two specific O-acetyltransferases, acetyl-CoA:N-acylneuraminate 4-O-acetyltransferase (EC 2.3.1.44) and acetyl-CoA: N-acylneuraminate 7(9)-O-acetyltransferase (EC 2.3.1.45, Fig. 9). While the 7(9)-O-acetyltransferase is wide-spread, as the occurrence of $Neu5,9Ac_2$, exhibiting highest activity observed so far in bovine submandibular glands, the 4-O-acetyltrans-

$Neu5,9Ac_2\alpha(2\text{->}8)Neu5Ac\alpha(2\text{->}3)Gal\beta(1\text{->}4)Glc\beta(1\text{->}1')Cer$

Fig. 8. Structure of O-acetylated GD3, an antigen specific for human melanoma cells (from Cheresh et al., 1984)

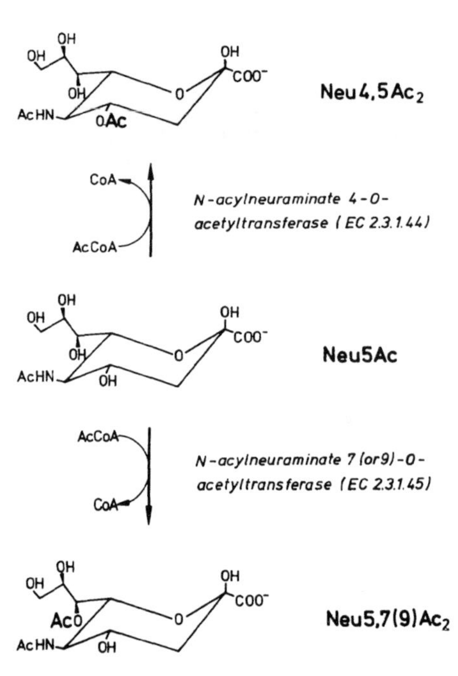

Fig. 9 Biosynthesis of O-acetylated sialic acids by the action of two specific O-acetyltransferases. For details see the text.

ferase, according to our present knowledge, is restricted to a few animals (horse, donkey and the Australian monotreme echidna (Tachyglossus aculeatus) (Schauer, 1986). The activity of the 4-O-acetyltransferase leads to O-acetylation of O-4 at the pyranose ring only, the 7(9)-O-acetyltransferase probably is responsible for O-acetylation of the whole glycerol side chain of sialic acids, by successive transfer of O-acetyl groups to O-7, followed by spontaneous migration of the ester groups to O-8 and O-9. These reactions are depicted in Fig. 10. The figure also shows the action of a sialic acid-specific esterase (N-acyl-O-acetylneuraminate O-acetylhydrolase (EC 3.1.1.53), which initiates the catabolic pathway of O-acetylated sialic acids (and consequently of glycoconjugates) by hydrolysis of the ester groups. As this enzyme, detected so far in many vertebrate tissues and isolated from horse liver and bovine brain (Schauer, 1986; Shukla *et al.*, unpublished results), only acts on O-acetyl groups at O-4 and O-9, such residues at O-7 and O-8 are only hydrolyzed after migration to O-9 (Fig. 10). Thus, three enzymes (two O-acetyltransferases and one esterase) are involved in the metabolism of O-acetylated sialic acids and correspondingly

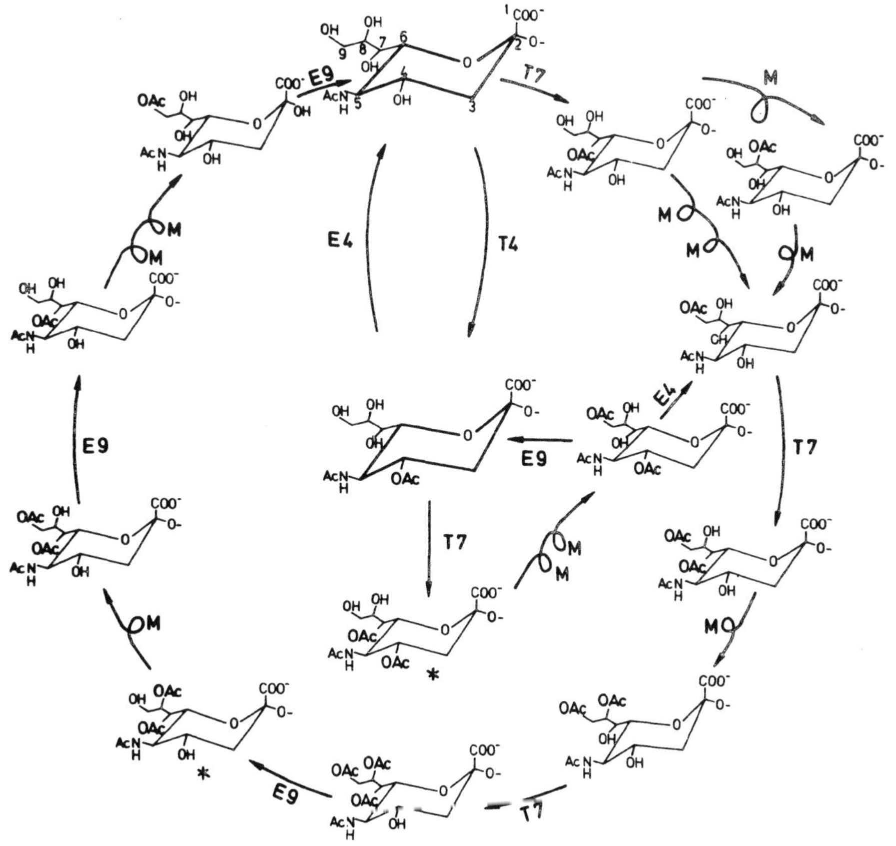

Fig. 10 Metabolism of O-acetylated sialic acids. Cooperative effects of two sialate-O-acetyltransferases, sialate-O-acetylesterase and non-enzymic migration of O-acetyl groups. T4, reaction catalyzed by N-acylneuraminate-4-O-acetyltransferase; T7, N-acylneuraminate-7(9)-O-acetyltransferase; E4, sialate-O-acetylesterase acting on 4-O-acetyl residues; and E9, the same esterase acting on 9-O-acetyl grups. Sialic acids marked with an asterisk have not yet been found, probably because they are very labile intermediates. For further details see the text.

in the control of O-acetylated glycoconjugates as antigens. The first chemical indication for the existence of O-acetylated sugars later identified as O-acetylated sialic acids was obtained in bovine submandibular gland 50 years ago (Blix, 1936).

Besides of their immunological role, sialic acid O-acetyl groups have other biological implications, as summarized in Scheme 2. Among the most important roles is their retarding effect on the degradation of glycoconjugates initiated by sialidase (Schauer, 1982, 1986). An *O*-acetyl residue at *O*-4 completely prevents the action of this enzyme obtained from animal and microbial sources, and *O*-acetyl groups at the sialic acid side chain strongly hinder enzymic release of sialic acids. As removal of sialic acids initiates further degradation of glycoconjugates by other glycosidases and proteases, the presence of *O*-acetyl groups extends the life-time and thus the biological role of glycoconjugates. This may be important for soluble glycoconjugates, e.g. serum glycoproteins, or for glycoconjugates immobilized in cell plasma membranes. Thus, the long life-time of erythrocytes from various mammalian species may be due to the presence of a high ratio of *O*-acetylated sialic acids in their cell membranes (Shukla & Schauer, 1982).

Furthermore, *O*-acetylated sialic acids have been found to modify the complement reactivity of mouse erythrocytes (Varki & Kornfeld, 1980), participate in environmental adaptation (e.g. in the dormouse the content of *O*-acetyl groups in ganglioside sialic acids is decreased during hibernation; Rahmann *et al*., 1983) and play an important, but dual role in the receptor specificity of influenza viruses. On the one hand, *O*-acetyl groups of $Neu5,9Ac_2$ transferred to erythrocytes prevent binding of various influenza virus isolates from human or animal origin (Higa *et al*., 1985), but on the other hand the same sialic acid functions as receptor for influenza C viruses (Herrler *et al*., 1985). After the attachment of influenza C virus a sialic acid-specific esterase as constituent of the virus hydrolyzes sialic acid *O*-acetyl groups, which leads to release of the virus from the erythrocytes or to destruction of the hemagglutination inhibitory potency of glycoconjugates bearing $Neu5,9Ac_2$. This viral esterase has properties similar to those of the mammalian sialic acid-specific esterase mentioned above. It represents, after sialidase, a second receptor destroying enzyme found in a virus species (Herrler *et al*., 1985) (Fig. 11).

Scheme 2. The data for these functions of *O*-acetylated sialic acids are taken from references cited in the text.

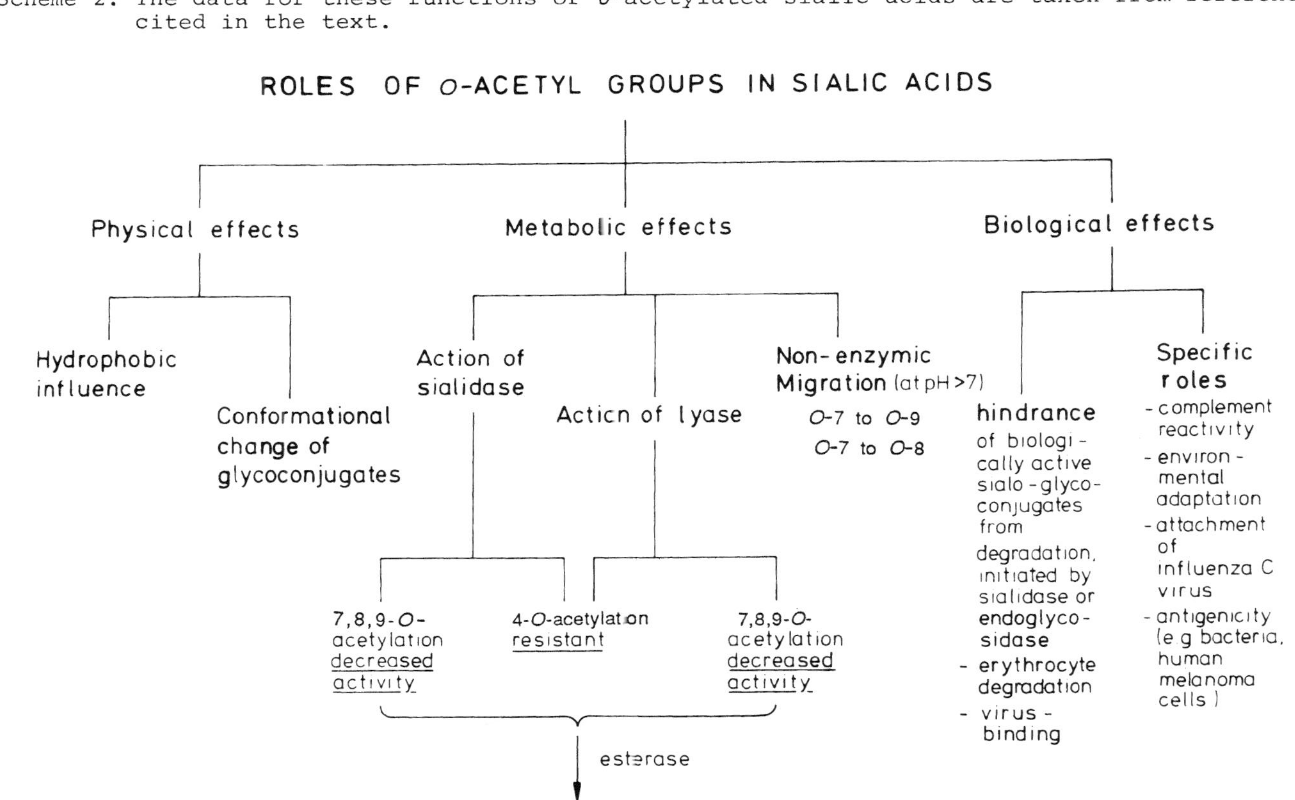

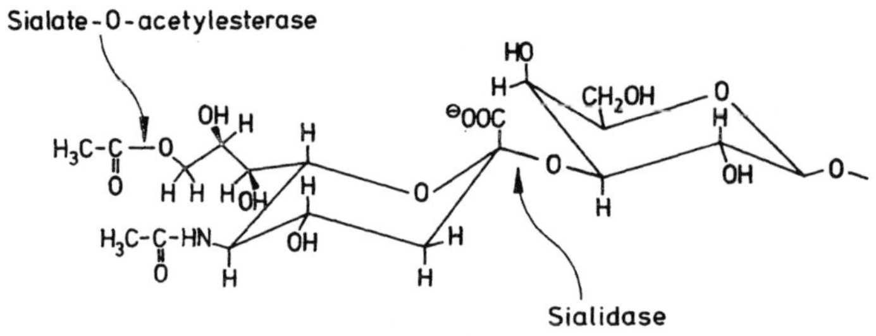

Fig. 11 The two receptor destroying enzymes (RDE) of influenza viruses (from Herrler et al., 1985). For details see the text.

MASKING OF ANTIGENIC AND OTHER RECOGNITION SITES BY SIALIC ACIDS

Although sialic acids represent antigens, even autoantigens that can lead e.g. to the production of cold agglutinins in blood, the main function of sialic acids seems to be the masking of antigenic and other recognition sites of soluble molecules and cells (Ashwell & Harford, 1982; Reutter et al., 1982; Schauer, 1982; Schauer et al., 1984b, Schauer, 1985). The chemical nature of these sites may be protein, carbohydrate or lipid. A selection of the most important and well-investigated examples for the masking effect of sialic acids is shown in Scheme 3.

A classical example for such a role (not schown in Scheme 3) is the masking by sialic acid of the T-antigen, the specificity of which is due to terminal galactose residues ß(1->3)-linked to oligosaccharide chains bound O-glycosidically to erythrocyte membrane glycoproteins (Anstee, 1981) (Fig. 12). Human erythrocytes exhibit T-reactivity only after sialidase treatment (Fischer & Poschmann, 1976). Furthermore, 3-fucosyl-N-acetyllactosamine antigen is expressed in human tissues only after the enzymic removal of sialic acid (Howie & Brown, 1985).

These are well-defined examples of the masking of antigenic sites by sialic acids. A variety of similar effects on

Scheme 3. The examples for this antirecognition function of sialic acids observed in various biological and pathological systems are taken from reviews by Ashwell + Harford, 1982; Reutter et al., 1982; Schauer, 1982; and Schauer *et al.*, 1984b. For more details see the text.

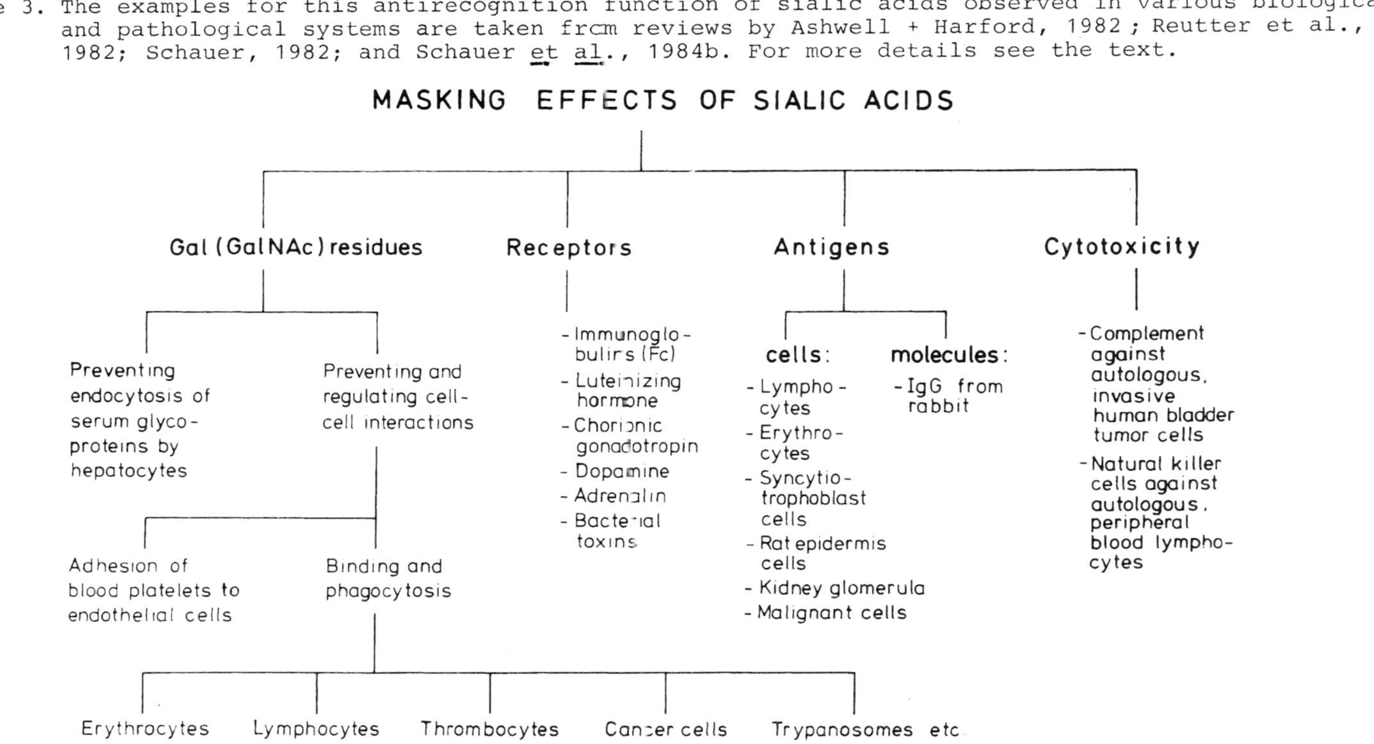

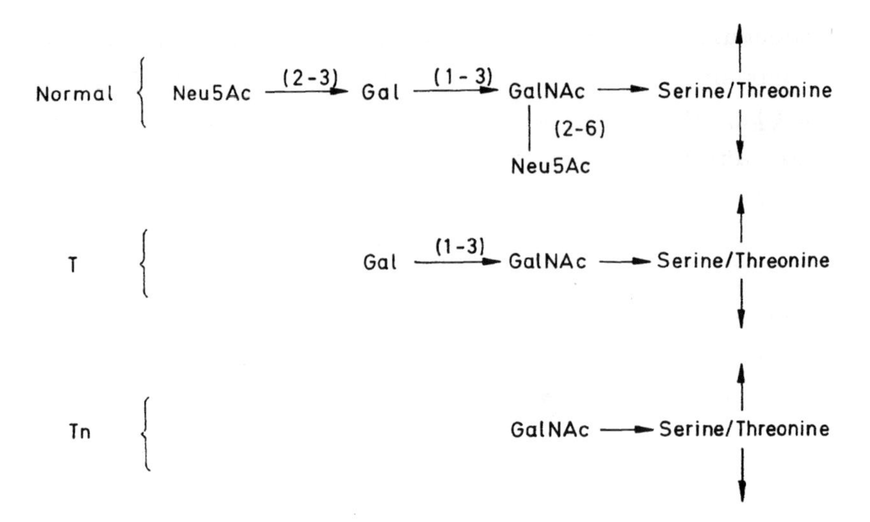

Fig. 12 Proposed structures of the major O-glycosidically linked oligosaccharides of the sialoglycoproteins in normal, T-, and Tn-exposed erythrocyte membranes (from Anstee, 1981)

the antigenicity of cells and macromolecules has become known since the last 20 years, although in most cases, as far as it was investigated, the antigenicity of the protein part of glycoconjugate molecules was found to be reduced or abolished by the presence of sialic acids. One of the most striking and important examples is the immunobarrier between mother and fetus due to the presence of highly sialylated glycoconjugate molecules on syncytiotrophoblast cells (Kelley *et al.*, 1979; Taylor *et al.*, 1979). These glycoconjugates prevent the formation of antibodies by the maternal organism against fetal cells, obviously by masking transplantation antigens. Antibodies were formed against mouse trophoblast cells by the maternal organism only after enzymic removal of sialic acids. These antibodies resulted in the rejection of mouse fetuses. Sialic acids have also been found to mask glycoproteins of kidney glomerular membranes (Blau & Haas, 1973). Enzymic removal of these sialic acid residues, which may occur during viral or bacterial infections, is considered to be related to immunological injuries of the glomeruli leading to chronic glomerulonephritis. It has generally been observed that cells and tissues rich in sialoglycoconjugates tend to be the cause of autoimmune disease (Reutter *et al.*, 1982). Evidence is also accumulating that sialic acid residues mask tumor antigens, summarized e.g. by Gahmberg and Andersson, 1982, Reutter *et al.*, 1982, and Schauer, 1982. These observations have lead to attempts to cure cancer by sialidase treatment of tumor cells and vaccination of the patient with these cells.

In the examples described removal of sialic acids results in recognition of cell- or tissue-specific antigens by cells of the immune-defense system. Macrophages may be involved in this process by recognition of immunoglobuline-loaded cells via their Fc-receptor, a mechanism which may also operate in the examples given above. Macrophages, however, can also phagocytoze cells in a non-immunological way by recognition of sugar residues on cells via lectins. Such a mechanism becomes active by the loss of sialic acids and exposure of galactose or *N*-acetylgalactosamine residues on cell membranes. It was shown that liver or peritoneal macrophages express a galactose-specific lectin which can bind partially desialylated blood and tumor cells (Scheme 3) (Kolb *et al.*, 1981; Müller *et al.*, 1983; Schauer et al., 1984b; Schauer, 1985). Binding and phagocytosis studies were mainly carried out with erythrocytes, showing that immunoglobulins and complement are not involved in this interaction, but only terminal sugar residues on the erythrocytes as ligands and galactose-specific lectin molecules on the macrophage surface as receptors, as outlined in Fig. 13. The lectin has been isolated from Kupffer cells and characterized (Roos *et al.*, 1985). This mechanism explains why the interaction can be inhibited

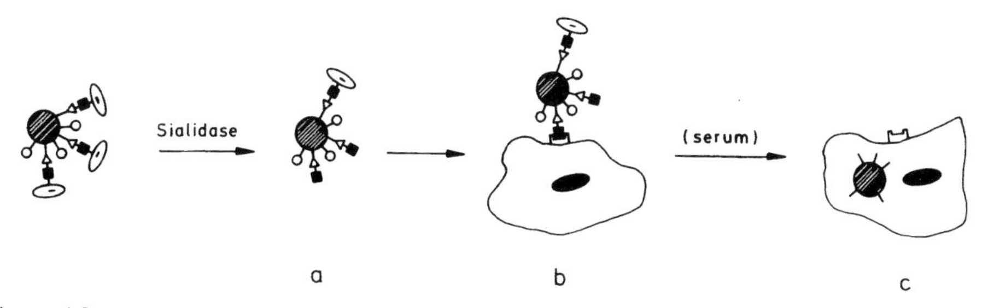

Fig. 13 Mechanism of binding (b) and phagocytosis (c) of sialidase-treated erythrocytes (a) by macrophages. ⊖, negatively charged sialic acid; ■, galactose residue; O, other membrane components; ┌─┐, galactose-specific lectin on the macrophage surface.

both in vivo and in vitro by free or glycosidically bound galactose. Galactose linked to glycoproteins was found to be a much stronger inhibitor than the free sugar or lactose, inhibiting the erythrocyte-macrophage interaction at µmolar concentration (Schauer et al., 1984b; Kelm et al., 1986).

With regard to the physiological significance of this phenomenon it is assumed that similar to hepatocytes and desialylated serum glycoproteins (Ashwell & Harford, 1982), macrophages are responsible for the removal of erythrocytes, lymphoyctes, thrombocytes and other, e.g. tumor cells from blood stream, as soon as a critical density of galactose or N-acetylgalactosamine residues has been exposed on these cells. This can be due to a loss of sialic acid in the course of ageing of these cells, or it can be the result of incomplete sialylation of e.g. tumor cells. The clearance of bacteria from blood stream may be mediated by the same lectin, or, as is assumed for fungi, e.g. yeast cells coated by mannan, by the mannose-specific receptor which is also present on macrophages (Stahl et al., 1984).

The pathophysiological importance of the galactose-specific macrophage lectin may be the sequestration of non-aged erythrocytes and other blood cells, which have lost sialic acid by viral or bacterial sialidases during infections. Several diseases (influenza virus, meningococcal, pneumococcal and clostridial (gas oedema) infections) are known which are accompanied by elevated serum sialidase levels, and which often develop light or severe anemic states (Schauer et al., 1984b, Schauer et al., 1985).

The frequent occurrence of galactose-specific lectins in vertebrates points to the possibility that cell-cell interactions as described here for e.g. macrophages and erythrocytes generally play a role in cellular interactions in tissues. As is shown in Fig. 14, such interactions may be regulated by reversible sialylation of galactose residues with the aid of sialyltransferases and sialidases. An example for such a mechanism might be the thymus, where immature lymphocytes known to contain only a small amount of sialic acid and to expose many galactose residues on their surface (Reisner *et al.*, 1976) are retained by macrophages exposing a galactose-specific lectin (unpublished observations). During maturation, the lymphocytes acquire more plasma membrane sialic acids which results in liberation of the cells into the blood stream. It certainly will be rewarding to search for more examples of sialic acids being involved in immunological and non-immunological recognition phenomena in mature and especially in developing organisms.

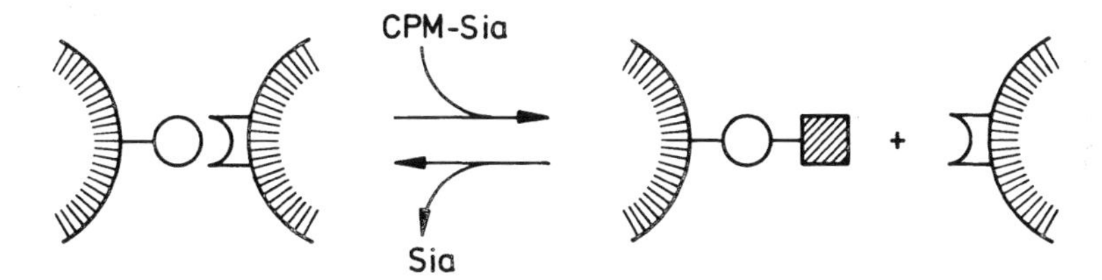

Fig. 14 Possible mechanism of regulation of cell aggregation by sialylation and desialylation.
○ , galactose; ▨ , sialic acid (Sia);
Ɒ , lectin molecule.

SUMMARY

Sialic acids, comprising a group of *N*- and *O*-acyl as well as *O*-methyl and *O*-sulfate derivatives of neuraminic acid, are differently linked to various sugars of glycoproteins and gangliosides. Thus, the great molecular variety already existing in the asialo-portion of glycoconjugates is much increased by the addition of sialic acids. This structural diversity may be mirrored by a high antigenic diversity of the oligosaccharide part of complex carbohydrates. However, not much is known of such an influence of the sialic acid moieties of glycane chains, although antibodies against many non-sialylated oligosaccharides are known. In most cases investigated so far, sialic acids reduce or even prevent antigenicity of oligosaccharide chains and of the protein or lipid parts of glycoconjugate molecules, respectively. Enzymic removal of sialic acid or reduction of its carboxylic group to an alcohol residue alters or significantly increases the antigenicity. In contrast to this masking effect, sialic acids have been found to act as antigenic determinants. N-Acetylneuraminic acid in different linkages to other sugars is an essential component of various antigens in human erythrocytes, and N-glycolylneuraminic acid determines the blood group specificity of East-Asian dogs. The latter sialic acid also plays a role in immunological processes in man. N-Glycolylneuraminic acid probably cannot be synthesized in man, but is derived from the foodstuffs and incorporated into tissue glycoconjugates in small quantities. Sialic acid O-acetyl groups are also involved in immunological processes by modifying the antigenicity of glycoproteins and polysaccharides or by contributing to the specificity of tumor antigens. Evidence for a role of the multiple forms of sialic acids as differentiation antigens is accumulating.

REFERENCES

Anstee D.J. (1981) The Blood Group MNSs-Active Sialoglycoproteins, Seminars Hematol. 18:13-31.

Ashwell G. and Harford J. (1982) Carbohydrate-Specific Receptors of the Liver, Ann. Rev. Biochem. 51:531-554.

Blau E.B. and Haas J.E. (1973) Glomerular Sialic Acid and Proteinuria in Human Renal Disease, Lab. Invest. 28:477-481.

Blix G. (1936) Über die Kohlenhydratgruppen des Submaxillarismucins, Hoppe-Seyler's Z. Physiol. Chem. 240:43-54.

Cheresh D.A., Varki A.P., Varki N.M., Stallcup W.B., Levine J. and Reisfeld R.A. (1984) A Monoclonal Antibody Recognizes an O-Acylated Sialic Acid in a Human Melanoma-associated Ganglioside, J. Biol. Chem. 259:7453-7459.

Corfield A.P. and Schauer R. (1982) Occurrence of Sialic Acids, in: Sialic Acids, Cell Biol. Monogr. Vol. 10, Schauer R, ed., Springer, Wien, pp. 5-50.

Deicher H. (1926) Über die Erzeugung heterospezifischer Hämagglutinine durch Infektion artfremden Serums, Z. Hyg. Infektionskr. 106:561-579.

Faillard H. and Cabezas J.A. (1963) Isolierung von N-Acetyl- und N-Glykolylneuraminsäure aus Kälber- und Hühnerserum, Hoppe-Seyler's Z. Physiol. Chem. 333:266-271.

Faillard H. and Schauer R. (1972) Glycoproteins as Lubricants, Protective Agents, Carriers, Structural Proteins and as Participants in Other Functions, in: Glycoproteins - Their Composition, Structure and Function, BBA Library, Vol. 5, Gottschalk A., ed., Elsevier, Amsterdam, pp. 1246-1267.

Feizi T. (1985) Demonstration by Monoclonal Antibodies that Carbohydrate Structures of Glycoproteins and Glycolipids are Onco-Developmental Antigens, Nature 314:53-57.

Finne J. (1982) Occurrence of Unique Polysialosyl Carbohydrate Units in Glycoproteins of Developing Brain, J. Biol. Chem. 257:11966-11970.

Fischer K., and Poschmann A. (1976) Neuraminidase-Induced Hemolytic Anemia - Diagnostic and Therapeutic Guidelines, Dtsch. Med. Wochenschr. 101:1731-1733.

Gahmberg C.G. and Andersson L.C. (1982) Surface Glycoproteins of Malignant Human Leukocytes, Biochim. Biophys. Acta 651:65-83.

Hakomori S., Patterson C.M., Nudelman E. and Sekiguchi K. (1983) A Monoclonal Antibody Directed to N-acetylneuraminosyl-α(2-6)-Galactosyl Residue in Gangliosides and Glycoproteins, J. Biol. Chem. 258:11819-11822.

Hanganutziu M. (1924) Hémagglutinines Hétérogénétiques après Injection de Sérum de Cheval, Compt. Rend. Soc. Biol. 91:1457-1459.

Herrler G.,Rott R., Klenk H.-D., Müller H.-P., Shukla A.K. and Schauer R. (1985) The Receptor-Destroying Enzyme of Influenza C Virus is Neuraminate-O-Acetylesterase, Embo J. 4:1503-1506.

Higa H.H., Rogers G.N. and Paulson J.C. (1985) Influenza Virus Hemagglutinins Differentiate between Receptor Determinants Bearing N-Acetyl-, N-Glycollyl-, and N,O-Diacetylneuraminic Acids, Virology 144:279-282.

Higashi H., Fukui Y., Ueda S., Kato S., Hirabayashi Y., Matsumoto M. and Naiki, M. (1984) Sensitive Enzyme-Immunostaining and Densitometric Determination on Thin-Layer Chromatography of N-Glycolylneuraminic Acid-Containing Glycosphingolipids, Hanganutziu-Deicher Antigens, J. Biochem. 95:1517-1520.

Higashi, H., Hirabayashi Y., Fukui Y., Naiki M., Matsumoto M., Ueda S. and Kato S. (1985) Characterization of N-Glycolylneuraminic Acid-containing Gangliosides as Tumor-associated Hanganutziu-DeicherAntigen in Human Colon Cancer, Cancer Res. 45:3796-3802.

Howie A.J. and Brown G. (1985) Effect of Neuraminidase on the Expression of the 3-Fucosyl-N-Acetyllactosamine Antigen in Human Tissues, J. Clin. Pathol. 38:409-416.

Jennings H.J., Katzenellenbogen E., Lugowski C., Michon F., Roy R. and Kasper D.L. (1984) Structure, Conformation and Immunology of Sialic Acid-Containing Polysaccharides of Human Pathogenic Bacteria, Pure Appl. Chem. 56:893-905.

Kasukawa R., Kano K., Bloom M.L. and Milgrom F. (1976) Heterophile Antibodies in Pathologic Human Sera Resembling Antibodies Stimulated by Foreign Species Sera, Clin. Exp. Immunol. 25:122-132.

Kelley L.K., King B.F., Johnson L.W. and Smith C.H. (1979) Protein Composition and Structure of Human Placental Microvillous Membrane, Exp. Cell Res. 123:167-176.

Kelm S., Jibril S., Lee H., Yoshino T. and Schauer R. (1986) The Galactose Receptor of Rat Peritoneal Macrophages: Binding of Sialidase-Treated Blood Cells and Glycoproteins, in: Lectins, Biology, Biochemistry, Clinical Biochemistry, Vol. 5, Bøg-Hansen T.C. and van Driessche E., eds., Walter de Gruyter, Berlin, pp. 221-228.

Kolb H., Vogt D. and Kolb-Bachofen V. (1981) Does the D-Galactose Receptor on Kupffer Cells Recycle? Biochem. J. 200:445-448.

Montreuil J. (1980) Primary Structure of Glycoprotein Glycans, Adv. Carbohydr. Chem. Biochem. 37:157-223.

Montreuil J., Debray H., Debeire P. and Delannoy P. (1983) Lectins as Oligosaccharide Receptors, in: Structural Carbohydrates in the Liver, Falk Symposium 34, Popper H., Reutter W., Köttgen E. and Gudat, F., eds., MTP Press Boston, pp. 239-258.

Müller E., Schröder C., Sharon N. and Schauer R. (1983) Binding and Phagocytosis of Sialidase-Treated Rat Erythrocytes by a Mechanism Independent of Opsonins, Hoppe-Seyler's Z. Physiol. Chem. 364:1410-1420.

Nishimaki T., Kano K. and Milgrom F. (1979) Hanganutziu-Deicher Antigen and Antibody in Pathologic Sera and Tissues, J. Immunol. 122:2314-2318.

Nöhle U. and Schauer R. (1984) Metabolism of Sialic Acids from Exogeneously Administered Sialyllactose and Mucin in Mouse and Rat, Hoppe-Seyler's Z. Physiol. Chem. 365:1457-1467.

Ørskov F., Ørskov I., Sutton A., Schneerson R., Lin W, Egan W., Hoff G.E. and Robbings J.B. (1979) Form variation in E. coli K1: Determined by O-Acetylation of the Capsular Polysaccharide. J. Exp. Med. 149:669-685.

Picard J.K., Loveday D. and Feizi T. (1985) Evidence for Sialylated Type 1 Blood Group Chains on Human Erythrocyte Membranes Revealed by Agglutination of Neuraminidase-Treated Erythrocytes with Waldenström's Macroglobulin IgMwoo and Hybridoma Antibody FC 10.2, Vox Sang. 48:26-33.

Rahmann H., Hilbig R., Probst W. and Mühleisen M. (1984) Brain Gangliosides and Thermal Adaptation in Vertebrates, in: Ganglioside Structure, Function and Biomedical Potential, Adv. Exp. Med. Biol., Vol. 174, Ledeen R.W., Yu R.K., Rapport M.M. and Suzuki K., eds., Plenum Press, New York, pp. 395-404.

Reisner Y., Linker-Israeli M. and Sharon N. (1976) Separation of Mouse Thymocytes into Two Subpopulations by the Use of Peanut Agglutinin, Cell. Immunol. 25:129-134.

Reutter W., Köttgen E., Bauer C. and Gerok W. (1982) Biological Significance of Sialic Acids, in Sialic Acids, Cell Biol. Monogr., Vol. 10, Schauer R., ed., Springer, Wien, pp. 263-305.

Roelcke D. (1984) Kälteagglutinine: Humane monoklonale Antikörper gegen Glykokonjugat-Antigene von Zelloberflächen, Funkt. Biol. Med. 3:106-127.

Roelcke D. (1985) Li Cold Agglutinin: A Further Antibody Recognizing Sialic Acid-Dependent Antigens Fully Expressed on Newborn Erythrocytes, Vox Sang. 48:181-183.

Roelcke D., Pruzanski W., Ebert W., Römer W., Fischer E., Lenhard V. and Rauterberg E. (1980) A New Human Monoclonal Cold Agglutinin Sa Recognizing Terminal N-Acetylneuraminyl Groups on the Cell Surface, Blood 55:677-681.

Roos P.H., Hartman H.-J., Schlepper-Schäfer J., Kolb H. and Kolb-Bachofen V. (1985) Galactose-Specific Receptors on Liver Cells. II. Characterization of the Purified Receptor from Macrophages Reveals no Structural Relationship to the Hepatocyte Receptor, Biochim. Biophys. Acta 847:115-121.

Schauer R. (1978) Biosynthesis of Sialic Acids, Methods Enzymol. 50:374-386.

Schauer R. (1982) Chemistry, Metabolism and Biological Functions of Sialic Acids, Adv. Carbohydr. Chem. Biochem. 40:131-234.

Schauer R. (1983) Sialic Acids as Potential Determinants on Differentiation Antigens, Biochem. Soc. Transact. 11:270-271.

Schauer R. (1985) Sialic Acids and their Roles as Biological Masks, Trends Biochem. Sci. 10:357-360.

Schauer R. (1986) Metabolism of O-Acetyl Groups of Sialic Acids, Methods Enzymol., in press.

Schauer R., Schröder C. and Shukla A.K. (1984a) New Techniques for the Investigation of Structure and Metabolism of Sialic Acids, in: Ganglioside Structure, Function and Biomedical Potential, Adv. Exp. Med. Biol. Vol. 174, Ledeen R.W., Yu R.K., Rapport M.M. and Suzuki, K. eds., Plenum Press, New York, pp. 75-86.

Schauer R, Shukla A.K., Schröder C. and Müller E. (1984b) The Anti-Recognition Function of Sialic Acids: Studies with Erythrocytes and Macrophages, Pure Appl. Chem. 56:907-921.

Schauer R., Sander-Wewer M., Gutschker-Gdaniec G.H.M., Roggentin P., Randow E.A, and Hobrecht R. (1985) Sialidase Activity in the Sera of Patients and Rabbits with Clostridial Myonecrosis, Clin. Chim. Acta 146:119-127.

Schröder C., Nöhle U., Shukla A.K. and Schauer R. (1983) Improved Methods for the Isolation and Structural Analysis of N-Glycolylneuraminic Acid in Man, in: Glycoconjugates, Proc. 7th Int. Symp., Chester M.A., Heinegård D, Lundblad A. and Svensson S, eds., Rahms, Lund, pp. 162-163.

Semino G.A., Restani P. and Cerletti P. (1985) Effect of Bound Carbohydrate on the Action of Trypsin on Lupin Seed Glycoproteins, J. Agricult. Food Chem. 33:196-199.

Shukla A.K. and Schauer R. (1982) Fluorimetric Determination of Unsubstituted and 9(8)-O-Acetylated Sialic Acids in Erythrocyte Membranes, Hoppe-Seyler's Z. Physiol. Chem. 363:255-262.

Smith D.F. and Ginsburg, V. (1980) Antibodies against Sialyloligosaccharides Coupled to Protein, J. Biol. Chem. 255:55-59.

Stahl P.D., Wileman T.E., Diment S. and Shepherd V.L. (1984) Manose-Specific Oligosaccharide Recognition by Mononuclear Phagocytes, Biol. Cell 51:215-218.

Staub C.A. (1985) Cold Reacting Antibodies Recognizing Antigens Dependent on N-Acetylneuraminic Acid, Transfusion 25:414-416 (1985)

Sugita M. (1979) Studies on the Glycosphingolipids of the Starfish, Asterina pectinifera. II. Isolation and Characterization of a Novel Ganglioside with an Internal Sialic Acid Residue, and III. Isolation and Structural Studies of Two Novel Gangliosides Containing Internal Sialic Acid Residues, J. Biochem. 86:289-300, 765-772, 1979.

Taylor P.V. and Hancock K.W. (1979) Effect of Neuraminidase on Immunogenicity of Early Mouse Trophoblast, Transplantation 28:256-257.

Thurin J., Herlyn M., Hindsgaul O., Strömberg N., Karlsson K.-A., Elder D., Steplewski Z. and Koprowski H. (1985) Proton NMR and Fast-Atom Bombardment Mass Spectrometry Analysis of the Melanoma-Associated Ganglioside 9-O-Acetyl-GD3, J. Biol. Chem. 260:14556-14563.

Tsai C.-M., Zopf D.A., Yu R.K., Wistar R.Jr. and Ginsburg V. (1977) A Waldenström macroglobulin that is both a cold agglutinin and a cryoglobulin because it binds N-acetylneuraminosyl residues, Proc. Natl. Acad. Sci. U.S.A. 74:4591-4594.

Van der Meer A., Kamerling J.P., Vliegenthart J.F.G., Schmid K. and Schauer R. (1983) Fucopyranosyl-(1->4)-N-Glycolylneuraminic Acid, A Constituent of Glycoproteins of the Cuvierian Tubules of the Sea Cucumber Holothuria forskali Della Chiaje, Biochim. Biophys. Acta 757:371-376.

Varki A. and Kornfeld S. (1980) Structural Studies of Phosphorylated High Mannose-type Oligosaccharides, J. Biol. Chem. 255:10847-10858.

Yasue S., Handa S., Miyagawa S., Inoue J., Hasegawa A. and Yamakawa T. (1978) Difference in Form of Sialic Acid in Red Blood Cell Glycolipids of Different Breeds of Dogs, J. Biochem. (Tokyo) 83:1101-1107.

BLOOD GROUP ACTIVE HAPTENS IN URINE AND FAECES

Arne Lundblad and M. Alan Chester

Department of Clinical Chemistry
University of Lund, Lund, Sweden

A number of oligosaccharides found in normal urine and faeces have been identified as blood group active haptens, particularly within the ABO and Lewis systems. The oligosaccharide composition in these materials is therefore dependent upon the ABO, H, Lewis and Secretor genotype of the individual.

The presence of A and B substances in urine was first demonstrated by Yosida in 1928 (1). Since then several investigations have been undertaken to isolate and characterize these antigens (2-5). It was soon recognized that the A and B antigens existed in both high and low molecular weight forms (6). The observation of Price Evans _et al_ (7) that secretors had a significantly higher excretion of fucose than non-secretors led to the discovery of a number of fucose-rich oligosaccharides present in the urine of secretors, but absent from the urine of non-secretors. Chemical and immunochemical studies of these oligosaccharides have shown that they are low molecular weight blood group active haptens in the ABO system (8).

Oligosaccharides isolated from human milk have given important information in the elucidation of the structures of the Lewis antigens (9). Lacto-_N_-fucopentaose II and Lacto-_N_-difucohexaose I were shown to be the most active Le^a and Le^b compounds respectively (9). These oligo-saccharides are excreted in the urine of pregnant and lactating women together with a large number of other milk-type oligosaccharides (10-12). On the other hand, A and B active oligosaccharides have not been isolated from human milk.

The urinary excretion of some of the A, B and H haptens is also dependent on the dietary intake of galactose (13,14). This finding led to further studies on the origin of the urinary compounds. Metabolic studies performed in a Rhesus monkey and by galactose loading experiments in man, strongly indicated that the intestine is the main site of biosynthesis (14,15).

These studies have now been extended to include analyses of faeces from breastfed infants. The results have revealed a new source of a number of oligosaccharides, including blood group active haptens. The intestine appears to be capable of an active biosynthesis of various compounds, in addition to its primary degradative role.

STRUCTURES

Tables 1-3 summarize all compounds with blood group specificity, that have been isolated from urine and faeces and characterized.

Table 1. H, Le^a or Le^b Blood Group Active Haptens Isolated from Urine in Various Physiological Conditions and/or Faeces of Breast Fed Infants.

Trivial Name	Structure	Hapten Origin*		Ref
O(H)-disaccharide	Gal |α1,2 Fuc	H	G	13,18
2-fucosyllactose	Galβ1-4Glc |α1,2 Fuc	H	P,F	11,22
Lactodifucotetraose	Galβ1-4Glc |α1,2 |α1,3 Fuc Fuc	(H)	N,P,F	11, 16-18, 22
Lacto-N-fucopentaose I	Galβ1-3GlcNAcβ1-3Galβ1-4Glc |α1,2 Fuc	H	P,F	11,22
Lacto-N-fucopentaose II	Galβ1-3GlcNAcβ1-3Galβ1-4Glc |α1,4 (on GlcNAc) Fuc	Le^a	P,F	10,22
Lacto-N-difucohexaose I	Galβ1-3GlcNAcβ1-3Galβ1-4Glc |α1,2 |α1,4 Fuc Fuc	Le^b,(H)	P,F	11,22
Lacto-N-difucohexaose IV	Galβ1-3GlcNAcβ1-3Galβ1-4Glc |α1,2 |α1,2 Fuc Fuc	H	P	11
Lacto-N-neodifucohexaose I	Galβ1-4GlcNAcβ1-3Galβ1-4Glc |α1,2 |α1,3 Fuc Fuc	H	P	11
Lacto-N-trifucoheptaose III	Galβ1-3GlcNAcβ1-3Galβ1-4Glc |α1,2 |α1,4 |α1,3 Fuc Fuc Fuc	Le^b(H)	N	17
Lacto-N-neotrifucoheptaose II	Galβ1-4GlcNAcβ1-3Galβ1-4Glc |α1,2 |α1,3 |α1,2 Fuc Fuc Fuc	H	P	11

(continued)

VI_H	Galβ1-3GlcNAcβ1-4Gal |α1,2 |α1,4 |α1,6 Fuc Fuc Fuc	Le^b(H)	G	18
Lacto-N-trifuco- sylated hexaose	Fuc |α1,3 Galβ1-4GlcNAc \β1,6 Galβ1-4Glc /β1,3 Galβ1-3GlcNAc |α1,2 |α1,4 Fuc Fuc	Le^b(H)	F	22

* N = normal urine
  G = urine after oral intake of galactose
  P = urine during pregnancy and lactation
  F = faeces of breast fed infants.

Table 2. Blood Group A Active Haptens Isolated from Urine in Various Physiological Conditions and/or Faeces of Breast Fed Infants.

Trivial Name	Structure	Origin*	Ref
A-tri	GalNAcα1-3Gal |α1,2 Fuc	G	13,18
A-tetra	GalNAcα1-3Galβ1-4Glc |α1,2 Fuc	N,F	20,21
A-penta	GalNAcα1-3Galβ1-4Glc |α1,2 |α1,3 Fuc Fuc	N,F	17-19,21
A-hexa	GalNAcα1-3Galβ1-3GlcNAcβ1-3Galβ1-4Glc |α1,2 Fuc	F	21
A-hepta	GalNAcα1-3Galβ1-3GlcNAcβ1-3Galβ1-4Glc |α1,2 |α1,4 Fuc Fuc	N,F	17,18,21
VII_A	GalNAcα1-3Galβ1-3GlcNAcβ1-4Gal |α1-2 |α1-4 |α1-6 Fuc Fuc Fuc	G	18

*See Table 1

Table 3. Blood Group B Active Haptens Isolated from Human Urine in Various Physiological Conditions

Trivial Name	Structure	Origin*	Ref
B-tri	Galα1-3Gal α1,2 Fuc	G	13,18
B-penta	Galα1-3Galβ1-4Glc α1,2 α1,3 Fuc Fuc	N	16-18
$VII_{B(a)}$	Galα1-3Galβ1-3GlcNAcβ1-3Galβ1-4Glc α1-2 α1-4 Fuc Fuc	N	17
VII_B	Galα1-3Galβ1-3GlcNAcαβ1-4Gal α1,2 α1,4 α1,6 Fuc Fuc Fuc	G	18

*See Table 1

URINE

O(H) secretors. Normal O(H) secretors have lactodifucotetraose as their most abundant urinary oligosaccharide (16) and in smaller amounts also lacto-N-trifucoheptaose III, a compound which also contains the Le^b hapten (17). When galactose is given per os either in the free form or as a glycoside, a biosynthetic process is induced which results in an excretion of 2-fucosylgalactose, which rapidly becomes the most predominant compound (13). In addition a trifucosylated hexasaccharide denoted VI_H was isolated in urine after galactose loading (18).

A secretors. One could anticipate that the compounds present in A secretor urine should be analogous to those found in O secretors, differing only in the addition of the terminal, non-reducing α(1-3)-linked N-acetylgalactosamine. This is only partially true (Table 2). As expected, the A-penta is the most abundant compound in normal urine and the A-tri predominates after galactose ingestion (13,19). The trifucosylated compounds VI_H and VII_A (18) also follow this pattern.

Normal A secretor urine contains in addition the A-tetra (20), but the O secretor equivalent, 2-fucosyllactose is not normally present in urine. Similarly the A-hepta occurs in normal urine (17,18) but the corresponding "milk" oligosaccharide, lacto-N-difucohexaose I has not been found.

B secretors. B-secretor urine contains the B-penta, the B-tri and small amounts of two heptasaccharides (Table 3), analogous to the A-oligosaccharides. No compound analogous to the A-tetra has been isolated.

Pregnancy and lactation. Human milk is a rich source of various fucosylated oligosaccharides and as early as in the 13th week of pregnancy, these "milk type" oligosaccharides start to appear in urine. Their maximal excretion is reached in the third trimester and first weeks of lactation. Lactose, lacto-Ntetraose and lacto-N-neotetraose are the most common core structures and most compounds in milk are fucosylated and sialylated derivatives of these. The fucosylation is again controlled by the H, Lewis and Secretor genes of the woman. Table 1 lists all milk type oligosaccharides with H, Le^a or Le^b haptens built into their structure which are excreted into urine during pregnancy and lactation (10-12).

In addition, myoinositol-containing oligosaccharides have been isolated; analogous to 2-fucosyllactose, lactodifucotetraose and A-penta; in which the glucose moiety is replaced by myoinositol (10-12).

FAECES

In order to obtain more information about the origin of the urinary oligosaccharides and also to learn to what extent the milk oligosaccharides survive passage through the intestine or how they are metabolized during this passage, faeces of breastfed infants with different blood groups has been studied.

Initially, one A secretor child was studied (21) and the results indicated a complete disappearance of all milk oligosaccharides. Instead a number of A-active oligosaccharides (Table 2) were excreted in quite large amounts (several hundred mg/24 h portions of faeces). Several other A secretor children have been studied and some have an excretion pattern similar to the first child studied, others give a more complex picture, excreting several milk type oligosaccharides as well as lower amounts of the A-active haptens.

One B secretor child was also studied in detail. In contrast to the A secretor children no B-active haptens could be identified, but several milk oligosaccharides were present including a new trifucosylated lacto-N-hexaose (22).

Faeces from children fed cow milk-derived formula (which does not contain fucosyl oligosaccharides) instead of breast milk was completely devoid of any of the oligosaccharides described above.

IMMUNOCHEMICAL STUDIES

Some of the oligosaccharides have been immunochemically characterized. For instance the activity of the A and B pentasaccharides was compared with that of mono- and difucosyl type 2 oligosaccharides of soluble blood group A and B substances (23).

The higher activity of the monofucosyl determinant was confirmed with most polyclonal antisera and lectins used, although with one anti-A serum the urinary difucosyl oligosaccharide was even more active than the monofucosyl derivative.

The A-tri, A-tetra, A-penta, A-hexa and A-hepta saccharides were compared in their ability to inhibit binding of polyclonal anti-A to A substance (21). The results are shown in Fig. 1. All these oligosaccharides were able to inhibit by at least 50% at amounts less than 50 nmoles under the conditions used. The A-hepta was the most active. The A-tetra was slightly better than the A-hexa.

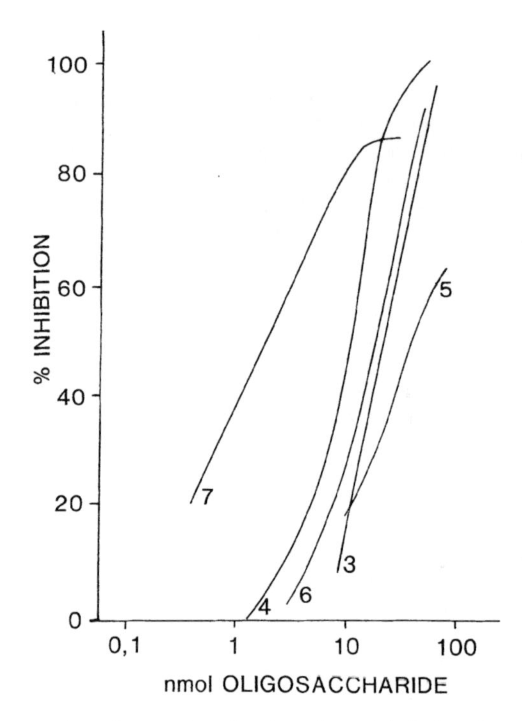

Fig.1. Inhibition of binding of polyclonal anti-A to A substance by A-tri (3), A-tetra (4), A-penta (5), A-hexa (6) and A-hepta (7).

In a study of various monoclonal antibodies it has been shown that these oligosaccharides are useful tools for the discrimination of antibodies reacting preferentially with type I and type II chains (24). This is illustrated in Table 4. Antibodies HT 29-36 and CB have broad A specificity and agglutinate all A cells. Antibody CLH6 on the other hand reacts preferentially with type I chains and does not therefore agglutinate A erythrocytes which contain predominantly type II chains.

Table 4. Inhibition of Monoclonal Antibody Binding to Blood Group A Substance in ELISA (nmoles required for 50% inhibition) (24).

Antibody	A-tetra	A-penta	A-hexa	A-hepta
HT 29-36	1.3	0.9	1.9	0.7
CB	5.8	4.0	6.1	2.2
CLH6	14.1	42.6	0.2	1.0

BIOSYNTHETIC STUDIES

Experiments with a Rhesus monkey indicate that at least the smallest ABH haptens are synthesized in the intestine (15).

^{14}C-Galactose was given either per os or via an intestinal vein to a blood group B Rhesus monkey. Non-radioactive galactose was given simultaneously per os in both experiments and the urinary excretion was examined. ^{14}C-Labelled B-tri was detected in urine only when the radioactive galactose had been administered per os. The results strongly suggest that the intestine is the site of biosynthesis of the B-tri, and by analogy the A-tri and H-disaccharide. Galactose given per os to human individuals also resulted in a much higher urinary excretion of the trisaccharides (or disaccharide in O secretors). Only slight increases in excretion were observed after intravenous administration (14).

The presence of A-active structures in the faeces of a blood group A child fed breast-milk (21), which are absent from faeces of children not ingesting human milk oligosaccharides clearly indicated that the milk oligosaccharides were converted to the corresponding A haptens during their passage through the gastrointestinal system. Subsequent examination of faeces from other blood group A babies has indicated that in some instances both A-haptens and unchanged milk oligosaccharides appear in the faeces. This variation is not a function of the childs A_1/A_2 blood group status.

Examination of faeces from a blood group B breast-fed child gave no indication of the presence of B-active haptens (22), although milk oligosaccharides were present. Since the blood group B glycosyltransferase requires a higher oligosaccharide acceptor concentration than the equivalent blood group A glycosyltransferase, one can speculate that the milk oligosaccharide concentration at the site of biosynthesis of these blood group active haptens is a critical factor. This is also suggested by the variations observed between the various blood group A individuals examined. Other, more physiological factors also presumably play an important role, such as intestinal absorption rates and flow through the intestine.

CONCLUSION

Until recently the major known human source of oligosaccharides has been milk and urine. Among the milk oligosaccharides are found haptens with H and Lewis activity. A- and B-active oligosaccharides are conspicuously absent. On the other hand urine is a rich source of ABH haptens but contains very small amounts of the Lewis structures. The oligosaccharide content of urine is also much more variable, being influenced by diet, pregnancy, and disease.

Recent investigations have shown that faeces is also a rich human source of oligosaccharides, the occurrence of which is also greatly dependent on diet and blood group. The oligosaccharide content of urine and faeces in an individual at any given time is the result of the complex interaction of biosynthetic and catabolic processes coupled to dietary and physiological parameters.

REFERENCES

1. Yosida, K. (1928). Uber die gruppenspezifischen Unterschiede der Transudate, Exsudate, Sekrete, Exkrete, Organextrakte und Organzellen des Menschen und ihre rechtsmedizinischen Anwendungen. Z. ges. explt. Med. 63:331-339.
2. Freudenberg, K. and Eichel, H. (1935). Uber spezifische Kohlenhydrate der Blutgruppen. Ann. Chem. 518:97-102.
3. Jorpes, E. and Norlin, G. (1934). Uber die blutgruppenmerkmale im Harn des Menschen. Acta path. microbiol. scand. 11:91-118.
4. Masamune, H., Hakomori, S., Masamune O. and Takase, S. (1956) Biochemical studies on carbohydrates. Group A, B and O substances from human urine. Tohoku. J. Exp. Med. 64:67-80.
5. Lundblad, A. and Berggård, I. (1967). Isolation and characterization of urinary A and B agglutinogen fractions. Biochim. Biophys. Acta 148:146-150.
6. Jorpes, E. and Thaning, T. (1945). On the inhomogeneity of the urinary human A-substance. J. Immunol. 51:221-225.
7. Price Evans, D.A., McConnel, R.B. and Donohoe, W.T.A. (1964) Fucose and agglutinogen contents of urine from patients with duodenal ulcers. J. Lab. Clin. Med. 64:581-593.
8. Lundblad, A. (1976). Urinary glycoproteins, glycopeptides and oligosaccharides. In "Mammalian Glycoproteins" (eds. M. Horowitz and W. Pigman). Acad. Press, New York, pp. 441-458.
9. Watkins, W.M. (1966). Blood group specific substances. "Glycoproteins", ed. Gottschalk, A. Elsevier, Amsterdam pp. 481-483.
10. Hallgren, P. and Lundblad, A. (1977). Structural analysis of urine oligosaccharides isolated from the urine of blood group O, non-secretor woman during pregnancy and lactation. J. Biol. Chem. 252:1014-1022.
11. Hallgren, P. and Lundblad, A. (1977). Structural analysis of oligosaccharides isolated from the urine of a blood group A, secretor woman during pregnancy and lactation. J. Biol. Chem. 252:1023-1033.
12. Hallgren, P., Lindberg, B.S. and Lundblad, A. (1977). Quantitation of some urinary oligosaccharides during pregnancy and lactation. J. Biol. Chem. 252:1034-1040.
13. Lundblad, A., Hallgren, P., Rudmark, A. and Svensson, S. (1973). Structures and serological activities of three oligosaccharides isolated from urines of nonstarved secretors and from secretors on lactose diet. Biochemistry 13:3341-3345.
14. Chester, M.A., Hallgren, P., Lundblad, A. and Messeter, L. (1979). Urinary oligosaccharide excretion induced by galactose given orally or intravenously. Eur. J. Biochem. 100:385-392.
15. Chester, M.A., Hallgren, P., Lindberg, B.S. and Lundblad, A. (1977). Biosynthesis of the blood group B-specific trisaccharide in a Rhesus monkey. Eur. J. Biochem. 77:87-91.
16. Björndal, H. and Lundblad, A. (1970). Structure of two urinary oligosaccharides characteristic of blood group O(H) and B-secretors. Biochim. Biophys. Acta 201:434-437.
17. Strecker, G. and Montreuil, J. (1973). Isolement et étude de la structure de 16 oligosaccharides isolés de l urine humaine. C.R. Acad. Sc. Paris 277:1393-1396.
18. Strecker, G., Trentesaux-Chauvet, C., Riazi-Farzad, T., Fournet, B., Bouquelet, S. and Montreuil, J. (1973). Mise en évidence d'une oligosaccharidosurie associée à diverses mélituries et détermination de la structure des oligosaccharides excrétés. Hypothèse concernant l' origine des oligosaccharides des liquides biologiques. C.R. Acad. Sc. Paris 277:1569-1572.

19. Lundblad, A. and Svensson, S. (1973). The structure of a urinary difucosyl pentasaccharide, characteristic of secretors with the blood group A gene. Carbohydr. Res. 30:187-189.
20. Derappe, C., Lundblad, A., Messeter, L. and Svensson, S. (1980). Isolation and characterization of a blood group A-specific urinary tetrasaccharide. FEBS Lett. 119:177-180.
21. Sabharwal, H., Nilsson, B., Chester, M.A., Sjöblad, S. and Lundblad, A. (1984). Blood group specific oligosaccharides from faeces of a blood group A breast-fed infant. Mol. Immunol. 21:1105-1112.
22. Sabharwal, H., Nilsson, B., Chester, M.A., Lindh, F., Grönberg, G. and Lundblad, A. (1985). Structures of the main oligosaccharides isolated from faeces of a blood group B, secretor, breast-fed infant. Proc. VIII Int. Symp. Glycoconj. Houston, Texas. 1985.
23. Lundblad, A. and Kabat, E.A. (1971). Immunochemical studies on blood groups. XLVIII Activity of fucose-containing oligosaccharides isolated from human urine and from human ovarian cyst A and B blood group substances. J. Immunol. 106:1572-1577.
24. Furukawa, K., Clausen, H., Hakomori, S., Sakamoto, J., Look, K., Lundblad, A., Mattes, M.J. and Lloyd, K.O. Analyses of the specificity of five murine anti-blood group A monoclonal antibodies, including one that identifies type 3 and type 4 A determinants. Biochemistry, in press.

THE SEPARATION OF IMMUNOCYTE SUBPOPULATIONS BY USE OF VARIOUS LECTINS

Toshiaki Osawa

Division of Chemical Toxicology and Immunochemistry
Faculty of Pharmaceutical Sciences
University of Toyko
Toyko, Japan

Recent development in immunology has revealed that immune responses can develop only after complicated cell-cell interactions between various subpopulations of immune-competent cells. Elucidation of the molecular mechanisms of these cell-cell interactions requires the separation of these cell subpopulations, and precise knowledge on their biological characteristics and on the factors secreted by them which mediate the cell-cell interactions. Furthermore, for the treatment of diseases which are caused by or at least related to a functional abnormality of certain subpopulation of immunocytes, adoptive transfer of the cell subpopulation in question may be effective.

On the other hand, it has also been revealed that various subpopulations of immunocytes, which differ in function and differentiation stage from each other, generally have different cell surface marker antigens. Since these cell surface marker antigens are complex-carbohydrates in many cases, lectins can be used for the mutual separation of immunocyte subpopulations.

In this article we describe mainly our attempts at the separation of functionally distinct immunocyte subpopulations by use of various purified lectins.

SEPARATION OF B CELLS FROM MOUSE SPLENOCYTES

In 1972, Schnebli and Dukor[1] suggested the use of wheat germ agglutinin (WGA) for the detection of B cells on the basis of the experimental fact that WGA can agglutinate splenocytes from nude mice but not cortisone-resistant mouse thymocytes or thoracic duct lymphocytes. This finding was later applied for the actual separation of T and B cells by Bourguignon et al.,[2] but the B cell population thus separated was found to be still contaminated by 10-15% T cells. However, Reisner et al.[3] were the first to attempt to separate B cells from mouse splenocytes. They treated mouse splenocytes with soybean agglutinin (SBA), and then the agglutinated cells (SBA^+) were separated from unagglutinated ones by sedimentation at unit gravity in 50% fetal calf serum. They found that 83% of the SBA^+ cells were cell surface immunoglobulin-positive and only a small percentage of the cells (5-8%) in this fraction was Thy-1^+ positive.

We treated mouse splenocytes with Bauhinia purpurea agglutinin (BPA), and the BPA^+ cells were separated by sedimentation at unit gravity in 50% fetal calf serum.[4] The BPA^+ fraction mainly consisted of cell surface IgM positive cells (68.4%) and Thy-1^+ positive cells (14%). BPA^+ cells responded to LPS more strongly than BPA^- cells did, whereas BPA^- cells showed better responses to concanavalin A (Con A) and pokeweed mitogen (PWM). Reisner et al.[3] deduced that the receptors for SBA on B cells are terminal GalNAc residues exposed on the cell surface, which are not exposed on mature T cells because they are possibly sialylated. We concluded, however, that the difference in agglutinability with BPA between T and B cells may be due to the difference in the total amount of cell-surface mucin-type sugar chains, since BPA binds strongly to a terminal GalNAc residue or a Galβ1-3GalNAc sugar sequence even when the sugar chain is sialylated.[5] So, the difference in the total amount of mucin-type sugar chains seems to more plausibly to explain the difference between T and B cells in the agglutinability with BPA than that in the extent of sialylation.

The B-cell-enriched fraction (BPA^+) was then treated with Lens culinaris agglutinin (LCA), and the LCA^+ and LCA^- cell populations were separated by sedimentation at unity gravity in 50% fetal calf serum. As shown in Table 1, almost the same percentage of cell surface IgM-positive cells was observed in each fraction, however, the percentage of surface IgD-positive cells in the LCA^- fraction was almost half that in the LCA^+ fraction and that in unfractionated B cells. Surface IgD is thought to be one of the differentiation markers on B cells, and it was claimed by Gronowicz and Coutinho[6] that the responses to polyclonal B cell mitogens, lipopolysaccharide (LPS), purified protein derivative (PPD) and dextran surface (DxS) vary according to the maturation stage of B-cell subsets. As shown in Table 2, both LCA^- and LCA^+ B cells responded almost equally to both LPS and PPD, but LCA^- B cells showed a weaker response to DxS than LCA^+ B cells. Thus the agglutinability with LCA may be assumed to reflect the maturational stage of B cell on the basis of the expression of surface IgD and of the response to DxS. These results are in accord with the fact that surface IgD has a stronger affinity to LCA than

Table 1. Surface IgM and IgD frequency in LcA-fractionated B cells

Fraction	Distribution, % of recovered cells[1]	Surface IgM^+ %[2]	Surface IgD^+ %[2]
Unfractionated B		58.2	65.5
Top	17.9	68.0	39.6
Bottom	54.5	64.8	75.2

[1] Remaining cells comprize agglutinated and non-agglutinated cells in the middle fraction.

[2] Percentages of surface IgM^+ or IgD^+ cells were determined from the immunofluorescence.

Table 2. Proliferative response of LcA-fractionated B cells[1]

Fraction	Mitogen[2]			
	None	LPS	PPD	DxS
Unfractionated B	4,100 ± 750	53,500 ± 300	38,500 ± 500	38,200 ± 6,000
Top	2,900 ± 200	57,300 ± 900	39,000 ± 1,800	29,700 ± 3,700
Bottom	7,800 ± 1,600	56,000 ± 3,300	33,100 ± 1,000	64,000 ± 3,800

[1] Data are expressed as ^{3}H-TdR incorporated (dpm/4 x 10^{5} cells). Each value represents the mean ± SD of triplicate cultures.

[2] Concentrations of mitogens: LPS, 20 μg/ml; PPD, 100 μg/ml; DxS, 50 μg/ml.

surface IgM,[7] and it may be possible that LCA recognizes surface IgD-positive B cells through the carbohydrate moieties of IgD molecules on the cell surface. The results described above show that staining with LCA may be a useful method for the determining the maturational stage of the B cell lineage.

INDUCTION OF SUPPRESSOR AND HELPER T CELLS BY LECTINS

Dutton[8] reported that helper and suppressor T cell activities could be induced when T lymphocytes were activated with mitogens such as Con A and phytohemagglutinin-P (PHA). We attempted to preferentially induce different T cell subsets with various lectins.[9] Mouse spleen cells were incubated with various amounts of a lectin for 40 hr at 37°C, and then the activated lymphocytes were separated by Ficoll-Urografin density gradient centrifugation followed by washing with a haptenic sugar solution to remove the lectins. The lectin-treated lymphocytes were then cultured with normal spleen cells and sheep red blood cells (SRBC). Five days later, the number of direct IgM producing plaque forming cells (PFC) was determined by the method of Cunningham and Szenberg.[10] As shown in Fig. 1, Con A and PHA induced suppressor cell activity, rather than helper cell activity, under our experimental conditions, whereas LCA and BPA were found to selectively induce helper activity. The results obtained using various lectins are summarized in Table 3. Raff[11] showed that T cells comprized two functionally distinct cell populations showing different maturation stages, i.e. T_1 and T_2 cells. T_1 cells are relatively short lived and immature. Gershon[12] indicated that T_1 cells might correspond to suppressor T cells, while T_2 cells might correspond to helper T cells. Erb _et al._[13] also showed that the helper cell precursors comprize a long lived T_2-cell pool. Although Con A, LCA and _Pisum sativum_ agglutinin (PSA), see Table 3, are mannose-binding lectins, they show subtle differences in their carbohydrate-binding specificities,[14,15] and, therefore, they may selectively activate distinct T cell subsets.

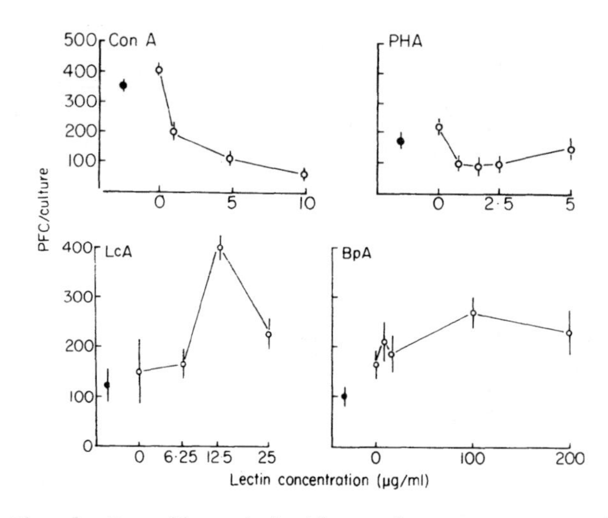

Fig. 1. In vitro induction of suppressor and helper cell activities with Con A, PHA, LCA or BpH. Spleen cells (10^7 cells/ml) were cultured for 40 hr with a lectin. The viable lymphocytes were purified and resuspended in RPMI 1640 medium. 2 x 10^5 lectin activated cells were added to 2 x 10^6 fresh spleen cells plus 10^6 SRBC followed by culturing for 5 days. (●) Without addition of lectin-activated cells. Vertical bars represent SEM for triplicate experiments.

Table 3. Lectins which induce suppressor and helper cells

Lectins which preferentially induce suppressor cells	Concanavalin A (Con A) Phaselous vulgaris agglutinin (PHA-P) Pisum sativum agglutinin (PSA) Wistaria floribunda mitogen (WFM)
Lectins which preferentially induce helper cells	Lens culinaris agglutinin (LCA) Bauhinia purpurea agglutinin (BPA)
Lectins which induce neither suppressor cells nor helper cells	Wheat germ agglutinin (WGA) Peanut (Arachis hypogaea) agglutinin (PNA) Limulus polyphemus agglutinin (LPA)

NATURAL THYMOCYTOTOXIC AUTOANTIBODY (NTA) AS A CARBOHYDRATE BINDING AUTOANTIBODY

Shirai and Mellors[16,17] found that a natural thymocytotoxic autoantibody (NTA), which exhibits cytotoxic activity toward thymocytes, is induced spontaneously in autoimmune-prone New Zealand mice within 3 months after birth. Furthermore, it was shown that this autoantibody has a selective cytotoxic effect on suppressor T cells,[18,19] and so it was considered to be an important pathologic factor for the development of autoimmune phenomena in New Zealand mice.

We demonstrated that NTA consists of two autoantibodies in terms of target cell specificity.[20] One of the autoantibodies, NTA-2, is only strongly cytotoxic toward desialized lymphocytes, whereas the other, NTA-1, is cytotoxic toward both intact thymocytes and asialolymphocytes. Since the cytotoxicity of NTA-2 was strongly inhibited by the asialo complex-type glycopeptide (unit B glycopeptide) of porcine thyroglobulin but not by the intact unit B glycopeptide, these two autoantibodies could be separated by affinity chromatography on an affinity adsorbent prepared by coupling aisalo-porcine thyroglobulin to Sepharose 4B. The purified NTA-2 was shown to be a carbohydrate-binding antibody, and, as shown in Fig. 2, oligosaccharides having a terminal β-galactosyl residue were potent inhibitors of NTA-2. On the other hand, NTA-1 was only slightly,

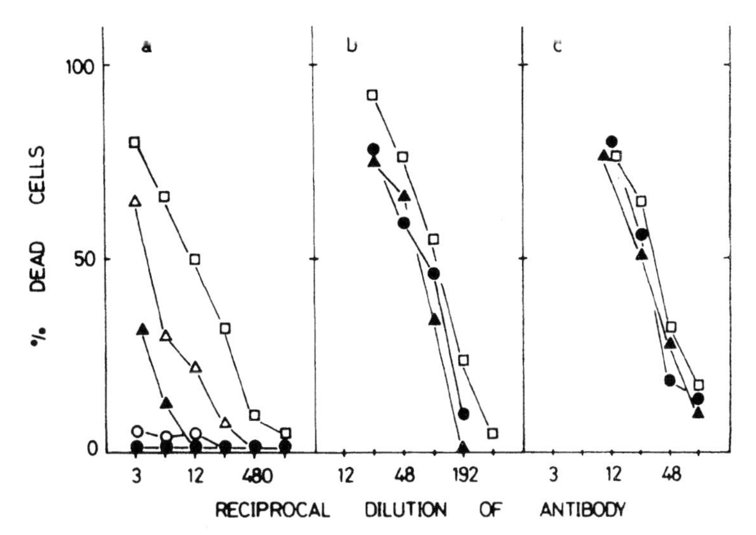

Fig. 2. Inhibition of cytotoxic activities of NTA-1 and NTA-2 with various sugars, a, NTA-2 against neuraminidase treated thymocytes; b, NTA-1 against neuraminidase-treated thymocytes; c, NTA-1 against intact thymocytes. Hapten inhibition assays were performed as described in the text in the presence of various sugars: without sugars (□), 180 mM lactose (●), 60 mM lactose (○), 180 mM melibiose (▲), 60 mM melibiose (△).

if at all, inhibited by these sugars. These results indicate that the carbohydrate-binding specificity of NTA-2 is quite similar to that of peanut agglutinin (PNA)[21,22] in that it has a strong affinity to terminal sialic acid-free β-galactosyl residues of sugar chains. Then, we examined the cytotoxic effect of the column-purified NTA fractions on the Con A-induced, antigen non-specific suppressor T cell function. The results presented in Table 4 show that Con A-induced suppressor cells treated with NTA-2 and complement (C) did not exhibit suppressive activity anymore. On the other hand, when the cells were treated with NTA-1 and C, their suppressor activity was affected to only a small extent. These results suggest that suppressor T cells are cells with abundant terminal β-galactosyl residues that are not masked by sialic acid residues on the cell surface. Furthermore, we also found[23] that injection of column-purified NTA-2 into normal mice led to several autoimmune phenomena in them. Actually, PNA^+ T cells amount Con A-activated spleen cells were found to markedly decrease with age in NZB mice, but not in non-autoimmune strains, by use of a fluorescence activated cell sorter (FACS).[24] Thus, NTA-2 possibly plays a role in the decrease of PNA suppressor T cells in autoimmune animals.

Table 4. Treatment of Con A-induced suppressor cells with NTA and C

Cells added	Anti-SRBC PFC/Culture[1]
Con A-induced suppressor cells	103 ± 25
Con A-induced suppressor cells	
After NTA-1 and C treatment	115 ± 10
After NTA-2 and C treatment	274 ± 78
Treated with C alone	102 ± 12
None	215 ± 49
Without SRBC	21 ± 8

[1] Numbers represent arithmetic means for five cultures and standard deviation.

SEPARATION OF MOUSE SUPPRESSOR AND HELPER T CELL SUBSETS WITH LECTINS

Since the carbohydrate-binding specificity of NTA-2 in autoimmune-prone animals indicated that suppressor T cells have abundant sialic acid-free β-galactosyl residues on the cell surface, we attempted to separate suppressor T cells by use of PNA.[25,26] Con A-activated mouse spleen cells were incubated with PNA for 10 min at room temperature. Then the cells were layered over 50% FCS. After 60 min at room temperature, the cells had separated into 3 fractions, top, middle and bottom, respectively. The effects of these fractions on the primary antibody response to SRBC were examined. The bottom fraction of the PNA-agglutinable cells showed a significant suppressor effect, whereas the nonagglutinable cells (top fraction) had no effect (Table 5). Some of the cell surface markers of these fractionated cells were analyzed and

Table 5. Effect of PNA^+ and PNA^- Con A activated cells on the primary PFC response to SRBC and their cell surface markers

	PFC/culture[1]	IgM^+ cells (%)	Thy-1^+ cells (%)	Lyt-1^+ cells (%)	Lyt-2^+ cells (%)	Recovery (%)
Control	177 ± 34					
Before fractionation[2]	20 ± 2					
PNA^-	135 ± 29	42.3	34.5	41.2	3.6	7.6
PNA^+	11 ± 4	13.7	67.0	10.1	74.6	14.8
PNA^+ + Anti-Thy 1.2 + C[3]	160 ± 15					
PNA^+ + Anti-Lyt 1.2 + C[3]	28 ± 18					
PNA^+ + Anti-Lyt 2.2 + C[3]	155 ± 20					

[1] Each value represents the mean ± S.E.M. for 3 cultures.

[2] 4 x 10^5 cells were added.

[3] 2 x 10^5 cells were added.

listed also in Table 5. The bottom fraction showed a much greater abundance of Thy-1$^+$ and Lyt-2$^+$ cells, and treatment of the bottom fraction with anti-Thy-1 or anti-Lyt-2 antiserum and C totally abolished the suppressor activity, indicating that the fraction contained suppressor T cells. Suppressor activity of PNA$^+$ cells amongst Con A-activated spleen cells was also seen in PNA$^+$ cells after fractionation with a fluorescence activated cell sorter.[25] However, the PNA-agglutinable cells obtained from normal spleen cells did not suppress the PFC response, indicating that we are dealing with Con A-inducible suppressor T cells.

The fact that suppressor T cells have abundant sialic acid-free β-galactosyl residues on the cell surface suggested that helper T cells may have abundant sialic acid residues on the cell surface. Reisner et al.[27] showed that the cell surface of mature T cells, namely T_2 cells, which are precursor of helper T cells,[13] is masked by sialic acid residues. Therefore, we attempted to separate LCA-induced helper cells with *Limulus polyhemus* agglutinin (LPA),[9,26] which specifically binds to sialyl residues on the cell surface.[28] LCA activated cells were incubated with LPA for 10 min at room temperature. Then the cells were layered over 50% fetal calf serum. After 60 min at room temperature, the cells had separated into three fractions, the top, middle and bottom fractions, respectively. The effects of these fractions on the primary antibody response to SRBC were tested. The LPA-agglutinable cells that settled at the bottom of the tube showed a significant helper effect, whereas the nonagglutinable cells which remained at the top of the tube showed no significant effect (Table 6). When the LPA-agglutinable cells were treated with anti-Thy-1 or anti-Lyt-1 antiserum and C, the stimulatory activity of the cells was abolished, while treatment with anti-Lyt-2 antiserum and C had no effect, indicating that the effectors in the bottom fraction were actually helper T cells.

Table 6. Effect of LPA$^+$ and LPA$^-$ LCA-activated cells on the primary PFC response to SRBC

	PFC/culture[1]
Control	152 ± 46
Before fractionation[2]	200 ± 20
LPA$^-$	166 ± 16
LPA$^+$	502 ± 28
LPA$^+$ + Anti-Thy 1.2 + C[2]	162 ± 23
LPA$^+$ + Anti-Lyt 1.2 + C[3]	126 ± 28
LPA$^+$ + Anti-Lyt 2.2 + C[3]	346 ± 17

[1] Each value represents the mean ± SEM for 3 cultures.

[2] 4 x 10^5 cells were added.

[3] 2 x 10^5 cells were added.

To further characterize the suppressor and helper T cells separated with lectins, the effects of treatment with various anti-glycolipid antibodies and C on the functional activities of PNA-agglutinable suppressor T cells and LPA-agglutinable helper T cells were investigated. As shown in Table 7, treatment of PNA-agglutinable cells with anti-asialo GM_1, anti-Forssman and anti-globoside plus C significantly decreased the suppressor activity. On the other hand, treatment of LPA-agglutinable cells with anti-GM_1 plus C abolished the helper activity, whereas treatment of LPA-agglutinable cells with other anti-glycolipid antibodies plus C did not significantly affect the helper activity. Anti-asialo GM_1 antiserum recognizes the terminal disaccharide structure of β-D-Gal-(1→3)-D-GalNAc,[29] which is also a determinant structure for the binding of PNA.[21,22] Thus, these results also indicate that asialo GM_1 or related sugar sequences, which are devoid of sialic acid residues, are abundant on the cell surface of suppressor T cells, while GM_1 or related sugar sequences containing sialic acid residues are abundant on the cell surface of helper T cells.

Table 7. Effect of the treatment of PNA^+ Con A-activated cells and LPA^+ LCA-activated cells with various anti-glycolipid antisera and C on the suppressor and helper activities of the cells

Treatment	PFC/culture[1]	
	PNA^+	LPA^+
Control	118 ± 11	119 ± 3
None	19 ± 6	227 ± 12
Anti-GM_1 + C	27 ± 4	106 ± 2
Anti-asialo GM_1 + C	67 ± 15	186 ± 15
Anti-Forssman + C	92 ± 12	246 ± 3
Anti-Globoside + C	97 ± 19	171 ± 1

[1] Each value represents the mean ± S.E.M. for 3 cultures. The number of cells added to the culture from each fraction was 2 x 10^5.

SEPARATION OF CYTOTOXIC T CELLS WITH LECTINS

Kimura et al.[30,31] reported the enrichment of cytotoxic T cells (CTL) induced in mixed lymphocyte cultures (MLC) with the use of an affinity adsorbent prepared by coupling *Vicia villosa* agglutinin (VVA) to Biogel P-2 or P-6. They found that VVA exhibited selective reactivity toward the T145 glycoprotein that is expressed specifically on the cell surface of CTL cells[32] and this glycoprotein could be isolated by specific binding to a VVA-Sepharose 4B column followed by elution with N-acetyl-D-galactosamine (GalNAc).[30] On the other hand, MacDonald et al.[33] treated MLC cells with VVA and then stained the VVA-treated cells with rabbit anti-VVA and FITC-goat anti-rabbit IgG antisera. After separation of the stained cells into VVA^+ and VVA^- cells with a FACS, they could not observe a significant difference in killer activity between VVA^+ and VVA^- cells. Lang et al.[34] also reported that VVA is not specific to CTL, and furthermore, Kaufmann et al.[35] did not observe a positive correlation between the expression of the T145 antigen on the cell surface and killer activity of the cells.

We treated MLC cells, derived from C57BL/6 spleen cells ($H\text{-}2^b$; responder cells) and mitomycin C-treated DBA/2 spleen cells ($H\text{-}2^d$; stimulator cells), with various GalNAc-binding lectins, and the agglutinated cells were separated from unagglutinated ones by sedimentation at unit gravity in 50% fetal calf serum.[36] Then, the agglutinated cells were dissociated into single cells by washing with a haptenic sugar-containing medium, and the cytolytic activity of the separated cells toward P815 cells ($H\text{-}2^d$) was investigated in vitro. The three lectins in Fig. 3 were most effective for the enrichment of specific CTL. Next we etablished the several CTL cell lines listed in Table 8 after long-term culturing of MLC cells in the presence of interleukin-2 (IL-2). These cell lines were stained with biotinyl Dolichos biflorus agglutinin (DBA) and FITC-avidin and then analyzed with a FACS. The same procedures were also performed using Lima bean (Phaselous limensis) agglutinin (LBA), Helix pomatia agglutinin (HPA) and PNA. Fig. 4 shows the results of staining with DBA. Normal spleen cells and normal T cells showed no significant fluorescence. However, 2° MLC (secondary mixed lymphocyte culture; primary MLC cells stimulated again with stimulator cells) cells and CTL cell lines showed brighter fluorescence, and the proportions of cells showing brighter fluorescence and the fluorescence intensities at the peaks in the fluorescence profiles increased almost parallel to the cytolytic activity. The relevance of lectin binding to the cytolytic activity of CTL cell lines was investigated by calculating Spearman

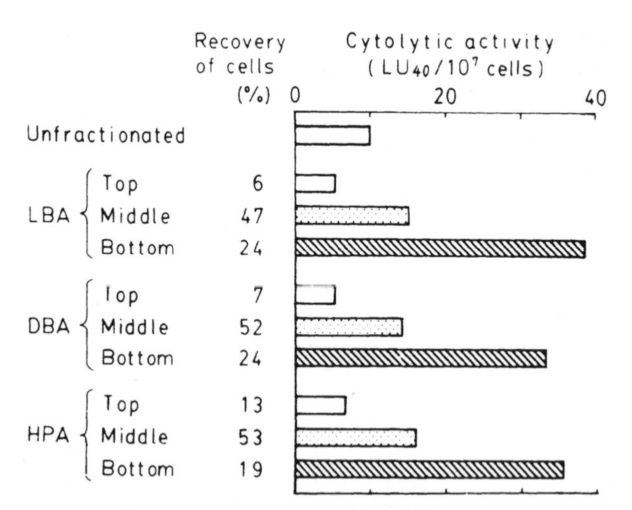

Fig. 3. Cytolytic activity of lectin-fractionated lymphocytes activated by primary MLC. One unit (LU_{40}) is defined as the number of lymphocytes required to lyse 40% of 2 x 10^4 P815 cells as target cells in a 4-hr, ^{51}Cr-release assay.

Table 8. Cytolytic activities of CTL populations

Cells	Cytolytic activity[1] ($LU/10^6$ cells)
Con A-activated spleen lymphocytes	0.036
Spleen lymphocytes primed in vivo	0.21
1°MLC cells	1.8
2°MLC cells	4.8
LC7 cells	6.9
LC4 cells	25
LC5	192
$LC5T^8$	152
$LC5B^5$	227

[1] One lytic unit (LU) is defined as the number of lymphocytes required to lyse 50% of 2×10^4 target cells within a 4 hr, ^{51}Cr-release assay, unless otherwise stated.

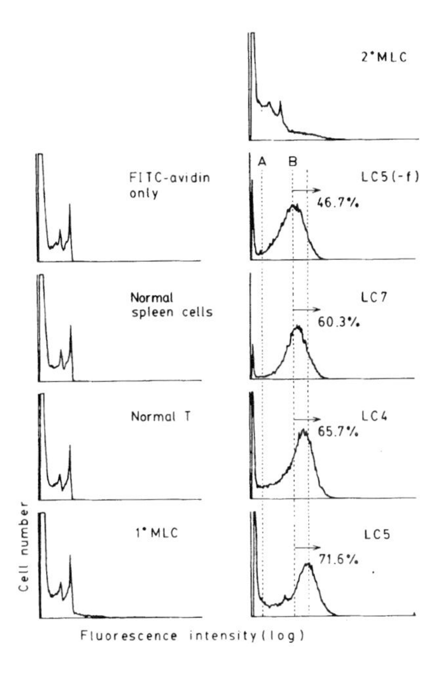

Fig. 4. Fluorescence profiles of normal cells and CTL cell-lines labeled with biotinyl DBA and FITC-avidin. The values shown as percentages represent the ratio of (the area having a higher fluorescence intensity than B in a peak) x 100/(the area having a higher fluorescence intensity than A in a peak).

rank-correlation coefficients. For DBA, strong rank-correlations (rs=0.979) between lectin binding and cytolytic activity were observed, but the correlations were weaker for HPA, LBA and PNA. Since the LC5 cell line was the strongest CTL cell line established, we fractionated LC5 cells by the agglutination-sedimentation method with DBA. The majoring (90%) of the cells was recovered in the agglutinated fraction, a much smaller part (3%) being found to be not agglutinated. Then, DBA-fractionation and reculturing were repeated several times for both cell populations to purity further the DBA^+ and DBA^- cells. The cytolytic activities of the unagglutinated cells ($LC5T^8$ cells; the selection procedure was repeated 8 times) and agglutinated cells ($LC5B^5$ cells; the selection procedure was repeated 5 times) thus obtained were not very different from that of the original LC5 cells (Table 8). These results suggest that not all CTL cells have receptors for DBA, but the majority of them do have DBA receptors. DBA is known to be the most specific lectin α-D-GalNAc residues, and its most potent inhibitor is a blood-group A active pentasaccharide,[37] α-D-GalpNAc-(1→3)-[α-L-Fucp-(1→2)]-β-D-Galp-(1→4)-β-GalpNAc-(1→6)-R. The other lectins used, HPA and LBA, appear to be less specific for α-D-GalNAc groups,[38,39] and may also react, to a more limited extent, with terminal α-D-Gal groups. It may, therefore, be reasonable to assume that this wider sugar-binding specificity of the latter two lectins is reflected in their lower selectivity for CTL. SBA is also a specific lectin for α-D-GalNAc residues,[40,41] but the α-L-fucosyl group in the blood group A active pentasaccharide was shown to greatly diminish the inhibitory activity of the oligosaccharide toward SBA. Since SBA is ineffective for the separation of CTL cells, most CTL cells may express a blood-group A antigen-like structure including an L-fucosyl group, DBA being able to bind most specifically to this sugar structure. In this context, it is also of interest to note that the blood-group A active pentasaccharide is much less active as an inhibitor of VVA than α-D-GalpNAc-(1→3)-D-Gal is.[42] Although VVA was not tested in this study owing to its unavailability, it may be less selective for CTL cell than DBA. Membrane receptors of CTLs for DBA may not exactly be a blood-group A antigen itself, because the immune, anti-human blood-group A serum used in this study did not react with CTL cells. Another possibility is that the anti-A antiserum recognizes the larger portion of the blood-group A antigen. Thus, the characterization of membrane receptors of CTL cells for DBA, even if they are not exclusive, differential cell-surface markers for CTL cells, is of interest.

IN VIVO TUMORICIDAL ACTIVITY OF DBA-SEPARATED SPLEEN CELLS OBTAINED FROM TUMOR-BEARING MICE[43]

Adoptive transfer of cytotoxic cells, which preferentially attack tumor cells, is one of the potential approaches to experimental tumor therapy in animals and may be clinically applicable to human cancers.

Based on our success in separating specific cytotoxic cells induced in allogeneic mixed lymphocyte cultures with the use of DBA, we attempted to separate tumor-specific cytotoxic cells from spleen cells of mice which had been inoculated with syngeneic tumor cells, and then to administer these cytotoxic cells to the same tumor cell-bearing mice. One of the biggest problems with this approach is the growth of the separated effector cells. However, culturing with IL-2 can potentially overcome this problem. Activated cytotoxic cells can proliferate in the presence of IL-2 without a stimulating antigen.[44-46] Recently, Rosenberg and coworkers demonstrated that intravenous injection of murine splenocytes activated by culturing with IL-2 mediate the regression of established pulmonary metastases of sarcomas.[47,48] We subcutaneously inoculated 1 x 10^6 X-5563 tumor cells ($H-2^k$) in to a C3H/He mouse, and after establishment of a tumor (11-12 days), spleen cells were isolated and then the cells (2 x 10^6 cells/ml) cultured with IL-2 for 4 days at 37°C. Then, the cells were separated into DBA^+ and DBA^- cells. When the in vitro cytotoxicity of the separated cells toward X-5563 tumor cells was tested (Fig. 5), DBA^+ cells showed strong cytotoxicity toward X-5563 cells, while the cytotoxicity of DBA^- cells was found to be weak. This cytotoxicity of DBA^+ cells was confirmed to be specific to X-5563 cells and, as shown in Fig. 6, it decreased on treatment with anti-Thy-1 antibody and complement, but was not affected by treatment with

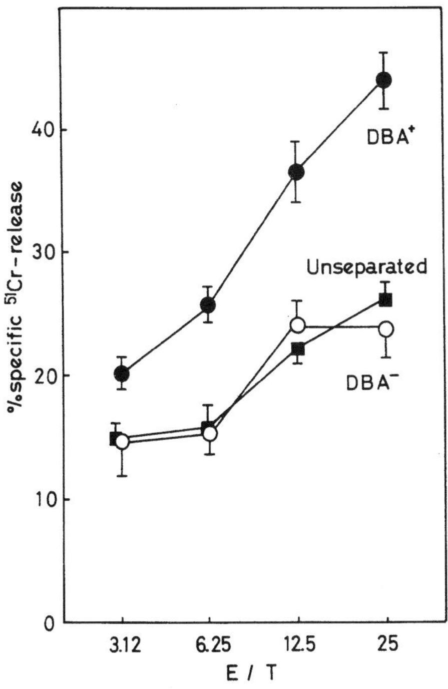

Fig. 5. Cytolytic activity of DBA-separated cells toward ^{51}Cr-labeled X-5563 cells in vitro.

anti-asialo GM_1 antibody and complement, suggesting that the effector cells were T cells and not natural killer cells. Then, DBA^+ or DBA^- cells (2×10^5) were mixed with 2×10^4 viable X-5563 cells, and the mixture was subcutaneously inoculated into a C3H/He mouse, and the growth of the tumor was investigated (Winn assay). Unexpectedly, DBA^- cells almost completely inhibited the growth of a tumor, while the inhibitory effect of DBA^+ cells was found to be much weaker (Fig. 7), contrary to that the latter cells showed a strong cytotoxic effect on X-5563 cells in vitro. A tumor growth inhibiting effect of DBA^- cells which were separated from X-5563-bearing C3H/He mice was also observed on another syngeneic tumor cell line, C3MC2 cells (Fig. 8). However, when DBA^+ and DBA^- cells were similarly separated from C3H/He mice which had not previously been exposed to tumor cells, the cells did not show cytotoxicity or tumor growth inhibiting activity in vitro or in vivo. Both DBA^+ and DBA^- cells were found to be $Lyt\text{-}1^-2^+$ cells. Actually, when DBA^- cells were treated with anti-Lyt-2 antiserum and complement, their

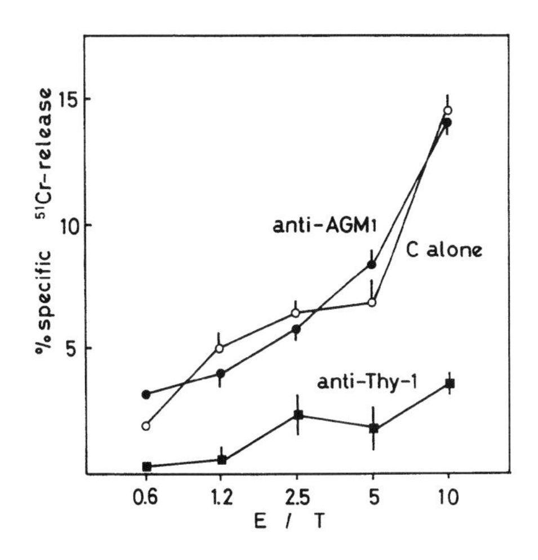

Fig. 6. Effect of treatment with various antibodies and complement on the in vitro cytolytic activity of DBA^+ cells toward ^{51}Cr-labeled X-5562 cells.

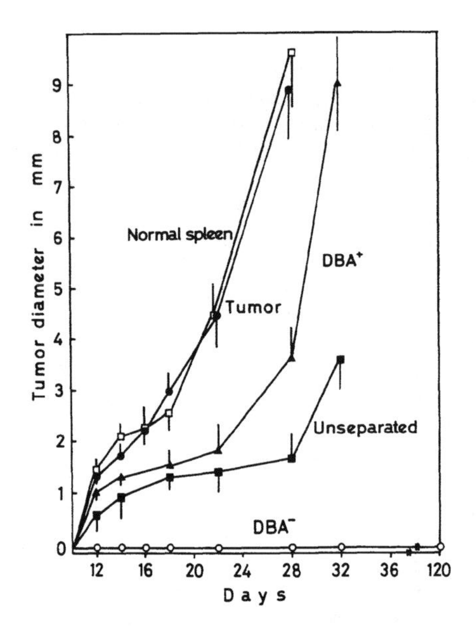

Fig. 7. Inhibitory activity toward tumor growth (X-5563 cells) of DBA⁻ cells in vivo (Winn assay).

tumor growth inhibiting activity in vivo was remarkably abolished. Thus, DBA separation can effectively enrich the effector cell population which plays a major role in mediating the regression of tumors in vivo.

FRACTIONATION OF NYLON-WOOL NONADHERENT MOUSE SPLEEN CELLS AND BONE MARROW CELLS BY USING BPA

(a) Enrichment of IL-2 producer T cells.

T cells can be divided into two groups, one can be induced to become IL-2 producing cells through activation with an antigen or a mitogen as well as interleukin-1 (IL-1) from macrophages, and the other express

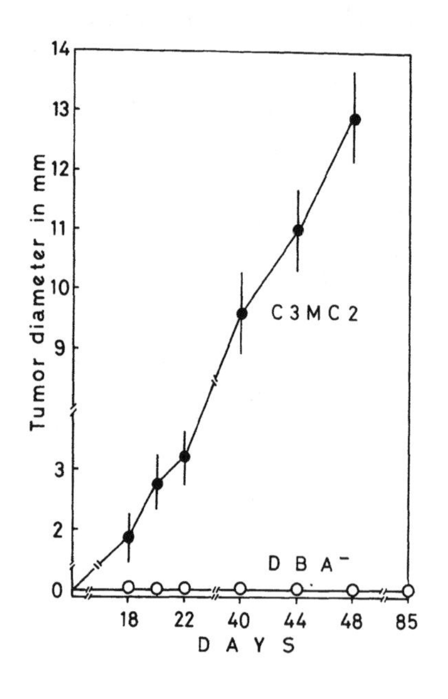

Fig. 8. Inhibitory activity toward tumor growth (C3MC2 cells) of DBA⁻ cells in vivo (Winn assay).

receptors for IL-2 depending on the activation with an antigen or a mitogen and subsequently continue growing only in the presence of IL-2. In a study on the mechanism of T cell activation with an antigen or a mitogen, it is desirable that these two T cell groups are separated and that the activation mechanism is investigated for each group of T cells separately.

Many attempts have been made to fractionate T cells into IL-2 producers and IL-2 responders in terms of cell surface phenotypes such as Lyt antigens. From the results of these studies, IL-2 was found to be produced by Lyt-1^+2^- T cells after stimulation with a mitogen such as Con A,[49-51] while Lyt-1^-2^+ T cells were found to be unable to produce IL-2.[51,52] With regard to the fractionation of IL-2 producers by means of lectins, Bödeker et al.[53] and Conlon et al.[54] reported that PNA^- thymocytes produced IL-2, whereas PNA^+ thymocytes did not, and that the latter were incapable of responding to Con A even in the presence of exogeneous IL-2.[51,54]

We attempted enrichment of IL-2 producing T cells from nylon-wool nonadherent mouse spleen cells rather than from thymocytes by use of BPA,[55,56] because the spleen is one of the major sites of IL-2 production and immune responses. We selected BPA, because Lyt-1^+2^- T cells have been shown to be IL-2 producers,[49-51] and because our previous studies indicated that Lyt-1^+2^- T cells have abundant sialic acid residues on the cell surface[26] and that BPA can recognize GalNAc or Gal residues on the cell surface regardless of whether they are masked with sialic acid residues or not[5,57] but PNA can recognize only sialic acid-free Gal residues,[21,22] expecting that a certain BPA^+ cell population could be the IL-2 producers.

Nylon wool-nonadherent mouse spleen cells were fractionated by the agglutination-sedimentation method with BPA, and the Con A-induced IL-2 production of the BPA^+ and BPA^- cells was compared. The distributions into the BPA^- and BPA^+ fractions were 66 ± 2% and 34 ± 2%, respectively, and the contents of Thy-1^+ cells in these fractions were 96% and 83%, respectively. IL-2 production from the BPA^- and BPA^+ fractions under serum-free conditions was assayed using an IL-2-dependent long-term cultured nonspecific suppressor T cell line[58] as an indicator. Representative results are shown in Fig. 9. Contrary to our expectation, the amount of IL-2 produced by the BPA^- fraction after Con A activation was greater than that produced by nylon-wool-nonadherent cells, whereas the ability of the BPA^+ fraction to produce IL-2 was considerably lower. Neither the control supernatant obtained from the BPA^- fraction nor that from the BPA^+ fraction without stimulation by Con A contained detectable IL-2 activity. Furthermore, it was confirmed that the low ability of the BPA^+ fraction to produce IL-2 was not due to the effect of suppressor cells or the difference in Ia-bearing accessory cells or interference by contaminating B cells. Then we tested the IL-2 dependence of the BPA^- and BPA^+ fractions for growth. Since Larsson[59] previously reported that a short (4 hr) pulse with Con A at 37°C induced IL-2 receptors on T cells, Con A pulsed cells obtained from the BPA^- and BPA^+ fractions were tested as to their responsiveness to IL-2. As shown in Table 9, the cells in both fractions were dependent upon IL-2, but the IL-2 requirement of BPA^+ cells was much greater than that of BPA^- cells.

Several laboratories have reported that autoimmune-prone mice such as the MRL/l strain show a decrease in the ability to produce IL-2 during the course of aging.[60,61] When nylon-wool-nonadherent spleen cells of 4-month-old female MRL/l mice were subjected to BPA fractionation, the level of BPA^+ cells was apparently increased in the spleens of MRL/l mice, and furthermore, BPA^- cells separated from the spleens of MRL/lpr

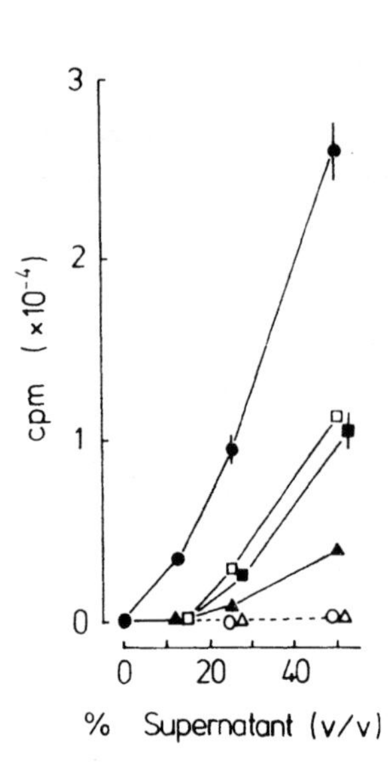

Fig. 9. IL-2 production by BPA-fractionated nylon-wool-non-adherent spleen cells from DBA/2 mice. Culture supernatants prepared under serum-free conditions were assayed for ^{3}H-thymidine incorporation of an IL-2-dependent nonspecific suppressor T-cell line (NSTL assay). Assays were performed with the indicated concentrations of supernatants from Con A-stimulated BPA$^-$ cells (●), BPA$^+$ cells (▲), nylon-wool-nonadherent cells (□), unfractionated spleen cells (■), from BPA$^-$ cells without Con A activation (○) or BPA$^+$ cells without Con A activation (△). Data are expressed as means ± range for duplicate cultures.

Table 9. Proliferation of T cells against BPA-fractionated cells pretreated with Con A in the presence of IL-2

Precultured[1]	Recultured[2]	^{3}H-TdR incorporation, cpm[3]	
		BPA$^-$ cells	BPA$^+$ cells
Con A	medium	615 ± 37	1,323 ± 277
Con A	Con A	37,415 ± 2,277	26,215 ± 7,015
Con A	50% IL-2[4]	32,154 ± 4,984	26,338 ± 4,708
Con A	25% IL-2	31,930 ± 2,333	17,908 ± 3,169
Con A	12.5% IL-2	32,308 ± 1,015	14,277 ± 2,738
Con A	1% IL-2	5,846 ± 831	not tested
Medium	50% IL-2	120 ± 41	4,492 ± 492

[1] BPA-nonagglutinated and BPA-agglutinated cells were stimulated with or without Con A (3 μg/ml) at 37°C for 10-11 hr.

[2] Cells precultured with or without Con A were washed with α-MM, and then cultured with IL-2 at 37°C for 60 hr.

[3] ^{3}H-TdR uptake was measured for the last 12 hr of the 3-day culture period. Values are expressed as mean cpm ± SD for triplicate cultures.

[4] The conditioned medium of an IL-2-producing MLA 144 cell line (MLA 144CM) was used as the source of IL-2.

mice were less potent producers of IL-2 than those separated from the spleens of MRL/n mice, a normal congenic MRL/l mouse strain. From these results we concluded that the deficiency of autoimmune MRL/l mice as to the production of IL-2 might be based not only on an intrinsic defect of the IL-2-producer T cells themselves but also on variations in the T cell subsets, for example an increase in the IL-2 low producer BPA^+ T cells.

(b) Enrichment of macrophage precursors.[56]

When BPA^+ cells were incubated with the culture supernatant of Con A-activated mouse spleen cells (Con A sup), which contains at most only 0.15 μg/ml free Con A, for 5 days, the appearance of long, macrophage-like cells was observed. Furthermore, the stimulation of BPA^+ cells with a high dose of Con A (50 μg/ml) also induced large, elongated (100-200 μm long) cells which were nonspecific esterase- and phagocytosis-positive, adherent to a plastic surface, and resistant to anti-BAT and C treatment. In contrast, no such cells were induced on incubation with Con A-sup and stimulation with a high dose of Con A. These results indicate that macrophage precursors were selectively enriched in the BPA^+ fraction (the $Thy\text{-}1^+$ cell content is approximately 80%). Then, we subjected mouse bone marrow cells, which are the richest source of macrophage precursor CFU-c cells, to BPA fractionation.[62] When bone marrow cells of C3H/He mice were fractionated with BPA, BPA^+ cells were much more responsive to a colony stimulating factor (CSF) in the L-cell conditioned medium as shown in Fig. 10, and the cells having macrophage morphology were found to be induced. Moreover, we found that the proportion of BPA^+ cells amongst bone marrow cells was greater in autoimmune-prone MRL/l mice than in normal C3H/He mice (Table 10, Fig. 11), and the response of BPA^+ cells from MRL/l mice to the L cell conditioned medium was found to be stronger than that of BPA^+ cells from normal mice (Fig. 10). Since CSF activity is much stronger in the serum of MRL/l mice than in that of MRL/n and normal mice,[62] these results of analyses of bone marrow cells using BPA suggest that abnormally intensive differentiation of CFU-c into macrophages is induced in autoimmune-prone MRL/l mice by the synergistic effect of the increased BPA+ precursors and the over-production of serum CSF.

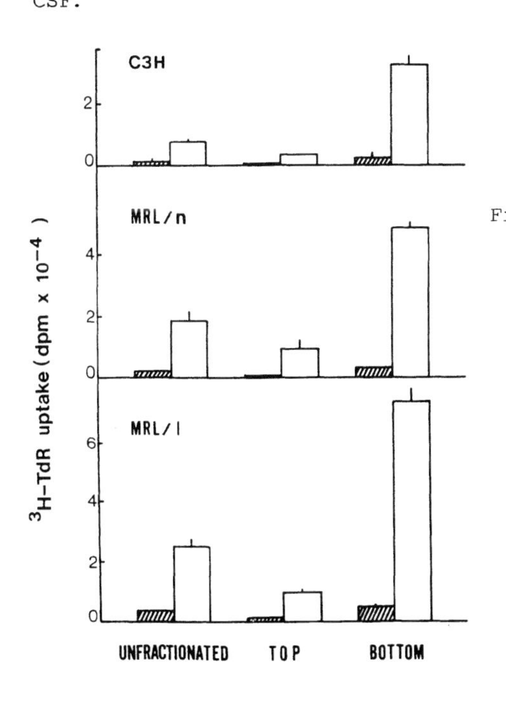

Fig. 10. Proliferation of BPA^+ (bottom) and BPA^- (top) bone marrow cells from normal (C3H and MRL/n) and autoimmune prone (MRL/l) mice. 5×10^4 cells were cultured with (open bars) or without (hatched bars) L cell conditioned medium (1/55 dilution).

Table 10. Fractionation of mouse bone marrow cells with BPA

Strain	Yield (%)	
	BPA^- (top fraction)	BAP^+ (bottom fraction)
C3H/He	20.9	7.1
MRL/n	37.8	15.8
MRL/l	36.0	22.2

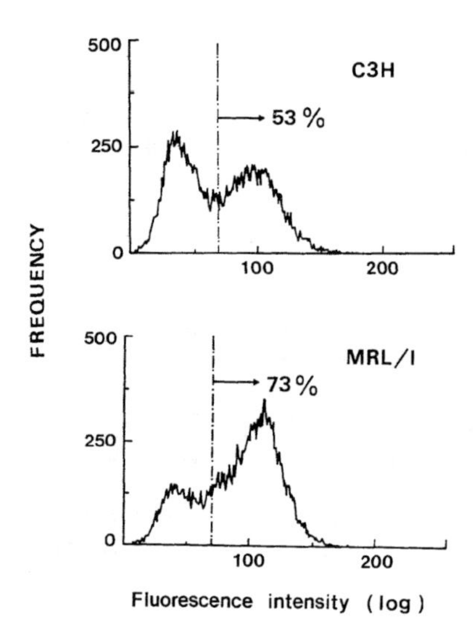

Fig. 11. Fluorescence profiles of bone marrow cells from normal (C3H) and autoimmune-prone (MRL/l) mice labeled with FITC-BPA.

CONCLUSION

Lectins are extremely useful tools for the enrichment of distinct immunocyte subpopulations. If they are used in combination, certain immunocyte subpopulations can be very effectively enriched. Although the final purification of a distinct immunocyte subpopulation should be preferably achieved with the use of a monoclonal antibody for the subpopulation, lectins can easily be obtained in relatively large amounts and can easily be removed from the cell surface with haptenic sugars. Thus, the use of lectins for the enrichment of immunocyte subpopulation is a promising approach for the medical application of certain functional immunocytes.

ABSTRACT

The use of lectins for the enrichment of various lymphocyte subpopulations was investigated. Bauhinia purpurea lectin (BPA) was found to be effective for the enrichment of B cells, and the B cells thus obtained were further fractionated with lentil lectin into subsets showing high and low responces to dextran sulfate. The ability of various lectins to selectively induce suppressor T cell activity or helper T cell activity was also examined. The suppressor T cells thus induced were enriched peanut lectin and, conversely, the helper T cells were enriched with Limulus polyhemus lectin. This method was applied to analysis of age-dependent changes in the levels of suppressor and helper T cells in autoimmune-prone mice. Cytotoxic T cells induced in an allogeneic mixed lymphocyte culture were enriched with Dolichos biflorus lectin (DBA). These cytotoxic T cells showed a specific killing effect in vitro. However, when spleen cells of tumor-bearing mice were fractionated by use of DBA, the DBA^- cells mediated the regression of the tumors in vivo. BPA was also found to be effective for the enrichment of the interleukin-2-producing T cell subset and macrophage precursor cells. Using this technique, bone marrow cells of autoimmune-prone MRL/l mice were found to be rich in macrophage precursor cells.

ACKNOWLEDGEMENTS

This work was supported by research grants from the Ministry of Education, Science and Culture of Japan, the Takeda Science Foundation, and the Nisshin Foundation.

REFERENCES

1. Schnebli, H.P. and Dukor, P. (1972). Plant agglutinins used to distinguish between different classes of mouse lymphocytes. Eur. J. Immunol. 2:607-609.
2. Bourguignon, L.Y.W., Rader, R.L. and McMahon, J.T. (1979). Rapid separation of mouse T and B lymphocytes using wheat germ agglutinin. J. Cell. Physiol. 99:95-100.
3. Reisner, Y., Ravid, A. and Sharon, N. (1976). Use of soybean agglutinin for the separation of mouse B and T lymphocytes. Biochem. Biophys. Res. Commun. 72:1585-1591.
4. Hishinuma, A., Imai, Y., Nakano, T. and Osawa, T. (1983). The reactivities of Bauhinia purpurea and Lens culinaris lectins to mouse B lymphocytes and their subsets. Int. Arch. Allergy & Appl. Immunol. 72:330-335.
5. Irimura, T., Kawaguchi, T., Terao, T. and Osawa, T. (1975). Carbohydrate binding specificity of the so-called galactose-specific phytohemagglutinins. Carbohyd. Res. 39:317-327.
6. Gronowicz, E. and Coutinho, A. (1975). Functional analysis of B cell heterogeneity. Transplant. Rev. 24:3-40.
7. Kobo, R.T., Yamaga, K.M. and Moore, R.E., Jr. (1979). Differential affinity of murine lymphocyte membrane IgM and IgD for the lentil lectin. Mol. Immunol. 16:107-112.
8. Dutton, R.W. (1972). Inhibitory and stimulatory effects of concanavalin A on the response of mouse spleen cell suspensions to antigen I. Characterization of the inhibitory cell activity. J. Exp. Med. 136:1445-1460.
9. Nakano, T., Oguchi, Y., Imai, Y. and Osawa, T. (1980). Induction and separation of mouse helper T cells by lectins. Immunology 40:217-222.
10. Cunngham, A.J. and Szenberg, A. (1968). Further improvements in the plaque technique for detecting single antibody-forming cells. Immunology 14:599-600.

11. Raff, M.C. (1971) Surface antigenic markers for distinguishing T and B lymphocytes in mice. Transplant. Rev. 6:52-80.
12. Gershon, R.K. (1974). T cell control of antibody production. In Contemporary Topics in Immunobiology (Edited by Cooper, M.D. and Warner, N.L.), Vol. 3, pp. 1-40. Plenum Publishing Corp., New York.
13. Erb, P., Meier, B., Krause, D., von Boehmer, H. and Feldmann, M. (1978). Nature of T cell-macrophage interaction in helper cell induction in vitro. I. Evidence for genetic restriction of T cell-macrophage interaction prior to T cell priming. Eur. J. Immunol. 8:786-792.
14. Yamamoto, K., Tsuji, T. and Osawa, T. (1982). Requirement of the core structure of a complex-type glycopeptide for the binding to immobilized lentil and pea lectins. Carbohyd. Res. 19:229-234.
15. Katagiri, Y., Yamamoto, K., Tsuji, T. and Osawa, T. (1984). Structural requirements for the binding of glycopeptides to immobilized Vicia faba (fava) lectin. Carbohyd. Res. 129:257-265.
16. Shirai, T. and Mellors, R.C. (1971). Natural thymocytotoxic autoantibody and reactive antigen in New Zealand Black and other mice. Proc. Natl. Acad. Sci. 68:1412-1415.
17. Shirai, T. and Mellors, R.C. (1972). Natural cytotoxic autoantibody against thymocytes in NZB mice. Clin. Exp. Immunol. 12:133-152.
18. Klassen, L.W., Krakauer, R.S. and Steinberg, A.D. (1977). Selective loss of suppressor cell function in New Zealand mice induced by NTA. J. Immunol. 119:830-837.
19. Shirai, T., Hayakawa, K., Okumura, D. and Tada, T. (1978). Differential cytotoxic effect of natural thymocytotoxic autoantibody of NZB mice on functional subsets of T cells. J. Immunol. 120:1924-1929.
20. Imai, Y., Nakano, T., Sawada, J. and Osawa, T. (1980). Specificity of natural thymocytotoxic autoantibody developed in New Zealand Black mice. J. Immunol. 124:1556-1561.
21. Terao, T., Irimura, T. and Osawa, T. (1975). Purification and characterization of a hemagglutinin from Arachis hypogaea. Hoppe-Seyler's Z. Physiol. Chem. 356:1685-1693.
22. Pereira, M.E.A., Kabat, E.A., Lotan, R. and Sharon, N. (1976). Immunochemical studies on the specificity of the peanut (Arachis hypogaea) agglutinin. Carbohyd. Res. 51:107-118.
23. Nakanishi, A., Imai, Y., Nakano, T. and Osawa, T. (1982). Induction of autoimmune phenomena in normal mice treated with natural thymocototoxic autoantibody. J. Immunol. 128, 2137-2141.
24. Imai, Y., Oguchi, Y., Nakano, T. and Osawa, T. (1980). Age-dependent changes of the levels of suppressor and helper t cells in New Zealand Black mice. Jap. J. Exp. Med. 50:415-421.
25. Imai, Y., Oguchi, Y., Nakano, T., Sawada, J. and Osawa, T. (1979). Separation of mouse T cell subsets by a fluorescent activated cell sorter using fluorescence-labeled peanut agglutinin. Immunological Communications 8:495-504.
26. Nakano, T., Imai, Y., Naiki, M. and Osawa, T. (1980). Characterization of mouse helper and suppressor T cell subsets separated by lectins. J. Immunol. 125:1928-1932.
27. Reisner, Y., Linker-Israeli, M. and Sharon, N. (1976). Separation of mouse thymocytes into two subpopulations by the use of peanut agglutinin. Cell. Immunol. 25:129-134.
28. Nowak, T.P. and Barondes, S.H. (1975). Agglutinin from Limulus polyphenus. Purification with formalinized horse erythrocytes as the affinity adsorbent. Biochim. Biophys. Acta. 393:115-123.
29. Naiki, M., Marcus, D.M. and Ledeen, R.J. (1974). Properties of antisera to ganglioside GM_1 and asialo GM_1. J. Immunol. 113:84-93.

30. Kimura, A., Wigzell, H., Holmquist, G., Ersson, B. and Carlsson, P. (1979). Selective affinity fractionation of murine cytotoxic T lymphocytes (CTL). Unique lectin specific binding of the CTL associated surface glycoprotein T 145. J. Exp. Med. 149:473-484.
31. Kimura, A., Örn, A. and Holmquist, G. (1979). Unique lectin-binding characteristics of cytotoxic T lymphocytes allowing their distinction from natural killer cells and K cells. Eur. J. Immunol. 9:575-578.
32. Kimura, A. and Wigzell, H. (1978). Cell surface glycoproteins of murine cytotoxic T lymphocytes I. T 145, a new cell surface glycoprotein selectively expressed on Ly 1⁻2⁺ cytotoxic T lymphocytes. J. Exp. Med. 147:1418-1434.
33. MacDonald, H.R., Mach, J.-P., Schreyer, M., Zaech, P. and Cerottini, J.-C. (1981). Flow cytofluorometric analysis of the binding of Vicia villosa lectin to T lymphoblasts: Lack of correlation with cytolytic function. J. Immunol. 126:883-886.
34. Lang, I., Banga, J.P., Varey, A.-M., Gunn, H., Cooke, A. and Roitt, I.M. (1982). Direct staining of mouse T lymphoblasts with fluoresceinated Vicia villosa lectin. Immunology 46:769-776.
35. Kaufmann, Y. and Berke, G. (1981). Cell surface glycoproteins of cytotoxic T lymphocytes induced in vivo and in vitro. J. Immunol. 126:1443-1446.
36. Yamazaki, T., Imai, Y., Oguchi, Y., Nakano, T. and Osawa, T. (1983). Separation of mouse cytotoxic T cells by lectins. Carbohyd. Res. 120:269-281.
37. Etzler, M.E. and Kabat, E.A. (1970). Purification and characterization of a lectin (plant hemagglutinin) with blood group A specificity from Dolichos biflorus. Biochemistry 9:869-877.
38. Hammerström. (1972). Snail (Helix pomatia) hemagglutinin. Methods Enzymol. 28 (part B):368-383.
39. Galbraith, W. and Goldstein, I.J. (1972). Lima bean (Phaselous lunatus) lectin. Methods Enzymol. 28 (part B):318-323.
40. Lis, H., Sela, B.A., Sachs, L. and Sharon, N. (1970). Specific inhibition by N-acetyl-D-galactosamine of the interaction between soybean agglutinin and animal cell surfaces. Biochim. Biophys. Acta 211:582-585.
41. Pereira, M.E.A., Kabat, E.A. and Sharon, N. (1974). Immunochemial studies on the specificity of soybean agglutinin. Carbohyd. Res. 37:89-102.
42. Kaladas, P.M., Kabat, E.A., Kimura, A. and Ersson, B. (1981). The specificity of the combining site of the lectin from Vicia villosa seeds which reacts with cytotoxic T-lymphoblasts. Mol. Immunol. 18:969-977.
43. Okada, T., Ezawa, K., Imai, Y. and Osawa, T.: unpublished results.
44. Gillis, S. and Smith, K.A. (1977). Long-term culture of tumor-specific cytotoxic T-cells. Nature (Lond.) 268:154-155.
45. Wagner, H. and Röllinghoff, M. (1978). T-T-cell interaction during in vitro cytotoxic allograft response. I. Soluble products from activated Lyl⁺ T cells trigger autonomously antigen-primed Ly23⁺ T cells to cell proliferation and cytolytic activity. J. Exp. Med. 148:1523-1538.
46. Rosenberg, S.A., Schwarz, S. and Spiess, P.J. (1978). In vitro growth of murine T cells. II. Growth of in vitro sensitized cells cytotoxic for alloantigens. J. Immunol. 121:1951-1955.
47. Mazumder, A and Rosenberg, S.A. (1984). Successful immunotherapy of natural killer-resistant established pulmonary melanoma metastases by the intravenous adoptive transfer of syngeneic lymphocytes activated in vitro by interleukin 2. J. Exp. Med. 159:495-507.
48. Mule, J.J., Shu, S., Scjwarz, S.L. and Rosenberg, S.A. (1984). Successful adoptive immunotherapy of established pulmonary metastases of multiple sarcomas with lymphokine-activated killer cells and recombinant interleukin-2. Science 225:1487-1489.

49. Peatkau, V., Show, J., Mills, G. and Caplan, B. (1980). Cellular origins and targets of costimulator (IL 2). Immunol. Rev. 51:157-175.
50. Shaw, J., Caplan, B., Peatkau, V., Pilarski, L.M., Delovitch, T.L. and McKenzie, I.F.C. (1980). Cellular origins of co-stimulator (IL 2) and its activity in cytotoxic T lymphocyte responses. J. Immunol. 124:2231-2239.
51. Wagner, H. and Röllinghoff, M. (1978). T-T cell interactions during in vitro cytotoxic allograft responses. I. Soluble production from activated Ly 1^+ T cells trigger autonomously antigen-primed Ly 23^+ T cells to cell proliferation and cytolytic activity. J. Exp. Med. 148:1523-1538.
52. Jandinski, J., Cantor, H., Tadakuma, T., Peavy, D.L. and Pierce, C.W. (1976). Separation of helper T cells from suppressor T cells expressing different Ly components. I. Polyclonal activation: Suppressor and helper activities are inherent properties of distinct T-cell subclasses. J. Exp. Med. 143:1382-1390.
53. Bödeker, B.G.D., Eijk, R.V.W. and Mühlradt, P.F. (1980). Mitogenic effects of partially purified interleukin 2 on thymocyte subpopulations and spleen T cells of mouse. Eur. J. Immunol. 10:702-707.
54. Conlon, P.J., Henney, C.S. and Gillis, S. (1982). Cytokine-dependent thymocyte responses: characterization of IL 1 and IL 2 target subpopulations and mechanism of action. J. Immunol. 128:797-801.
55. Imai, Y. and Osawa, T. (1983). Enrichment of IL 2 producer T cells from mouse spleen by use of Bauhinia purpurea lectin. Scand. J. Immunol. 18:217-224.
56. Yamaguchi, N. and Osawa, T. (1984). Separation and characterization of macrophage precursors and of IL 2-responding cells from nylon wool-nonadherent murine spleen cells by using Bauhinia purpurea agglutinin. Int. Archs. Allergy appl. Immunol. 75:309-316.
57. Osawa, T., Irimura, T. and Kawaguchi, T. (1978). Bauhinia pupurea hemagglutinin. Methods Enzymol. 50:367-372.
58. Ezawa, K., Toyoshima, S. and Osawa, T. (1983). Establishment and characterization of continuous murine non-specific suppressor T cell lines. Jap. J. Exp. Med. 53:139-146.
59. Larsson, E.-L. (1981). Mechanism of T cell activation. II. Antigen- and lectin-dependent acquisition of responsiveness to TCGF is a nonmitogenic, active response of resting T cells. J. Immunol. 126:1323-1326.
60. Altmann, A., Theofilopoulos, A.N., Weiner, R., Katz, D.H. and Dixon, F.J. (1981). Analysis of T cell function in autoimmune murine strains. Defects in production of and responsiveness to IL 2. J. Exp. Med. 154:791-808.
61. Wofsy, D., Murphy, E.D., Roths, J.B., Dauphinee, M.J., Kipper, S.B. and Talal, N. (1981). Deficient interleukin 2-activity in MRL/Mp and C57BL/6J mice bearing the lpr gene. J. Exp. Med. 154:1671-1680.
62. Nakanishi, A., Imai, Y. and Osawa, T.: unpublished results.

Addendum

Current concepts of the combining sites of lectins tested are also presented in the appendix of this proceedings book, in an article entitled "A guide for carbohydrate specificities of lectins".

editor

MANNOSE-BINDING PROTEINS OF ANIMAL ORIGIN

Y.C. Lee

Biology Department,
The Johns Hopkins University
Charles and 34th Street
Baltimore, Maryland 21218 USA

More than ninety years ago, Stillmark (1888) discovered that castor-bean extracts caused agglutination of human red cells. Boyd and Shapeleigh (1954) invented the term "lectin" (from legere, to select) for this group of plant proteins which are capable of hemagglutination. It soon became apparent that lectins are carbohydrate-binding proteins. Modern nomenclature defines lectins as carbohydrate-binding proteins which are not immunoglobulins or enzymes.

Since the discovery of the Gal/GalNAc-binding protein in mammalian livers (Ashwell and Morell, 1974), there have been many reports of the detection and isolation of carbohydrate-binding proteins in animal tissues and fluids. Table I summarizes some of the more extensively characterized carbohydrate-binding systems of animal origin. Most carbohydrate-binding proteins fit the above definition and thus can be called animal lectins. Vertebrate soluble lectins have been reviewed recently by Barondes (1984). However, some carbohydrate-binding proteins of animal origin are clearly of immunological nature (see <u>chicken serum and egg yolks Man-binding protein</u> below) and may be disqualified as lectins by the above definition.

Table I. Some carbohydrate-binding systems of animal origin.

Animal	Source	Specificity	Calcium required	Sub-units	Cellular activity		Ref.
MAMMALIAN (rat, rabbit, human)				(kDa)	Surface	Inside	
	Liver	GalNAc, Gal	+	40-50	+	+	a
	Lung macrophage	Man, L-Fuc	+	?	+	+	b
	Liver non-parenchymal cells	Man, GlcNAc	+	?	-	+	a
	Liver parenchymal cells	Man, GlcNAc ManNAc	+	31	-	+	c
	Ubiquitous	Man-6-P	+/-	120,45	+	+	d
MAMMALIAN AND AVIAN							
	Ubiquitous	Gal	-	12-29	-	+	e
AVIAN AND REPTILIAN							
	Liver	GlcNAc, Man	+	26	+	+	a

[a]Ashwell and Harford (1982); [b]Shepherd *et al.* (1981); [c]Maynard and Baenziger (1982); [d]Sly and Fischer (1982); [e]Barondes, S.H. (1984).

Baynes and Wold (1976) showed that ribonuclease B, which has a Man-terminated oligosaccharide chain, has a very short half-life in blood circulation. Achord *et al.* (1977) also reported that lysosomal enzymes (possessing exposed Man or GlcNAc residues) are rapidly cleared from the circulation. A large portion of the cleared Man- or GlcNAc-terminated glycoproteins is deposited in the liver, which we know now is due to the Man-binding activities of hepatic non-parenchymal cells. Another locus for deposition of the cleared Man-ligands is lung macrophages (Stahl *et al.*, 1978). In addition to these Man-binding proteins, a "mannan-binding protein" from rabbit liver was isolated by Kawasaki *et al.* (1978).

There are other Man-binding proteins found in animal sources. Some of the better characterized Man-binding proteins are listed in Table II. Animal Man-binding proteins studied so far appear to have different specificity patterns compared to those of plant origin. Most of Man-binding activities in animals can bind GlcNAc-containing ligands quite well, and for some, GlcNAc is as good or even a better recognition marker than Man. For example, chicken GlcNAc-binding protein shows substantial binding of Man-containing ligands (Kuhlenschmidt and Lee, 1984). Man-specific plant lectins react with Glc, but not GlcNAc, and GlcNAc-specific plant lectins do not react with Man. In this review, only those animal binding proteins which are primarily specific for Man are considered.

MANNOSE-BINDING PROTEIN IN MAMMALIAN ALVEOLAR MACROPHAGES

Stahl *et al.* (1976) first discovered that rapid clearance of lysosomal enzymes from the circulation can be suppressed by periodate oxidation of the enzymes, which suggested a possible link between the carbohydrate chains of the enzymes and the uptake mechanism. Unambiguous determination of the sugar specificity of binding by macrophages was obtained (Schlesinger *et al.*, 1980) using a series of neoglycoproteins prepared with bovine serum albumin (BSA). Although the macrophage

Table II. Animal Man-binding systems.

	Calcium required	Molecular weight Native	Subunit	Specificity hierarchy	Ref.
		(kDa)			
MAMMALIAN (rat, rabbit, human)					
Lung Macrophages	+	?		Man = L-Fuc > GlcNAC	a
Liver Nonparenchymal cells	+	?		ManNAc = Man = L-Fuc > GlcNAc	b
Parenchymal cells	+	190	32	ManNAc = Man = GlcNAc	c
Lymphoid tissues	+	?	32,58	ManNAc = GlcNAc = Man > $ManNH_2$	d
Serum	+	500	32	ManNAc = Man = GlcNAc	e
AVIAN					
Chicken Liver	+	280-740	32	GlcNAc > Man = L-Fuc	f
Serum	-		27,72	Man = L-Fuc > GlcNAc >> ManNAc	g

[a]Shepherd *et al.* (1981); [b]Ashwell and Harford (1982); [c]Mori *et al.* (1983); [d]Kawasaki *et al.* (1980); [e]Kozutsumi *et al.* (1980); [f]Kuhlenschmidt and Lee (1984); [g]Wang *et al.* (1985).

carbohydrate-binding protein is specific for ligands containing D-Man, it also reacts with ligands containing L-fucose. This can be easily explained by some stereochemical features which are shared by these two sugars (Lee and Lee, 1982). The corollary of this interpretation is that the substituents of C-6 of D-Man (or L-Fuc) and the aglycon linked to the anomeric carbon of D-Man (or L-Fuc) are not important for binding. This Man-binding protein exist not only on the surface of the macrophages but also intracellularly. After solubilization of the cells with a detergent such as Triton X-100, the total Man-binding activity is several-fold greater than the activity measured on the surface. This is analogous to the carbohydrate binding proteins of mammalian and chicken hepatocytes. Like the Gal/GalNAc-binding protein of mammalian hepato-cytes (Ashwell and Harford, 1982) or the GlcNAc-binding protein of chicken hepatocytes (Kuhlenschmidt et al., 1984), the Man-binding protein of the macrophages catalyzes the endocytosis of the appropriate sugar-containing solutes. It has been postulated that the Man-binding system of alveolar macrophages can also mediate phagocytosis of yeast cells (Warr, 1980). The binding and phagocytosis of zymosan by mouse peritoneal macrophages was shown to be inhibited by mannan suggesting that the Man-binding protein may be able to mediate phagocytosis of insoluble materials (Sung et al., 1983).

The Man-binding protein on the cell surface, upon complexing with Man-containing ligands, is internalized, and the ligands are eventually degraded in the lysosomes. However, the receptor does recycle back to the cell surface (also analogous to the hepatic carbohydrate-binding proteins), and this process can be inhibited with chloroquine or ammonia (Tietze et al., 1980, Tietze et al., 1982). These investigators also proposed the presence of two functionally distinct intracellular pools of the Man-binding protein.

Studies from our laboratory showed some interesting features of the Man-binding system of rabbit alveolar macrophages. (i) The rabbit alveolar Man-binding system is more active than many comparable carbohydrate-binding systems of other cells with respect to the rate of endocytosis and also the rate of eventual degradation of internalized ligands (Hoppe and Lee, 1983). This is understandable, since the macrophages are "professional phagocytes" of great efficiency. (ii) Unlike the Gal/GalNAc-binding system of mammalian hepatocytes or the GlcNAc-binding system of chicken hepatocytes, in which surface-bound ligands can be displaced with 0.1 M GalNAc or GlcNAc, respectively, alveolar macrophages show enhancement of binding when exposed to 0.1 M Man. The apparent number of binding sites remain nearly the same after such treatment, but the binding affinity increases (Hoppe and Lee, 1982).

Unfortunately, the isolation and purification of this Man-binding protein has been very difficult and thus molecular information on this particular animal lectin is rather scarce.

MANNALIAN LIVER MANNOSE-BINDING PROTEINS

The presence of Man-binding activity was first discovered in rabbit liver by Kawasaki et al. (1978). The binding protein was isolated with an affinity column of Sepharose 4B modified with yeast mannan, and thus the name of "mannan-binding" protein was coined. Using sequentially deglycosylated human orosomucoid derivatives (which possess Gal, GlcNAc, and Man, respectively, at the non-reducing terminal) and some BSA-based neoglycoproteins containing these sugars, the binding was confirmed to be specific for Man and GlcNAc. Neoglycoproteins containing 30-40 mol of Man or GlcNAc per mol of BSA bind more avidly than yeast mannan on a weight basis. Surprisingly, among the numerous monosaccharides tested, ManNAc shows the strongest inhibition of binding of mannan, Man-BSA or GlcNAc-BSA to the liver Man-binding protein. This is in contrast to the Man-binding proteins from alveolar macrophages and chicken serum, in which case ManNAc is much less potent than Man in inhibition of binding (Mizuno et al., 1981). This Man-binding protein requires calcium for binding activity, and consists of subunits of 31-32 kDa which exist as oligomers in aqueous solution. The molecular weight estimated by gel filtration is 250 kDa. The binding affinity (K_a) for mannan and GlcNAc-BSA is on the order of 10^8 to 10^9 M.

Maynard and Baenziger (1981) reported the isolation of a Man-binding protein from rat hepatocytes. This protein is similar to the rabbit liver Man-binding protein isolated by Mizuno et al. (1981), and is also similar to that found in reticuloendothelial cells (see below). While the reticuloendothelial binding protein is responsible for the binding and uptake of Man/GlcNAc-terminated glycoproteins, the liver Man-binding protein resides intracellularly and does not engage in endocytic activity. In addition, this protein is reported to recognize the oligomannoside core portion of N-linked oligosaccharides, even when the mannosyl residues are masked by other sugars, and this is called "core-specific lectin". Studies on the biosynthesis of the "core-specific" Man-binding protein showed a unique kinetics of the biosynthesis and secretion of this protein, which may be related to post-translational glycosylation and subunit assembly processes, has been noted (Brownell et al., 1984, Colley and Baenziger, 1985).

Kawasaki et al. (1985) isolated endogenous ligands for the rat liver Man-binding protein and found that they all possess a high mannose-type structure. These ligands are believed to be intermediates produced during processing of secretory glycoproteins. They speculated that the liver Man-binding protein can be a translocation protein associated with glycoprotein processing.

Similar Man-binding proteins have been found in other tissues and other animals. In fact, in spleen, lung, and lymph nodes, specific activities of this protein are even higher than that in liver, although the liver contains the largest amount of the Man-binding activity (Kawasaki, 1979).

It is to be noted that Tolleshaug et al. (1984) reported that rat hepatocytes can bind, internalize, and degrade yeast invertase (M_r 270 kDa, 50% mannan), while ribonuclease B was not recognized by these cells. Therefore a new "mannan-binding" protein capable of catalyzing endocytosis for mammalian hepatocytes seems to exist, but its isolation and characterization is not yet accomplished.

CHICKEN LIVER MANNOSE-BINDING PROTEIN

Two types of Man-binding proteins were isolated from chicken liver by affinity chromatography on a column of mannan-Sepharose (Oka *et al.*, 1985). The neutral Man-binding protein is quite similar to the Man-binding protein of mammalian liver. Both have molecular weight ranging from 280 to 740 kDa in native form and consist of subunits of 28 and 32 kDa. The acidic Man-binding protein is estimated to be ca. 640 kDa in its native state, and is comprised of 41 kDa subunits. Both Man-binding proteins are distinct from the GlcNAc-binding protein of chicken hepatocytes. Like the Man-binding protein from the mammalian liver, this Man-binding protein also does not take part in endocytosis.

MANNOSE-BINDING PROTEIN IN RABBIT SERUM

A Man-binding protein has been found in rabbit serum (Kozutsumi *et al.*, 1980). Immunochemically, this protein is indistinguishable from the rabbit liver Man-binding protein, but differs from it in its inhibition pattern by haptenic monosaccharides and in its much larger size (ca. 500 kDa by gel filtration). Kawasaki *et al.* (1983) reported that human serum also contains a "mannan-binding protein" of ca. 600 kDa, which consists of subunits of 31 kDa. They demonstrated that this binding protein is distinct from human C-reactive protein and amyloid P-component, both known to bind certain polysaccharides in the presence of calcium.

Table III. Comparison of chicken Man-binding proteins.[a]

	CS-MBP	CY-MBP	CL-MBP I	CL-MBP II	CL-GnBP
M.W. (kDa)	200	200	740	640	250
Subunits	27 75	27 75	28 32	41	27
Anti chicken IgG	+	+	-	-	-
Ca for Requirement	-	-	+	+	+
Binding pH	5-9	5-9			7-9
Specificity	Man > GlcNAc	Man > GlcNAc	Man, ManNAc GlcNAc		GlcNAc > Man
pI	6.5	6.5	?		?
References	b	c	d		e

[a]CS-, chicken serum; CY-, chicken egg yolk; CL-, liver; BP, binding protein; M, mannose; Gn, GlcNAc; [b]Wang *et al.* (1985); [c]Lee, Y.C., unpublished results; [d]Oka *et al.* (1985); [e]Kuhlenschmidt and Lee (1984).

MANNOSE-BINDING PROTEIN OF HEPATIC NONPARENCHYMAL CELLS

The Man-binding protein isolated from the whole rat liver (Townsend and Stahl, 1981) was initially thought to be the Man-binding protein responsible for the binding and uptake of Man-containing ligands by the non-parenchymal cells. However, as described above, later work (Maynard and Baenziger, 1981; Mori et al., 1983) revealed that there is a potent Man-binding protein in the hepatocytes which does not possess endocytic activity. Mori et al. (1983) noted that the procedure which was used to prepare Man-binding protein from the whole liver could be used to isolate Man-binding protein from the rat hepatocytes. However, the same procedure failed to yield a significant amount (less than 5% of the amount obtainable from the hepatocytes) of Man-binding protein from non-parenchymal cells. Therefore, it is clear that the Man-binding protein isolated from the whole liver is mostly derived from the parenchymal cells rather than the nonparenchymal cells.

Mori et al. (1983) partially purified the Man-binding protein from the rat liver non-parenchymal cells and reported its properties. They have also studied the binding and uptake of Man- and GlcNAc-terminated ligands by the non-parenchymal cells. Although the binding proteins from parenchymal and non-parenchymal cells both require calcium for binding, there are several important differences between the Man-binding proteins from the parenchymal and non-parenchymal cells. (i) While the antibody against the hepatocyte Man-binding protein completely eliminates its Man-binding activity, neither the IgG or Fab fragment prepared from this antibody affects the binding and uptake of Man- or GlcNAc-containing ligands by the non-parenchymal cells. (ii) ManNAc is a more potent inhibitor of binding by the hepatocyte Man-binding protein than Man, but Man is more potent in the non-parenchymal cells. (iii) The non-parenchymal cells show greater activity for GlcNAc-BSA or Man-BSA than mannan, but the hepatocyte Man-binding protein binds them with equal avidity. Although the Man-binding protein from the hepatocytes is termed "core-specific" (Maynard and Baenziger, 1981), the results of Mori et al. (1983) do not agree with this contention.

MANNOSE-BINDING PROTEIN IN RAT LYMPHOID TISSUES

A Man-binding protein from rat lymphoid tissues was isolated by Kawasaki et al. (1980). This protein shares many common features with the liver Man-binding protein, but differs from the liver protein in its antigenicity and in the inhibition of binding by $ManNH_2$ as well as ManNAc.

CHICKEN SERUM AND EGG YOLK MANNOSE-BINDING PROTEIN

Shortly after the report of the isolation of "mannan-binding protein" from rabbit serum by Kozutsumi et al. (1980), we reported the presence of Man-binding protein in chicken serum (1982). This is a protein distinct from the well-studied GlcNAc-binding protein of the chicken liver (Kuhlenschmidt and Lee, 1984, Kuhlenschmidt et al., 1984), although the GlcNAc-binding protein from chicken liver does react with Man. Purification of this protein is simply achieved by affinity chromatography on a column of Man-BSA-Sepharose, followed by gel filtration through an agarose column. The total quantity of the Man-binding protein present in the chicken serum is comparable to the GlcNAc-binding protein in the liver of a chicken.

Sera from chicken of various ages and from both sexes were found to contain this Man-binding protein (Lee et al., unpublished results). Therefore, it was undertaken to identify the Man-binding activity in chicken eggs (Wang et al., unpublished results). Both egg white and egg yolk contain the Man-binding activity, but the apparent total activity is greater in the yolk than in the white. Purification of the Man-binding protein from chicken egg yolk requires a more involved procedure due to the abundance of lipids, but the final product isolated is apparently indistinguishable from that isolated from serum. Unlike other Man-binding proteins described above, this binding protein does not require calcium for activity. It consists of two subunits, 72 kDa and 27 kDa, in a nearly equimolar ratio. In its native state, the molecular weight appears to be 200-230 kDa, suggesting that the molecule is comprised of two large and two small subunits. This notion led us to postulate a IgG-type structure for this protein, and indeed, as described below, we now believe this Man-binding protein is a subset of chicken IgG's.

Immunochemically, the chicken serum Man-binding protein is distinct from the chicken liver GlcNAc-binding protein. It also does not cross-react with the Man-binding protein from rabbit serum. However, the antisera raised against the chicken serum Man-binding protein reacts distinctly with the Man-binding protein from egg yolk. Most interestingly, the antisera against the serum Man-binding protein cross-reacts strongly with chicken serum IgG. Conversely, antisera against chicken IgG cross-reacts with Man-binding proteins from both egg yolk and serum. Protein A, which is known not to react with chicken IgG, did not react with the chicken Man-binding protein. Summing all the evidence presented above, it is reasonable to conclude that the Man-binding proteins from chicken serum and egg yolk belong to a subset of chicken serum IgG.

The fact that the same type of protein exists both in serum and egg is not a unique phenomenon. There is a known immunoglobulin transport system which mediates the deposition of serum IgG to egg white (Loeken and Roth, 1983). Riboflavin-binding protein exists in both chicken serum and chicken egg (white and yolk) in which case, some differences in amino acid and sugar composition have been noted. Whether such a subtle difference exists between the Man-binding proteins from serum and egg awaits further studies. Table III compares some properties of Man-binding proteins from chicken.

Yeast mannans are good ligands to these Man-binding proteins. This fact tends to suggest that the function of the chicken Man-binding proteins may be for anti-microbial invasion.

MANNOSE-SPECIFIC ADHESION OF RABBIT ALVEOLAR MACROPHAGES

Polyacrylamide gels modified with sugar derivatives (see, for example, Pless et al., 1983) have been very useful in demonstrating sugar-specific cell adhesion to solid surfaces. Largent *et al.* (1984) reported that rabbit alveolar macrophages specifically adhere to gel layers containing aminohexyl mannosides. The adhesion is inhibited by those sugars which are potent in inhibiting macrophage binding of Man-containing ligands (*i.e.*, D-mannose and L-fucose).

DELIVERY BY TARGETING

Obviously, the Man-binding system which has very high affinity for certain Man-containing ligands could be exploited for the specific delivery of chemicals and drugs to cells and tissues which possess such binding activity. Systematic studies on Man-modified di-, tri- and oligolysines showed that these compounds are potent inhibitors of the uptake of radiolabeled Man-BSA by rat alveolar macrophages (Ponpipom *et al.*, 1981). It was concluded from this study that the chirality of the peptide backbone is unimportant, but the number of the Man residues and the length of the side chain to which the Man derivatives are attached are important. In agreement with these generalizations was the observation that β-glucuronidase modified with Man_3Lys_2 shows enhanced uptake by reticuloendothelial cells when injected into rats (Doebber *et al.*, 1982). In a recent report, Roche *et al.* (1985) showed that specific delivery of muraminyl dipeptide (macrophage activator) chemically conjugated to Man-BSA to monocytes caused sufficient activation of these cells to manifest effective tumoricidicity *in vivo*.

ENZYME SECRETION

Zymosan is known to induce the secretion of β-hexosaminidase from peritoneal macrophages. Rodmer and Dean (1983) reported that this secretion is inhibited by mannose. The effect of zymosan is a long-term effect. If the macrophages are exposed to zymosan and then subsequently maintained in medium free of zymosan, the secretion of the enzyme continues for some time. However, mannose does not inhibit the secretion previously induced by endocytosis of zymosan. Zymosan-mediated secretion of another lysosomal enzyme, β-galactosidase, is apparently different from β-galactosidase secretion induced by methylamine (Riches *et al.*, 1983).

CONCLUSION

There are now a large number of Man-binding proteins reported to be present in vertebrates. They seem to have diverse structural organization and different apparent functions, which may range from endocytosis (mammalian macrophages), intracellular translocation (mammalian hepatocytes), or anti-microbial invasion (chicken serum and egg). Future studies are expected to shed more light on our understanding of these carbohydrate-binding proteins.

ACKNOWLEGEMENTS

The author wishes to thank Drs. Theresa Kuhlenschmidt, Ke-Yi Wang, Pradip K. Datta, and Ms. Amy Fogelstrom for their work on chicken Man-binding proteins. The author is also indebted to Ms. Nancy Stults and Dr. Craig Hoppe for many valuable suggestions. The author acknowledges research supports from National Science Foundation Research Grant DCB 8408680 and National Institutes of Health Research Grant CA21901.

REFERENCES

Achord, D., Brot, F., and Sly, W. (1977). "Inhibition of the rat clearance system for agalacto-orosomucoid by yeast mannan and by mannose." Biochem. Biophys. Res. Commun. 77:409-415.

Ashwell, G. and Harford, J. (1982). "Carbohydrate-specific receptors of the liver." Ann. Rev. Biochem. 51:531-554.

Ashwell, G. and Morell, A. (1974). "The role of surface carbohydrates in the hapatic recognition and transport of circulating glycoproteins." Adv. Enzymol. Relat. Areas Mol. Biol. 41:99-128.

Barondes, S.H. (1984). "Soluble lectins: a new class of extracellular proteins." Science 223:1259-1264.

Baynes, J., and Wold, F. (1976). "Effect of glycosylation on the in vivo circulating half-life of ribonuclease." J. Biol. Chem. 261: 6016-6124.

Bodmer, J. L. and Dean, R. T. (1983). "Does the induction of macrophage lysosomal enzyme secretion by zymosan involve the mannose receptor?" Biochem. Biophys. Res. Commun. 113, 192-198.

Boyd, W. C. and Shapeleigh, E. (1954). "Specific precipitating activity of plant agglutinins (Lectins)." Science 119:419.

Brownell, M. D., Colley, K. J., and Baenziger, J. U. (1984). "Synthesis, processing, and secretion of the core-specific lectin by rat hepatocytes and hepatpoma cells." J. Biol. Chem. 259:3925-3932.

Colley, K. J. and Baenziger, J. U. (1985). "Relationship of post-translational modification to complex assembly and secretion of the rat core-specific lectin." Proc. VIIIth Interntl. Symp. on Glycoconjugates (Houston, TX), Praeger, NY, p. 353.

Doebber, T. W., Wu, M. S., Bugianess, R. L., Ponpipom, M. M., Furbish, F. S., Barranger, J., Brady, R.O., and Shen, T. Y. (1982). "Enhanced macrophage uptake of synthetically glycosylated human placenatal p-glucoscerebrosidase." J. Biol. Chem. 257:2193-2199.

Hoppe, C. A. and Lee, Y. C. (1982). "Stimulation of mannose-binding activity in the rabbit alveolar macrophage by simple sugars." J. Biol. Chem. 257:12831-12834.

Hoppe, C. A. and Lee, Y. C. (1983). "The binding and processing of mannose-binding serum albumin derivatives by rabbit alveolar macrophages." J. Biol. Chem. 258:14193-14199.

Kawasaki, T. (1979). "Mannan-binding protein - a receptor for lysosomal enzymes." Seikagaku, 51:447-451.

Kawasaki, T., Etoh, R., and Yamashina, I. (1978). "Isolation and characterization of a mannan-binding protein from rabbit liver." Biochem. Biophys. Res. Commun. 81:1018-1024.

Kawasaki, N., Kawasaki, T., and Yamashina, I. (1983). "Isolation and charactercation of mannan-binding protein from human serum." J. Biochem. (Tokyo) 94:937-947.

Kawasaki, T., Mizuno, Y., Masuda, T., and Yamashina, I. (1980). "Mannan-binding protein in lymphoid tissues of rats." J. Biochem. (Tokyo) 88:1891-1894.

Kawasaki, T., Mori, K., Oka, S., Ikeda, K., and Yamashina, I (1985) Proc. VIIIth Internatl. Symposium on Glycoconjugates, (Houston, TX), Praeger, N.Y., p.146.

Kozutsumi, Y., Kawasaki, T., and Yamashina, I. (1980). "Isolation and characterization of a mannan-binding protein from rabbit serum." Biochem. Biophys. Res. Commun. 95:658-664.
Kuhlenschmidt, T. and Lee, Y. C. (1982). "Isolation of a mannose-binding protein from chicken serum." Fed. Proc. 41:1033.
Kuhlenschmidt, T. and Lee, Y. C. (1984). "Specificity of chicken liver carbohydrate binding protein." Biochemistry 23:3569-3575.
Kuhlenschmidt, T., Kuhlenschmidt, M., Roseman, S., and Lee, Y. C. (1984) Biochemistry 23:437-6444.
Largent, B. L., Walton, K. M., Hoppe, C. A., Lee, Y. C., and Schnaar, R. L. (1984). "Carbohydrate-specific adhesion of alveolar macrophages to mannose-derivatized surfaces." J. Biol. Chem. 259:1764-1769.
Lee, R.T. and Lee, Y.C. (1982). "Neoglycoproteins as probes for binding and cellular uptake of glycoconjugates." in "The Glycoconjugates", Vol. 4 (Ed. M. Horowitz), Acad. Press, N.Y., p.57-83.
Loeken, M. R. and Roth, T. F. (1983). "Analysis of maternal IgG subpopulations which are transported into the chicken oocyte." Immunology 49:21-28.
Maynard, Y. and Baenziger, J. U. (1981). "Oligosaccharide-specific endocytosis by isolated rate hepatic reticuloendothelial cells." J. Biol. Chem. 256:8063-8068.
Mizuno, Y., Kozutsumi, Y., Kawasaki, T., Yamashina, I. (1981). Isolation and characterization of a mannan-binding protein from rat liver." J. Biol. Chem. 256:4247-4252.
Mori, K., Kawasaki, T., Yamashina, I. (1983). "Identification of the mannan-binding protein from rat livers as a hepatocyte protein distinct from the mannan receptor on sinusoidal cells." Arch. Biochem. Biophys. 222:542-552.
Oka, S., Kawasaki, T., and Yamashina, I. (1985). "Isolation and characterization of mannan-binding proteins from chicken liver." Arch. Biochem. Biophys. 241:95-105.
Pless, D.D., Lee, Y.C., Roseman, S., and Schnaar, R.L. (1983). "Specific cell adhesion to immobilized glycoproteins demonstrated using new reagents for protein and glycoprotein immobilization." J. Chem. 258:2340-2349.
Ponpipom, M. M., Bugianesi, R. L., Robbins, J. C., Doebber, T. W., and Shen, T.Y. (1981). "Cell-specific ligands for selective drug delivery to tissues and organs." J. Med. Chem. 24:1388-1395.
Riches, D. W., Watkins, J. L., and Stanworth, D. R. (1983). "Biochemical differences in the mechanism of macrophage lysosomal exocytosis initiated by zymosan particles and weak bases." Biochem. J. 212: 869-874.
Roche, A. C., Midoux, P., Mayer, R., and Monsigny, M. (1985). "Activation of murine macrophages and human monocytes by membrane lectin-mediated endocytosis of activators-bound to neoglycoproteins." Proc. VIIIth International Symposium on Glycoconjugates, Houston, TX, Praeger, N.Y., p.569.
Schlesinger, P. H., Rodman, J. S., Doebber, T. W., Stahl, P., Lee, Y. C., Stowell, C. P., and Kuhlenshmidt, T. (1980). "The role of extra-hepatic tissues in the receptor-mediated plasma of glycoproteins terminated by mannose or N-acetylglucosamine." Proc. Natl. Acad. Sci. U.S.A. 192:597-606.

Shepherd, V. L., Lee, Y. C., Schlesinger, P. H., and Stahl, P. D. (1981). "L-Fucose-terminated glycoconjugates are recognized by pinocytosis receptors on macrophages." Proc. Natl. Acad. Sci. (USA) 78:1019-1022.

Sly, W.S. and Fischer, H.D. (1982). "The phosphomannosyl recognition system for intracellular and intercellular transport of lysosomal enzymes." J. Cell. Biochem. 18:67-85..

Stahl, P., Six, H., Rodman, J. S., Schlesinger, P., Tulsiani, D. R. P., and Touster, O. (1976). "Evidence for specific recognition sites mediating clearance of lysosomal enzymes in vivo." Proc. Natl. Acad. Sci. U.S.A. 73:4045.

Stahl, P., Rodman, J. S., Miller, M. J., and Schlesinger, P. H. (1978). "Evidence for receptor-mediated binding of glycoproteins, glycoconjugates, and lysosomal glycosidases by alveolar macrophages." Proc. Natl. Acad. Sci U.S.A. 73:1399-1403.

Stillmark, H. (1888) Inaug. Diss., Dorpat., as cited in Goldstein and Hayes (1978).

Sung, S.J., Nelson, R.S., and Silverstein, S.C. (1983). "Yeast mannans inhibit binding and phagocytosis of zymosan by mouse peritoneal macrophages." J. Cell. Biol. 96:160-166.

Tietze, C., Schlesinger, P. H., and Stahl, P. (1980). Chloroquine and ammonium ion inhibit receptor-mediated endocytosis of mannose-glycoconjugates by macrophages: apparent inhibition of receptor recycling." Biochem. Biophys. Res. Commun. 93:1-8.

Tietze, C., Schlesinger, P. H., and Stahl, P. (1982). "Mannose-specific endocytosis receptor of alveolar macrophages: Demonstration of two functionally distinct intracellular pools of receptor and their roles in receptor recycling." J. Cell Biol. 92:417-424.

Tolleshaug, H., Berg, T., and Blomhoff, R. (1984). "Uptake of mannose-terminated glycoproteins in isolated rat liver cells. Evidence for receptor-mediated endocytosis in hepatocytes." Biochem. J. 223: 151-160.

Townsend, R. and Stahl, P. (1981). Isolation and characterization of a mannose/N-acetylglucosamine/fucose-binding protein from rat liver." Biochem. J. 194:209-214.

Wang, K. Y., Kuhlenschmidt, T. B., and Lee, Y. C. (1985). "Isolation and characterization of the major mannose-binding protein in chicken serum."

Warr, G. (1980). "A macrophage receptor for (mannose/glucosamine)-glycoproteins of potential importance in phagoytic activity." Biochem. Biophys. Res. Commun. 93:737-745.

SIALIC ACID AND N-ACETYLGALACTOSAMINE SPECIFIC BACTERIAL LECTINS OF ENTEROTOXIGENIC *ESCHERICHIA COLI* (ETEC)

Mats Lindahl[1], Reinhard Brossmer[2],
And Torkel Wadström[1]

[1]*Department of Veterinary Microbiology, Swedish University of Agricultural Sciences Uppsala, Sweden, and* [2]*Department of Biochemistry 2 University of Heidelberg, Heidelberg, West-Germany*

Many pathogenic and nonpathogenic bacteria have the ability to produce proteins that mediate adhesion to surfaces. These "adhesins" are usually seen in the electron microscope as filaments, protruding from the cell surface. The filaments are called fimbriae (or pili) and in general they have the ability to bind to carbohydrates and cause hemagglutination. Their ability to mediate adhesion to cells has been correlated with the pathogenicity of bacteria.

E.COLI TYPE-1 FIMBRIAE

Hemagglutinating properties were discovered by Guyot in 1908. He found that E.coli cells agglutinated erythrocytes of different origin and that the agglutinability was different for a number of E.coli strains using different erythrocytes. When this subject was re-investigated by Duguid et al. in 1955, filamentous structures were detected on the surface of hemagglutinating E.coli by electron microscopy. The filaments were suggested to be responsible for the hemagglutinating properties and were given the name fimbriae. Morphologically they appeared, in the electron microscope, to be about 10 nm in diameter and 0,3-1,0 µm long. Hemagglutinating E.coli strains were divided into groups I through IV, based on their ability to agglutinate erythrocytes from a number of different animal species.

Collier and de Miranda (1955) found that the hemagglutination with some E.coli strains could be specifically inhibited with D-mannose. These E.coli strains were found to fit into group I of the grouping system developed by Duguid. The fimbriae responsible for mannose specific (or mannose sensitive - MS) hemagglutination was subsequently named type I fimbriae.

D-Mannose specific fimbriae have been identified, during the last three decades, on a number of different bacterial species. It seems as if D-mannose specificity is a general property of most gram-negative bacteria. Furthermore, the gene coding for this protein has been located on the chromosome.

Type I fimbriae appear to be homopolymeric proteins. The subunits (Mw 17 kD) were found, by X-ray diffraction, to be arranged in a helix having 3,125 subunits per turn (Brinton, 1965). When purified fimbriae were depolymerized, the subunits retained their carbohydrate specific binding property (Eshdat et al., 1981).

For a number of years, E.coli type I fimbriae, were believed to be involved in colonization and thus responsible for ascending infections in the urinary tract of humans (Svanborg-Edén and Hansson, 1978). However, type I fimbriae are now believed to be less important, since other types of fimbriae coexisting with type I fimbriae have been shown to have a more significant association with pyelonephritis (Leffler and Svanborg-Edén, 1980).

FIMBRIAE OF ENTEROPATHOGENIC E.COLI

Up to now, a number of different fimbriae have been discovered, both on E.coli and on other species of enterobacteria. D-Mannose specificity is the dominant feature but specificities for other carbohydrates have been found (Table 1). In general, fimbriae are non-glycosylated, homopolymeric proteins with subunits having molecular weights ranging from 8 kD to 34 kD.

The ability of bacteria to hemagglutinate with specific

Table 1. Carbohydrate specificities of bacterial lectins

Carbohydrate Specificity	Bacterial Species
D-Mannose	Aeromonas hydrophila
	Aeromonas salmonicida
	Citrobacter diversus
	Citrobacter freundii
	Corynebacterium parvum
	Enterobacter agglomerans
	Enterobacter cloacae
	Escherichia coli
	Klebsiella aerogenes
	Klebsiella pneumoniae
	Myxococcus xanthus
	Proteus morganii
	Pseudomonas echinoides
	Salmonella enteritidis
	Salmonella typhi
	Salmonella typhimurium
	Serratia marcescens
	Shigella flexneri
	Vibrio cholerae
L-Fucose	Aeromonas hydrophila
	Aeromonas salmonicida
	Vibrio cholerae
D-Galactose	Actinomyces naeslundii
	Actinomyces viscosus
	Aeromonas hydrophila
	Escherichia coli
	Fusobacterium nucleatum
	Pseudomonas aeruginosa
	Streptococcus pyogenes
N-Acetyl D-galactosamine	Escherichia coli
	Myxococcus xanthus
N-Acetyl D-glucosamine	Escherichia coli
N-Acetylneuraminic acid	Escherichia coli
	Mycoplasma pneumoniae
	Pseudomonas aeruginosa
	Streptococcus sanguis

binding to carbohydrates other than D-mannose is often reported to be associated with virulence. In some cases this concept has been proved, e.g. for K88 fimbriae. It was shown that an enterotoxin producing E.coli, isolated from piglets suffering from diarrhea, agglutinated guinea-pig erythrocytes in the presence of D-mannose (Jones and Rutter, 1972). It was found that the hemagglutinin was a fimbrium and it was named K88. When the wild type strain was deprived of the ability to produce K88 fimbriae, it could not colonize and cause diarrhea in piglets (Smith and Linggood, 1971). Furthermore, it was found that the susceptibility of pigs to this E.coli strain was determined genetically and that the susceptible pigs produced a receptor for K88 while the non-susceptible pigs did not produce the receptor (Rutter et al., 1975).

In 1975 a new antigen (K99), associated with diarrhea in calves and lambs, was discovered (Ørskov et al., 1975). Both K88 and K99 were originally believed to be capsular antigens, which is the explanation for the K in their names. Both have been found to be proteins and fimbriae. 987p was the third fimbrium discovered on enterotoxigenic E.coli. This fimbrium was also isolated from piglets having diarrhea. However, unlike the other two, 987p did not cause agglutination of erythrocytes but was found to cause adherence to epithelial cells from the small intestine of pigs (Isaacson and Richter, 1981). A fourth E.coli fimbriae (F41) associated with piglet diarrhea has recently been described (de Graaf and Roorda, 1982). However, it was previously described in 1980 with the name "anionic K99" (Morris et al., 1980).

MULTIPLE FIMBRIATION

Any E.coli strain is capable of producing type I fimbriae. If there is a plasmid coding for another fimbrium, both type I and a second fimbrium can be produced by one and the same cell. E.coli strains often have plasmids coding for different fimbriae. Thus a number of different fimbriae may be produced by a single strain, e.g. Neisseria gonorrhoeae (Lambden, 1982), uropathogenic (Nowicki et al., 1984) and enterotoxigenic E.coli (Morris et al., 1980) and Aeromonas hydrophila (Atkinson and Trust, 1980).

The presence of different fimbriae complicates the hemagglutination test, especially when specificity is studied using different monosaccharides as inhibitors. Therefore, it is necessary to use strains producing only one single type of fimbriae for the hemagglutination inhibition studies. The inconvenience of the presence of type I fimbriae can be circumvented by addition of D-mannose or αD-methylmannoside.

CULTIVATION OF BACTERIA FOR EXPRESSION AND DETECTION OF FIMBRIAE

Cultivation of bacteria for fimbriae production is a difficult task, especially if nothing is known about the demand on medium and growth temperature. Generally a nutrient poor solid medium and a temperature of 37°C is most suitable for the production of plasmid mediated fimbriae. Although, there are exceptions, e.g. K88 are plasmid mediated E.coli fimbriae which are produced in nutrient rich cultures at 37°C. The production of some fimbrial types, e.g. E.coli type I, is enhanced in a nutrient rich static broth (Duguid et al., 1955). Other fimbriae may only be produced when cultivation is performed below 33°C, as shown for Yersinia enterocolitica (Faris et al., 1983).

The classical method for fimbriae detection is hemagglutination on glass slides or on porcelain depression tiles, using a set of erythrocytes of different origin (Duguid et al., 1955). The procedure is to add one drop of 3% (vol/vol) erythrocyte suspension to a drop of bacterial suspension (10^{10} cells/ml) and mix for a few minutes at room temperature. Sometimes the mixture needs cooling to give visible agglutination. This method is not very sensitive and may give false negative results.

Another, more sensitive, hemagglutination test has proved useful. Bacterial suspensions (50µl) are added to microtiter wells (V-shaped), mixed with erythrocyte suspensions (0,5-1% vol/vol) and incubated, without shaking, at room temperature or at 4°C for 4 hours. A red dot of erythrocytes in the middle of the wells are considered as negative hemagglutination, while positive reactions appear with a thin layer of erythrocytes completely covering the bottom of wells (Arai and

Munoz, 1979). This method is also suitable for hemagglutination inhibition tests (Smit et al., 1984).

Hemagglutination can also be performed in a cell aggregometer (Firon et al., 1983). This method is very sensitive and is suitable for inhibition tests since the reaction can be measured and recorded continuously. Erythrocyte suspensions ($5x10^6$ cells/ml) in 0.01M sodium phosphate buffered saline, pH 6.8, containing α D-methylmannoside, are mixed with fimbriated bacterial cells ($4x10^7$ cells/ml) with stirring at 20°C. The amount of light transmitted through the mixture, per minute is measured. An increase in transmitted light is a measure of a decreasing number of particles (i.e. erythrocytes or clumps of erythrocytes) scattering light, as the hemagglutination reaction is completed. Relative inhibiting potency of inhibitors can be calculated from the decrease of the slope value. If the slope values are plotted against the log concentration of inhibitors, the concentration causing 50% inhibition can be calculated.

Some fimbrial types may not cause hemagglutination with any type of erythrocytes since receptors can be either lacking or not properly exposed on the erythrocyte surface. These fimbriae are referred to as non-hemagglutinating fimbriae (Rohde et al., 1975). To be able to detect these fimbriae, it is preferable to use adhesion experiments where epithelial cells from the relevant organ of the host is incubated with the bacterial suspension and observed by light microscopy (Isaacson et al., 1978).

The presence of fimbriae is readily confirmed by transmission electron microscopy. Bacteria are allowed to settle on formvar coated grids and negatively stained with phosphotungstic acid or uranyl acetate (Brinton, 1965) and the grids examined at a magnification of about 12.000 X.

SIALIC ACID SPECIFIC FIMBRIAE

Colonization factor antigen I (CFA/I) of enterotoxigenic _E.coli_ (ETEC), which had been isolated from humans with diarrhea, was the first fimbrium proven to be sialic acid specific (Evans et al., 1979). So far, four different sialic acid specific fimbriae of _E.coli_ have been discovered (Evans et al., 1979; Lindahl and Wadström, 1984; Parkkinen et al., 1983 and Wadström et al., 1985). Other bacterial species have also been shown to have sialic acid specific fimbriae. Bacteria with sialic acid specific fimbriae have been associated with different diseases e.g. diarrhea, pneumonia, caries and meningitis (Evans et al., 1979; Lindahl and Wadström, 1984; Parkkinen et al., 1983; Sobeslavsky et al., 1968; Wadström et al., 1985 and Murray et al., 1982).

COLONIZATION

Fimbriae enable bacteria to adhere to a mucosal surface of the host where they can multiply and cause infection. Thus the bacterium is specialized for the environment at the site of colonization and has an organelle that enables adhesion to this site. To cause infection toxins or enzymes that affect the host are usually needed. Colonization is not necessarily linked to infection since bacteria living in symbiosis with the host also uses adhesive factors, e.g. fimbriae, to colonize (Savage and Blumershine, 1974), and pathogens which cause infection in the urinary tract may live and multiply in the intestines of the host without causing clinical symptoms. Even if the appropriate receptor promoting adhesion is present, the environment may not fit the nutritional requirements for bacterial multiplication.

In studies on E.coli causing diarrhea in animals, it has been found that susceptibility to infection can be correlated with the age of the host.

E.coli strain RDEC-1 causing diarrhea in rabbits was shown to bind to brush border membranes from rabbits older than 21 days and that binding could be correlated to susceptibility to infection (Cheney and Boedeker, 1984).

K99 fimbriated enterotoxigenic E.coli were found to adhere to epithelial cells from the small intestine of newborn piglets and calves. As the piglets and calves aged, adhesion to their epithelial cells was significantly reduced (Runnels et al., 1980). In contrast, K88 fimbriated bacteria were found to bind to epithelial cells independent of the age of the animal. These findings are consistant with the susceptibility of the animal, since diarrhea caused by K99 producing strains occurs in newborn piglets but is very rare in 6 weeks old piglets. K88 strains on the other hand, causes diarrhea both in the newborn and in older piglets.

THE MUCOSAL SURFACE

Enterotoxigenic E.coli adhere to the small intestinal mucosa of the host. The epithelial cells of the intestinal mucosa are covered by a mucous film (mucus). A small fraction of the epithelial cells, goblet cells, produce the mucus. The function of the mucus is to protect epithelial cells, to lubricate and keep the epithelium hydrated. Its main component, mucin, is composed of a linear protein backbone covered by oligosaccharide chains with a carbohydrate content up to 80% dry weight (Forstner, 1978). Porcine and human small intestinal mucins have been found to be composed of Gal, Fuc, GalNAc, GlcNAc and sialic acid residues; Glc and Man were virtually absent (Mantle and Allen, 1981; Forstner et al., 1979). So far no sequences of oligosaccharides from small intestinal mucins have been published.

Culling et al. (1977) suggested that sialic acids in the small intestinal mucin of adult humans are O-acetylated on the tricarbon tail, since the mucin was insensitive to V.cholerae sialidase. Mucins from adults show a lower sensitivity to sialidase as compared to fetal mucins, despite the fact that 4-O-acetyl-NeuAc was not detected (Lev, 1968). This implies that O-acetylation of sialic acids in the small intestinal mucin may increase with age. However, sialic acids in bovine small intestinal mucin were not O-acetylated in the adult animal (Nishiura, 1984).

THE HOST

Young animals, e.g. piglets, calves and lambs are susceptible to intestinal infections of enterotoxigenic E.coli. The newborn may be infected immediately following birth and die after only 3-4 days due to dehydration. Since the neonatal animal's immunological defence system is immature, it is dependent on immunoglobulins in the maternal milk. Thorough investigations of this problem have been done. If the mother has been vaccinated or otherwise exposed to the pathogen, the milk will supply sufficient protection (Rutter et al., 1976). However, other components in the maternal milk may also protect against infection.

It has been shown that the protein content in cow and pig colostrum decreases during the first week after parturition (Perrin, 1955). The sialic acid content in cow and human milk also decreases during this period (Hirano et al., 1966). Thus it is possible that the high levels of protein and sialic acid in colostrum may play a role as inhibitors of bacterial infections in the gastrointestinal tract during this initial period.

Gibbons et al. (1975) showed that glycoproteins and oligosaccharides from sow colostrum could inhibit hemagglutination of K88 fimbriated E.coli , but the immunoglobulin fraction was found to be the most efficient inhibitor.

Fat globule membranes (FGM) from milk contains membrane glycoproteins from the apical plasma membrane of the secretory cells of the mammary gland. K99-, K88- and CFA/I-fimbriated organisms have been found to adhere to FGMs from bovine, porcine and human milk respectively (Reiter, 1981). However, secretory IgA has also been found on porcine FGMs and may perhaps be the major factor providing adhesion of bacterial cells (Honkanen-Buzalski and Sandholm, 1981), even though there are other components in FGMs with the ability to inhibit K88 hemagglutinin (Atroshi et al., 1983).

Furthermore, skimmed milk of human origin has been shown to bind heat - labile enterotoxin from E.coli (Otnaess and Halvorsen, 1980). These results were further supported by the fact that receptors (GM_1 ganglioside) for the enterotoxin were purified from the milk fractions (Kolstø Otnaess et al., 1983).

Still, it is clear that only passive immunization by transfer of immunoglobulins to the newborn gives sufficient protection towards infection in vivo. Whether the non-immunoglobulin fractions of milk have any effect at all before or during infection is still to be elucidated.

SIALIC ACID SPECIFIC E.COLI LECTINS

Evans et al. (1979) claimed that hemagglutination by CFA/I fimbriated enterotoxigenic E.coli could be inhibited with sialic acid. A year later it was shown that hemagglutination by CFA/I or K99 fimbriated cells was weaker with sialidase treated than with untreated erythrocytes (Faris et al., 1980). Hemagglutination inhibition with GM_2 ganglioside was detected, while neutral glycolipids did not inhibit CFA/I or K99. Thus it was concluded that sialic acid must be in the structure of the receptor for these fimbriae. Further evidence for the sialic acid specificity of CFA/I was supplied when sialic acid monosaccharides and bovine submaxillary mucin were found to inhibit hemagglutination while poly α2-8-NeuAc (colominic acid) could not (Lindahl et al., 1982). The lack of inhibitory activity of colominic acid might be explained by the fact that colominic acid can be O-acetylated in the tricarbon tail (Ørskov et al., 1979).

K99 hemagglutination was also demonstrated to be sensitive to sialic acid monosaccharides as well as to a few sialoglycoproteins (Lindahl and Wadström, 1984). Human glycophorin and Fab fragments of rabbit anti-N-antigen antibodies were also found to inhibit hemagglutination.

A third type of sialic acid specific fimbriae was found on an E.coli strain from a child with meningitis (Parkkinen et al., 1983). It was shown that the hemagglutination with this strain was inhibited by sialyllactoses and that the NeuAcα2-3D-Galß1-4D-Glc and NeuAcα2-3D-Galß1-4D-GlcNAc were the most potent inhibitors.

The CS 2 antigen found on enterotoxigenic E.coli of the O6 and O8-serogroups isolated from human diarrhea has recently been shown to be specific for sialic acid (Wadström et al., 1985). This fimbrium was found to be more specific for NeuGc than for NeuAc, twice as much NeuAc as compared to NeuGc was needed to give 50% hemagglutination inhibition. Hemagglutination occurs with bovine but not with human erythrocytes. Whether this reflects the specificity towards carbohydrates is still unknown.

The specificity of K99 fimbriae has been subjected to more detailed studies than any other sialic acid specific fimbriae. By using the aggregometer assay, the sensitivity of

Table 2. Inhibition of K99 hemagglutination with monosaccharides

Monosaccharide	50 % inhibition (mM)
NeuGc	6.8
NeuAc	11.8
Gal	121
Fuc	160
GalNAc	232
GlcNAc	299
Glc	532

Inhibition data obtained by using the aggregometer assay

K99 hemagglutination to NeuAc was found to be 10 times higher than to Gal and 15, 20 and 25 times higher than to Fuc, GalNAc and GlcNAc respectively (Table 2). Sialic acid derivatives lacking the carboxyl group (2α-benzyl-nonulosamine, 2α-benzyl-NeuAc-methylester and 2α-benzyl- NeuAc-amide) had lower inhibiting activity on hemagglutination (Table 3). Hence, the carboxyl group must be involved in the protein-carbohydrate interaction which could explain why a lowered ionic strength increases the hemagglutinating activity. The position of the carboxyl group was shown to be important since 2ß-methyl-NeuAc was less inhibitory than 2α-methyl-NeuAc (Figure 1). NeuGc was found to be a stronger inhibitor than NeuAc, which implies that the additional hydroxyl group, in NeuGc, is likely to form a hydrogen bond. Acetylation of the hydroxyl groups on the tricarbon tail decreased the inhibition activity. Thus it is likely that the hydroxyl groups at C_7 and C_9 are involved in binding. An additional hydroxyl group at C_3 enhances binding (Table 3). Both 4-epi-NeuAc, with an axially orientated hydroxyl group at C_4, and 4-O-acetyl-NeuAc were better inhibitors than NeuAc. This might be due to a new possibility of interaction with the protein, where the

Table 3. Inhibition of K99 hemagglutination with sialic acid derivatives

Derivative	50 % inhibition (mM)
2-α-Benzyl-NeuAc	0,4
NeuAcα2-8NeuAcα2-3galß1-4glc	0,8
NeuAcα2-6galß1-4glc	3,1
2-α-Benzyl-9-O-acetylNeuAc	3,2
4-epi-NeuAc	3,5
NeuAcα2-3galß1-4glc	3,5
2-α-Methyl-NeuAc	3,7
4-O-Acetyl-NeuAc	4,1
2-α-Benzyl-nonulosamine	5,2
NeuGc	6,8
3-Hydroxy-NeuAc	6,9
NeuAc	11,8
2-ß-Methyl-NeuAc	13,5
2-α-Benzyl-8,9-isopropylidene-NeuAc	18,0
2-α-Benzyl-NeuAc-methylester	20,4
2-α-Benzyl-NeuAc-amide	115
4,7-O-diacetyl-NeuAc	2,5[a]
9-O-Acetyl-NeuAc	3,0[a]
2-α-Benzyl-5-noracetyl-NeuAc	6,0[a]

Inhibition data obtained by using the aggregometer assay.
[a] No inhibition at this concentration, hemagglutination reaction appeared as the control.

inverted position of the hydroxyl or the addition of a carbonyl group can enhance binding to the protein. The acetamido group was found to be involved in binding, since a derivative lacking this group, and 2α-benzyl-5-noracetyl-NeuAc, was a very weak inhibitor. It was also found that 2α-benzylated derivatives were more efficient inhibitors. 2α-benzyl-NeuAc was a 10 times stronger inhibitor than 2-α-methyl-NeuAc. Similar findings have been obtained with D-mannose derivatives for E.coli type I fimbriae, where the

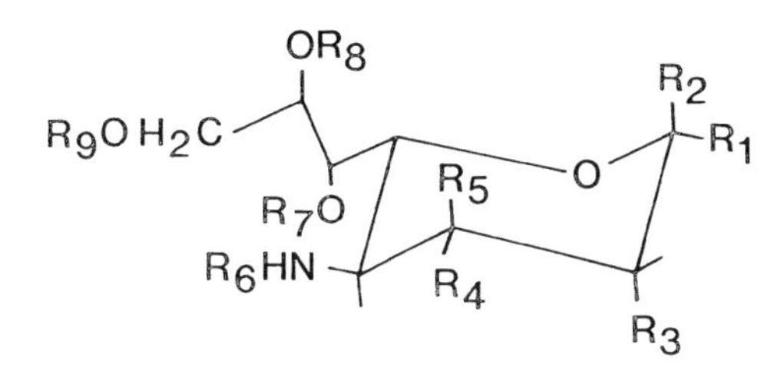

Fig. 1.

2-α-Benzyl-NeuAc
$R_1=OCH_2C_6H_5$, $R_2=COOH$, $R_3=H$, $R_4=OH$, $R_5=H$, $R_6=COCH_3$, $R_7=R_8=R_9=H$

2-α-Benzyl-9-O-acetyl-NeuAc
$R_1=OCH_2C_6H_5$, $R_2=COOH$, $R_3=H$, $R_4=OH$, $R_5=H$, $R_6=COCH_3$, $R_7=R_8=H$, $R_9=COCH_3$

4-epi-NeuAc
$R_1=COOH$, $R_2=OH$, $R_3=H$, $R_4=H$, $R_5=OH$, $R_6=COCH_3$, $R_5=OH$, $R_6=COCH_3$, $R_7=R_8=R_9=H$

2-α-Methyl-NeuAc
$R_1=OCH_3$, $R_2=COOH$, $R_3=H$, $R_4=OH$, $R_5=H$, $R_6=COCH_3$, $R_7=R_8=R_9=H$

4-O-Acetyl-NeuAc
$R_1=COOH$, $R_2=OH$, $R_3=H$, $R_4=OCOCH_3$, $R_5=H$, $R_6=COCH_3$, $R_7=R_8=R_9=H$

2 α-Benzylnonulosamine
$R_1=OCH_2C_6H_5$, $R_2=CH_2OH$, $R_3=H$, $R_4=OH$, $R_5=H$, $R_6=COCH_3$, $R_7=R_8=R_9=H$

NeuGc
$R_1=COOH$, $R_2=OH$, $R_3=H$, $R_4=OH$, $R_5=H$, $R_6=COCH_2OH$, $R_7=R_8=R_9=H$

3-Hydroxy-NeuAc
$R_1=COOH$, $R_2=OH$, $R_3=OH$, $R_4=OH$, $R_5=H$, $R_6=COCH_3$, $R_7=R_8=R_9=H$

NeuAc
$R_1=COOH$, $R_2=OH$, $R_3=H$, $R_4=OH$, $R_5=H$, $R_6=COCH_3$, $R_7=R_8=R_9=H$

2-β-Methyl-NeuAc
$R_1=COOH$, $R_2=OCH_3$, $R_3=H$, $R_4=OH$, $R_5=H$, $R_6=COCH_3$, $R_7=R_8=R_9=H$

2-α-Benzyl-8,9-isopropylidene-NeuAc
$R_1=OCH_2C_6H_5$, $R_2=COOH$, $R_3=H$, $R_4=OH$, $R_5=H$, $R_6=COCH_3$, $R_7=H$, R_8, R_9=isoprop

2-α-Benzyl-NeuAc-methylester
$R_1=OCH_2C_6H_5$, $R_2=COOCH_3$, $R_3=H$, $R_4=OH$, $R_5=H$, $R_6=COCH_3$, $R_7=R_8=R_9=H$

2-α-Benzyl-NeuAc-amide
$R_1=OCH_2C_6H_5$, $R_2=CONH_2$, $R_3=H$, $R_4=OH$, $R_5=H$, $R_6=COCH_3$, $R_7=R_8=R_9=H$

4,7-Di-O-acetyl-NeuAc
$R_1=COOH$, $R_2=OH$, $R_3=H$, $R_4=OCOCH_3$, $R_5=H$, $R_6=R_7=COCH_3$, $R_8=R_9=H$

9-O-Acetyl-NeuAc
$R_1=COOH$, $R_2=OH$, $R_3=H$, $R_4=OH$, $R_5=H$, $R_6=COCH_3$, $R_7=R_8=H$, $R_9=COCH_3$

2-α- Benzyl-5-nor-acetyl-NeuAc
$R_1=OCH_2C_6H_5$, $R_2=COOH$, $R_3=H$, $R_4=OH$, $R_6=R_5=R_7=R_8=R_9=H$

p-nitrophenyl-α-D-mannoside was a 30 times better inhibitor than D-mannose (Firon et al., 1983). Sialyllactoses were stronger inhibitors, for K99 hemagglutination, than NeuAc monosaccharides (Table 3).

NeuAcα2-8NeuAcα2-3Galβ1-4Glc was a four times stronger inhibitor than the monosialyllactoses. Since K99 is a basic protein, it is possible that the additional sialic acid in the disialyllactose forms an electrostatic bond to the protein which strengthens binding.

OTHER SIALIC ACID SPECIFIC LECTINS

A number of sialic acid specific lectins have been found in organisms other than bacteria (Table 4). Among these, the specificity has been thoroughly investigated only with Limulin.

Binding of Limulin to sialic acid was found to be highly dependent on the presence of the acetamido group (Roche and Monsigny, 1979), which is the only similarity to K99. Changes in the tricarbon tail did not affect binding and subsequently it was not believed to be involved in binding. Furthermore, the 4-O-acetylated sialic acid derivative was not bound by Limulin (Maget-Dana et al., 1979), in contrast to K99.

The sialic acid specific lectin from the slug *Limax flavus* is highly specific for NeuAc. In a hemagglutination inhibition experiment, concentrations of 0.13mM NeuAc, 0.90mM NeuGc, 23mM GlcNAc or 24mM GalNAc resulted in 50% inhibition (Miller, 1982).

Carcinoscorpin, from Indian Horseshoe crab, shows a fifty times higher specificity to NeuGc than to NeuAc (Mohan et al., 1982). However, it binds glucuronic acid and 3-deoxy-2-oxo-octonate (KDO) slightly better than NeuAc.

The specificity of Edulin, from the mussel *Mytilus edulis*, has not been fully investigated, although inhibition of hemagglutination has been detected with NeuAc (Hardy et al., 1976).

Table 4. Sialic acid specific lectins

Lectin	Source
Limulin	Limulus polyphemus
Carcinoscorpin	Carcinoscorpus rotunda cauda
Edulin	Mytilus edulis
Lobster agglutinin 1	Homarus americanus
	Cepaea hortensis
	Limax flavus
	Crassostrea gigas
Achatinin	Achatina fulica

Sialic acid specificity has also been claimed for a lectin from the snail Cepaea hortensis (Wagner, 1982), but its interaction with other monosaccharides has not yet been studied.

A lectin purified from the lobster Homarus americanus, LAg1, was found to be specific for NeuAc (Hall and Rowlands, 1974). However, the protein could be eluted from a Sepharose-mucin column with ManNAc (Hartman et al., 1978).

Hemagglutination with a lectin purified from the snail Achatina fulica has been shown to be sensitive to NeuAc. Other N-acetyl sugars were weaker inhibitors than NeuAc, while NeuGc and non-acetylated neutral sugars were not inhibitory (Iguchi et al., 1983).

The interactions of these lectins with sialic acid seem to be rather different. Hopefully, their specific interactions with sialic acid will be studied in detail for future use in biological assays.

E.COLI FIMBRIAE WITH GALNAC SPECIFICITY

F41 is so far, the only E.coli fimbrium with a specificity for GalNAc. However, this fimbrium has an interesting history. It was first purified from cells giving agglutination with anti-K99 sera. Unfortunately both K99 and F41 fimbriae were present (Morris et al., 1980). Subsequently both were called K99, but since they had isoelectric points at pH 9,5 and pH 4,6 respectively, they were discriminated by the prefixes cationic- and anionic- (Isaacson, 1977). Hemagglutination with anionic K99 was found, by the conventional hemagglutination technique on glass slides, to be inhibited with GalNAc but not with any other monosaccharide (Morris et al., 1977). In 1982, anionic K99 was finally named F41 (de Graaf and Roorda, 1982).

By using the aggregometer assay, GalNAc was found to be an 11 times stronger inhibitor than GlcNAc and 20 times stronger than Gal. Glc and NeuAc were not inhibitory and Fuc gave a very weak inhibition (Table 5). These results implied that the acetamido group has a strong influence on the protein-carbohydrate interaction, and that the hydroxyl group at C_4 is important for binding (Lindahl and Wadström, in manuscript). In this respect the interaction of F41 with GalNAc is similar to other GalNAc specific lectins. However, hydrophobic interactions are likely to be involved in the protein-carbohydrate interaction, since the hemagglutination reaction was enhanced when the ionic strength was increased (Lindahl and Wadström, 1983).

Table 5. Inhibition of F41 hemagglutination with monosaccharides

Monosaccharide	50 % inhibition (mM)
GalNAc	14,8
GlcNAc	162
Gal	355
Fuc	1100
Glc	1100[a]
NeuAc	1100[a]

Inhibition data obtained by using the aggregometer assay.
[a] No inhibition of hemagglutination reaction was detected at this concentration.

GALNAC SPECIFIC LECTINS

The number of known GalNAc specific lectins exceeds the number of sialic acid specific lectins. Furthermore, their carbohydrate specificities have been extensively studied. The acetamido group and the hydroxyl groups at C_4 and C_6 have been proven to be prerequisites for binding of lectins from Phaseolus lunatus (Galbraith and Goldstein, 1972), Sophora japonica (Wu et al., 1981), Maclura pomifera (Wu et al., 1982), Phaseolus vulgaris (Dupuis and Leclair, 1982) and Wistaria floribunda (Sugii and Kabat, 1980). Disaccharides and p-nitrophenyl derivatives generally bind more tightly to the lectins (Sarkar et al., 1981). However, the subtle structural differences between hexoses determines the restricted interaction with the acetamido group and the C_4 and C_6 hydroxyl groups, for GalNAc specific lectins.

BACTERIAL BINDING TO INTESTINAL MUCIN

Since the small intestinal epithelium is covered with a mucin layer, which is highly glycosylated, it is plausible that a number of receptors for different enteropathogens will be discovered in this layer.

Crude preparations of small intestinal mucins, from a three weeks old calf and a human adult, were found to inhibit the hemagglutination of K99 and F41, at a concentration of 15µg/ml. Human jejunal mucin was a strong inhibitor for both fimbrial types, while bovine duodenal mucin was a weak inhibitor for K99 but strong for F41. Bovine jejunal mucin was not inhibitory (Lindahl and Wadström, 1983).

Hemagglutination with K99 and F41 was also found to be inhibited with purified mucin glycopeptides (Mw 4×10^5 daltons) from the small intestines of pigs (I. Carlstedt, in preparation). By using I^{125}-labelled pig mucin glycopeptides, we could detect binding to K99 fimbriated bacterial cells. Binding was specific since it could be inhibited with unlabelled glycopeptides (Fig. 2) and the dissociation constant was calculated to 3×10^{-7}M (Lindahl et al., 1985). The number of binding sites per cell was calculated to 1.7×10^3. However, the number of fimbriae per cell is usually found by electron microscopy to be 300-400. It is likely that this number is too low since many fimbriae may not be visible if they are covered by, or cover the cells. The mucin glycopeptides used in this study were probably heterogenous in their oligosaccharide chains. Sulphomucins were probably present to some extent which would result in an underestimation of the number of binding sites per cell.

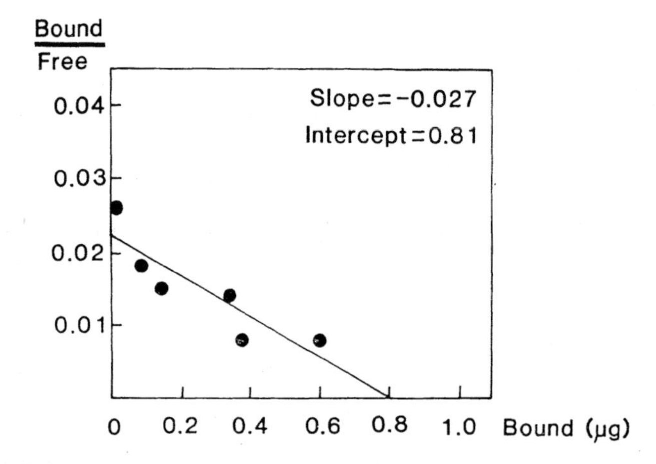

Fig. 2. Scatchard plot of mucus glycopeptide binding to K99 producing bacterial cells

In each sample $7x10^8$ fimbriated bacteria were incubated for one hour with I^{125} labelled mucus glycopeptides and different amounts of unlabelled mucin glycopeptides ranging from 0,25 μg to 80 μg.

BINDING OF K99 AND F41 FIMBRIAE TO ERYTHROCYTE MEMBRANE GLYCOPROTEINS

Electrophoretic transblots of the electrophoretic patterns of porcine, bovine, equine and human erythrocyte ghosts were immersed into solutions of K99 or F41 fimbriae. Binding of fimbriae onto the transblots was detected with specific rabbit IgG and peroxidase-conjugated goat anti-rabbit IgG.

K99 and F41 were both found to bind human PAS-1, PAS-2 and PAS-3 (Fig. 3). Glycophorins from porcine, bovine, equine and human erythrocytes have been found to have NeuAc or NeuGc and GalNAc residues (Kawashima et al. 1982; Fukuda et al. 1982, 1980; Lisowska et al. 1980). The 40 kD proteins of porcine, bovine and human erythrocytes were intensively stained due to binding of F41 (Fig. 3). The binding of K99 to these proteins was much weaker. Similar observations were made for the 32,5 kD and 30 kD proteins of bovine erythrocytes indicating occurrence of a higher GalNAc/NeuGc ratio for these proteins. A variety of sialoglycoproteins was found on porcine and bovine erythrocytes (Fig. 3). Twice as many proteins were stained after K99 binding to porcine and bovine erythrocytes, as compared to equine and human erythrocytes.

The binding of K99 to glycoproteins of the human erythrocyte membrane corresponds well with the PAS components described by Liljas (1978). It seems as if very little PAS-3 monomers were present in the SDS-PAG electrophoretic pattern of human erythrocytes (Fig. 3). This was confirmed by using the double-stain technique of Dzandu et al.(1984). Binding of K99 to glycoproteins of bovine erythrocytes was also demonstrated for the 34 kD protein which was characterized by Fletcher (1982) and for the 26 kD and 16 kD proteins which were characterized by Merrick (1977). The main PAS bands of bovine, human, porcine and equine erythrocytes have been reported as 285 kD, 89 kD, 78 kD and 56 kD proteins respectively (Kobylka et al., 1972). However, the sialic acid specific K99 binds predominantly to the 84 kD, 82 kD, 80 kD and 78 kD proteins of bovine, human, porcine and equine erythrocytes respectively (Fig. 3).

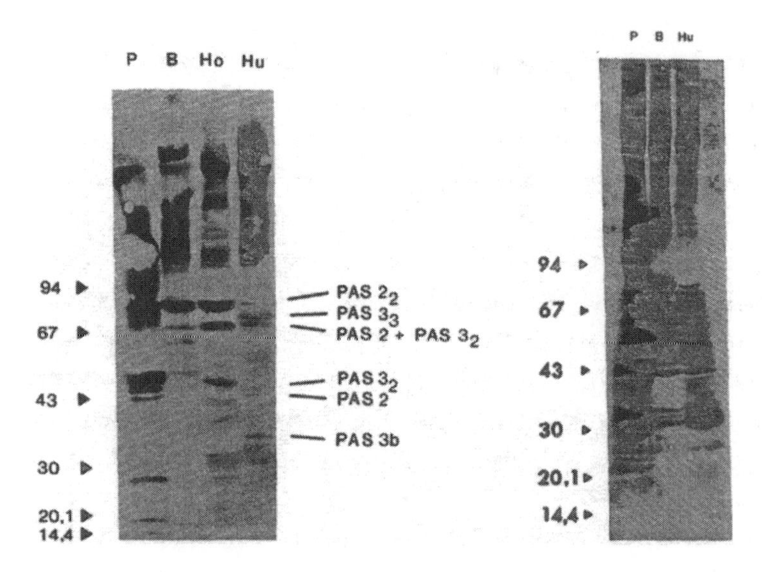

Fig. 3. Binding of K99 and F41 fimbriae to erythrocyte membrane glycoproteins.

Erythrocyte membrane proteins (60 µg) were separated on SDS-PAGE (5-15 % gradient). Electrophoretic transblots were immersed in fimbrial solutions containing, K99 (left) and F41 (right), 1 µg protein per ml. Molecular weight markers were obtained from Pharmacia Fine Chemicals, Sweden. P=porcine, B=bovine, Ho=equine and Hu=human erythrocyte membranes. PAS denotation is according to Liljas (1978).

It is evident that the glycophorins serve as prominent receptors for both K99 and F41. The importance of gangliosides, as proposed by Smit et al. (1984), is only speculative since hemagglutination occurs both with human and equine erythrocytes even though human erythrocytes have very low amounts of gangliosides (Yamakawa et al. 1960).

BINDING OF K99 AND F41 FIMBRIAE TO MILK GLYCOPROTEINS

Colostral milk collected from a cow on the day of parturition and from a sow on the third day after farrowing, was skimmed by centrifugation and the casein fractions were separated by isoelectric precipitation. Fat globule membranes (FGMs) were prepared and solubilized with Triton X-100 (1%) as described by Fischer et al. (1984).

K99 and F41 binding was almost identical to nitrocellulose transfers of the electrophoretic patterns of the milk fractions. FGMs and whey fractions of sow colostrum were found to be rich in glycoprotein receptors for K99 and F41 (Fig. 4). The similar binding patterns to sow FGMs and whey suggests a similar protein composition. The Commassie brilliant blue stained gel also showed a resemblance between the protein composition of these fractions, however a major component of 43 kD was only present in the FGM fraction. The FGM fraction also contains a number of additional glycoproteins to which K99 and F41 can bind. Bovine colostrum has less sialoglycoprotein components than porcine colostrum. The 120 kD protein of bovine FGMs might be identical to the membrane mucoprotein described by Jackson et al.(1962). Similar components may be present in porcine FGMs (e.g. 105 kD and 117 kD components in Fig. 4). The 53 kD component in bovine FGMs is likely to be Glycoprotein B described by Basch et al.(1976). This glycoprotein has been found to have 7,8% (by weight) sialic acid. Another sialoglycoprotein, Glycoprotein 2 (Snow et al.,1977) has an apparent molecular weight of 70 kD and could be identical to the 73 kD protein from bovine FGMs (Fig. 4). Alkaline phosphatase is a sialoglycoprotein with a molecular weight of 85 kD and has been purified from bovine FGMs by Linden et al.(1976). The 83 kD protein (shown in fig. 4) from bovine FGMs can possibly correspond to alkaline phosphatase. Bovine FGMs have been suggested to have secretory IgA on the surface, with the secretory component embedded in the membrane (Honkanen-Buzalski and Sandholm, 1981). The secretory component, which is a sialoglycoprotein, may correspond to the 79 kD band of bovine FGMs.

The casein fractions contained only small amounts of glycoprotein with receptors for K99 and F41. Bovine κ-casein appears as a 29 kD protein and the 30 kD protein found in the porcine casein fraction is likely to be κ-casein. κ-casein has been described by Jolles and Fiat(1979) as a sialoglycoprotein. Sialylated whey proteins present in bovine and porcine colostrum are ß-lactoglobulin (18 kD), IgG heavy chain (50 kD) and lactoferrin (86 kD). The 24 kD and 25 kD bands in figure 4 represents porcine and bovine ß-lactoglobulin and the bands near 50 kD are probably IgG heavy chains. Lactoferrin was absent in the Commassie stained gel.

The presence of glycoproteins in FGM fractions with receptors for K88, K99 and F41 fimbriae raises the question; do these components together with other milk glycoproteins have a prophylactic or therapeutic function against neonatal

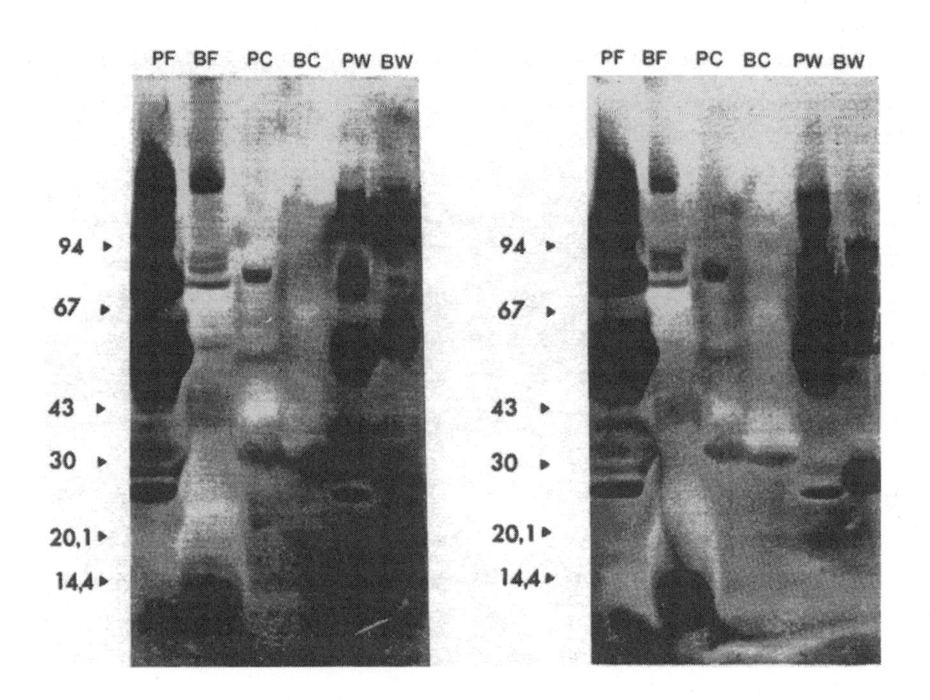

Fig. 4. Binding of K99 and F41 fimbriae to porcine and bovine colostrum fraction.

Whey and casein fractions (120 µg) and fat globule membrane extracts from 4 mg cream were separated on SDS-PAGE (5-15 % gradient). Molecular weight markers were obtained from Pharmacia Fine Chemicals, Sweden. Electrophoretic transblots were immersed in fimbrial solutions containing, K99 (left) and F41 (right), 1 µg protein per ml.
PF=porcine FGM, BF=bovine FGM, PC=porcine casein, BC=bovine casein, PW=porcine whey and BW=bovine whey.

diarrhea? Experimental infections have proved the opposite, since colostrum from non-vaccinated sows failed to prevent diarrhea (Rutter et al., 1976). On the other hand, very high challenge doses were used in the infection experiments (10^{10} cells). At lower challenge doses milk glycoproteins might be deterrent on the development of neonatal diarrhea, however, this has yet to be investigated.

CONCLUSIONS

The ability of a microbe to bind to a receptor molecule on the mucosal surface of an organ is a prerequisite for colonization. Environmental factors, e.g. temperature, redox potential, nutrients and osmotic pressure, are of course important for the multiplication of bacteria and probably do play a role in the restrictive colonization of a specific organ or host. The significance of fimbrial specificity to a monosaccharide, with respect to selective colonization of an organ or host, is low. K99 which is specific for sialic acid is likely to adhere to a variety of cells and to the small intestinal mucosa of a number of animals, since sialic acids are abundant in the mucus layer and on epithelial cells. However, the specificity of K99 is higher for sialyllactoses than for sialic acid monosaccharides, suggesting that a host or organ specificity may be possible if the natural receptor is an oligosaccharide. Moreover, the high affinity for 4-O-acetyl-NeuAc and the low affinity for sialic acids with an O-acetylated tricarbon tail could be another possibility for K99 fimbriated E.coli cells to achieve host or organ specificity. At present there are no results available on the lowest required affinity an adhesin should have to be able to cause colonization.

Neonates are very sensitive to enteric infections due to lack of an autogenous immune defence system, but they also lack a stable population of commensal bacteria to protect against the "invader" (Freter 1984). Both pathogenic and commensal enteric bacteria have been found to associated with the mucosal surface (Nagy et al.; 1976, Bhat et al., 1980). However, the search for receptors has been restricted to epithelial cell membrane proteins or lipids (Kearns and

Gibbons 1979; Leffler and Svanborg-Edén 1980; Cheney and Boedeker 1984). Mucus has merely been studied in the sense of nutrient source for enteric bacteria (Hoskins and Boulding 1981; Roberton and Stanley 1982). The complexity of the carbohydrate structures present in mucus offers favourable possibilities for adhesion. Our finding that glycopeptides from pig small intestinal mucin binds to K99 fimbriated E.coli cells stresses the importance of mucin as a site for colonization. Furthermore, the adhesin of S.sanguis has been found to be specific for NeuAc α 2,3Galß1,3GalNAcol present in human salivary mucin (Murray et al., 1982). Thus the general concept of mucus has to be altered to include the function of mucus as an important anchorage for bacterial colonization.

In neonatal diarrhea pathogens overcome the immune defence, either by outnumbering the colostral immunoglobulins, or by lacking antigens for the immunoglobulins. In addition to the systemic immune response colostral glycoconjugates may act as receptor analogues and deter colonization.

ACKNOWLEDGEMENTS

We are indebted to Dr. I. Carlstedt for supplying us with pig small intestinal mucin. This work was supported by a grant from the Swedish Medical Research Council (16X0473).

REFERENCES

Arai, H. and Munoz, J. J. (1979) Fimbrial hemagglutinin in stationary and shake cultures of Bordetella pertussis, Infect.Immun., 25: 764-767.

Atkinson, H. M. and Trust, T. J. (1980) Hemagglutination properties and adherence ability of Aeromonas hydrophila, Infect.Immun., 27: 938-946.

Atroshi, F., Schilt, R. and Sandholm, M. (1983) K88-mediated adhesion of E.coli inhibited by fractions in sow milk, Zbt.Vet.Med.B, 30: 425-433.

Basch, J. J., Farrell Jr., H. M. and Greenberg, R. (1976) Identification of the milk fat globule membrane proteins. I. Isolation and partial characterization of glycoprotein B, Biochim.Biophys.Acta, 448: 589-598.

Bhat, P., Albert, M.J., Rajan, D., Ponniah, J., Methan, V.I. and Baker, S.J. (1980) Bacterial flora of the jejunum: a comparison of luminal aspirate and mucosal biopsy, J.Med.Microbiol., 13: 247-256.

Brinton, C. C. (1965) The structure, function, synthesis and genetic control of bacterial pili and a molecular model for DNA and RNA transport in gram negative bacteria, Trans. N.Y.Acad.Sci., 27: 1003-1054.
Cheney, C. P. and Boedeker, E. C. (1984) Rabbit mucosal receptors for an enteropathogenic Escherichia coli strain: appearance of bacterial receptor activity at weaning, Gastroenterology, 87: 821-826.
Collier, W. A. and de Miranda, J. C. (1955) Bacterien-haemagglutination. III. Die hemmung der coli-haemagglutination durch mannose, Antoine van Leeuwenhoek, J.Microbiol.Serol., 21: 135-140.
Culling, C. F. A., Reid, P. E., Dunn, W. L. and Clay, M. G. (1977) Histochemical comparison of the epithelial mucins in the ileum in Crohn's disease and in normal controls, J.Clin. Path., 30: 1063-1067.
Duguid, J. P., Smith, I. W., Dempster, G. and Edmunds, P. N. (1955) Nonflagellar filamentous appendages ("fimbriae") and haemagglutinating activity in bacterium coli, J.Path.Bact., 70: 335-348.
Dupuis, G. and Leclair, B. (1982) Structural requirements for simple sugars to inhibit the agglutination of human group A erythrocytes, FEBS Lett., 144: 29-32.
Dzandu, J. K., Deh, M. E., Barratt, D. L. and Wise, G. E. (1984) Detection of erythrocyte membrane proteins, sialoglycoproteins and lipids in the same polyacrylamide gel using a double-staining technique, Proc.Natl. Acad.Sci., 81: 1733-1737.
Eshdat, Y., Silverblatt, F. J. and Sharon, N. (1981) Dissociation and reassembly of Escherichia coli type 1 pili, J.Bact., 148: 308-314.
Evans, D. G., Evans, D.J., Clegg, S. and Pauley, J. A. (1979) Purification and characterization of the CFA/I antigen of enterotoxigenic Escherichia coli, Infect.Immun., 25: 738-748.
Faris, A., Lindahl, M. and Wadström, T. (1980) GM-like glycoconjugate as possible receptor for CFA/I and K99 haemagglutinins on enterotoxigenic Escherichia coli, FEMS Microbiol.Lett., 7: 265-269.
Faris, A., Lindahl, M., Ljungh, Å., Old, D. C. and Wadström, T. (1983) Autoaggregating Yersinia enterocolitica express surface fimbriae with high surface hydrophobicity, J. Appl.Bact., 55: 97-100.
Firon, N., Ofek, I. and Sharon, N. (1983) Carbohydrate specificity of the surface lectins of Escherichia coli, Klebsiella pneumoniae and Salmonella typhimurium, Carbohydr.Res., 120: 235-249.
Fischer, J., Klein, P.-J., Farrar, G. H., Hanisch, F.-G. and Uhlenbruck, G. (1984) Isolation and chemical and immunochemical characterization of the peanut-lectin-binding glycoprotein from human milk-fat-globule membranes, Biochem.J., 224: 581-589.
Fletcher, M. A., Brunschwig, J. P., Lo, H., Caldwell, K. E. and Lo, T. M. (1982) Biochemical and morphological properties of bovine erythrocyte membrane glycoproteins, J.Cell. Biol., 19: 157-170.
Forstner, J. F. (1978) Intestinal mucins in health and disease, Digestion, 17: 234-263.
Forstner, J. F., Jabbal, I., Qureshi, R., Kells, D. I. C. and Forstner, G. G. (1979) The role of disulphide bonds in human intestinal mucin, Biochem. J., 181: 725-732.

Freter, R. (1984) Interdependence of mechanisms that control bacterial colonization of the large intestine. Microecology and Therapy, 14: 89-96.

Fukuda, K., Tomita, M. and Hamada, A. (1980) Isolation and characterization of alkali-labile oligosaccharide units from horse glycophorin, J.Biochem., 87: 687-693.

Fukuda, K., Kawashima, I., Tomita, M. and Hamada, A. (1982) Structural studies of the acidic oligosaccharide units from bovine glycophorin, Biochim. Biophys.Acta, 717: 278-288.

Galbraith, W. and Goldstein, I. J. (1972) Phytohemagglutinin of the Lima bea (Phaseolus lunatus). Isolation, characterization and interaction with type A blood-group substance, Biochem., 11: 3976-3984.

Gibbons, R. A., Jones, G. W. and Sellwood, R. (1975) An attempt to identify the intestinal receptor for the K88 adhesin by means of a haemagglutination inhibition test using glycoproteins and fractions from sow colostrum, J.Gen.Microbiol., 86: 228-240.

de Graaf, F. K. and Roorda, I. (1982) Production, purification, and characte rization of the fimbrial adhesive antigen F41 isolated from calf enteropathogenic Escherichia coli strain B41M, Infect.Immun. 36:751-758.

Guyot, G. (1908) Ueber die bakterielle hämagglutination (bakterio-hemagglutination), Zbt.Bakt.I.Abt.Orig., 47: 640-653.

Hall, J. L. and Rowlands Jr, D. T. (1974) Heterogeniety of lobster agglutinins. II. Specificity of agglutinin-erythrocyte binding, Biochem., 13: 828-832.

Hardy, S. W., Fletcher, T. C. and Gerrie, L. M. (1976) Factors in haemolymph of the mussel Mytilus edulis L., of possible significance as defence mechanisms, Biochem. Soc.Trans., 4: 473-475.

Hartman, A.L., Campbell, P.A. and Abel, C.A. (1978) An improved method for the isolation of lobster lectins, Develop. and Comp. Immunology, 2:617-625.

Hirano, S., Hayashi, H., Masuda, F. and Onodera, K. (1966) Mucopolysaccharides isolated from human and cow colostrums, Agr.Biol.Chem., 30: 212-219.

Honkanen-Buzalski, T. and Sandholm, M. (1981) Association of bovine secretory immunoglobulins with milk fat globule membranes, Comp.Immun.Microbiol. Infect.Dis., 4: 329-342.

Hoskins, L. C. and Boulding, E. T. (1981) Mucin degradation in human colon ecosystems: evidence and role of bacterial subpopulations producing glycosidases as extracellular enzymes, J.Clin.Invest., 67: 163-172.

Iguchi, S. M. M., Egawa, K. and Momi, T. (1983) Isolation and characterization of a novel sialic acid-specific lectin (Achatinin) in the mucus of Achatina fulica, in: "Glycoconjugates", Chester, M. A., Heinegård, D., Lundblad, A. and Svensson, S. eds., Proceedings of the 7th international symposium on glycoconjugates, Lund-Ronneby, 304-305.

Isaacson, R. E. (1977) K99 surface antigen of Escherichia coli: purification and partial characterization, Infect.Immun., 15: 272-279.

Isaacson, R. E., Fusco, P. C., Brinton, C. C. and Moon, H. W. (1978) In vitro adhesion of Escherichia coli to porcine small intestinal epithelial cells: pili as adhesive factors, Infect.Immun., 21: 392-397.

Isaacson, R. E. and Richter, P. (1981) Escherichia coli 987P pilus: purification and partial characterization, J.Bact., 146: 784-789.
Jackson, R. H., Coulson, E. J. and Clark, W. R. (1962) The mucoprotein of the fat/plasma interface of cow's milk. I. Chemical and physical characte rization, Arch.Biochem. Biophys., 97: 373-377.
Jollès, P. and Fiat, A.-M. (1979) The carbohydrate portions of milk glycoproteins, J.Dairy Res., 46: 187-191.
Jones, G. W. and Rutter, J. M. (1972) Role of the K88 antigen in the pathogenesis of neonatal diarrhoea caused by Escherichia coli in piglets, Infect.Immun., 6: 918-927.
Kawashima, I., Fukuda, K., Tomita, M. and Hamada, A. (1982) Isolation and characterization of alkali-labile oligosaccharide units from porcine erythrocyte glycophorins, J.Biochem. 93: 865-874.
Kearns, M.J. and Gibbons, R.A. (1979) The possible nature of the pig intestinal receptor for the K88 antigen of Escherichia coli, FEMS Microbiol.Lett. 6: 165-168.
Kobylka, D., Khetty, A., Shin, B.C. and Carraway, K.L. (1972) Proteins and glycoproteins of the erythrocyte membrane, Arch.Biochem.Biophys. 148: 475-487.
Kolstø Otnaess, A.-B., Laegreid, A. and Ertresvåg, K. (1983) Inhibition of enterotoxin from Escherichia coli and Vibrio cholerae by gangliosides from human milk, Infect.Immun. 40: 563-569.
Lambden, P.R. (1982) Biochemical comparison of pili from variants of Neisseria gonorrhoeae P9, J.Gen.Microbiol., 128: 2105-2111.
Leffler, H. and Svanborg-Edén, C. (1980) Chemical identification of a glycosphingolipid receptor for Escherichia coli attaching to human urinary tract epithelial cells and agglutinating human erythrocytes, FEMS Microbiol. Lett., 8: 127-134.
Lev, R. (1968) A histochemical study of glycogen and mucin in developing human foetal epithelia, Histochem.J., 1:152-165.
Liljas, L. (1978) Aggregates of human erythrocyte membrane sialoglycoproteins in the presence of deoxycholate and dodecyl sulphate, Biochim.Biophys.Acta., 532: 347-353.
Lindahl, M., Faris, A. and Wadström, T. (1982) Colonization factor antigen on enterotoxigenic Escherichia coli is a sialic- specific lectin, Lancet, 2: 280.
Lindahl, M. and Wadström, T. (1983) Comparative properties of K99 and 3P-(F41); two sialic acid specific hemagglutinins of enterotoxigenic E.coli (ETEC), in "Proceedings of fourth international symposium on neonatal diarrhea, Acres, S.D., ed., Veterinary infectious disease organization (VIDO), Saskatchewan, 225-234.
Lindahl, M. and Wadström, T. (1984) K99 surface haemagglutinin of enterotoxigenic E.coli recognize terminal N-acetylgalactosamine and sialic acid residues of glycophorin and other complex glycoconjugates, Vet.Microbiol., 9: 249-257.
Lindahl, M., Carlstedt, I. and Wadström, T. (1985) Binding of K99 fimbriae to pig small intestinal mucin and erythrocyte membrane glycoproteins, in: "Molecular biology of microbial pathogenicity. Role of protein-carbohydrate interactions", Fullerlove, G., ed. Academic Press, London.

Linden, G. and Alais, C. (1976) Phosphatase alkaline du lait de vache II. Structure sous-unitaire, nature metalloproteique et para-metres cinetique, Biochim.Biophys. Acta, 429: 205-213.

Lisowska, E., Duk, M. and Dahr, W. (1980) Comparison of alkali- labile oligosaccharide chains of M and N blood-group glycopeptides from human erythrocyte membrane, Carbohydr.Res., 79: 103-113.

Maget-Dana, R., Roche, A.-C. and Monsigny, M. (1979) Ganglioside- Limulin interactions, in: "Progress in clinical and biological research vol. 29", Cohen E., ed., Alan R. Liss, Inc., New York, 567-578.

Mantle, M. and Allen, A. (1981) Isolation and characterization of the native glycoprotein from pig small-intestinal mucus, Biochem.J., 195: 267-275.

Merrick, J.M., Schifferle, R., Zadarlik, K., Kano, K. and Milgrom, F. (1977) Isolation and partial characterization of the heterophile antigen of infectious mononucleosis from bovine erythrocytes, J.Supramol.Struct., 6: 275-290.

Miller, R.L. (1982) A sialic acid-specific lectin from the slug Limax flavus, J.Invertebr.Pathol., 39: 210-214.

Mohan, S., Thambi, Dorai, D., Srimal, S. and Bachawat, B.K. (1982) Binding studies of a sialic acid-specific lectin from horseshoe crab Carcinos corpus rotunda cauda with various silaoglycoproteins, Biochem.J., 203: 253-261.

Morris, J. A., Stevens, A. E. and Sojka, W. J. (1977) Preliminary characterization of cell-free K99 antigen isolated from Escherichia coli B41, J.Gen.Microbiol., 99: 353-357.

Morris, J. A., Thorns, C. J. and Sojka, W. J. (1980) Evidence for two adhesive antigens on the K99 reference strain Escherichia coli B41, J.Gen.Microbiol., 118: 107-113.

Murray, P. A., Levine, M. J., Tabak, L. A. and Reddy, M. S. (1982) Specificity of salivary-bacterial interactions: II. Evidence for a lectin on Streptococcus sanguis with specificity for a NeuAcα2,3Galß1,3GalNAc sequence, Biochem. Biophys.Res.Commun., 106: 390-396.

Nagy, B., Moon, H. W. and Isaacson, R. E. (1976) Colonization of porcine small intestine by Escherichia coli: ileal colonization and adhesion by pig enteropathogens that lack K88 antigen and by acapsular mutants, Infect. Immun., 13: 1214-1220.

Nishiura, M. (1984) Analysis of sialic acids in bovine intestinal mucins, Bull.Tokyo Med.Dent.Univ., 31: 1-11.

Nowicki, B., Rhen, M., Väisenen-Rhen, V., Pere, A. and Korhonen, T. K. (1984) Immunofluorescence study of fimbrial phase variation in Escherichia coli KS 71, J.Bact., 160: 691-695.

Ørskov, I., Ørskov, F., Smith, H. W. and Sojka, W. J. (1975) The establishment of K99, a thermolabile, transmissible Escherichia coli K antigen, previously called "Kco", possessed by calf and lamb enteropathogenic strains, Acta Pathol.Microbiol.Scand.Sect.B, 83: 31-36.

Ørskov, F., Ørskov, I., Sutton, A., Schreerson, R., Lin, W., Egan, W., Hoff, G. E. and Robbins, J. B. (1979) Form variation in Escherichia coli K1: determined by O-acetylation of the capsular polysaccharide, J.Exp. Med., 149: 669-685.

Otnaess, A.-B. and Halvorsen, S. (1980) Non-antibody components in human milk inhibit Escherichia coli heat labile enterotoxin measured by an enzyme- linked immunosorbent assay, Acta Path.Microbiol.Scand.Sect.C, 88: 247-253.

Parkkinen, J., Finne, J., Achtman, M., Väisenen, V. and Korhonen, T. K. (1983) Escherichia coli strains binding neuraminyl α2-3 galactosides, Biochem. Biophys.Res. Commun., 111: 456-461.

Perrin, D. R. (1955) The chemical composition of the colostrum and milk of the sow, J.Dairy Res., 22: 103-107.

Reiter, B. (1981) The contribution of milk to resistance to intestinal infection in the newborn, in: "Immunological aspects of infection in the fetus and newborn", Lambert, H. P. and Wood, C. B. S., eds., Academic Press, London, 155-193.

Roberton, A. M. and Stanley, R. A. (1982) In vitro utilization of mucin by Bacteroides fragilis, Appl.Environ. Microbiol., 43: 325-330.

Roche, A.-C. and Monsigny, M. (1979) Limulin (Limulus polyphemus lectin). Isolation, physicochemical properties, sugar specificity and mitogenic activity, in: "Progress in clinical and biological research vol. 29", Cohen, E., ed., Alan R. Liss, Inc., New York, 603-616.

Rohde, R., Aleksic, S., Müller, G., Plavsic, S. and Aleksic, V. (1975) Profuse fimbriae conferring O-inagglutin-ability to several strain of S.typhimurium and S.enteritidis isolated from pasta products cultural, morphological, and serological experiments, Zbl.Bakt. Hyg.I.Abt.Orig.A, 230: 38-50.

Runnels, P. L., Moon, H. W. and Schneider, R. A. (1980) Development of resistence with host age to adhesion of K99+ Escherichia coli to isolated intestinal epithelial cells, Infect.Immun., 28: 298-300.

Rutter, J. M., Burrows, M. R., Sellwood, R. and Gibbons, R. A. (1975) A genetic basis for resistance to enteric disease caused by E.coli, Nature, 257: 135-136.

Rutter, J. M., Jones, G. W., Brown, G. T. H., Burrows, M. R. and Luther, P. D. (1976) Antibacterial activity in colostrum and milk associated with protection of piglets against enteric disease caused by K88-positive Escherichia coli, Infect.Immun., 13: 667-676.

Sarkar, M., Wu, A. M. and Kabat, E. A. (1981) Immunochemical studies on the carbohydrate specificity of Maclura pomifera lectin, Arch.Biochem.Biophys., 209: 204-218.

Savage, D. C. and Blumershine, R. V. H. (1974) Surface-surface associations in microbial communities populating epithelial habitats in the murine gastrointestinal ecosystem: scanning electron microscopy, Infect.Immun. 10: 240-250.

Smit, H., Gaastra, W., Kamerling, J. P., Vliegenhart, F. G. and de Graaf, F. K. (1984) Isolation and structural characterization of the equine erythrocyte receptor for enterotoxigenic Escherichia coli K99 fimbrial adhesin, Infect. Immun., 46: 578-584.

Smith, H. W. and Linggood, M. A. (1971) Observations on the pathogenic properties of the K88, Hly and Ent plasmids of Escherichia coli with particular reference to porcine diarrhoea, J.Med.Microbiol., 4: 467-485.

Snow, L. D., Colton, D. G. and Carraway, K. L. (1977) Purification and properties of the major sialoglycoprotein of the milk fat globule membrane, Arch.Biochem.Biophys., 179: 690-697.

Sobeslavsky, O., Prescott, B. and Chanock, R. M. (1968) Adsorption of Mycoplasma pneumoniae to neuraminic acid receptors of various cells and possible role in virulence, J.Bact., 96: 695-705.

Sugii, S. and Kabat, E. A. (1980) Immunochemical specificity of the combining site of Wistaria floribunda hemagglutinin, Biochem., 19: 1192-1199.

Svanborg-Edén, C. and Hansson H. A. (1978) Escherichia coli pili as mediators of attachment to human urinary tract epithelial cells, Infect.Immun., 21: 229-237.

Wadström, T., Sjöberg, P.-O. and Lindahl, M. (1985) Sialic acid specific lectins of enterotoxigenic E.coli, in: "Lectins, Biology, Biochemistry, Clinical Biochemistry" vol.4, Bøg-Hansen, T. C. and Breborowicz, J., eds., W. de Gruyter, Berlin, 417-424.

Wagner, M. (1982) Agglutination of bacteria by a sialic acid-specific lectin of the snail Cepaea hortensis, Acta Histochem., 71: 35-39.

Wu, A. M., Kabat, E. A., Gruezo, F. G. and Poretz, R. D. (1981) Immunochemical studies on the reactivities and combining sites of D-galacto-pyranose- and 2-acetamido-2-deoxy-D- galacto-pyranose-specific lectin purified from Sophora japonica seeds, Arch.Biochem.Biophys., 209: 191-203.

Wu, A. M., Kabat, E. A., Pereira, M. E. A., Gruezo, F. G. and Liao, J. (1982) Immunochemical studies on blood groups: internal structure and immunological properties of water-soluble human blood group A substance studied by Smith degradation, liberation and fractionation of oligosaccharides and reaction with lectins, Arch.Biochem. Biophys., 215: 390-404.

Yamakawa, T., Irie, R. and Iwanaga, M. (1960) Chemistry of the lipids of posthemolytic residue or stroma of erythrocytes. Silicic acid chromatography of mammalian stroma glycolipides, J.Biochem., 48: 490-507.

INTERACTION OF VIRUSES, BACTERIA AND BACTERIAL TOXINS WITH HOST CELL SURFACE GLYCOLIPIDS. ASPECTS ON RECEPTOR IDENTIFICATION AND DISSECTION OF BINDING EPITOPES

Klaus Bock[a], Karl-Anders Karlsson, Nicklas Strömberg, and Susann Teneberg*

[a]*Department of Organic Chemistry, The Technical University of Denmark, DK-2800 Lyngby, Denmark and Department of Medical Biochemistry, University of Göteborg P.O. Box 33031, S-400 33 Göteborg, Sweden*

Specific attachment of viruses and bacteria to carbohydrate has long been known, although it is only recently that the subject is being treated more extensively. The work of Burnet and co-workers in the 1940's revealed an enzyme activity, "receptor destroying enzyme", RDE, responsible for specific inhibition of influenza virus binding to cell surfaces.[1] Later, Faillard in Klenk's laboratory identified the substance being split off as NeuAc.[2] At about the same time it was shown that several bacteria specifically recognized Man.[3] About 15 years after, van Heyningen and co-workers opened up[4] an intense period of studies on the NeuAc-containing glycolipid, the ganglioside GM1, as a specific receptor for cholera toxin.[5] Concerning glycolipid receptors**, Haywood demonstrated[6] a specific binding of Sendai virus to liposomes containing brain gangliosides. Since then, its specificity and biological relevance have been further investigated.[7,8] More recently, two independent groups[9,10]

*Authors appear in alphabetical order
**The designation receptor is used throughout in the more general meaning of attachment site.

demonstrated a specific binding of *E. coli* to Galα1→4Gal-containing glycolipids causing urinary tract infection of man. This result was the impetus for a more systematic study of glycolipids as receptors for viruses and bacteria as is now being performed in our laboratory. Recent reviews summarize the receptor knowledge in the case of viruses,[11] bacteria[12] and bacterial toxins.[5]

The purpose of this paper is to discuss some current molecular aspects of the interaction of microbiological ligands with animal cell surface carbohydrate, primarily glycolipids (as these are more accessible technically). Although there is old background knowledge as indicated above, this field is only in its infancy at the chemical level. Thus no bacterial lectin has up to now been isolated in a pure form, although sequence information is becoming available at a nucleotide level in some instances.[13] Bacterial toxins have, on the other hand, been isolated[5] but the receptor-binding site is not yet known in any case. One exception is the unique crystal conformation of the influenza virus hemagglutinin[14] where the NeuAc-binding pocket has been localized. Furthermore, receptor-binding variants were identified[15] where a single amino acid substitution in the pocket may explain the interaction with NeuAcα2→3 and NeuAcα2→6, respectively. By NMR analysis and computer-based calculation (see also below) of the receptor conformation in these cases[16] it should later be possible to construct the protein-carbohydrate interface to find out atomic interactions. Attempts to crystallize a complex of the hemagglutinin and sialyllactose[15] have so far been unsuccessful, possibly due to the low-affinity binding (J.C. Paulson, personal communication). In this regard, the crystal structure[17] of the complex between L-Ara and the high-affinity L-Ara-binding transport protein of *E. coli* is of general interest concerning precise protein-carbohydrate interactions in microbiological systems.

This presentation will be limited to selected aspects on detection and characterization of receptors for microbiological ligands, and in part on the general molecular character of the receptor sites. More biologically relevant data, which are covered in recent receptor reviews,[5,11,12] will not be considered.

CARBOHYDRATE AT THE ANIMAL CELL SURFACE AND RECEPTOR AVAILABILITY

The cell surface is abundant in carbohydrate. This may be one reason why microbiological ligands appear to have primarily selected carbohydrates as specific attachment sites for colonization or invasion. This carbohydrate may exist in strictly membrane-bound form as glycoproteins,[18] glycolipids[19] and proteoglycans,[20] and in more loosely associated form as secreted glycoproteins[18] or mucins[18,21] and proteoglycans.[22] This means that carbohydrate is the most probable first collision partner for a virus, bacterium or a toxin molecule that approaches the host cell to infect or produce a toxin effect.

The majority of cell surface glycoconjugates may have mainly protective and structural roles for cell integrity. They do not differ much in oligosaccharide structure between different origins, as is mostly the case for the major N-linked Man-containing chains. On the other hand, the number of identified surface oligosaccharides is rapidly increasing and the complexity is impressive. Especially O-linked and lipid-linked oligosaccharides show distinct variations among species, individuals, tissues and cells. This has been most clearly demonstrated for glycolipids in the gastrointestinal tract,[23-26] where there is a variation not only in terminal sequences, as in blood group determinants of the ABO, Lewis, P and other systems, but also in the core oligosaccharides (see also contribution by G.C. Hansson in this volume). A second well-studied object is the red blood cell, which may be used for hemagglutination assaying and which differs characteristically in glycoconjugates between species and individuals.[19] The characteristic tissue tropism for infections (only certain species or cells are susceptible to infection) may therefore in part be explained at the receptor level. Good experimental evidence for this is still mostly lacking. In the case of cholera toxin[5] and Sendai virus[8] it has, however, been possible to induce sensitivity to the toxin effect or infection by the coating of primarily resistant cells with the appropriate glycolipid receptors.

One important issue in relation to interactions at the cell surface is the molecular topology, or relative arrangement of surface molecules. This may be of crucial importance not only in particle attachment and invasion (see for instance the process of virus infection[27]), but also for the recognition of cells by various antibodies. This is an experimentally complex problem, in part explaining our scant knowledge.[28] Also, there is a dynamic situation at the cell surface including lateral diffusion and position exchange of molecules anchored in the membrane bilayer, and this may vary with experimental conditions and the nature of the ligand. In the pioneering studies on NIL cells[29,30] using galactose oxidase and tritiated $NaBH_4$ for labeling of surface carbohydrate, it was found that shorter-chain glycolipids such as lactosylceramide and glucosylceramide were practically not labeled. In contrast, the specific activity of these glycolipids in polyomatransformed NIL cells was considerably higher, indicating a better accessibility on the tumor cell surface. This was explained by decreased masking from more complex glycoproteins and glycolipids on the transformed, compared with the non-transformed, cells. An analogous situation may explain the appearance of the interesting tumor-associated ganglioside GD3 (NeuAcα2→ 8NeuAcα2→3Galß1→4Glcß-Cer), where specific monoclonal antibodies[31] are able to bind to melanoma cells but not to normal cells, although the antigen may be shown by chemical methods to be present in both cases. In more recent experiments on galactose oxidase labeling of human erythrocytes[32], lactosylceramide was cryptic while globoside (no. 3 of Table 1) was labeled 40-70%. The fact that globoside was only partially labeled in spite of optimized experimental conditions may be due to different topological forms, one being more intimately associated with the other surface molecules, such as proteins, and therefore not accessible at the terminal GalNAc for the enzyme.

These scattered data on receptor accessibility on the

cell surface pose some central questions on viral, bacterial and toxin interactions with host cells. In the case of cholera toxin[5] the receptor requirement appears to be the complete pentasaccharide, Galß1→3GalNAcß1→4(NeuAcα2→3) Galß1→4Glc, which is known only in glycolipid form as the ganglioside GM1. In the target cells for cholera diarrhea, the epithelial cells of the small intestine of man, the receptor glycolipid is present in very low concentration; only 10^4 per cell.[33] In one of the actual models for toxin action,[5] the toxin induces cAMP formation after binding to the receptor and inserting into or penetrating the cell membrane. How is the toxin molecule (85 kD compared to 12 kD for galactose oxidase) able to gain access to the bilayer-close receptor site on the complex surface of the epithelial cell? *In vivo*, is it explained by an enzymatic action from the bacterium, *Vibrio cholerae*, producing the toxin, and stripping of surface substances? In case of the Sendai virus, for example, there exists no good explanation[27] of how the close proximity necessary for membrane-membrane fusion is achieved by primary binding to NeuAc-containing receptors. Is there a stepwise approach from more distant protein-bound receptor epitope to bilayer-close lipid-bound receptor? Finally, the finding to be reported below, that several bacteria recognize lactosylceramide may imply that the bacterial lectins are able to reach this bilayer-close epitope which is not normally accessible for galactose oxidase (see discussion above). In this regard, a virus or bacterium, in contrast to a single molecule, may be able to gain access to a bilayer-close epitope by induction of lateral diffusion of masking surface components. This possibility is included in a recent theory of cell adhesion.[34]

The conclusion from these few examples of molecule or particle interaction with bilayer-close binding sites is that many more experiments are required to explain host cell surface molecular topology and dynamics. It remains to be shown if an infectious particle colliding with the host cell surface is fundamentally superior to a free molecule (galactose oxidase, antibody) in attaching to a bilayer-close epitope.

THE RELEVANCE OF SPECIFIC ATTACHMENT TO CARBOHYDRATE IN CASE OF INFECTIONS

There is general agreement as to the necessity, in most cases, for a primary attachment or adhesion to the host cell in the infectious process.[5,11,12,35,36] This may be based on a rather unspecific association, often of non-polar nature, or a molecularly specific lock-in-key interaction. Although there is also evidence for specific protein-protein interactions in pathogen-host relationships,[35,36] the vast majority of chemically defined associations appear to be based on a protein on the ligand specifically recognizing carbohydrate on the host cell.[5,11,12] As noted above, the selection of carbohydrate receptors on the host cell surface may be due to abundance and distinct differences between species and tissues, in part explaining cases of tissue tropism for infections. Also of importance may be the great structural variability of carbohydrate compared to protein (three amino acids combine into six tripeptides while three different monosaccharides produce more than one thousand variant trisaccharides), allowing a large variability within a small mass. This could lead to receptor shifts through limited mutations (see also discussion below).

ASSAYING OF ATTACHMENT AND ADHESION

In the case of bacteria, one method of analysing the specific binding to tissues and cells is the direct observation and counting of particles under the microscope.[35] A simple and convenient assay applicable also for viruses is hemagglutination.[11,12,35,36] Evidence for carbohydrate as attachment sites may be obtained by chemical modification of the target cells by proteolytic enzymes, neuraminidase, periodate, or by inhibition of binding by incubation with various oligosaccharides and glycoconjugates.[11,12,35,36] A powerful and informative approach for NeuAc receptors is being used by J.C. Paulson and collaborators mainly for viruses[15] but also for bacteria.[37] After neuraminidase treatment of erythrocytes and abolishment of hemagglutination, various well characterized sialyl transferases are used to add back NeuAc in specific positions. In this way, two receptor-binding variants of influenza virus were shown to specifically recognize NeuAcα2→3 and NeuAcα2→6, respectively.[15]

Inhibition experiments were successful in the classical case of Man-binding bacteria[3,35] with effective inhibitory concentrations of soluble Man derivatives in the order of 0.001-0.1 mM.[38] Also, inhibition studies with glycolipids revealed the receptor for cholera toxin.[4] A series of carbohydrate receptors primarily for bacteria have been discovered using the inhibition approach[12,35] (see also contribution by T. Wadström et al. in this volume). From our own recent experience, however, the method of using soluble univalent oligosaccharides for inhibition of binding in hemagglutination or against relevant target cells may be of value only in the case of binding sites with relatively high affinity. When of lower affinity, the experiment may produce a falsely negative result, as the inhibitory substance is unable to compete with the multivalency of receptors on the cell surface. One example may serve to illustrate this.[39] The specific receptor for Shiga toxin produced by *Shigella dysenteriae type 1* was picked up as Galα1→4Gal using the solid-phase overlay assay based on glycolipids as described below. This multivalent cell-surface-similar assay efficiently provided the specificity for this disaccharide placed in terminal or internal position of the oligosaccharide chain (see further below), and there was no binding to similar sequences like Galα1→3Gal. Attempts to inhibit toxin action (inhibition of protein synthesis) in a cell assay by use of the free disaccharide were, however, unsuccessful (up to a concentration of the order of 2-5 mg/ml or about 5 mM). In contrast, the disaccharide coupled multivalently to bovine serum albumin (BSA) gave a 50% inhibition at a level of 0.1 mg/ml in a cell suspension, or 0.001 mg/ml in a monolayer assay.[39] Lactose coupled to BSA was inactive when tested up to 2 mg/ml. The conclusion from these experiments is that the specific receptor for Shiga toxin is Galα1→4Gal, but that the low-affinity binding sites of the toxin (5-7 per molecule) requires multivalent interaction for efficient binding. We have several other examples of microbiological ligands where the univalent free oligosaccharide receptor is unable to inhibit in routine inhibition tests (unpublished results).

Therefore, a screening for carbohydrate receptors using various natural and synthetic oligosaccharides including those released from the target cell surfaces may be efficient only in cases with relatively high-affinity binders. Solubilization of amphipathic glycoconjugates (glycoproteins and glycolipids) may be used as in the case of cholera toxin.[4] However, the difficulties in controlling micellar behaviour may cause falsely positive or negative results depending on exposition of hydrophobic surfaces or masking of binding epitopes. An alternative approach of value in certain cases may be to coat primarily negative cells with receptor glycolipids thus inducing binding and infection.[8,10] For toxins and Sendai virus, a hydrophobic coating of purified gangliosides from water dispersions to Petri dishes has been used.[7,40] Presently the most efficient approach, avoiding most of the mentioned drawbacks, is the solid-phase overlay assay to be described below.

A NOVEL SOLID-PHASE OVERLAY ASSAY BASED ON A THIN-LAYER CHROMATOGRAM WITH SEPARATED GLYCOLIPIDS

The success in binding antibodies to glycolipids separated on a thin-layer plate[41] and finding of glycolipids being specifically recognized[10] by uropathogenic *E. coli* stimulated us, on the basis of our experience with monoclonal antibodies,[42] to develop an analogous assay for viruses[43] and bacteria.[44] The essential ingredient is a plastic coating of the developed chromatogram with separated glycolipids, followed by treatment with serum albumin, overlay of ligand suspension and detection of bound ligand after extensive washings. The detailed conditions of various steps were recently described and discussed.[45] Fig. 1 is an illustration of the binding of *E. coli* causing urinary tract infection in man to Galα1→4Gal-containing glycolipids.[9,10,46] There is a selective binding to some glycolipid bands, with no background staining or binding to major rapid- or slow-moving species. The nature of the glycolipid receptors will be discussed

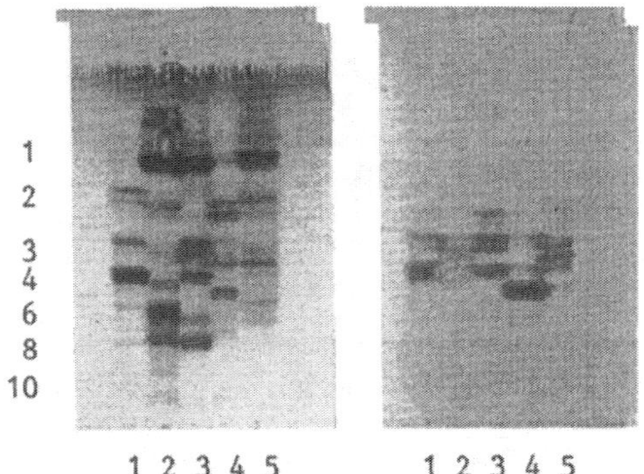

Fig. 1. Thin-layer chromatogram detected with anisaldehyde (left) and autoradiogram (right) after binding of 35 S-labeled uropathogenic *E.coli* (typical for strain J96) to non-acid glycolipids of the following sources: human erythrocytes(1), human meconium (2), intestine from *Macaca cynomolgus* (3), dog small intestine (4), and rabbit small intestine(5). Numbers to the left indicate the approximate number of sugars. The solvent was chloroform-methanol-water 60:35:8, by volume. Autoradiography for 24 h. The binding specificity is for Galα1→4Gal in terminal or internal position of the oligosaccharide chain (compare Table 1).

below. The detection in this case was done by autoradiography after metabolic labeling. Alternatively one may use external labeling, detection by specific antibody, or overlayering with erythrocytes.[45]

This approach has several advantages over traditional assaying. One is the multivalent presentation of the lipid-linked oligosaccharides in the separated spots, allowing low-affinity binders to attach (see Shiga toxin referred to above[39]). It is also probably similar to the presentation at the biological surface membrane. The effect of the plastic coating may be to anchor the glycolipids hydrophobically but also to make the albumin blocking of the silica gel more efficient and avoid background staining. The sensitivity of detection of a receptor glycolipid depends on the avidity of the ligand; for *E. coli* (Fig. 1) it is in the picomole range allowing a few nanograms of receptor to be detected.[46] An obvious advantage is the possibility of efficient screening for receptors using mixtures of glycolipids of target tissues and cells and other sources. This is a unique capability compared to conventional assays.

One central question is the biological relevance of the binding specificities revealed by the assay. In the detailed study of uropathogenic *E. coli*,[46] the recognition in the assay of only Galα1→4Gal-containing sequences is in agreement with binding to living cells and inhibition with receptor analogues. In most of the large number of other cases with binding specificities to glycolipids as demonstrated by this overlay assay (mostly unpublished), we have not yet proven identical or similar interactions at the natural cell membrane. The unexpected preliminary finding[43] that Sendai virus was unable to bind to brain gangliosides, which is in contrast to earlier results[7,40] may be due to differences in assay technology. On the other hand, our finding[43,47] that the lacto series of gangliosides (human erythrocytes), which was not tested by the earlier workers[7,40] was preferred before the ganglio series has been confirmed by a Japanese group using liposome technology.[48] Our theory is that there may be subtle differences in the technology used resulting in different presentations of ceramide-close binding epitopes. One should also consider the possible existence of receptor-binding variants of viruses[15] (see contribution by Y. Suzuki et al. in this volume) which may explain such differences.[47]

In bacteria, a large number of species appear to recognize lactosylceramide.[44,47,49] The peculiar dependence of binding on the ceramide structure is nicely reproducible in the overlay assay. Most species prefer lactosylceramide with hydroxy fatty acid, while some prefer binding to lactosyl-ceramide with non-hydroxy fatty acid. We have interpreted this as a lactose recognition, so that different ceramide components may have a different effect on the presentation of the disaccharide bound directly to ceramide. Variants of bacteria recognizing slightly different lactose epitopes may thus prefer separate molecular species with regard to ceramide. This appears to be true at the overlay assay surface but we have not yet proven analogous bindings to natural membranes, although it is known that epithelial cells, where most of these bacteria[47,49] colonize, have hydroxy fatty acids in their sphingolipids.[23] We have discussed elsewhere[44,45] that one- and two-sugar glycolipids may sometimes give a falsely negative binding, probably due to overcoating with the plastic.

One obvious drawback of the assay is the present limitation to lipid-linked oligosaccharides (hydrophobic anchoring essential). A similar approach for glycoproteins after separation, such as by electrophoresis, is of course possible. However, one particular glycoprotein may carry several heterogeneous oligosaccharide chains linked to the same peptide, excluding a direct conclusion from an overlay on a specific sequence being the binder. An interesting alternative is the release of free oligosaccharide, separation on a thin-layer plate, and chemical coupling of the reducing end to the derivatized gel.[50] This allows assaying of Man-containing oligosaccharides, which are not present in glycolipids of mammals. One should, however, remember that with the exception of Man-containing sequences, which are abundant on cell surfaces and appear to show little variation with animal species and tissues,[18] the majority of known oligosaccharides are included among glycolipids.[19] This refers to most O-linked glycoprotein sequences including the important mucins,[18,21] which lack Man and are assumed to carry the same sequences as glycolipids, although the link monosaccharide is GalNAcα instead of Glcß. Although the present assay is limited to natural and synthetic glycolipids, it covers a wide range of cell surface oligosaccharides of relevance as receptors for microbiological ligands. To also include peptide-linked oligosaccharide in the overlay assay, we are working on a technique which uses a release of reducing oligosaccharide and coupling to paraffin chains. This allows handling of all receptor candidates with glycolipid technology, including assaying, isolation and structural analysis. To estimate the avidity of binding to various glycolipids one may simply use the overlay assay and a dilution series of glycolipids.[46] We prefer to use coating of glycolipid dilutions in microtiter wells with conditions analogous to antibody binding.[45]

RECEPTOR ISOLATION AND STRUCTURAL ANALYSIS

A detailed discussion on this aspect is outside the scope of the present review. Therefore a few comments will be made in reference to the problem of receptor identification related to the overlay assay described above.

Preparation of Mixtures of Glycolipids of Various Origins for Receptor Screening

The first step in screening for a receptor for a particular ligand may be to detect or find a positive binding in the overlay assay. This is most efficiently done using mixtures of glycolipids of various origins to cover as wide a range of receptor candidates as possible. Thus the first panel to be analysed in a screening may not necessarily contain the target tissue for infection. As illustrated in Fig. 1, the patterns of non-acid glycolipids may differ much between the same tissue as in small intestine of different animal species, but this is also the case between tissues of the same individual.[19,23,24,26] This means that a glycolipid in a mixture outside the target tissue may show a positive binding and be a major component which dramatically helps to identify the receptor after isolation. A nice illustration is one of the target cells for urinary tract infection with Galα1→4Gal recognizing *E. coli*. In Fig. 2, the overlay result, using total non-acid glycolipids of a mucosal scraping of the upper ureter, shows a series of very minor components which are binding. This documents a binding to glycolipids extracted from the surface of the target organ. However, it is practically impossible to obtain enough material for isolation and structure determination using this tissue. Instead, globoside of human erythrocytes was used, which is the major glycolipid of this cell (Fig. 1), and easily available through experiments on blood group glycolipids.

Although glycolipids may be major lipids in the surface monolayer of a plasma membrane,[23] they are minor components of a lipid extract of whole cells or tissues, where lipids of other membranes dominate. We have worked out an improved

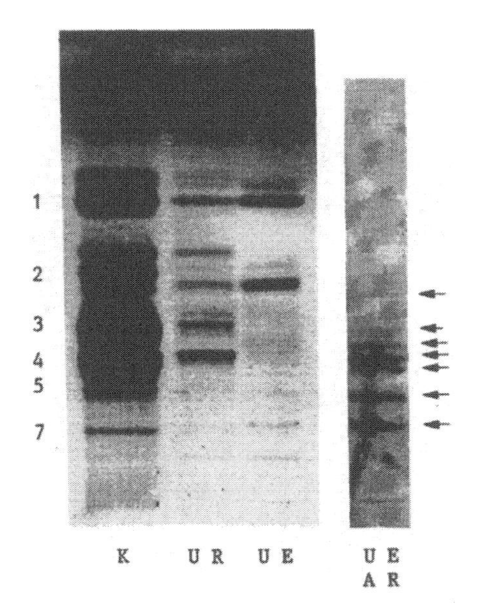

Fig. 2. Thin-layer chromatogram detected with anisaldehyde (left) and autoradiogram (right) after binding of 14-C-labeled uropathogenic *E. coli* (strain 36692). The lanes contain non-acid glycolipids of human kidney (K), ureter (U), epithelial scraping (E) and residue after scraping(R). Solvent as for Fig. 1. Autoradiography for 4 days. As shown,the ureter epithelial part (UE) has a very simple glycolipid composition with practically only 1- and 2-sugar glycolipids as detected with anisaldehyde.[70] On detection with bacteria (UE,AR) several bands appear which are not seen with anisaldehyde. These glycolipids most probably contain Galα1→4Gal and some of them may be identical to species gathered in Table 1.

procedure for isolation of acid[51] and non-acid[52] glycolipids free of non-glycolipid contaminants. The different fractions of Figs. 1 and 2 were prepared in this way.[52] One advantage of this, compared to using a total lipid extract, is that a moderate amount is added to the plate to optimize separation, without need to overload with phospholipids which may also produce trailing and unspecific binding. Also the pattern from autoradiography (Figs. 1 and 2) may be directly compared with the same mixture after chemical spray detection. An additional advantage is the ability to perform structural analysis of the mixture by mass spectrometry (see below).

Isolation and Identification of a Glycolipid Receptor Detected by the Overlay Assay

Frequently the composition of the mixtures used in the panel for screening has already been carefully characterized by us or by others.[19,23-26] In the simplest situation the detected binding is to a known glycolipid or a group of glycolipids from which the common binding denominator may be directly concluded. The actual glycolipids may already be available in pure form to allow more quantitative binding studies in, for example, microtiter wells.[45] However, when there exists a series of isoreceptors which are required for detailed analysis of the binding specificity (see further below) or when the receptor structure is unknown, the substance has to be isolated, which may be the most laborious step of the procedure, and structurally analysed. This includes high-technology separation methods and analysis by mass spectrometry and NMR spectroscopy as recently reviewed in detail concerning glycolipids.[19]

Since our first presentation of mass spectrometry for the direct sequencing of oligosaccharide using various derivatives,[53] we have also found the procedure increasingly valuable for depicting sequences of a mixture.[25,54,55] The derivatized mixture of glycolipids, prepared free of non-glycolipid contaminants as discussed above, is distilled off from the direct inlet probe by a programmed heating allowing a continuous recording of mass spectra for sequences of up to 10-15 sugars successively evaporating. This very informative primary screening for sequences of a mixture may be directly compared with the same mixture separated by thin-layer chromatography as the evaporation sequence and mobility on the plate roughly follow each other.[25,54] As outlined in Fig. 3, this may allow a direct comparison of an overlay result with mass spectrometry. A very interesting further development of this is direct scanning of the thin-layer plate with argon atoms using a fast atom bombardment (FAB) ion source.[56] The limit for complete sequence resolution by mass spectrometry is at present twelve sugars[57], which is within access for most of the interesting receptor glycolipids for microbiological ligands.

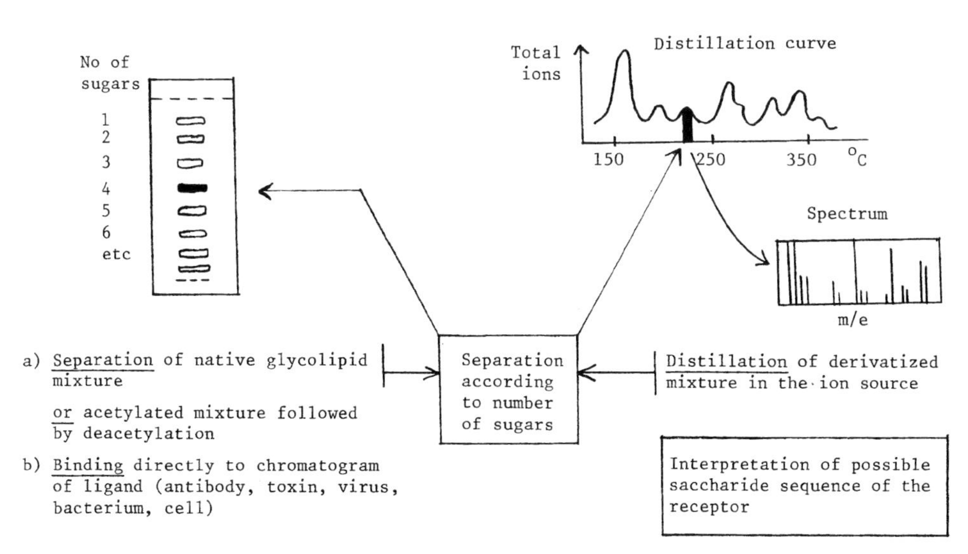

Fig. 3. Illustration of the possible comparison of a mixture of glycolipids by overlay thin-layer chromatography and direct inlet mass spectrometry. Information on the sequence of a receptor glycolipid binding a particular ligand (here shown as a 4-sugar band on the overlay assay to the left) may be obtained by distillation-mass spectrometry of the same glycolipid mixture in suitably derivatized form (right). As the separation in both systems is mainly according to the number of sugars in the glycolipid, interpreted sequences along the distillation may be directly compared with the mobility of the active band on the thin-layer plate.

NATURE OF RECEPTOR GLYCOLIPIDS AND RECEPTOR INTERACTIONS

From about three years experience of receptor screening, the following characteristics of carbohydrate receptors for microbiological ligands have become evident. The picture is in part novel and different from what is known concerning protein-carbohydrate interactions in immunology, which is presently the best known field (see contribution by E.A. Kabat in this volume).

Table 1 contains several glycolipids shown to bind *E. coli*,[46] the bacterium which is frequently found in urine of

Table 1. Examples of glycolipids tested for binding of uropathogenic *E. coli* and the Shiga toxin. Different binding preferences of the two ligands to internally placed Galα1→4Gal indicate separate binding epitopes on the disaccharide

No.		*E. coli*	Shiga toxin
1	Galα1→4GalβCer	+	+
2	Galα1→4Galβ1→4GlcβCer	+	+
3	GalNAcβ1→3Galα1→4Galβ1→4GlcβCer	+	(+)
4	GalNAcβ1→3GalNAcβ1→3Galα1→4Galβ1→4GlcβCer	+	–
5	GalNAcα1→3GalNAcβ1→3Galα1→4Galβ1→4GlcβCer	+	–
6	Galβ1→3GalNAcβ1→3Galα1→4Galβ1→4GlcβCer	+	–
7	Fucα1→2Galβ1→3GalNAcβ1→3Galα1→4Galβ1→4GlcβCer	+	–
8	GalNAcα1→3(Fucα1→2)Galβ1→3GalNAcβ1→3Galα1→4Galβ1→4GlcβCer	+	–
9	NeuAcα2→3Galβ1→3GalNAcβ1→3Galα1→4Galβ1→4GlcβCer	+	–
10	NeuAcα2→6(NeuAcα2→3)Galβ1→3GalNAcβ1→3Galα1→4Galβ1→4GlcβCer	+	–
11	Galα1→3Galα1→4Galβ1→4GlcβCer	+	+
12	$(Gal\alpha1{\rightarrow}3)_{2-5}$Galα1→4Galβ1→4GlcβCer	+	–
13	GalNAcβ1→3$(Gal\alpha1{\rightarrow}3)_{1-5}$Galα1→4Galβ1→4GlcβCer	+	–
14	Galα1→3Galβ1→4GlcβCer	–	–
15	Galα1→4Galβ1→4GlcNAcβ1→3Galβ1→4GlcβCer	+	+

patients with urinary tract infection (compare also Figs. 1 and 2). The common denominator for positive binders is Galα1→4Galß which confirms the proposals done in the original contributions.[9,10] No glycolipids shown to lack this sequence are able to bind the bacteria using overlay assay. However, it is known that most of the strains isolated from infected urine also carry Man-recognizing lectins,[9,10,35,38] which escape detection using natural mammalian glycolipids lacking Man.[19] We have also used in our panel for screening synthetic Man-containing glycolipid[58] and a mixture of Man-containing glycolipids from the invertebrate *Hyriopsis schlegelii,* kindly donated by Dr. Hori.[59] Tested in this supplementary way, the actual bacteria present two separate binding properties which are unrelated concerning recognized sugar.

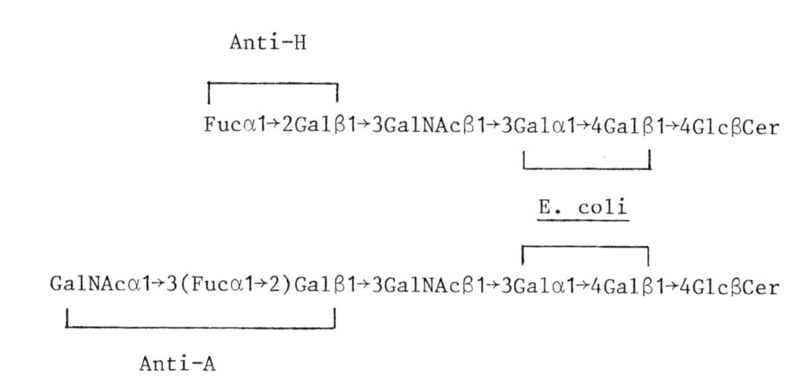

Fig. 4. Different locations in the saccharide chain of binding epitopes for antibodies and uropathogenic *E.coli*. Immunodominant groups are most often terminal sequences, differing between individuals (self and non-self). Microbiological ligands have, on the other hand, selected internal core sequences, which sometimes differ between organs but not between individuals.

One conclusion from Table 1 is the existence of a series of glycolipids which are isoreceptors. The pattern of these varies with tissue origin, as easily demonstrated by the overlay approach (see Figs. 1 and 2). Interestingly, there is a similar binding strength of all positive binders,[46] regardless of whether the binding sequence is placed terminally or internally in the chain. As noted elsewhere,[47,60] this differs from antibodies directed against the same glycolipids or other short sequences, which mostly recognize terminal parts (see examples in Fig. 4). To assure adhesion,

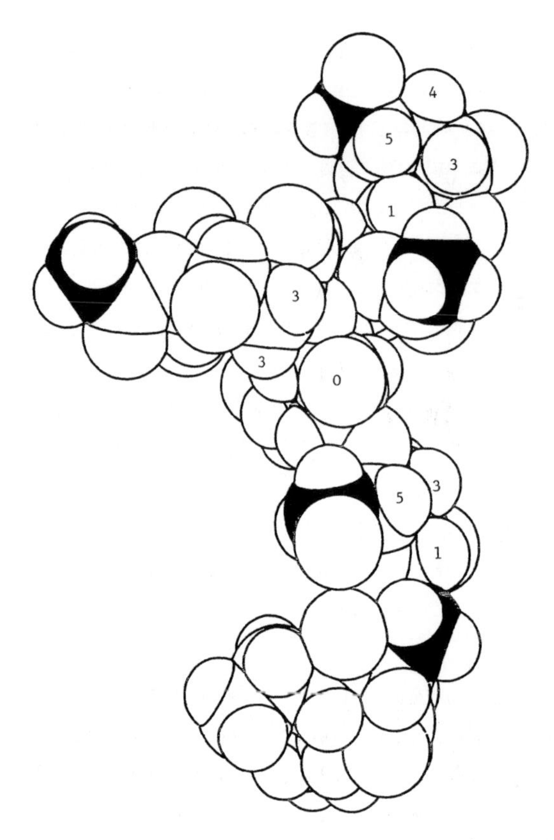

Fig. 5. Molecular model of the conformation of GM1 ganglioside calculated by HSEA and analysed by NMR.[61] The purpose is to visualize a probable limitation in the extension of the binding epitope for cholera toxin, although all five sugars are involved: Galβ1→3GalNAcβ1→4(NeuAcα2→3)Galβ1→4GlcβCer. Although the binding site of the toxin and the binding epitope on GM1 are not yet known, available receptor data[5,71-72] (unpublished) together make it likely that the projected side carries the binding epitope. The two cut paraffin chains (only six carbons) of the fatty acid and long-chain base are shown below (C-1 of the base shaded). The ring hydrogens on the non-polar side of Glcβ are numbered (H-1, H-3 and H-5), as are those on the non-polar side of Galβ (H-1, H-3, H-4 and H-5). The H-3's of NeuAc are numbered. The methyl carbons are shaded on the acetamido groups on NeuAc (left) and GalNAc. C-6 of Glc and terminal Gal are shaded and one of the carboxyl oxygens of NeuAc is labeled O. A probable binding epitope may extend from the non-polar side of Gal (top) over the GalNAc acetamido group and NeuAc carboxyl to the non-polar side of Glc. The terminal Gal may be substituted in various positions without complete blocking of the binding: Fucα1→2, NeuAcα2→3, GalNAcβ1→4 (ref. 71 and unpublished). Similarly, substitution of NeuAc with NeuAcα2→8 (on the back side of this projection) gives a binding similar to GM1 without the terminal Gal (GM2), which is about 50x less strong than GM1 in a liposome assay.[71] GM1 containing N-glycoloylneuraminic acid and GM1 oxidized by galactose oxidase are as effective as GM1.[72]

this fundamental property may have been selected to avoid differences often appearing in terminal parts, as in blood group determinants. We have found such internal recognitions for several other microbiological ligands[47] (mostly unpublished), which may reflect a common shape of the binding site of the lectin-like proteins to adopt a continuous chain. In Kabat's definition of antibody binding sites for carbohydrate, he formulates cavity-type sites for binding of short terminal parts and groove-type sites for longer and internal sequences (see Kabat's contribution in this volume). A groove-type site would be the most probable for microbiological lectins. It is of interest in this regard that the crystal conformation of the L-Ara-binding transport protein of *E. coli* shows the sugar-binding part in a cleft between two major domains of the protein.[17]

Antibodies may recognize from one up to seven monosaccharides in a sequence (see Kabat's contribution in this volume). In our experience so far, the microbiological lectins are in most cases limited to one or two sugars. As an exception, cholera toxin is said to require the whole pentasaccharide receptor for optimal binding,[5] although NeuAc is a branch, possibly giving a total length of the epitope corresponding to 2-3 sugars in a linear sequence as outlined below. Fig. 5 shows the molecular model of the GM1 ganglioside as obtained by HSEA calculation and NMR analysis of the released pentasaccharide[61] (see also below). Although the binding site of the toxin is not yet known, one may visualize the most probable binding surface of the receptor as shown in the model, based on the scarce data available on modified GM1 molecules.[5] We are at present analysing binding preferences for cholera toxin and *E. coli* heat-labile toxin (LT)[5] to various natural and chemically modified gangliosides to obtain further information on the most probable binding epitope (see general approach described below). One conclusion allowed so far is that cholera toxin tolerates oligosaccharide extensions on GM1 (to be published), thus recognizing an internal part analogous to *E. coli* and Shiga toxin (Table 1) and other ligands (see discussion above). A binding epitope on GM1 is probably limited in size although involving as many as five sugars. The borders (Fig. 5) may be part of the non-polar surfaces of the terminal Galß (ring hydrogens 1,3,4 and 5 are numbered) and the ceramide-linked Glcß (ring hydrogens 1,3, and 5 are numbered). In between these, the carboxyl group and C3 and C4 of NeuAc, and the acetamido groups of GalNAc may be part of the epitope. The overall extension of the binding site of the toxin may correspond to only 2-3 sugars in total length. Thus the binding site of the protein may be groove-type and of limited size. It may therefore be a rule that microbiological ligands carry groove-type binding sites capable of recognizing a limited size of internally placed epitopes. However, in the only case so far where the conformation of the binding site has been worked out (crystal structure of influenza virus hemagglutinin[14,15]), the site appears more of the cavity-type and may bind a terminal sequence. We are at present analysing the binding preferences of the X-31 and X-31 HS influenza viruses to various gangliosides to find out in more detail the binding epitope[15,16] (collaboration with J.C. Paulson and co-workers).

Data are accumulating on a number of receptor-binding variants analogous to the two influenza viruses recently interpreted to recognize NeuAcα2→3 and NeuAcα2→6, respectively.[15] Two closely related species of *Propionibacterium* were concluded to bind different epitopes on lactose,[44,47,49] and uropathogenic *E. coli*[46] and the Shiga toxin[39,62] both recognize Galα1→4Gal, but probably slightly different epitopes, as will be discussed below. There is some preliminary evidence for earlier unknown variants of Sendai virus with different binding preferences to various gangliosides.[47] In the case of the two influenza virus variants, it was shown[15] that a single amino acid substitution in the binding site of the hemagglutin, from Gln to Leu, probably is related to the shift in binding specificity from NeuAcα2→3 to NeuAcα2→6. In analogy, the variants noted above with a slight shift in binding epitope on the same disaccharide may be explained by point mutations in the binding sites of the lectin-like proteins. The effect of such a mutation may be greater, and therefore more easily discovered when the binding epitope is placed internally in an oligosaccharide chain.

As noted above, several ligands bind in the low-affinity mode, including the Shiga toxin[39,62] and a large number of lactosylceramide-binding bacteria.[44,47,49] This is based on the inability of free oligosaccharide to inhibit binding. In other cases, as with Man-binders,[35] the cholera toxin[5] and Galα1→4Gal-recognizing *E. coli*,[46] free saccharides in moderate concentrations are active as inhibitors. As proven in the case of Shiga toxin[39] the presence of fairly high levels (see discussion above) of free Galα1→4Gal has no effect on toxin binding and toxin effect (inhibition of protein synthesis) on target cells. The disaccharide multivalently linked to serum albumin is an effective inhibitor. The probable multivalent binding of the toxin to the target cell surface cannot be inhibited unless the receptor analogue is multivalently presented. A low-affinity binding may have been selected in some ecological niches to avoid inhibition by free oligosaccharides present at the target cell surface as a result of physiological enzymatic degradations. One may note that influenza virus is not inhibitable by simple NeuAc derivatives. As discussed in the methodological section above, such low-affinity interactions, which in our experience may be a rule rather than an exception, have important consequences for the assay approach when screening for receptors.

In some cases glycolipids have been selected before glycoproteins as specific receptors. The classical example is cholera toxin and the ganglioside GM1 carrying a pentasaccharide which has not yet been found in glycoproteins. This may offer two advantages. Firstly, the absence of binding sites in secreted glycoproteins means absence of soluble inhibitors for the ligand to attach to the strictly membrane-bound receptor. Secondly, the glycolipid receptor creates a proximity to the membrane bilayer which is assumed to be important for membrane penetration.[5] In the case of the large number of lactosylceramide binders one should note that lactose is not known in glycoproteins, again an advantage to remain adhered to true membrane-bound receptor. (Free lactose is inactive as inhibitor in moderate concentrations. The physiological importance of the high level of free lactose in human milk, about 60 mg/ml, is at present not known. At this very high level there may be some influence at the receptor level on the colonization of intestinal bacteria in the newborn. However, evidence for this is still lacking.)

When considering various membrane-bound and secreted glycoconjugates and their role as binding sites for bacteria, it is of some interest that apparently the most commonly recognized sugars by bacteria, namely Man[35] and lactose,[44,47,49] are not present in the most important of the secreted glycoconjugates, the mucins.[18,21] One reason for selecting these sites may then be to avoid interference from mucins and optimize adhesion to cells.

FINE DISSECTION OF BINDING EPITOPES ON GLYCOLIPID RECEPTORS

In pioneering studies Kabat and Lemieux and co-workers have defined binding epitopes for a series of antibodies and plant lectins using a laborious synthesis of systematically modified sugars.[63,64] An attractive generalization has been formulated on basis of these data.[65] The protein interacts with a tight cluster of two or three hydroxyl groups which may originate in one or two sugar units. This strongly polar interaction largely determines the specificity of the binding. In addition to this, there is a relatively more extensive surface which is non-polar in character and extends from the polar grouping. The interaction of this surface with a complementary non-polar surface of the protein is the main contribution to the stability of the complex. In the following we will illustrate two aspects on receptor epitopes of glycolipids. As a first point, the Lemieux formulation as given above appears to be valid also for microbial interactions. Secondly, the property of microbiological ligands to recognize epitopes placed internally in an oligosaccharide chain is of considerable help in defining a limited epitope before the need of laborious organic synthesis.

To illustrate this we have chosen the Galα1$\rightarrow$4Gal-recognition by *E.coli* and the Shiga toxin (also see Figs. 1

and 2, Table 1, and discussion above). In both cases the biological relevance of the analyzed binding to thin-layer plates and microtiter wells has been documented by inhibition of adhesion and toxicity, respectively, by receptor analogues. A careful comparison of the two ligands has revealed different preferences for the internally placed disaccharide, see Table 1. As already noted, *E. coli* is able to recognize Galα1→4Gal equally well regardless of the nature of the extension of the chain, whether it is a rather bulky blood group determinant or a disialyl branch. The Shiga toxin on the other hand tolerates some, but not other, extensions of Galα1→4Gal. Therefore the two binding sites of the proteins differ slightly and recognize different epitopes on Galα1→4Gal.[60]

To analyse this further we have applied computer based molecular modeling to several of the actual glycolipids.[46] This is based on HSEA calculations[66] which have been shown in several cases to be in agreement with NMR analysis in solution.[61] Fig. 6 shows the model of the partial structure of the Forssman glycolipid (no. 5 of Table 1), which binds the bacteria well but not the toxin (Table 1). At about 90 degrees there appears a bend, or knee in the oligosaccharide chain at the Galα1→4Gal. The fact that the bacterium binds all Galα1→4Gal-containing glycolipids (Table 1) makes it likely that the binding epitope should reside on the convex, more accessible side of the bend, since various protrusions on other sides may be expected to cause sterical hindrance of the binding. This bend exposes a continuous non-polar surface of the disaccharide, made up of the α side of Galß (H-1, H-3, H-4 and H-5 marked in the Fig.) and H-1 and H-2 of Galα (also marked). This non-polar surface is surrounded by several polar groupings some of which may be key polar groupings according to the Lemieux concept (see above). In this regard it was recently shown[67] by inhibition studies with bacteria and synthetic substances that the 6-OHs of both Gals are essential, since neither D-Fucα1→4Gal nor Galα1→4D-Fuc were active. In this particular case, the epitope character falls within the generalization of Lemieux (see above).

The toxin binding epitope should differ only slightly from that of *E. coli*. Several extensions on Galα block access

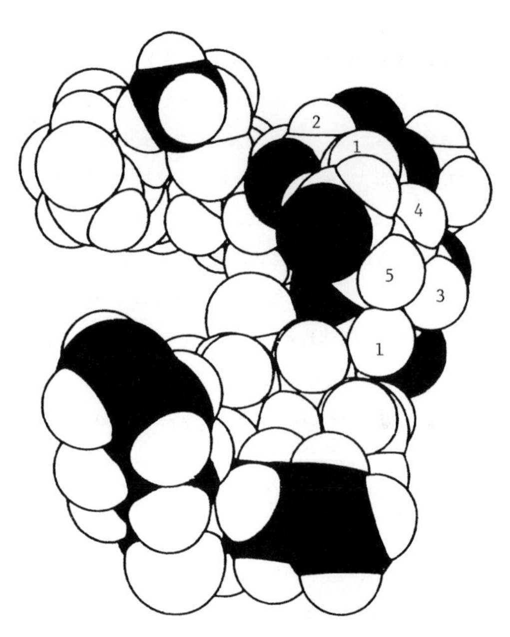

Fig. 6. Molecular model of the HSEA-calculated conformation of Forssman glycolipid (no. 5 of Table 1). The model is projected to visualize Galα1→4Gal forming a bend or knee in the chain (numbered ring hydrogens, H-1 and H-2 of Galα and H-1, H-3, H-4 and H-5 of Galß). In the bottom the two cut chains (first six carbons only) of the fatty acid (left) and the long-chain base are seen projecting towards the viewer (shaded carbons). There is a continuous non-polar surface over the convex side of Galα1→4Gal (numbered hydrogens) surrounded by polar oxygens (several shaded). *E. coli* gains access to this epitope while the Shiga toxin does not, probably due to a steric hindrance from the acetamido group (methyl carbon shaded) of GalNAcß.

for the toxin but not for the bacterium. Galα1→3 (substance 11 of Table 1) is tolerated but GalNAcß1→3 and extensions on this interfere. This may indicate that the binding epitope for the toxin is shifted more to the Galα side of the disaccharide where the acetamido group may induce a steric hindrance. This may also mean a smaller surface of non-polar interaction with Galß, and this is supported by a 5-10 times weaker binding of the toxin than the bacterium as measured in microtiter wells, with coated glycolipid having the disaccharide in terminal position. To definitely establish this postulate of epitope difference for the bacterium and the toxin, organic synthesis is required of receptor analogues with selected changes primarily along the convex side of the disaccharide.

We have several other examples to be reported where binding preferences of ligands to internally placed epitopes in combination with molecular modelling have allowed encircling of analogous binding epitopes. A probable binding side of the GM1 receptor for cholera toxin was discussed above (Fig. 5) with both polar and non-polar character. This specific approach is of limited value in case of terminally placed epitopes as for most antibodies (see discussion above), since all natural oligosaccharide extensions block the binding and do not provide further information. The knowledge gained from assaying microbiological ligands against a large number of natural isoreceptors may drastically reduce the starting alternatives for synthesis of receptor analogues to definitely establish the epitope. This potential advantage in the case of viruses, bacteria and bacterial toxins encourages design of efficient receptor analogues for various biomedical applications.

Another potent approach we have at present is a biochemical or chemical modification of the natural receptor glycolipids to discover effects on binding in relation to molecular modelling. This is much more easily achieved than total organic synthesis. It also provides information about possible effects of neighbouring groups on binding. It is known, for example, that p-nitrophenylαMan is over a hundred times more active as inhibitor than methylαMan in case of Man-binding bacteria.[35] Thus a modification of neighbouring groups to the binding epitope may produce a more potent receptor.

The property among microbiological ligands to recognize internally placed epitopes of an oligosaccharide chain has several interesting consequences. Firstly, binding preferences to a large number of natural isoreceptors may reveal closely related variants of ligands recognizing slightly different epitopes (see Table 1). This is not expected to the same degree if the epitope is placed terminally, as seen in substances no. 1,2 and 15 for the two ligands of Table 1. (Different core oligosaccharides of a terminally placed epitope, however, influence binding in some cases.) Secondly, this property is of considerable help when trying to encircle a binding epitope before laborious organic synthesis has to begin. Thirdly, chemical neighbouring-group modification may improve the binding and direct the synthesis of receptor analogues for various applications. One may finally note the reasons for selecting internally placed epitopes. One probable part of the selection pressure was discussed above-to avoid differences such as blood groups, between individuals, residing mostly in terminal sequences. Another may be a more efficient "receptor shift" upon mutations. Consider the system of Table 1 and target cells which lack terminally placed Galα1→4Gal. Then the presence on these cells of, for example, substance no. 5, allows binding of bacteria but not toxin. A point mutation in the binding site of the ligand may therefore in some cases be expected to shift specificity from one target cell to another.

OUTLINE OF A GENERAL APPROACH FOR SCREENING AND CHARACTERIZATION OF GLYCOLIPID RECEPTORS

The solid-phase overlay assay with current supplementations is a very *efficient way of detecting a specific binding* of viruses, bacteria and bacterial toxins to lipid-linked oligosaccharides. These may be natural glycolipids or oligosaccharides released from peptide and linked synthetically to paraffin chains. New ways of preparing natural glycolipids free of non-glycolipid contaminants facilitate the interpretation of binding patterns within a panel of glycolipid mixtures of various origins. Using this approach most membrane-linked and secreted glycoconjugates are included.

In the simplest case, the detected receptor substance is known in its structure and may be available in pure form. Alternatively, a laborious *isolation and structural analysis* with high-technology methods may be required before more detailed quantitative binding studies will take place.

Analysis of binding preferences to a large number of natural glycolipids with the binding site placed internally in the oligosaccharide chain gives partial *information on the binding epitope* when supplemented with *molecular modeling*.

This information decisively facilitates the planning of *organic synthesis* of receptor analogues to definitely find out the binding epitope and for detailed *studies of protein-carbohydrate interactions with various physicochemical methods.*

A final step may be to prove the biological relevance of the binding as characterized chemically. This complex problem may sometimes require more sophisticated approaches. Tests for the presence of the receptor structure in the target cells are straightforward using, for example, the overlay assay and extracted glycolipids (see Fig. 2). In cases of low-affinity binding sites the obligate inhibition studies may have to be done with soluble synthetic analogues with multivalently presented epitope. However, inhibition of binding in a bioassay may be due to agglutination of the ligand by the multivalent reagent and therefore not necessarily proving the presence of the receptor on the target cell. In cases of several binding specificities being carried by a ligand, the proof of the biological relevance of these may be the most difficult problem of the whole procedure.

PERSPECTIVES

The field of carbohydrate receptors for microbiological ligands is only beginning to develop, and has interesting promises for the future. Although evidence for carbohydrate as essential attachment sites on the host cell surface has existed for some time,[35,36] the development has been relatively slow. One reason for this has been a lack of technology for efficient handling of often minor amphipathic membrane glycoconjugates. In our view two things have contributed importantly to a rapid progress at present. One is the improvement of carbohydrate technology in general including isolation, high-technology mass spectrometry for sequence analysis and NMR spectroscopy for primary structure and conformation in conjunction with conformation calculations. A second decisive improvement is in the assay technology. The solid-phase overlay analysis using defined mixtures of lipid-linked oligosaccharides of diverse origins holds great promises for picking up receptors for a large number of viruses, bacteria and bacterial toxins. When supplemented with peptide-linked oligosaccharides (see discussion above) and with improved quantitation and inhibition assays,[45] the great majority of animal cell surface oligosaccharides will be within range of efficient analysis. This includes both a detection of binding in the assay and isolation and structural characterization, as well as epitope dissection through conformation analysis. Good progress is also being made in the chemical synthesis of carbohydrates, although this is at present probably the most difficult and laborious task within preparative organic chemistry.

One reason why the vast majority of essential attachment sites on the host cell surface is carbohydrate may be the abundance of surface glycoconjugates imposing a selection pressure for a sometimes rapidly mutating ligand. Although the normal or physiological function of the complexity and variability of cell surface carbohydrate is still essentially unknown, (possibly recognition in cell-cell interactions), the fact that many micro-organisms in host-pathogen and host-normal flora relations produce lectin-like proteins with often high specificity is of potential biomedical interest. The characteristic tissue tropism[35,36] for colonization and infection may in part be explained at the receptor level. Clinical diagnosis and therapy could be supplemented by the knowledge of such specific receptors. Synthetic receptor analogues with optimized binding property may be linked to solid supports for a selective pick up and detection of a ligand from infected urine, feces, blood, sputum, etc., to speed up diagnosis in relation to actual several-days cultivation procedures. A combination of this isolation step and DNA-probe analysis may be a realistic approach.

A most spectacular application is in clinical therapy. Although this principle has not yet been convincingly proven in the laboratory,[68,69] use of potent receptor analogues to inhibit ligand attachment to mucous membranes and thereby interfere with colonization and infection is a very attractive approach. Especially in viral infections, with lack of antibiotic treatment and where vaccination therapy often has limited value due to a rapid antigenic drift, the highly conserved receptor-binding property may be utilized. Due to the complex formal procedure of drug testing in the human, use in veterinary medicine may be the first choice to prove the principles.

Use of receptor analogues in biotechnology is another possibility. Affinity purification, solid-phase attachment of catalytic bacteria in continuous biomass processings, preparation of lectins for vaccination or for use as reagents or in affinity separations of molecules and cells are attractive applications.

SUMMARY

An overview and perspective is presented on animal cell surface carbohydrate (primarily lipid-linked oligosaccharides) as specific receptors for viruses, bacteria and bacterial toxins. Although carbohydrate has been known for many years to be specific attachment sites for these ligands, it is only in very recent time that carbohydrate technology and receptor assays in combination afford a rational approach.

One generalization from present experience is the property of microbiological ligands to recognize sequences placed internally in an oligosaccharide chain which differs from antibody recognition of short sequences which most often involves terminally placed determinants. This is of both biological and technical importance. Biologically it may assure attachment by avoiding differences between host individuals often residing in terminal parts (e.g. blood group determinants), and may also make a shift of target cells by mutations more efficient. Technically this property is an important help when dissecting narrow binding epitopes, and for disclosing receptor-binding variants with only slight differences in binding epitopes (e.g. different epitopes on the same disaccharide). Such variants representing a kind of "epitope drift" are probably a consequence of point mutations in the binding site of the lectin-like proteins to select a proper host environment.

Current technology allows an efficient screening for carbohydrate receptors with interesting consequences for applications within medicine (diagnosis and therapy) and biotechnology.

ACKNOWLEDGEMENT

The work reported herein was supported by a grant from the Swedish Medical Research Council (no. 3967). We are indebted to B.-E. Uhlin for the labeled bacteria used for Fig. 1, to C. Svanborg Edén for the labeled bacteria used for Fig. 2, and to M.E. Breimer, G.C. Hansson and H. Leffler for help with glycolipid samples of Fig. 2.[70]

REFERENCES

1. Burnet, F.M. and Stone, J.D. (1947) The receptor-destroying enzyme of *Vibrio cholerae*, Austr. J. Exp. Biol. Med. Sci., 25:227-233.
2. Faillard, H. (1957) Uber die Abspaltung von N-Acetylneuraminsäure aus Mucinen durch das "Receptor-Destroying-Enzyme" aus *Vibrio cholerae*, Z. Physiol. Chem., 307:62-86.
3. Collier, W.A. and DeMiranda, J.C. (1955) Bakterien-Hämagglutination. III. Die Hemmung der Coli-Hämagglutination durch Mannose, Antonie van Leeuwenhoek J. Microbiol. Serol., 21:135-188.
4. van Heyningen, W.E., Carpenter, C.C.J., Pierce, N.F. and Greenough, W.B. (1971) Deactivation of cholera toxin by ganglioside, J. Infect. Dis., 124:415-418.
5. Eidels, L., Proia, R.L. and Hart, D.A. (1983) Membrane receptors for bacterial toxins, Microbiol. Rev.47:596-620.
6. Haywood, A. M. (1974) Characteristics of Sendai virus receptors in a model membrane, J. Mol. Biol., 83:427-436.
7. Holmgren, J., Svennerholm, L., Elwing, H., Fredman, P. and Strånnegard,Ö. (1980) Sendai virus receptor: Proposed recognition structure based on binding to plastic-adsorbed gangliosides, Proc. Natl. Acad. Sci. USA, 77:1947-1950.
8. Markwell, M.A., Svennerholm, L. and Paulson, J.C. (1981) Specific gangliosides function as host cell receptors for Sendai virus, Proc. Natl. Acad. Sci. USA, 78:5406-5410.
9. Källenius, G., Möllby, R., Svenson, S.B., Winberg, J., Lundblad, A., Svensson, S. and Cedergren, B. (1980) The P^k antigen as receptor for the haemagglutinin of pyelonephritic *Escherichia coli*, FEMS Microbiol. Lett., 7:297-302.
10. Leffler, H. and Svanborg Edén, C. (1980) Chemical identification of a glycosphingolipid receptor for *Escherichia coli* attaching to human erythrocytes, FEMS Microbiol. Lett., 8:127-134.
11. Dimmock, N.J. (1982) Initial stages in infection with animal viruses, J. Gen. Virol., 59:1-22.
12. Jones, G.W. and Isaacson, R.E. (1983) Proteinaceous bacterial adhesins and their receptors, CRC Crit. Rev. Microbiol., 10:229-260.
13. Lund, B., Lindberg, F.P., Båga, M. and Normark, S. (1985) Globoside-specific adhesins of uropathogenic *Escherichia coli* are encoded by similar trans-complementable gene clusters, J. Bacteriol., 162:1293-1301.

14. Wilson, I.A., Skehel, J.J. and Wiley, D.C. (1981) Structure of the haemagglutinin membrane glycoprotein of influenza virus at 3Å resolution, Nature, 289:366-373.
15. Rogers, G.N., Paulson, J.C., Daniels, R.S. Skehel, J.J., Wilson, I.A. and Wiley, D.C. (1983) Single amino acid substitutions in influenza haemagglutinin change receptor binding specificity, Nature, 304:76-68.
16. Sabesan, S., Bock, K. and Paulson, J.C. (1985) Synthesis of sialyloligosaccharides and the determination of their conformational properties based on HSEA calculations and nuclear magnetic resonance spectroscopy, in: "Glycoconjugates", E.A. Davidson, J.C. Williams and N.M. Di Ferrante, eds., Praeger Publishers, New York.
17. Quiocho, F.A. and Vyas, N.K. (1984) Novel sterospecificity of the L-arabinose-binding protein, Nature, 310:381-386.
18. Berger, E.G., Buddecke, E., Kamerling, J.P., Kobata, A., Paulson, J.Ç. and Vliegenthart, J.F.G. (1982) Structure, biosynthesis and functions of glycoprotein glycans, Experientia, 38:1129-1162.
19. Hakomori, S.-i. (1983) Chemistry of Glycosphingolipids, in: "Sphingolipid Biochemistry", J. Kanfer and S.-i Hakomori, eds. Plenum Press, New York.
20. Höök, M., Kjellén, L. Johansson, S. and Robinson, J.(1984) Cell-surface glycosaminoglycans, Ann. Rev. Biochem., 53:847-869.
21. Podolsky, D.K. (1985) Oligosaccharide structures of human colonic mucin, J. Biol. Chem.,260:8262-8271.
22. Hascall, V.C. (1981) Proteoglycans: Structure and function, in: "Biology of Carbohydrates", Vol. 1, V. Ginsburg and P. Robbins, eds., John Wiley and Sons, New York.
23. Karlsson, K.-A. (1982) Glycosphingolipids and surface membranes, in: "Biological Membranes", Vol. 4, D. Chapman, ed., Academic Press, London.
24. Björk, S., Breimer, M.E., Hansson, G.C., Karlsson, K.-A., Larson, G. and Leffler, H.(1983) Blood group glycosphingolipids of human gastrointestinal tissues, in: "Red Cell Membrane Glycoconjugates and Related Genetic Markers", J.P. Cartron, P. Rouger and C. Salmon, eds., Librairie Arnette, Paris.
25. Breimer, M.E., Hansson, G.C., Karlsson, K.-A. and Leffler, H. (1981) Blood group type glycosphingolipids from the small intestine of different animals analysed by mass spectrometry and thin-layer chromatography. A note on species diversity, J. Biochem., 90:589-609.
26. Breimer, M.E., Hansson, G.C., Karlsson, K.-A. and Leffler, H. (1982) Glycosphingolipids of rat tissue. Different composition of epithelial and nonepithelial cells of small intestine, J. Biol. Chem., 257:557-568.
27. White, J., Kielian, M. and Helenius, A. (1983) Membrane fusion proteins of enveloped animal viruses, Quart. Rev. Biophys., 16:151-195.
28. Gahmberg, C.G. (1977) Cell surface proteins: changes during cell growth and malignant transformation, in: "Cell Surface Reviews", Vol. 3, G. Poste and G.L. Nicolson, eds., North-Holland Publishing Company, Amsterdam.
29. Gahmberg, C.G. and Hakomori, S.-i. (1975) Surface carbohydrates of hamster fibroblasts. I. Chemical characterization of surface-labeled glycosphingolipids and a ceramide tetrasaccharide specific for transformants. J. Biol. Chem., 250:2438-2446.

30. Gahmberg, C.G. and Hakomori, S.-i. (1975) Surface carbohydrates of hamster fibroblasts. II. Interaction of hamster NIL cell surfaces with *Ricinus communis* lectin and concanavalin A as revealed by surface galactosyl label. J. Biol. Chem., 250:2447-2451.
31. Hellström, I., Brankovan, V. and Hellström, K.E., (1985) Strong anti-tumor activities of IgG3 antibodies to a human melanoma-associated ganglioside, Proc. Natl. Acad. Sci. USA, 82:1499-1502.
32. Lampio, A., Finne, J., Homer D. and Gahmberg, C.G (1984) Exposure of the major human red-cell glycolipid, globoside, to galactose oxidase, Eur. J. Biochem., 145:77-82.
33. Breimer, M.E., Hansson, G.C., Karlsson, K.-A. Larson, G. and Leffler, H. (1986) Glycosphingolipid composition and surface carbohydrate architecture of epithelial cells of human small intestine including the density of cholera toxin receptor, submitted for publication.
34. Bell, G.I., Dembo, M. and Bongrand, P. (1984) Cell adhesion: Competition betwen nonspecific repulsion and specific bonding. Biophys. J., 45:1051-1064.
35. Beachey, E.H. (1980) "Bacterial Adherence", Receptors and Recognition, series B, Vol. 6, Chapman and Hall, London.
36. Lonberg-Holm, K. and Philipson, L. (1981) "Virus Receptors. Part 2. Animal Viruses", Receptors and Recognition, Series B, Vol.8, Chapman and Hall, London.
37. Loomes, L.M., Uemura, K.-i., Childs, R.A., Paulson, J.C., Rogers, G.N., Scudder, P.R., Michalski, J.-C., Hounsell, E.F., Taylor-Robinson, D. and Feizi, T(1984) Erythrocyte receptors for *Mycoplasma pneumoniae* are sialylated oligosaccharides of the Ii antigen type, Nature, 307:560-563.
38. Firon, N., Ofek, I. and Sharon, N. (1984) Carbohydrate-binding sites of the mannose-specific fimbrial lectins of Enterobacteria, Infect. Immun. 43:1088-1090.
39. Lindberg, A.A., Schultz, J.E., Westling, M., Brown, J.E., Rothman, S.W., Karlsson, K.-A. and Strömberg, N. (1986) Identification of the receptor glycolipid for Shiga toxin produced by *Shigella dysenteriae* type 1, in: "Protein-Carbohydrate Interactions in Biological Systems. The Molecular Biology of Microbial Pathogenicity", D.L.Lark, ed., Academic Press, pp. 439-446.
40. Holmgren, J., Elwing, H., Fredman, P., Strannegård, Ö. (1980) Gangliosides as receptors for bacterial toxins and Sendai virus, Adv. Exp. Med. Biol., 125:453-470.
41. Magnani, J.L., Brockhaus, M., Smith, D.F., Ginsburg, V., Blaszczyk, M., Mitchell, K.F.Steplewski, Z. and Koprowski, H (1981) A monosialoganglioside is a monoclonal antibody-defined antigen of colon carcinoma, Science, 212:55-56.
42. Hansson, G.C., Karlsson, K.-A., Larson, G., McKibbin, J.M., Blaszczyk, M,, Herlyn, M. Steplewski, Z. and Koprowski, H. (1983) Mouse monoclonal antibodies against human cancer cell lines with specificities for blood group and related antigens. Characterization by antibody binding to glycosphingolipids in a chromatogram binding assay, J. Biol. Chem., 258:4091-4097.
43. Hansson, G.C., Karlsson, K.-A., Larson, G., Strömberg, N., Thurin, J., Örvell, C. and Norrby, E. (1984) A novel approach to the study of glycolipid receptors for viruses. Binding of Sendai virus to thin-layer chromatograms, FEBS Lett., 170:15-18.
44. Hansson, G.C., Karlsson, K.-A., Larson, G., Strömberg, N. and Thurin, J. (1985) Carbohydrate-specific adhesion of

bacteria to thin-layer chromatograms: a rationalized approach to the study of host cell glycolipid receptors, Anal. Biochem., 146:158-163.
45. Karlsson, K.-A. and Strömberg, N. (1986) Overlay and solid-phase analysis of glycolipid receptors for bacteria and viruses, Meth. Enzymol., in press.
46. Bock, K., Breimer, M.E., Brignole, A., Hansson, G.C., Karlsson, K.-A., Larson G., Leffler, H., Samuelsson, B.E., Strömberg, N., Svanborg Edén, C. and Thurin, J. (1985) Specificity of binding of a strain of uropathogenic *Escherichia coli* to Galα1→4Gal-containing glycosphingolipids, J. Biol. Chem., 260:8545-8551.
47. Holgersson, J., Karlsson, K.-A., Karlsson, P., Norrby, E., Örvell, C. and Strömberg, N. (1985) Approaches to the study of receptors, in: "World's Debt to Pasteur", H. Koprowski and S.A. Plotkin, eds., Alan Liss, New York.
48. Umeda, M., Nojima, S. and Inoue, K. (1984) Activity of human erythrocyte gangliosides as a receptor to HVJ, Virol., 133:172-182.
49. Hansson, G.C., Karlsson, K.-A., Larson, G., Lindberg, A.A., Strömberg, N. and Thurin, J. (1983) Lactosylceramide is the probable adhesion site for major indigenous bacteria of the gastrointestinal tract, in: "Glycoconjugates", M.A. Chester, D. Heinegård, A. Lundblad and S. Svensson, eds., Rahms i Lund, Lund, Sweden.
50. Magnani, J.L. (1985) Immunostaining free oligosaccharides directly on thin-layer chromatograms, Anal. Biochem., 150:13-17.
51. Breimer, M.E., Hansson, G.C., Karlsson, K.-A. and Leffler, H. (1983) The preparative separation of sialic acid containing lipids from sulphate group-containing glycolipids from small intestine of different animals, J. Biochem., 93:1473-1485.
52. Karlsson, K.-A.(1986) Preparation of total non-acid glycolipids for overlay analysis of receptors for bacteria and viruses and for other studies, Meth. Enzymol., in press.
53. Karlsson, K.-A. (1973) Carbohydrate composition and sequence analysis of cell surface components by mass spectrometry. Characterization of the major monosialoganglioside of brain, FEBS Lett., 32:317-320.
54. Breimer, M.E., Hansson, G.C., Karlsson, K.-A., Leffler, H., Pimlott, W. and Samuelsson, B.E. (1979) Selected ion monitoring of glycosphingolipid mixtures. Identification of several blood group type glycolipids in the small intestine of an individual rabbit, Biomed. Mass Spectrom., 6:231-241.
55. Samuelsson, B.E. (1986) Investigation of the structures of large molecules with special reference to glycoconjugates, Adv. Mass Spectrom., 10:225-241.
56. Kushi, Y. and Handa, S. (1985) Direct analysis of lipids on thin-layer plates by matrix-assisted secondary ion mass spectrometry, J. Biochem., 98:265-268.
57. Breimer, M.E., Hansson, G.C., Karlsson, K.-A., Leffler, H., Pimlott, W. and Samuelsson, B.E. (1981) Sequencing of oligosaccharides by mass spectrometry applied on a 12-sugar glycolipid, FEBS Let., 124:299-303.
58. Dahmén, J., Frejd, T., Grönberg, G., Lave, T., Magnusson, G. and Noori, G (1983) 2-Bromoethylglycosides: applications in the synthesis of spacer-arm glycosides, Carb. Res., 118:292-301.

59. Sugita, M., Nakae, H., Yamamura, T., Takamiya, Y., Itasaka, O. and Hori, T. (1985) The occurence of glycosphingolipids containing mannose in the sea-water bivalve, *Meretrix lusoria* (Hamaguri), J. Biochem., 98:27-34.
60. Bock, K., Karlsson, K.-A., Strömberg, N. and Tenenberg, S. (1986) Fine dissection of binding epitopes on carbohydrate receptors for microbiological ligands, in: "Protein-Carbohydrate Interaction in Biological Systems. The Molecular Biology of Microbial Pathogenicity", D.L. Lark, ed., Academic Press, pp. 207-213.
61. Sabesan, S., Bock, K. and Lemieux, R.U. (1984) The conformational properties of the gangliosides GM2 and GM1 based on ^{1}H and ^{13}C nuclear magnetic resonance studies, Can. J. Chem., 62:1034-1045.
62. Brown, J.E., Karlsson, K.-A.., Lindberg, A., Strömberg, N. and Thurin, J. (1983) Identification of the receptor glycolipid for the toxin of *Shigella dysenteriae*, in: "Glycoconjugates", M.A. Chester, D. Heinegård, A. Lundblad and S. Svensson, eds., Rahms i Lund, Lund, Sweden.
63. Lemieux, R.U., Wong, T.C., Liao, J. and Kabat E.A. (1984) The combining site of anti-I-Ma (Group 1), Mol. Immunol., 21:751-759.
64. Lemieux, R.U., Venot, A.P., Spohr, U., Bird, P., Mandal, G., Morishima, N., Hindsgaul, O. and Bundle, D.R. (1985) Molecular recognition. V. The binding of the B human blood group determinant by hybridoma monoclonal antibodies, Can. J. Chem, 63:2664-2668.
65. Lemieux, R.U. (1985) The hydrated polar-group 'gate' effect on the specificity and strength of the binding of oligosaccharides by protein receptor sites, manuscript.
66. Thøgersen, H., Lemieux, R.U., Bock, K. and Meyer, B. (1982) Further justification for the *exo*-anomeric effect. Conformational analysis based on nuclear magnetic resonance spectroscopy of oligosaccharides, Can. J. Chem., 60:44-57.
67. Svensson, S.B., Janson, P.E., Källenius, G., Leontein, K., Oscarson, S., Nilsson, M. and Norberg, T. (1986) Identification, structural and conformational studies of carbohydrate receptor structures in fimbriae mediated binding, in: "Protein-Carbohydrate Interactions in Biological Systems. The Molecular Biology of Microbial Pathogenicity", D.L. Lark, ed., Academic Press, in press.
68. Svanborg Edén, C., Freter, R., Hagberg, L., Hull, R., Hull, S., Leffler, H. and Schoolnik, G.(1982) Inhibition of experimental ascending urinary tract infection by an epithelial cell-surface receptor analogue, Nature, 298:560-562.
69. Källenius, G., Jacobson, S.H., Tullus, K. and Svenson, S.B. (1985) P-fimbriae studies on the diagnosis and prevention of acute pyelonephritis, Infection, 13:159-162.
70. Breimer, M.E., Hansson, G.C. and Leffler, H. (1985) The specific glycosphingolipid composition of human ureteral epithelial cells, J. Biochem., 98:1169-1180.
71. Iwamori, M., Shimomura, J. and Nagai, Y. (1985) Specific binding of cholera toxin to rat erythrocytes revealed by analysis with a fluorescence-activated cell sorter, J. Biochem., 97:729-735.
72. Fishman, P.H., Pacuszka, T., Hom, B. and Moss, J. (1980) Modification of ganglioside GM1. Effect of lipid moiety on choleragen action, J. Biol. Chem., 255:7657-7664.

MEMBRANE GLYCOPROTEINS AND PLANT AND ANIMAL PROTEINS WITH LECTIN OR LECTIN-LIKE PROPERTIES

Zhao-Wen Shen

Shanghai Institute of Biochemistry
Chinese Academy of Sciences, Shanghai, China

This article is a resume′ of our studies on complex carbohydrates carried out at the Shanghai Institute of Biochemistry. Some of the work may not be directly related to conventional immunochemistry, but may be of interest to most of the attendants of this symposium. In the past five years, we carried out investigations on membrane glycoproteins, isolation and characterization of lectins and lectin-like substances from various plant and animal sources, proteins exhibiting biological activity from plant sources traditionally used in medical treatment, and cell surface proteins from certain specialized cells or cancer cells. These topics will be presented in the order mentioned.

STUDIES ON THE FAT CELL MEMBRANE

The fat cell was chosen as the object of study from the supposition that its membrane may not be very complicated. Indeed, our preliminary results of gel electrophoresis on a Triton X-100 extract of the membrane showed (staining with Coomassie blue) twenty or more bands, corroborating our supposition. Although we used fat pads from the rat epididymis initially as a source of fat cells, logistic reasons drove us to turn to the greater omentum of the pig.

First, a reproducible method for preparing cell membranes from adipose tissue had to be established. After a few trials we settled upon the technique of suspending the washed intact cells in hypertonic sodium chloride solution 15:1 (v/v) (15% for rat epididymal fat cells and 20% for fat cells of the greater omentum of the pig), stirring at 37° C for 2 hours, centrifuging and the "cell ghosts" collected. These were washed several times with Tris-HCl buffer (pH 7.4) and stored at -30° C.

The fat cells assumed various shapes after hypertonic treatment, but after about 1 hour, took the form of spherical globules for cells from rat epididymis or irregular forms for cells from pig greater omentum. These were examined for the presence of "marker" enzymes, i.e., alkaline phosphatase, succinic dehydrogenase, and NADPH-cytochrome c reductase. The last two enzymes showed very low activity, indicating absence of significant contamination of the membranes with intracellular components, while the alkaline phosphatase as membrane marker showed activity approximately 3.5 fold that of intact cells. The "ghosts" were also shown to be practically free of cell nuclei (DNA assay and hematoxylin staining) and extremely low in residual fat (Sudan III staining).[1]

Next, we proceeded to examine the periodate-Schiff positive bands revealed by gel electrophoresis of Triton X-100 extracts of fat cell membranes. The existence of at least five main membrane glycoproteins was indicated by the stains. Washed "ghosts" (50g) were suspended in an equal volume of 0.02M Tris-HCl buffer, pH 7.4, containing 2% Triton X-100, homogenized, incubated for 2 hours, centrifuged at 30,000 g for 30 minutes and the supernatant solution applied to a column of Con A-Sapharose 4B. The glycoproteins which were adsorbed onto the column were eluted with methyl-α-D-mannopyranoside (0.2M) in the same buffer. The material eluted with mannose showed 4 main bands which stained with periodate-Schiff reagent when subjected to SDS-PAGE. The apparent molecular weights were estimated on the gel using standard proteins as reference. They are 74, 79, 88, and >100 kilodaltons (Fig. 1,2).

The fractions eluted from the Con A-Sepharose 4B column were pooled and after extensive dialysis against 0.03M Tris-HCl buffer, pH 7.0, containing 0.1% Triton X-100, were loaded onto a DEAE-cellulose column. The fraction which was not adsorbed was concentrated and once more subjected to SDS-PAGE; only one band of molecular weight 74,000 was seen (Fig. 3). This 74 kDa material was found to contain 7% neutral sugars,

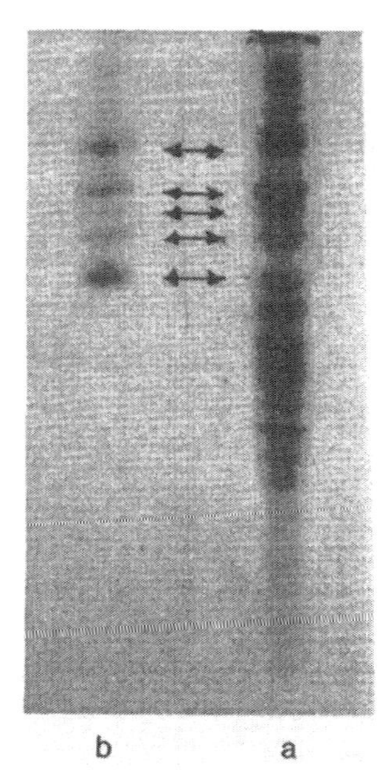

Fig. 1. Fat cell plasma membrane proteins of pig greater omentum separated into over 20 components, of which at least 5 are Schiff-positive glycoproteins, in SDS-PAGE. (a) Stained with Coomassie blue for protein components. (b) Stained with periodic acid-Schiff's reagent for glycoprotein components. Electrophoresis was performed in 5-10% linear gradients of polyacrylamide gels.

5% hexosamines and 2.6% sialic acid. It is abundant in glycine, serine, alanine, and also glutamic and aspartic acids, but deficient in sulfur-containing amino acids[2].

During our studies of fat cell membranes we came across certain phenomena which led to the discovery of a lectin or lectin-like component, existing presumably as a membrane component, which we describe tentatively as porcine adipocyte lectin (**PAL**).

The fat cell membranes were "solubilized" as usual with Triton X-100 (1%) containing 0.4M KCl, but when SM-2 (a styrene-divinyl benzene copolymer) was added to the extract to remove most of the Triton X-100, 35 to 45% of the proteins on

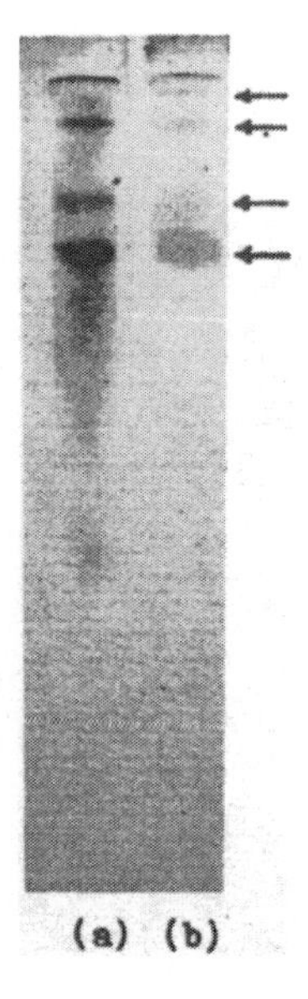

Fig. 2. Upon gel electrophoresis in SDS, the Con A-binding material separated into 4 protein components, all of which are Schiff-reactive glycoproteins. No apparent nonspecific adsorption was observed. The apparent molecular weights of the glycoproteins are estimated to be 74,000, 79,000, 88,000, and >100,000 daltons. (a) Stained with Coomassie blue for protein components (b) Stained with periodic acid-Schiff's reagent for glycoprotein components.

the extract were also removed. When the SM-2 from this operation was suspended in 0.01M Tris-HCl, pH 7.5, stirred and centrifuged, the adsorbed proteins dissociated from the copolymer and remained in solution. Affinity chromatography on a maltose-starch gel column followed by elution with Tris-HCl, pH 2, removed the lectin (Fig. 4). SDS-PAGE gave two bands with apparent molecular weights of 89,000 and 112,000, Pharmacia high molecular weight proteins being used as reference material. Since these two bands also stain with periodate-Schiff reagent, we conclude that both are glycoproteins.

(a) (b)

Fig. 3. SDS gel electrophoretic examination of the unadsorbed material which passes through the DEAE-cellulose column, reveals that this fraction consists of only one glycoprotein component migrating as a single band on SDS gel, with an apparent molecular weight of 74,000 daltons. (a) Stained with Coomassie blue for protein. (b) Stained with periodic acid-Schiff's reagent for glycoprotein.

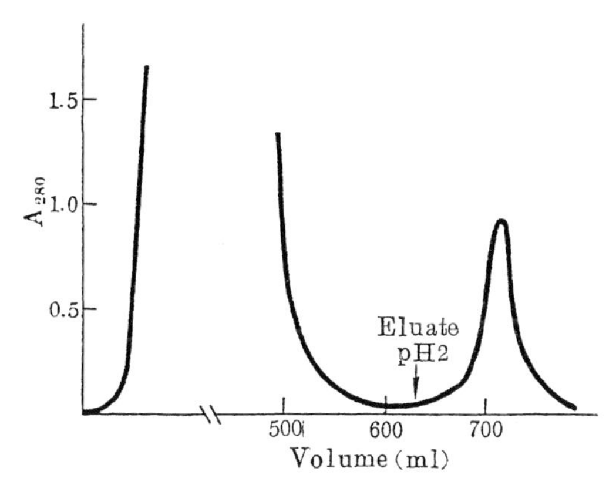

Fig. 4. Chromatography on a column of starch gel covalently coupled to maltose.

Triton X-100 extracts after SM-2 indicated a hemagglutination activity of 142$(mg/ml)^{-1}$ (Table 1). Hapten inhibition was strongest for maltose and slightly weaker for mannose and arabinose.

However, mannan, thyroglobulin, porcine submaxillary gland protein, ovalbumin, as well as hepatic glycogen and dextran all showed more or less inhibition of hemagglutination.

PAL seems to cause adhesion between cells of different types, based presumably upon specific recognition among cells. The cell adhesion ratio is expressed as the number of rosettes/number of fat cells counted. Calcium ions are required for expression of adhesion. Treatment of erythrocytes with neuraminidase raised the cell adhesion ratio for fat cells from 13% to 70%, while if the fat cells were similarly treated, the ratio rose to 90%. When fat cells were treated with trypsin, there was a great fall in cell adhesion (to 18.6%). These findings could be explained by a drop in negative charges on the cell surface, fostering closer contact and by the existence of some intimate relationship between the structure of fat cell protein, especially the sugar binding site, and cell adhesion.

Table 1. PURIFICATION OF PORCINE ADIPOCYTE LECTIN (**PAL**)

Fraction	Total Protein (mg)	Activity of PAL (mg/ml)-1	Total Activity	Activity Recovery (%)
Extract of rat cell membrane by Triton X-100	40			
Triton X-100 removed	38	142	$5.4x10^3$	100
PAL obtained by chromatography on maltose-starch column	1.7	1670	$2.8x10^3$	52

FURTHER STUDIES ON LECTINS

This leads quite naturally to another phase of our research: various studies on lectins, both old and new. In collaboration with our neighboring Institute of Plant Physiology, we carried out studies on the purification of a

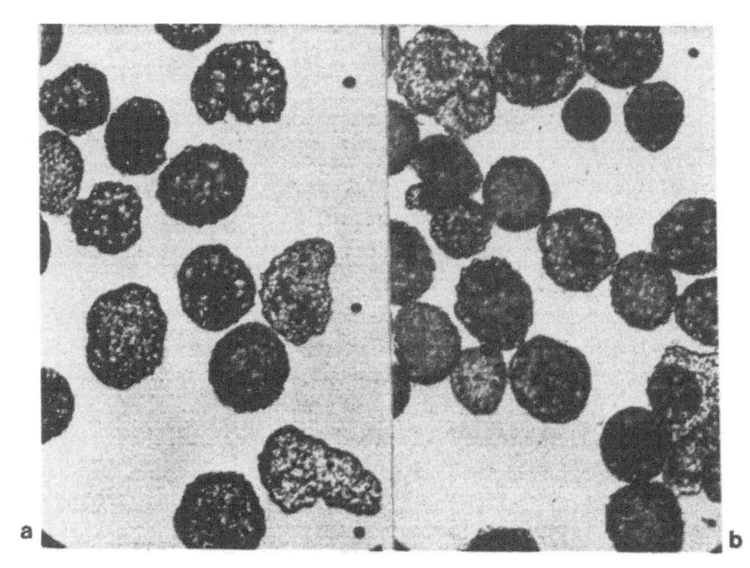

Fig. 5. Aggregation of adipocytes induced by PAL. (a) Control: well dispersed adipocytes remain in suspension. (b) Adipocytes in suspension aggregate after addition of PAL.

lectin from rice germ, its properties and distribution in various parts of the rice plant and also its possible biological functions. By extraction of rice germ with acidified saline, ammonium sulfate precipitation, and affinity adsorption on chitin, an electrophoretically homogenous protein was obtained, with a molecular weight of 23,000 as determined by gel filtration on Sephadex G-75. SDS-polyacrylamide electrophoresis with 1% mercaptoethanol gave 3 bands, corresponding to 11,300, 13,700, and 19,000 kDa, respectively.

The intact lectin was found to show cell agglutination activity on erythrocytes and lymphocytes (Fig.6), sperm cells, sweet potato mesophyll cells, and rice callus cells. Rice Germ Lectin (RGL) agglutinated rice embryo and suspended callus cells, also isolated cells from the leaves of rice seedlings. N-acetylglucosamine specifically inhibited these agglutinations. The lectin also showed mitogenic activity when tested on human peripheral lymphocytes and mouse spleen lymphocytes. The protein has two N-terminal residues: glycine and lysine and an isoelectric point of 6.5 to 6.8.

PAL was also able to bring about aggregation of dispersed fat cells, as shown by a fat cell suspension in Hank's buffer when incubated at 37° C. for one hour. The suspension with added PAL showed 81% aggregation as compared to 15% in the control (Fig. 5).

We venture to put forward a hypothesis that porcine adipocyte lectin is able to recognize the specific oligosaccharide structure of the red cell membrane and that the sugar binding properties are retained after undergoing fairly drastic operations like detergent solubilization, SM-2 treatment, and affinity chromatography.[3]

It is of interest to note that rice germ lectin distributes itself among all tissues of the young rice plant 4 days after germination. RGL content was found to increase in the rice embryos all through its differentiation and development. From this and also from evidence of the presence of lectin in early nonembryonic tissue, we postulate the existence of an intimate relationship between early nonembryonic tissues and embryo differentiation[4].

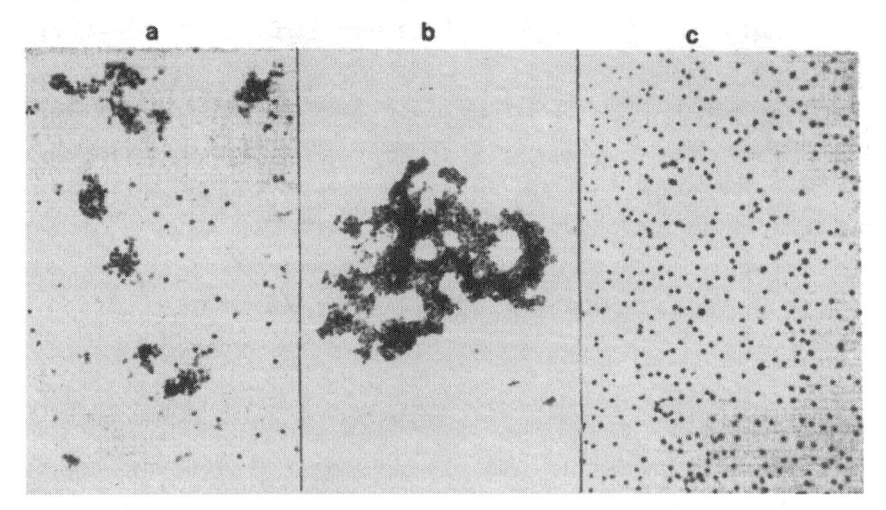

Fig. 6. RGL-induced agglutination of mouse spleen lymphocytes.(a) 0.83 μg/ml; (b) 33.3 μg/ml: (c) same as (b), except that the incubation medium contains 0.5M GlcNAc. (x300).

We also studied various species of *Allium*, which to the best of our knowledge, do not seem to have been studied with respect to their content of lectins. Of the several species studied, i.e., *Allium sativum, A. odorum, A. fistulorum, A. ledeborianum, A. cepa,* and *A. cepa yellow,* extracts of leaves and bulbs all showed hemagglutinating activity on rabbit and human erythrocytes. The sugar inhibitor for the agglutination by all species was mannose and compounds containing mannose, except for *A. cepa yellow* which was specifically inhibited by lactose and L-fucose. The *A. sativum* lectin was purified by affinity chromatography on mannose-Sapharose 4B coupled by divinyl sulfone. Upon elution by methyl alpha-D-mannoside followed by discontinuous polyacrylamide gel electrophoresis, 4 major bands were detected. Since DEAE-cellulose chromatography yielded 4 components and since each component possesses hemagglutinating activity, it was concluded that they were isolectins. This is evidenced by the fact that the methyl alpha-D-mannoside eluate, when subjected to SDS-PAGE and stained with Coomassie blue, gave only two bands either in the presence or absence of mercaptoethanol. The apparent molecular weights of the two nonidentical subunits were 47,500 and 26,500 and the isoelectric points: 6.0, 5.8, and 5.1. It was low in sulfur-containing amino acids and the N-terminals were alanine and glycine.

We also detected the presence of a lectin in the hemolymph of the silkworm, *Philosamia cynthia ricini,* which feeds on the leaves of the castor plant. It was detected during the fifth instar and early pupal stage, agglutinated rat and rabbit erythrocytes, but showed no effect on human cells, irrespective of blood type. The hemagglutinating activity of the lectin is specifically inhibited by galactose and lactose.[5]

The loquat, *Eriobotrya japonica,* a plant bearing edible fruits and indigenous to China was also found to contain a lectin in its leaves. It is a glycoprotein of molecular weight of 12,500, with 15% hexose, 2.7% hexosamine and no detectable amounts of sialic acid. Its isoelectric point is 4.5, N-terminal amino acid is alanine, and it is high in glycine and glutamate but lacks arginine and histidine. It agglutinates erythrocytes from a variety of animals including man, cattle and rabbits. From studies on the inhibition of hemagglutination by sugars, the lectin was found to specifically bind mannan, porcine thyroglobulin, and ovalbumin. Pinellin (see below) and loquat lectin are the only known proteins which bind mannose specifically without also binding to glucose.

The lectin showed a strong tendency to aggregate in acidic and neutral solution to form a homopolymer of molecular weight of around 500,000. It was also quite stable to high temperature: scarcely losing hemagglutinating capacity after heating for 15 minutes at 100° C.[6]

Since we needed various lectins in our research, we prepared them when we had access to the proper material. With some of these, a modified purification method was adopted. For example, peanut lectin was prepared by affinity chromatography on lactose-starch gel cross linked through epichlorhydrin. The modification combines the advantages of simple manipulation and cheaper source material[7].

From sesame seeds, *Sesame indicum L.*, a lectin was isolated in pure form by affinity chromatography on a chitin column. It turned out to be a glycoprotein of molecular weight 56,000 which yielded 2 nonidentical subunits of 30,000 and 26,000 upon treatment with 2-mercaptoethanol. N-acetyl-glucosamine is the specific sugar inhibitor for this lectin, in contrast to the report by Tomita, in which the specific sugar for a sesame seed lectin was claimed to be D-glucose.[8]

STUDIES ON A PROTEIN FROM *Pinella ternata*

One of our traditional medicinal herbs used for ages as a remedy for coughing is *Pinella ternata*. Our protein laboratory succeeded in isolating pinellin, a crystalline protein from the plant, exhibiting cell agglutinating and mitogenic activity. Erythrocytes from the sheep, dog, cat, rabbit, guinea pig, rat, mouse, and pigeon were agglutinated, while human erythrocytes were not agglutinated as were also those of the monkey, pig, chicken, goose, duck, tortoise, toad, and eel. Thus there seems to be rather strong donor species specificity. Of the sugars tested for hapten inhibition of hemagglutination, only mannan and thyroglobulin which has an oligomannoside core in its molecule, were found to inhibit agglutination of rabbit erythrocytes by pinellin. These observations suggest that pinellin binds only to some oligomannoside structures and that the binding site should be much larger than a single mannose molecule. Pinellin is also unique in that, unlike concanavalin A and other lectins, which bind mannose, no binding to glucose could be demonstrated.

In addition to erythrocytes, pinellin also agglutinated splenocytes, Ehrlich ascite cells (Fig. 7), and Hep A hepatoma

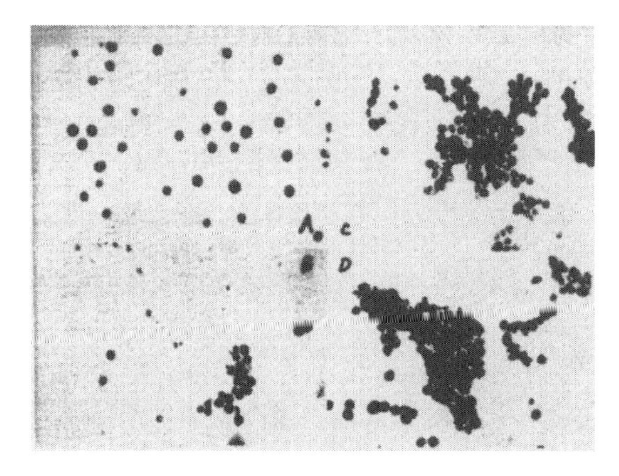

Fig. 7. Agglutination of Ehrlich ascites cells by pinellin. (x 1680) Concentration (μg/ml) of pinellin in reaction media: (a) control; (b) 8.3; (c) 33.3; (d) 166.7.

ascitic cells from the mouse and cultured human hepatoma cells (Fig. 8), but it does not agglutinate fat cells of rat epididymis and those of pig greater omentum, though it binds to them. Pinellin is also a mitogen for rabbit peripheral lymphocytes, showing species specificity as well as cell type specificity[9].

Binding of pinellin by fat cells from rat epididymal pads was studied using ^{125}I labelled pinellin (2.2 μCi/microgram). It was concluded that the binding is rather specific, as the iodo-pinellin may be completely replaced by the unlabelled native protein (Fig. 9). The binding is also a saturable

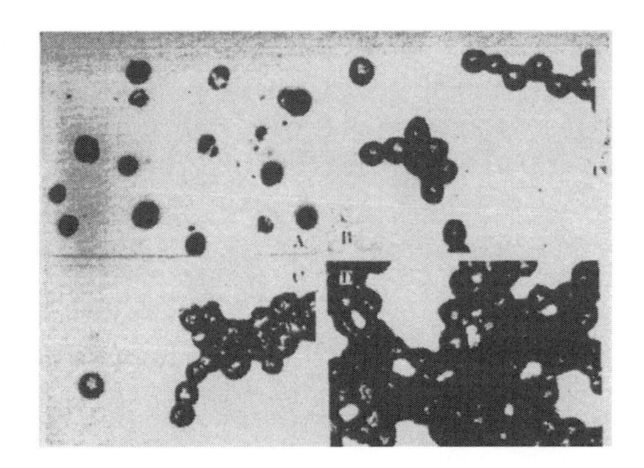

Fig. 8. Agglutination of cultured human hepatoma cells. QGY 7703-3 by pinellin (x 1920). Concentration (μg/ml) of pinellin in reaction media: (a) control; (b) 8; (c) 80; (d) 160.

process with respect to the protein with a single fat cell binding with a maximum of $6\text{-}7 \times 10^7$ molecules of pinellin (Fig. 10). The binding capacity decreased with increasing temperature, binding at 37° C. being only 10% of the binding at 0° C. Binding of pinellin to the receptor attained equilibrium quickly both at 0° C and at a higher temperature, but at 37° C. the complex dissociated at a rapid rate. Methyl alpha-D-Mannoside, even at 0.1M, does not cause dissocoation of the complex, nor does mannan at a concentration of 2 mg/ml. A Scatchard plot of ^{125}I-pinellin binding to its receptor reveals the rather complex nature of the binding, not amenable to simple analysis. To study the binding in somewhat greater detail, concanavalin A was added to the pinellin receptor complex which responded by rapid dissociation to attain a new equilibriuim. From the similarity in sugar specificity between Con A and pinellin, we think it may be possible that they both bind at the same site or sites. Low concentrations of Con A (2.5×10^{-9}M) showed little effect; higher concentrations (5×10^{-8}M) inhibit pinellin receptor complex formation. However, complete inhibition was not attained, even with 2.5×10^{-5}M Con A, showing that the Con A receptor and pinellin receptor on the fat cell could not be totally identical. Insulin at 2.5×10^{-5}M did not show any effect on pinellin and its binding with receptors. The interrelationship among complex formation between each of the proteins and its corresponding receptor appears to present a rather intricate phenomena which warrants further investigation.[10]

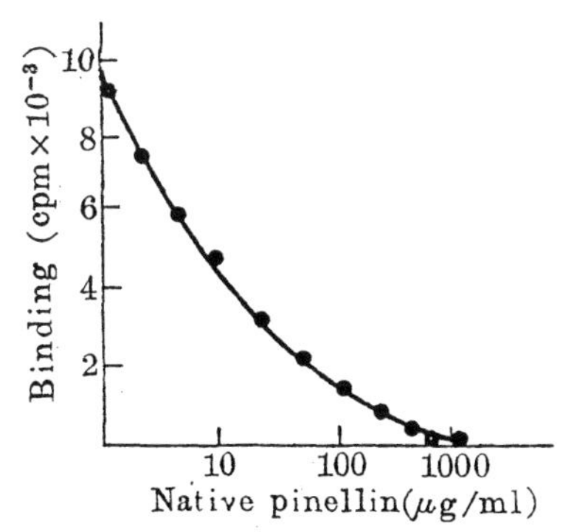

Fig. 9. The effect of unlabelled pinellin on binding of ^{125}I-pinellin with its specific receptor. Cells ($3.1x10^4$) were suspended in Hank's solution, 0.5% bovine serum albumin, and the prescribed amounts of unlabelled pinellin, and then incubated at 24° C. for 15 min., after which ^{125}I-pinellin ($7.2x10^4$ cpm) was added, and further incubation for 50 min. was completed after which measurements were taken. Total volume: 0.4 ml.

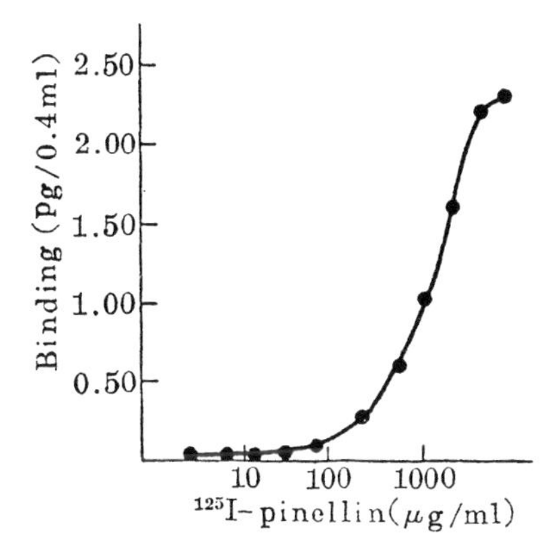

Fig. 10. The specific binding of ^{125}I-pinellin to its receptor. Fat cells ($2.2x10^4$) were suspended in Hank's solution containing 0.5% bovine serum albumin, and the prescribed amounts of ^{125}I-pinellin. These were incubated at 24° C. for 60 min. (logarithmic coordinates). Total volume: 0.4 ml.

Pinellin has been shown to have a strong anti-implantation activity in rabbits. Studies by the method of blastocyst transfer showed that pinellin primarily affects the normal function of the uterine endometrium (Fig. 11). By using the immunoenzyme technique, the membranous surface of the glandular epithelium in the endometrium was found to be the binding site of pinellin. The results suggest that the anti-implantation effect of pinellin may be due to its interference of normal interactions between the endometrial and the embryonic cells.[11]

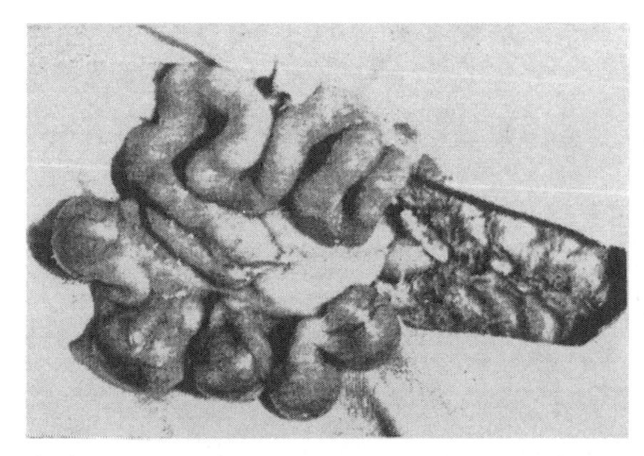

Fig. 11. Anti-implantation effect of pinellin in the rabbit. No blastocyst was found in the uterine horn on the 11th day of pregnancy, into which 500 μg of pinellin was injected 6 days before. Five embryos developed in the control horn.

PRELIMINARY STUDIES ON REACTIONS OF LECTINS WITH THEIR RECEPTORS

(1) Lectin receptors on ram spermatozoa. Wheat germ agglutinin, concanavalin A, *Ricinus communis* agglutinin, and peanut agglutinin have been separately tested on ram spermatozoa collected by the pseudo-vagina method. Wheat germ agglutinin was found to be the strongest lectin tested; the minimum concentration required followed the order: WGA, 1.4 μg/ml; Con A, 5.5 μg/ml; RCA, 5.5 μg/ml; PNA, 21.3 μg/ml. The agglutination by these lectins was inhibited by N-acetyl-glucosamine, methyl alpha-D-mannoside, D-galactose, and lactose, respectively. Ram spermatozoa prefixed with glutar-aldehyde were incubated with Con A, subsequently with horseradish peroxidase and then stained with 3-3'-diamino-benzidine tetrahydrochloride. Con A receptors were found to be uniformly distributed on the head of the spermatozoa, with only a few on the flagellum and middle piece. Under the electron microscope, electron-dense material was found to

cover the head portion. Methyl alpha-D-mannoside inhibits the binding of Con A to these spermatozoa.[12]

Mention should also be made of the inhibitory effect of L-fucose on the formatiion of rosettes by ram spermatozoa binding with neuraminidase-treated fixed sheep erythrocytes (Fig.12).[13]

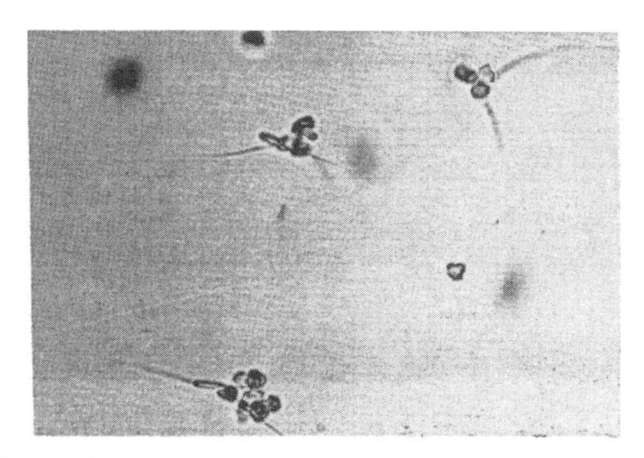

Fig. 12. Test for rosette formation. Rosettes formed from ram spermatozoa acting on neuraminidase-treated sheep erythrocytes.

(2) The distribution of Con A receptors has also been observed in esophageal epithelial cells, normal as well as cancerous, by means of the Con A-HRP immunoenzyme localization technique. Under the light microscope, the Con-A receptor is mainly localized on the plasma membrane of the superficial and prickle cells. The receptor increases markedly on carcinoma cells and appears on the nuclear membrane, showing a tendency to increase with development of the carcinoma. It appears that the Con A receptor on the nuclear membrane might be related to the degree of differentiation of cancer cells.[14]

SUMMARY

The fat cell, chiefly from porcine greater omentum, was treated with hypertonic sodium chloride and membranes left over from exuded fat were extracted with Triton X-100 and the extract subjected to gel electrophoresis. Of the 20 or more bands which showed on the gel, at least five were periodate-

Schiff positive. By the treatment of the glycoproteins with Con A, followed by DEAE-cellulose, one with an apparent molecular weight of 74 kDA was obtained pure. The other components are not yet purified. However, two glycoproteins of apparent molecular weights of 89 and 112 kDa were obtained from the fat cell membrane by affinity chromatography. These were apparently lectins, since they possessed hemagglutinating activity and specificity towards maltose and caused adhesion among cells (erythrocytes) and aggregation of fat cells.

The properties of lectins from various sources have also been studied: rice germ lectin, *Allium sativum* lectin, a lectin from a species of silkworm and one from loquat leaves.

A crystalline protein called pinellin was isolated from *Pinella ternata.* Since it has been shown to exhibit cell aggregation and mitogenic activity, it is also a lectin, showing species as well as cell type specificity. Preliminary studies have also been made of the binding of pinellin to its receptor on fat cells. The protein has also been shown to possess an anti-implantation effect in rabbits.

Studies have been made of membrane proteins on ram spermatozoa and Con A receptors on squamous epithelial cancer cells.

REFERENCES

1. Xu, Ji-hua, Wu, Bo-liang, Sun, Ce, Shen, and Zhao-wen (1982). Preparation of fat cell ghosts (in Chinese with abstract in English). Acta Biochim. Biophys. Sinica, 14:251-255.
2. Mo, Hanqing, Zhu, Chou-rong, Sun, Ce, Shen, and Zhao-wen (1983). Purification of a glycoprotein for porcine adipocyte plasma membranes (in English). Kexue Tong-bao, 28:544-548.
3. Zhu, Zheng, Sun, Ce, Shen, and Zhao-wen (1985). Preliminary studies on a lectin on the surface of adipocytes (in English). Scientia Sinica, B, 28:152-159.
4. Shen, Zhao-wen, Sun, Ce, Zhu, Zheng, Tang, Xi-hua, Shen, Rui-juan (1984). Purification and properties of rice germ lectin. Can. J. Biochem. Cell. Biol., 62:1027-1032.
5. Mo, Hanqing, Sun, and Ce (1983). A lectin from the hemolymph of *Philosamia cynthia ricini* (in Chinese with abstract in English).Acta Biochim. Biophys. Sinica, 15:383-384.

6. Wu, Chaun-yue, Shen, and Zhao-wen (1986). Kexue Tongbao. in press.
7. Sun, Ce, Ding, Chang-rong, Shen, and Zhao-wen (1984). Purification of peanut agglutinin by affinity chromatography on lactose-starch gel (in English). Kexue Tongbao, 29:823-827.
8. Zhu, Zheng, Sun, Ce (1982). Purification and properties of sesame lectin (in Chinese with abstract in English). Acta Biochim. Biophys. Sinica, 14:91-93
9. Sun, Ce, Xu, Ji-hua, Zhai, Shi-kang, Tao, Zong-jin, Yau, Tai-ying, Zhu, Zheng, Shen, Zhao-wen(1983). Some biological porperties of pinellin (in Chinese with abstract in English). Acta Biochim. Biophys. Sinica, 15:333-338.
10. Sun, Ce, Xu, Ji-hua, Tao, Zong-jin, Shen, Zhao-wen (1983). Pinellin receptors on the adipose cell (in Chinese with abstract in English). Acta Biochim. Biophys. Sinica, 15:1-8.
11. Chen, Hui-ling, Song, Jin-fen, Tao, Zong-jin (1984). Anti-implantation effect of pinellin in rabbits (in Chinese with abstract in English). Acta Physiologica Sinica, 36:388-392.
12. Guo, Hsiao-hui, Liu, Yan-ying, Lu, Zhao-qi, Sun, Ce, Zhu, Zheng,(1985). Lectin receptors on ram spermatozoa (in Chinese with abstract in English). Acta Biochim. Biophys. Sinica, 17:736-742.
13. Zhu, Zheng, Sun, Ce, Lin, Yun-ying, Lu Zhao-qi (1984) A carbohydrate binding protein in ram sperm (in Chinese, with abstract in English). Acta Biochim. Biophys. Sinica, 16:102-10
14. Sun, Ce, Zhang, Jun, Zhu, Zheng, Wang, Ya-lin, Liu, Yan-ying (1983). Con A receptors on esophageal epithelial squamous cells (in Chinese with abstract in English). Chinese Med. J., 62:486-489.

DIFFERENTIAL BINDING PROPERTIES OF GalNAc AND/OR Gal SPECIFIC LECTINS*[1]

Albert M. Wu[2] and Shunji Sugii[+]

Department of Veterinary Pathology, College of Veterinary Medicine, Texas A & M University, College Station, TX 77843, USA [1] and Department of Serology and Immunology, School of Medical Technology, Kitasato University, 1-15-1 Kitasato, Sagamihara, Kanagawa, 228, Japan [+]

The binding properties of lectins have been used to study the structural and functional role of cell surface carbohydrates (1-6), to detect sugar moieties on normal and neoplastic cell surfaces (1), to isolate mutants resistant to the cytotoxic action of some lectins (7-10), and to isolate and characterize glycoconjugates (5,11-13A). This group of lectins generally recognize D-pyranose sugars, and require configurational and structural complementarity of sugars for interaction to occur. All lectin molecules have more than two carbohydrate binding sites, a property resulting in their ability to agglutinate cells or to precipitate complex carbohydrates (1,4,14,15). Until the early seventies, the carbohydrate specificities of lectins, were determined by the ability of monosaccharides or their glycosides to inhibit lectin-induced hemagglutination (2,14,17,18). In the early eighties, lectins of the same apparent monosaccharide specificity were found to demonstrate different reactivities toward different oligosaccharide chains, and differential affinities to animal cells and glycoproteins, implying that they have their own binding specificity extending beyond the monosaccharide (1-4,14,18).

During the past decade, the concept of the combining

---

*This paper is dedicated to Dr. A. Herp, who has assisted my research for a decade. [1]Some of the content in this review article, especially the section related to Galβ1→3GalNAc specific lectins, has been published in Mol. Cell. Biochem., 61:131-141, 1984 and is reproduced with permission by Martinus Nijhoff publisher, Boston.
[2]Corresponding author. [3]Abbreviations, see next page.

sizes of many lectins was extended to five or more sugars, and many of the multi-branched oligosaccharides exhibit a significant increment in their lectin-binding reactivities as compared with the linear counterparts. For example, *Datura stramonium* (TAL, Thorn apple) shows over 480-fold higher specificity for a biantennary Penta-2,6-saccharide, which contains two branches of Galβ1→4GlcNAcβ1→6Man and Galβ1→4GlcNAcβ1→2Man at the nonreducing ends, than one of its branched trisaccharides. Some of the lectins, having broad spectra of binding properties, possess dual or multiple affinities to various disaccharides. For example, *Wistaria floribunda* is specific for GalNAcα1→3GalNAc, GalNAcα1→3Gal, GalNAcα1→Ser(Thr), and Galβ1→4(3)GlcNAc residues; *Bauhinia purpurea alba,* for Galβ1→3GalNAc, GalNAcα1→Ser(Thr) and Galβ1→4(3)GlcNAc residues, and *Datura stramonium,* for *N*-acetyllactosamine and *N*-acetylchitobiose (2,16). Differentiating among lectin affinities shall aid in elucidating the binding properties of cell surface carbohydrates at both the molecular and structural levels. The purpose of this review article is to list the binding affinities of well studied GalNAc and/or Gal specific lectins - the lectins which are specific for Mäkelä's Group II carbohydrates (17). The article is divided into five sections. In **I.** Gal and/or GalNAc specific lectins are further grouped into six classes; their general properties are summarized in **II.;** binding properties of lectins among classes are compared in **III.;** in **IV.**, their

---

[3a]Abbreviations. Gal, D-galactopyranose; Glc, D-glucopyranose; Man, D-mannopyranose; LFuc, L-fucopyranose; GalNAc, 2-acetamido-2-deoxy-D-galactopyranose; GlcNAc, 2-acetamido-2-deoxy-D-glucopyranose; (GlcNAc β1→4)n repeating unit of GlcNAcβ1→4GlcNAc; NeuAc, *N*-acetylneuraminic acid; R, carbohydrate residues; Melibiose Galα1→6Glc; Raffinose, Galα1→6Glcβ1→2Fruf, and Stachyose Galα1→6Galα1→6Glcβ1→2Fruf; BGS, blood group active glycoproteins (substances); QPA, Quantitative precipitin assay; QPIA, Quantitative precipitin-inhibition assay.

[3b]Abbreviation for lectin classification: Bold letters of F, A, Af, Tn, I (II), and B symbolize classification of subgroups of lectins and the regular forms of these letters represent part or all of the structure of the lectin determinants. **F**, GalNAcα1→3GalNAc; **A**, GalNAcα1→3Gal; **Af**, GalNAcα1→3[LFucα1→2]Gal related; **T**, Galβ1→3GalNAcα1→Ser(Thr) of protein core or Galβ1→3GalNAcβ1→ at the terminal nonreducing end of the ganglioside; **Tn**, GalNAcα1→Ser(Thr) of the peptide chain; **I(II)**, Human blood groups type I and II carbohydrate sequence, Galβ1→3(4)GlcNAc at the nonreducing end of the carbohydrate chain (17A); and **B**, Galα1→3Gal at the nonreducing end of the carbohydrate chain.

[3c]Abbreviations of lectins are listed in the section on lectin classification (next page).

differential and unique affinities are illustrated; finally in **V.**, a strategy of application of the lectins is given.

I. CLASSIFICATION OF GalNAC AND/OR Gal SPECIFIC LECTINS

Inhibition of hemagglutination or precipitin reactions with known oligosaccharides and glycopeptides can provide important information about structural differences in binding specificities of lectins. Based on the information available, the lectins selected in this review article are divided into six classes according to their specificity for GalNAcα1→Ser(Thr) of the peptide backbone (lectin T determinant) and disaccharides (lectin F, A, I(II), T, Tn and B determinants). The source and structural relationship of the lectin determinants are shown in Fig. 1. Most of the lectin F determinants are found in glycosphingolipids and the other five determinants can be found in human blood group A, B, H, Le^a, Le^b, and Ii active glycoproteins prepared from human ovarian cyst fluid. The reactivities of lectin determinants (lectin active disaccharides) represent a combination of binding of two individual sugars. The contribution of each sugar in the binding is not necessarily equal, and is different among lectins. For example, both *Maclura pomifera* (MPL) and peanut (PNA) are Galβ1→3GalNAc specific, however, the inhibitory profile of the monosaccharides with these two lectins is quite different; GalNAc>Gal in MPL and Gal>>GalNAc (inactive) in PNA. The classification scheme utilized here provides better understanding of the lectins, and serves as a useful guide for application-minded novices. A scheme of classification is shown as follows.

A GalNAc-specific lectins (Section IIIA)

1. **F** specific lectins (GalNAcα1→3GalNAc, **F**orssman specific disaccharide) - *Dolichos biflorus* (DBL), *Helix pomatia* (HPL), Hog peanut (ABrL, *Amphicarpaea bracteata*), and *Wistaria floribunda* (WFL) lectins.
2. **A** specific lectins (GalNAcα1→3Gal, Blood Group A specific disaccharide) - *Griffonia(Bandeiraea) simplicifolia*-A_4 (GSI-A_4), Lima bean (LBL), soy bean (SBL), *Vicia villosa (VVL)*, *Wistaria floribunda* (WFL) lectins, *Dolichos biflorus* (DBL) and *Helix pomatia (HPL)*.
3. **Tn** specific lectins [GalNAcα1→to Ser(Thr) of the protein core]- *Vicia villosa* B_4 (VVL-B_4), *Salvia sclarea* (SSL), *Maclura pomifera* (MPL), *Bauhinia purpurea alba* (BPL), HPL and WFL.

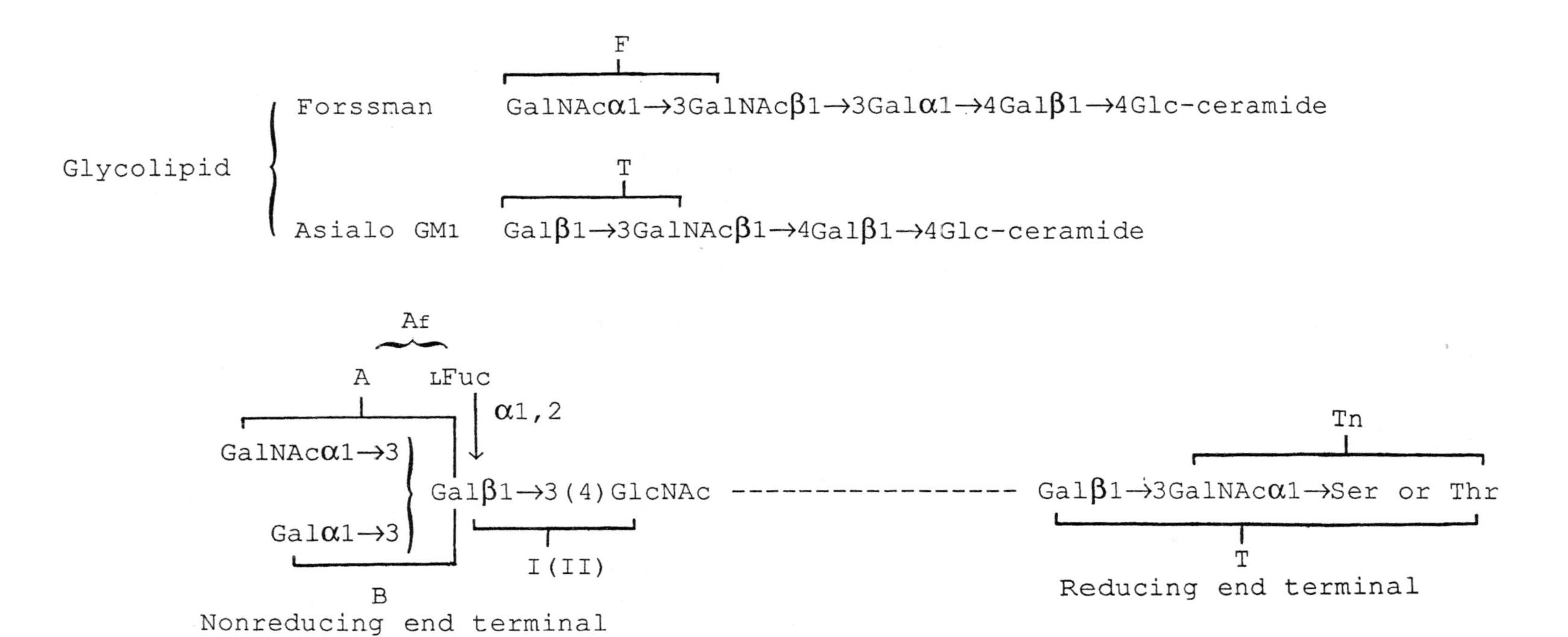

Fig. 1. Proposed Gal and GalNAc specific lectin determinants. Lectin F determinant can be found in glycosphingolipid (the principal glycolipid of mammalian tissues) of the tissues of the guinea pig, horse, cat, and chicken. It can also be found on the surface of some bacteria, viruses, human gastric carcinoma and colon tumors (76-77). Five out of six (A/A_f,B,I/II,T,and Tn) determinants can be found in human blood group A,B,H,Le^a,Le^b, and Ii active glycoproteins, prepared from human ovarian cyst fluid. Galβ1→3GalNAcβ1→ linked at the nonreducing end of asialo GM_1 is also considered as Lectin T determinant. I and II lectin determinants defined in this article are equivalent to human blood group type I (Lacto-N-biose, Galβ1→3GlcNAc) and type II (*N*-acetyllactosamine) carbohydrate sequences.

B Gal-specific lectins (Section IIIB)

1. **T** specific lectins (Galβ1→3GalNAc, the mucin type sugar sequences on human erythrocyte membrane, T antigen or the disaccharides at the terminal nonreducing end of the gangliosides) - Peanut (PNA), *Bauhinia purpurea alba* (BPL), *Maclura pomifera* (MPL), *Sophora japonica* (SJL), *Artocarpus integrifolia* (Jacalin, AIL), and *Artocarpus lakoocha* (Artocarpin) lectins.
2. Type **I** and **II** specific lectins (Galβ1→3(4)GlcNAc - the disaccharide residues at nonreducing end of the carbohydrate chains derived from either *N*-glycosidic or *O*-glycosidic linkages) - *Ricinus communis* agglutinin (RCA_1), *Datura stramonium* (TAL, Thorn apple), *Erythrina cristagalli* (ECL, Coral tree), and *Geodia cydonium* (GCL).
3. **B** specific lectin (Galα1→3Galβ1, Human blood group B specific disaccharide) - *Griffonia(Bandeiraea) simplicifolia* B_4(GSI-B_4).

II. GENERAL PROPERTIES OF GalNAc- AND Gal-SPECIFIC LECTINS

The general properties of GalNAc- and Gal-specific lectins are summarized in Table I. These lectins are usually purified from the seed extract or tissue extract by adsorption onto insoluble blood group substances, or by affinity chromatography on melibiose-BioGel, galactose derivatized-Sepharose, or acid-treated Sepharose and by subsequent elution with GalNAc, Gal or lactose (References listed in Table I). Further purification and fractionation to isolate isolectins or other components are performed by gel filtration, ion-exchange, or ConA Sepharose chromatography. With a few exceptions, the purified lectins are glycoproteins containing up to 37% carbohydrate, but most of them have less than 10% carbohydrate (Table I). Either Mn^{2+} or Ca^{2+} is required for reactivity of DBL, SBA, GSI-A_4, LBL, SJL, ECL, and GSI-B_4. MPL, AIL, ALL, ECL, GCL, and TAL are mitogenic. A code for classification, molecular weights, subunits, isoelectric points and sedimentation coefficients ($S_{20,W}$) of the purified lectins is also included in Table I.

II-A. GalNAc-specific lectins

DBL, *Dolichos biflorus* (Horse Gram) agglutinates human type A_1 cells strongly and type A_2 cells weakly (19,20). The removal of divalent cations from the medium by EDTA abolishes the hemagglutinating and precipitating abilities of these lectins (21). The A isolectin binds to ConA-Sepharose column

Table I. Molecular properties of GalNAc- and Gal-specific lectins

Lectin	Monosaccharide Specificity	Class[b]	Human erythrocytes agglutinated	MW $\times 10^3$ ($S_{20,W}$)	Carbohydrate content (%)	No. of subunits	pI	Metal ion required for reactivity	Mitogenicity or Toxicity
DBL[a] (19-25)	GalNAc only	**F**	$A_1>A_{2(10x)}$	110.0	3.8-4.7	4[c]		Ca^{++}, Mn^{++}	
HPL (26-28)	GalNAc only	**F**	A	79.0(5.28)	8.0	6			
ArL (2,28A)	GalNAc	**F**	A_1	135.0	Yes	4[c]			
WFL (29-31)	GalNAc>Gal	**F**	A, B, O	68.0	3.2	2	5.0,5.5		
SBA (32-34)	GalNAc>Gal	**A**	A>0, B	120.0	7.0	4	6.0	Ca^{++}, Mn^{++}	
GSI-A4 (35-38)	GalNAc	**A**	A	114.0	9.0	$4(A_4)$	4.9,5.1	Ca^{++}, Mn^{++}	
LBL-II (39-44)	GalNAc	**Af**	A>>B	269.0(9.60)	3.0-5.0	8		Ca^{++}, Mn^{++}	
LBL-III (39-44)	GalNAc	**Af**	A>>B	138.0(6.00)	3.0-5.0	4		Ca^{++}, Mn^{++}	
VVL-A4 (45,46)	GalNAc	**A**	A_1	134.0	6.7	$4(A_4)$		none	
VVL-B4 (45,46)	GalNAc>>Gal	**Tn**	Tn	108.3	9.8	$4(B_4)$		none	
SSL (47)	GalNAc>>Gal	**T**	Tn	50.0	15.0	2	8.8		
MPL (48-50)	GalNAc>Gal	**T**	A, B, O	40.0(4.10)	trace	$2(\alpha,\beta)$[cd]	4.3	none	Mitogen
BPL (51-53)	GalNAc>Gal	**T**	A, B, O, M, N	195.0(7.50)	11.0	4		none	
SJL (54-57)	GalNAc>Gal	**T**	A, B, O	132.8(6.80)	5.90	$4(\alpha_2\beta_2)$	5.5	Ca^{++}, Mn^{++}	
RCA_2 (58-63)	Gal, GalNAc	**T**	A, B, O	60.0	4.5	$2(\alpha,\beta)$	7.1		Toxic
PNA (64-67)	Gal only	**Tn**	A[e], B[e], O[e]	110.0	None	$4(\alpha_4)$			
AIL (68-69A)	Gal>GalNAc	**T**	A, B, O, etc.	40.0	3.0	$4(\alpha_4)$	4	none	Mitogen
ALL (69B,69C)	Gal, GalNAc	**T**	A[e], B[e], O[e], etc.	70.0	11.7	4		none	Mitogen
RCA_1 (58-63)	Gal only	**I(II)**	A, B, O	120.0	4.5	$4(\alpha_2\beta_2)$	7.80		
ECL (70,71)	GalNAc	**I(II)**	A, B, O	56.8(3.90)	4.5	$2(\alpha,\beta)$[c]	4.50	none[f]	Mitogen
GCL (72)	GalNAC	**I(II)**	A, B, O	60.0	14.0	4	4.40		Mitogen
TAL (73-75)	GlcNAc	**I(II)**	A, B, O	86.0	37.0	$2(\alpha,\beta)$			Mitogen
GSI-B4 (2,38)	Gal	**B**	B	114.0	9.0	$4(B_4)$		Ca^{++}, Mn^{++}	

[a]Two forms, Major A and Minor B with same activity (2,22,23); [b]Structures represented are shown in abbreviation; [c]Subunits are of different molecular weights; [d]Three subunits were reported in reference 48; [e]the neuraminidase-treated erythrocytes. [f]ECL contains tightly bound Mn^{++} and Ca^{++} and its reactivity is not affected by EDTA.

whereas the B isolectin does not (22,23). The molecular weights of A (major) and B (minor) isolectins are 113,000 and 109,000, respectively. Both are composed of four subunits with molecular weights of 26,500 and 26,000 for the A and B isolectins, respectively (22,23). The carbohydrate specificity and hemagglutinating activity of both DBL A and B isolectins are identical.

HPL, *Helix pomatia* (Edible snail) contains 18-half cysteine residues and consists of 6 identical subunits (26,27). Two subunits are linked by an intrachain S-S bond to form subunit dimers with a molecular weight of 26,000 and three dimers hold together by noncovalent interactions (27,28). The lectin specifically agglutinates human type A but not types B or O erythrocytes (26). The optimal pH of the lectin for precipitation is between 5.4 and 9.5 (26).

ABrL, *Amphicarpaea bracteata* (Hog-Peanut) shows anti-blood A activity. It has a molecular weight of 1.35×10^5, and consists of four heterogeneous subunits (M_r.3.7 to 3.2×10^4), all of which contain carbohydrates. No cysteine and only a small amount of methionine is found (2,28A).

WFL, *Wistaria floribunda* has a M_r of 6.8×10^4, composed of two identical subunits with molecular weights of 32,000 linked together by an S-S bond (29-31). The lectin agglutinates human erythrocytes of all types. A tetrameric molecular form with a molecular weight of 125,000 is sometimes obtained by gel filtration.

SBA, Soy bean (*Glycine max*) has a molecule weight of 1.2 x 10^5 and consists of 4 identical subunits with molecular weights of 30,000 each and has a pI of 6.0-6.1 (32-34). The lectin agglutinates rabbit and human erythrocytes (A>O>B). Trypsinizing the erythrocytes greatly increases the agglutinating activity.

GSI-A$_4$, *Griffonia*(*Bandeiraea*) *simplicifolia* I-A$_4$. The native mixture of GSI isolectins, agglutinates human types B and AB erythrocytes strongly, A$_1$ weakly and A$_2$ and O cells not at all (36,37). GSI consists of five isolectins: tetrameric forms composed of two subunits (A and B) in different proportions. All are glycoproteins and designated as A$_4$, A$_3$B, A$_2$B$_2$, AB$_3$ and B$_4$ (2,3,38). The subunits have similar molecular weights and immunological reactivity but differ in isoelectric points. GSI-A$_4$ consists of 4 identical subunits with molecular weights of 32,000 each (2). GSI-A_4 agglutinates human type A but not type B erythrocytes. The optimal pH of the lectin mixture for the precipitation reaction lies between 6 and 10.

LBL, Lima bean (*Phaseolus lunatus*) contains two components (II and III)(39-42). Both components are composed of identical subunits with molecular weights of 31,000 each. Two of these are linked by an interchain S-S bond to form a larger subunit with a molecular weight of 62,000. Component II contains 4 dimers and component III two dimers. Both components agglutinate human type A erythrocytes strongly but type B cells weakly. The carbohydrate binding properties of the lectin are strongly dependent on the integrity of the free SH-groups (41). The removal of Mn^{2+} reduces the hemagglutinating activity. EDTA completely inhibits precipitation and the optimal pH for precipitation is between 4.5 and 8.5 (39).

VVL, *Vicia villosa* seeds contain at least three lectins. These isolectins are composed of two different subunits with M_r=35,900 (subunit B) and 33,600 (subunit A), and designated as B_4, A_4, and A_2B_2. The purified B_4 and A_4 lectins contain 6.7-9.8% carbohydrate by weight. Both are rich in the acidic and hydroxylic amino acids and lack cysteine and methionine. The A_4 lectin (minor fraction) agglutinates A_1 erythrocytes specifically. The predominant lectin in *V. villosa* seeds, B_4, does not agglutinate A, B, or O erythrocytes (45,46).

SSL, *Salvia sclarea* lectin, is a glycoprotein of M_r= 5.0×10^4 composed of two identical subunits of M_r=3.5×10^4 linked together by disulfide bonds. The purified lectin agglutinates specifically Tn erythrocytes and at high concentrations also agglutinates Cad erythrocytes. Native A, B, or O red blood cells are not agglutinated by the lectin (47).

II-B. Gal-specific lectins

MPL, *Maclura pomifera*, has a molecular weight of 4.0-4.3 x 10^4. It is composed of three nonidentical subunits with molecular weights of 12,000, 13,500, and 14,500 and has a pI of 4.3 (48,49). The lectin agglutinates human erythrocytes of all types and shows a weak mitogenic activity to rat splenocytes (50).

BPL, *Bauhinia purpurea alba*, is a glycoprotein of M_r. 1.95 x 10^5, which is composed of 4 identical subunits with molecular weights of 44,000 linked by S-S bonds (51-53). The lectin shows nonspecific agglutination of human erythrocytes.

SJL, *Sophora japonica,* has a molecular weight of 132,800 by gel filtration and consists of two dissimilar subunits joined together by noncovalent bonds (54-57). Each subunit is composed of two identical disulfide-linked polypeptides (56). Its optimal pH for hemagglutination and precipitation is between 6.7 and 9.5. Ca^{++} or Mn^{++} is required for interaction (57).

RCA2 (Ricin), *Ricinus communis* agglutinin II,(58-63) is a highly toxic nonagglutinating protein with a molecular weight of 60,000 consisting of one A chain (M_r=29,500) and one B chain (M_r=34,000). It shows hemagglutination when it is in dimeric form.

PNA, Peanut(*Arachis hypogaea*), a carbohydrate free protein with a M_r of 1.1 x 10^5, is made up of 4 identical subunits with molecular weights of 27,000 (64-67). It does not agglutinate untreated human erythrocytes at all but agglutinates neuraminidase-treated human erythrocytes of all types. Its mitogenic activity to animal peripheral lymphocytes is not well characterized.

AIL, *Artocarpus integrifolia* (Jacalin, jack fruit), is a glycoprotein of M_r 39,500. and is composed of four identical subunits of M_r 10,500. Interaction of jacalin with serum or colostrum results in precipitation of IgA (68-69).

ALL, *Artocarpus lakoocha* lectin (Artocarpin), has a molecular weight of 70,000 as determined by Sephadex gel filtration. The lectin is a tetramer composed of four apparently identical subunits. ALL is a glycoprotein, and contains 11.7% carbohydrate in which D-xylose (6%) is the main sugar. It contains high proportions of acidic and hydroxylic amino acids and a paucity of sulfur-containing amino acids (69B).

RCA1, *Ricinus communis* agglutinin I, is a hemagglutinin with a molecular weight of 120,000 consisting of two A chains (M_r=29,500) and two B chains (M_r=37,000) (58-63).

ECL, *Erythrina cristagalli,* is composed of two nonidentical subunits with molecular weights of 28,000 and 26,000 (70-71). The lectin agglutinates human erythrocytes of all types and rabbit erythrocytes. It is mitogenic for human peripheral T lymphocytes but not mitogenic for mouse thymocytes.

GCL, *Geodia cydonium,* has a M_r of 6.0 x 10^4 and consists of 4 identical subunits with molecular weights of 15,000 each (72). It has a pI of 4.4 and agglutinates human erythrocytes of all types. It is a mitogen.

TAL, *Datura stramonium* (DSL - Thorn apple), has a M_r of 8.6 x 10^4 and is made up of two nonidentical subunits with molecular weights of 46,000 and 40,000 (73 to 75). It non-specifically agglutinates human erythrocytes. The aggregated TAL shows mitogenic activity to human peripheral lymphocytes.

GSI-B$_4$, *Griffonia* (*Bandeiraea*) *simplicifolia* I-B$_4$, like GSI-A$_4$, consists of 4 identical subunits with a molecular weight of 28,500. GSI-B$_4$ agglutinated human type B but not type A erythrocytes (2,38).

III. COMPARISON OF BINDING PROPERTIES AMONG SUBGROUPS OF GALNAC/GAL SPECIFIC LECTINS

III-A. GalNAc specific Lectins

Eleven well characterized lectins are classified into three subgroups: **(1)** GalNAcα1→3GalNAc as part of the Forssman determinant and are coded **F**, **(2)** GalNAcα1→3Gal as part of the human blood group A determinant, coded **A**, while the structure α1→2 fucosylated at the subterminal Gal is coded **A**$_f$ and **(3)** GalNAcα1→Ser(Thr) of the protein core as a Tn determinant and are coded **Tn**. Source and structural relationships of these lectins are shown in Fig. 1 Many of them have dual or multiple specificities. A summary of their specificities and their code is given in Table II.

(1) Lectins specific for **F** determinants

The terminal disaccharide (GalNAcα1→3GalNAc, coded **F**) appears to be the main serological determinant of Forssman antigens. It is a heterogenetic determinant and found in many animal species such as in tissues of the guinea pig, horse, cat, and chicken, but not rat and rabbit (76, 76A). The Forssman-like determinant has also been found to occur on the surface of some bacteria, influenza and Rous sarcoma viruses, and in human gastric carcinomas and colon tumors (77). Four well characterized lectins are considered in this group; DBL,

Table II. Comparison of the reactivities of GalNAc-specific lectins[a]

Carbohydrate residues capable of reacting	Class[b]	DBL (21;25)	HPL (21;25)	ABrL (2,28A)	WFL[c] ([illegible],29)	SBA (25,32)	LBL[c] (43,44)	GSI[c] (A4) (35)	VVL (A&B) (78)	VVL (B4) (46)	MPL (48)	BPL (51)	SSL[c] (47)
GalNAcα1→3GalNAc	F	+++	+++	++++	++	ND	-	ND	++	ND	-	-	ND
GalNAcα1→3GalNAcβ1→3Galα1→4 Galβ1→4Glc	F	++++	++++	ND	+++	ND	ND	ND	ND	ND	-	-	ND
GalNAcα1→3Gal	A	+	ND	+++	+++	+++	-[e]	+++	++++	ND	-	++	ND
GalNAcα1→3Galβ1→3GlcNAc	A	+	++	ND	++-	++++	-[e]	+++	++	ND	-	+++	ND
GalNAcα1→3Galβ1→4GlcNAcβ1→6R[d] ↑α1,2 LFuc	Af	++	-	ND	+	+	++	+	++	ND	-	+++	ND
Galβ1→3GalNAc(α1→Ser or Thr)[f]	T	ND	ND	ND	ND	ND	ND	ND	ND	+	++++	++++	ND
GalNAcα1→(to Ser or Thr)[g]	Tn	-	+	+	+++	++	ND	ND	++	++++	+++	+++	++[d]
Galβ1→3GlcNAc-R	I	-	-	ND	ND	+	ND	ND	ND	ND	-	+++	ND
Galβ1→4GlcNAc-R	II	-	-	ND	+	+	ND	ND	ND	ND	-	+++	ND

[a]Based mainly on the summarized results obtained from precipitin inhibition assays. ++++, best inhibitor; +++, about 1/2 of the best; ++, about 1/10 of the best; +, about 1/100 of the best; -, less than 1/100 or no inhibition; ND, not determined.

[b]Representive structure is shown in footnote 3b abbreviation.

[c]The best inhibitors are GalNAcα1→6Gal and o,p-nitrophenyl αGalNAc for WFL and GSI(A4), respectively; GalNAcα1→Ser, GalNAcα1→Thr for SSL; and GalNAcα1→3(LFucα1→2)Galβ1→3GlcNAcβ1→3Galβ1→4Glc, an A active hexasaccharide (43) for LBL.

[d]R; related carbohydrate residues.

[e]From the results that GalNAcα1→3[LFucα1→2]Galβ1→4GlcNAc(A active, tetra-), +++ and Galα1→3[LFucα1→2]Gal, ++, it can be concluded that LFuc is important for binding (43).

[f]Values obtained from Galβ1→3GalNAc related inhibitors.

[g]Values obtained from GalNAcα1 glycosides and also interact well with desialized ovine salivary mucus glycoprotein, and mild acid treated armadillo salivary glycoprotein [GalNAcα1→Ser(Thr) of the protein core] - (unpublished data).

HPL, ABrL, and WFL. Knowledge of VVL reactivity is still hampered by the paucity of its A active isolectin VVL-A_4 and requires further study.

For decades, DBL and HPL were recognized as being specific for human blood group A active oligosaccharides (25). However, in the early eighties, these two lectins were found to be more specific for the Forssman-active pentasaccharide, GalNAcα1→3GalNAcβ1→3Galα1→4Galβ1→4Glc, than for human blood group A (GalNAcα1→3Gal) related oligosaccharides (21). WFL exhibits equal reactivities toward the above two types of oligosaccharides. Recently, it was also reported that hog-peanut lectin (*Amphicarpaea bracteata* lectin, ABrL) is also specific for the Forssman and human blood group A specific disaccharides GalNAcα1→3GalNAc and GalNAcα1→3Gal (2,28A). All Forssman specific lectins show dual or multiple specificities (which will be discussed in Section IV) toward human blood group A active oligosaccharides and GalNAcα1→ Ser(Thr) of the protein core (Tn antigen). Their differential reactivities can be demonstrated by variation in binding to monosaccharides and corresponding glycosides (Table IV) and also by their precipitating profile of lectins with various complex carbohydrates (Table V).

(2) Lectins specific for **A** determinants

Nine out of eleven GalNAc-specific lectins show to various extents specificity for the human blood group A-active determinant oligosaccharides - GalNAcα1→3Gal, GalNAcα1→3Galβ1→3GlcNAc, and GalNAcα1→3[LFucα1→2]Galβ1→3GalNAc, etc. (Table II). This group of lectins is coded **A**, and the α1→2 fucosylated A at the subterminal Gal of the disaccharide as $\mathbf{A}_f$. VVL-A_4, a minor component in this isolectin mixture, is the blood group A active fraction. VVL (a mixture of A_4, A_2B_2 and B_4), GSI-A_4, SBA, and WFL are the most specific lectins for the A-active disaccharide (78,35,32,29,25) and trisaccharide GalNAcα1→3Galβ1→3GlcNAc, but are less active to the A-active fucosyl pentasaccharide (Table II). LBL reacts 20 times better with a synthetic A-active fucosyl trisaccharide than with the A-active disaccharide (44). These results imply that the LFucα1→ blocks the binding to VVL-A_4, GSI-A_4, SBA, WFL, and HPL but provides a more favorable conformation for the binding of

DBL and LBL. While DBL and LBL are precipitated strongly by only the A-active substances (19,39), SBA, HPL, and WFL react with a large variety of blood group substances (26,29, 32,44,78)(See Section IV-C(1) or Table V). DBL does not bind the first Smith degraded product of the A-active substance, whereas HPL, WFL, and SBA react well because this treatment splits off the terminal GalNAc and LFuc and exposes Galβ1→3(4) linked terminal residues (human blood group type I and type II structure) as well as GalNAcα1→Ser(Thr) of the core protein (lectin Tn determinant) (21,32,79,108). From these results, it can be assumed that DBL and LBL are more specific for fucosyl A-active than for non-fucose containing A determinants.

As seen in Table II, VVL-A_4 in a mixture of VVL reacts with some Forssman-active disaccharides (78) while LBL is inactive. WFL and SBA show some reactivity for Galβ1→3(4)GlcNAc as one of their multiple specificities (21,29,32).

(3) Lectins specific for **Tn** determinant

The Tn determinant infers a structure of GalNAcα1→Ser(Thr) of the peptide core, the most widely encountered *O*-glycosidic linkage in glycoproteins. The Tn transformation indicates an acquired disorder characterized by the exposure, at the red cell surface, of normally cryptic GalNAc residues α→linked to the hydroxyl of Ser or Thr on membrane sialoglycoproteins (80). It is the result of a selective deficiency of the 3-β-D-galactosyltransferase involved in the biosynthesis of the T structure: Galβ1→3GalNAcα→Ser(Thr) (81). Tn antigen can be detected at the cell surface of erythrocytes, granulocytes, platelets, and B and T lymphocytes of patients presenting the Tn syndrome (82). This antigen has also been proposed as a marker of cancerous tissues (83,84).

Two lectins (VVL-B_4 and SSL) are GalNAc-specific lectins but do not agglutinate human types A, B, and O erythrocytes (45-47). They bind to the Tn exposed erythrocytes, i.e. they react well with GalNAcα1→ linked to the Ser or Thr of glycophorin A (46). VVL-B_4 is less reactive with glycoproteins or glycopeptides possessing Galβ1→3GalNAcα1→Ser(Thr) [T antigenic determinant] while no information is available on SSL. MPL, BPL, WFL, HPL, and SBA are also specific for the residue of GalNAcα1→ linked to the protein core. Fraction A of the desialized armadillo salivary glycoprotein (ASG-A), which is composed of GalNAcα1→Ser or Thr is one of the best reagents to test for the specificity of this group of lectins.

III-B. Gal-specific Lectins

Eleven selected lectins are classified into three classes according to their specificity for (1) Galβ1→3GalNAc groups, a mucin-type sequence on human erythrocyte membrane or Galβ1→3GalNAcβ1→ at the terminal nonreducing end of the ganglioside and are coded **T**; (2) Galβ1→3(4)GlcNAc sequence, as the human blood group type I and type II structure and coded **I** and **II**; and (3) Galα1→3Gal as part of the human blood group B determinant and coded **B**.

(1) Lectins specific for **T** antigen

Galβ1→3GalNAcα1→Ser(Thr) of the core protein is the mucin type sequence of the glycoprotein on the human erythrocyte membrane. It is the antigenic determinant of the Thomsen-Friedenreich antigen, which is a tumor-associated antigen of non-oncofetal origin and is one of the few chemically well defined antigens with a proven link to malignancy; therefore, anti-T probes are potentially very useful in cancer research (84-87).

Up to now, the combining sites of BPL (*Bauhinia purpurea alba*), MPL (*Maclura pomifera*), PNA (peanut, *Arachis hypogaea*), SJL (*Sophora japonica*), AIL (*Artocarpus integrifolia*) lectins, ALL (*Artocarpus lakoocha*, Artocarpin), and ricin (RCA_2) are found to be the most complementary to the T determinant disaccharide [Galβ1→3GalNAcα1→Ser(Thr) of the protein core]. All of them show dual or triple specificities. For example BPL, MPL, SJL, and RCA_2 are also specific for Tn (GalNAcα1→*O* to the Ser or Thr of the protein core) and PNA, BPL, and RCA_2, for Type I and II sequences (Galβ1→3(4)GlcNAc).

The binding affinity of RCA_2 is dependent upon the number and location of this disaccharide moiety along the peptide backbone (58). RCA_2 also reacts well with the tri-branched antennary oligosaccharides containing Galβ1→4GlcNAc (58,88).

Table III. Comparison of the reactivities of Galβ1→ linked-specific lectins[a]

		Galβ1→3GalNAc							Galβ1→3(4) GlcNAc				Galα1→3Gal
Carbohydrate residues capable of reacting	Class of Lectin[b]	PNA (67)	MPL (48)	BPL (51)	SJL[c] (57)	RCA2 (58)	AIL[e] (68)	ALL (69C)	RCA1 (89)	ECL (70)	GCL (72)	TAL (75)	GSI-B4 (35)
Galβ1→3GalNAcα1→*O*	**T**	++++	++++	++++	++++	++++[e]	++++[e]	++++	ND	ND	ND	ND	ND
GalNAcα1→*O* to Ser(Thr)[d]	**Tn**	-	++++	+++	?	++[e]	++[e]	++	ND	+	++	ND	-
Galβ1→4GlcNAcβ1→R	**II**	++	-	+++	++	++[e]	±[e] -	++++	+++	+++	+[f]	ND	ND
Galβ1→3GlcNAcβ1→R	**I**	+	-	+++	++	ND	±[e] -	+++	++	++++	ND	ND	ND
Galα1→3Gal	**B**	+	-	++	-	ND	ND	ND	++	+	ND	ND	++++
GalNAcα1→3Galβ1→R	**A**	-	-	++	-	ND	ND	ND	ND	ND	ND	ND	-

[a]Based mainly on the summarized results obtained from precipitin inhibition assays. ++++, best inhibitors; +++, about 1/2 the best; ++, about 1/10 of the best; +, about 1/100 of the best; ±, less than 1/100; –, no inhibition; ND, not determined.

[b]Representive structures shown in footnote 3b. [c]The best inhibitor for SJL is Galβ1→3GalNAcβ-*N*-tosyl-L-serine.

[d]The specificity for GalNAcα1→ to Ser or Thr is estimated from the results of precipitin inhibition assays: the inhibitory potency of GalNAc for PNA, that of GalNAcα-linked glycosides for MPL and BPL, and that of methyl α-GalNAc for ECL. The reactivity of SJL towards this residue is questionable. [e]Based on association constant (68).

[f]The best inhibitor for TAL was a biantennary saccharide with two Galβ1→4GlcNAcβ1→ residues (Penta-2,6) at the nonreducing end (75) as ++++.

(2) Lectins specific for Galβ1→3(4)GlcNAc(**I/II**)

Geodia cydonium lectin (GCL), *Ricinus communis* agglutinin (RCA_1), *Erythrina cristagalli* lectin (ECL), and Thorn apple lectin (TAL) are most specific for three monosaccharides, i.e. GalNAc, GalNAc, Gal, and GlcNAc, respectively; but based on a disaccharide structure, they are most specific for Galβ1→4(3)GlcNAcβ1→ linked residues (72,89,70,95). Galβ1→4(3)GlcNAc occurs in many glycoproteins of normal and malignant cell surfaces as the human blood group type **I** and type **II** carbohydrate sequence (78,90). *N*-acetyllactosamine at the nonreducing end of a branched complex carbohydrate is found in many *N*-glycosidically linked oligosaccharides or polysaccharides and these are known as bi-, tri-, and/or tetraantennary *N*-glycosidically linked complex carbohydrates. These branched oligosaccharides are ideal reagents for defining the differential binding properties of type I(II) specific lectins. Tetraantennary oligosaccharides containing *N*-acetyllactosamine units are most specific for ECL (70), whereas the biantennary one is most specific for TAL and RCA_1 (75,88,89). These lectins will be further discussed in Section IV-B. The combining site of GCL is also specific for Galβ1→4(3)GlcNAc but it has not yet been well characterized as compared with that of other lectins. PNA, BPL, and RCA_2 are also specific for Galβ1→4(3)GlcNAc (67,51,88). The strength of their affinities toward branched complex oligosaccharides containing *N*-acetyllactosamine at their nonreducing ends has not yet been determined. Recently, the carbohydrate specificity of *Phaseolus vulgaris* leukoagglutinin (L_4) has been characterized and shown to be most specific for the pentasaccharide containing *N*-acetyllactosamine (91,92).

```
Galβ1→4GlcNAcβ1,2↓
                  Man
Galβ1→4GlcNAcβ1,6↑
```

This structural unit occurs in tetraantennary *N*-acetyllactosamine-type glycoproteins (92).

(3) Lectin specific for Galα1→3Gal (**B**)

Only GSI-B_4 is available in this group. It reacts well with oligosaccharides with terminal α1→linked Gal, with Galα1→3Gal being the best. Substitution at the subterminal residues, such as LFucα1→2, affects the binding substantially indicating that the B_4 site involves at least the subterminal α1→3 linked Gal and is strictly B specific (35). BPL and RCA_1 are weakly specific for this residue.

IV. DIFFERENTIAL AND UNIQUE AFFINITIES OF GalNAc-AND Gal-SPECIFIC LECTINS

Differential and unique binding properties of lectins can be demonstrated in three ways as described in the following section. (**A**) Variation in the ratio of binding to monosaccharides and their glycosides (Table IV). (**B**) Differences in inhibition profile with branched or linear oligosaccharides and/or glycopeptides. Inhibition profiles of dozens of lectins with branched oligosaccharides and glycopeptides have been extensively studied by Debray *et al.*(88), ECL by Kaladas *et al.*(70), TAL by Crowley *et al.*(75), VVL-B_4 by Tollefsen and Kornfeld (46), and RCA_1 and RCA_2 (Ricin) by Baenziger and Fiete (58). In these studies, detailed information about the carbohydrate specificities of many lectins was obtained by reaction with multiple determinant oligosaccharides. (**C**) Comparison of precipitation profiles of lectins with various well characterized complex carbohydrates (Table V to VII). It is one of the valuable methods to evaluate the overall reactivity between complex carbohydrates and lectins.

IV-A. Unique Binding Properties of Lectins

The individual binding characteristics of Gal and GalNAc specific lectins can be differentiated by their specificity for methyl glycosides, hydrophobicity, and GalNAc/Gal. As shown in Table IV, no two lectins share the same binding properties and three lectins (WFL, ECL, and VVL-B_4) show reverse anomeric specificity between the methyl and *p*-nitrophenyl glycosides (29,70,46). With the methyl glycosides, WFL binds methyl α GalNAc best while *p*-nitrophenyl β GalNAc is a better inhibitor than its α anomer (29). It is

Table IV. Variations in binding to monosaccharides and their glycosides of some GalNAc- and Gal-specific lectins[a].

Lectin	Specificity for *p*-nitrophenyl glycosides	Specificity for methyl glycosides	Hydrophobicity (*p*-nitrophenyl vs methyl glycosides)	GalNAc/Gal (RP[b])
DBL (19,25)	$\alpha > \beta$	NA	yes	NA[c]
WFL (29)	$\beta > \alpha$	$\alpha > \beta$	yes	50.0
SBA (25,32)	$\alpha > \beta$	$\alpha > \beta$	yes	25.0
LBL (25,44)	$\alpha > \beta$	$\alpha > \beta$	none	17.6
GSI-A4 (35)	$\alpha > \beta$	$\alpha > \beta$	yes	17.1
VVL (78)	$\alpha > \beta$	$\alpha > \beta$	yes	143.0
VVL-B4 (46)	$\beta \geq \alpha$	$\alpha >> \beta$	none	200.0
PNA (67)	$\alpha \geq \beta$	$\alpha \geq \beta$	none	NA[d]
MPL (48)	$\alpha >> \beta$	$\alpha >> \beta$	yes	3.3
BPL (51)	$\alpha > \beta$	$\alpha > \beta$	none	4.6
SJL (57)	$\beta >> \alpha$	$\alpha = \beta$	yes	5.5
ECL (70)	$\beta > \alpha$	$\alpha > \beta$	yes	1.5
RCA1 (89)	$\beta > \alpha$	$\beta > \alpha$	yes	NA[d]
GSI-B4 (35)	$\beta > \alpha$	$\alpha >> \beta$	none	NA[d]

[a]Values are calculated from moles required for 50% inhibition; no data is available for HPL and GCL; TAL is not shown due to ambiguity of the results; NA not available due to inactivity of sugars tested or not done; [b]expressed as reciprocal of relative potency; [c]Gal is inactive; [d]GalNAc is inactive.

important, therefore, that the *p*-nitrophenyl group is able to interact with that portion of the lectin combining site that is complementary to the subterminal *N*-acetylhexosamine moiety as the subterminal sugar is often an acetylated amino sugar in glycoproteins reacting with lectins. It is also important to clarify lectin reactivity more extensively than just on the basis of their specificities for monosaccharides or methyl glycosides. Thus, ECL and RCA_2 were inappropriately classified as GalNAc-specific lectins since their best inhibitory sugar structure is a linear oligosaccharide with Galβ1→ at the nonreducing end.

IV-B. Comparison of Reactivities of GalNAc- and Gal-Specific Lectins With Various Oligosaccharides

The reactivities of lectins with monosaccharides and their glycosides provide some information concerning hydrophobic interactions and the inhibition profile of GalNAc/Gal to demonstrate unique binding characteristics. However, using oligosaccharides and glycopeptides with well known structure, more important information about the fine specificity of individual lectins can be obtained, and this in turn can be used to explain their different reactivities with glycoproteins at the structural level. In this section, the specificities of GalNAc and Gal specific lectins for different oligosaccharides are compared.

(1) Lectins specific for GalNAcα1→3GalNAc

For DBL and HPL, Forssman-active oligosaccharides are much more specific than A-active oligosaccharides (21). The most reactive oligosaccharide with DBL and HPL is the Forssman-active pentasaccharide GalNAcα1→3GalNAcβ1→3Galα1→4Galβ1→4Glc. The Forssman-active synthetic trisaccharide GalNAcα1→3GalNAcβ1→3Gal-*O*-$(CH_2)_8COOCH_3$ is slightly more potent towards both lectins than the corresponding disaccharide GalNAcα1→3GalNAc, which in turn is much more active than the A-active disaccharide GalNAcα1→3Gal. This suggests that the penultimate GalNAc unit of the Forssman determinant is also involved in the binding and the third sugar, Gal, from the nonreducing end is not as important as the terminal and penultimate ones. DBL reacts better with blood group A active fucosyl pentasaccharides than with blood group A non-fucosylated tri- or disaccharides; but in the case of HPL, GalNAcα1→3Gal is somewhat inferior to GalNAcα1→3Galβ1→3GlcNAc, which however exhibits higher specificity than the corresponding fucosylated pentasaccharide (26). This suggests that the Fucα1→2 linkage stabilizes the conformation of the GalNAcα1→ linkage with respect to the rest of the molecule by facilitating entry into DBL but blocks sterically the reactive surface of the terminal GalNAc to HPL.

With WFL, the order of the reactivities toward various oligosaccharides is: a) GalNAcα1→6Gal > b) GalNAcα1→3Gal, GalNAcα1→3GalNAcβ1→3Galα1→4Galβ1→4Glc, (Forssman specific pentasaccharide) and GalNAcα1→3Galβ1→3GlcNAc > c) GalNAcα1→3Galβ1→4Galβ1→4Glc (Globoside) > d) GalNAcα1→3GalNAc > GalNAc, and e) GalNAcα1→3[LFucα1→2]Galβ1→4GlcNAcβ1→6R > f) Galβ1→4GlcNAc. The Forssman-active pentasaccharide is more potent than GalNAc but less active than the best inhibitor GalNAcα1→6Gal (21,29). Globosides with β-linked GalNAc and GalNAcα1→3Gal have almost the same potency and both structural features are more reactive than GalNAc (21). Galβ1→4GlcNAc is only about one third as active as GalNAc but the branched oligosaccharide with Galβ1→4GlcNAc residues is expected to be a potent inhibitor. This is confirmed by the fact that WFL reacted strongly with asialo-fetuin in which three Galβ1→3GalNAc and nine Galβ1→4GlcNAc residues per mole are present at the terminal nonreducing ends (93-95). Recently, it was found that one of the triantennary branches (III) (See Table V, footnote "g") in asialofetuin may also be Galβ1→3GalNAc linked (95A). Thus, the affinity of the oligosaccharide to WFL caused by such a structural difference needs to be further studied. Mild acid hydrolysis of blood group B substance to expose internal Galβ1→4(3)GlcNAc activity and GalNAcα1→Ser(Thr) of the protein core (96) also increases its reactivity to WFL. From the data of Sugii and Baker (21,29) it can be concluded that WFL has a broad range of affinities.

At the present time, only limited information is available for ABrL. The best information so far shows this lectin to be most specific for GalNAcα1→3GalNAc and GalNAcα1→3Gal, which are 50- to 100-fold more potent as agglutinating agents than melibiose or lactose (2,28A).

(2) Lectins specific for GalNAcα1→3Gal

The reactivity of SBA towards various oligosaccharides decreases in the following order: Phenyl αGalNAc > GalNAcα1→3Galβ1→3GlcNAc > GalNAcα1→3Gal >> GalNAcα1→3[LFucα1→2]Galβ1→4GlcNAcβ1→6R. This profile indicates that SBA exhibits greatest affinity for phenyl αGalNAc, the A active trisaccharide (GalNAcα1→3Galα1→3GlcNAc), and the A active disaccharide (GalNAcα1→3Gal), which implies that the GalNAcα1→ linkage confers a favorable conformation for binding. Substitution of blood group A-active oligosaccharides (i.e. GalNAcα1→3Gal sequence) by an L-fucosyl group at the subterminal Gal residue greatly diminishes their binding capacity (2,32).

GSI-A_4 is most specific for terminal nonreducing α-linked GalNAc but reacts moderately with oligosaccharides having a terminal nonreducing α-linked Gal, this accounts for its good reactivity with blood group A, B, and precursor I substances (35). GSI-A_4 reacts best with the A-active di- and trisaccharides (GalNAcα1→3Gal and GalNAcα1→3 Galβ1→3GlcNAc). Blood group A-active fucosyl oligosaccharides (GalNAcα1→3[LFucα1→2]Galβ1→4GlcNAcβ1→6Gal-ol, R_{IMS}, 2.56) and (GalNAcα1→3Galβ1→3GlcNAcβ1→6-3-hexenetriols, MSS, AR_L, 0.52) are respectively 40 and 26 times less active than the A-active disaccharide. This indicates that the substitution of Lfucose at the penultimate residue interferes substantially with lectin-binding (35).

GalNAcα1→3Gal for VVL (a mixture of A_4, A_2B_2, and B_4, with B_4 as a dominant fraction) is nearly as potent as the best inhibitor, methyl αGalNAc (78). The A-active trisaccharide GalNAcα1→3Galβ1→3GlcNAc is 7 times less active than the corresponding disaccharide but showed almost the same potency as the Forssman-active disaccharide GalNAcα1→3GalNAc. Again, the A-active fucosyl pentasaccharide is a much weaker inhibitor than the corresponding trisaccharide. Thus, the Fucα1→2 of the pentasaccharide reduces the interaction of the subterminal Gal with the combining site.

LBL is most specific for fucose-containing oligosaccharides having the A trisaccharide determinant, GalNAcα1→3[LFucα1→2]Gal. The best inhibitor, an A-specific hexasaccharide, GalNAcα1→3[LFucα1→2]Galβ1→3GlcNAcβ1→3Galβ1→4Glc, is about 11 times more active than the A trisaccharide (GalNAcα1→3[LFucα1→2]Gal). A difucosyl oligosaccharide having a second fucosyl residue linked α1→3 to the GlcNAc is even less active; and fucose linked α1→4 to GlcNAc is completely inactive. These results suggest that the combining sites of this lectin involves at least the last four and perhaps five sugars of the nonreducing terminal in the hexasaccharide. This would be in the upper range of antibody combining site sizes (43).

(3) Lectins specific for GalNAcα1→Ser(Thr)

Only VVL-B_4 and SSL have been studied well. A fetuin glycopeptide containing two GalNAcα1→*O* to Ser (Thr) of the peptide chain (i.e. two lectin Tn determinants) is the most potent determinant for VVL-B_4. The order of decreasing strength is two Tn in clusters >> single Tn >> a fetuin glycopeptide containing one or two Galβ1→3GalNAcα1→Ser(Thr) of the peptide chain (lectin T structure) > NeuNAcα2→3Galβ1→3GalNAcα1→Ser (Thr) of the peptide chain (46).

The carbohydrate binding specificity of SSL to various glycoproteins is: glycopeptides containing two Tn structures > glycopeptides containing one Tn structure or glycopeptides containing three sequential Tn structures (47). The precise difference in specificity between these two lectins has to be further studied.

(4) Lectins specific for Galβ1→3GalNAc

Based on the reactivity of monosaccharides, the best precipitating inhibitor for four T specific lectins is GalNAc for BPL, MPL, and SJL and Gal for PNA, respectively. However, all four lectins are specific for Galβ1→3GalNAc residues. The inhibitory profile of SJL is similar to that of BPL (Table III). The type I and II disaccharides (Galβ1→3GlcNAc and Galβ1→4GlcNAc) are much less active than Galβ1→3GalNAc. The di- and oligosaccharides containing type I carbohydrate sequence (Galβ1→3GlcNAc), such as blood group A-active type I trisaccharide and lacto-*N*-tetraose are more specific for SJL and BPL than the di- and trisaccharides containing type II carbohydrate sequence (Galβ1→4GlcNAc). On the contrary, type

II sugar sequence is more active to PNA than type I (51,57,67). Galβ1→4(3)GalNAc related oligosaccharides are inactive to MPL. MPL precipitates well with the first Smith degraded and mild acid hydrolyzed product of A- and B-active substances (48). This is explained by the ability of MPL to react well with both Galβ1→3GalNAcα1→Ser(Thr) and GalNAcα1→Ser(Thr) of the protein core, i.e. both lectin T and Tn determinants. This is further confirmed by a good precipitation of MPL with the antifreeze glycoproteins containing repeating units of Galβ1→3GalNAc as well as armadillo salivary glycoprotein (Fraction A), which has GalNAcα1→Ser(Thr) as carbohydrate side chains.

The reactivity of PNA with the antifreeze glycoproteins and with the Smith degraded and mild acid hydrolyzed products of blood group substances revealed that PNA specificity is directed toward Galβ1→3GalNAc and Galβ1→4(3)GlcNAc (67).

As mentioned above, the inhibitory profile of SJL is similar to that of BPL. However, the reactivity of SJL with glycoproteins differs from that of PNA, BPL, and MPL (see Section C). The broad reactivity of BPL with glycoproteins is supported by its specificity with various oligosaccharides (51). Pneumococcus type XIV polysaccharide with I Ma activity is precipitated strongly by BPL but not at all by SJL (51,57). Both BPL and SJL react equally well with the I Ma-active trisaccharide Galβ1→4GlcNAcβ1→6Gal and *N*-acetyllactosamine. This indicates that the reactivity of these lectins with the I Ma-active trisaccharide mainly involves the *N*-acetyllactosamine moiety of the I Ma determinant. The non- or weak reactivity of SJL with the antifreeze glycoproteins, the Smith degraded, and mild acid hydrolyzed products of A- and B-active substances still remains unexplained (Table VI). It is possible that some unknown factor(s) such as LFucα1→2 linked to the subterminal Gal of the sugar chain is(are) also involved in precipitation of SJL by glycoproteins.

RCA_2 is most specific for GalNAc among the monosaccharides tested (58A,88), but it is most reactive with glycopeptides containing Galβ1→3GalNAc (58). The decreasing order of the carbohydrate specifities is: (a) [Galβ1→3GalNAcα1→3Ser or Thr]$_4$ glycopeptides >> (b) Triantennary oligosaccharides containing Galβ1→4GlcNAc at the non-reducing end > (c) [Galβ1→3GalNAcα1→Ser or Thr]$_2$ glycopeptides with spacing between carbohydrate side chains, Galβ1→3GalNAcα1→Ser >> (d) GalNAcα1→Ser or Thr. The number and location of the Galβ1→3GalNAc units along the peptide backbone is very important to the degree of binding. Removal of Gal residues from the glycopeptides significantly decreases lectin binding ability (58). RCA_2 also reacts relatively strongly with Galβ1→4GlcNAc (type II sequence), especially when occurring within bi- and tri-branched oligosaccharides (58,88). A comparison of the reactivities between Galβ1→3GalNAc and Galβ1→4GlcNAc residues by QPA has not yet been studied.

AIL, *Artocarpus integrifolia* agglutinin, binds the T-antigenic disaccharide very strongly but binds extremely poorly to other disaccharides such as lactose, *N*-acetyl-lactosamine, and Galβ1→3GlcNAc, which are topographically similar to Galβ1→3GalNAc (68).

ALL, *Artocarpus lakoocha* lectin, Galβ1→3GalNAcα1→*O*$(CH_2)_2NHCO(CH_2)_7COOCH_3$, is the best and about twice as potent as only the disaccharide itself. The order of decreasing potency is Galβ1→3GalNAcα1→*O*$(CH_2)_2NHCO(CH_2)_7COOCH_3$ > Galβ1→3GalNAc > Melibiose > Raffinose > Gal, GalNAc (69B,69C and personal communication).

(5) Lectins specific for Galβ1→3(4)GlcNAc

ECL and TAL are specific for GalNAc and β1→4 linked oligomers of GlcNAc, respectively. When the blood group type II sequence (Galβ1→4GlcNAc, *N*-acetyllactosamine) is used as inhibitor, it is 35 times more active than GalNAc for ECL and 1.7 times more specific than *N,N'*-diacetylchitobiose for TAL (70,75,88). This accounts for the high reactivity of ECL with blood group I substances exposing the terminal Galβ1→4GlcNAc sequence (70). LFuc linked to Galβ1→4GlcNAc greatly diminishes inhibitory activity, consistent with the relative inability of H substances to precipitate these lectins.

The inhibitory abilities of linear GlcNAcβ1→4 oligosaccharides to TAL increase with the number of sugar residues up to tetrasaccharide (*N,N',N'',N'''*-tetraacetylchitotetraose)(75). Branched oligosaccharides containing *N*-acetyllactosamine units are the most complementary to the combining sites of both ECL and TAL (70,75). The best inhibitors to ECL range from bi- to tetra-antennary oligosaccharides containing two to four Galβ1→4GlcNAc residues at the nonreducing end (70), while the best one to TAL is a biantennary penta-2,6[4] (75). Unlike ECL, TAL shows a remarkable difference in inhibitory potency between Galβ1→4GlcNAcβ1→2Man, and Galβ1→4GlcNAcβ1→6Man (70,75).

Among the various branched oligosaccharides tested, the tri- and tetraantennary oligosaccharides Nona II[4] and Undeca[4] which contain a Penta-2,6 moiety are about 1/2 as active to TAL as Penta-2,6[4] itself and 200 times more active than *N,N'*-diacetylchitobiose (75). Other branched oligosaccharides are much less active than Penta-2,6. These results indicate that a biantennary structure of Penta-2,6 is the most specific for the combining site of TAL (89).

The carbohydrate specificities of TAL and ECL can be summarized in decreasing order as follows:

TAL, Penta-2,6 >> Penta-2,4 > [GlcNAcβ1→4]$_4$ > [GlcNAcβ1→4]$_3$ > Galβ1→4GlcNAc (*N*-acetyllactosamine), GlcNAcβ1→4GlcNAc and Galβ1→4GlcNAcβ1→6Man > Manα1→4Manβ1→2GlcNAc > GlcNAcβ1→4Glc, More information is available in reference 75.

ECL, Tetra and tri-antennary oligosaccharides containing Galβ1→4GlcNAc linked units at nonreducing ends > Galβ1→4GlcNAcβ1→6Gal (Blood group I Ma specific) > Galβ1→4GlcNAc > Galβ1→3GlcNAc >> Lactose >>> Galβ1→6Gal and, Galα1→3Gal > Raffinose (70).

When comparing the binding affinities of these two lectins, they show differences and TAL behaves with dual binding properties,i.e. specificities for Galβ1→4GlcNAc and [GlcNAcβ1→4]$_4$.

4

Penta-2,6

Galβ1→4GlcNAcβ1,2↓
Man
Galβ1→4GlcNAcβ1,6↑

Nona II

Galβ1→4GlcNAcβ1→4Manα1,3↓
Man
Galβ1→4GlcNAcβ1→6Manα1,6↑
Galβ1→4GlcNAcβ1,2↑

Undeca

Galβ1→4GlcNAcβ1,4↓
Galβ1→4GlcNAcβ1→2Manα1,3↓
Galβ1→4GlcNAcβ1,6↓ Man
Manα1,6↑
Galβ1→4GlcNAcβ1,2↑

Alternatively, the specificity of TAL for Galβ1→4GlcNAc could be mainly for GlcNAc.

RCA_1 is most specific for Galβ1→4GlcNAcβ1→6Gal (human blood group I Ma-active trisaccharide); branched oligosaccharides with Galβ1→4GlcNAc at the nonreducing end are more potent than a linear human blood group I Ma-active trisaccharide or *N*-acetyllactosamine (89).

RCA_1 also reacts with Galβ1→3GalNAcα1→Ser(Thr) of the peptide chain. However, the difference in reactivity between the lectin T determinant and lectin I and II determinants remains to be established.

Detailed information about differential binding properties of RCA_1 and RCA_2 can be obtained from references 58 and 88.

The most reactive monosaccharide with GCL is GalNAc (72), but Galβ1→4GlcNAc and Galβ1→3GlcNAc are more reactive (72). Lactose (Galβ1→4Glc) is less active than these two disaccharides, indicating that the *N*-acetylamide group in C-2 is important for binding.

(6) Lectins specific for Galα1→3Gal

Only GSI-B_4 has been thoroughly studied. The combining site of B_4 was found to be strictly blood group B determinant specific. The most active disaccharide is Galα1→3Gal. Substitution of an αLFuc group on the subterminal Gal in a tetrasaccharide of B sequence (Galα1→3[LFucα1→2]Galβ1→4GlcNAcβ1→R) decreased the binding affinity about 3.5 fold (2,35).

IV-C. Reactivities of GalNAc-and Gal-specific lectins with glycoproteins

Although the combining sites of lectins can be elucidated by quantitative precipitin-inhibition assays, the actual lectin glycoprotein interaction is more complicated. The factors in interactions not only involve the combining sites, but also other parts of the molecule. Therefore, studies on the mode of interaction between lectins and various well-defined glycoproteins, polysaccharides, and human blood group active glycoproteins should provide further information about the binding affinities of lectins such as steric and/or charge effects on the lectin-glycoconjugate complex as well as the solubility of the complex.

Binding characteristics of lectins as demonstrated by quantitative precipitin assays are shown in Table V to VII. More than two dozen glycoproteins or polysaccharides ranging from simple to complex carbohydrates as well as their chemically modified products have been tested. Among these, the carbohydrate moieties of five samples are well established - fetuin, active and inactive antifreeze glycoproteins, and ovine and porcine salivary glycoproteins. All of these contribute important clues for grouping and differentiating various lectins. The antifreeze glycoproteins, which are composed of repeating units of the diglycosyl-tripeptide (Galβ1→3GalNAcα1→Thr-Ala-Ala), are unique reagents for detecting the T antigen structure (samples 23 and 24 in Table V). The desialized ovine and armadillo salivary glycoproteins are ideal reagents for studying the binding properties of Tn structure (Table V, sample 19). The asialo-porcine salivary glycoprotein has a suitable structure for testing three lectin determinants - A, T, and Tn structures (Table V, #20). Mild acid hydrolyzed fetuin is an ideal reagent for T and I(II) lectin determinants (Table V, sample #22).

More than twenty samples of blood groups A, B, H, Le^a, Le^b, and I active glycoproteins purified from human ovarian cyst fluid[5] (98-103), saliva, and other animal sources (104-106), as well as their mild acid hydrolyzed or Smith degraded products (79,96,97) have also been chosen to examine reaction profiles. These complex glycoproteins are suitable ligands for understanding the interaction of lectins with a broad

---

[5] Regardless of A, B, H, or Le^a, Le^b and I activity, water-soluble blood group substances purified from human ovarian cyst fluid have a similar overall structure. They are polydispersed macromolecules (Mr 3.0×10^5 to 1.0×10^6) of similar composition (75 to 85% carbohydrate, 15 to 20% protein): they all consist of multiple heterosaccharide (branched) chains attached by an *O*-glycosidic linkage at their internal reducing ends to serine or threonine of the polypeptide backbone. These carbohydrate side chains may range in size from one to more than twelve sugars (Figs. 2 and 3). More than two thirds of these side chain residues are smaller than twelve sugars (Fig. 3), which can result from incomplete biosynthesis or by degradation in the cyst cavity. With a few exceptions, the carbohydrate binding determinants of glycoprotein are located at the non-reducing ends of carbohydrate side chains (18). See lectin determinant in Fig. 1.

range of lectin determinants - A, A_f, B, I/II, T, and Tn (Figs. 2 and 3). The carbohydrate chains of these glycoproteins are all *O*-glycosidically linked through the GalNAc to the peptidyl Ser or Thr in the protein core. The carbohydrate side chains comprise a heterogenous population varying in size from one to more than twelve sugars although most are less than six units in length (Fig. 3). With a few exceptions, the lectin binding determinants are located at the nonreducing end of the carbohydrate side chains. A representive branched core carbohydrate moiety of human ovarian cyst fluid glycoproteins is illustrated in Fig. 2. It represents an over-all structure derived from its smaller oligosaccharide elements (107,108). This core structure is found in precursor blood group glycoproteins and is easily prepared by either Smith degradation (79) or mild acid hydrolysis (96,97). Galβ1→4GlcNAcβ1→ linked residues at the non-reducing ends of branched chains (II, III, and IV in Fig. 2) react with both antisera to type XIV pneumococcal capsular polysaccharide (sample 5 in Table VI) and human blood group I Ma serum. The residues responsible for A, B, H, Le^a, and Le^b activities are demonstrated by adding a particular sugar to the appropriate location. The blood group active glycoprotein from other sources (as listed in Table V #9 and Table VI #13, 18, and 23) have a similar schematic structure as glycoproteins purified from human ovarian cyst fluid, but probably different linkages in the carbohydrate core structure and/or the branches (13).

(1) Lectins specific for GalNAcα1→3GalNAc (**F**-sequence)

Almost all of the glycoconjugates containing the disaccharide GalNAcα1→3GalNAc as the terminal sugar sequence at the nonreducing ends are found in glycolipids. Therefore, the quantitative precipitin data of **F**-specific lectins with Forssman active complex cannot be obtained. However, all of these lectins also react with the human blood group A structure, which is available in large quantities from a variety of sources (98). As a result, blood group A active glycoproteins, especially A_1 type, are major probes for studying differential lectin-binding affinities.

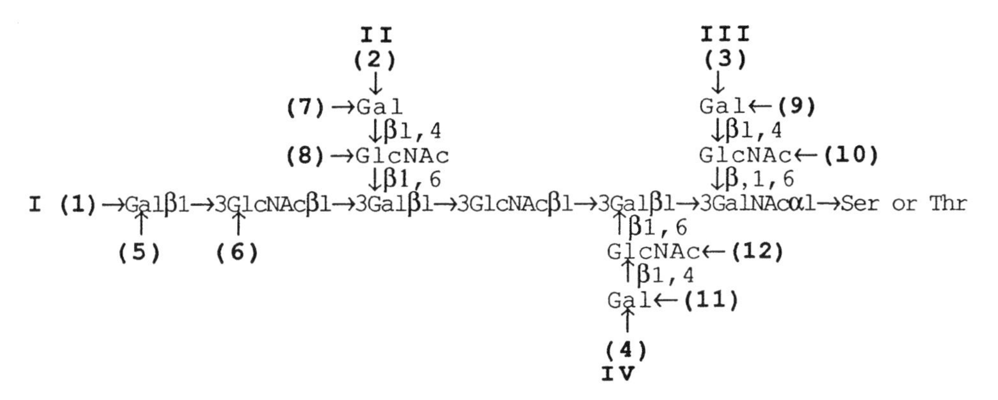

Proposed internal portion of carbohydrate side chains of blood group active glycoproteins.

Samples in Table V	Blood group active glyco-protein purified from human ovarian cyst fluid	Human blood group determinant present	Sugar added	Site of Addition
2	MSS 1st Smith			
8	Beach phenol insoluble P1	Ii		
10	Tij II phenol insoluble	pneumococcus type		
11	Tij II 20% 2X	XIV polysaccharide		
12	Tij II 10% 2X	antigenic determinant	None	
14	JS 1st Smith			
18	OG 20%			
17	N-1 phenol insoluble	Le^a	LFucα1→4	(6)
		unknown	LFucα1→3	(8), (10) & possible (12)
13	JS phenol insoluble	H, Le^b	LFucα1→2	(5), (7), (9),
16	Tighe phenol insoluble		as in sample 17	(6), (8), (10) (11), and (12)
1	MSS, native		GalNAcα1→3	(1), (2), (3)
4	MSM	A	and as in	and/or (4)
5	Cyst 9		samples 16	(5), (7), (9)
6	Cyst 14		and 17	and (12)
7	Beach phenol insoluble		Galα1→3 as	(1), (2), (3)
10	Tij II phenol insoluble		in samples	and/or (4)
11	Tij II 20% 2X	B	16 and 17	(5), (7), (9)
12	Tij II 10% 2X			(6), (8), (10) and (12)

Fig 2. Structural relationship of blood group antigenic determinants. The structure shown is the proposed internal portion of the carabohydrate moiety of blood group substances to which the residues responsible for A, B, H, Le^a, and Le^b activities are attached. This structure represents precursor blood group active glycoproteins (102) and can be prepared by Smith degradation of A, B, H active glycoproteins, purified from human ovarian cyst fluids (79). They are the sources of lectin A/Af, B, I/II, T, and Tn determinants (See Fig. 1). The residues responsible for A, B, H, Le^a, Le^b, and I activities are numbered in parentheses. An intact megalosaccharide of twenty-four sugars has not been isolated. However, most of the carbohydrate chains isolated are parts of this structure. The designation of blood group substances is described in Footnote b of Table V (modified from Mol. Cell. Biochem. 61: 136, Fig. 2., 1984).

Size and Mole %[a]	Structure of carbohydrate chains	Lectin Determinants
Monosaccharide[b] (21%)	1. GalNAcα1→Ser(Thr) of the protein core	Tn
Disaccharide (15.4%)	2. Galβ1→3GalNAcα1→Ser(Thr)	T
	3. GlcNAcβ1,6↓ (major) GalNAcα1→Ser(Thr)	
Trisaccharides (24.9%)	4. GlcNAcβ1,6↓ Galβ1→3GalNAcα1→Ser(Thr)	T(?)
	5. Galβ1→4GlcNAcβ1,6↓ GalNAcα1→Ser(Thr)	II
	6. GlcNAcβ1→3Galβ1→3GalNAcα1→Ser(Thr)	
	7. Galβ1→3GlcNAcβ1,6(3)↓ GalNAcα1→Ser(Thr)	I
	8. Galβ1→3GlcNAcβ1,6↓ R[c]→GalNAcα1→Ser(Thr)	I
Tetrasaccharide (18.0%)	9. Galβ1→3GlcNAcβ1→3Galβ1→3GalNAcα1→Ser(Thr) (major)	I
	10. Galβ1→4GlcNAcβ1,6↓ Galβ1→3GalNAcα1→*O* to Ser(Thr)	II and T (?)
	11. GlcNAβ1→3Galβ1→4GlcNAcβ1,6↓ GalNAcα1→Ser(Thr)	
Pentasaccharides (4.2 mole%)	12. Galβ1→3GlcNAcβ1→3Galβ1→4GlcNAcβ1,6↓ GalNAcα1→Ser(Thr)	I
Hexasaccharides	13. Galβ1→3GlcNAcβ1→3Galβ1→3GlcNAcβ1→3Galβ1→3GalNAcα1→Ser(Thr)	I
(6.1%)	14. Galβ1→3GlcNAcβ1→3Galβ1→3GalNAcα1→Ser(Thr) Galβ1→4GlcNAcβ1,6↑	I&II
	15. Galβ1→4GlcNAcβ1,6↓ Galβ1→3GlcNAcβ1→3Galβ1→3GalNAcα1→Ser(Thr)	I&II

Fig. 3. Heterogeneity and structure of major carbohydrate side chains of MSS 1st Smith degraded blood group A active glycoprotein, purified from human ovarian cyst fluid. This fraction represents 77.1% of total carbohydrate residues and its chain size ranges from monosaccharide (Tn sequence) up to hexasaccharides (79,108,108A). Three kinds of lectin receptors are available - Tn, T, and I/II (Galβ1→3(4)GlcNAcβ1→ linked residues) which play an important role in the lectin-carbohydrate interaction. Human blood group Ii active glycoproteins (102) and mild acid hydrolyzed (pH 1.5, 100°C for 2 hours) blood group A and B active glycoproteins prepared from human ovarian cyst fluids (96,97) give a similar heterogeneity profile.

[a]About 10% is chromogen. [b]Chromogen was excluded.
[c]R, carbohydrate residues, cleaved by peeling reaction.

Table V. Comparison of precipitating activities of GalNAc-specific lectins with blood group-active substances and glycoproteins[a]

#	Glycoprotein	DBL (19)	HPL (26,28)	WFL (29)	SBA (32)	LBL (43)	GSI-A4 (35)	VVL (78)	MPL (48)	BPL (51)
1.	Cyst MSS, native (A_1)[b]	4+	5+	4+	3+	3+	5+	2+	±	+
2.	1st Smith degraded[c]	–	2+	4+	3+	ND	ND	2+	4+	4+
3.	2nd Smith degraded[c]	–	4+	3+	ND	ND	ND	2+	3+	3+
4.	Cyst MSM (A_1)[b]	4+	4+	4+	4+	3+	ND	ND	ND	3+
5.	Cyst 9 (A_1)[b]	5+	5+	4+	3+	2+	5+	2+	±	±
6.	Cyst 14 (A_2)[b]	3+	5+	4+	2+	+	3+	3+	2+	4+
7.	Cyst Beach phenol insoluble (B)[b]	–	5+	2+	+	±	5+	2+	2+	±
8.	P1 (mild acid hydrolyzed[d])	±	5+	4+	3+	ND	ND	ND	4+	+
9.	Horse 4 25% (B)	ND	ND	4+	+	ND	4+	ND	3+	
10.	Cyst Tij II phenol insoluble (B, I)[b]	ND	ND	3+	ND	ND	2+	2+	3+	4+
11.	Cyst Tij II 20% 2X (B,I)[b]	ND	ND	4+	ND	ND	4+	4+	3+	4+
12.	Cyst Tij II 10% 2X (B,I)[b]	ND	ND	ND	ND	ND	ND	3+	3+	4+
13.	Cyst JS phenol insoluble (Le^b, H)[b]	±	5+	2+	±	–	–	+	3+	3+
14.	1st Smith degraded[c]	–	ND	ND	±	ND	ND	ND	4+	2+
15.	2nd Smith degraded[c]	–	ND	ND	2+	ND	ND	ND	3+	ND
16.	Cyst Tighe phenol insoluble (H)[b]	–	5+	2+	–	ND	±	ND	–	4+
17.	Cyst N-1 phenol insoluble (Le^a)[b]	ND	5+	4+	3+	ND	±	±	±	4+
18.	Cyst OG 20% from 10% (I)[b]	±	5+	4+	ND	ND	5+	5+	4+	5+
19.	Asialo-ovine salivary glycoprotein[e]	±	4+	3+	2+[i]	ND	ND	5+	4+	5+
20.	Asialo-porcine salivary glycoprotein[f]	3+	5+	5+	ND	ND	ND	4+	4+	ND
21.	Fetuin[g]	–	–	–	ND	ND	ND	–	–	±
22.	Asialo-fetuin[g]	±	±	5+	ND	ND	ND	±	2+	5+
23.	Active antifreeze glycoprotein[h]	–	+	±	ND	ND	ND	–	4+	4+
24.	Inactive antifreeze[h]	–	–	–	ND	ND	ND	–	4+	3+

[a]6.6 μgN of DBL; 6.5 μgN of HPL; 6.6 μgN of WFL; 9.1 μgN of SBA; 5.3 μgN of GSI-A4; 6.0 μgN of VVL; 6.0 μg of MPL and 6.1 μg of BPL were used for the quantitative precipitin assays. Glycoproteins were tested up to 60 μg. –, less than 3% of lectin precipitated; ±, less than 15% lectin precipitated; +, between 16-25% of lectin precipitated; ++, between 26-50% of lectin precipitated; +++, between 51-75% of lectin precipitated; ++++, between 76-95% of lectin precipitated; +++++, completely precipitated; ND, not determined.

[b] to [i] shown on next two pages

(continued)

Legend to Table V (continued)

[b]The blood group substances were purified from human ovarian cyst fluid by digestion with pepsin and precipitation with ethanol; the dried ethanol precipitates were extracted with 90% phenol, the insoluble fraction being named after the corresponding blood group substance (e.g. Cyst Beach phenol insoluble). The supernatant was fractionally precipitated by addition of 50% ethanol in 90% phenol to the indicated concentrations. The designation 10 or 20% (ppt) denotes a fraction precipitated from phenol at an ethanol concentration of 10 or 20%; 2X signifies that a second phenol extraction and ethanol precipitation were carried out (e.g., Cyst OG 20% 2X). The A, B, H, Le^a, Le^b, and I in parentheses indicate blood group activities. General structural concepts of blood group active glycoproteins are described in Fig. 2.

[c]The terminal sugars at the nonreducing ends were removed by Smith degradation (79).

[d]P1 fractions represent the non-dialyzable portion of the blood group substance after mild acid hydrolysis at pH 1.5 to 2.0 at 100°C for 2 hr, which removed most of the LFucα1→ linked end groups as well as some blood group A and B active oligosaccharide side chains (96,97).

[e]Ovine salivary glycoprotein (110) with a molecular weight of 343,000 is composed of 40% protein, 30% sialic acid, and 22% GalNAc with small amounts of Gal and LFuc. The carbohydrate moiety of this glycoprotein is *O*-glycosidically linked through GalNAc at the reducing end of the carbohydrate side chain to Ser or Thr of the protein core. The carbohydrate side chain of ovine salivary glycoprotein is over 75% to 94% NeuNAcα2→6GalNAc and some larger oligosaccharides containing GalNAc, Gal, Fuc, and NeuNAc (110). The carbohydrate structure of the desialized ovine salivary mucus glycoproteins is mostly GalNAcα1→*O* to the Ser/Thr of the protein core.

[f]Porcine salivary glycoprotein with a molecular weight of 6.3×10^5 is composed of 35% protein, 15% sialic acid, 7.3% LFuc , 10% Gal and 26% HexN. The carbohydrate side chains of this glycoprotein are also *O*-glycosidically linked through GalNAc at the reducing end to Ser or Thr of the protein core. Twelve kinds of carbohydrate side chains have been isolated, composed of one to five sugar residues with Galβ1→3GalNAcα1→Thr or Ser as the carbohydrate core region. Substitution by NeuAc, NeuGc, LFuc, or GalNAc exhibits blood group A and H activities, respectively. The mild acid treated mucin contains mainly Galβ1→3GalNAcα1→ Ser or Thr (lectin T determinant) and GalNAcα1→ Ser or Thr (lectin Tn determinant)(110).

[g]Fetuin, the major glycoprotein in fetal calf serum (93) with a molecular weight of 48,400, is composed of 78% amino acids, 8.7% sialic acid, 6.3% hexosamine, and 8.3% neutral sugar (94). It has three oligosaccharide side chains with two different structures *O*-glycosidically linked to Ser or Thr of the protein core as well as three identical carbohydrate side chains per molecule *N*-glycosidically linked

(Legend to Table V continued on next page)

Legend to Table V (continued)

to asparagine. The structures of the carbohydrate side chains are (95):

$$\begin{array}{l} \text{NeuAc}\alpha 2\rightarrow 3\text{Gal}\beta 1\rightarrow 3\text{GalNAc}\alpha 1\rightarrow \text{OH of Ser of Thr,} \\ \qquad\qquad \text{NeuNAc}\alpha 2,6\uparrow \end{array}$$

$$\text{NeuAc}\alpha 2\rightarrow 3\text{Gal}\beta 1\rightarrow 3\text{GalNAc}\alpha 1\rightarrow \text{OH of Ser of Thr,}$$

and

$$\begin{array}{ll} \text{I} & \text{NeuAc}\alpha 2\rightarrow 3\text{Gal}\beta 1\rightarrow 4\text{GlcNAc}\beta 1\rightarrow 2\text{Man}\alpha 1,6\downarrow \\ \text{II} & \text{NeuAc}\alpha 2\rightarrow 6\text{Gal}\beta 1\rightarrow 4\text{GlcNAc}\beta 1\rightarrow 2\text{Man}\alpha 1\rightarrow 3\text{Man}\beta 1\rightarrow 4\text{GlcNAc}\beta 1\rightarrow 4\text{GlcNAc}\beta 1\text{-}N\text{-Asn} \\ \text{III} & \text{NeuAc}\alpha 2\rightarrow 3\text{Gal}\beta 1\rightarrow 4(3)^{*}\text{GlcNAc}\beta 1,4\uparrow \end{array}$$

Recently, it was reported that Branch III in the *N*-linked triantennary carbohydrate chain can also be Galβ1→3*GalNAc linkage(95A) and an O-glycosidically linked hexasaccharide (NeuAcα2→3Galβ1→3[NeuAcα2→3Galβ1→4GlcNAcβ1→6]GalAc-ol) was also found in this glycoprotein. Carbohydrate side chains of mild acid treated (desialized) fetuin contain no NeuAc in the above structures (95B).

[h]The active antifreeze glycoproteins from the Antarctic fish *Trematomus borchgrevinki* are composed of repeating units of diglycosyl-tripeptide, namely,

$$\left[\begin{array}{l} \text{-Ala-Ala-Thr-} \\ \qquad\qquad \uparrow\alpha 1 \\ \text{Gal}p\beta 1\rightarrow 3\text{GalNAc}p \end{array}\right]_n$$

The molecular weights of these glycoproteins vary from 10,500 to 21,000. The inactive antifreeze glycoproteins have a molecular weight ranging from 2600 to 3800 (111). Their structures are similar to the active preparations except that proline is present in the peptide chain (111). The inactive glycoproteins were initially thought not to lower the freezing point, but in later work they were reported to have some antifreeze activity (112).

[i]Unpublished data.

---

[6]Group C streptococcal polysaccharides have terminal nonreducing GalNAc residues linked to rhamnose backbone (109).

[7]The primary structure of classes A, B, BF, C, and CF carbohydrate units of glycosylation site of human plasma α_1-acidic glycoprotein (113) is indicated in the following structure for asialo-orosomucoid.

$$\begin{array}{l} \qquad\qquad\qquad\qquad \text{LFuc} \\ \qquad\qquad\qquad\qquad \downarrow\alpha 1,3 \\ \qquad\qquad \text{Gal}\beta 1\rightarrow 4\text{GlcNAc} \\ \qquad\qquad\qquad\qquad \downarrow\beta 1,4 \\ \text{Gal}\beta 1\rightarrow 4\text{GlcNAc}\beta 1\rightarrow 2\text{Man} \\ \qquad\qquad\qquad\qquad \downarrow\alpha 1,3 \\ \qquad\qquad\qquad\qquad \text{Man}\beta 1\rightarrow 4\text{GlcNAc}\beta 1\rightarrow 4\text{GlcNAc}\beta 1\text{-Asn} \\ \qquad\qquad\qquad\qquad \uparrow\alpha 1,6 \\ \text{Gal}\beta 1\rightarrow 4\text{GlcNAc}\beta 1\rightarrow 2\text{Man} \\ \qquad\qquad\qquad\qquad \uparrow\beta 1,6 \\ \qquad\qquad \text{Gal}\beta 1\rightarrow 4\text{GlcNAc} \end{array}$$

(Brackets on the left group the units as CF, C, BF, B, A.)

The carbohydrate units of this asialo-glycoprotein can be grouped into compounds with biantennary, triantennary and tetraanternary structures with fucose residue. The structural variability of the carbohydrate units of glycosylation site I and also of glycosylation sites II and V (114) accounts largely for the microhetrogeneity of α_1-acid glycoprotein.

DBL and HPL are the most specific lectins for human type A erythrocytes but they differ greatly in their reactivities to glycoproteins and polysaccharides (19,26,28). DBL reacts strongly only with glycoconjugates possessing a terminal GalNAc residue such as blood group A substances, asialo-porcine salivary glycoprotein with A activity, and group C polysaccharide[6]. However, it reacts neither with their Smith degraded products (in which the A determinants GalNAcα1→ as the nonreducing end are abolished) nor with other blood group substances and glycoproteins having a different terminal sugar (Table V). On the other hand, HPL shows broad specificities and reacts with a variety of complex carbohydrates such as blood group A, B, H, Le^a, and precursor I substances, the mild acid hydrolyzed product of B-active glycoprotein, desialyzed ovine and porcine salivary glycoproteins, guaran, group C polysaccharide, and staphylococcal teichoic acids (20), which indicates that HPL reacts with both A and Tn types of GalNAcα1→ linked residues. Both lectins are unreactive towards active and inactive antifreeze glycoproteins, as well as native and asialo-fetuin (#23 in Table V). However, substantial differences exist between their reactivities with the asialo-ovine salivary glycoprotein (21). This desialized glyco-protein, which contains a substantial proportion of GalNAcα1→Ser(Thr) of the protein core (110), reacts strongly with HPL but not at all with DBL. This data indicates that DBL recognizes only the terminal GalNAc at the nonreducing end(s) (A and F structure) but does not react with internal GalNAc residues linked to the protein core (Tn determinant) that react with HPL. This accounts for the broad specificities of HPL for various glycoproteins.

Like HPL, WFL strongly reacts with all blood group A, B, H, Le^a, and precursor I substances tested and with asialo-glycoproteins such as ovine and porcine salivary glyco-proteins. Asialo-fetuin is one of the best reagents to differentiate the carbohydrate specificities of WFL and HPL as its lectin determinants (T and II) have been well characterized (93-95). Asialo-fetuin is precipitated well by WFL but not at all by HPL (Table V). This asialo-glycoprotein is composed of three disaccharide side chains of Galβ1→3GalNAc residues per molecule (*O*-glycosidically linked to Ser or Thr of the protein core), as well as three identical carbohydrate side chains of three *N*-acetyllactosamine residues at the nonreducing end per molecule (*N*-glycosidically linked to asparagine)(93-95). WFL also reacts well with asialo-orosomucoid[7] which has multiple *N*-acetyllactosamine residues at the nonreducing ends of complex carbohydrate chains. The strong precipitability of the two desialized glycoproteins by WFL is attributed to the lectin I and II determinants.

Based on the data obtained from QPA (Table V) and QPIA (Table II), the binding properties of DBL, HPL, and WFL, both in terms of active carbohydrate sequences (lectin determinants) and active glycoproteins, are summarized here.

Active Carbohydrate Sequences (Lectin determinants)	Active Glycoproteins
DBL	
1. GalNAcα1→3GalNAc (Forrsman, **F**)	
2. GalNAcα1→3Gal (Human blood group A determinant) at nonreducing end of carbohydrate chain, especially fucosyl A determinant (A_f)	a. BGS(Blood group substances) with blood group A_1 (A) activity only.
HPL	
1. GalNAcα1→3GalNAc (Forssman, **F**)	
2. GalNAcα1→3Gal (A determinant)	a. and
also	b. BGS with BI, H, I
3. GalNAcα1→*O* to Ser or Thr of protein core (Tn determinant)	c. Smith degraded BGS and mild acid treated BGS (P-1 substances)
WFL	
1 to 3 (F,A,Tn) and also Galβ1→3(4)GlcNAcβ1→ (Type I and II structure)	a, b, c and also d. asialo fetuin and asialo-orosomucoid

(2) Lectins specific for GalNAcα1→3Gal (**A** sequence)

WFL, SBA, LBL, GSI-A_4, and VVL are most specific for human A determinant carbohydrate sequences containing GalNAcα1→3Gal, GalNAcα1→3Galβ1→3GlcNAc, or GalNAcα1→3[Fucα1→2]Galβ1→6R, etc. and are defined as GalNAcα1→3Gal specific (**A**) lectins and GalNAcα1→3[LFucα1→2]Gal coded $\mathbf{A_f}$. The reaction profile of WFL has been demonstrated in a previous section. Therefore, only SBA, LBL, GSI-A_4, and VVL will be considered here.

SBA, which shows nonspecificity for human erythrocytes (A>B,O), reacts to various extents with blood group A, Le^a, and precursor I substances, the mild acid hydrolyzed product of B substance, porcine salivary glycoprotein, and group C polysaccharide but not with B and H substances (32). As shown in Table V, its precipitin patterns with blood group substances are very similar to those of WFL (29) but even the most reactive substances precipitated less than 80% of SBA used (32). The best precipitinogens to SBA were the A- and I-active substances and the mild acid hydrolyzed product of B substances possessing a terminal GalNAc or Galβ1→4GlcNAc. QPA and QPIA results suggest that SBA has some affinity for the terminal α1-linked GalNAc (lectin A determinant), Galβ1→4(3)GlcNAc linked residues (lectin I/II determinant) and GalNAcα1→Ser(Thr) of the protein core (lectin Tn determinant).

LBL precipitates best with blood group A_1 substances from hog gastric mucosa (45) and human ovarian cyst and moderately with A_2 substances. B substances precipitate it very poorly and H, Le^a, Le^b, and precursor I substances do not react (Table V). Blood group A_1 and A_2 substances react with LBL to varying extents, the differences being attributable to heterogeneity resulting from incomplete biosynthesis of the carbohydrate chains. The lectin agglutinates human type A erythrocytes strongly but type B erythrocytes weakly, which is similar to the precipitation profile observed with DBL (43). The precipitating abilities of LBL with blood group substances are quite similar to its hemagglutinating potency which indicates that LBL is specific for the terminal GalNAcα1→R linked sequences at the nonreducing end (A_f structure mainly).

GSI-A_4 precipitates well with human blood groups A_1, A_2, B, and precursor I substances, but blood group A_2 glycoproteins are less efficient than are A_1 glycoproteins. H, Le^a, and Le^b glycoproteins do not react (35). From the reaction profile, it is concluded that GSI-A_4 recognizes most specifically GalNAcα1→ at the nonreducing end (i.e. lectin A determinant). It reacts with terminal nonreducing α-linked Gal (lectin B determinant) moderately.

The precipitin reaction profile of VVL-A_4 has not been studied as the availability of this isolectin is quite limited. The quantitative precipitin reaction profile used (78) assumes that interaction results between glycoproteins with a mixture of the isolectins A_4, B_4, and A_2B_2, with the isolectin B_4 being the dominant component in this mixture.

Thus, only some properties can be assigned to VVL-A_4 based on this profile.

The lectin reacts strongly with both asialo-ovine and porcine salivary glycoproteins. It precipitates to different extents with blood group A_1, A_2, H, B, and precursor I substances from saliva and ovarian cysts and even with different blood group substances of the same blood group specificity indicating the major lectin determinant is not a blood group determinant site, which is in good agreement with the properties of the isolectin mixtures, containing B_4 (lectin Tn determinant) as a major component and A_4 (lectin A determinant) as a minor one.

From QPA data (Table V), it can be concluded that all GalNAcα1→3GalNAcβ1→ and GalNAcα1→3Gal specific lectins (DBL, HPL, WFL, SBA, GSI-A_4, and VVL) react well with cyst MSS, cyst MSM, and Cyst 9 blood group A-active glycoproteins and to various extents with other glycoproteins. The range of the reactivities is dependent upon the number of lectin determinants available as well as the presence of a "help" or "blocking" factor (LFuc?).

The binding properties of GalNAcα1→3Gal specific lectins, based on the active carbohydrate sequences (lectin determinants), are illustrated as follows:

SBA (A, I/II and Tn)

a. GalNAcα1→3Gal (**A** determinant at nonreducing end with no LFucα1→2 substituted on subterminal Gal) > its fucosylated derivative (A_f).

b. Galβ1→3(4)GlcNAc residue (Type **I** and **II** sequences) at the nonreducing end of the carbohydrate chain.

c. GalNAcα1→*O* to Ser or Thr of the protein core (Tn sequence).

GSI-A_4 (A > A_f >> B)

a. GalNAcα1→3Gal (**A** determinant at nonreducing end with no LFucα1→2 substituted on subterminal Gal) > its fucosylated derivative (A_f).

d. Galα1→3Gal (Blood group B active structure) at the nonreducing end of the carbohydrate chain.

LBL (A_f mainly)

GalNAcα1→3(LFucα1→2)Galβ1→3GlcNAcβ1→3Galβ1→4Glc (Blood group A active hexasaccharide, Hexa-A_f) >
GalNAcα1→3(LFucα1→2)Galβ1→4GlcN (Tetra-A_f) >
GalNAcα1→3(LFucα1→2)Gal (A_f) >> GalNAcα1→3Galβ1→4GlcNAc (A)
LFucα1→2 to subterminal Gal residue is very important for binding.

VVL (A > A_f and Tn)
(A mixture of three isolectins A_4, B_4, and A_2B_2)

a. GalNAcα1→3Gal (lectin A determinant) and

c. GalNAcα1→Ser(Thr) of the protein core (lectin T determinant)

(3) Tn specific lectins

VVL-B_4 and SSL have been demonstrated to be most specific for GalNAcα1→ Ser or Thr (Tn determinant)(46,47), but their reactivities with other glycoproteins have not been studied extensively. Four other lectins, WFL, HPL, MPL, and BPL, also show some specificity for the Tn determinant. The reactivities of WFL and HPL have been described in the sections on F and A specific lectins. MPL and BPL possess dual and triple specificities, respectively, which will be discussed in the following paragraph. The Tn specific lectins are expected to react well with the desialized ovine and armadillo salivary mucus glycoproteins as the carbohydrate side chains of both glycoproteins are mainly of the Tn structure (110).

(4) Galβ1→3GalNAc specific lectins
(**T** sequence)

The binding characteristics of PNA, MPL, SJL, and BPL have been intensively investigated by the quantitative precipitin assay (48,51,57,67). Among the simple compounds with well defined structures tested, the antifreeze glycoproteins (samples 1-2 in Table VI), which are composed of repeating units of diglycosyltripeptide (111,112) containing Galβ1→3GalNAc residues as carbohydrate side chains, react strongly with PNA, MPL, and BPL, but do not react with SJL. It is assumed that some other factors are required for the precipitation of SJL. In native fetuin (93-95, Table V, Footnote g) both types of lectin determinants, Galβ1→3GalNAcα1→ linked and Galβ1→4GlcNAcβ1→ linked, are masked by sialic acid. Therefore, access of their sugar determinants to the combining site of the lectin is hindered so that the resulting binding affinity is weak (sample 3 in Table VI). However, the desialized product (sample 4) shows reasonably good reactivity. The antigenic determinant of Type XIV pneumococcal polysaccharide (sample 5), which resides in the Galβ1→4 linked branches to GlcNAc residues of the carbohydrate

Table VI. Comparison of precipitating activities of Galβ1→3GalNAc specific lectins with various glycoproteins and blood group-active substances[a]

Sample tested		% of lectin precipitate[c]			
#	Glycoprotein	PNA (67)	MPL (48)	BPL (51)	SJL (59)
1.	Active antifreeze glycoprotein[h] (M_r 10,500-21,000)	5+	4+	4+	-
2.	Inactive antifreeze glycoprotein[h] (M_r 2,600-3,600)	4+	4+	3+	-
3.	Fetuin[g]	-	-	±	-
4.	Asialo-fetuin[g]	2+	2+	5+	-
5.	Pneumococcus type XIV polysaccharide[i]	-	-	4+	-
6.	Cyst MSS, native (A1)[b]	±	±	+	2+
7.	1st Smith degraded[c]	5+	4+	4+	±
8.	2nd Smith degraded[c]	5+	3+	3+	±
9.	Cyst Mcdon, native (A1)[b]	ND	2+	3+	3+
10.	P1 (mild acid hydrolyzed)[d]	ND	3+	4+	+
11.	Cyst 9 (A1)[b]	-	±	2+	3+
12.	Cyst 14 (A2)[b]	4+	2+	4+	-
13.	Salivary WG phenol insoluble (A2)	±	±	4+	-
14.	Cyst Beach phenol insoluble (B)[b]	±	2+	±	5+
15.	P1 (mild acid hydrolyzed)[d]	5+	4+	5+	2+
16.	Salivary FM phenol insoluble (B)	2+	+	4+	4+
17.	P1 (mild acid hydrolyzed)[d]	5+	3+	4+	±
18.	Cow 21, native (B, I)	5+	_	4+	2+
19.	P1 (mild acid hydrolyzed)d	5+	4+	4+	-
20.	Cyst Tij II phenol insoluble (B, I)[b]	2+	2+	4+	4+
21.	Cyst Tighe phenol insoluble (H)[b]	-	-	4+	-
22.	Cyst JS phenol insoluble (Le[b], H)[b]	2+	+	3+	-
23.	Hog 14 (A + H)	±	±	4+	2+
24.	Cyst N-1 10% 2X (Le[a])[b]	5+	±	4+	2+

[a]6.1 μgN of BPL, 6.0 μgN of MPL, 7.0 μgN of SJL & 6.0 μgN of PNA were used for quantitative precipitin assays. Glycoproteins were tested up to 60 mg. -, less than 3% of lectin precipitated; ±, less than 15% lectin precipitated; +, between 16-25% of lectin precipitated; ++, between 26-50% of lectin precipitated; +++, between 51-75% of lectin precipitated; ++++, between 76-95% of lectin precipitated; +++++, completely precipitated; ND, not determined.

[b,c,d,g,h] are described in Legend for Table V.

[i]Type XIV pneumococcal polysaccharide is composed of the following repeating unit (115).

$$\left[\begin{array}{l} 6\text{GlcNAc}\beta1\rightarrow3\text{Gal}\beta1\rightarrow4\text{Glcp}\beta1\rightarrow \\ \quad\uparrow\beta1,4 \\ \text{Galp} \end{array}\right]_n$$

in which Gal as branches and β1→4 linked to the carbohydrate core structure.

core structure (115) and also shares some human blood group I Ma activity, reacts only with BPL.

From the precipitating profiles of the samples tested, the differential affinity of Galβ1→3GalNAc specific lectins is evident. As summarized in Table VI, PNA, MPL, and BPL share similar precipitin profiles as described below: (i) they precipitate well with active and inactive glycoproteins as well as with the mild acid hydrolyzed product of salivary PM glycoprotein (P1), (ii) they react poorly or weakly with cyst MSS (Blood group A_1 active) glycoprotein, cyst Beach ØOH insoluble (B active) glycoprotein, and native fetuin, and (iii) all three lectins precipitate strongly with sequential Smith degraded products of MSS and with Beach P1 glycoprotein. The precipitating profile of SJL is found to be different from these three lectins. Among the blood group active glycoproteins tested, cyst Beach glycoprotein having blood group B activity is the most active and precipitated over 95% of the lectin. SJL showed substantial differences in precipitin reactions as compared with other B active glycoproteins. Saliva PM and cyst Tij II react strongly and precipitated over 85% of the lectin but Cow 21 gave intermediate reactions. The lectin precipitated moderately with A_1 substances (Cyst 9, MSS, and Mcdon) but weakly with A_2 blood group glycoproteins (Cyst 14 and WG saliva). The first and second stages of the Smith degraded products of MSS and the mild acid hydrolyzed products of blood group A or B glycoproteins (P1 substances) substantially decreased precipitation of the lectin indicating that a mild acid-sensitive compound near the terminal group at the nonreducing end plays an important role in this reaction. SJL shows its unique binding character by its reactivity profile towards cyst Beach phenol insoluble and salivary PM glycoproteins and their mild acid hydrolyzed products (P1), as well as MSS and its Smith products. BPL has a broad range of affinities; its special binding character is demonstrated by its ability to react strongly with pneumococcus Type XIV polysaccharides, salivary WG glycoprotein (A_2) Cyst Tighe (H), and Hog 14 (A+H) glycoproteins. The binding properties of these four Galβ1→3GalNAc specific lectins can also be differentiated by their affinity to the following glycoproteins. Both PNA and BPL, but not MPL, react well with N-1 10% 2X (blood group Le^a active) glycoprotein and Cyst 14 (blood group

A_2 active) glycoproteins. Desialized fetuin completely precipitates BPL, while less than one third of PNA and MPL is precipitated. Native salivary PM glycoproteins react strongly with BPL and SJL, but react only moderately or weakly with PNA and MPL, respectively. These findings, together with differences in their inhibitory profiles (Tables III and IV), further confirm that each lectin has its own characteristic specific binding sites.

BPL shows a broad range of precipitability with glycoproteins (Table VI) (51). Thus, BPL is the lectin of choice to screen for complex carbohydrate chains with Galβ1→3GalNAc, Galβ1→4GlcNAc and GalNAcα1→ linked to peptidyl Ser or Thr (i.e. lectin T, I/II and Tn determinants) whereas MPL is the best probe to verify the Galβ1→3GalNAcα→ and GalNAcα→ linked sequences (lectin T and Tn determinants).

From the combination of the data available from both QPA (Table VI) and QPIA (Table III), the binding properties of PNA, MPL, BPL, and SJL are expressed by both active carbohydrate sequences and active glycoproteins as follows:

Active Carbohydrate Sequences (Lectin determinants)	Active Glycoproteins (In Table VI)
PNA	
1. T sequence and 2. Type I/II structure	a. Active and inactive antifreeze glycoproteins (#1 and #2). b. Smith degraded and mild acid hydrolyzed blood group substances (P-1) (#7, #8, #15, and #19).
MPL	
1. T sequence and 3. Tn structure	a. and b. c. Asialo glycoproteins, prepared from submandibular glands of armadillo, ovine and porcine (unpublished data; for structures, see reference 110).
BPL	
Structures 1 to 3 i.e.lectin T, I/II, and Tn determinants	a. to c. d. Pneumococcus type XIV polysaccharides (#5). e. Cyst Mcdon native (#9); salivary PM phenol insoluble (#16), Cyst Tij II phenol insoluble (#20).
SJL	
Structures 1 and 2 (i.e.lectin T and I/II determinants)*	e. and f. Cyst Beach phenol insoluble (#14), unique glycoprotein for SJL.

*Some unknown mild acid sensitive compounds near the terminal group at the nonreducing end play an important role for binding or precipitating this lectin.

(5) Galβ1→3(4)GlcNAc specific lectins (Lectin **I/II** determinants)

At least four lectins (RCA_1, GCL, ECL, and TAL) are recognized as being the best ones for these disaccharides, especially TAL for Penta-2,6^4 type oligosaccharides and another five lectins (WFL, SBA, BPL, PNA, and SJL) cross-react with these residues (Tables II and III). The precipitability of TAL with glycoproteins has not been as well studied and has been shown to react with asialo-fetuin and asialo-orosomucoid, carcinoembryonic antigen (CEA), and pneumococcus type XIV polysaccharide; the latter is composed of repeating units of a tetrasaccharide with branched Galβ1→4 linked to the GlcNAc of the core structure (Footnote i in Table VI). Therefore, the precipitin patterns of only three lectins (RCA_1, GCL, and ECL) are compared. Of 31 complex carbohydrates tested for RCA_1, all except active and inactive antifreeze glycoproteins, *Streptococcus* group C polysaccharide, and native rat salivary glycoprotein, reacted strongly; 22 (only fourteen are shown in Table VII) completely precipitated the lectin, indicating that RCA_1 has both a broad range of affinities and that its carbohydrate bound complex is of low solubility (89). As shown in Table VII, GCL shares the same reaction profile with RCA_1 except some of them are about 3/5 to 4/5 as active as those of RCA_1. ECL precipitated most strongly with two fractions of a precursor human ovarian cyst blood group substance with I and i activities - Cyst OG 20% from 10% (#17) and Cyst OG 10% from 20% (35). A_1, A_2, B, H, Le^a, and Le^b blood group substances precipitate poorly to moderately and substances of the same blood group activity precipitate the lectin to varying extents. These differences are attributable to heterogeneity resulting from incomplete biosynthesis of carbohydrate chains. ECL precipitates strongly with 1st Smith cycle degraded MSS (#2) due to unmasking of the I/II determinants. The precipitability of this lectin is strongly affected by the volume of the reaction mixture. The reaction profiles among these three lectins with various glycoproteins and polysaccharides demonstrate that each has its own binding properties; especially RCA_1 and ECL can be used for differential purposes.

From the data available from both QPA (Table VII) and QPIA (Table III), the binding properties of RCA_1, ECL, and GCL expressed by both active carbohydrate sequences (lectin determinants) and active glycoproteins, are given as follows.

Active Carbohydrate Sequences (Lectin determinant)	Active Glycoproteins (# in Table VII)
ECL	
1. Type I(II) structure > 2. B determinant and 3. Tn sequence	a. Cyst OG 20% from 10% (#17) b. 1st Smith degraded Cyst MSS (#2)
GCL	
1. and 3. i.e. I/II and Tn determinants	a., b., and c. Cyst MSS, native (#1) d. Pneumococcus type XIV (#20)
RCA_1	
1.to 3., i.e. lectin I(II), B, and Tn determinants and 4. T structure	a. to d. and e. Native and asialo-fetuin (#18, #19) f. Desialized porcine salivary glycoprotein g. Bovine and ovine salivary glycoproteins

(6) Galα1$\rightarrow$3Gal specific lectin

As shown in Table VII, GSI-B_4 precipitates extensively with human blood group B active glycoproteins and to a lesser extent with a precursor human blood group substance, while A_1, A_2, H_1, Le^a, and Le^b substances are inactive. From the QPA profile, it is clear that this lectin has its own specific binding property (with blood group B determinant).

V. Selecting Lectins for Carbohydrate Studies - a strategy for application

The common specificities (Gal/GalNAc lectin determinants) of lectins as well as their individual binding characteristics allows them to be used not only as special reagents for cell markers, but also as fine tools for elucidating changes in mammalian cell surface structures during their differentiation (1-6). Lectins can be conjugated with fluorescein isothiocyanate (FITC), rhodamine, ferritin, peroxidase, or radioisotopes (116-120) for histochemical use. They can also be immobilized on agarose or acrylamide and used for isolating and separating glycoproteins, glycolipids, polysaccharides, subcellular particles, and cells and for purifying detergent-

Table VII. Comparison of precipitating activities of Galβ1→4(3)GlcNAc and Galα1→3Gal-specific lectins with blood group-active substances and glycoproteins[a]

Glycoprotein	Galβ1→4(3)GlcNAc RCA1 (89)	ECL (70)	GCL (72)	Galα1→3Gal GSI-B4 (35)
1. Cyst MSS, native (A1)[b]	5+	+	4+	-
2. 1st Smith degraded[c]	5+	4+	5+	ND
3. 2nd Smith degraded[c]	5+	-	3+	ND
4. Cyst 9 (A1)[b]	5+	-	3+	-
5. Cyst 14 (A2)[b]	5+	3+	5+	-
6. Cyst Beach phenol insoluble (B)[b]	5+	2+	4+	4+
7. P1 (mild acid hydrolyzed)[d]	5+	ND	4+	-
8. Horse 4 25% (B)	5+	-	ND	4+
9. Cyst Tij II phenol insoluble (B,I)[b]	ND	3+	ND	4+
10. Cyst Tij II 20% 2X (B,I)[b]	ND	3+	ND	2+
11. Cyst Tij II 10% 2X (B, I)[b]	ND	3+	ND	ND
12. Cyst JS phenol insoluble (Le^b, H)[b]	4+	4+	3+	-
13. 1st Smith degraded[c]	ND	-	4+	ND
14. 2nd Smith degraded[c]	ND	ND	4+	ND
15. Cyst Tighe phenol insoluble (H)[b]	5+	3+	ND	-
16. Cyst N-1 phenol insoluble (Le^a)[b]	5+	4+	4+	-
17. Cyst OG 20% from 10% (I)[b]	5+	5+	5+	2+
18. Fetuin[g]	4+	-	ND	ND
19. Asialo-fetuin[g]	5+	3+	ND	ND
20. Pneumococcus type XIV polysaccharide[i]	5+	ND	3+	ND

[a]5.4 µgN of RCA1; 5.9 µgN of ECL; 6.2 µgN of GCL and 5.04 µgN of GSI-B4 were used for the quantitative precipitin assays. Glycoproteins were.tested up to 60 µg; -, less than 3% of lectin precipitated; ±, less than 15% lectin precipitated; +, between 16-25% of lectin precipitated; ++, between 26-50% of lectin precipitated; +++, between 51-75% of lectin precipitated; ++++, between 76-95% of lectin precipitated; +++++, completely precipitated; ND, not determined.

[b,c,g] are described in Legend to Table V; [i] is described in footnote of Table VI.

solubilized cell membrane components. Therefore, selecting lectins for specific applications will be the state of the art in lectinology.

A scheme using a combination of Gal/GalNAc specific lectins as tools for differentiating carbohydrate structures on cell surfaces and glycoconjugates in general is proposed in Fig. 4. RCA_1, BPL, WFL, and HPL possess broad affinities (Table V to VII) and reasonable stability. Thus, they are ideal reagents for screening purposes. When BPL, RCA_1, and WFL bind to carbohydrate residues of mammalian cells or precipitate glycoproteins and/or proteoglycans, this indicates that the cell surface or the carbohydrate side chains of these glycoproteins must contain one or several determinants such as Galβ1→3GalNAcα1→Ser(Thr) of the protein moiety (lectin T determinant), GalNAcα1→Ser(Thr) (lectin Tn structure) and Galβ1→ 4(3)GlcNAc (lectin I/II sequences) residues at the terminal nonreducing ends of long carbohydrate side chains. A positive reaction with PNA and a negative test with $VVL-B_4$ narrows the range of possible receptors, which can be Galβ1→3GalNAcα1→linked to Ser or Thr of the core protein (lectin T determinant) and/or Galβ1→4GlcNAc residues at terminal nonreducing ends of glycoconjugates. Positive of PNA and a negative result with MPL and $VVL-B_4$ limits the receptor to Galβ1→3(4)GlcNAc (lectin I/II determinant) residues at the terminal nonreducing ends of the glycoproteins (Table VIII). BPL or PNA binding weakly or not at all to cells indicates that the receptors are either shielded by sugars or that no such receptors are present on the cell surface. This can be confirmed by glycosidase treatment. If after glycosidase treatment the lectins bind to the cells, it is obvious that the determinants (receptors) are substituted by the sugars corresponding to the glycosidases utilized.

The presence of GalNAcα1→3GalNAc(Gal) at the nonreducing end of carbohydrate chains of glycoconjugates can be confirmed by the use of DBL and LBL. As shown in Table IX, DBL, HPL, WFL positive, and RCA_1 and LBL negative results indicate that the receptor is most likely GalNAcα1→3GalNAc. Positive DBL and LBL, and negative RCA_1 tests imply that the receptor is a GalNAcα1→3[LFucα1→2]Gal(Af) related oligosaccharide at the nonreducing end.

The same approach can also be used for the detection of other carbohydrate (GlcNAc, Man and LFuc) specificities to

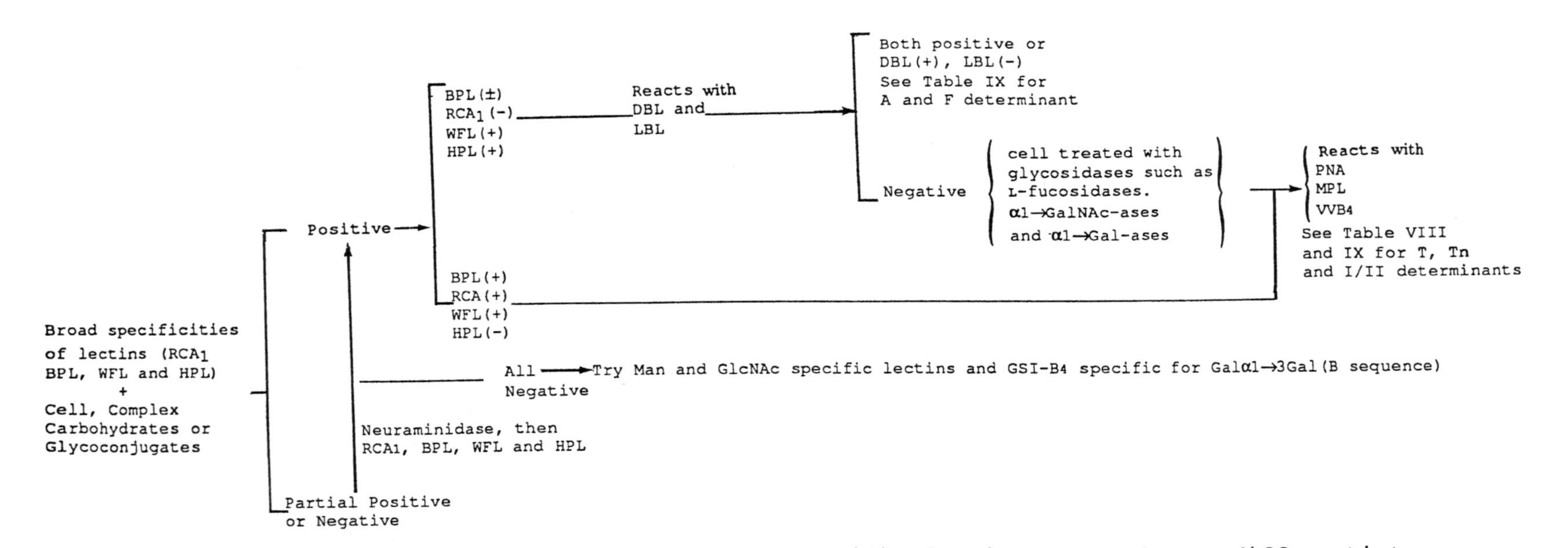

Fig. 4. Diagram of using Galβ1→ and GalNAcα1→ specific lectins as tools to differentiate binding properties and structure of carbohydrate side chain residues of complex carbohydrates or glycoconjugates.

Table VIII. Differential Simple Lectin Receptors (T, Tn and I/II) of Complex Carbohydrates

Lectin tested[a]					Possible oligosaccharides at the terminal non-reducing end of carbohydrate side chains as lectin determinants[b]
PNA	BPL	MPL	RCA_1	$VVL\text{-}B_4$	
+	+	+	+[c]	+	T, and/or Tn and/or A
+	+	−[c]	+	−[c]	Type I & II carbohydrate sequences predominately
−[c]	+	+	+[c]	±	T
+	+	+	+	±	T and Tn when HPL (+), T and I/II when HPL(−)
−	+	+	−	+	Tn

[a] +: Positive reaction: lectin binding to receptors, agglutinating cells or precipitating either glycoprotein or polysaccharides; −: no binding or precipitating. ±, binding upon the complexity of glycoconjugates, and specificity of lectin.

[b] Positive reaction after glycosidase treatment indicates determinants (or receptors) substituted by the sugars corresponding to the glycosidases used. T, Galβ1→3GalNAcα1→Ser(Thr) of protein core; Tn, GalNAcα1→Ser (Thr) of peptide chain; I/II, Galβ1→3(4)GlcNAc at non-reducing end of carbohydrate chains; A, GalNAcα1→3Gal(blood group A determinant).

[c] The positive results of GCL and TAL confirm further that Galβ1→4GlcNAc and/or Galβ1→3GlcNAc residues are present at the non-reducing end of complex carbohydrate (72,75).

[d] Negative results before and after glycosidase treatment.

Table IX. Differential Complex Lectin Receptors (F, A, A_f and I/II) of Complex Carbohydrates

Lectin tested[a]										Possible oligosaccharides at the terminal non-reducing end of carbohydrate side chains as lectin determinants[b,e]
RCA_1	PNA	BPL	VVB_4	MPL	SBA	WFL	HPL[d]	LBL	DBL	
–	–	–	–	–	NA[c]	+	+	–	+	F
–	–	–	–	–	+[d]	+[d]	+[d]	+	+	A and A_f
–	+	+	–	–	+	+	–	–	–	I/II
–	–	+	+	+	+	+	+	NA[c]	–	Tn
–	+	+	±	+	±	+	–	–	–	T and/or probably I/II

[a] "+": Positive reaction: lectin binding to receptors, agglutinating cells or precipitating either glycoprotein or polysaccharide; "-", no binding or precipitating; "±", depending upon the complexity of carbohydrate side chains and the carbohydrate affinity of lectin.

[b] Positive reaction after glycosidase treatment indicates determinants (or receptors) substituted by the sugars corresponding to the glycosidases used.

[c] NA, Not available.

[d] Positive after αLFucose glycosidase treatment indicates A_f determinant.

[e] Abbreviation; **F**, GalNAcα1→3GalNAc; A_f, GalNAcα1→3[LFucα1→2]Gal related; A, T, Tn, and I/II shown in Table VIII.

detect other types of receptors. Recently, some of the differential affinities of Man and LFuc specific lectins have also been reported (16). However, characterization of combining (receptor) sites on lectins is still hampered by the lack of suitable specific competitive inhibitory reagents (oligosaccharides). Using lectins as tools for differential purposes, a general principle is that better characterized lectins yield more precise information about structural differences.

Summary

Grouping of lectin binding properties, based on determinant structure rather than monosaccharide inhibition pattern, should facilitate the selection of lectins as structural probes for glycans as well as for the interpretation of the distribution and the properties of the carbohydrate chains on the cell surface.

Based on the binding specificities studied with glycan by precipitin-inhibition, competitive-binding and hemagglutinin-inhibition assays, twenty Gal and/or GalNAc specific lectins have been divided into six classes according to their specificity for the disaccharide as all or part of the determinants and GalNAcα1→Ser(Thr) of the peptide chain. The differential affinities of these lectins were characterized by quantitative precipitin assay. Abbreviation of the following six lectin determinants can also be used to classify these lectins.

(1) **F** determinant (GalNAcα1→3GalNAc, **F**orssman specific disaccharide).

(2) **A** (Af) determinant (GalNAcα1→3Gal, Human blood group **A** specific disaccharide; Af, fucosylated A, (GalNAcα1→3[LFucα1→2]Gal).

(3) **Tn** determinant (GalNAcα1→*O* to Ser(Thr) of the protein core, Tn antigen).

(4) **T** determinant (T antigen, Galβ1→3GalNAcα1→*O* to Ser(Thr) of the protein core, the mucin type sugar sequence on the human erythrocyte membrane or Galβ1→3GalNAcβ1→ at the nonreducing end of ganglioside).

(5) **I** and **II** determinants (human blood group type I and II

carbohydrate sequences). Most of the lectins reactive to Galβ1→4GlcNAc(II) are also reactive to Galβ1→3GlcNAc(I). Lectin I(II) determinants (i.e.Galβ1→3(4)GlcNAc residues) can be found at the nonreducing end of the carbohydrate chains derived from either *N*-glycosidic or *O*-glycosidic linkages.

(6) **B** determinant (Galα1→3Gal,Human blood group **B** specific disaccharide).

Their carbohydrate specificities are classified as following:

Gal/GalNAc Specific Lectins	Lectin Determinants (Active carbohydrate sequence)
F specific lectins	
Dolichos biflorus (DBL)	F > Af^1 > A
Helix pomatia (HPL)	F > A (>Af^2) ≥ Tn
Hog peanut (ABL, *Amphicarpaea bracteata*)	F > A > Tn
Wistaria floribunda (WFL)	A(>Af^2), F > Tn, I(II)
A specific lectins	
Lima bean (LBL)	Hexa-Af^1 > Af^1 >> A
Griffonia (Bandeiraea) simplicifolia-A4 (GSI-A4)	A > Af^2 >> B
Soy bean (SBL)	A(>Af^2), Tn and I(II)
Vicia villosa (VVL) (A mixture of A4, A2B2, and B4)	A(>Af^2) and Tn mainly
Tn specific lectins	
Vicia villosa B4 (VVL-B4)	Two Tn >> one Tn >> one or two T
Salvia sclarea (SSL)	Two Tn > single or three sequential Tn structures
T specific lectins	
Peanut (*Arachis hypogaea*, PNA)	T >> I(II)
Maclura pomifera (MPL)	T > Tn
Bauhinia purpurea alba (BPL)	T > I(II) and Tn
Sophora japonica (SJL)[3]	T and I(II)
Artocarpus integrifolia (Jacalin, AIL)	T and I(II)
Ricinus communis toxin (Ricin, RCA2)	T > I(II) and Tn
I(II) specific lectins	
Ricinus communis agglutinin (RCA1)	I(II) > T and B
Datura stramonium (TAL, Thorn apple)	Biantennary I(II) (Penta-2,6) >> C^4
Erythrina cristagalli (Coral tree, ECL)	Multiple antennary I(II)
Geodia cydonium (GLC)	I(II) and Tn

B specific lectins

Griffonia(Bandeiraea) simplicifolia-B4 (GSI-B4) B

[1]Substitution of LFucα1→2 to subterminal Gal is an important factor for binding. [2]Substitution of LFucα1→2 to subterminal Gal blocks binding.
[3]Some helping factor is required for precipitation of the lectin.
[4]C, chitin oligosaccharide

The differential binding properties of lectins can be defined from comparisons of their carbohydrate specificities listed above.

Acknowledgements

[1]This work was aided by Grants from the Texas Agricultural Experiment Station (TAES H6194), USDA/SEA Formula Animal Health Funds (Project 6648) and the Robert J. and Helen C. Kleberg Foundation.
[2]We thank Doretha A. Jones and Kay Sanders for their secretarial assistance in preparing this manuscript.

REFERENCES

1. Nicolson, G.L. (1974) The interaction of lectins with animal cell surfaces, Int. Rev. Cytol., 39:89-190.
2. Goldstein, I.J. and Poretz, R.D. Isolation, physiochemical characterization, and carbohydrate binding specificity of lectins. *In*: "The Lectins, Properties, Functions and Applications in Biology and Medicine," (Liener, I.E., Sharon, N., and Goldstein, I.J. eds.), Academic Press, pp. 33-247.
3. Lis, H. and Sharon, N. (1974) Lectins: their chemistry and application to immunology, *In*: "The Antigens", (Sela, M., ed.), Academic Press, New York, Vol 4., pp. 429-529.
4. Lis, H. and Sharon, N. (1986) Lectins as molecules and as tools. Ann. Rev. Biochem., 55:35-67.
5. Pereira, M.E.A. and Kabat, E.A. (1979) Immunochemical studies on lectins and their application to the fractionation of blood group substances and cells. Crit. Rev. Immunol., 1:33-78.
6. Kimura, A., Wigzell, H., Holmquist, G., Ersson, B., and Carlson, G. (1979) Selective affinity fractionation of murine cytotoxic T lymphocyte (CTL) - Unique lectin specific binding of the CTL associated surface glycoproteins, T 145., J. Exp. Med., 149:473-484.
7. Stanley, P. (1980) Surface carbohydrate alterations of mutant mammalian cells selected for resistance to plant lectins. *In*: "The Biochemistry of Glycoproteins and Proteoglycans," (Lennarz, W.J., ed.), Plenum Press, New York and London, pp. 161-189.

8. Stanley, P.(1983) Lectin-resistant CHO cells: Selection of new mutant phenotypes. Somatic Cell Genetics, 9:593-608.
9. Stanley, W.S., Peters, B.P., Blake, D.A., Yep, D., Chu, E.H.Y., and Goldstein, I.J. (1979) Interaction of wild-type and variant mouse 3T3 cells with lectins from *Bandeiraea simplicifolia* seeds. Proc. Natl. Acad. Sci. U.S.A., 76:303-307.
10. Briles, E.B., Li, E., and Kornfeld, S. (1977) Isolation of wheat germ agglutinin-resistant clones of Chinese hamster ovary cells deficient in membrane sialic acid and galactose. J. Biol. Chem., 252:1107-1116.
11. Kristiansen, T. (1974) Group-specific separation of glycoproteins. Meth. Enzymol., 34(B):331-341.
12. Pereira, M.E.A. and Kabat, E.A. (1976) Immunochemical studies on blood groups. LXII. Fractionation of hog and human A, H, and AH blood group active substance on insoluble immunoadsorbents of Dolichos and Lotus lectins. J. Exp. Med., 143:422-436.
13. Watkins, W.M. (1972) Blood-group specific substances *In*: Glycoproteins (Gottschalk, A., ed.) 2nd ed., Part B, pp. 830-891. Elsevier, Amsterdam.
13A. Merkle, R.K. and Cumming, R.D. (1987) Lectin affinity chromatography of glycopeptides. Meth. Enzymol., 138:232-259.
14. Gallagher, J.T. (1984) Carbohydrate-binding properties of lectins: a possible approach to lectin nomenclature and classification. Biosci. Rep., 4:621-632.
15. Kabat, E.A. (1976) Structural Concepts in Immunology and Immunochemistry, 2nd ed., Holt, Rinehart and Winston, New York.
16. Wu, A.M., Sugii, S., and Herp, A. (1987) A guide for carbohydrate specificities of lectins *In*: "The Molecular Immunology of Complex Carbohydrates" (Wu, A.M., ed.), Plenum Press, New York and London. Appendix 1.
17. Mäkelä, O. (1957) Studies in hemagglutinins of legumiosae seeds. Ann. Med. Exp. Biol. Fenn. Suppl., 11,35:1-156.
17A. Watkin, W.M. (1980) Biochemistry and Genetics of the ABO, Lewis, and P Blood Group System. Advan. Hum. Genet. 10:1-136.
18. Wu, A.M. (1984) Differential binding characteristics and application of DGalβ1→3DGalNAc specific lectin, Mol. Cell. Biochem., 61:131-141.
19. Etzler, M.E. and Kabat, E.A. (1970) Purification and characterization of a lectin (plant hemagglutinin) with blood group A specificity from *Dolichos biflorus*. Biochemistry, 9: 869-877.
20. Etzler, M.E. (1972) Horse gram (*Dolichos biflorus*) lectin. Meth. Enzymol., 28:340-344.
21. Baker, D.A., Sugii, S., Kabat, E.A., Ratcliffe, R.M., Hermentin, P., and Lemieux, R.U. (1983) Immunochemical studies on the combining sites of Forssman hapten reactive hemagglutinins from *Dolichos biflorus*, *Helix pomatia* and *Wistaria floribunda*. Biochemistry, 22:2741-2750.
22. Carter, W.G. and Etzler, M.E. (1975) Isolation and characterization and subunit structure of multiple forms of *Dolichos biflorus* lectin. J. Biol. Chem., 250:2756-2762.
23. Carter, W.G. and Etzler, M.E. (1975) Isolation and characterization of subunits from the predominant form of *Dolichos biflorus* lectin. Biochemistry, 14:2685-2689.

24. Kocourek, J., Jamieson, G.A., Votruba, T., and Hořejší, V. (1977). Studies on Phytohemagglutinins: 1. Some properties of the lectins of horse gram seeds (*Dolichos Biflorus* L.) Biochim. Biophys. Acta 500, 344-360.
25. Hammarström, S., Murphy, L.A., Goldstein, I.J., and Etzler, M.E. (1977) Carbohydrate binding specificity of four *N*-acetyl-D-galactosamine-"specific" lectins: *Helix pomatia* A hemagglutinin, soybean agglutinin, lima bean lectin and *Dolichos biflorus* lectin Biochemistry, 16:2750-2755.
26. Hammarström, S. and Kabat, E.A. (1969) Purification and characterization of a blood-group A reactive hemagglutinin from the snail *Helix pomatia* and a study of its combining site. Biochemistry, 8:2696-2705.
27. Hammarström, S., Westöö, A., and Bjork, I. (1972) Subunit structure of *Helix pomatia* A hemagglutinin. Scand. J. Immunol., 1:295-309.
28. Hammarström, S. (1974) Structure, specificity, binding properties and some biological activities of a blood group A-reactive hemagglutinin from the snail *Helix pomatia*. Ann. N.Y. Acad. Sci., 234:183-197.
28A Maliarik, M.J., Roberts, D.D., and Goldstein, I.J. (1987) Properties of the lectin from the Hog Peanut (*Amphicarpaea bracteata*) Arch. Biochem. Biophys., 255:194-200.
29. Sugii, S. and Kabat, E.A. (1980) Immunochemical specificity of the combining site of *Wistaria floribunda* hemagglutinin. Biochemistry, 19:1192-1199.
30. Kurokawa, T., Tsuda, M. and Sugino, Y. (1976) Purification and characterization of a lectin from *Wistaria floribunda* seeds. J. Biol Chem., 251:5686-5693.
31. Cheung, G., Haratz, A., Katar, M., Skrokov, R., and Poretz, R.D. (1979) Purification and properties of the hemagglutinin from *Wistaria floribunda* seeds. Biochemistry, 18:1646-1650.
32. Pereira, M.E.A., Kabat, E.A., and Sharon, N. (1974) Immunochemical studies on the specificity of soybean agglutinin. Carbohydr. Res., 37:89-102.
33. Gordon, J.A., Blumberg, S., Lis, H., and Sharon N. (1972) Method II. Purification of SBA by affinity chromatography. Meth. Enzymol., 28:365-368.
34. Lotan, R., Siegelman, H.W., Lis, H., and Sharon, N. (1974) Subunit structure of soybean agglutinin. J. Biol. Chem., 249:1219-1224.
35. Wood, C., Kabat, E.A., Murphy, L.A., and Goldstein, I.J. (1979) Immunochemical studies on the combining sites of the two isolectins, A_4 and B_4, isolated from *Bandeiraea simplicifolia*. Arch. Biochem. Biophys., 198:1-11.
36. Hayes, C.E. and Goldstein, I.J. (1974) An αD-galactosyl-binding lectin from *Bandeiraea simplicifolia* seeds. Isolation by affinity chromatography and characterization. J. Biol. Chem., 249:1904-1914.
37. Murphy, L.A. and Goldstein, I.J. (1978) *Bandeiraea simplicifolia* I isolectins. Meth. Enzymol., 50:345-349.
38. Murphy, L.A. and Goldstein, I.J. (1977) Five α-Dgalactopyranosyl-binding isolectins from *Bandeiraea simplicifolia* seeds. J. Biol. Chem., 252:4739-4742.
39. Galbraith, W. and Goldstein, I.J. (1979) Phytohemagglutinin of the lima bean (*Phaseolus lunatus*). Isolation, characterization and interaction with type A blood-group substance. Biochemistry, 11:3976-3984.

40. Galbraith, W. and Goldstein, I.J. (1970) Phytohemagglutinins: a new class of metalloproteins. Isolation, purification and some properties of the lectin from *Phaseolus lunatus*. FEBS Lett., 9:197-201.
41. Gould N.R. and Scheinberg, S.L.(1970) Isolation and partial characterization of two anti-A hemagglutinins from *P. lunatus*. Arch. Biochem. Biophys., 137:1-11.
42. Galbraith, W. and Goldstein, I.J. (1972) Lima bean (*Phaseolus lunatus*) lectin. Meth. Enzymol., 28:318-323.
43. Sikder, S.K., Kabat, E.A., Roberts, D.D., and Goldstein, I.J. (1986) Immunochemical studies on the combining site of the blood group A-specific lima bean lectin. Carbohydr. Res., 151:247-260.
44. Roberts, D.D. and Goldstein, I.J. (1984) Effect of carbohydrate and metal ion binding on the reactivity of the essential thiol groups of lima bean lectin. J. Biol. Chem., 259:903-908.
45. Tollefsen, S.E. and Kornfeld, R. (1983) Isolation and characterization of lectins from *Vicia villosa*. Two distinct carbohydrate binding activities are present in seed extracts. J. Biol. Chem., 258:5166-5171.
46. Tollefsen, S.E. and Kornfeld, R. (1983) The B_4 lectin from *Vicia villosa* seeds interacts with *N*-acetylgalactosamine residues α-linked to serine or threonine residues in cell surface glycoproteins. J. Biol. Chem., 258:5172-5176.
47. Piller, V., Piller, F., and Cartron, J.P. (1986) Isolation and characterization of an *N*-acetylgalactosamine specific lectin from *Salvia sclarea* seeds. J. Biol. Chem., 261:14069-14075.
48. Sarker, M., Wu, A.M., and Kabat, E.A. (1981) Immunochemical studies on the carbohydrate specificity of *Maclura pomifera* lectin. Arch. Biochem. Biophys., 209:204-218.
49. Bausch, J.N. and Poretz, R.D. (1977) Purification and properties of the hemagglutinin from *Maclura pomifera* seeds. Biochemistry, 16:5790-5794.
50. Jones, J.M. and Feldman, J.D. (1973) Binding of *Maclura pomifera* lectin to rat lymphoid cells and erythrocytes. J. Immunol., 111:1765-1770.
51. Wu, A.M., Kabat, E.A., Gruezo, F.G., and Allen, H.J. (1980) Immunochemical studies on the combining site of the D-galactopyranose and 2-acetamido-2-deoxy-D-galactopyranose-specific lectin isolated from *Bauhinia purpurea alba* seeds. Arch. Biochem. Biophys., 204:622-639.
52. Irimura, T. and Osawa, T. (1972) Studies on a hemagglutinin from *Bauhinia purpurea alba* seeds. Arch. Biochem. Biophys., 151:475-482.
53. Osawa, T., Irimura, T., and Kawaguchi, T. (1978) *Bauhinia purpurea* agglutinin. Meth. Enzymol., 50:367-372.
54. Poretz, R.D.(1972) *Sophora japonica* hemagglutinin. Meth. Enzymol., 28:349-354.
55. Poretz, R.D., Riss, H., Timberlake, J.W., and Chien, S.M. (1974) Purification and properties of the hemagglutinin from *Sophora japonica* seeds. Biochemistry, 13:250-256.
56. Timberlake, J.W., Won, R.B.C., and Poretz, R.D.(1980) Properties and subunit characterization of affinity purified *Sophora japonica* lectin. Prep. Biochem., 10:173-190.
57. Wu, A.M., Kabat, E.A., Gruezo, F.G., and Poretz, R.D. (1981) Immunochemical studies on the reactivities and

combining sites of the D-galactopyranose and 2-acetamido-2-deoxy-D-galactopyranose specific lectin purified from *Sophora japonica* seeds. Arch. Biochem. Biophys., 209:191-203.

58. Baenziger, J.U. and Fiete, D. (1979) Structural determinants of *Ricinus communis* agglutinin and toxin specificity for oligosaccharides. J. Biol. Chem., 254:9795-9799.
59. Olsnes, S., Saltvedt, E., and Pihl, A. (1974) Isolation and comparison of galactose-binding lectin from *Abrus precatorius* and *Ricinus communis*. J. Biol. Chem., 249:803-810.
60. Gürtler, L.G. and Horstmann, H.J. (1973) Subunits of toxin and agglutinin of *Ricinus communis*. J. Biophys. Acta, 295:582-594.
61. Nicolson, G.L., Blaustein, J., and Etzler, M.D. (1974) Characterization of two plant lectins from *Ricinus communis* and their quantitative interaction with murine lymphoma. Biochemistry, 13:196-204.
62. Olsnes, S., Refsnes, K., Christensen, T.B., and Phil, A. (1975) Studies on the structure and properties of the lectins from *Abrus precatorius* and *Ricinus communis*. Biochim. Biophys. Acta., 405:1-10.
63. Olsnes, S. (1978) Ricin and ricinus agglutinin, toxic lectins from castor bean. Meth. Enzymol., 50:330-335.
64. Lotan, R., Skutelsky, E., Danon, D., and Sharon, N. (1975) The purification, composition and specificity of the anti-T lectin from peanut (*Arachis hypogaea*). J. Biol. Chem., 250:8518-8523.
65. Terao, T., Irimura, T., and Osawa, T. (1975) Purification and characterization of a hemagglutinin from *Arachis hypogaea*. Hoppe-Seyler's Z. Physiol. Chem., 356:1685-1692.
66. Lotan, R. and Sharon, N. (1978) Peanut (*Arachis hypogaea*) agglutinin. Meth. Enzymol., 50:361-367.
67. Pereira, M.E.A., Kabat, E.A., Lotan, R., and Sharon N. (1976) Immunochemical studies on the specificity of the peanut (*Arachis hypogaea*) agglutinin. Carbohydr. Res., 51:107-118.
68. Sastry, M.V.K., Banarjee, P. Patanjali, S.R., Swamy, M.J., Swarnalatha, G.V., and Surolia, A. (1986) Analysis of saccharide binding to *Artocarpus integrifolia* lectin reveals specific recognition of T-antigen [Galβ1→3GalNAc]. J. Biol. Chem., 261:11726-11733.
69. Kumar, G.S., Appukuttan, P.S., and Baus, D.(1982) α-*D*-Galactose-specific lectin from jack fruit (*Artocarpus integrifolia*) seeds. J. Biosci., 4:257-261.

69A. Chatterrjee, B.P. and Uhlenbruck, G. (1982) Occurrence of anti-Thomsen-Freidenreich (TF) like lectin in the seeds of jack fruit (*Artocarpus integrifolia*) reacting with ant (Oecophylla smargdina fabri) egg glycoprotein. Experientia, 38:1225-1226.

69B. Chowdhury, S., Ahmed, H., and Chatterjee, B.P. (1987) Purification and characterization of an α-D-galactosyl-binding lectin from *Artocarpus lakoocha* seeds. Carbohydr. Res., 159:137-148.

69C. Chatterjee, B.P., Chowdhury, S., and Ahmed, H. Further characterization of *Artocarpus lakoocha* lectin purified by rivanol. *In*: Abstract B-8, 4th European Carbohydrate Symposium, July 12-17, 1987, Darmstadt, FRG.

70. Kaladas, P.M., Kabat, E.A., Iglesias, J.L., Lis, H., and Sharon, N. (1982) Immunochemical studies on the combining site of the D-galactose/*N*-acetyl-D-galactosamine specific lectin from *Erythrina cristagalli* seeds. Arch. Biochem. Biophys., 217:624-637.
71. Iglesias, J.L., Lis, H., and Sharon, N.(1982) Purification and properties of a D-galactose/*N*-acetyl-D-galactosamine-specific lectin from *Erythrina cristagalli*. Eur. J. Biochem., 123:247-252.
72. Bretting, H., Phillips, S.G., Klumpart, H.J., and Kabat, E.A. (1981) A mitogenic lactose-binding lectin from the sponge *Geodia cydonium*. J. Immmunol., 127:1652-1658.
73. Horejsí, V. and Kocourek, J. (1978) Studies on lectins. XXXVII. Isolation and characterization of the lectin from jimson-weed seeds (*Datura stramonium*). Biochim. Biophys. Acta., 532:92-97.
74. Crowley, J.F. and Goldstein, I.J.(1981) *Datura stramonium* lectin: isolation and characterization of the homogeneous lectin. FEBS Lett., 130:149-152.
75. Crowley, J.F., Goldstein, I.J., Arnarp, J., and Lönngren, J. (1984) Carbohydrate binding studies on the lectin from *Datura stramonium* seeds. Arch. Biochem. Biophys., 231: 524-533.
76. Springer, G.F. (1971) Blood group and Forssman antigenic determinants shared between microbes and mammalian cells. Prog. Allergy, 15:9-77.
76A. Makita, A., Suzuki, C., and Yohizawa, Z. (1966) Chemical and immunological characterization of the Forssman hapten isolated from equine organs. J. Biochem., 60:502-513.
77. Hakomori, S-I. and Kannagi, R. (1983) Glycosphingolipids as tumor-associated and differentiation markers. J. Nat. Cancer Inst., 71:231-351.
78. Kaladas, P.M., Kabat, E.A., Kimura, A., and Ersson, B. (1981) The specificity of the combining site of the lectin from *Vicia villosa* seeds which reacts with cytotoxic T-lymphocytes. Mol. Immunol., 18:969-977.
79. Wu, A.M., Kabat, E.A., Pereira, M.E.A., Gruezo, F.G., and Liao, J. (1982) Immunochemical studies on blood groups: the internal structure and immunological properties of water-soluble human blood group A substance studied by Smith degradation, liberation and fractionation of oligosaccharides and reaction with lectins. Arch. Biochem. Biophys., 215:390-404.
80. Dahr, W., Uhlenbruck, G., and Bird, G.W.G. (1974) Cryptic A-like receptor sites in human erythrocyte glycoproteins: proposed nature of Tn-antigen. Vox. Sang., 27:29-42.
81. Cartron, J.-P., Andreu, G., Cartron, J., Bird, G.W.G., Salmon, C., and Gerbal, A. (1978) Demonstration of T-transferase deficiency in Tn polyagglutinable blood samples. Eur. J. Biochem., 92:111-119.
82. Cartron, J.-P., Blanchard, D., Nurden, A., Cartron, J., Rahuel, C. Lee, D., Vainchenker, W., Testa, U., and Rochant, H. (1982) *In*: "Blood Groups and Other Cell Surface Markers in Health and Disease." (Salmon, C., ed.) pp. 39-54, Masson Publishing USA, Inc., New York.
83. Springer, G.F.(1984) T and Tn, general carcinoma autoantigens. Science, 224:1198-1206.
84. Hirohashi, S., Clausen, H., Yamada, T., Shimosato, Y., and Hakomori, S.-I. (1985) Blood group A cross-reacting epitope defined by monoclonal antibodies NCC-LU-35 and -81 expressed in cancer of blood group O or B individuals: Its identification as Tn antigen. *Proc. Natl. Acad. Sci.* U.S.A., 82:7039-7043.

85. Springer, G.F., Desai, P.R., and Banatwala, I. (1975) Blood group MN antigens and precursors in normal and malignant human breast glandular tissue. J. Natl. Cancer Inst., 54:335-339.
86. Anglin, J.H., Jr., Lerner, M.P., and Nordquist, R.E. (1977) Blood group-like activity released by human mammary carcinoma cells in culture. Nature, 269:254-255.
87. Springer, G.F. and Desai, P.R. (1975) Human blood group MN and precursor specificities: structural and biological aspects. Carbohydr. Res., 40:183-192.
88. Debray, H., Decout, D., Strecker, G., Spik, G., and Montreuil, J. (1981) Specificity of twelve lectins towards oligosaccharides and glycopeptides related to *N*-glycosylproteins. Eur. J. Biochem., 117:41-55.
89. Wu, A.M., Sugii, S., Gruezo, F.G., and Kabat, E.A. (1987) Immunochemical studies on the Galβ1→4(3)GlcNAc specificity of *Ricinus communis* agglutinin. Carbohydr. Res., (in press).
90. Feizi, T. (1985) Demonstration by monoclonal antibodies that carbohydrate structures of glycoproteins and glycolipids are onco-developmental antigens. Nature,314:53-57.
91. Cummings, R.D. and Kornfeld, S. (1982) Characterization of the structural determinants required for the high affinity interaction of asparagine-linked oligosaccharides with immobilized *Phaseolus vulgaris*. leukoagglutinating and erythroagglutinating lectins. J. Biol. Chem., 257:11230-11234.
92. Hammarström, S., Hammarström, M.L., Sundblad, G., Arnarp, J., and Lonngren, J. (1982) Mitogenic leukoagglutinin from *Phaseolus vulgaris* binds to a pentasaccharide unit in *N* acetyllactosamine-type glycoprotein glycans. Proc. Natl. Acad. Sci., 79:1611-1615.
93. Spiro, R.G. and Bhoyroo, V.D. (1974) Structure of the O-glycosidically linked carbohydrate units of fetuin. J. Biol. Chem., 249:5704-5717.
94. Graham, E.R.B. (1972) Fetuin. *In*: Glycoproteins (Gottschalk, A., ed.), 2nd ed., pp. 717-731, Elsvier, Amsterdam.
95. Nilsson, B., Norden, N.E., and Svensson, S. (1979) Structural studies on the carbohydrate portion of fetuin. J. Biol. Chem., 254:4545-4553.
95A. Townsend, R.R., Hardy, M.R., Wong, T.C., and Lee, Y.C. (1986) Binding of *N*-linked bovine fetuin glycopeptides to isolated rabbit hepatocytes: Gal/GalNAc hepatic lectin discrimination between Galβ1→4GlcNAc and Galβ1→3GlcNAc in a triantennary structure. Biochemistry, 25:5916-5725.
95B. Edge, A.S.B. and Spiro, R.G. (1987) Presence of an O-Glycosidically Linked Hexasaccharide in Fetuin. J. Biol. Chem. 262:16135-16141.
96. Leskowitz, S. and Kabat, E.A. (1954) Immunochemical studies on blood groups. XV. The effect of mild acid hydrolysis on the glucosamine and galactosamine in blood group substances. J. Amer. Chem. Soc., 76:5060-5065.
97. Allen, P.Z. and Kabat, E.A. (1959) Immunochemical studies on blood groups. XXII. Immunochemical studies on the nondialyzable residue from partially hydrolyzed blood group A, B, and O(H) substances (P1 fraction). J. Immunol., 82:340-357.
98. Kabat, E.A. (1956) *In*: "Blood group substances. Their Chemistry and Immunochemistry." Academic Press, New York pp. 135-139.

99. Beiser, S.M. and Kabat, E.A. (1952) Immunochemical studies on blood groups. XII. Preparation of blood group substances from bovine stomach linings and a comparison of their chemical and immunochemical properties with those of blood group substances from other species. J. Immunol., 68:19-40
100. Lloyd, K.O. and Kabat, E.A. (1968) Immunochemical studies on blood groups. XLI. Proposed structures for the carbohydrate portions of blood group A, B, H, Lewis^a, and Lewis^b substances. Proc. Natl Acad. Sci., 61:1470-1477.
101. Vicari, G. and Kabat, E.A. (1969) Immunochemical studies on blood groups. XVII. Isolation and characterization from ovarian cyst fluid of a blood group substances lacking A, B, H, Le^a, and Le^b specificity. J. Immunol., 102:821-825.
102. Maisonrouge-McAuliffe, F. and Kabat, E.A. (1976) Immunochemical studies on blood groups. Fractionation, heterogeneity, and chemical and immunochemical properties of blood group substances with B, I, and i activities purified from human ovarian cyst fluid. Arch. Biochem. Biophys., 175:71-80.
103. Kabat, E.A., Bendich, A., Bezer, A.E., and Knaub, V. (1948) Immunochemical studies on blood groups. VI. The cross-reaction between type XIV antipneumococcal horse serum and purified blood group A, B, and O substances from hog and human sources. J. Exp. Med., 87:295-300.
104. Newman, W. and Kabat, E.A. (1976) Immunochemical studies on blood groups. Purification, chemical and immunochemical properties of blood group-active glycoproteins from horse gastric mucosae. Arch. Biochem. Biophys., 172:510-523.
105. Kabat, E.A., Bendich, A., Bezer, A.E., and Beiser, S.M. (1947) Immunochemical studies on blood groups. IV. Preparation of blood group A substances from human sources and comparison of their chemical and immunochemical properties with those of the blood group A substance from hog stomach. J. Exp. Med., 85:685-699.
106. Baer, H., Kabat, E.A., and Knaub, V. (1950) Immunochemical studies on blood groups. X. The preparation of blood group A and B substances and an active substance from individual horse stomachs and of blood group B substance from human saliva. J. Exp. Med., 91:105-114.
107. Rovis, L., Anderson, B., Kabat, E.A., Gruezo, F., and Liao, J. (1973) Structures of oligosaccharides produced by base-borohydride degradation of human ovarian cyst blood group H, Le^b, and Le^a active glycoproteins. Biochemistry, 12:5340-5354.
108. Wu, A.M., Kabat, E.A., Nilsson, B., Zopf., D.A. Gruezo, F.G., and Liao, J. (1984) Immunochemical studies on blood groups. Purification and characterization of radioactive ^{3}H-reduced di- to hexasaccharides produced by alkaline β-elimination-borohydride ^{3}H reduction of Smith degraded blood group A active glycoproteins. J. Biol. Chem., 259:7178-7186.
108A Mutsaers, J.H.G.M., Van Halbeek, H., Vliegenthart, J.F.G., Wu, A.M., and Kabat, E.A. (1986) Typing of core and backbone domains of mucin-type oligosaccharides from human ovarian-cyst glycoproteins by 500-MHz ^{1}H-NMR spectroscopy. Eur. J. Biochem., 157:139-146.

109. Krause, R.M. and McCarthy, M. (1962) Studies on the chemical structure of the streptococcal cell wall. II. The composition of group C cell walls and chemical basis for serologic specificity of the carbohydrate moiety. J. Exp. Med., 115:49-62.
110. Herp, A., Borelli, C., and Wu, A.M. (1987) Biochemistry and lectin binding properties of mammalian salivary mucous glycoprotein. *In*: "The Molecular Immunology of Complex Carbohydrates." (Wu, A.M., ed.) Plenum Press, New York and London. pp. II-6.
111. DeVries, A.L., Komatsu, S.K., and Feeney, R.E. (1970) Chemical and physical properties of freezing point-depressing glycoproteins from antarctic fishes. J. Biol. Chem., 245:2901-2908.
112. Lin, Y., Duman, J.G., and DeVries, A.L. (1972) Studies on the structure and activity of low molecular weight glycoproteins from an antarctic fish. Biochem. Biophys. Res. Commun., 46:87-92.
113. Schmid, K., Binette, J.P., Dorland, L., Vliegenthart, J.F.G., Fournet, B., and Montreuil, J. (1979) The primary structure of the asialo-carbohydrate units of the first glycosylation site of human plasma α_1-acid glycoprotein. Biochim. Biophys. Acta., 581:356-359.
114. Fournet, B., Montreuil, J., Strecker, G., Dorland, L., Haverkamp, J., Vliegenthart, J.F.G., Binette, J.P., and Schmid, K. (1978) Determination of primary structures of 16 asialo-carbohydrate units derived from human plasma α_1-acid glycoprotein by 360-MHz ^{1}H-NMR spectroscopy and permethylation analysis. Biochemistry, 17:5206-5214.
115. Lindberg, B., Lönngren, J., and Powell, D.A. (1977) Structural studies on the specific for type-14 pneumococcal polysaccharide. Carbohydr. Res., 58:177-186.
116. Singer, S.J. and Nicolson, G.L., (1972) The fluid mosaic model of the structure of cell membranes. Science 175: 720-731.
117. Nicolson, G.L. (1973) Cis- and trans-membrane control of cell surface topography. J. Supramol. Struct. 1:410-416.
118. Nicolson, G.L. (1974) Factors influencing the dynamic display of lectin binding sites on normal and transformed cell surfaces. *In*: Control of Proliferation in Animal Cells (Clarkson, B. and Baserga, R., eds.), pp. 251-270. Spring Harbor, New York.
119. Tanner, M.J.A. and Anstee, D.J. (1976) A method for the direct demonstration of the lectin-binding components of the human erythrocyte membrane. Biochem. J., 153:265-270.
120. Gros, D., Obrenovitch, A., Challice, C.E., Monsigny, M., and Schrevel, J. (1977) Ultrastructural visualization of cellular carbohydrate components by means of lectins on ultrathin glycol methacrylate sections. J. Histochem. Cytochem., 25:104-114.

ANTIGENIC PROPERTIES OF HUMAN ERYTHROCYTE GLYCOPHORINS

Elwira Lisowska

Department of Immunochemistry
Institute of Immunology and
Experimental Therapy
Polish Academy of Sciences
53-114 Wroclaw, Poland

The name of glycophorin was given by Marchesi et al. (1972) to the major sialoglycophorin of human erythrocyte membranes, known earlier as the glycoprotein carrying blood group M and N determinants and receptors for agglutinins of influenza viruses (Baranowski et al., 1959; Romanowska 1959; Klenk & Uhlenbruck, 1960; Kathan et al., 1961; Springer et al., 1966). Fractionation of the erythrocyte membranes by sodium dodecylsulfate-polyacrylamide gel electrophoresis (SDS-PAGE) and visualization of sialoglycoproteins with periodic acid-Schiff (PAS) reagent give a complex pattern of bands which may differ in details, depending on electrophoretic conditions. It has been now accepted that there are at least four distinct sialoglycoproteins in human erythrocyte membranes. Furthmayr et al. (1975) designated three of them as glycophorin A, B and C, in order of their decreasing amount in the membrane. Anstee et al. (1979) denoted them glycoprotein α, β, γ and δ, in order of their decreasing molecular weight. Dahr et al. (1978c) used other designations. The more numerous bands seen in SDS-PAGE correspond to monomers of the sialoglycoproteins and to homo- and heterodimers (and higher oligomers) formed by the most abundant glycophorins A and B (Fig. 1).

Human erythrocyte glycophorins are complex glycoprotein antigens suitable to study the relation between the structure and immunological properties. They can be solubilized from red cell membranes by various relatively simple procedures (reviewed by Lisowska, 1977) that enabled their chemical characterization. The primary structures of glycophorin A and C have been totally elucidated, and that of glycophorin B is almost totally known (see the next section). Glycophorins represent a high degree of polymorphism which is reflected in their antigenic properties. Glycophorins A and B are genetically differentiated into blood group MN and Ss antigens,

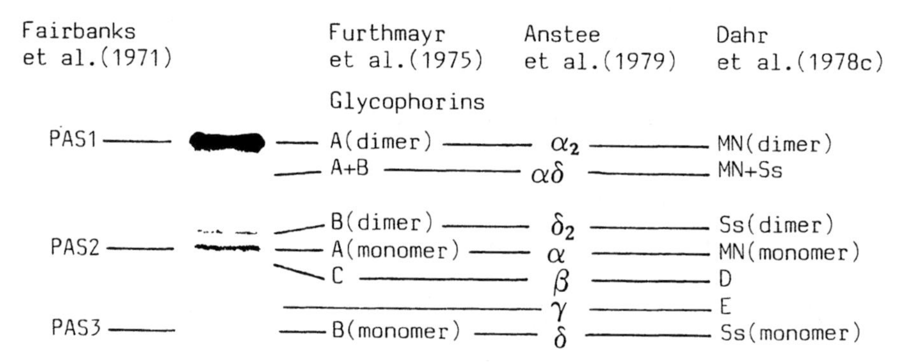

Fig. 1. Electrophoretic pattern (SDS-PAGE, PAS-staining) and nomenclatures used for human erythrocytemembrane sialoglycoproteins.

respectively. Various rare human allo- and autoantibodies directed against blood group determinants and common antigenic sites of glycophorins have been found and it is also relatively easy to obtain polyclonal immune sera and monoclonal antibodies against these determinants. Moreover, many rare antigenically distinct variant forms of glycophorins have been detected by means of human antibodies or immune animal sera. Collectively this indicates that immunochemical studies on glycophorins are attractive and relatively advanced.

STRUCTURE OF GLYCOPHORINS AND THEIR BLOOD GROUP-RELATED POLYMORPHISM

Glycophorins solubilized from red cell membranes strongly aggregate in water solution and can be effectively fractionated only in the presence of detergents or other dissociating agents (reviewed by Lisowska, 1977). So far, glycophorins A, B and C have been isolated and structurally characterized, but the most extensive studies have been carried out on glycophorin A.

Glycophorin A has the molecular weight of about 30,000 (Morawiecki, 1964; Tomita & Marchesi, 1975) and spans the erythrocyte membrane (for refs. see Lisowska, 1977). Its polypeptide chain consists of 131 amino acid residues of known sequence (Tomita & Marchesi, 1975; Tomita _et al_. 1978). Three distinct domains of the molecule can be distinguished: the glycosylated NH_2-terminal portion located outside erythrocyte surface, intramembranous hydrophobic domain and cytoplasmic COOH-terminal fragment. In the intramembranous domain (amino acid residues approx. 70-100) the hydrophobic amino acids comprise almost a half of all amino acid residues. The glycosylated domain, rich in serine and threonine residues, contains about 15 _O_-glycosidic oligosaccharide chains of the structure elucidated by Thomas and Winzler (1969) and one _N_-glycosidic biantennary chain with bisecting _N_-acetylglucosamine residue (Yoshima _et al_., 1980; Irimura _et al_., 1981). The structure of glycophorin A and its oligosaccharide chains shown in Fig. 2 is simplified, because it does not reflect the microheterogeneity in the carbohydrate portion of this glycoprotein. Some of the oligosaccharide chains are incompletely sialylated and other structural variations may occur in the minor portion of oligosaccharides (Takasaki & Kobata, 1976; Petryniak _et al_., 1980; Moulds, 1983). Moreover, there is an incomplete glycosylation of some serine and threonine residues that affects the susceptibility of proximate peptide bonds to proteolytic degradation. Treatment of glycophorin A with trypsin, chymotrypsin (Tomita _et al_., 1978) or pronase (Wasniowska & Lisowska, 1981) reveals heterogeneity in degradation of the glycosylated portion and two NH_2-terminal glycopeptides (shorter and longer) are produced by each enzyme in approximately equal amounts. It cannot be ruled out that double bands of dimer and monomer of glycophorin A frequently seen in SDS-PAGE (Lisowska, 1981) correspond to less and more glycosylated molecules.

The structural difference between glycophorin A of blood group M and N was not known for a long time. Inactivation of M and N antigens by desialylation (Makela & Cantell, 1958; Springer & Ansell, 1958; Romanowska, 1959; Klenk & Uhlenbruck, 1960; Lisowska & Duk, 1976) suggested that anti-M and anti-N antibodies react with oligosaccharide chains which are different in M and N glycoprotein (Huprikar & Springer, 1970; Sringer & Desai, 1975, 1982) On the other hand, it has been

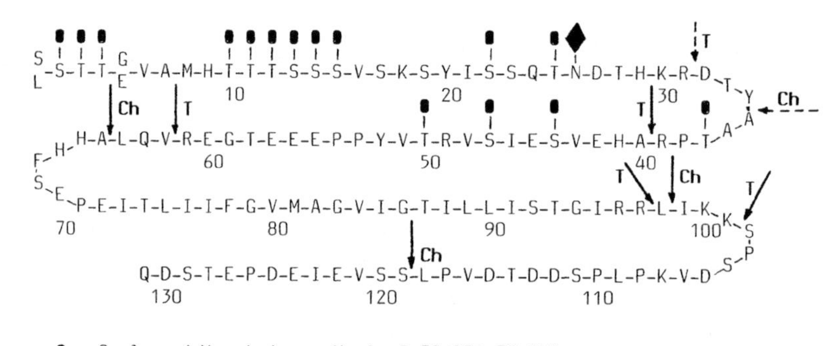

- O-glycosidic chains: NeuAcα2-3Galβ1-3GalNAcα- (6|2 NeuAcα)

- N-glycosidic chain:

NeuAcα2-6Galβ1-4GlcNAcβ1-2Manα1-6
GlcNAcβ1-4 Manβ1-4GlcNAcβ1-4GlcNAc- (Fucα1-6)
NeuAcα2-6Galβ1-4GlcNAcβ1-2Manα1-3

Fig.2. Stucture of glycophorin A (Tomita et al., 1978) and its oligosaccharide chains (Thomas & Winzler, 1969; Yoshima et al., 1980; Irimura et al., 1981). Arrows indicate sites of total (solid line) or partial (broken line) cleavage with trypsin (T) and chymotrypsin (Ch). Changes to the structure suggested by Dahr et al. (1980b) are shown in Fig. 3.

found that modification of amino groups also destroys blood M and N activity (Lisowska & Morawiecki, 1967; Ebert et al., 1972; Lisowska & Duk, 1975a, 1975b; Uhlenbruck et al., 1976) and makes glycoprotein M reactive with blood group N-specific *Vicia graminea* lectin (Lisowska & Duk, 1975a, 1975b; Dahr et al., 1975a). These findings suggested that M and N antigens differ in the structure of their polypeptide chains. After Tomita and Marchesi (1975) reported the amino acid sequence of glycophorin A from pooled blood and found two different amino acid residues at the positions 1 and 5 of the polypeptide chain (see Fig.2), it was independently shown in four laboratories that this heterogeneity represents blood group MN-related polymorphism (Wasniowska et al., 1977; Dahr et al., 1977a; Furthmayr, 1978; Blumenfeld & Adamany, 1978). Blood group M glycoprotein contains serine and glycine residues, whereas N glycoprotein has leucine and glutamic acid residues at the positions 1 and 5 respectively. It is the only difference found between the both antigens. The comparative studies on the oligosaccharides released from M and N glycopeptides by alkaline-borohydride degradation revealed neither qualitative nor quantitative differences (Lisowska et al., 1980).

All human erythrocytes, including homozygous M, contain "cryptic" N receptors which are exposed after tryptic digestion of red cells (for refs, see Issitt, 1981) or glycoproteins isolated from M or N erythrocytes (Baranowski & Lisowska, 1963; Lisowska & Jeanloz, 1973). Serological analysis of red cell sialoglycoproteins fractionated in the presence of detergents showed that "cryptic" N determinants are located in glycophorin B, which also carries blood group S and s antigenic sites (Fujita & Cleve, 1975; Anstee & Tanner, 1975; Dahr et al., 1975d, 1977a; Dahr & Uhelnbruck, 1978). The blood groups Ss are inherited with MN: the extensive family studies showed the existence of gene complexes Ns, Ms, MS and NS (Issitt, 1981). Location of blood group Ss determinants in glycophorin B was confirmed by finding that membranes of rare erythrocytes lacking Ss antigens (phenotype S-s-) do not show in electrophoretic analysis the PAS-positive bands corresponding to glycophorin B (Dahr et al., 1975c; Tanner et al., 1977). Furthmayr (1978) determined amino acid sequence of NH_2-terminal portion (residues 1-35) of glycophorin B and surprisingly found that it is identical to glycophorin A of blood group N up to the 26th amino acid residue. The results of Furthmayr did not give any information about a possible location of Ss determinants. Dahr et al. (1980a, 1980b) analyzed NH_2-terminal glycopeptides of glycophorin B obtained from homozygous S or s red blood cells. They confirmed the results of Furthmayr (1978), except the positions 11, 17, 28, 29 and 31 and they also pointed out an incomplete glycosylation of some serine and threonine residues (Fig.3). The important finding of Dahr et al. (1980a, 1980b) was that blood group Ss-related polymorphism of glycophorin B consists of the presence of different amino acid residues at the position 29 of the polypeptide chain, namely methionine in S and threonine in s glycoprotein. Testing the activity of the modified glycopeptides, they showed that these amino acid residues are essential components of blood group S or s antigenic sites. Dahr et al. (1980b) also reinvestigated the sequence of amino acid residues 9-30 of glycophorin A and found the residues at the positions 11 and 17 to be identical with those found by them in glycophorin B. In conclusion, Dahr et al. confirmed the identity of glycophorins A and B up to the 26th amino acid residue, but they suggested the corrections to their structures at these two positions. The corrected structure of glycophorin A has been recently confirmed by isolation and sequencing cDNA encoding glycophorin A (Siebert & Fukuda, 1986). Recently, Dahr *et al*.

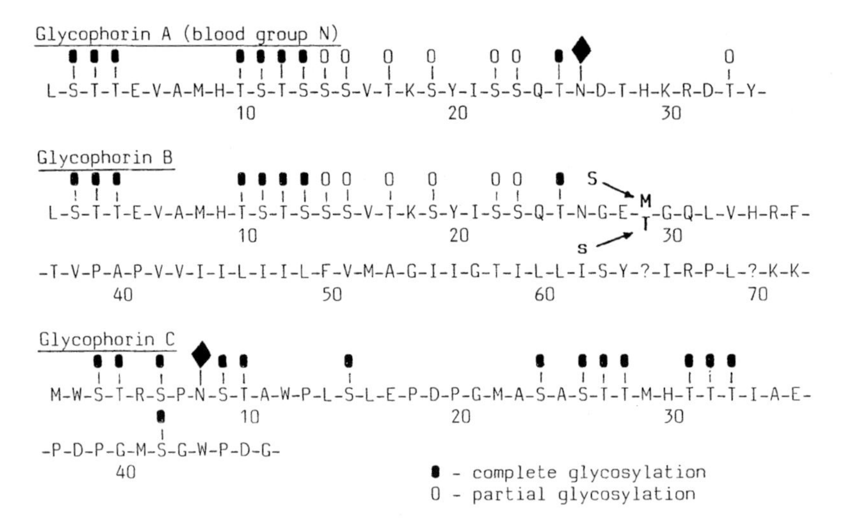

Fig.3. Structure of the NH_2-terminal portions of glycophorins A,B and C (Dahr *et al*., 1980b, 1982, 1985c, Dahr and Beyreuther, 1985, Colin *et al*, 1986).

(1985c) reported that polypeptide chain of glycophorin B is composed of approx. 97 amino acids and they extended the sequence determination up to its 72nd amino acid residue (Fig.3). They found a high degree of homology between the hydrophobic portions of glycophorin A (residues 58-101) and B (residues 29-72). Glycophorin B has no N-glycosidic chain. In view of the well established rule glycosylation of asparagine residue requires the sequence -Asn-X-Ser- or -Asn-X-Thr-; therefore, the 26th Asn residue, which is glycosylated in glycophorin A, cannot be glycosylated in glycophorin B due to the lack of serine or threonine residue at the position 28. The O-glycosidic chains of glycophorin B are the same as those carried by glycophorin A, since determination of the structure of the O-glycosidic chains was performed by Thomas and Winzler (1969) on a crude glycoprotein preparation containing all glycophorins. Moreover, glycophorins A and B react identically with lectins and antibodies directed against the O-glycosidic chains. Glycophorin B, in contrast to A, is resistant to trypsin treatment at the erythrocyte surface, but is more sensitive to chymotrypsin digestion.

Glycophorin C, present in erythrocytes in smaller amount than glycophorin B, has the polypeptide chain composed of 128 amino acid residues. Dahr et al. (1982, 1985c) and Dahr and Beyreuther (1985) determined the sequence of the NH_2-terminal glycosylated fragment (amino acid residues 1-47) of glycophorin C and a partial sequence of 45 amino acid residues in its hydrophobic intramembranous peptide. Recently a full sequence of glycophorin C has been deducted from the sequence of respective cDNA (Colin et al. 1986). The structure of glycophorin C is different from that of glycophorins A and B (Fig.3). Glycophorin C and another minor red cell membrane glycoprotein (γ according to Anstee et al., see Fig.1) probably carry the common erythrocyte antigens Gerbich (Ge) and, in contrast to glycophorins A and B, are associated with the membrane cytoskeleton (Mueller & Morrison, 1981; Anstee et al., 1984a, 1984b; Dahr et al., 1985a).

VARIANT FORMS OF GLYCOPHORINS

The occurrence of structural variations of the human red cell sialoglycoproteins has been suggested by serological detection of many variant antigens related to the MNSs blood group system (Issitt, 1981). These variant antigens, occurring usually with low incidence, have been defined by antibodies found in humans (in some cases antibodies to rare antigens are quite common) or in animal immune sera. On the other hand, rare individuals with erythrocytes lacking glycophorins or containing defective glycophorins usually have in the sera alloantibodies directed against common antigens absent in their red cells. Only few of the known variants have been structurally characterized and these are shortly described here. Generally, the variant forms of glycophorins fall into three categories: (1) with changes in the oligosaccharide chains, (2) with amino acid replacements in the polypeptide chain and (3) hybrid molecules.

Glycophorins with changed oligosaccharide chains

It is shown in Fig.4 that variant O-glycosidic chains of glycophorins may be either defective or more complex than the regular chains. Desialylation of erythrocytes, which can be easily performed *in vitro*, uncovers "cryptic" antigenic determinants recognized by antibodies commonly present in human and animal sera which agglutinate human and most of animal asialoerythrocytes. These determinants, known as Thomsen-Friedenreich (TF or T) receptors, were identified as desialylated O-glycosidic chains of glycophorins (for review see Uhlenbruck, 1981). These disaccharide chains, usually present in glycoproteins in a "cryptic" form, are strongly immunogenic. Besides the existence of the common antibodies mentioned above, it is possible to obtain polyclonal antisera or monoclonal antibodies with anti-TF specificity, using asialoerythrocytes or asialoglycophorin as an immunogen (Hanisch et al., 1983; Rahman & Longenecker, 1982; Longenecker et al., 1984; Steuden et al., 1985) However, asialoglycophorins are not real variant forms, because they may be formed *in vivo* only transiently, by the action of viral or bacterial neuraminidases. Another type of polyagglutination (Tn) was found in some indivivuals as an acquired long-lasting syndrome resulting from the somatic mutation in the precursor cells. Therefore, usually only a part of erythrocytes (descendants of the mutated cells) of a given individual shows the Tn characteristics. Tn erythrocytes (found to be different from TF by means of lectins) contain glycophorins with defective O-glycosidic chains, lacking sialic acid and galactose residues

(Dahr et al., 1975b) This defect is caused by the lack of galactosyltransferase involved in biosynthesis of these chains (Berger & Kozdrowski, 1978). The glycophorin variant with more complex oligosaccharide chains is present in erythrocytes carrying a rare antigen Cad. Cad is inherited as an autosomal dominant character and was first recognized by an unexpected strong reactivity of blood group O or B erythrocytes with

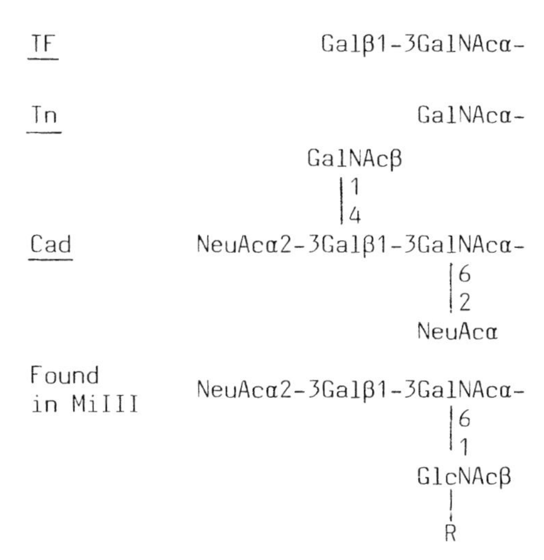

Fig.4. Variant O-glycosidic chains of human erythrocyte glycophorins (Vaith & Uhlenbruck, 1978; Dahr et al., 1975b; Blanchard et al., 1983; Adamany et al., 1983).

blood group A-specific *Dolichos biflorus* lectin. Glycophorins A and B in erythrocytes Cad have over 80% of their O-glycosidic chains modified by an additional N-acetylgalactosamine residue linked to carbon 4 of galactose residue (Blanchard et al., 1983; Herkt et al., 1985) Another example of more complex O-glycosidic chains of glycophorins was reported by Adamany et al. (1983). They found that MiIII erythrocytes (carrying type III antigen of Miltenberger system, discussed later in this article) contained glycophorin A with a significant proportion of the O-glycosidic chains in which N-acetylgalactosamine residue is substituted at carbon 6 with N-acetylglucosamine residue, instead of sialic acid. This N-acetylglucosamine residue may be in turn substituted with further galactose and N-acetylglucosamine residues, forming linear or branched structures. However, the GlcNAc-containing O-glycosidic chains of glycophorin A are not related to the

MiIII antigen, since they were not found in another MiIII individual (Adamany *et al*., 1983). Moulds (1983) reported a common presence of a small amount of GlcNAc-containing *O*-glycosidic chains in human glycophorins and an increased amount of such chains in M1, Tm and Can phenotypes which are more frequent in black donors than in Caucasians.

Glycophorins with changes in the MN-specific region of the polypeptide chain

Variant glycophorins with altered polypeptide chain are formed by exchange of one or more amino acid residues. If this alteration occurs within the NH_2-terminal blood group *M*- or *N*-specific portion of the molecule, the blood group activity is lost or changed. For example, the rare M^c erythrocytes are serologically identified as a blood group type intermediate between M and N, because they are agglutinated by most anti-M and by a minority of anti-N sera. Elucidation of the structure of M^c determinants (Fig.5) gave a reasonable explanation of the serological observations. Glycophorin A in M^c erythrocytes has the blood group M-typical residue (Ser) at the position 1 and blood group N-characteristic one (Glu) at the position 5 of the polypeptide chain (Dahr *et al*., 1981b; Furthmayr *et al*., 1981) So far, no antibody with anti-M^c specificity has been found or obtained. Another genetic variation at MN locus leads to the formation of M^g antigen. This extremely rare antigen is defined by anti-M^g antibodies that are common in the sera of normal individuals and also can be produced by immunization of animals with M^g erythrocytes (Issitt, 1981) The structural studies showed that M^g erythrocytes contain an altered glycophorin A of blood group N in which the glycosylated threonine residue at the position 4 is replaced with asparagine residue (Fig. 5). This change affects in turn the post-translational glycosylation of neighboring serine and threonine residues (Blumenfeld *et al*., 1981; Dahr *et al*.,1981a; Furthmayr *et al*.,1981) The lack of glycosylation in the antigenic site may be the major reason that M^g does not crossreact with anti-M or anti-N sera. However, it was found to react with two monoclonal anti-M antibodies which is rather surprising (see the section on

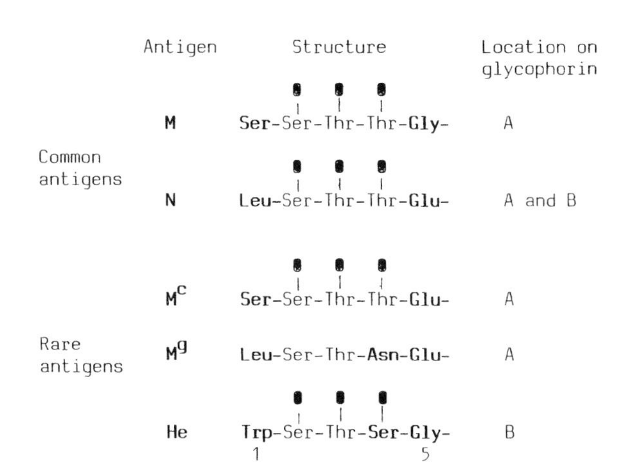

Fig. 5. Common and rare variant forms of glycophorins A and B with changes in the NH_2-terminal portion of the polypeptide chain (Dahr et al., 1977a, 1981a, 1981b, 1984a; Wasniowska et al., 1977; Furthmayr, 1978; Furthmayr et al., 1981; Blumenfeld & Adamany, 1978; Blumenfeld et al., 1981).

monoclonal antibodies). The second immunologically distinct antigen (He) associated with MN locus was found first in erythrocytes of Mr. Henshaw and then in about 3% of blacks, but only rarely in Caucasians (Issitt, 1981) Anti-He antibodies can be produced by immunization of animals with He erythrocytes and occur rarely in humans. Moreover, the He antigen reacts with some rabbit or human anti-M sera which agglutinate M-N+He+ red cells and are designated as anti-M^e (Wiener & Rosenfield, 1961; McDougall & Jenkins, 1981; Levene et al., 1984). Dahr et al. (1984a) found that He erythrocytes contain normal glycophorin A, but glycophorin B is altered at three positions of the polypeptide chain: leucine(1), threonine(4) and glutamic acid (5) are replaced with tryptophan, serine and glycine residues, respectively, without affecting the glycosylation in this region of the molecule (Fig. 5). The NH_2-terminal tryptophan and sialic acid residues were required for the reaction with human anti-He serum tested by Dahr et al. (1984a) The structure of He antigen suggests that rare anti-M^e sera are those of anti-M which have predominance of antibodies specific for an epitope with glycine residue at the position 5, present in M and He glycoproteins.

The variant glycophorins described above are products of altered genes. Interestingly, the genes encoding M^c and M^g antigens can be formed from the normal counterparts by a single DNA base mutation, whereas formation of the gene coding for He antigen must have required an exchange of several bases, or involvement of other mechanisms (Dahr et al., 1984a).

Glycophorins with changes in the internal portion of the polypeptide chain

Some low incidence antigens genetically related to the MN blood group system are serologically identified in red cells with unaltered MN antigens. It suggests that variant antigens of this type are not located in the region of blood group M or N determinants, but may be present in another portion of glycophorin A or B. Such antigens are exemplified by those belonging to the Miltenberger system. Initially, the Miltenberger system included erythrocytes reacting with anti-Mi^a sera (first found in Mrs. Miltenberger) and subdivided into 4 classes (MiI - MiIV), dependently on their reactivity with 4 other antisera: Verweyst (Vw), Hutchinson (Hut), Murrell (Mur), and Hill (Hil) (Cleghorn, 1966). Later, the system was extended to 8 classes, after new variant red cells and 3 additional antisera had been found (Dybkjaer et al., 1981). As is shown in Table 1, erythrcytes of some new classes do not react with anti-Mi^a sera and therefore, the initial definition of the system is not valid any more.

Table 1. The eight Miltenberger classes of erythrocytes (Dybkjaer et al., 1981).

Cell classes	Typing sera							
	Mi^a	Vw	Hut	Mur	Hil	Anek	Raddon	Lane
MiI	+	+	-	-	-	-	-	-
MiII	+	-	+	-	-	-	-	-
MiIII	+	-	-	+	+	-	-	-
MiIV	+	-	-	+	-	+	(-)	-
MiV	-	-	-	-	+	-	-	-
MiVI	+	-	-	+	+	+	(-)	-
MiVII	-	-	-	-	-	(-)	+	+
MiVIII	-	-	-	-	-	+	+	+

+ or -: positive or negative reaction, respectively;
(-): a weak positive reaction

The structural studies of Dahr et al. (1984b) showed that MiI and MiII erythrocytes carry an altered glycophorin A in which the 28th threonine residue is exchanged for methionine in MiI and for lysine residue in MiII red cells (Fig. 6). Both variants may result from a single DNA base exchange in the gene encoding the normal glycophorin A. The MiI and MiII specific glycophorin A molecules do not have the N-glycosidic oligosaccharide chain attached to the 26th asparagine residue, since similarly as glycophorin B, they do not have threonine or serine residue at position 28 of the polypeptide chain. It explains why the MiI and MiII glycoproteins show in SDS-PAGE a lower apparent molecular weight than normal glycophorin A. The reactivity of MiI glycoprotein with anti-Vw serum and of MiII with anti-Mi^{a} serum was impaired after treating the glycoproteins with methionine- or lysine-modifying reagents, respectively (Dahr et al., 1984b). It confirmed that the amino replacements described contribute to the novel epitopes of glycophorin A in MiI and MiII red cells.

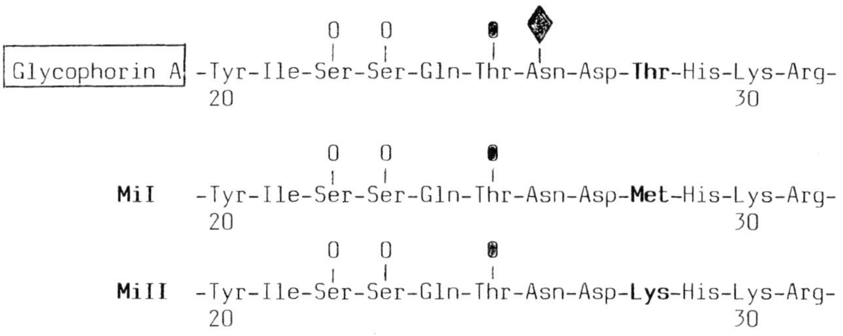

```
                     O    O          ●   ◆
                     |    |          |   |
Glycophorin A -Tyr-Ile-Ser-Ser-Gln-Thr-Asn-Asp-Thr-His-Lys-Arg-
               20                                  30

                     O    O          ●
                     |    |          |
MiI           -Tyr-Ile-Ser-Ser-Gln-Thr-Asn-Asp-Met-His-Lys-Arg-
               20                                  30
                     O    O          ●
                     |    |          |
MiII          -Tyr-Ile-Ser-Ser-Gln-Thr-Asn-Asp-Lys-His-Lys-Arg-
               20                                  30
```

Fig. 6. Structural changes in the MiI and MiII variants of glycophorin A (Dahr et al., 1984b).

The structural and antigenic relationship between MiI, MiII and other classes of the Miltenberger system and the complexity of crossreactions of the antisera defining this system will be fully understood after the structure of the other variant antigens is elucidated. Including these antigens into one system may appear fortuitous. Presently it is known that erythrocytes of MiIII, MiIV and MiVI classes must have altered glycophorin B, because it has an increased apparent molecular weight in SDS-PAGE (Dahr et al., 1978b, 1984b; Anstee et al., 1979). The red cells of MiV class carry a hybrid glycophorin which is described in the next section.

Hybrid glycophorins

The hybrid red cell membrane sialoglycoproteins characterized so far contain portions of glycophorins A and B. These hybrid molecules probably result from chromosomal misalignment and unequal crossing-over between the genes encoding glycophorins A and B (with elimination of normal genes), involving the mechanism similar to that proposed for the formation of Lepore and anti-Lepore hemoglobins (Anstee et al., 1982b). If the hybridization occurs within homologous amino-terminal portion of glycophorins A and B (amino acid residues 1-26), the hybrid molecules may be indistinguishable from normal glycophorin A or B. Such a hybrid was detected in En(a-) erythrocytes which lack glycophorin A, but have apparently normal glycophorin B. However, the En(a-) red cells of members of an English family were found to contain glycophorin B which expressed the blood group M activity and had amino-terminal serine residue (Dahr et al., 1978c). It suggested that this glycoprotein is a hybrid molecule composed of NH_2-terminal portion of glycophorin A of blood group M and COOH-terminal portion of glycophorin B. If the hybridization occurs in other regions of the molecules, the hybrids have molecular weights intermediate between those of glycophorins A and B. Besides the presence of antigenic determinants characteristic for the both glycophorins, such hybrids usually carry a novel antigenic determinant typical for the hybrid and most probably formed at the region of connection of the polypeptide chains derived from glycophorin A and B. Erythrocytes carrying the hybrid glycoproteins are characterized by the lack (homozygotes) or reduced content (heterozygotes) of normal glycophorins A and B. Two types of hybrid glycophorins are known, with NH_2-terminal portion of glycophorin A and COOH-terminal portion of glycophorin B, and vice versa.

The electrophoretic pattern of sialoglycoproteins from MiV erythrocytes shows the presence of several unusual bands representing a sialoglycoprotein migrating similarly to glycophorin C and its homo- and hetero- dimers, whereas the bands corresponding to glycophorins A and B are either less intense (MiV heterozygotes) or absent (MiV homozygotes). The

Table 2. Hybrid glycophorins

Composition	Designation	Antigenic characteristics
NH_2-terminal portion of glycophorin A (α) and COOH-terminal portion of glycophorin B (δ)	$(\alpha\text{-}\delta)^{MiV}$	M or N, s, Wr(b-), Hil+
	(α-δ) found in J.R.	M, S, Wr(b-), Hil-
NH_2-terminal portion of glycophorin B (δ) and COOH-terminal portion of glycophorin A (α)	$(\delta\text{-}\alpha)^{Dantu}$	N, s, Wr(b-), St(a-), Dantu+
	$(\delta\text{-}\alpha)^{St^a}$	N, Wr(b+), St(a+), Dantu-

References are given in the text.

MiV glycoprotein carries the blood group M or N and s determinants, some other epitopes of glycophorin A, and Hil antigenic determinant characteristic for III, V and VI classes of the Miltenberger system. The MiV glycoprotein, similarly to glycophorin A, is digested by trypsin at the erythrocyte surface. These findings led to the conclusion that MiV glycoprotein is a hybrid containing NH2-terminal portion of glycophorin A and COOH-terminal portion either of glycophorin B of blood group s, or of MiIII variant of glycophorin B carrying s and Hil antigens (Anstee & Tanner, 1978; Dahr et al., 1978b; Anstee et al., 1979, 1982b; Vengelen-Tyler et al., 1981; Judd et al., 1983). An apparently similar hybrid glycoprotein was found in erythrocytes of J.R. individual, which do not express any antigens of the Miltenberger system (Langley et al., 1981). The hybrid glycoprotein of J.R. carries blood group M and S antigenic determinants. Although the J.R. and MiV hybrid glycoproteins are in some respects indistinguishable, they are not identical. Besides the S antigen present in J.R. hybrid instead of s present in all known MiV glycoproteins, the former one does not express the Hil antigen. It shows that fragments of different glycophorin B molecules are involved in formation of J.R. and MiV hybrid glycoproteins, and it cannot be ruled out either that both hybrids contain different NH_2-terminal fragments derived from glycophorin A.

Another type of hybrid molecules represents glycoproteins containing NH_2-terminal portion of glycophorin B and COOH-terminal portion of glycophorin A (Tanner et al., 1980; Mawby et al., 1981; Anstee et al., 1982a; Blanchard et al., 1982). These hybrid sialoglycoproteins show in SDS-PAGE an apparent molecular weight similar to that of the minor glycoprotein γ and also are involved in formation of homo- and heterodimers. They express blood group N determinants and are resistant to proteolytic digestion on the erythrocyte surface. Despite many common features, these hybrid glycoproteins can be subdivided into two distinct types (Table 2), dependent on the presence of Stones (St^a) and Dantu antigens. These antigens were defined on rare erythrocyte samples by occasionally found human antibodies (Cleghorn, 1962; Contreras et al., 1984). It has been recently found that St^a and Dantu antigens are present on erythrocytes carrying δ-α type of hybrid glycoprotein, and therefore, they may be located at the junction site of glycophorin B and A fragments. One type of the hybrid glycoprotein is Dantu+, St(a-) and Wr(b-) (Wr^b is a common antigenic determinant discussed later in this article), whereas the second type is Dantu-, St(a+) and Wr(b+) (Anstee et al., 1982a; Ridgwell et al., 1984; Dahr et al., 1985d). Moreover, St(a+) hybrid does not express blood group Ss activity, and St(a+) heterozygotes have an approx. 50% dose of normal glycophorins A and B. Dantu+ erythrocytes have about half a dose of glycophorin A, but do not have glycophorin B, and the hybrid glycoprotein shows the presence of weak s. Recently, the complete amino acid sequence of Dantu glycoprotein has been reported (Dahr et al., 1985d). The hybrid molecule is composed of 99 amino acid residues and has the residues 1-39 of blood group S specific glycophorin B and the residues 72-131 of glycophorin A. The structure of other hybrid glycophorins and the location and size of the antigenic sites characteristic for the hybrids have not yet been elucidated.

ERYTHROCYTES LACKING GLYCOPHORINS

Besides the existence of many variant forms of glycophorins, some rare erythrocytes have been found in which one or even two glycophorins are totally absent. These rare individuals who lack glycophorin A have in their sera alloantibodies directed against common epitopes of this glycoprotein. The total absence of glycophorin A was revealed in En(a-) erythrocytes (Dahr *et al.*, 1976a, 1976b; Gahmberg *et al.*, 1976; Tanner *et al.*, 1976). These deficient red cells and alloantibodies present in the sera of En(a-) persons are discussed in detail in the next section. A more profound defect has been disclosed in M^k erythrocytes. The M^k red cells, initially known in heterozygous form, showed an approximate 50% decrease in the content of glycophorins A and B (Dahr *et al.*, 1977b, 1978b; Anstee & Tanner, 1978). The first two cases of homozygous M^k erythrocytes, reported by Tokunaga *et al.* (1979), showed a complete absence of glycophorins A and B. The sera of these M^k homozygous persons contained antibodies directed against the glycosylated portions of glycophorins A and B. A selective absence of glycophorin B has been mentioned earlier. The red cells lacking Ss antigens (phenotype S-s-) were shown to contain no glycophorin B demonstrable in SDS-PAGE (Dahr *et al.*, 1975c, 1978a; Tanner *et al.*, 1977). These cells contain an abnormal component which is not glycosylated and shows a slightly higher apparent molecular weight than glycophorin B, but its relation to glycophorin B has not been fully established (Tanner *et al.*, 1977; Anstee *et al.*, 1979) The S-s- phenotype is quite common (up to 30%) in negroid populations (ref. in Anstee *et al.*, 1982b) Glycophorin C, and minor glycoproteins γ and β_1 are absent from erythrocytes which do not have common Gerbich (Ge) antigens (Anstee *et al.*, 1984a, 1984b; Dahr *et al.*, 1985a) and from red cells with homozygous hereditary elliptocytosis, which were not typed for Ge antigens (Alloisio *et al.*, 1985). It should be noted that Ge-negative erythrocytes of two persons studied by Anstee *et al.* (1984a) also were elliptocytic. Other Ge-negative red cells with normal shape contain abnormal sialoglycoproteins, immunologically related to glycophorin C (Anstee et al., 1984b; Dahr et al., 1985a).

It is surprising that lack of the major membrane sialoglycoprotein, glycophorin A (approx. 10^6 copies are present in one cell) does not affect physical properties or function of erythrocytes. It is possible that the still unrecognized role of glycophorin A is taken over by other membrane components in the deficient red cells. An increased glycosylation of band 3 (the major protein of red cell membranes) observed in En(a-) erythrocytes (Tanner et al., 1976; Gahmberg et al., 1976) may be connected with this possibility, but there is no evidence whether this fact has any functional significance. On the other hand, the minor membrane components, glycophorin C and glycoprotein γ, may play a role in maintaining the discoid shape of erythrocytes. Both glycoproteins are associated with the membrane cyto-skeleton (Owens et al., 1980; Anstee et al., 1984a; Alloisio et al., 1985) and the term glycoconnectin was suggested for glycophorin C (Mueller & Morrison, 1981). Anstee et al. (1984b) implied that lack of glycophorin C and glycoproteins β_1 and γ is connected with ellipocytosis and that this deficiency may be compensated by the presence of structurally related variant glycoproteins.

COMMON ANTIGENIC DETERMINANTS OF GLYCOPHORIN A DEFINED BY HUMAN ANTIBODIES

En^a and Wr^b antigens

Darnborough et al. (1969) and Furuhjelm et al. (1969) described an English (M.E.P.) and Finnish (V.B.) individual, respectively, whose sera contained antibodies against a common erythrocyte antigen which was absent in red cells of the propositi. The antigen was called En^a, and the deficient erythrocytes were designated as En(a-) The depressed expression of MN antigens and decreased amount of sialic acid in En(a-) erythrocytes suggested that En^a may be related to MN sialoglycoproteins (glycophorin A). In 1976 three research groups reported independently that membranes of En(a-) red cells demonstrate a lack of glycophorin A in SDS-PAGE and an increased glycosylation of band 3 (Dahr *et al.*, 1976a, 1976b; Gahmberg *et al.*, 1976; Tanner *et al.*, 1976). The results of the electrophoretic studies and immunological properties of En(a-) erythrocytes indicated that these cells lack the whole molecules of glycophorin A. A strong support for this conclusion was provided by finding that En(a-) membrane extracts fail to react with antibodies against the COOH-terminal portion of glycophorin A (Furthmayr, 1978). A heterozygous En(a-) condition, with approx. 50% of normal glycophorin A content in erythrocytes, has been shown for several members of families of the homozygous En(a-) individuals. Despite the total lack of glycophorin A, the En(a-) erythrocytes expressed weakly M or N antigens. It was shown to result from the unchanged content of glycophorin B (N, S or s), or glycophorin B-like hybrid glycoprotein (M, S), detected in the English family (Dahr *et al.*, 1976b, 1978c; Anstee *et al.*, 1977). The findings described above strongly suggested that anti-En^a antibodies are directed against common epitopes of glycophorin A. However, agglutination of erythrocytes by the anti-En^a sera was not inhibited by glycophorin A. This confusing result, which gave rise to various speculations, has been partly explained after more anti-En^a sera were found. The En(a-) condition is extremely rare, the third case (R.L.) was found in Canada by Taliano *et al.* (1980). However, anti-En^a were found to be present as alloantibodies in homozygotes with hybrid glycophorins of α-δ type(the antibodies are directed against a lacking fragment of glycophorin A) and as autoantibodies in patients suffering from autoimmune hemolytic anemia (Issitt, 1981; Dahr *et al.*, 1985b).

The detailed serological studies on a larger number anti-En^a sera showed that the term En^a includes various epitopes located on different fragments of glycophorin A polypeptide chain, and anti-En^a denotes a broad group of antibodies with different specificities (Issitt, 1981). A diversity among anti-En^a antibodies was shown by comparing their reactivity with untreated and protease-treated red cells. Some of the antibodies recognize epitopes removed (or destroyed) by trypsin, the other ones react with determinants resistant to trypsin treatment, but removed by papain or ficin, and antibodies of the third category react with determinants left on red cells after all enzyme treatments. Issitt et al. (1981) proposed to subdivide the En^a antigens (and antibodies defining them) into 3 groups: En^aTS (trypsin-sensitive), located nearer the NH_2-terminus of glycophorin A; En^aFS (ficin-sensitive), located intermediately; En^aFR (ficin-resistant), located near the membrane lipid bilayer. Anti-En^a sera may contain antibodies of all categories which can be separated by absorption on the protease-treated erythrocytes. Moreover, the antibodies in each group are heterogenous in respect to specificity. Issitt (1981) pointed out that the proposed terminology of anti-En^a antibodies may be replaced by the more precise one after their exact specificity is elucidated. Interestingly, the lack of inhibition of anti-En^a antibodies by glycophorin A was found to concern the anti-En^aFR only, whereas anti-En^aTS and anti-En^aFS are inhibitable (Pavone et al., 1981; Issitt, 1981) which facilitates studies on their specificity. Dahr et al. (1985b) tested the inhibition of several anti-En^aTS and anti-En^aFS allo- and autoantibodies by various untreated and modified fragments of glycophorin A. The six anti-En^aFS sera studied are directed against the residues approx. 46-56 of the polypeptide chain and five of them require a sialooligosaccharide chain attached to the 50th threonine residue. The three anti-En^aTS sera tested were found to react with overlapping regions of glycophorin A, comprising amino acid residues approx. 27-33, 31-39 and 36-42. These studies confirmed the heterogeneity in glycosylation of glycophorin A and showed that two of the anti-En^aTS sera (specific for 27-33 or 31-39 residues) react with those glycophorin A molecules or fragments only in which the 33rd threonine residue is not glycosylated. The knowledge of specificity of anti-En^a antibodies makes them useful tools to study the structure of hybrid glycophorins and other glycophorin A-related glycoproteins.

It is more difficult to characterize determinants for anti-En^aFR sera since they are not inhibited by glycophorins. The same concerns a related Wr^b antigen, belonging to a still hypothetical blood group Wright (Wr) system. The story of the

Wr system is shortly as follows. Many human sera contain antibodies defining a rare antigen called Wr^a which segregates independently of blood group MN antigens (Issitt, 1981). The serum of one Wr(a+) individual (M.Fr.) was found to contain an antibody which reacts with a common antigen present in all erythrocytes tested, but absent from M.Fr. red cells (Adams et al., 1971) It suggested that the antibody defines the antigen antithetical to Wr^a and therefore termed Wr^b. The subsequent serological studies with the use of unique Wr(a+b-) erythrocytes of M.Fr. allowed to identify anti-Wr^b antibody in more sera: it is present as alloantibody in anti-En^aFR sera and as autoantibody in some patients with autoimmune hemolytic anemia (Issitt et al., 1976; Pavone et al., 1978; Issitt, 1981). Moreover, mouse monoclonal antibodies with anti-Wr^b specificity were obtained (see the section on monoclonal antibodies). Although erythrocytes of M.Fr. do not show any defect or alteration in MN antigens and membrane sialoglycoproteins, there are several data suggesting the connection of Wr^b antigen with glycophorin A. The Wr^b antigen is absent in En(a-) erythrocytes lacking glycophorin A and in homozygous erythrocytes carrying a hybrid glycophorin of α-δ type, lacking a part of glycophorin A molecule (Issitt et al., 1975, 1975, 1976; Issitt, 1981). Moreover, monoclonal anti-Wr^b antibodies bind specifically glycophorin A from Wr(b+) erythrocytes that was shown by immunoprecipitation assay (Ridgwell et al., 1983). Employing the same procedure, Ridgwell et al. (1984) showed the presence of Wr^b antigen in (δ-α)St^a and its absence in (δ-α)Dantu hybrid glycoprotein. All these findings and resistance of Wr^b antigen on erythrocytes to ficin treatment suggest that Wr^b is one of the En^aFR antigenic determinants which is selectively modified in apparently normal glycophorin A of M.Fr. red cells. One of the possible explanations of the failure of glycophorin A to inhibit anti-Wr^b and anti-En^aFR antibodies is the location of the epitopes in the α-helical polypeptide chain portion which is denatured during isolation procedures (Schulte & Marchesi, 1979; Ridgwell et al., 1983). Involvement of phospholipids in En^aFR and Wr^b determinants also is possible. Recently, Rearden (1985) reported that treatment of erythrocytes or their membranes with phospholipase A_2 distinctly decreases the binding of two anti-Wr^b monoclonal antibodies, being without any effect on several other red cell antigens. Probably both, the proper conformation and interaction with lipids are required for the expression of Wr^b and En^aFR antigenic determinants (Dahr, 1983).

Similar to En^aFR, labile common antigens in the internal portion of glycophorin B are defined by anti-U and anti-Duclos human antibodies (Issitt, 1981). Limited immunochemical studies performed so far indicate that these antigenic sites are located closer to the membrane lipid bilayer than blood group Ss determinants (Dahr, 1983).

Pr antigens

Various cold agglutinins detected in human sera (reviewed by Roelcke, 1984) are usually monoclonal immunoglobulins which occur in patients with chronic lymphoproliferation and recognize carbohydrate antigens. Those cold agglutinins which react with protease-sensitive antigens of red cells are named anti-Pr (Roelcke & Uhlenbruck, 1969). Anti-Pr agglutinins, except anti-Pr^a with still unclear specificity, fail to react with neuraminidase-treated erythrocytes. They show a strong reduction in agglutination of En(a-) red cells, and are inhibited by glycophorin A and its different glycosylated fragments (Merz & Roelcke, 1971; Roelcke et al., 1971; Roelcke, 1981, 1984; Relcke & Kreft, 1984). It shows that anti-Pr are directed against the sialylated O-glycosidic chains present on glycophorin A and other glycophorins, and for that reason they agglutinate weakly En(a-) erythrocytes. Anti-Pr agglutinins are a differentiated group of antibodies, most frequently of IgM class. Although desialylated glycophorins fail to inhibit anti-Pr antibodies, the effects of modifications of antigen sialic acid residues are not uniform and became the basis for dividing anti-Pr into 3 major subspecificities: anti-Pr_1; -Pr_2; and -Pr_3. Modifications used include periodate oxidation that transforms sialic acid residue into 7-carbon derivative, or blocking the carboxyl groups of sialic acid with carbodiimide compounds (Suttajit & Winzler, 1971). Both modifications abolish the reaction of glycophorins with anti-Pr_1 antibodies. Anti-Pr_2 show strongly enhanced reactivity with periodate-oxidized glycoproteins and they do not react with carbodiimide-treated ones. In contrast to anti-Pr_2, the reactivity of anti-Pr_3 antibodies with the periodate-oxidized glycoproteins is abolished, and with carbodiimide-treated ones is enhanced (Lisowska & Roelcke, 1973; Roelcke et al., 1976; Roelcke, 1981, 1984; Roelcke & Kreft, 1984). Actually, more subspecificities of anti-Pr antibodies exist, since some of them do not fit to any of the 3 categories, and anti-Pr_1 and anti-Pr_3 are further subdivided, dependent on their reactivity with animal erythrocytes (Roelcke 1981, 1984; Roelcke & Kreft, 1984). Some anti-En^aTS antibodies resemble anti-Pr, since they also are dependent on antigen sialic acid residues and react with glycosylated determinants removed from red cells by proteases (Dahr et al., 1985b). Therefore, classification of a new antibody to anti-Pr or anti-En^a category may encounter difficulties. The basic difference between anti-En^a and anti-Pr antibodies is that anti-En^a are specific for a definite peptide fragment of glycophorin A, located usually in the less glycosylated or nonglycosylated portion, whereas anti-Pr recognize predominantly oligosaccharide chains and react with heavily glycosylated portions of glycophorins and possibly other sialoglycoproteins. However, the exact specificity of anti-Pr antibodies has not been fully elucidated yet and there are many unsettled problems, like in the case of anti-TF antibodies, directed against desialylated O-glycosidic chains of glycophorins.

'NEOANTIGENS' FORMED BY CHEMICAL MODIFICATIONS OF BLOOD GROUP M AND N DETERMINANTS

The properties of blood group M and N related antigenic determinants are discussed in detail in the section on monoclonal antibodies. Here are described products of their chemical modifications and antibodies specific for the modified determinants. Oligosaccharide chains contribute to antigenicity of M and N determinants and release of sialic acid abolishes the reactivity of the antigens with most anti-M and anti-N sera. However, the glycoproteins submitted to mild acid hydrolysis (desialylation) or several steps of Smith degradation (deglycosylation) elicit the production of antibodies which do not react with native antigen, but are specific for desialylated or deglycosylated blood group M or N determinants, respectively (Lisowska & Kordowicz, 1977a, 1977b; Kordowicz & Lisowska, 1978). Another type of antigenically changed blood group M and N determinants can be obtained by modifications of amino groups of the NH_2-terminal amino acid residues (Ser of Leu) of glycophorin A. Especially interesting are the reactions of these amino groups with glucose and formaldehyde, since they occur also *in vivo* and human antibodies against such modified M and N determinants are known.

Some human sera agglutinate erythrocytes pretreated with glucose and frequently this agglutination is specific for blood group M or N red cells (Drzeniek et al., 1981; Morel et al., 1981; Reid et al., 1981). The 'activation' of erythrocytes by incubation in glucose-containing solution has the features of chemical reaction: it is dependent on time, temperature, glucose concentration and pH (the reaction occurs at neutral and better at slightly alkaline pH). It is well known that sugars react with amino groups of proteins, this reaction can be performed *in vitro* and occurs *in vivo*, being enhanced in diabetic patients (Bunn & Higgins, 1981). It was suggested that the antibodies mentioned above recognize blood group M or N determinants modified by the reaction of glucose with amino groups of the NH_2-terminal serine or leucine residue, respectively (M_{glc} or N_{glc} determinants). The results of inhibition studies confirmed this suggestion. The serum Mar, containing anti-M_{glc} antibodies (Drzeniek et al., 1981), was inhibited by 0.1 M glucose and by the low concentration solution of M glycoprotein (40-200 µg/ml), pretreated with glucose and then separated from free glucose. Untreated M and N glycoproteins and glucose-treated N glycoprotein were inactive. Out of many other monosugars tested, only mannose and N-acetylglucosamine (epimers of glucose) gave the same effects as glucose. The other anti-M_{glc} and anti-N_{glc} antibodies also were weakly inhibited by glucose and some other monosugars, not necessarily epimers of glucose (Morel et al., 1981; Reid et al., 1981). The weak affinity of antibodies in the Mar serum to glucose was used for their purification by affinity chromatography on Sephadex G-25 that gave the preparation containing IgG and IgM (Drzeniek, 1983). Interestingly, M-positive erythrocytes of diabetic patients are agglutinated by anti-M_{glc} sera without prior incubation in glucose solution (Reid et al., 1981; our unpublished results). Probably the same result would be obtained with anti-N_{glc} sera, but they were not tested with diabetic red cells. The modification by glucose did not affect the agglutination of erythrocytes by rabbit and human anti-M and anti-N sera (Morel et al., 1981), but glucose-pretreated M glycoprotein showed a slightly decreased inhibition of rabbit anti-M (Drzeniek et al., 1981). It suggests that M and N determinants modified by glucose fail to react with amino group-dependent anti-M and anti-N sera but the extent of the modification is too low to give a pronounced reduction of activity.

The origin of antibodies specific for glucated M or N determinants is unknown. One antiserum was found in a diabetic patient (Reid et al., 1981) but the remaining ones were derived from apparently healthy untransfused blood donors (Morel et al., 1981; Drzeniek et al., 1981). The existence of antibodies against glucated red cell determinants has practical implications. Such antibodies may be useful for diagnostic purposes. Moreover, attention should be paid to this problem when sera are screened for the presence of anti-erythrocyte antibodies, since test red cells are frequently stored in glucose-containing solutions.

The antibodies against glucated M or N determinants are usually compared with anti-N-like antibodies found in some dialysis patients. The latter antibodies are formed only if dialyzers are sterilized with formaldehyde and they are directed against blood group N determinants modified by this reagent, highly reactive with amino groups (Dahr & Moulds, 1981; Lynen et al., 1983). However, the problem of immunization process and of specificity of these antibodies is rather complicated. They are formed independently of MN blood group type of the patient and their properties change at different stages of immunization. The antibodies react best with formaldehyde-treated blood group N glycoprotein, but they also react with similarly treated M and with untreated N glycoprotein. Desialylation of the antigens abolishes their reaction with dialysis-related antibodies, whereas it is without any effect on the reaction with antibodies against glucated M and N determinants. Therefore, both kinds of antibodies are different, except that they are specific for blood group M or N determinants with modified amino groups.

MOUSE HYBRIDOMA MONOCLONAL ANTIBODIES IN STUDIES ON GLYCOPHORINS

Blood group M and N determinants and anti-M and anti-N monoclonal antibodies

Blood group M and N types of glycophorin A are defined by amino acid residues at the positions 1 and 5 of the polypeptide chain, where M glycoprotein contains serine and glycine, and N glycoprotein has leucine and glutamic acid residues, respectively; the amino acid residues 2,3 and 4 carry the oligosaccharide chains (Figs. 2 & 5). The polyclonal anti-M and anti-N reagents widely used so far are either rabbit antisera, usually obtained by immunization with homozygous M or N erythrocytes and cross-absorption, or occasionally found human allo- or autoantibodies (Issitt, 1981). The question arises what is the exact size and location of epitopes involved directly in the reaction with anti-M and anti-N antibodies. The contribution of both, carbohydrates and amino acid residues to the antigenicity of M and N determinants has been demonstrated by early findings that M and N antigens are inactivated by desialylation or modification of amino groups (for refs. see the section on the structure of glycophorins). Not only removal, but also the modifications of sialic acid residues by periodate oxidation-$NaBH_4$ reduction or carbodiimide treatment abolish blood group M and N activity (Lisowska & Duk, 1972; Ebert et al., 1972; Liao et al., 1973; Dahr et al., 1975d). It has been established that inactivation of M and N antigens by amino group-modifying reagents occurs due to the modification of α-amino group of the NH_2-terminal amino acid residue (Lisowska & Wasniowska, 1978; Prohaska et al., 1981). Moreover, the removal of the NH_2-terminal amino acid residue by Edman degradation, or elongation of the peptide chain by chemical attachment of an additional amino acid residue also destroy the M and N activity (Dahr & Uhlenbruck, 1978; Lisowska _et al._, 1979). These results show that sialic acid residues and NH_2-terminal amino acid residue with free amino group are essential components of blood group M and N determinants. However, a great diversity has been observed among anti-M and anti-N sera and the results described above do not pertain to all of them.

Although most anti-M and anti-N sera require sialic acid residues in the antigens, the antisera which react equally well with untreated and neuraminidase-treated erythrocytes are not rare (Judd _et al._, 1979; Issitt & Wilkinson, 1983). Moreover, the sialic acid-dependent antisera are not uniform either, since for some of them the presence of one sialic acid residue (attached to galactose residue) in _O_-glycosidic chains is sufficient for maximal hemagglutination, whereas the other ones require both sialic acid residues, attached to galactose and _N_-acetylgalactosamine residue, respectively (Sadler _et al._, 1979). Similar differences concern the effect of amino group modification. In our studies we came across several rabbit anti-M sera which, in contrast to others, reacted with _N_-acetylated M glycoprotein (Lisowska & Duk, 1975b; and unpublished results). Human and rabbit anti-M and anti-N sera can also be subdivided dependent on their reactivity with variant M^c and He antigens. As mentioned earlier, glycophorin A of M^c type has the amino acid residues typical for M and N blood group at the positions 1(Ser) and 5(Glu), respectively, and reacts with most of anti-M and minority of anti-N sera. Some anti-M sera (termed anti-M^e) react with He antigen, which is the glycophorin B variant similar to M glycoprotein in that it contains glycine residue at the position 5 (see Fig.5 and the section on variant glycophorins). It indicates that those anti-M and anti-N sera which are reactive with variant M^c and He antigens require the presence of only one out of the two blood group M or N specific amino acid residues. The antisera which do not react with these variant antigens may be directed either against an epitope containing the amino acid residue lacking in the variant, or against a more complex determinant which includes amino acid residues at the position 1 and 5 as well. All the above-listed differences in the serological properties of polyclonal anti-M and anti-N sera must result from the presence of different predominant antibodies in these sera. It indicates in turn that a variety of anti-M and anti-N antibodies may be produced which react with different and perhaps overlapping portions of the MN-specific region of glycophorin A.

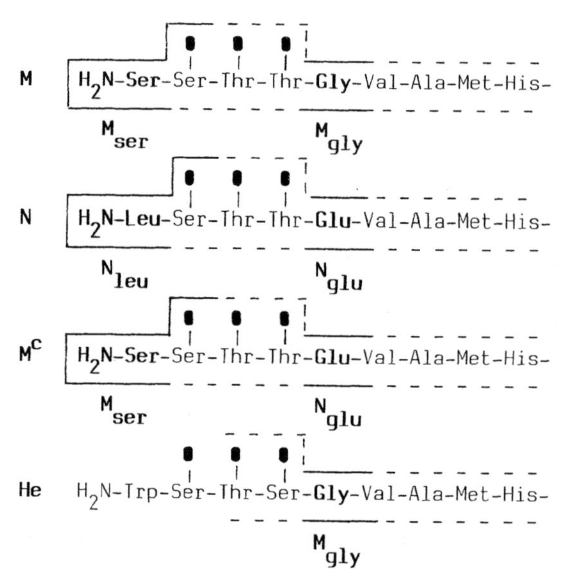

Fig. 7. Suggested subspecificities within blood group M and N antigenic determinants.

The differences in reactivity of anti-M and anti-N sera with M^c and He antigens suggest that two major epitopes can be distinguished within the blood group specific NH_2-terminal portion of M or N glycoprotein. These epitopes, located in the region of the 1st and 5th amino acid residue (Fig.7), are provisionally designated as M_{ser}, M_{gly}, N_{leu} and N_{glu} (Wasniowska et al., 1985). The average size of an antigenic epitope is in the range of several amino acid or carbohydrate residues (see Dr. E. A. Kabat, this book). Therefore, the NH_2-terminal epitopes M_{ser} or N_{leu} should include not only the NH_2-terminal serine or leucine, but also at least one adjacent glycosylated amino acid residue. It increases the probability that carbohydrates are directly involved in the antigenic site and that respective antibodies are dependent on antigen sialic acid residues and amino group. On the other hand, the epitopes of M_{gly} or N_{glu} type may be extended in both directions and may include glycosylated amino acid residue(s) and(or) nonglycosylated portion of the polypeptide chain (Fig.7). Antibodies against these epitopes are more likely to be independent or only partly dependent on antigen sialic acid residues and independent of amino groups.

The suggested blood group M and N subspecificities may represent an oversimplified view, especially due to the lack of knowledge of a spatial arrangement of the NH_2-terminal region of glycophorin A. The nuclear magnetic resonance studies showed that M and N glycopeptides have very similar, if not identical secondary structure (Prohaska et al., 1981; Carter et al., 1984). It supports the view that anti-M and anti-N antibodies directly recognize the differences in their primary structure, i.e. different amino acid residues in M and N antigens. However, it does not rule out the possibility of existence of conformational blood group M and N antigenic determinants.

A convenient tool to study antigenic properties of blood group M and N determinants are mouse hybridoma monoclonal antibodies. Many anti-M and anti-N monoclonal antibodies have already been obtained and some of them described in available literature, are listed in Table 3. They represent different Ig classes and subclasses, and there is no apparent relation between the antigen used for immunization (erythrocytes or isolated glycoprotein), Ig isotype and other properties of the antibodies. All anti-N monoclonal antibodies agglutinate not only N-positive erythrocytes, but also in lower degree homozygous M red cells. Including into the tests S-s-U-, En(a-) and protease-treated erythrocytes showed that this cross-reactivity occurs due to interaction of the antibodies with blood group N determinants of glycophorin B. Interestingly, the anti-N antibodies 35/5F and 179K agglutinate more strongly M,S+s- erythrocytes than M,S-s+ ones. It confirms earlier findings that homozygous S erythrocytes have more glycophorin B than homozygous s red cells (Dahr, 1981). A high anti-N specificity of the antibodies NN3, NN4, NN5, 179K, 35/5F and iB5 was additionally demonstrated by other methods, including quantitative immunoflourescence studies, electrophoretic fractionation of erythrocyte membranes

Table 3. Monoclonal antibodies with anti-M and anti-N specificity

Immunogen used[1]	Antibody		Reaction with desialylated antigen[2]	Type of pH-dependence[3]	References
			Anti-M		
E+sgp	IgG1	6A7	-		Bigbee et al., 1983, 1984
	IgG1	9A3	-		
E	IgG1	BS38	+++	(1)	Sonneborn et al., 1984; Fraser et al., 1985
		BS39	+		
	IgG1	Bs44	+		
E	IgG2b	G8	-		Nichols et al., 1985
	IgG1	E3	++		
sgp	IgG1	LM110/140	-	(1)	Fraser et al., 1985
	IgG2b	LM110/149	-	(1)	
sgp	IgM	425/2B	+	(1)	Lisowska et al., unpublished
			Anti-N		
sgp	IgG2b	LM17/19	+	(2)	Fraser et al., 1982
	IgG1	LM17/20	-	(2)	
E+sgp	IgG1	8A2	-		Bigbee et al., 1983, 1984
	IgM	NN3	-		
	IgG2a	NN4	-		
	IgG1	NN5	-		
E	IgM	2/23	-	(3)	Fletcher & Harbour, 1984
E	IgG1	HD4-D10			Allen et al.,1984
E		BS40	-	(4)	Sonneborn et al., 1984
		BS41	-		
E(cord)	IgG1	iB5			Edelman et al., 1984
E	IgM	179K	-	(1)	Waśniowska et al., 1985
	IgM	35/5F	-	(1)	
sgp	IgG	N/61	-	(3)	Lisowska et al., unpublished
	IgG	N/92	++	(2)	

[1]E: erythrocytes, sgp: sialoglycoprotein; [2]+++, ++, +, -: an enhanced, unchanged, decreased or negative reaction, respectively, with desialylated erythrocytes or glycoproteins, in comparison with the untreated ones; [3]four different patterns of pH-dependence are described in the text.

followed by immunoblot technique, immunoprecipitation, microplate ELISA (enzyme-linked immunosorbent assay), and inhibition studies with the use of highly purified red cell sialoglycoproteins and glycopeptides. Most anti-N monoclonal antibodies fail to react with desialylated antigen. An exception is the antibody N/92, obtained and investigated presently in our laboratory, which agglutinates equally well untreated and desialylated erythrocytes. Monoclonal antibodies with anti-M specificity show more differentiated properties. Except the potent 6A7 and weaker BS44 which are shown to be highly specific for M antigen, the other antibodies described crossreact with homozygous N erythrocytes or with purified N glycoprotein. Only 5 out of 10 anti-M antibodies are totally nonreactive with desialylated antigens, the other ones show decreased, unchanged or even enhanced reaction (Table 3).

Despite the crossreactivity of most anti-M and anti-N monoclonal antibodies in agglutination of M and N red cells, the antibodies may serve as specific blood group-typing reagents after dilution or after selecting the proper pH for testing. The pH-dependence of polyclonal anti-M and anti-N sera is known (Issitt, 1981). However, the effect of pH on monoclonal anti-M and anti-N antibodies is more strongly pronounced and highly diverse. The monoclonal antibodies tested showed a decrease of activity at pH above 8.5 and below 6, except anti-M BS38 which agglutinates M and N erythrocytes more strongly at pH 4.5 than 6. The most interesting differences in the effect of pH are, however, observed around neutral condition, since monoclonal anti-M and anti-N antibodies show at least 4 different patterns of dependence of hemagglutination on pH at the range of 6 - 8.5:

(1) The agglutination of M and N red cells by anti-M and anti-N antibodies, respectively, is unaffected or only slightly decreased at higher pH, whereas agglutination of crossreacting erythrocytes N and M, respectively, is strongly decreased and usually disappears at pH 8 - 8.5.

(2) No agglutination occurs at pH 6, then it strongly increases, up to maximum at pH 8.5; the effect is similar for M and N erythrocytes.

(3) No distinct effect of pH is seen over the range 6 - 8.5.

(4) Maximal agglutination occurs in pH range 6 - 7 and it distinctly decreases in the range 7 - 8.5, similarly for M and N erythrocytes.

The effect of pH of type (1) or (2) can be used for selecting the conditions for the most specific agglutination of M or N erythrocytes. It is difficult at the moment to explain the complexity of pH-dependence of anti-M and anti-N antibodies, especially that both the antibodies and amino group-containing M and N antigenic determinants may be affected by pH changes. However, it seems to be important to check the pH-dependence of monoclonal anti-M and anti-N antibodies, since not only the hemagglutination, but also other assays (e.g. binding to glycoproteins determined by microplate ELISA or immuno-blotting) may give different results, dependent on pH of the assay (our unpublished results).

The diversity of anti-M and anti-N monoclonal antibodies also is pronounced in their dependence on antigen amino groups and in reactivity with variant antigens. Unfortunately, only a few of the antibodies were tested in regard to these properties. Two highly specific and sialic acid-dependent anti-N antibodies 179K and 35/5F, reacting exclusively with N glycoprotein and glycophorin B, completely fail to react with the N-acetylated glycoproteins. Therefore, we classified them as anti-N_{leu}, since they seem to require the free amino group of the NH_2-terminal leucine residue. This specificity is probably very common among anti-N antibodies, since another one tested in this regard (8A2) also was amino group-dependent and none of anti-N studied (8A2, NN3, NN4, NN5 and iB5) reacted with M^c antigen, suggesting their dependence on leucine residue. The anti-M antibody 425/2B studied in our laboratory has different properties: it crossreacts in pH dependent manner with glycoprotein N and glycophorin B; is independent of antigen amino groups and is only partly dependent on sialic acid residues. This antibody would fit to anti-M_{gly} category that was confirmed by its reactivity with He antigen. The reactions of some other anti-M antibodies with variant antigens are shown in Table 4. Four antibodies tested with M^c and He erythrocytes showed the expected interrelation of both reactions: three reacted with M^c and not with He; and one reacted with He and gave a trace reaction only with M^c, that suggested anti-M_{ser}, and anti-M_{gly} specificities, respectively. However, only two out of the three presumably

Table 4. Reactions of monoclonal anti-M antibodies with variant antigens

Antibody	Antigens tested M^c	M^g	He	References
6A7	-			Bigbee et al., 1984
9A3	+			
BS38	+	-	-	Fraser et al., 1985
BS44	(-)	+	+	
LM110/140	+	-	-	
LM110/149	+	-	-	
G8		-		Nichols et al., 1985
E3		+		

+: positive reaction; - or (-): negative or trace reaction

anti-M_{ser} antibodies (LM110/149) are sialic acid-dependent, whereas the third one (BS38) gives an enhanced agglutination of desialylated erythrocytes (Table 3). Two other anti-M antibodies (6A7 and 9A3) studied by Bigbee et al. (1983, 1984) and Langlois et al. (1985) require sialylated antigen, but are independent of its amino groups and, therefore, are candidates for anti-M_{gly} specificity. However, one of them (9A3) reacts with M^c antigen, which suggests that its specificity is defined by serine, and not by glycine residue. The properties of BS38 and 9A3 antibodies indicate that some anti-M antibodies with specificity dependent on the NH_2-terminal serine residue may not require free amino group or sialic acid residues in the antigen. The most puzzling result, which cannot be explained in any reasonable way, is the cross-reactivity of two anti-M antibodies (BS44 and E3) with M^g antigen (Table 4), which is structurally unrelated to blood group M antigen (Fig. 5).

The studies on monoclonal antibodies performed so far confirm that a variety of anti-M and anti-N antibodies may be produced which react with different portions or sides of the blood group M or N specific domain of glycophorin A. It is noteworthy that anti-M antibodies seem to be more differentiated than anti-N. Further detailed studies on the fine specificity of a greater number of monoclonal antibodies should shed more light on the character of blood group M- and N-related antigenic epitopes.

Monoclonal antibodies against common antigenic epitopes of glycophorins

The blood group-unrelated monoclonal antibodies against glycophorin A are easily obtained by immunization of mice either with intact erythrocytes or with isolated glycophorin fraction. If erythrocytes are used for immunization, the anti-glycophorin A specificity of the antibodies is usually assessed by finding that they fail to react with En(a-) red cells, or, more directly, by specific immunoprecipitation of glycophorin A from the extract of labeled erythrocyte membranes. The monoclonal antibodies directed against glycophorin A have specificities comparable with those of human antibodies of anti-En^a or anti-Pr type (Table 5). The antibodies which do not require sialic acid residues in glycophorin A, or those which react with trypsin-treated erythrocytes and do not crossreact with glycophorin B, can be regarded as anti-En^a. This type of specificity is frequent among anti-glycophorin A antibodies. Some of such antibodies (R7, BRIC13, BBRIC14, BRIC15, MoAb36 and MoAb145) have anti-Wr^b specificity, because in addition to En(a-) erythrocytes, they also do not react with Wr(b-) red cells of M.Fr. Similarly to human anti-Wr^b, they recognize epitopes resistant at erythrocyte surface not only to trypsin but also to ficin, and they are either not inhibited or only weakly inhibited by isolated glycophorin A (Anstee & Edwards, 1982; Ridgewell et al., 1983: Rearden et al., 1983, 1984). The antibodies which do not react with desialylated glycophorin A or erythrocytes and crossreact with glycophorin B are likely to be specific for the heavily glycosylated NH_2-terminal domain identical in glycophorins A and B and therefore may have anti-Pr specificity. The antibody R1.3 has such properties (Table 5).

Table 5. Monoclonal antibodies against common antigenic determinants of glycophorin A

Antibody		Type of specificity	Sensitivity of antigens to neuraminidase[1]	trypsin[2]	References
IgG1	R1.3	anti-Pr	NS	TR	
IgG1	R7	anti-Wr^b	NR	TR	Edwards, 1980;
IgG1	R10	anti-En^a	NR	TS	Anstee & Edwards,
IgG2b	R18	anti-En^a	NR	TR	1982
IgG1	R23	anti-En^a	NR	TS	
	BRIC13	anti-Wr^b	NR	TR	Ridgwell et al.,
	BRIC14	anti-Wr^b	NR	TR	1983
	Bric15	anti-Wr^b	NR	TR	
IgG2a	B22A	anti-En^a	NR	TR	
IgG3	D22	anti-En^a	NS	TR	Ochiai et al.,
IgG1	E11B	anti-En^a	?	TR	1983
IgM	F11	anti-En^a	NS	TR	
IgG1	J11A	anti-Pr(?)[3]	NS	TS	
IgM	VieG4	?	NS		Liszka et al., 1983
IgG1	10F7	anti-En^a	NR	TS	Bigbee et al.,
IgM	MN1	anti-En^a	NR	TS	1983, 1984
IgG1	MoAb31	?	NS	TR	Rearden et al.,
IgG1	MoAb36	anti-Wr^b	NR	TR	1983, 1984, 1985
IgM	MoAb145	anti-Wr^b	NR	TR	
IgG1	6B5	(anti-CNBr3[4])	NS		
IgG1	8F10	(anti-CNBr1)	NS		
IgG2b	9C3		NS		
IgG3	1B4		NS		
IgG1	4C3	anti-Pr	NS		
IgG2b	4E7	(anti-CNBr1	NS		Barsoum et al.,
IgG1	7B10	& CNBr3)	NS		1985
IgG2b	7C11		NS		
IgG1	9D6		NS		
IgG1	3D2	(probably			
IgG1	4C6	anti-CNBr2)			

[1]NS: neuraminidase-sensitive, NR: neuraminidase-resistant; [2]concerns red cell-bound antigen, TR: trypsin-resistant, TS: trypsin-sensitive; [3]J11A shows preference for blood group M antigen; [4]CNBr1, CNBr2, CNBr3: glycophorin A fragments obtained by cyanogen bromide degradation, containing amino acid residues 9-81, 82-131 and 1-8, respectively.

The reactivity of R1.3 with trypsin-treated red cells and a weak reactivity with En(a-) erythrocytes result most probably from the reaction with trypsin-resistant glycophorin B, because the antibody does not react with trypsin-treated S-s-U- red cells, lacking glycophorin B (Anstee & Edwards, 1982). The MoAb31 shows similar properties, but in contrast to R1.3 it does react with trypsin-treated S-s-U- erythrocytes (Rearden et al., 1985). Several antibodies with evident anti-Pr specificity were obtained by Barsoum et al. (1985). These antibodies react with glycoslylated CNBr1 and CNBr3 fragments of glycophorin A (amino acid residues 9-81 and 1-8, respectively), and most of them do not recognize the difference between CNBr1 and CNBr3 (Table 5). Generally, the structures (fragments of glycophorin A) recognized by the monoclonal antibodies have not been precisely defined yet.

Besides the antibodies against glycophorin A, two monoclonal antibodies (BRIC4 and BRIC10) specific for glycophorin C (Glycoprotein β) were described (Anstee et al., 1984a, 1984b). These antibodies fail to react with neuraminidase-treated or trypsin-treated erythrocytes. In immunoprecipitation assay they bind specifically glycophorin C and a newly detected glycoprotein β_1, located with SDS-PAGE between glycophorin C and glycoprotein γ BRIC4 and BRIC10 do not precipitate these components from extracts of Gerbich (Ge)-negative erythrocyte membranes. As mentioned earlier, the glycoproteins missing in Ge-negative red cells are involved in maintaining the normal shape of erythrocytes. Those Ge-negative erythrocytes which have the normal shape show the presence of abnormal glycoproteins precipitated by BRIC4 and BRIC10. The immunological relationship between glycophorin C and the variant glycoproteins, shown by means of monoclonal antibodies, supports the suggestion that the variant glycoproteins may replace glycophorin C in its function (Anstee et al., 1984b).

The monoclonal antibodies are very useful reagents for the studies on normal and variant glycophorins. As mentioned above and earlier in this article, monoclonal antibodies served for characterization of Ge-negative red cells and for identification of Wr^b antigen in normal glycophorin A and in the hybrid $(\delta-\alpha)$ St^a glycoprotein. Merry et al. (1984). demonstrated the potential of using monoclonal antibodies for the quantification of antigenic sites on erythrocyte surface. Moreover, monoclonal antibodies have already been used for identification and characterization of glycophorin A or glycophorin-like glycoproteins in erythroid precursor, leukemic and other cancer cells. The advantages resulting from using monoclonal antibodies will obviously increase after the specificity of the antibodies is more precisely characterized.

Use of monoclonal antibodies against glycophorin A in studies on leukemic and other cells

Glycophorin A was detected in established human leukemic cell lines K562 (Gahmberg et al., 1979) and HEL (Papayannopoulou et al., 1982). This finding opened the possibility to study the biosynthesis of glycophorin A and alterations occurring in this glycoprotein during the cell differentiation. Gahmberg et al. (1984) reported that glycophorin A in K562 cells (and in erythroid precursor cells at early stages of erythropoiesis) is less glycosylated than in mature erythrocytes, and that degree of the glycoslylation is increased after hemin-induced differentiation of K562 cells. This difference in glycosylation does not cause a distinct difference in the reaction of glycophorin A with polyclonal antiserum. However, hemin-induced K562 and HEL cells show increased binding of monoclonal antiglycophorin A antibodies R10 and R18, and the reaction of glycophorin A with R10 is demonstrable in immunoprecipitation assay only after the induced differentiation of K562 cells (Gahmberg et al., 1984). The antibodies R10 and R18 are specific for nonglycosylated epitopes of glycophorin A (Anstee & Edwards, 1982). The weaker reactivity of these antibodies with less glycosylated glycophorin A from immature cells suggests that glycosylation contributes to conformation or exposure of the nonglycosylated antigenic sites (Gahmberg et al., 1984). Other authors also demonstrated the binding of several anti-glycophorin A monoclonal antibodies to K562 and HEL cells, using cytofluorometric method (Liszka et al., 1983; Rimmer & Horton, 1984; Tabilio et al., 1984; Rearden et al., 1985). The results obtained with individual antibodies differed in respect to the intensity of flourescence and proportion of reactive cells. It indicates that exposure of various antigenic sites of glycophorin A in K562 and HEL cells is not uniform and changes not only during the induced cell differentiation, but also during the normal cell cycle. It is noteworthy that three antibodies with anti-Wr^b specificity (MoAb36, MoAb145 and R7) fail to bind to K562 and HEL cells (Rearden et al., 1985).

The monoclonal anti-glycophorin A antibodies are also used in attempt to detect glycophorin in other cell lines and in leukocytes from patients with different types of leukemia. The presence of glycophorin A in the malignant blast cells from 3 out of 15 acute leukemia cases was shown by Anderson et al. (1979) by means of polyclonal rabbit antiserum. The monoclonal antibodies seem to be more selective in reaction with nonerythroleukemic cells. For example, the antibody VieG4 is reactive with K562 and bone marrow erythroid precursor cells and with erythroleukemic cells from patients, but it did not bind either to 11 other hemopoietic cell lines, or to the blast cells of 227 patients diagnosed as nonerythroleukemia (Liszka et al., 1983). Similarly, the monoclonal antibodies tested by Rearden et al. (1985) demonstrated the presence of glycophorin A on erythroid cells only. The antibody R10 was bound to malignant cells from 20 of 724 nonerythroleukemia cases and is considered as a suitable reagent for detection of 'cryptic' erythroleukemias (Greaves et al., 1983). In contrast to the examples given above, the group of monoclonal antibodies reacting with glycosylated CNBr-fragments of glycophorin A (see Table 5) show significant binding to many malignant and nonmalignant cells, including erythroleukemia, melanoma, carcinomas of breast, cervix, larynx and colon, normal fibroblasts and mammary cells (Barsoum et al., 1984). Most probably these antibodies are directed against the carbohydrate portion of glycophorin A (anti-Pr specificity) and therefore, they may crossreact with sialoglycoproteins typical for other cells and not related to glycophorin A in tests with polyclonal antisera (Ogata & Lloyd, 1982: Carlsson & Fukuda, 1985). MoAb145, which has anti-Wr^b specificity, does not react with K562 and HEL cells, but is bound to the cells of human bladder epithelium which is lost in some cases of transitional cell carcinoma of the bladder (Rearden et al., 1983).

The studies cited above demonstrate the advantages and limitations resulting from using monoclonal antibodies against glycophorin A. Since the specificity of these antibodies is not exactly known yet, and glycophorins in immature cells and mature erythrocytes appear to be different, the conclusions from such studies must be drawn with caution. However, the history of monoclonal antibodies against glycophorin A is only 3 years-old and their usefulness will increase concomitantly with increasing knowledge in this field.

GENERAL CONCLUSION

Human erythrocyte glycophorins represent a group of complex glycoprotein antigens which have relatively strong immunogenic properties. They elicit formation of antibodies in animals and of allo- and autoantibodies in humans. Moreover, some human alloantibodies are found in persons which were never pregnant, immunized or transfused, and the origin of such antibodies is not explained. Due to primary interest in red cell surface antigens, most of known anti-glycophorin antibodies are directed against their external NH_2-terminal domains. Both, the polypeptide chain and carbohydrates contribute to antigenic properties of glycophorins, since antibodies specific for peptide fragments, oligosaccharides, or composite determinants - including the oligosaccharide chain(s) and amino acid residues - are known. The most immunogenic portions of the outer domains of glycophorins A and B are the NH_2-terminal blood group M or N-specific region and less glycosylated internal portion, located nearer to the membrane lipid bilayer. Alterations in the structure of the polypeptide chain or oligosaccharide chains of glycophorins cause the change of their antigenic properties and production of antibodies specific for the altered regions of these glycoproteins.

Studies on glycophorins illustrate how immunological findings facilitate the biochemical investigation. The discovery of MN and Ss blood groups in erythrocytes gave the impulse to isolate and characterize the red cell components carrying these blood group antigens (glycophorins A and B). These studies showed that MN and Ss genes code for the polypeptide chains of glycophorin A and B and that two forms of each of these glycoproteins exist. The serological detection of many variant antigens immunologically and/or genetically related to the MNSs blood group system suggested, in turn, that other structural alterations occur in glycophorin A or B in some individuals. The biochemical studies performed on some of the variant red cells enabled the identification of glycophorins with exchanged amino acid residues or hybrid glycoproteins. The absence of glycophorins in some erythrocytes also was detected due to previous serological observations. Since the structural changes or absence of glycophorins are inherited, they demonstrate the possibilities of mutation, crossing-over and deletion or inactivation of genes encoding the polypeptide chains of glycophorins. Glycophorins are probably not exceptional in respect to a great number of existing genetic variants. In many other cases the variations may be not detected if they have no functional or immunological consequences, or if the specificity of respective antibodies cannot be identified. Immunological methods also enabled the detection of glycophorin A in normal and malignant erythroid precursor cells and following the structural changes in glycophorin during the cell differentiation. Despite the fact that biological function of glycophorins A and B is still unknown, the significant recent progress in immunochemical studies on glycophorins has supplied many interesting data. The results obtained are important not only from the point of view of basic biochemical, immunologic and genetic research, but also have an impact on applied research, in such fields as transfusiology, legal medicine, hematology and other clinical studies.

REFERENCES

Adamany, A.M., Blumenfeld, O.O., Sabo, B. and McCreary, J. (1983): A carbohydrate structural variant of MM glycoprotein (glycophorin A). J. Biol. Chem. 258:11537-11545.

Adams, J., Broviac, M., Brooks, W., Johnson, N.R. and Issitt, P.D. (1971): An antibody in the serum of a Wr(a+) individual reacting with an antigen of very high frequency. Transfusion 11:290-291.

Allen, R.W., Nunley, N., Kimmeth, M.E. Wallhermfechtel, M., and Vengelen-Tyler, V (1984): Isolation and serological characterization of a monoclonal antibody recognizing the N blood group antigen. Transfusion 24:136-140.

Alloisio, N., Morlé, L., Bachir, D., Guertani, D., Colonna, P., and Delaunay J. (1985): Red cell membrane sialoglycoprotein β in homozygous and heterozygous 4.1(-) hereditary elliptocytosis. Biochem. Biophys. Acta 816:57-62.

Anderson, L.C., Gahmberg, C.G., Teerenhovi, L. and Vuopio, P. (1979): Glycophorin A as a cell surface marker of early erythroid differentiation in acute leukemia. Int. J. Cancer 24:717-720.

Anstee, D.J. and Edwards, P.A.W. (1982): Monoclonal antibodies to human erythrocytes. Eur. J. Immunol. 12:228-232.

Anstee, D.J. and Tanner, M.J.A. (1975): Separation of ABH, I, Ss antigenic activity from the MN-active sialoglycoprotein of the human erythrocyte membrane. Vox Sang. 29:378-389.

Anstee, D.J. and Tanner, M.J.A. (1978): Genetic variants involving the major membrane sialoglycoprotein of human erythrocytes. Studies on erythrocytes of type M^k, Mitenberger class V and M^g. Biochem. J. 175:149-157.

Anstee, D.J., Barker, D.M., Judson, P.A. and Tanner, M.J.A. (1977): Inherited sialoglycoprotein deficiencies in human erythrocytes of type En(a-). Brit. J. Haemat. 35:309-320.

Anstee, D.J., Mawby, W.J. and Tanner, M.J.A. (1979): Abnormal blood group Ss-active sialoglycoproteins in the membrane of Miltenberger class III, IV and V human erythrocytes. Biochem. J. 183: 193-203.

Anstee, D.J., Mawby, W.J., Parsons, S.F., Tanner, M.J.A. and Giles, C.M. (1982a): A novel hybrid sialoglycoprotein in St^a-positive human erythrocytes. J. Immunogenet. 9:51-55.

Anstee, D.J., Mawby, W.J. and Tanner, M.J.A. (1982b): Structural variation in human erythrocyte sialoglycoproteins. *Membranes and Transport*, Martonosi, A.N., ed., Plenum Press, New York, NY, pp. 427-433.

Anstee, D.J., Parsons, S.F., Ridgwell, K., Tanner, M.J.A., Merry, A.H., Thomson, E.E., Judson, P.A., Johnson, P., Bates, S. and Fraser, I.D.(1984a): Two individuals with elliptocytic red cells apparently lack three minor erythrocyte membrane sialoglycoproteins. Biochem. J. 218:615-619.

Anstee, D.J., Ridgwell, K., Tanner, M.J.A., Daniels, G.L. and Parsons, S.F. (1984b): Individuals lacking the Gerbich blood group antigen have alterations in the human erythrocyte membrane sialoglycoproteins β and γ. Biochem. J. 221:97-104.

Baranowski, T. and Lisowska, E. (1963): Studies on blood group antigens M and N. VIII. Fractionation of the products of proteolytic digestion. Arch. Immunol. Ther. Exp. 11:631-640.

Baranowski, T., Lisowska, E., Morawiecki, A., Romanowska, E. and Strozecka, K. (1959): Studies on blood group antigens

M and N. III. Chemical composition of purified antigens. Arch. Immunol. Ther. Exp. 7:15-27.

Barsoum, A.L., Czuczman, M.S., Bhavanandan, V.P. and Davidson, E.A. (1984): Epitopes immunologically related to glycophorin A on human malignant and nonmalignant cells in culture. Int. J. Cancer 34:789-795.

Barsoum, A.L., Bhavanandan, V.P. and Davidson, E.A. (1985): Monoclonal antibodies to cyanogen bromide fragments of glycophorin A. Mol. Immunol. 22:361-367.

Berger, E.G. and Kozdrowski, I. (1978): Permanent mixed-field polyagglutinable erythrocytes lack galactosyltransferase activity. FEBS Lett. 93:105-108.

Bigbee, W.L., Vanderlaan, M., Fong, S.S.N. and Jensen, R.H. (1983): Monoclonal antibodies specific for the M- and N-forms of human glycophorin A. Mol. Immunol. 20:1353-1362.

Bigbee, W.L., Langlois, R.G., Vanderlaan, M. and Jensen, R.H. (1984): Binding specificities of eight monoclonal antibodies to human glycophorin A - studies with M^cM, and $M^kEn(UK)$ variant human erythrocytes and M- and MN^V-type chimpanzee erythrocytes. J. Immunol. 133:3149-3155.

Blanchard, D., Cartron, J.P., Rouger, P. and Salmon, C. (1982): Pj variant, a new hybrid MNSs glycoprotein of the human red cell membrane. Biochem. J. 203:419-426.

Blanchard, D., Cartron, J.P., Fournet, B., Montreuil, J., van Halbeek, H. and Vliegenthart, J.F.G. (1983): Primary structure of the oligosaccharide determinant of blood group Cad specificity. J. Biol. Chem. 258:7691-7695.

Blumenfeld, O.O. and Adamany, A.M. (1978): Structural polymorphism within the amino-terminal region of MM, NN and MN glycoproteins (glycophorins) of the human erythrocyte membrane. Proc. Natl. Acad. Sci. USA 75:2727-2731.

Blumenfeld, O.O., Adamany, A.M. and Puglia, K.V. (1981): Amino acid and carbohydrate structural variants of glycoprotein products (M-N glycoproteins) of the M-N allelic locus. Proc. Natl. Acad. Sci. USA 78:747-751.

Bunn, H.F. and Higgins, P.J. (1981): Reaction of monosaccharides with proteins: possible evolutionary significance. Science 213:222-224.

Carlsson, S.R. and Fukuda, M. (1985): Isolation and characterization of leukosialin, the major cell surface sialoglycoprotein on human leukocytes, and demonstration of its cell specific glycosylation. *Glycoconjugates*, Proc. VIIIth Intern. Symp. (Houston, TX.), pp. 540-541.

Carter, D.R., Hardy, E.R., Lannom, H.K., Dill, K., Ferrari, B. and Pavia, A.A. (1984): ^{13}C n.m.r. study of the pH behavior of N-methylated peptides related to the NH_2-terminus of glycophorins. Int. J. Biol. Macromol. 6:348-352.

Cleghorn, T.E. (1962): Two human blood group antigens St^a(Stones) and Ri^a(Ridley) closely related to the MNSs blood group system. Nature (London) 195:895.

Cleghorn, T.E. (1966): A memorandum on the Miltenberger blood groups. Vox Sang. 11:219-222.

Colin, Y., Rahuel, C., London, J., Romeo, P.H., d'Auriol, L., Galibert, F. and Cartron, J.P. (1986) Isolation of cDNA clones and complete amino acid sequence of human erythrocyte glycophorin C. J. Biol. Chem. 261: 229-233.

Contreras, M., Green, C., Humphreys, J., Tippett, P., Daniels, G., Teesdale, P., Armitage, S. and Lubenko, A. (1984): Serology and genetics of MNSs-associated antigen Dantu. Vox Sang. 46:377-386.

Dahr, W. (1981): Serology, genetics and chemistry of the MNSs blood group system. Blood Transf. Immunohaemat. 24:85-95.

Dahr, W. (1983): Biochemical studies of erythrocyte membrane glycoprotein variants. In *Red Cell Membrane Glycoconjugates and Related Genetic Markers,* Cartron, J.P., Rouger, P. and Salmon, C., eds., Librairie Arnette, Paris, pp. 27-36.

Dahr, W. and Beyreuther, K. (1985) A revison of the N-terminal structure of sialoglycoprotein D (glycophorin C) from human erythrocyte membranes. Biol. Chem. Hoppe Seyler 366: 1067.

Dahr, W. and Moulds, J. (1981): An immunochemical study on anti-N antibodies from dialysis patients. Immunol. Commun. 10:173-183.

Dahr, W. and Uhlenbruck, G. (1978): Structural properties of the human M and N blood group system antigen receptor sites. Hoppe-Seyler's Z. Physiol. Chem. 359:835-843.

Dahr, W., Uhlenbruck, G. and Bird, G.W.G. (1975a): Influence of free amino and carboxyl groups on the specificity of plant anti-N. Vox Sang. 28:389-391.

Dahr, W., Uhlenbruck, G., Gunson, H.H. and van der Hart, M. (1975b): Molecular basis of Tn polyagglutinability. Vox Sang. 29:36-50.

Dahr, W., Uhlenbruck, G., Issitt, P. and Allen, F.H. (1975c): SDS-polyacrylamide gel electrophoretic analysis of the membrane glycoproteins from S-s-U- erythrocytes. J. Immunogenet. 2:249-251.

Dahr, W., Uhlenbruck, G. and Knott, H. (1975d): Immunochemical aspects of the MNSs blood group system. J. Immunogenet. 2:87-100.

Dahr, W., Uhlenbruck, G., Leikola, J., Wagstaff, W. and Landfried, K. (1976a): Studies on the membrane glycoprotein defect of En(a-) erythrocytes. I. Biochemical aspects. J. Immunogenet. 3:329-346.

Dahr, W., Uhlenbruck, G., Wagstaff, W. and Leikola, J. (1976b): Studies on the membrane glycoprotein defect of En(a-) erythrocytes. II. MN antigenic properties of En(a-) erythrocytes. J. Immunogenet. 3:383-394.

Dahr, W., Uhlenbruck, G., Janssen, E. and Schmalisch R. (1977a): Different N-terminal amino acids in the MN glycoprotein from MM and NN erythrocytes. Human Genet. 35:335-343.

Dahr, W., Uhlenbruck, G. and Knott, H. (1977b): The defect of Mk erythrocytes as revealed by sodium dodecysulfate polyacrylamide gel electrophoresis. J. Immunogenet. 4:191-200.

Dahr, W., Issitt, P., Moulds, J. and Pavone, B. (1978a): Further studies on the membrane glycoprotein defects of S-s- and En(a-) erythrocytes. Hoppe-Seyler's Z. Physiol. Chem. 359:1217-1224.

Dahr, W., Longster, G., Uhlenbruck, G. and Schumacher, K. (1978b): Studies on Miltenberger class III, V, M^v and M^k red cells. I. Sodium dodecylsulfate polyacrylamide gel electrophoretic investigations. Blut 37:129-138.

Dahr, W., Uhlenbruck, G., Leikola, J. and Wagstaff, W. (1978c): Studies on the membrane glycoprotein defect of En(a-) erythrocytes. III. N-terminal amino acids of sialoglycoproteins from normal and En(a-) red cells. J. Immunogenet. 5:117-127.

Dahr, W., Gielen, W., Beyreuther, K. and Krüger, J. (1980a): Structure of the Ss blood group antigens. I. Isolation of Ss-active glycopeptides and differentiation of the antigens by modification of methionine. Hoppe-Seyler's Z. Physiol. Chem. 361:145-152.

Dahr, W., Beyreuther, K., Steinbach, H., Gielen, W. and

Krüger, J. (1980b): Structure of the Ss blood group antigens. II. A methionine/threonine polymorphism within the N-terminal sequence of the Ss glycoprotein. Hoppe-Seyler's Z. Physiol. Chem. 361:895-906.

Dahr, W., Beyreuther, K., Gallasch, E., Krüger, J. and Morel, P. (1981a): Amino acid sequence of the blood group M^g-specific major human erythrocyte membrane sialoglycoprotein. Hoppe-Seyler's Z. Physiol. Chem. 362:81-85.

Dahr, W., Kordowicz, M., Beyreuther, K. and Krüger, J. (1981b): The amino acid sequence of the M^c-specific major red cell membrane sialoglycoprotein- an intermediate of the blood group M- and N-active molecules. Hoppe-Seyler's Z. Physiol. Chem. 362:363-366.

Dahr, W., Beyreuther, K., Kordowicz, M. and Krüger, J. (1982): N-terminal amino acid sequence of siologlycoprotein D. (glycophorin C) from human erythrocyte membranes. Eur. J. Biochem. 125:57-62.

Dahr, W., Kordowicz, M., Judd, W.J., Moulds, J., Beyreuther, K. and Krüger, J. (1984a): Structural analysis of the Ss sialoglycoprotein specific for Henshaw blood group from human erythrocyte membranes. Eur. J. Biochem. 141:51-55.

Dahr, W., Newman, R.A., Contreras, M., Kordowicz, M., Teesdale, P., Beyreuther, K. and Krüger, J. (1984b): Structures of Miltenberger class I and II specific human erythrocyte membrane glycoproteins. Eur. J. Biochem. 138:259-265.

Dahr, W., Moulds, J., Baumeister, G., Moulds, M., Kiedrowski, S. and Hummel, M. (1985a): Altered membrane sialoglycoproteins in human erythrocytes lacking the Gerbich blood group antigens. Biol. Chem. Hoppe-Seyler 366:201-211.

Dahr, W., Müller, T., Moulds, J., Baumeister, G., Issitt, P.D., Wilkinson, S. and Garratty, G. (1985b): High frequency antigens of human erythrocyte membrane sialoglycoproteins. I. En^a receptors in the glycosylated domain of the MN sialoglycoprotein. Biol. Chem. Hoppe-Seyler 366:41-51.

Dahr, W., Blanchard, D., Hummel, M., Cartron, J.P. and Beyreuther, K. (1985c): Preparative purification of the Ss and D sialoglycoproteins (glycophorins B and C) from human erythrocyte membranes by high performance liquid chromatography and amino acid sequence analysis of their intramembranous domains. Glycoconjugates, Proc. VIIIth Intern. Symp. (Houston, TX.), pp.21-22.

Dahr, W., Moulds, J., Unger, D., Blanchard, D. and Cartron, J.P. (1985d): Structural analysis and properties of a blood group Dantu-active hybrid sialoglycoprotein (glycophorin) from human red cell membranes. Glycoconjugates, Proc. VIIIth Intern. Symp. (Houston, TX.), pp. 543-544.

Darnborough, H., Dunsford, I. and Wallace, J.A. (1969): The En^a antigen and antibody. A genetical modification of human red cells affecting their blood grouping reactions. Vox Sang. 17:241-255.

Drzeniek, Z. (1983): Purification of antibody reacting with nonenzymatically glycosylated blood group M determinants. Immunol. Lett. 6:179-183.

Drzeniek, Z., Kusnierz, G. and Lisowska, E. (1981): A human antiserum reacting with modified blood group M determinants. Immunol. Commun. 10:185-197.

Dybkjaer, E., Poole, J. and Giles, C.M. (1981): A new Miltenberger class detected by a second example of Anek type serum. Vox Sang. 41:302-305.

Ebert, W., Metz, J. and Roelcke, D. (1972): Modification of N-acetylneuraminic acid and their influence on the antigen reactivity of erythrocyte glycoproteins. Eur. J. Biochem. 27:470-472.

Edelman, L., Blanchard, D., Rouger, P., Doinel, C., Cartron, J.P., Salmon, C. and Reviron, J. (1984): A monoclonal antibody directed against the homologous N-terminal domain of glycophorin A and B. Exp. Clin. Immunogenet. 1:129-139.

Edwards, P.A.W. (1980): Monoclonal antibodies that bind to the human erythrocyte membrane glycoproteins glycophorin A and Band 3. Biochem. Soc. Trans. 8:334-335.

Fairbanks, G., Steck, T.L. and Wallach, D.F.H. (1971): Electrophoretic analysis of human erythrocyte membrane. Biochemistry 10:2606-2616.

Fletcher, A. and Harbour, C. (1984): An interesting monoclonal anti-N produced following immunization with human group O, NN erythrocytes. J. Immunogenet. 11:121-126.

Fraser, R.H., Munro, A.C., Williamson, A.R., Barrie, E.K., Hamilton, E.A. and Mitchell, R. (1982): Mouse monoclonal anti-N. I. Production and serological characterization. II. Physicochemical characterization and assessment for routine blood grouping. J. Immunogenet. 9:295-309.

Fraser, R.H., Inglis, G., Mackie, A., Munro, A.C., Alla, E.K., Mitchell, R., Sonneborn, H.H. and Uthemann, H. (1985): Mouse monoclonal antibodies reacting with M blood group-related antigens. Transfusion 25:261-266.

Fujita, S. and Cleve, H. (1975): Isolation and partial characterization of two minor glycoproteins from human erythrocyte membranes. Biochim. Biophys. Acta 382:172-180.

Furthmayr, H. (1978): Structural comparison of glycophorins and immunochemical analysis of genetic variants. Nature (London) 271:519-524.

Furthmayr, H., Tomita, M. and Marchesi, V.T. (1975): Fractionation of the major sialoglycopeptides of the human red cell membrane. Biochem. Biophys. Res. Commun. 65:113-121.

Furthmayr, H., Metaxas, M.N. and Metaxas-Bühler, M. (1981): M^g and M^c: mutations within the amino terminal region of glycophorin A. Proc. Natl. Acad. Sci. USA 78:631-635.

Furuhjelm, U., Meyllyla, G., Nevanlinna, H.R., Nordling, S., Pirkola, A., Gavin, J., Gooch, A., Sanger, R. and Tippett, P. (1969): The red cell phenotype En(a-) and anti-En^a: serological and physicochemical aspects. Vox Sang. 17:256-278.

Gahmberg, C.G., Myllyla, G., Leikola, J., Pirkola, A. and Nordling, S. (1976): Absence of the major sialoglycoprotein in the membrane of human En(a-) erythrocytes and increased glycosylation of Band 3. J. Biol. Chem. 251:6108-6116.

Gahmberg, C.G., Jokinen, M. and Anderson, L.C. (1979): Expression of the major red cell sialoglycoprotein, glycophorin A, in the human leukemic cell line K562. J. Biol. Chem. 254:7442-7448.

Gahmberg, C.G., Ekblom, M. and Anderson, L.C. (1984): Differentiation of human erythroid cells is associated with increased glycosylation of the major sialoglycoprotein, glycophorin A. Proc. Natl. Acad. Sci. USA 81:6752-6756.

Greaves, M.F., Sieff, C. and Edwards, P.A.W. (1983): Monoclonal antiglycophorin as a probe for erythroleukemias. Blood 61:645-651.

Hanisch, F.G., Farrar, G.H., Schmalisch, R. and Uhlenbruck, G.

(1983): Immunochemistry of O-glycosidically-linked Gal(β1-3) GalNAc on fragments of human glycophorin A. Immunobiology 165:147-160.

Herkt, F., Parente, J.P., Leroy, Y., Fournet, B., Blanchard, D., Cartron, J.P., van Halbeek, H. and Vliegenthart, J.F.G. (1985): Structure determination of oligosaccharides isolated from Cad erythrocyte membranes by permethylation analysis and 500 MHZ ^{1}H-NMR spectroscopy. Eur. J.Biochem. 146:125-129.

Huprikar, S.V. and Springer, G.F. (1970): Structural aspects of human blood group M and N specificity. In *Blood and Tissue Antigens,* Aminoff, D., ed., Academic Press, New York-London, PP. 327-335.

Irimura, T., Tsuji, T., Tagami, S., Yamamoto, K. and Osawa, T. (1981): Structure of a complex type sugar chain of human glycophorin A. Biochemistry 20:560-566.

Issitt, P.D. (1981): *The MN blood group system,* Montgomery Scientific Publications, Cincinnati, OH.

Issitt, P.D. and Wilkinson, S.L. (1983): Further studies on the dependence of some examples of anti-M and anti-N on the presence of red cell-borne sialic acid. Transfusion 23:117-119.

Issitt, P.D., Pavone, B.G., Goldfinger, D. and Zwicker, H. (1975): An En(a-) red cell sample that types as Wr(a-b-). Transfusion 15:353-355.

Issitt, P.D., Pavone, B.G., Wagstaff, W. and Goldfinger, D. (1976): The phenotypes En(a-), Wr(a-b-) and En(a+), Wr(a+b+), and further studies on the Wright and En blood group systems. Transfusion 15:396-407.

Issitt, P.D., Daniels, G. and Tippet, P. (1981): Proposed new terminology for En^{a}. Transfusion 21:473.

Judd, W.J., Issitt, P.D., Pavone, B.G., Anderson, J. and Aminoff, D. (1979): Antibodies that define NANA-independent MN-system antigens. Transfusion 10:12-18.

Judd, W.J., Geisland, J.R., Issitt, P.D., Wilkinson, S.L., Anstee, D.J., Shin, C. and Glidden, H. (1983): Studies on the blood of an Mi^{V}/M^{k} proposita and her family. Transfusion 23:33-36.

Kathan, R.H., Winzler, R.J. and Johnson, C.A. (1961): Preparation of an inhibitor of viral hemagglutination from human erythrocytes. J. Exp. Med. 113:37-45.

Klenk, E. and Uhlenbruck, G. (1960): Uber neuraminsäure-haltige Mucoide aus Menschenerythrozytenstroma: ein Bietrag zur Chemie der Agglutinogene. Hoppe-Seyler's Z. Physiol. Chem. 319:151-160.

Kordowicz, M. and Lisowska, E. (1978): Antisera against polypeptide chains derived from M and N blood group glycoproteins. Arch. Immunol. Ther. Exp. 26:145-149.

Langley, J.W., Issitt, P.D., Anstee, D.J., McMahan, M., Smith, N., Pavone, B.G., Tessel, J.A. and Carlin, M.A. (1981): Another individual (J.R.) whose red blood cells appear to carry a hybrid MNSs sialoglycoprotein. Transfusion 21:15-24.

Langlois, R.G., Bigbee, W.L. and Jensen, R.H. (1985): Flow cytometric characterization of normal and variant cells with monoclonal antibodies specific for glycophorin A. J. Immunol. 134:4009-4017.

Levene, C., Sela, R., Lacser, M., BarShany, S., Giles, C.M. and Poole, J. (1984): Further examples of human anti-M^{e} found in sera of Israeli donors. Vox Sang. 46:207-210.

Liao, T.H., Gallop, P.M. and Glumenfeld, O.O. (1973): Modification of sialyl residues of sialoglycoprotein(s) of the human erythrocyte surface. J. Biol. Chem. 258:8247-8253.

Lisowska, E. (1977): Chemistry of M and N blood group antigens. In *Clinical Laboratory Science*, Seligson, D., ed., Section D: *Blood Banking*, Vol. I, Greenwalt, T. J. and Steane, E.A., eds., CRC Press Inc. , Cleveland, OH. pp. 281-299.

Lisowska, E. (1981): Biochemistry of M and N blood group specificities. Blood Transf. Immunohaemat. 24:75-84.

Lisowska, E. and Duk, M. (1972): The reaction of products of sequential periodate oxidation of human erythrocyte glycoproteins with hemagglutinin from *Helix pomatia*. Arch. Immunol. Ther. Exp. 20:869-875.

Lisowska, E. and Duk, M. (1975a): Effect of modification of amino groups of human erythrocytes on M, N and N_{Vg} blood group specificities. Vox Sang. 28:392-397.

Lisowska, E. and Duk, M. (1975b): Modification of amino groups of human erythrocyte glycoproteins and the new concept on the structural basis of M and N blood group specificity. Eur. J. Biochem. 54:469-474.

Lisowska, E. and Duk, M. (1976): The effect of gradual mild acid hydrolysis on serological activities of glycoproteins from human erythrocytes. Arch. Immunol. Ther. Exp. 24:39-45.

Lisowska, E. and Jeanloz, R.W. (1973): Composition and distribution of carbohydrate chains in glycoproteins of human erythrocyte membrane. Carbohyd. Res. 29:181-191.

Lisowska, E. and Kordowicz. (1977a): Specific antibodies for desialized M and N blood group antigens. Vox Sang. 33:164-169.

Lisowska, E. and Kordowicz, M. (1977b): Immunochemical properties of M and N blood group antigens and their degradation products. In *Human Blood Groups*, Mohn, F., ed., Karger, Basel, pp. 188-196.

Lisowska, E. and Morawiecki, A. (1967): The role of free amino groups in the blood group activity of M and N mucoids. Eur. J. Biochem. 3:237-241.

Lisowska, E. and Roelcke, D. (1973): Differentiation of anti-Pr_1 and anti-Pr_2 sera with periodate-oxidized erythrocyte glycoproteins. Blut 26:339-341.

Lisowska, E. and Wasniowska, K. (1978): Immunochemical characterization of cyanogen bromide degradation products of M and N blood group glycopeptides. Eur. J. Biochem. 88:247-252.

Lisowska, E., Drzeniek, Z. and Klis, W. (1979): Structural requirements for the interaction of M and N glycopeptides with anti-M and anti-N antibodies and *Vicia graminea* lectin. Glycoconjugates, Proc. Vth Intern. Symp. (Kiel), pp. 512-513.

Lisowska, E., Duk, M. and Dahr, W. (1980): Comparison of alkali-labile oligosaccharide chains of M and N blood group glycopeptides from human erythrocyte membrane. Carbohydr. Res. 78:103-113.

Liszka, K., Majdic, O., Bettelheim, P. and Knapp, W. (1983): Glycophorin A expression in malignant hematopoiesis. Amer. J. Hemat. 15:219-226.

Longenecker, B.M., Rahman, A.F.R., Leigh, J.B., Purser, R.A., Greenberg, A.H., Willans, D.J., Keller, O. Petrik, P.K., Thay, T.Y., Suresh, M.R. and Noujaim, A.A. (1984): Monoclonal antibody against a cryptic carbohydrate antigen of murine and human lymphocytes. I. Antigen expression in non-cryptic or unsubstituted form on certain murine lymphomas, on a spontaneous murine mammary carcinoma, and on several human adenocarcinomas. Int. J. Cancer 33:123-129.

Lynen, R., Rothe, M. and Gallasch, E. (1983): Characterization of formaldehyde-related antibodies encountered in hemodialysis patients at different stages of immunization. Vox Sang. 44:81-89.

Mäkelä, O. and Cantell, K. (1958): Destruction of M and N blood group receptors of human red cells by some influenza viruses. Ann. Med. Biol. Fenn. 36:366-374.

Marchesi, V.T., Tillack, T.W., Jackson, R.L., Segrest, J.P. and Scott, R.E. (1972): Chemical characterization and surface orientation of the major glycoprotein of the human erythrocyte membrane. Proc. Natl. Acad, Sci. USA 69:1445-1449.

Mawby, W.J., Anstee, D.J. and Tanner, M.J.A. (1981): Immunochemical evidence for hybrid sialoglycoproteins of human erythrocyte. Nature (London) 291:161-162.

McDougall, D.C.J. and Jenkins, W.J. (1981): The first human example of anti-M^{e}. Vox Sang. 40:412-415.

Merry, A.H., Thomson, E.E., Anstee, D.J. and Stratton, F. (1984): The quantification of erythrocyte antigen sites with monoclonal antibodies. Immunology 51:793-800.

Merz, W. and Roelcke, D. (1971): Biochemische Differenzierung der Pr_1/Pr_2 determinierenden von der MN-determinierenden N-acetyl-Neuraminsäure durch Acetylierungversuche mit Erythrocytenglykoproteinen. Eur. J. Biochem. 23:30-35.

Morawiecki, A. (1964): Dissociation of M and N-group mucoproteins into sub-units in detergent solution. Biochim. Biophys. Acta 83:339-347.

Morel, P.A., Bergren, M.O., Hill, V., Garratty, G. and Perkins, H.A. (1981): M and N specific agglutinins of human erythrocytes stored in glucose solutions. Transfusion 21:652-662.

Moulds, J.J. (1983): Immunochemistry of the MNSs blood group system. In *Red Cell Membrane Glycoconjugates and Related Genetic Markers,* Cartron, J.P., Rouger, P. and Salmon, C., eds., Librairie Arnette, Paris, pp. 17-26.

Mueller, T.J. and Morrison, M. (1981): Glycoconnectin (PAS-2), a membrane attachment site for the human erythrocyte cytoskeleton., In *Erythrocyte Membranes, 2. Recent Clinical and Experimental Advances,* Kruckenberg, W.C., Eaton, J.E. and Brewer, G.J., Eds., A.R. Liss Inc., New York, NY., pp. 95-112.

Nichols, M.E., Rosenfield, R.E. and Rubinstein, P. (1985): Two blood group M epitopes disclosed by monoclonal antibodies. Vox Sang. 49:138-143.

Ochiai, T., Furthmayr, H. and Marcus, D.M. (1983): Diverse in specificities of five monoclonal antibodies reactive with glycophorin A of human erythrocytes. J. Immunol. 131:864-868.

Ogata, S. and Lloyd, K.O. (1982): Gp110 - a major sialoglycoprotein of human cells: Isolation and partial characterization from a malignant melanoma cell line. Arch. Biochem. Biophys. 217:665-673.

Owens, J.W., Meuller, T.J. and Morrison, M. (1980): A minor sialoglycoprotein of the human erythrocyte membrane. Arch. Biochem. Biophys. 204:247-254.

Papayannopoulou, T., Yokochi, T. and Martin, P. (1983): The surface antigen profile of HEL cells. In *Globin Gene Expression and Hemopoietic Differentiation,* Stamatoyannopoulos, G. and Nienhuis, A.W., eds., Alan R. Liss, New York, NY., pp. 277-292.

Pavone, B.G., Pirkola, A., Nevanlinna, H.R. and Issitt, P.D.

(1978): Demonstration of anti-Wr^b in a second serum containing anti-En^a. Transfusion 18:155-159.
Pavone, B.G., Billman, R., Bryant, J., Sniecinski, I. and Issitt, P.D. (1981): An auto-anti-En^a, inhibitable by MN sialoglycoprotein. Transfusion 21:25-31.
Petryniak, J., Petryniak, B., Wasniowska, K. and Krotkiweski, H. (1980): Isolation and immunochemical characterization of the *Euonymus europaeus* lectin receptor from the major sialoglycoprotein of human O erythrocytes. Eur. J. Biochem. 105:335-341.
Prohaska, R., Koerner, T.A.W., Armitage, I.M. and Furthmayr. H. (1981): Chemical and carbon-13 nuclear magnetic resonance studies of the blood group M and N active sialoglycopeptides from human glycophorin A. J. Biol. Chem. 256:5781-5791.
Rahman, A.F.R. and Logenecker, B.M. (1982): A monoclonal antbody specific for the Thomsen-Friedenreich cryptic T antigen. J. Immunol. 129:2021-2024.
Rearden, A. (1985): Phospholipid dependence of Wr^b antigen expression in human erythrocyte membranes. Vox Sang. 49:3466-353.
Rearden, A., Nachtscheim, D.A., Frisman, D.M., Chiu, P., Elmajian, D.A. and Baird, S.M. (1983): Altered cell surface antigen expression in bladder carcinoma detected by a new hemagglutinating monoclonal antibody. J. Immunol. 131:3073-3077.
Rearden, A., Elmajian, D.A. and Baird, S.M. (1984): Comparison of human and siamang ABH and MN blood groups using monoclonal antibodies. J. Med. Primatol. 13:315-325.
Rearden, A., Taetle, R., Elmajian, D.A., Majda, J.A. and Baird, S.M. (1985): Glycophorin A on normal and leukemia cells detected by monoclonal antibodies, including a new monoclonal antibody reactive with glycophorins A and B. Mol. Immunol. 22:369-378.
Reid, M.E., Ellisor, S.S., Barker, J.M., Lewis, T. and Avoy, D.R. (1981): Characteristics of an antibody causing agglutination of M-positive non-enzymatically glycosylated human red cells. Vox Sang. 41:85-90.
Ridgwell, K., Tanner, M.J.A. and Anstee, D.J. (1983): The Wr^b antigen, a receptor for *Plasmodium falciparum* malaria, is located on a helical region of the major membrane sialoglycoprotein of human red blood cells. Biochem. J. 209:273-276.
Ridgwell, K., Tanner, M.J.A. and Anstee, D.J. (1984): The Wr^b antigen in St^a positive and Dantu-positive human erythrocytes. J. Immunogenet. 11:365-370.
Rimmer, E.F. and Horton, M.A. (1984): Expression of myeloid-specific antigens on two human erythroleukemia cell lines, HEL and K562. Leukemia Res. 8:207-211.
Roelcke, D. (1981): Pr and Gd antigens. Blood Transf. Immunohaemat. 24:27-36.
Roelcke, D. (1984): Kälteagglutinine: humane monoclonale Antikörper gegen Glykokonjugat-Antigene von Zelloberflächen. Funkt. Biol. Med. 3:106:127.
Roelcke, D. and Kreft, H. (1984): Characterization of various anti-Pr cold agglutinins. Transfusion 24:210-213.
Roelcke, D. and Uhlenbruck, G. (1969): Letter to the editor. Vox Sang. 18:478-479.
Roelcke, D., Ebert, W., Metz, J. and Weicker, H. (1971): I-, MN- and Pr_1/Pr_2-activity of human erythrocyte glycoprotein fractions obtained by ficin treatment. Vox Sang. 21:352-361.

Roelcke, D., Ebert, W. and Geisen, H.P. (1976): Anti-Pr3: serological and immunochemical identification of a new anti-Pr specificity. Vox Sang. 30:122-133.

Romanowska, E. (1959): Studies on blood group antigens M and N. IV. Action of influenza virus enzyme on the blood group substances M and N. V. The M and N blood group substances as inhibitors of influenza virus haemagglutination. Arch. Immunol. Ther. Exp. 7:749-764.

Salder, J.E., Paulson, J.C. and Hill, R.L. (1979): The role of sialic acid in the expression of human MN blood group antigens. J. Biol. Chem. 254:2112-2119.

Schulte, T.H. and Marchesi, V.T. (1979): Conformation of human erythrocyte glycophorin A and its constituent peptides. Biochemistry 18:275-280.

Siebert,, P.D. and Fukuda, M. (1986): Isolation and characterization of human glycophorin A cDNA clones by a synthetic oligonucleotide approach: nucleotide sequence and mRNA structure. Proc. Natl. Acad. Sci. USA 83: 1665-1669.

Sonneborn, H.H., Uthemann, H., Munro, A.C., Bruce, M., Fraser, R.H.and Inglis, G. (1984): Reactivity of monoclonal antibodies directed against blood group antigens M and N. Develop. Biol. Standard. 57:61-68.

Springer, G.F. and Ansell, N. (1958): Inactivation of human erythrocyte agglutinogens M and N by influenza viruses and receptor-destroying enzyme. Proc. Natl. Acad. Sci. USA 44:182-189.

Springer, G.F. and Desai, P.R. (1975): Human blood group MN and precursor specificities: structural and biological aspects. Carbohyd. Res. 40:183-192.

Springer, G.F. and Desai, P.R. (1982): Extent of desialylation of blood group MM, NN and MN antigens required for reactivity with human anti-T antibody and *Arachis hypogea* lectin. J. Biol. Chem. 257:2744-2746.

Springer G.F., Nagai, Y. and Tegtmeyer, H. (1966): Isolation and properties of human blood group NN and meconium V_g antigens. Biochemistry 5:3254-3272.

Steuden, I., Duk, M., Czerwinski, M., Radzikowski, C. and Lisowska, E. (1985): The monoclonal antibody anti-asialoglycophorin from human erythrocytes specific for β-D-Gal-(1-3)-αD-GalNAc- chains (Thomsen-Friedenreich receptors). Glycoconjugate J. 2:303-314.

Suttajit, M. and Winzler, R.J. (1971): Effect of modification of N-acetyl-neuraminic acid on the binding of glycoproteins to influenza virus and on susceptibility to cleavage by neuraminidase. J. Biol. Chem. 246:3398-3402.

Tabilio, A., Rosa, J.P., Testa, U., Kieffer, N., Nurden, A.T., Del Canizo, M.C., Breton-Gorius, J. and Vainchenker, W. (1984): Expression of platelet membrane glycoproteins and α-granule proteins by a human erythroleukemia cell line (HEL). EMBO J. 3:453-459.

Takasaki, S. and Kobata, A.(1976): Chemical charactarerization and distribution of ABO blood group active glycoprotein in human erythrocyte membrane. J. Biol. Chem 251:3610-3615.

Taliano, V., Guévin R.M., Hébert, D., Daniels, G.L., Tippett, P., Anstee, D.J., Mawby, W.J. and Tanner, M.J.A. (1980): The rare phenotype En(a-) in a French canadien family. Vox Sang. 38:87-93.

Tanner, M.J.A. and Anstee, D.J. (1976): The membrane change in En(a-) human erythrocytes. Absence of the major sialoglycoprotein. Biochem J. 153:271-277.

Tanner, M.J.A., Jenkins, R.E., Anstee, D.J. and Clamp, J.R. (1976): Abnormal carbohydrate composition of the major penetrating membrane protein of En(a-) erythrocytes. Biochem. J. 155:701-703.

Tanner, M.J.A., Anstee, D.J. and Judson, P.A. (1977): A carbohydrate-deficient membrane glycoprotein in human erythrocytes of phenotype S-s-. Biochem. J. 165:157-161.

Tanner, M.J.A., Anstee, D.J. and Mawby, W.J. (1980): A new human erythrocyte variant (Ph) containing an abnormal sialoglycoprotein. Biochem. J. 187:493-500.

Thomas, D.B. and Winzler, R.J. (1969): Structural studies on human erythrocyte glycoproteins. Alkali-labile oligosaccharides. J. Biol. Chem. 244:5943-5946.

Tokunaga, E., Sasakawa, S., Tamaka, K., Kawamata, H., Giles, C.M., Ikin, E.W., Poole, J., Anstee, D.J., Mawby, W.J. and Tanner M.J.A. (1979): Two apparently healthy Japanese individuals of type M^kM^k have erythrocytes which lack both the blood group MN and Ss-active sialoglycoproteins. J. Immungenet. 6:383-390.

Tomita, M. and Marchesi, V.T. (1975): Amino acid sequence and oligosaccharide attachment sites of human erythrocyte glycophorin. Proc. Natl. Acad. Sci. USA 72:2964-2968.

Tomita, M., Furthmayr, H. and Marchesi, V.T. (1978): Primary structure of human erythrocyte glycophorin A. Isolation and characterization of peptides and complete amino acid sequence. Biochemistry 17:4756-4769.

Uhlenbruck, G. (1981): The Thomsen-Friedenreich (TF) receptors - an old history with new mystery. Immunol. Commun. 10:251-264.

Uhlenbruck, G., Dahr, W., Schmalisch, R. and Janssen, E. (1976): Studies on the receptors of the MNSs blood group system. Blut 32:163-170.

Vaith, P. and Uhlenbruck, G. (1977): The Thomsen agglutination phenomenon: a discovery revisited 50 years later. Z. Immun. -Forsch. 154:1-14.

Vengelen-Tyler, V., Anstee, D.J., Issitt, P.D., Pavone, B.G., Ferguson, S.J., Mawby, W.J., Tanner, M.J.A., Blajchman, M.A. and Lorque, P. (1981): Studies on the blood of an Mi^V homozygote. Transfusion 21:1-14.

Wasniowska, K. and Lisowska, E. (1981): Products of digestion of tryptic M and N blood group glycopeptides with pronase. Arch. Immunol. Ther. Exp. 29:551-558.

Wasniowska, K., Drzeniek, Z. and Lisowska, E. (1977): The amino acids of M and N blood group glycopeptides are different. Biochem. Biophys. Res. Commun. 76:385-390.

Wasniowska, K., Reichert, C.M., McGinniss, M.H., Schroer, K.R., Zopf, D., Lisowska, E., Messeter, L. and Lundblad, A. (1985): Two monoclonal antibodies highly specific for the blood group N determinant. Glycoconjugate J. 2:163-176.

Wiener, A.S. and Rosenfield, R.E. (1961): M^e, a blood group factor common to the antigenic properties M and He. J. Immunol. 87:376-378.

Yoshima, H., Furthmayr, H. and Kobata, A. (1980): Structures of the asparagine-linked sugar chains of glycophorin A. J. Biol. Chem. 255:9713-9718.

CARBOHYDRATE STRUCTURES AS ONCO-DEVELOPMENTAL ANTIGENS AND COMPONENTS OF RECEPTOR SYSTEMS

Ten Feizi

Applied Immunochemistry Research Group
Clinical Research Centre, Watford Road
Harrow, Middlesex HA1 3UJ. U.K.

In this article I shall review our research activities which have been focused on the antigenicities and biological roles of carbohydrate structures of glycoproteins and glycolipids. I shall first discuss examples of carbohydrate structures which behave as onco-developmental antigens recognized by monoclonal antibodies. This will be followed by a discussion of the possible involvement of carbohydrates in the functions of growth factor receptors and in the interactions of embryonic cells. The last section will be concerned with observations on the carbohydrate receptors for an infective agent, *Mycoplasma pneumoniae*, and new clues to the triggering of autoantibodies following human infection with this agent.

CARBOHYDRATE STRUCTURES AS ONCO-DEVELOPMENTAL ANTIGENS RECOGNIZED BY MONOCLONAL ANTIBODIES

The term onco-developmental antigens is used here to refer to antigens which appear or disappear during successive stages of embryogenesis and cell differentiation, and may 're-appear' in certain tumours. It is now clear that many cell surface antigens of this category recognized by monoclonal antibodies are carbohydrate structures of glycoproteins and glycolipids (Feizi, 1985). This concept first arose as a result of work with human monoclonal autoantibodies known as anti-I and anti-i. These antibodies occur in the sera of patients with an autoimmune haemolytic disorder known as cold haemagglutinin syndrome (Feizi, 1981a; Feizi and Hadler, 1983). There is a persistent form of this syndrome of unknown aetiology, and a transient form which frequently arises following infection with *Mycoplasma pneumoniae* (anti-I) and Epstein Barr virus infection (anti-i). The antigens recognized by these autoantibodies, I and i, are developmentally-regulated on human erythrocytes (Marsh, 1961). I antigen is strongly expressed on erythrocytes of adults and weakly on those of the fetus and neonate; i antigen is weakly expressed on adult erythrocytes, but is the predominant antigen in the fetus and neonate. The monoclonal antibodies in the sera of patients with cold agglutinin syndrome have been remarkable in the way that they have set a precedent for subsequent studies with hybridomas. Our studies in Dr. Kabat's laboratory established that both antigens are expressed on carbohydrate backbones of the major blood group antigens (Feizi *et al.*, 1971 a,b); we showed that there are differences in the fine specificities of the monoclonal anti-I antibodies in the sera of different patients, and that the anti-I of one patient (Ma) recognizes the sequence

Galβ1–4GlcNAcβ1\
6Gal(GalNAc)

which commonly occurs on branched backbone structures. Our subsequent collaborative studies with Dr. Hakomori's group established that the i antigenic determinants are expressed on linear backbones of poly-*N*-acetyllactosamine type while the majority of I determinants are expressed on branched poly-*N*-acetyllactosamine backbones as shown on the glycolipids below:

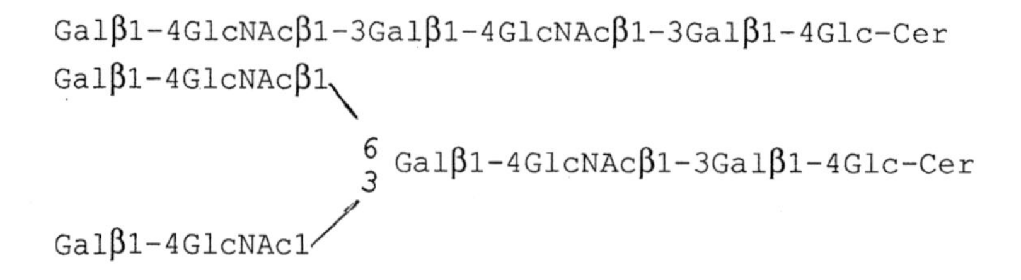

Thus the change in antigenicity of erythrocytes from i to I during the first year of life could be attributed to an increase in branched carbohydrate sequences on glycoproteins and glycolipids. This was later confirmed by biochemical studies on band 3 protein (Fukuda *et al.*, 1979, 1984 a and b).

Carbohydrate backbones of I and i antigen type occur on a variety of glycoproteins and glycolipids where they are frequently further glycosylated and converted into blood group A, B and H active structures (Kabat, 1973; Watkins, 1980) or they may be sialylated (Hakomori, 1981). In the presence of the blood group monosaccharides, the I and i antigen activities are masked but may be revealed by removal of these monosaccharides (Feizi *et al.*, 1971 b). The knowledge that there is a lack of expression of genetically predicted A, B and H antigens in certain epithelial tumours (Masamume *et al.*, 1958; Davidsohn *et al.*,1971) stimulated us to use monoclonal anti-I and anti-i antibodies to look for inappropriately expressed I and i antigen activities on glycoproteins extracted from gastrointestinal tumours. Glycoprotein extracts from metastatic (but not primary) colon tumours were found to express these antigens (Feizi *et al.*,1975). Later studies showed that the IMa determinant behaves as a tumour-associated antigen on glycoprotein extracts of gastric carcinomas derived from secretors (Picard *et al.*, 1978). We also used monoclonal anti-I and i antibodies to study the antigenic changes during mouse embryogenesis, and observed that the I and i antigens behave as 'stage specific' antigens during early embryonic development (Kapadia *et al.*, 1981). The earliest embryos from single cell stage onwards were found to express I antigen. We now believe that this may be the product of maternal genes derived from the oocyte. The embryo cells continue to express I as a major antigen until differentiation begins in the inner cell mass and the primary endoderm is formed. At this stage i antigen appears for the first time. Thus i antigen heralds the onset of differentiation in the early mouse embryo.

When we became aware of a hybridoma antibody produced by Solter and Knowles (1978) against a stage specific embryonic antigen (SSEA-1) which specifically appears in the mouse at the 8 cell stage, we were eager to investigate whether this antigen was in any way related to I antigen. We (Gooi *et al.*,

1981) established that the SSEA-1 determinant involves a fucose residue α1-3 linked to *N*-acetyllactosamine:

$$\begin{array}{l} \mathrm{Gal}\beta 1\text{-}4\mathrm{GlcNAc} \\ \quad\quad\quad | \alpha 1,3 \\ \quad\quad\quad \mathrm{Fuc} \end{array}$$

N-acetyllactosamine is the terminal disaccharide unit of the I and i antigens, so it was envisaged that these antigens could be readily converted into SSEA-1 by a single fucosylation step (Fig. 1).

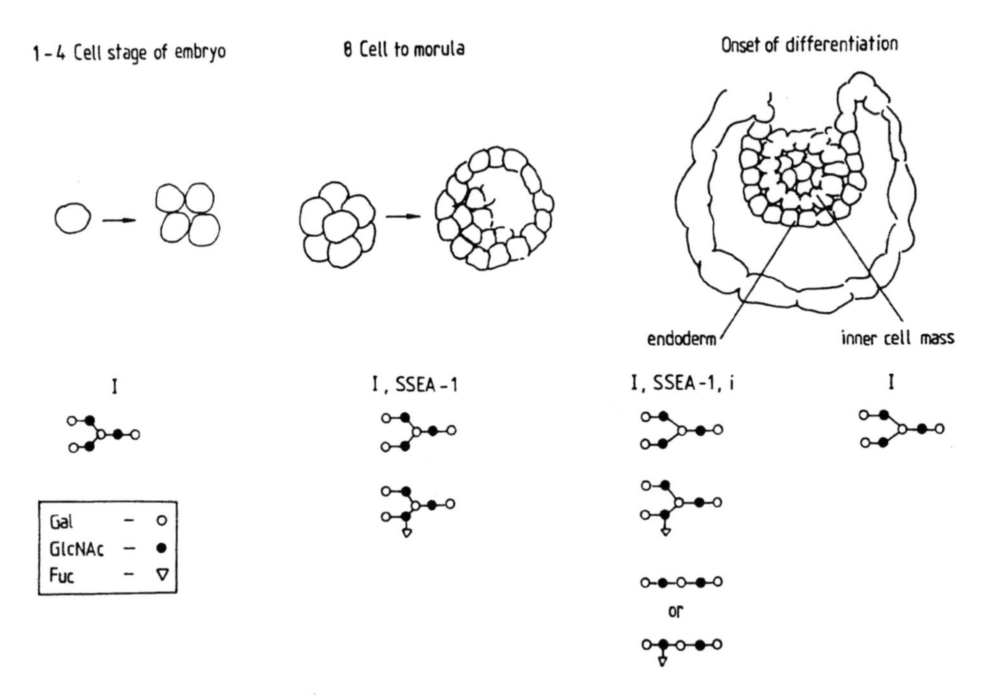

Fig. 1. Schematic presentation of the sequential changes in antigen expression on poly-*N*-acetyllactosamine chains during early embryogenesis in the mouse. Reproduced from Feizi (1983) with the publishers' permission.

Thus, it immediately became clear that sequential changes in antigenicity during embryonic development and oncogenesis could arise from sequential changes in the branching patterns of the backbone structures, and the addition (or deletion) of peripheral monosaccharides (Feizi, 1981 b; Fig. 2). Indeed, this has turned out to be a general phenomenon, borne out by studies from this and other laboratories. Extensive reviews of the subject have been made recently (Feizi, 1984 and 1985;

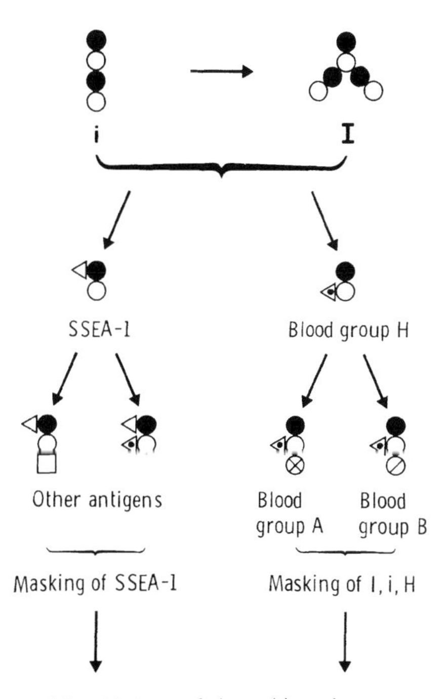

Fig. 2. Schematic presentation of interrelation of I, i, SSEA-1 and the blood group A, B, H antigens. Symbols for monosaccharides: ○ , Galβ1-4; ● , GlcNAcβ1-3 (or 6 at branch point); △ , Fucα1,3; ◬ , Fucα1,2; ⊘ , Galα1-3; ⊗ GalNAcα1,3; □ , sialic acid. Reproduced from Feizi, 1981b with the publishers'permission.

Hakomori, 1985); suffice it to say that among cell surface antigens of the onco-developmental variety detected using hybridoma antibodies, almost all that have been characterized thus far at the epitope level have been carbohydrate structures. Moreover, no unique embryonic or tumour-associated antigens have been identified. For example, a tumour-associated antigen in one cell type occurs as a normal antigen in other cell types. This knowledge is important and essential for rationalising the use of monoclonal antibodies in cell biology and for clinical purposes, such as tumour diagnosis and immunotherapy. The result is that a large array of monoclonal antibodies to diverse carbohydrate structures is now available as exquisitely specific reagents. These can be used for (a) detecting these genetically determined structures in individual cells and organized tissues, (b) for investigating the biochemical events which lead to their altered expression during cell differentiation and oncogenesis and (c) for investigating the functions of individual carbohydrate structures in normal and neoplastic cells and, in particular, for understanding the roles of these structures in growth regulation and cell interactions.

CARBOHYDRATE STRUCTURES AS ENDOGENOUS RECEPTORS?

The precise functions of individual carbohydrate sequences synthesized by cells is a subject of great interest. As reviewed recently (Feizi, 1985), information is gradually accumulating as a result of biochemical and immunochemical studies. I shall briefly discuss two immunochemical observations which may ultimately enable us to understand the roles of blood group-related carbohydrate antigens of the onco-developmental family. The first is the occurrence of this family of carbohydrate structures on the glycoprotein receptor for epidermal growth factor (EGF), and the second is the appearance of SSEA-1 antigen at the 8 cell stage of the mouse embryo which is the time of onset of compaction (tight apposition) of the embryonic cells.

Immunoblotting of the receptor for EGF isolated from A431 cells has shown that this glycoprotein expresses blood group A and other carbohydrate antigens of onco-developmental family (Childs *et al.*, 1984). In fact, the majority of monoclonal antibodies to this receptor glycoprotein are directed against

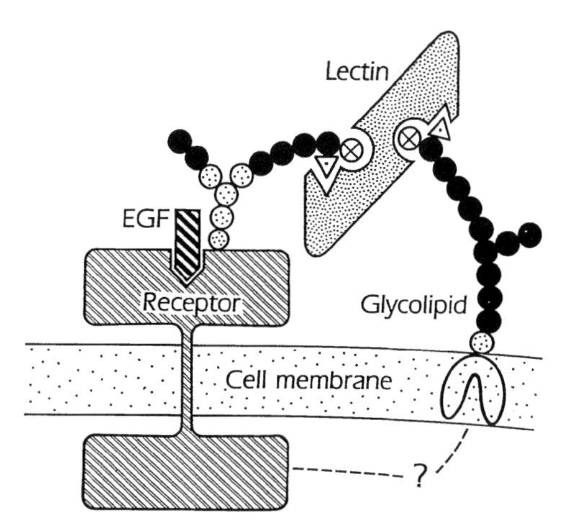

Fig. 3. Diagram of a cell membrane depicting the glycoprotein receptor for epidermal growth factor, carrying a *N*-glycosidically linked carbohydrate chain which contains backbone and peripheral region sequences that may be shared by other membrane glycoproteins and glycolipids. Thus, through interactions with carbohydrate binding proteins (here an hypothetical endogenous lectin with blood group A-related specificity is shown) the receptor glycoprotein may be part of a large complex through which it could be linked to other extra- and intracellular signalling systems and regulators. Symbols: ⊙, monosaccharides in the core region; ●, monosaccharides in the backbone regions; △, and ⊗, monosaccharides in the peripheral regions (depicting fucose and *N*-acetylgalactosamine,respectively, which constitute the blood group A determinant). Adapted from Feizi and Childs, 1985,with the publishers' permission.

its carbohydrate moiety (reviewed by Gooi *et al.*, 1985). There is no evidence to indicate that the carbohydrate moiety is involved in the interaction of the receptor with EGF. Moreover, it is reasonable to predict that the carbohydrate structures will be different in EGF receptors isolated from different cell types. On the other hand, certain lectins inhibit EGF binding, and at least two of the EGF receptor antibodies with carbohydrate specificities have been shown to elicit a biological response (reviewed by Childs *et al.*, 1984). These effects may represent allosteric influences on the EGF binding site, and we have raised the possibility that the carbohydrate chains may be receptors for other endogenous ligands (endogenous lectins) distinct from EGF, which regulate the cellular response to the growth factor (Feizi & Childs, 1985). Figure 3 was designed to stimulate some thoughts on the way the carbohydrate structures which are shared by the receptor and other glycoproteins and glycolipids of the cell surface may interact with endogenous lectins to form a macromolecular complex through which the receptor glycoprotein may be linked to other extra- and intracellular signalling systems. In fact, there are potential reactive sites among the oligosaccharides of the EGF receptor for a widely distributed soluble β-galactoside binding lectin of animal tissues. Thus it will be interesting to investigate whether this lectin or related proteins whose levels change during embryonic development (Barondes, 1984) and lymphocyte transformation (Carding *et al.*, 1985) exert growth regulatory effects through interactions with growth factor receptors.

When we observed that SSEA-1, which appears just at the time of embryonic compaction, consists of the 3-fucosyl-*N*-acetyllactosamine sequence, we suggested that this structure may be involved in this cell interaction event which is crucial for the further development of the embryo (Gooi *et al.*, 1981). Evidence in favour of this concept has been presented by Bird and Kimber (1984) and Fenderson *et al.* (1984). Our own experiments (Rastan *et al.*, 1985) in which endo-β-galactosidase was used to treat experimentally decompacted embryos suggest strongly that carbohydrate binding proteins with high affinities for structures based on poly-*N*-acetyllactosamine sequences [SSEA-1 is one among them (Childs *et al.*, 1983)] participate in the recompaction process. However, our impression is that further work is required to establish the precise recognition system operating at compaction stage (for a detailed review see Rastan *et al.*, 1985).

CARBOHYDRATE RECEPTORS FOR *MYCOPLASMA PNEUMONIAE*

One of the triggers for the production of anti-I autoantibodies in man is infection with *Mycoplasma pneumoniae*. This agent causes epidemics and sporadic cases of atypical pneumonia and some 30% of the patients develop transiently-occurring anti-I autoantibodies which are mono- or oligoclonal (reviewed by Feizi and Hadler, 1983). Our recent studies have shown that the erythrocyte receptors for this agent are sialyl oligosaccharides of Ii antigen type with terminal α2,3-linked

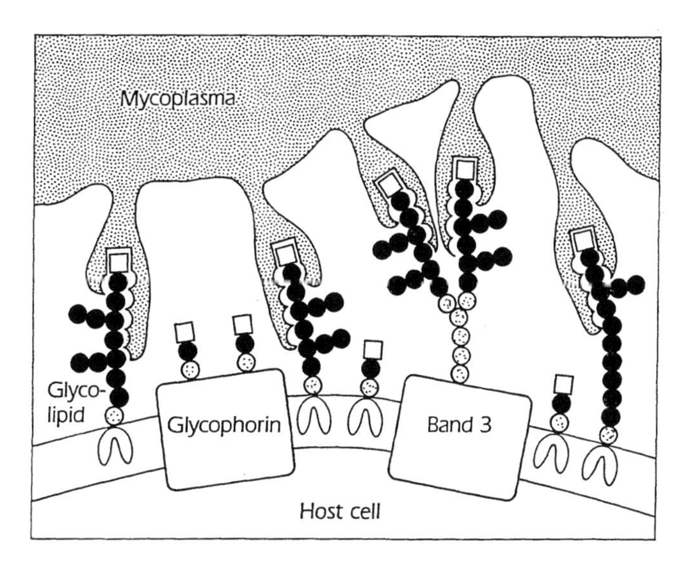

Fig. 4. Schematic presentation of the human erythrocyte membrane depicting the interaction of *M. pneumoniae* with long carbohydrate chains of I antigen type, such as those of band 3 protein and certain glycolipids, but not with the short chain structures. Symbols: □ sialic acid; ● ,monosaccharides in the backbone regions; ⊙, monosaccharides in the core regions of the carbohydrate chains. The long, branched backbone regions constitute the I antigen. Adapted from Feizi and Childs, 1985, with the publishers' permission.

sialic acid (Loomes *et al.*, 1984 and 1985) of the type occurring on band 3 protein and long chain glycolipids of erythrocyte membranes. Thus, we envisage, as schematically shown in Figure 4, that the mycoplasma adheres to such long chain structures in preference to short sialyl oligosaccharides, and somehow, following this interaction the host produces autoantibodies to the I antigen structure, i.e. to the backbone region of the receptor sequence. These developments are a sequel to our original studies (Feizi *et al.*, 1969) which showed that anti-I antibodies could be elicited in rabbits immunized with human erythrocytes that had been treated with *M. pneumoniae*.

It is our belief that the complexing of infective agents to specific saccharides of host-cell membranes may be an important mechanism for eliciting autoantibodies, and that the same principle may apply to autoimmune diseases with anti-receptor antibodies where the causal agents are not yet known but infective agents are by no means ruled out.

CONCLUDING REMARKS

In this article I have attempted to highlight the wide relevance of carbohydrate structures of glycoproteins and glycolipids in the fields of developmental and tumour biology, the biochemistry of receptors and disorders of immunity. The enormous heterogeneity and limited amounts of carbohydrate materials from mammalian cells have posed great problems for the elucidation of functions of specific carbohydrate sequences. However, the availability of monoclonal antibodies and newly developed techniques for microscale analysis of the antigenicities and receptor functions of the carbohydrate moieties of glycolipids (Magnani *et al.*, 1980) and glycoproteins (Tang *et al.*, 1985) should greatly facilitate future research in this important field.

ACKNOWLEDGEMENTS

I am grateful to Dr. R.A. Childs for helpful criticisms and to Mrs. M. Moriarty for the preparation of this manuscript.

REFERENCES

Barondes, S.H. (1984). Soluble lectins: A new class of extracellular proteins. Science 223:1259-1264.

Bird, J.M. and Kimber, S.J. (1984) Oligosaccharides containing fucose linked α(1-3) and α(1-4) to N-acetylglucosamine cause decompaction of mouse morulae. Devel. Biol. 104:449-460.

Carding, S.R., Thorpe, S.J., Thorpe, R. and Feizi, T. (1985). Transformation and growth related changes in levels of nuclear and cytoplasmic proteins antigenically related to mammalian β-galactoside-binding lectin. Biochem. Biophys. Res. Commun. 127:680-686.

Childs, R.A., Pennington, J., Uemura, K., Scudder, P., Goodfellow, P.N., Evans, M.J. and Feizi, T. (1983). High molecular weight glycoproteins are the major carriers of the carbohydrate differentiation antigens I, i and SSEA-1 of mouse teratocarcinoma cells. Biochem. J. 215:491-503.

Childs, R.A., Gregoriou, M., Scudder, P., Thorpe, S.J., Rees, A.R. and Feizi, T. (1984). Blood group active carbohydrate chains on the receptor for epidermal growth factor of A431 cells. EMBO. J. 3:2227-2233.

Davidsohn, I., Kovarik, S. and Lee, C.L. (1966) A, B, and O substances in gastrointestinal carcinoma. Arch. Path. 81:381-390.

Feizi, T., Taylor-Robinson, D., Shields, M.D. and Carter, R.A. (1969). Production of cold agglutinins in rabbits immunized with human erythrocytes treated with Mycoplasma pneumoniae. Nature 222:1253-1256.

Feizi, T. (1981a). The blood group Ii system: a carbohydrate antigen system defined by naturally monoclonal or oligoclonal autoantibodies of man. Immunol. Commun. 10:127-156.

Feizi, T. (1981b). Carbohydrate differentiation antigens. Trends in Biochemical Sciences 6:333-335.

Feizi, T. (1983). In: Fetal antigens and cancer. Ciba Foundation Symposium 96. p.216-221. Ed. D. Evered & J. Whelan, Pitman.

Feizi, T., Gooi, H.C., Childs, R.A., Picard, J.K., Uemura, K., Loomes, L.M., Thorpe, S.J. and Hounsell, E.F. (1984). Tumour-associated and differentiation antigens on the carbohydrate moieties of mucin-type glycoproteins. Biochem. Soc. Transact. 12:591-596.

Feizi, T. (1985). Demonstration by monoclonal antibodies that carbohydrate structures of glycoproteins and glycolipids are onco-developmental antigens. Nature 314:53-57.

Feizi, T., Kabat, E.A., Vicari, G., Anderson, B. and Marsh, W.L. (1971a). Immunochemical studies on blood groups. XLVII. The I antigen complex - Precursors in the A, B, H, Le^a, and Le^b blood group system-Hemagglutination inhibition studies. J. Exp. Med. 133:39-52.

Feizi, T., Kabat, E.A., Vicari, G., Anderson, B. and Marsh, W.L. (1971b). Immunochemical studies on blood groups XLIX. The I antigen complex: specificity differences among anti-I sera revealed by quantitative precipitin studies; partial structure of the I determinant specific for one anti-I serum. J. Immunol. 106:1578-1592.

Feizi, T., Turberville, C., Westwood, J.H. (1975). Blood-group precursors and cancer-related antigens. Lancet, ii:391-393.

Feizi, T. and Hadler, N. (1983). Autoantibodies and disease. In: Biochemical Aspects of Human Disease. Eds: Elkeles, R.S. and Tavill, A.S. Blackwell Scientific Publications. pp. 656-692.

Feizi, T. and Childs, R.A. (1985). Carbohydrate structures of glycoproteins and glycolipids as differentiation antigens, tumour-associated antigens and components of receptor systems. Trends in Biochem. Sci. 10:24-29.

Fenderson, B.A., Zehavi, U. and Hakomori, S-I. (1984). A multivalent lacto-N-fucopentaose III - Lysyllysine conjugate decompacts preimplantation-stage mouse embryos while the free oligosaccharide is ineffective. J. Exp. Med. 160: 1591-1596.

Fukuda, M.N., Fukuda, M. and Hakomori, S. (1979). Cell surface modification by endo-β-galactosidase. Change of blood group activities and release of oligosaccharides from glycoproteins and glycosphingolipids of human erythrocytes. J. Biol. Chem. 254:5458-5465.

Fukuda, M., Dell, A. and Fukuda, M.N. (1984a). Structure of fetal lactosaminoglycan. The carbohydrate moiety of band 3 isolated from human umbilical cord erythrocytes. J. Biol. Chem. 259:4782-4791.

Fukuda, M., Dell, A., Oates, J.E. and Fukuda, M.N. (1984b). Structure of branched lactosaminoglycan, the carbohydrate moiety of band 3 isolated from adult human erythrocytes. J. Biol. Chem. 259:8260-8273.

Gooi, H.C., Feizi, T., Kapadia, A., Knowles, B.B., Solter, D. and Evans, M.J. (1981). Stage specific embryonic antigen SSEA-1 involves α1-3 fucosylated Type 2 blood group chains. Nature 292:156-158.

Gooi, H.C., Hounsell, E.F., Lax, I., Kris, R.M., Libermann, T.A., Schlessinger, J., Sato, J.D., Kawamoto, T., Mendelsohn, J. and Feizi, T. (1985). The carbohydrate specificities of the monoclonal antibodies 29.1, 455 and 3C1B12 to the epidermal growth factor receptor of A431 cells. Bioscience Reps. 5:83-94.

Hakomori, S. (1981). Blood group ABH and Ii antigens of human erythrocytes: chemistry, polymorphism and their developmental change. Semin. Hematol. 18:39-62.

Hakomori, S. (1984). Tumor-associated carbohydrate antigens. Ann. Rev. Immunol. 2:103-126.

Kabat, E.A. (1973). Immunochemical studies on the carbohydrate moiety of water soluble blood group A, B, H, Le^a and Le^b substances and their precursor I antigens. In: Carbohydrates in Solution. Advances in Chemistry Series 117. American Chemical Society Washington, 334-361.

Kapadia, A., Feizi, T. and Evans, M.J. (1981). Changes in the expression and polarization of blood group I and i antigens in post-implantation embryos and teratocarcinomas of mouse associated with cell differentiation. Exp. Cell Res. 131:185-195.

Loomes, L.M., Uemura, K-I., Childs, R.A., Paulson, J.C., Rogers, G.N., Scudder, P.R., Michalski, J-C., Hounsell, E.F., Taylor-Robinson, D. and Feizi, T. (1984). Erythrocyte receptors for *Mycoplasma pneumoniae* are sialylated oligosaccharides of Ii antigen type. Nature 307:560-563.

Loomes, L.M., Uemura, K-I. and Feizi, T. (1985). Interaction of *Mycoplasma pneumoniae* with antigen erythrocyte glycolipids of I and i antigen types. Infect. Immun. 47:15-20.

Magnani, J.L., Smith, D.F. and Ginsburg, V. (1980). Detection of gangliosides that bind cholera toxin: direct binding of ^{125}I-labelled toxin to thin-layer chromatograms. Anal. Biochem. 109:399-402.

Marsh, W.L. (1961). Anti-i : a cold antibody defining the Ii relationship in human red cells. Brit. J. Haematol. 7:200-209.

Masamune, H., Kawasaki, H., Abe, S., Oyama, K. and Yamaguchi, Y. (1958). Molisch positive mucopolysaccharides of gastric cancers as compared with the corresponding components of gastric mucosa. Tohoku J. exp. Med. 68:81-91.

Picard, J., Waldron, E.D. and Feizi, T. (1978). Changes in the expression of the blood group A, B, H, Le[a] and Le[b] antigens and the blood group precursor associated I(Ma) antigen in glycoprotein-rich extracts of gastric carcinomas. J. Clin. Lab. Immunol. 1:119-128.

Rastan, S., Thorpe, S.J., Scudder, P., Brown, S., Gooi, H.C. and Feizi, T. (1985). Cell interactions in preimplantation embryos: evidence for involvement of saccharides of the poly-N-acetyllactosamine series. J. Embryol. Exp. Morph. 87:115-128.

Solter, D. and Knowles, B.B. (1978). Monoclonal antibody defining a stage-specific mouse embryonic antigen (SSEA-1). Proc. Natl. Acad. Sci. USA 75:5565-5569.

Tang, P.W., Gooi, H.C., Hardy, M., Lee, Y.C. and Feizi, T. (1985). Novel approach to the study of the antigenicities and receptor functions of carbohydrate chains of glycoproteins. Biochem. Biophys. Res. Commun. 132:474-480.

Watkins, W.M. (1980). Biochemistry and genetics of the ABO, Lewis and P blood group systems. In: Advances in Human Genetics. Vol. 10 (Edited by Harris, H. and Hirschhorn, K.) pp. 1-136, 379-385. Plenum Publishing Co.

CHEMISTRY OF HUMAN ERYTHROCYTE POLYLACTOSAMINE GLYCOPEPTIDES (ERYTHROGLYCANS) AS RELATED TO ABH BLOOD GROUP ANTIGENIC DETERMINANTS

Roger A. Laine[1] and Jeffrey S. Rush[2]

[1]Department of Biochemistry, Louisiana State University, Baton Rouge, LA 70803 USA and [2]Department of Biochemistry, University of Kentucky-Medical, Lexington, KY 40536 USA

Twenty-four years after the first discovery in 1900 of the blood group antigens on erythrocytes by Landsteiner and others (1-3) several researchers successfully extracted small amounts of ABH substances from red cell membranes using 95% ethanol (4-7) as an elutrient. These findings that ethanol could extract some ABH substance were confirmed by others in subsequent years leading to the idea that lipids carried the ABH antigens on erythrocytes. The main proponents of this idea were Masamune and his students Hirata, Tokura, and Satoh (8-10) in the 1950's. Yamakawa, a pioneer in this area, also isolated at the same time a "globoside" fraction containing ABH activity from erythrocyte stroma using methanol-ether (11, 12). Modern knowledge suggests that these substances were probably the shorter, 5-20 sugar lactosaminyl ceramides having terminal ABH groups. Masamune's student, Hakomori, has performed the most careful and detailed work on the latter class of lipids and has developed methods for their purification and characterization (13).

Many attempts were made by several researchers to isolate small quantities of dilute alchohol-water or water-soluble ABH substances. Many of these studies were controversial and unrepeatable in competing laboratories (reviewed in Kabat,14). Stepanov (15), for example, using aqueous/alcoholic solvents found about 1.2mg of A substance per unit (250ml) of A cells and 0.6mg of B substance per unit of B cells (our calculations from data in reference 15). From recent work by Koscielak and colleagues (16) Stepanov's substances were probably "megaloglycolipids" or "poly(glycosyl)ceramides". These compounds are long, (up to DP50+) polylactosamine glycolipids which may be partially soluble in the more aqueous-dilute alcohol extractants used by these workers. Howe (17) probably came closest to identifying the true location of the majority of the ABH substance when he reported that Blood Groups A and B were associated with a fraction representing the bulk of the stroma protein which was sedimentd at alkaline pH at 25,000 x G. After protease digestion he obtained saccharidic material with ABH activity. Unfortunately the latter work was apparently not published (18). Reports describing the isolation of ABH substances from other sources (ovarian cyst, mecomium, seminal fluid) and the release of soluble substances and ABH activity from these sources with protease treatment have been thoroughly reviewed by Kabat (14)

Early evidence for the saccharidic nature of the ABH antigenic substance on human erythrocytes was suggested by many of these studies (14) chiefly because isolated, active complex carbohydrates alone would inhibit hemagglutination by appropriate antibodies. The first work defining the molecular basis of the ABH binding sites was suggested by Watkins and Morgan who found that monosaccharides would inhibit antibody hemagglutination (19) or lectin-specific hemagglutination (20). They also found that monosaccharides could inhibit the degradative action of enzymes from *Trichomonas foetus* on ABH substances, suggesting that the enzymes were glycosidases (21). Much further work ensued (14) but the ABH substances were never successfully isolated from erythrocytes for structural determination. Instead, Kabat and his associates turned to ABH-active oligosaccharidic substances from human ovarian cyst fluid and rigorously characterized the molecular domains of the A, B, and H antigens (14, review).

Other interesting observations to locate the ABH substances on erythrocytes were published. Pinto da Silva (22) showed that the ABH substances were associated with certain intramembranous particles, thought to contain major erythrocyte proteins. The molecular location on erythrocytes of the large amount of immunologically detectable ABH substance remained a mystery.

Hakomori and his colleagues isolated many forms and have published many reports of extremely minor glycosphingolipid components, which carried ABH activity, from erythrocytes (23-35). These substances, however, comprised only about 100 micrograms per 250ml unit of red cells, about 1/250 of the ABH-bearing saccharide obtainable from Band 3 polypeptide as noted below. Karls Anders Karlsson's group in Göteborg used in-beam electron impact mass spectrometry to help characterize size and branching in ABH glycosphingolipids (36-41).

Recently, FAB mass spectrometry has been used to assist in determining sequence and branching structure (42). The FAB technique, for glycosphingolipids, although useful, has not shown significant improvement over Karlssons pioneering electron-impact methods.

Koscielak's group reported somewhat higher amounts of "megaloglycolipids" or "poly(glycosyl)lipids" which were putatively larger forms of the glycosphingolipids which Hakomori's group had studied. These substances may comprise 1.5mg of saccharidic substance as reported by Koscielak (43) and Sharon and his colleagues (44) and probably also carry some ABH activity.

ABH SUBSTANCE ON BAND 3 GLYCOPROTEIN

In 1977, our research group found that more than a third of the carbohydrates on human erythrocyte membranes consisted of *N*-linked long chain, 40-60 sugar branched, mannose-containing polylactosamine glycopeptide (45) with the repeating unit

$$\begin{array}{c} \text{--Gal}\beta 1\rightarrow 4\text{GlcNAc} \\ \downarrow \beta 1,6 \\ \text{--GlcNAc}\beta 1\rightarrow 3\text{Gal}\beta 1\rightarrow 4\text{GlcNAc}\beta 1\rightarrow 3\text{Gal}\beta 1\rightarrow 4\text{GlcNAc}\beta 1\rightarrow 3\text{--} \end{array}$$

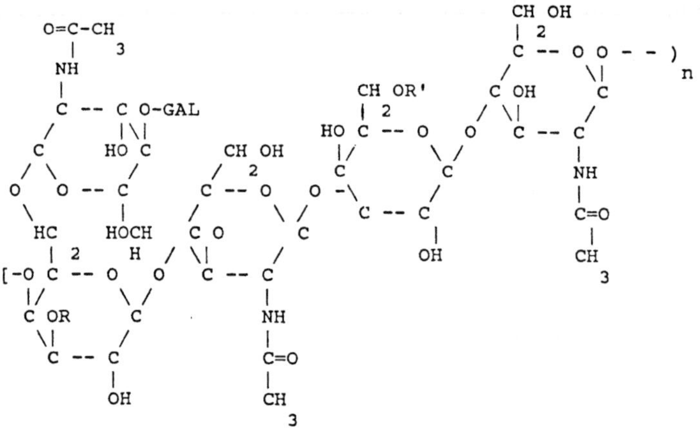

```
                                                        CH OH
 O=C-CH                                                 | 2
    |   3                                               C -- O O - - )
   NH                                                  /      \|       n
    |                                   CH OR'        C OH     C
    C -- C O-GAL                         | 2         / \|     /
   /     |\|                         HO C -- O    O   C -- C
  C      HO C         CH OH           |/       \ /         |
 / \       /         /  2             C         C          NH
O   O -- C          C -- O   O -\            /             |
 \       |         /       \ /    C -- C                   C=O
  HC    HOCH     C O        C          |                   |
  | 2    H  / \|        /             OH                   CH
[-O C -- O   O   C -- C                                      3
  |/      \ /         |
  C OR     C          NH
  \|      /           |
   C -- C            C=O
        |             |
        OH            CH
                        3
```

This repeating structure, according to our methylation linkage analysis, partial endo-β-galactosidase degradation and nitrous acid deamination, contained lactosamine oligomers whose alternate galactoses were further substituted on the 6-position (45). This architecture produced a densely arborized oligosaccharide which we called "erythroglycan" (45). In the same year, reports from Finland confirmed our structure, using other methods (46) and proposed that some ABH activity was found in these fractions (47). Previously,in other unrelated studies, Gahmberg and his co-workers and proposed that the oligosaccharide on Band 3 protein could be as large as 10,000 in molecular weight (48). It was also reported that Ii antigens occurred in this structure (49) (see also ref. 50, a review). Others later confirmed our findings (51).

Fig. 1 shows the profile of (protein-released) oligosaccharides found on human erythrocytes as arrayed by gel permeation chromatogaphy on Sephadex G-50 (45). The pooled fraction in the high molecular weight region comprises the element that we termed "erythroglycan". This fraction eluted with a similar mobiltiy to a T-10 dextran standard (10,000u) (Pharmacia), and shows a broad peak ranging from 5,000 to

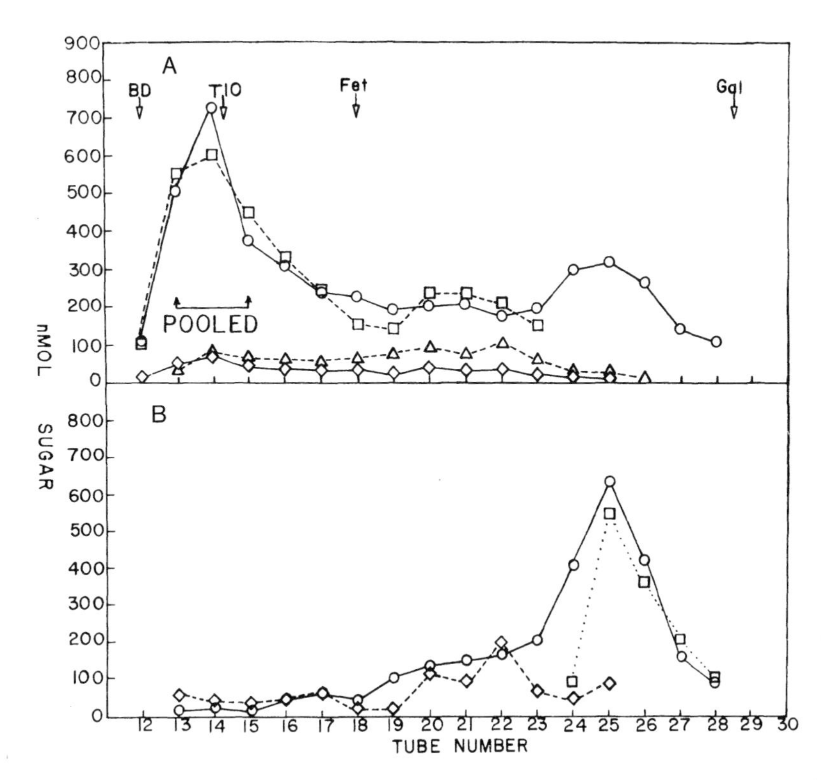

Fig. 1. Sephadex G-50 Chromatography of total glycopeptides from human erythrocytes. Molecular size standards are Blue Dextran,exclusion volume,(BD); Parmacia Dextran T10, 10,000 da-(T10); Fetuin triantennary glycopeptide, 3500 da,-(Fet); Galactose, 180 da,-(Gal). Quantity of sugars was estimated by gas-liquid chromatography of the trimethylsilyl methyl glycosides of the monosaccharides after methanolysis and N-acetylation. PANEL A: (triangle, mannose; circle, galactose; square, GlcNAc; diamond, Fucose); PANEL B: (circle, sialic acid; square, GalNAc). The pooled polylactosamine fraction near 10,000 in molecular weight was termed "Erythroglycan". [adopted from J. Biol. Chem. *253*:8006-8009 (1978)]

12,000 in molecular weight. From the sugar composition, as determined by gas-liquid chromatography of the trimethylsilyl ethers of methyl glycosides, we estimated well over a third of the erythrocyte carbohydrates to be in this fraction, and to be comprised chiefly of galactose (Gal) and 2-acetamido-2-deoxy glucose (GlcNAc) (45).

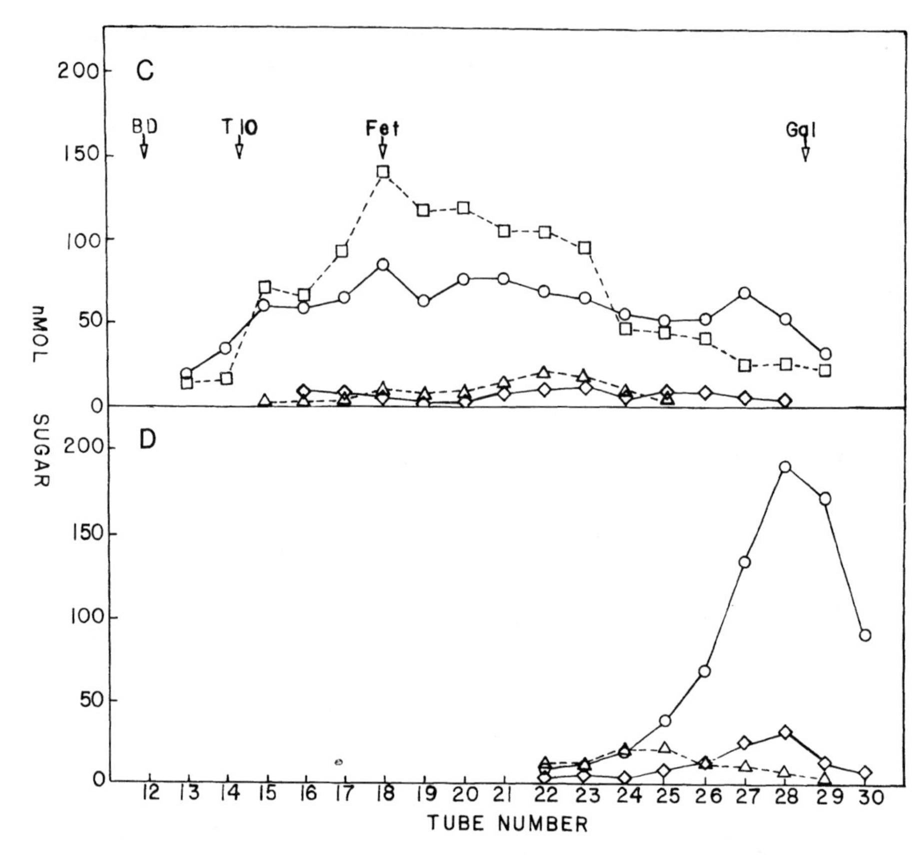

Fig. 2. Upper Panel (C): The pooled fraction from Fig. 1 was treated with endo-β-galactosidase from *E. Freundii* and re-chromatographed on Sephadex G-50. In 2 Lower Panel (D): the pooled fraction was deacylated with anhydrous hydrazine and deaminated with nitrous acid before rechromatography on Sephadex G-50. PANEL C: (square, GlcNAc; circle, Galactose; triangle, Mannose) PANEL D: (circle, Galactose; triangle, Mannose; diamond, Fucose) [adopted from J. Biol. Chem. *253*:8006-8009 (1978)]

As shown in Figure 2C, we found, in collaboration with Professors Y.T. Li and S.C. Li (45) that the enzyme *endo*-β-galactosidase (endogal) from *E. freundii*, an enzyme with a specificity directed toward keratan-like structures, was capable of degrading this oligomer. The polylactosamine-like structures, however, were somewhat resistant to total digestion due to branching. Methylation linkage analysis showed that 37% of the galactoses were 3-linked and another 37% were 3,6-disubstituted, with about 11% as terminal (45) This predicted that the main chains were linear arrays of β1→3Galβ1→4GlcNAc type, and that every second galactose was also substituted on the 6-position, confering an arborized architecture. This explained the partial resistance to the *E. freundii* endo-gal digestion since the enzyme does not work efficiently on branched galactoses (52).

POLYMERIC STRUCTURE AND MOLECULAR WEIGHT

To prove the repeating nature of the polylactosamine oligomers and to see whether any hexosamine-free sequences were present in the molecule, and to determine the identity of the sugars inserted as the branches to galactose, we de-acetylated (GlcNAc becomes GlcN) and performed a nitrous acid de-amination experiment which causes a rearrangement of the glucosamine to 2,5 anhydromannose (2,5-AHM), after loss of nitrogen, and cleavage of the glycosidic linkage (45). The predicted products from this type of degradation are shown below:

1.) If the unknown substance contained a series of galactoses, and a series of Glucosamines giving the 1:1 ratio of Gal:GlcN:

Gal-Gal-Gal-Gal-GlcN-GlcN-GlcN-GlcN-

↓ Nitrous acid

Gal-Gal-Gal-Gal-2,5-AHM + three 2,4-AHM

2.) If the substance had alternating Gal-GlcN:

Gal-GlcN-Gal-GlcN-Gal-GlcN-Gal-GlcN

↓ Nitrous acid

Gal-2,5-AHM Gal-2,5-AHM (total 4)

Condition 2.) would lead to a main product consisting of a disaccharide, with the mannose core structure surviving as well as any larger substances with a non-aminosugar substituted galactose. Fig. 2D shows the results of this experiment: All of the erythroglycan was degraded to low molecular weight oligosaccharides, indicating a polymeric structure made up of polylactosamines (45). The major peak in the low molecular weight inclusion area of the column was pooled and re-chromatographed on Bio-Gel P-2 for greater resolution. Fig. 3 shows the sugar composition as determined by gas-liquid chromatography, of the P-2 profile (33). The

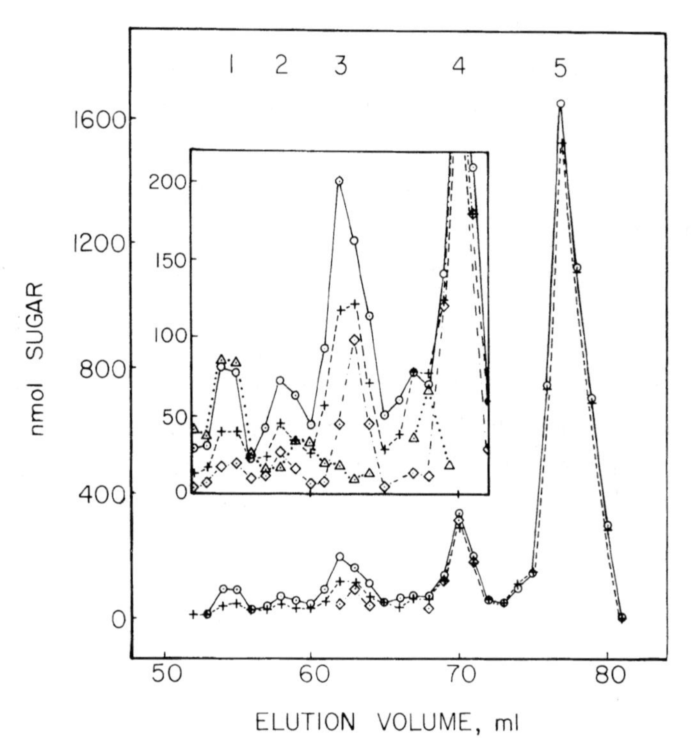

Fig. 3. Chromatography of the products of hydrazinolysis-deamination from Fig. 2D on Bio-Gel P-2. Sugar peaks were quantified by gas-liquid chromatography of the trimethylsilyl methyl glycosides of the monosaccharides after methanolysis: (triangle, mannose; circle, galactose; cross, 2-5-anhydro-mannitol; diamond, fucose). Inset is magnified 5 fold. Peak 5 is Galβ1→4[2,5-anhydromannitol], and stoichiometrically represents 18 moles of disaccharide. Peak 4 is Fucα1→2Galβ→4 [2,5-anhydromannitol], and represents 2 moles of trisaccharide. The shoulder to peak 4 is Manα1→3[Manα1→6]Manβ1→4[2,5-anhydromannitol], the remnant of the trimannosyl chitobiose core structure. Peak 3 consists of two unresolved oligosaccharides, and peaks 1 and 2 are uncharacterized.

major peak is Galβ1→4-2,5-AHM. Relative to the mannose core tetra-saccharide described below, this peak represents about 18 disaccharides (interchain mass of the lactosamine unit = 365u) with a molecular weight contribution of 365 x 18 = 6,570 daltons. The small peak preceding the major peak contained an oligosaccharide with molar ratio Fucose:Gal:2,5-AHM of 1:1:1. Integration showed there were 2 of these trisaccharides per each 18 disaccharides (53).

Methylation linkage analysis showed the Fucose to be terminal and the Galactose to be substituted on the 2-position (mass spectrum shown in Fig. 4).[53] The structure was therefore Fucα1→2Galβ1→4-2,5-AHM (the H-substance determinant). Comparing the integration of this peak with the disaccharide peak from the polylactosamine sequence, *there are two H-determinant trisaccharides per 18 disaccharide subunits,* adding another 1000 to the molecular weight. The single mole of expected mannose core substance consisting of 3 mannoses and a 2,5-AHM remnant is visible as a shoulder to the H-trisaccharide peak. These four sugars add another 650 to the mass, bringing the total to 8222μ. With the addition of an average of 1 sialic acid and 3 aminoacids the mass is brought to around 8800 and molar amounts of another surviving oligosaccharide near 1000 in molecular weight brings the total sugars to near the gel-permeation chromatography-estimated weight of 10,000.

BAND THREE PROTEIN ON ERYTHROCYTES CARRIES 2 X 10^6 ABH SITES

Each erythrocyte has been estimated to contain about 1,000,000 copies of Band 3 glycoprotein by coomassie blue staining (54). Each Band 3 molecule contains at least 1 large saccharide (48). This seems likely since the molecular weight of Band 3 is 93,000 and it is about 9% carbohydrate, therefore its oligosaccharide should be about 9,000 in mass. Each large saccharide (erythroglycan) contains an average of 2 possible H-active sites as we have determined chemically. Thus, each erythrocyte would contain about 2,000,000 ABH-type sites carried by Band 3 Oligosaccharide (erythroglycan).

If each erythrocyte contains 2,000,000 oligosaccharides of 10K weight, the total per erythrocyte would be 2 x 10 E 10 daltons. If each ml of packed cells contains 10E9 erythrocytes, this would be 2 x 10E19 daltons. A unit of erythrocytes containing 250ml of packed cells would then contain 5 x 10E21 daltons = 2.5*10E-2 g or 25mg.

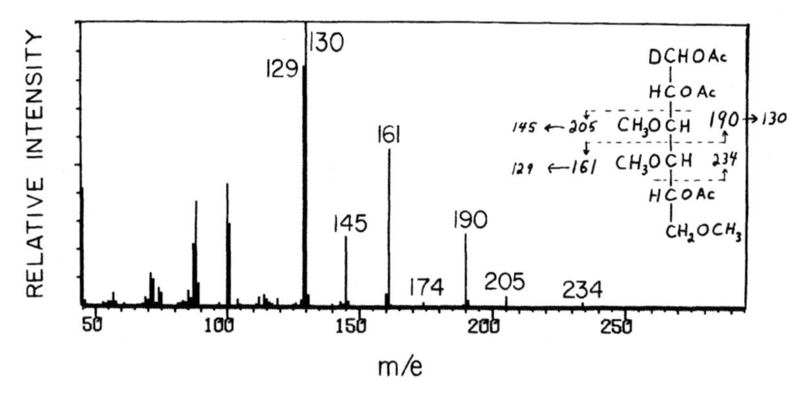

Fig. 4. The electron-impact mass spectrum of the galactose moiety obtained from peak 4 after methylation, hydrolysis, reduction with sodium borodeuteride, acetylation and gas-chromatographic separation. The galactose was 2-linked as shown by the primary and secondary fragments of the spectrum, identifying the derivative as 1,2,5-tri-O-acetyl-3,4,6-tri-O-methyl-1-deuterio-galactitol. The Fucose was terminal (not shown) and the 2,5-anhydromannitol was 4-linked (not shown). This data supports the H-blood group structure Fucα1→2Galβ1→4 GlcNAc in the original erythroglycan.

Another approach to this calculation would be that 1 mole of erythroglycan would weight 10,000g. A milliliter of erythrocytes would contain 2 x 10E15 molecules and 250ml would contain 5 x 10E17 molecules. Since Avagadro's number is 2.06 x 10E23, 250ml would contain 2.5 x 10E-6 moles (x 10,000g) = 0.025g. Thus, about 25mg of polylactosamine saccharide, bearing ABH substances occurs on Band 3 of erythrocytes. Our yields have been from 13 to 20mg/unit of cells (45,53). The highest estimates for "megaloglycolipids" of the polylactosamine type is aobut 5mg per unit of erythrocytes (44) although this figure has not been reported by other investigators. The shorter chain glycosphingolipids contain only 100-200μg/unit (Hakomori, S, Personal communication). Thus, at least 80% of the ABH substance is probably carried by Band 3.

Discussion

Finne, in 1979 (56) used *Bandeiraea simplicifolia* and *Lotus tetragonolobus* lectin binding to erythrocyte proteins in polyacrylamide gels to depict the major glycoprotein carriers of ABH-type substances. The highest molecular weight band at 105Kda bound 125-I *Lotus* lectin which is specific for "type 2" H determinants. This band was higher in its molecular weight centroid at (105,000) than the centroid of the Band 3 coomassie blue-stained band at 93,000 (54). The difference is that the lectin staining band is 12,000 daltons higher in apparent molecular weight than the center of the coomassie staining band 3. The entire *Lotus*-stained band appears above the coomassie blue stained Band 3 (56). It is a well known phenomenon that glycophorin A (PAS 1 and 2) does not stain well with coomassie due to its high carbohydrate content (54). We would like to suggest that a large portion of more highly glycosylated Band 3 also does not stain with coomassie and therefore Band 3 is much more heterogeneous than depicted by coomassie staining alone. The larger forms of Band 3 apparently contain bigger oligosaccharides. The larger forms appear to carry the majority of ABH immunological activity. This must mean that the broad coomassie-blue staining "Band 3" contains only the medium-sized to shorter, "incomplete" forms of erythroglycan.

The *N*-linked glycosylation system in the Golgi must therefore be able to recognize the size or some other parameter of the structure of the growing polylactosamine chain which triggers the attachment of the ABH determinants to only the larger forms of erythroglycan. Our estimation of the "average of 2 ABH sites per erythroglycan molecule" therefore is a probable reflection of a large number of ABH sites on the larger examples of Band 3 and few or none on the smaller species. It has been reported that in AB-type blood, the A-type determinants are on different erythroglycan molecules than the B-type groups(57). Thus, the A-type determinants would be on different Band 3 molecules than the B-type. Are these on different erythrocytes? This is an enigmatic result, since the glycoproteins must presumably travel through the same Golgi apparatus if they are on the same cells.

In Finne's work (56). the *Lotus* lectin also stained a group of 7 bands in the 40K to 80K region of the gel (termed "Band 4.5" (54)). Whether these are breakdown products of Band 3 or other proteins which bear the erythroglycan chain, or both, are as yet undetermined. Rothstein's group has shown that DIDS and NAP-taurine, specific affinity probes for extracytoplasmic and cytoplasmic domains of the Band 3 anion transporter also label proteins in the "Band 4" region of polyacrylamide gel electrophoretograms (55). Either these are smaller proteins with nearly the same ion transport properties as Band 3, or they are functional breakdown products. The latter is more likely. It has also been reported that the glucose transport protein (within Band 4.5) is glycosylated with a polylactosamine-type chain (58) but contamination with Band 3 fragments such as those suggested by Rothsteins group (55) has not been ruled out.

Thus, the final answer to Landsteiner's original observation of ABH blood types is that the ABH antigenic structures, so carefully worked out by Kabat and his co-workers, reside on a single polylactosamine carbohydrate chain on the major transmembrane protein of human erythrocytes. This glycoprotein is Band 3, the anion transporter. Minor amounts of ABH substances are borne by glycosphingolipids and perhaps a few other proteins containing polylactosamine chains. Band 3 is the most abundant glycoprotein on the erythrocyte surface, while its carbohydrate is polylactosamine and resistant to periodate stains. Band 3 and its "erythroglycan" carbohydrate, bearing ABH groups on its arborized structure, is the major topological determinant on the human erythrocyte. Unlike glycophorins, it seems to be essential to life, since no genetic variants have been uncovered who are missing the protein or its carbohydrate.

SUMMARY

Human erythrocytes bear carbohydrates linked to both proteins and lipids. The majority of the carbohydrates is carried on two proteins: 1) Band 3 (which carries a high molecular weight polylactosamine, variously termed "Erythroglycan", "poly(glycosyl)peptide" or "lactosaminoglycan" and 2) Glycophorin A (which carries 15 *O*-linked tetrasaccharides and 1 triantennary *N*-linked structure). The remainder of carbohydrates are carried mainly by a few other glycoproteins (glycophorins B,C, the glucose transporter and others) with a minor amount carried by glycosphingolipids. This report concerns the Band 3 carbohydrate and its content of potential ABH-active sites. We have determined that an average number of two [Fuc1→2Gal1→4GlcNAc] sequences are carried by each "erythroglycan", polylactosamine *N*-linked oligosaccharide. One such large oligosaccharide occurs on each molecule of Band 3 polypeptide of which there are 1,000,000 copies per erythrocyte. Therefore, about 2,000,000 possible ABH sites are borne by Band 3 on each erythrocyte. This approximates the number of immunologically estimated ABH sites on human erythrocytes. Thus, Band 3 carbohydrate probably carries the majority of ABH substance on human red cells, while other glycoproteins and glycosphingolipids carry a minor fraction.

REFERENCES

1. K. Landsteiner. (1900). Zent. Bakteriol Parasitenk 27:361.
2. K. Landsteiner, K. (1901). Wein. Klin. Woch-Scr. 14:1132.
3. A. von Decastello, A. Sturli. (1902). Munch. Med. Wochschr. 49:1090.
4. F. Schiff, L.Z. Adelsberger. (1924). Blood group specific antibodies and antigens. Z. Immunitatsforsch. 19:115.
5. S. Hesser. (1924). Does Moss' grouping of human blood with respect to isoagglutinins apply also to isohemolysins ? Serological studies with human red corpuscles. Acta. Med. Scand. Suppl. 9.
6. K. Landsteiner, J. van der Scheer, D.H. Whitt. (1924). Specificity of agglutinins and precipitins. Proc. Soc. Exptl. Biol. Med. 22:289.

7. K. Landsteiner, J. van der Scheer. (1925). Antigens of red corpuscles. II. Flocculation reaction with alcoholic extractions of erythrocytes. J. Exptl. Med. 42:123.
8. H. Masamune, H. Hirata. (1952). Chemistry and biology of lipides (IX group lipide from hog lung). Tohoku J. Exptl. Med. 55:233-241.
9. H. Masamune, M. Tokura, R. Satoh. (1952). Chemistry and biology of lipides (XIII group lipide of pig pancreas). Tohoku J. Exptl. Med. 57:2.
10. M. Tokura. (1952). Lipides (XI) Group A lipide (XII) Group B lipide (XIV) Group lipoid of stomach mucosa. Tohoku J. Exptl. Med. 56:299-307.
11. T. Yamakawa, T. Iida. (1953). Immunochemical study of red blood cells. Japan J. Exptl. Med. 23:327.
12. T. Yamakawa, S. Suzuki. (1952). Lipides of posthemolytic residue on stroma of erythrocytes. J. Biochem. 39:393.
13. S.-I. Hakomori, (1978). Biosynthesis of glycolipidis in normal NRK line cells. Methods Enzymol. 50:207-211.
14. E. A. Kabat. (1956). Blood Group Substances, Academic Press 15. A.V. Stepanov, A. Kuzin, Z. Makaeva, P.N. Kosyakov. (1940). Specific polysaccharides of blood. Biokhimiya 5:547.
16. J. Koscielak, H. Miller-Podraza, E. Zdebska. (1978). Isolation of poly(glycosyl)ceramides with A, B, H, and I blood group activities. Methods Enzymol. 50:211-216.
17. C. Howe. (1951). Influenza virus receptor and blood group antigens of human erythrocyte stroma. J. Immunol. 66:9.
18. C. Howe. (unpublished data as reported in ref. 14)
19. W. M. Watkins, and W.T.J. Morgan. (1952). Neutralization of the anti-H agglutinin in eel serum by simple sugars. Nature 169:852.
20. W. T.J. Morgan, and W.M. Watkins. (1953). Inactivation of the H-receptor on human erythrocytes by an enzyme obtained from Trichomonas foetus. Brit. J. Exptl. Pathol. 34:94.
21. W. M. Watkins, and W.T.J. Morgan. (1955). Inhibition by simple sugars of enzymes which decompose the blood-group substances. Nature 175:676.
22. P. Pinto da Silva, S.D. Douglas, D. Branton. (1971). Localization of A antigen sites on human erythrocyte ghosts. Nature 232:194-196.
23. S. I. Hakomori. (1970). Glycosphingolipids having blood groups ABH and Lewis specificities. Chem. Phys. Lipids 5:96-115.
24. S. I. Hakomori, G.D. Strycharz. (1968). Cellular blood group substances. I. Isolation and chemical composition of blood groups ABH and Leb isoantigens of sphingoglycolipid nature. Biochemistry 7:1279-1286.
25. S. I. Hakomori. (1978). Isolation of blood groups ABH-active glycolipids from human erythrocyte membranes. Methods Enzymol. 50:207-211.
26. K. Watanabe, R.A. Laine, S.I. Hakomori. (1975). Glycosphingolipids with blood groups A, H, and I activity: their status in group A1 and A2 erythrocytes and their changes associated with ontogeny and oncogeny. Biochemistry 14:2725-2733.

27. S. I. Hakomori, B. Siddiqui. (1974). Isolation and characterization of glycosphingolipid from animal cells and their membranes. Methods Enzymol. 32:345-367.

28. K. Stellner, K. Watanabe, S. Hakomori. (1973). Cellular blood-group substances. III. Isolation and characterization of glycosphingolipids with blood-group H specificity from membranes of human erythrocytes. Biochemistry 12:3047-3051.

29. H. Clausen, E. Holmes, S. Hakomori. (1986). Novel blood group H glycolipid antigens exclusively expressed in blood group A and AB erythrocytes(type three chain H). Differential conversion of different H substances substrates by A1 and A2 enzymes, and type 3 chair H expression in relation to secretor status. J. Biol. Chem. 261:1388-1392.

30. H. Clausen, S.B. Levery, R. Kannagi, S. Hakomori. (1986). Novel blood-group H glycolipid antigens exclusively expressed in blood-group A and AB erythrocytes (type 3 chain H) J. Biol. Chem. 261:1380-1387.

31. H. B. Clausen, S.B. Levery, J.M. McKibbin, S. Hakomori. (1985). Repetitive A epitope (type 3 chain A) defined by blood group A1-specific monoclonal antibody TH-1: chemical basis of qualitative A1 and A2 distinction. Biochemistry 24:3578-3586.

32. H. B. Clausen, K. Watanabe, R. Kannagi, S.B. Levery, E. Nudelman, Y. Arão-Tomumu, S. Hakomori (1985) Blood group A glycolipid (Ax) with globo-series structure which is specific for blood group A1 erythrocytes: one of the chemical bases for A1 and A2 distinction. Biochem. Biophys. Res. Commun. 124:523-529.

33. R. Kannagi, S.B. Levery, S. Hakomori. (1984). Hybrid type glycolipids (lacto-ganglio series) with a novel branched structure. Their presence in undifferentiated murine leukemia cells and their dependence on differentiation. Blood group H antigen with globo-series structure. Isolation and characterization from human blood group O erythrocytes. FEBS Lett. 175:397-401.

34. M. N. Fukuda, S. Hakomori. (1982). Structure of branched blood group A-active glycosphingolipids in human erythrocytes and polymorphism of A-and H-glycolipids in A1 and A2 subgroups. J. Biol. Chem. 257:446-455.

35. S. Hakomori. (1981). Blood group ABH and Ii antigens of human erythrocytes: chemistry, polymorphism, and their developmental change. Seimin. Hematol. 18:39-62.

36. M. E. Breimer, G.C. Hansson, K.A. Karlsson, and H. Leffler. (1980). Mass spectrometry of glycolipids-fingerprinting of lipid-linked oligosaccarides by mass spectrometry. FEBS Lett. 114:51-56.

37. M. E. Breimer, G.C. Hansson, K.A. Karlsson, and H. Leffler. (1978). Structure determination of blood group type glycolipids of cat small intestine by mass fragmentography. FEBS Lett. 124:299-303.

38. M. E. Breimer, G.C. Hansson, K.A. Karlsson, and H. Leffler. (1981). Blood group type glycosphingolipids from the small intestine of different animals

analyzed by mass spectrometry and thin-layer chromatography. A note on species diversity. J. Biochem. 90:589-609.

39. M. E. Breimer, K.A. Karlsson, B.E. Samuelsson. (1981). Characterization of a blood group H type pentaglycosylceramide of human pancreas based on a type 1 carbohydrate chain. J. Biol. Chem. 256:3810-3816.

40. M. E. Breimer, K.E. Falk, G.L. Hansson, K.A. Karlsson. (1982). Structural identification of two 10-sugar branched-chain glycosphingolipids of blood group H type present in epithelial cells of rat small intestine. J. Biol. Chem. 257:50-59.

41. M. E. Breimer, K.A. Karlsson, B.E. Samuelsson. (1982). Characterization of human intestinal difucosyl heptaglycosylceramide with a blood group B determinant and a type 1 carbohydrate chain. J. Biol. Chem. 257:1079-1085.

42. P. Hänfland, M. Kordowicz, H. Niermann, H. Egge, U. Dabrowski, J. Peter-Katalinic, J. Dabrowski. (1984). Purification and structure of branched blood group type B active glycosphingolipids from human erythrocyte membranes. Eur. J. Biochem. 145:531-542.

43. J. Koscielak, H. Miller-Podraza, R. Krauze, A. Piasek. (1976). Isolation and characterization of poly(glycosyl)ceramides (megalogylcolipids) with A, H, and I blood group activities. Eur. J. Biochem. 71:9-18.

44. M. Dejter-Juszynski, N. Harpaz, H.M. Flowers, N. Sharon. (1978). Blood group ABH specific macroglycolipids of human erythrocytes: isolation in high yield from a crude membrane glycoprotein fraction. Eur. J. Biochem. 83:363-373.

45. J. Jarnefelt, S.J. Rush, Y.T. Li, R.A. Laine. (1978). Erythroglycan, a high molecular weight glycopeptide with the repeating structure [galactosyl(1->4)-2-deoxy-2-acetamidoglucosyl(1->3)] comprising more than one-third of the protein bound carbohydrate of human erythrocyte stroma. J. Biol. Chem. 253:8006-8009.

46. T. Krusius, J. Finne, H. Rauvala. (1978). The poly(glycosyl) chains of glycoproteins. The characterization of a novel type of glycoprotein saccharides from human erythrocyte membrane. Eur. J. Biochem. 92:289-300.

47. J. Finne, T. Krusius, H. Rauvala, R. Kekomaki, G. Myllyla. (1978). Alkali-stable blood group A and B active poly(glycosyl) peptides from human erythrocyte membranes. FEBS Lett. 89:111-115.

48. C. G. Gahmberg, G. Myllyla, J. Leikola, A. Pirkola, S. Nordling. (1976). Absence of the major sialogylcoprotein in the membrane of human En(a-) erythrocythes and increased glycosylation of band 3. J. Biol. Chem. 251:6108-6116.

49. R. A. Childs, T. Feizi, M. Fukuda, S. Hakomori. (1978). Blood group I activity associated with band 3, the major intrinsic membrane protein of human erythrocytes. Biochem. J. 173:33-336.

50. T. Feizi. (1981). The blood group Ii system: a carbohydrate antigen system defined by naturally monoclonal or oligoclonal autoantibodies of man. Immunol. Comm. 10:127-156.
51. M. Fukuda, M.N. Fukuda, and S. Hakomori. (1979). Developmental change and genetic defect in the carbohydrate structure of band 3 glycoprotein of human erythrocyte membrane. J. Biol. Chem. 254:3700-3703.
52. P. Scudder, P. Hanfland, K. Uemura, T. Feizi. (1984). Endo-B-D-galactosidases of Bacteroides fragilis and Escherichia freundii hydrolyze linear but not branched oligosaccharide domains of glycolipids of the neolacto series. J. Biol. Chem. 259:6586-6592.
53. J. Rush (with R.A. Laine) (1981) Ph.D. Thesis; University of Kentucky-Medical, Dept. of Biochemistry, Lexington, KY.
54. T. L. Steck, J. Yu. (1973). Selective solublization of proteins from red blood cell membranes by protein perturbants. J. Supramol. Struct. 1:220-232.
55. S. Grinstein, L. McCulloch, A. Rothstein. (1978). Transmembrane effects of irreversible inhibitors of anion transport in red blood cells. Evidence for mobile transport sites. J. Gen. Physiol. 73:493-514.
56. J. Finne. (1980). Identification of the blood group ABH active glycoprotein components of human erythrocyte membrane. Eur. J. Biochem. 104:181-189.
57. J. Viitala, K.K. Karhi, C.G. Gahmberg, J. Finne, J. Jarnefell, O. Myllyla, T. Krusius. (1981). Blood group A and B determinants are located in different polyglycosyl peptides isolated from human erythrocytes of blood group AB. Eur. J. Biochem. 105:259-265.
58. G. Froman, P. Lundhal, F. Acevedo. (1981). The stereospecific d-glucose transport protein in cholate extracts of human erythrocyte membranes. Molecular sieve chromatography and estimation of molecular weight. FEBS Lett. 129:100-104.

ABSTRACT

BLOOD GROUP ANTIGENS AND THE ENZYMES INVOLVED IN THEIR SYNTHESIS: PAST AND PRESENT

W.M. WATKINS

Division of Immunochemical Genetics, MRC Clinical Research Centre, Harrow, Middlesex (IK)

Chemical studies on the antigens associated with the ABO system were initiated in the early years of the century but progress was limited by the paucity of techniques then available for the study of complex macromolecules. By the 1940s the secreted blood group substances were known to be polysaccharide-amino acid complexes but it was not until the 1950s that the nature of the precise differences which give rise to ABH and Lewis blood group specificities began to be recognized (1). Three main lines of attack were particularly useful in providing information about the specific structures. The first involved indirect methods of inhibition of haemagglutination and precipitation, and of the enzymatic inactivation of the specific substances, by simple sugars and oligosaccharides of known structure. The second method involved the examination of the chemical changes that occur when the serological activity is destroyed by a specific enzyme. The third method was the direct isolation and identification of serologically active fragments from the products of partial degradation of the macromolecules (2). A combination of these methods allowed the now well established carbohydrate structures of the A,B,H,Le^a and Le^b determinants to be elucidated. Enzymic degradation of the blood group substances revealed that loss of a single monosaccharide destroyed one blood group specificity and revealed another (3). These findings led to the prediction that the determinant structures are built up by the sequential addition of monosaccharides to a precursor carbohydrate chain. The antigenic determinants are therefore secondary gene products and the structural alleles at the ABO, Hh and Lewis loci are thought to encode glycosyltransferases that catalyse the addition of the immunodominant sugars to complete the determinants. Enzymes that conform to these predictions have been identified in individuals of the appropriate blood groups (3). Subgroups and rare ABO variants have been found to arise from mutant forms of the glycosyltransferases that are more or less efficient in transferring the appropriate sugars than are the products of the major A or B alleles (3). The enzymes specified by the A and B genes both have the inherent capacity to synthesize A or B structures. This overlapping specifity is not normally manifested *in vivo* but may explain the aberrant blood group expression which sometimes occurs in malignant tissues (4). The precursor of A and B, namely H, is expressed in epithelial tissues only when individuals have a secretor gene Se; the role of Se has still to be conclusively established.

References: 1) Reviewed in Kabat, E.A., Blood group substances. Academic Press, New York (1956). 2) Reviewed in Morgan, W.T.J. Proc. Roy. Soc., Ser.B., 151,308 (1960). 3) Reviewed in Watkins, W.M., Adv. Hum. Genet. 10,1 (1980). 4) Yates, A.D., Greenwell, P. and Watkins, W.M. Biochem. Soc. Trans., 10,445 (1984).

STRUCTURAL CONCEPTS OF THE HUMAN BLOOD GROUP A, B, H, Le^a, Le^b, I AND i ACTIVE GLYCOPROTEINS PURIFIED FROM HUMAN OVARIAN CYST FLUID[1-4]

Albert M. Wu
Department of Veterinary Pathology
Texas A&M University, College Station
Texas, 77843 USA

The chemical structures of the blood group A, B, H, Le^a, Le^b, I and i determinants in water soluble blood group substances isolated from human ovarian cyst fluid have been studied for over four decades and are well established (1-10). However, the internal structure of the carbohydrate moiety had not been approached until base was applied to cleave the *O*-glycosidic linkage between the GalNAc of the carbohydrate moiety and Thr and Ser of the protein core (8,11-15). The proposed composite structure of the carbohydrate side chains of blood group A, B, H, Le^a and Le^b substances (5,16,17) and of precursor substances with I and i determinants (18) shown in Fig. 1 were inferred from the mechanism of the alkaline β-elimination and peeling reactions, together with the structures of the oligosaccharide fragments isolated (16). More evidence is needed to confirm these carbohydrate internal structures. Alkaline β-elimination of intact blood group substances

---

[1] The author did some of this work in Dr. E.A. Kabat's laboratory between 1976-1981 at the Department of Microbiology, the College of Physicians and Surgeons, Columbia University, New York, New York, (17, 22, 23) and thanks him for his kind guidance.

[2] The abbreviations used are: GalNAc, 2-acetamido-2-deoxy-D-galacto-pyranose; Gal, D-galactopyranose; Fuc or LFuc, L-fucopyranose; GlcNAc, 2-acetamido-2-deoxy-D-glucopyranose; GalNAc-ol, 2-acetamido-2-deoxy-D-galactitol; Gal-ol, galactitol; GLC-MS, gas-liquid chromatography-mass spectrometry; HPLC, high pressure liquid chromatography; FAB, fast atom bombardment; NOE, nuclear Overhauser enchancement.

[3] For abbreviations of lectins, see Table IV in this text. All of the glycoproteins used for structural studies were purified from human ovarian cyst fluid. Therefore, it is the source if no indication is given otherwise; blood group active glycoprotein = blood group active substance; cyst glycoprotein = human blood group active glycoprotein, purified from ovarian cyst fluid.

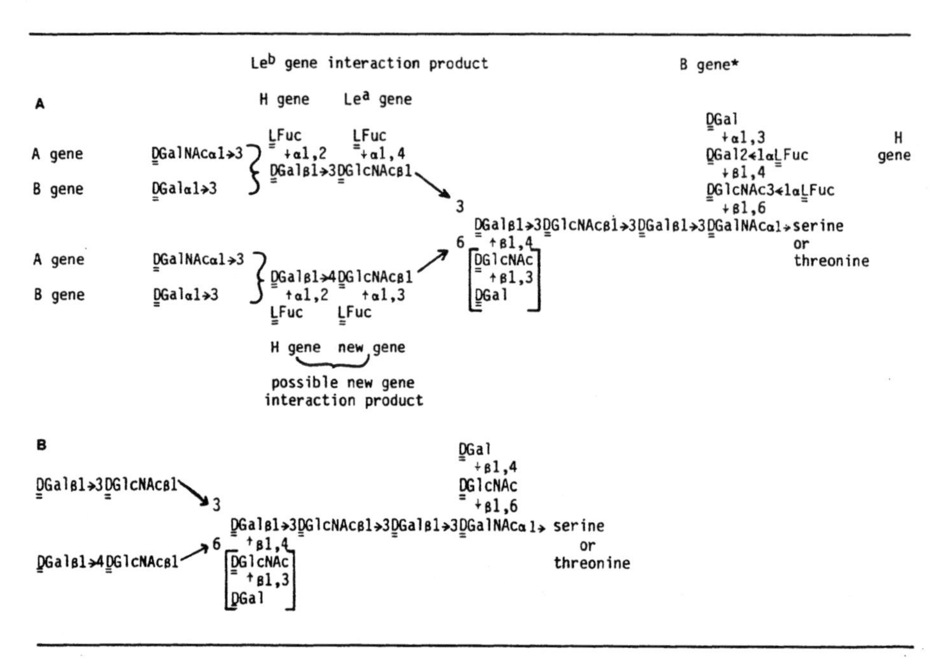

Fig. 1 (A) Proposed composite oligosaccharide structure (2,16,19) showing the relationship of the various blood group determinants and genes involved. Asterisk denotes that A substances have not been examined for oligosaccharides in this region (17). (B) Proposed overall composite structure for OG oligosaccharide moiety (18). The structure presented is based on the earlier composite structure for the A, B, H, Le^{a}, and Le^{b} substances (16,19), and shows the relationships of the type I and type II determinants upon which these antigens are built. The proposed structure is subject to all of the limitations considered earlier (15,16,19). As in the earlier studies, incomplete chains are present and can result from incomplete biosynthesis or from degradation within the cyst cavity. Moreover, the bracketed substitution on carbon 4 of the 3,4,6-linked Gal could be a Gal; whatever the residue, it must be a sequence capable of giving Gal-ol on peeling. Some side chains may be larger than the largest shown as indicated from the data presented. (Reproduced with permission from Wu *et al.*, Arch. Biochem. Biophys., 215:390-404,1982, Fig. 1).

liberates large amounts of nondialyzable oligosaccharides and glycopeptides, which are difficult to identify. Smith degradation of blood group substances prior to alkaline borohydride treatment decreases the size of the more highly substituted nondialyzable carbohydrate chains by removal of terminal blood-group active sugars whose structures are already known (1-10,12-16,19). As a result, the internal portion of longer chains becomes available for structural analysis. Moreover, the presence of radioactive [^{3}H] borohydride during the alkaline β-elimination-borohydride reduction would yield oligosaccharides with [^{3}H] at their reduced ends thus providing evidence that such oligosaccharides

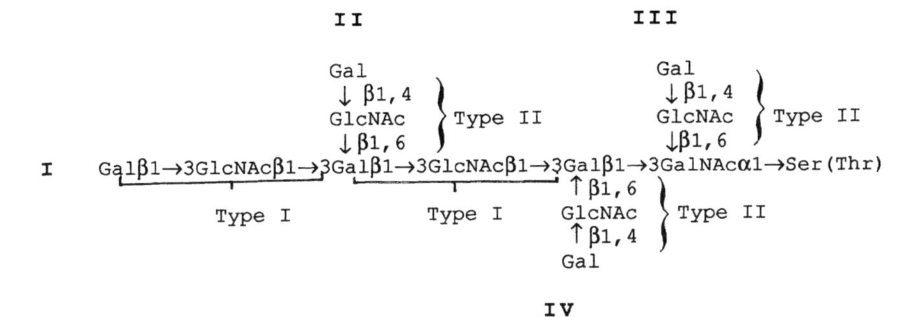

Fig. 2. Current concepts of the internal structure of carbohydrate chains, prepared from human ovarian cyst glycoproteins.

were not degraded by peeling (20,21,21A).

In the late seventies, a blood group A_1 active glycoprotein, MSS[5] 10% 2X, was chosen for studying size distribution and the internal structure of carbohydrate chains, in which the size of carbohydrate chains was reduced by Smith degradation and the carbohydrate chains were labeled with [^{3}H] at their reducing ends (17). These results (17,22,23), combined with other work (5,24-27), provide up-to-date information for constructing a general structure of the human blood group A, B, H, Le^a, Le^b, I and i active carbohydrate chains. The generalized internal structure of the carbohydrate chain is shown in Fig. 2. In this structure, a new branch (**IV**) is added to the subterminal Gal at the reducing end. Furthermore, a structural unit system of three disaccharide units and three key sugars involving specific linkages is used to elucidate the chemical structure of human cyst blood group active carbohydrate chains.

As shown in Table I, the complete composite structure is considered to be the core structure with four branches. The key sugars which specify blood group determinants are attached at the appropriate location along the branches. The core

---

[4] Previous carbohydrate "core" structure is equivalent to internal structure (core structure and branches) in this text; and the carbohydrate chains prepared by alkaline β-elimination and borohydride reduction are expressed as either R-GalNAc-ol or R-GalNAcα1→Ser(Thr) where R represents carbohydrate residues,but not for the peeling degraded products.

[5] MSS cyst, MSS cyst glycoprotein, or MSS 2X is a fraction of blood group A_1 active glycoprotein, purified from human ovarian cyst fluid (17).

Table I Structural Units of Human Blood Group Active Carbohydrate Chains, Prepared from Human Ovarian Cyst Glycoproteins.

Structural Units	Unit Location
Type I chain	Galβ1→3GlcNAc(Lacto-*N*-biose)β1→3 as found in Branch **I**, and sometimes as in Branch **III**. It is also part of the core structure.
Type II chain	Galβ1→4GlcNAc(*N*-acetyllactosamine)β1→6 as found in Branches **II** and **IV**, and frequently in Branch **III**.
T structure	Galβ1→3GalNAcα1→Ser(Thr)
Complete core structure	Galβ1→3GlcNAcβ1→3Galβ1→3GalNAcα1→Ser(Thr) (Galβ1→3GlcNAcβ1→3: Type I chain; Galβ1→3GalNAcα1→Ser: T determinant) The complete core structure is composed of one unit of Type I chain and a T determinant.
Branches	Four branches of two types of structures have been found - Type I chain (Branch **I** and sometimes Branch **III**) and Type II chain (Branches **II** and **IV**, and very often Branch **III**).
Elongated Branch **III**	Branch **III** can be longer, such as one unit of Type I chain added to Type II chain.
Key sugars for blood group determinants	GalNAcα1→3 linked to Gal at the nonreducing end for A activity. Galα1→3 linked to Gal at the nonreducing end for B activity. Fucα1→2 linked to Gal at the nonreducing end for H determinants. Fucα1→4 linked to subterminal GlcNAc of Branch **I** for Le^a determinant. Fucα1→3 linked to subterminal GlcNAc of Branch **II** for unknown determinant.

structure is a tetrasaccharide, composed of one unit of Type I chain at the nonreducing end and the T determinant sequence[Galβ1→3GalNAcα1→Ser(Thr)] at the other end. Branch **I** is a Type I chain and Branch **II** is a Type II chain. They are linked to Gal at the nonreducing end of the core structure. Branch **III** is mainly a Type II chain, but sometimes it is found as a Type I chain. It is linked to GalNAc at the reducing end. The length of Branch **III** can be increased by adding one or more sugars of the Type I chain sequence, such as Galβ1→3GlcNAcβ1→3Galβ1→4GlcNAcβ1→6GalNAcα1→Ser(Thr), a combination of Type I chain and Type II chain. A new Branch **IV** is a Type II chain, which is linked to the terminal Gal of the T determinant.

In this review article, the method to fractionate the glycoproteins (Section I), the procedure used to prepare and to purify the carbohydrate chains (section IIA to IIC), and the methods employed to identify the chains isolated (Section II-D) will be briefly described. The evidence for constructing current structural concepts of carbohydrate side chains will be given (Section III), and finally, the binding properties of the glycoproteins with lectins and human blood group anti-Ma serum before and after Smith degradation will be compared (Section IV).

I. Isolation and Purification of Human Ovarian Cyst Glycoproteins

Blood group active glycoproteins from cyst fluids have been most commonly purified by a fractionation involving precipitation of the 90% phenol soluble extract with ethanol(4,9). The products are highly active in serological tests and are largely free from contaminating proteins. An example of the method for the isolation of the blood group glycoprotein from human ovarian cyst fluid is given in Fig. 3(28). This procedure involves digestion with pepsin, precipitation by 90% phenol, and fractional precipitation with ethanol from the phenol extract. There was a progressive decrease in Fuc and Gal and in B activity and an increase in I and i activities in fractions requiring more ethanol for precipitation from phenol, indicating that the carbohydrate chains are becoming smaller. When dried ethanol precipitates were extracted with 90% phenol, the insoluble fraction was named after the corresponding blood group substance (e.g. Cyst Beach phenol insoluble). The supernatant was fractionally

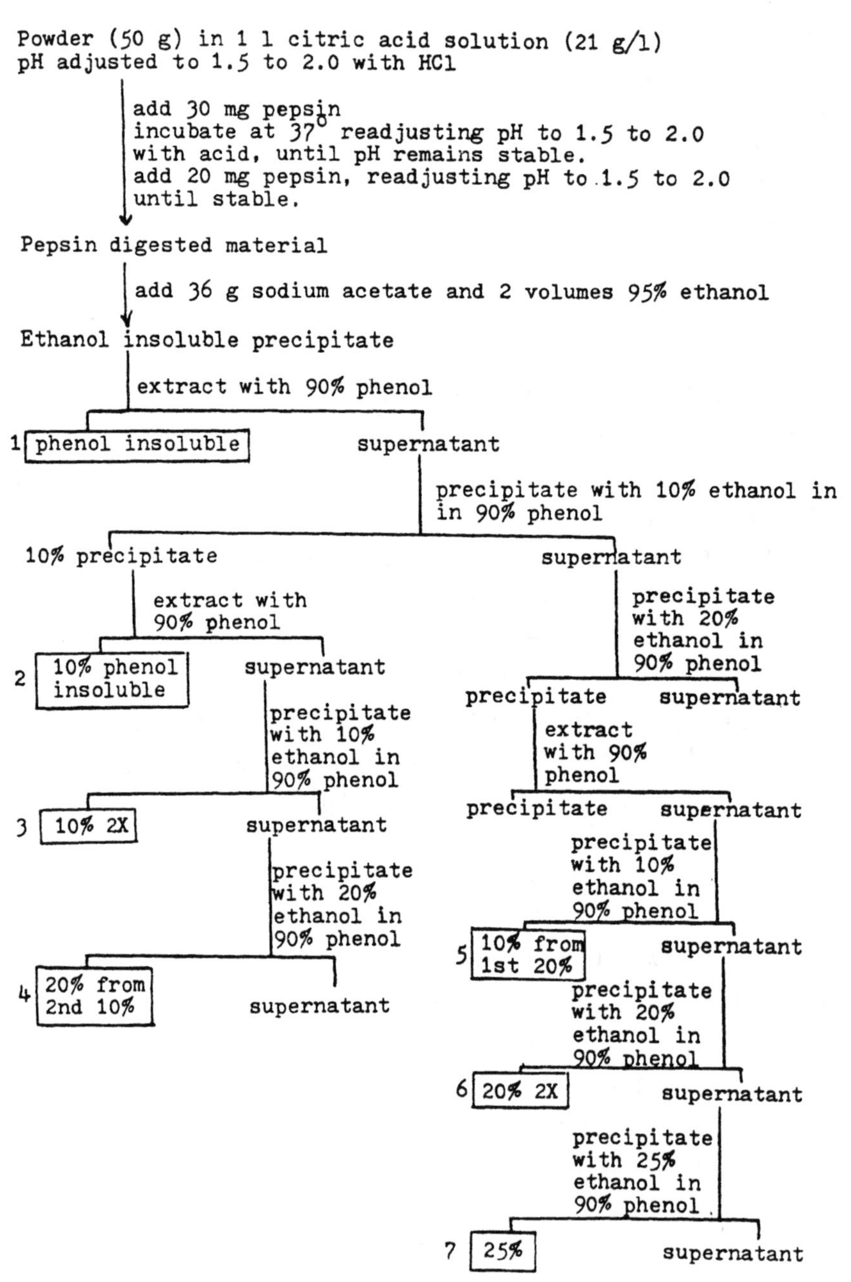

Fig 3. Flow diagram for the purification of blood group glycoproteins (of a Nigerian,Tij). Fractions mentioned in the blood group substances from Dr. E.A. Kabat's laboratory are in rectangles (Fig.12) and are numbered in order of increasing concentration of ethanol needed for precipitation.[Reproduced with permission from Maisonrouge-McAuliffe,F. and Kabat, E.A., Arch. Biochem. Biophys., 175:71-80, 1976, Fig.1].

precipitated by addition of 50% ethanol in 90% phenol to the indicated concentrations. The designation 10 or 20% (ppt) denotes a fraction precipitated from phenol at an ethanol concentration of 10 or 20%, and 2X signifies that a second phenol extraction and ethanol precipitation were carried out [as Cyst Tij 10% 2X (B,I) in Fig.3]. The A, B, H, Le^a, Le^b, and I in parentheses indicate blood group activities. Some of the cyst glycoproteins prepared in E.A. Kabat's laboratory and their structural relationship to blood group antigenic determinants are given in Fig. 12.

II. Preparation and Characterization of the Carbohydrate Chains

During the past four decades, the methods used to prepare the carbohydrate moiety of cyst glycoproteins can be divided into three stages. Until the early sixties (1st stage), mild acid hydrolysis had been the most popular method used to isolate blood group related oligosaccharides such as blood group A and B trisaccharides, Type I (Lacto-*N*-biose, Galβ1→3GlcNAc) and/or Type II (*N*-acetyllactosamine, Galβ1→4GlcNAc) blood group disaccharides (further discussion in Section III)(1-4,9,20). In the sixties (2nd stage), the alkaline β-elimination and peeling reactions were used to isolate oligosaccharide fragments to construct a generalized structure of cyst carbohydrate chains(Fig. 1A)(1-4,9,15,16,19-20). In the seventies (3rd stage), alkaline β-elimination-borohydride reduction was used to isolate intact carbohydrate chains (5,24) and in the late seventies, Smith degradation was applied to reduce the size of the carbohydrate moiety of the glycoproteins and radioactive [^{3}H] borohydride was added to the alkaline β-elimination-borohydride reduction system to create oligosaccharides with [^{3}H] at their reducing ends, which provided direct evidence that such oligosaccharides were not degraded by peeling (17,21A,22). In this Section, the mechanisms of the cleavage of carbohydrate chains from the polypeptide chain are illustrated in Part A, and the method (Smith degradation) used to reduce the size of the carbohydrate chains is briefly discussed in Part B. Examples of the methods used to purify oligosaccharides are described in Part C, and current methods for structural analysis are given in Part D.

II-A. Alkaline β-Elimination/Peeling Reaction and Borohydride Reduction

The carbohydrate chains of cyst glycoproteins are *O*-glycosidically linked through GalNAc to Ser or Thr of the protein moiety of the molecule. This linkage is characterized by its alkaline lability (21,29). The reaction in which the carbohydrate chains are separated from the polypeptide

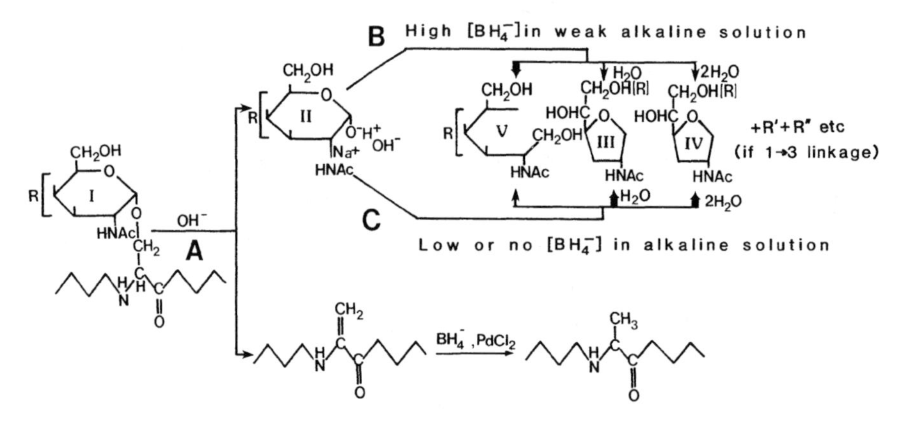

Fig 4. Effects of alkaline borohydride on the *O*-glycosidic linkage between *N*-acetylgalactosamine of the carbohydrate moiety and Ser or Thr of the protein core in the mucin type glycoproteins. Alkali treatment of the *O*-glycosidic linkage of cyst glycoprotein results in the formation of GalNAc as the reducing sugar (Compound II). This product is not stable in an alkaline environment. The direction of further reaction is dependent upon the concentration of $[OH^-]$ and $[BH4^-]$. In the absence of or with low $[BH4^-]$ in alkaline solution, the reaction is susceptible to peeling (Reaction **C**), and most of the carbohydrate chains are further degraded yielding carbohydrate (Oligosaccharide) fragments (R' and/or R") and a large amount of chromogen (III, and/or IV). With high $[BH4^-]$ in weak alkaline solution, borohydride reduction predominates (Reaction **B**), and the major products will be intact carbohydrate chains. In the above reaction direction, the yield is proportional to the extent of reaction shown by the size of the arrow "↓". R', R": Carbohydrate fragments result from peeling degradation, [Modified from Herp *et al.* Mol. Cell. Biochem., 23:27-43, Fig. 5, 1979 and Zinn *et al.*, (1977) in The Glycoconjugates Vol I., p. 77, Fig. 5, Academic Press].

chain in the presence of diluted alkali is defined as alkaline β-elimination, i.e. the reaction (**A**) from Compound I to Compound II in Fig. 4. The core structure and Branch **I** of the cyst carbohydrate chain is made up of alternating Gal and GlcNAc residues in β1→3 linkages (Fig. 2). Such 3-*O*-substituted reducing sugars are particularly unstable and susceptible to base degradation. In the absence of or at low concentration of borohydride (reaction direction **C** in Fig. 4), the reducing oligosaccharide chains released from the protein core are subject to further degradation by a base-catalyzed reaction. Successive degradation reactions will continue until a relatively alkali-stable glycosidic linkage remains. This type of degradation is termed the "peeling reaction" (19,21,29). In the early sixties, a reaction condition (such as Reaction **B** in Fig. 4.) to prevent the peeling reaction had not been established. Therefore, the megalooligosaccharides of the cyst glycoproteins were subjected to further degradation (peeling reaction) and the resulting fragments became feasible for purification and identification (16). The results of these conditions provided valuable information for our current understanding of the structure of cyst carbohydrate side chains. With a high concentration of borohydride in a low concentration of base (such as 1 M $NaBH_4$ in 0.05 N NaOH), the aldehydic group released initially is reduced with sodium borohydride, and further degradation is stopped (Reaction **B** in Fig. 4). This step of the reaction is known as borohydride reduction. When the cyst type glycoproteins are treated with 0.05 M NaOH and 1 M $NaBH_4$ at 45°C-50°C for 12-30 hours, intact carbohydrate chains are released from the protein moiety without suffering serious peeling degradation (5,21,22,24-27,29). Today, this condition is being widely used to separate intact carbohydrate chains from the protein core, and is known as the Carlson degradation (17,22).

II-B Smith Degradation

Smith degradation has been used to remove the terminal sugars at nonreducing ends, to study the blood group active determinants (16,30), as well as to eluciate the carbohydrate core structure by the differences in carbohydrate composition (16). It has also been applied to decrease the size of the carbohydrate moiety and to simplify isolation and characterization of core oligosaccharides(17,22). In the following , the stepwise reactions are given and the chemical composition of cyst glycoprotein before and after Smith degradation is compared. The resulting structures of the

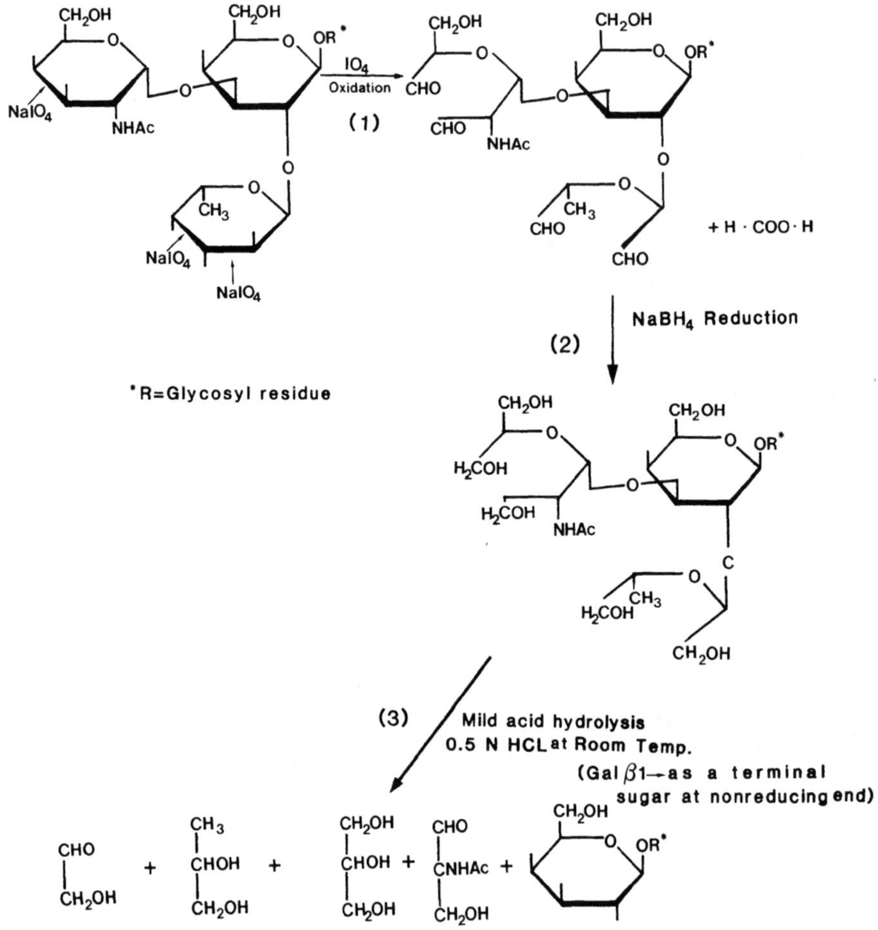

Fig. 5. Smith degradation of human blood group A determinants of cyst carbohydrate chains. Three stepwise reactions are involved in this degradation. (**1**) Periodate oxidation to cleave the linkage between carbon atoms carrying vicinal hydroxyl groups; (**2**) the reduction of the aldehydic groups of the periodate-oxidized carbohydrate chain with sodium borohydride to the polyhydroxyl compound; (**3**) removal of the resulting alcoholic derivatives by mild acid. The final product has Galβ1→ as the sugar at the nonreducing end. [Revised, Danishefsky, I. *et al.*, (1970) Introduction to polysaccharide chemisty in The Carbohydrates: chemistry and biochemistry. (W. Pigman, D. Horton, and A. Herp, eds.) Vol IIA p.406, Fig. 9. Academic Press, and Marshall, R.D. and Neuberger, A. (1972) Structural Analysis of the Carbohydrate Groups in Glycoproteins in Glycoproteins, Part A (Gottschalk, A. ed.) 2nd ed. p. 338, Fig.2., Elsevier, Amsterdam].

carbohydrate chains are illustrated in detail in Section III-A and Fig. 10.

Smith degradation includes three stepwise reactions (Fig. 5)(31,32) - Reaction **1** involves periodate oxidation to cleave the linkage between carbon atoms carrying vicinal hydroxyl groups. In the case of cyst carbohydrate chains, such groups are only available at the terminal nonreducing sugars such as GalNAc and LFuc as shown in Fig.5. Reaction **2** is the reduction of the aldehydic groups of the periodate-oxidized carbohydrate chain with sodium borohydride to the polyhydroxyl compound. In Reaction **3** there is hydrolysis with very dilute acid of the resulting alcoholic derivative, which is a true acetal and therefore susceptible to hydrolysis with very dilute acid. On the other hand, the glycosidic linkage of a residue that is not oxidized by periodate is comparatively stable to acid.

Table II

Chemical Composition of Native and Smith Degraded Products of a Human Ovarian Cyst Fluid Blood Group A Active Glycoprotein

	g/100g dry wt.		
Glycoprotein	MSS 10% ppt	MSS 1st Smith degradation	MSS 2nd Smith degradation
Lowry phenol color reaction	6.2	12.5	27.2
Total N	5.1	5.5	6.5
HexN[a]	32.3	36.1	37.8
HexNAc[b]	26.1	32.9	35.0
GalN[c]	21.2	14.7	15.0
GalN/GlcN[d]	1/.93	1/2.0	1/3.0
LFuc[e]	15.3	Trace	0.1
Gal[f]	20.8	29.3	19.1

[a]Determined by the method of Elson-Morgan; [b]Determined by the method of Morgan-Elson as described by Reissig; [c]Determined by the method of Ludowieg and Benmaman; [d]Determined by amino acid analyzer; [e]Analyzed by the Dische & Shettles method; and [f]Estimated by the orcinol method and corrected for fucose present. (Reproduced with permission from Wu *et al.*, Arch. Biochem. Biophys., 215:390-404, 1982, Table I).

When a cyst blood group A active glycoprotein was Smith degraded to reduce the length of the carbohydrate chain, its size was reduced to about 50% by one and then to about 60% by a second Smith degradation(16,17). These results indicate that about 50% of the carbohydrate residues in the human ovarian cyst blood group active glycoproteins have to be at the nonreducing ends of the carbohydrate side chains. Since the carbohydrate structure contains alternating *N*-acetylated hexosamine residues linked 1→3 or 1→4 which were not susceptible to periodate oxidation, the original structure must be highly branched. The chemical composition of human ovarian cyst fluid glycoprotein MSS before and after Smith degradations is shown in Table II. The original blood group A substance contained 15% LFuc and considerable amounts of *N*-acetyl-D-galactosamine. All of the LFuc as well as much of the DGalNAc was removed by the first Smith degradation. HexNAc/HexN were 0.81 and 0.91 for the native and first Smith degraded products as compared to 1.00 and 0.33 for GlcNAc and GalNAc, respectively. The ratios of GalNAc/GlcNAc calculated from the amino acid analytical data were 1/0.93, 1/2.0, and 1/3.0. With bovine serum albumin as reference, the Lowry phenol values of the first and second Smith degraded products increased by 100% and 440% respectively, indicating that the protein content was not affected by the Smith degradations and that the carbohydrate side chains blocked the Lowry phenol reaction in the intact blood group A active substance, as previously reported (33).

II-C. Purification of Cyst Glycoprotein Carbohydrate Chains

Once the carbohydrate chains are released from the protein core of the cyst glycoproteins and/or their chemically degraded products by alkaline β-elimination and borohydride-reduction, the oligosaccharide mixtures isolated can be separated by procedures involving charcoal column, gel filtration, paper chromatography, and HPLC. A scheme of the procedure used to purify carbohydrate chains is shown in Fig. 6. This procedure includes most of the conventional methods used for several decades. Once the chain mixtures have been fractionated by gel filtration and eluted from charcoal and paper chromatography. Further purification of each fraction by HPLC is recommended. This method offers improved resolution and speed. An example of the NMR characterization of hexa-isomers separated by HPLC is given in Fig. 7. Application of different types of chemical bonded phases by HPLC results in expanded chromatographic flexibility. Different stationary phases can be used to separate complex mixtures of oligosaccharides.

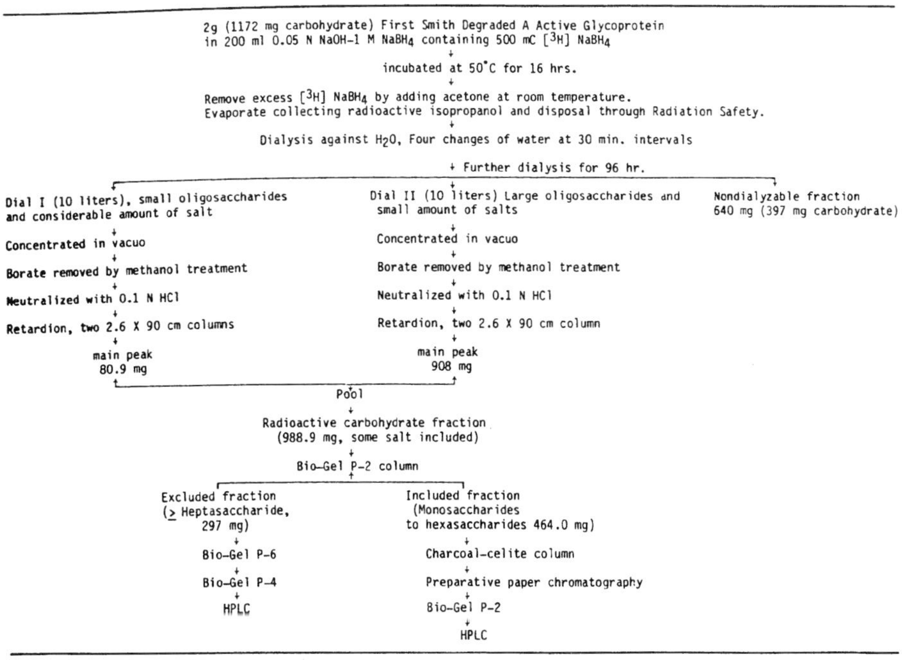

2g (1172 mg carbohydrate) First Smith Degraded A Active Glycoprotein
in 200 ml 0.05 N NaOH-1 M $NaBH_4$ containing 500 mC [^{3}H] $NaBH_4$
↓
incubated at 50°C for 16 hrs.
↓
Remove excess [^{3}H] $NaBH_4$ by adding acetone at room temperature.
Evaporate collecting radioactive isopropanol and disposal through Radiation Safety.
↓
Dialysis against H_2O, Four changes of water at 30 min. intervals
↓ Further dialysis for 96 hr.

Dial I (10 liters), small oligosaccharides and considerable amount of salt	Dial II (10 liters) Large oligosaccharides and small amount of salts	Nondialyzable fraction 640 mg (397 mg carbohydrate)
↓ Concentrated in vacuo	↓ Concentrated in vacuo	
↓ Borate removed by methanol treatment	↓ Borate removed by methanol treatment	
↓ Neutralized with 0.1 N HCl	↓ Neutralized with 0.1 N HCl	
↓ Retardion, two 2.6 X 90 cm columns	↓ Retardion, two 2.6 X 90 cm column	
↓ main peak 80.9 mg	↓ main peak 908 mg	

Pool
↓
Radioactive carbohydrate fraction
(988.9 mg, some salt included)
↓
Bio-Gel P-2 column

Excluded fraction (≥ Heptasaccharide, 297 mg)	Included fraction (Monosaccharides to hexasaccharides 464.0 mg)
↓ Bio-Gel P-6	↓ Charcoal-celite column
↓ Bio-Gel P-4	↓ Preparative paper chromatography
↓ HPLC	↓ Bio-Gel P-2
	↓ HPLC

Fig. 6. Flow diagram of alkaline β-elmination and radioactive ^{3}H-borotritide reduction of the first Smith degraded A active glycoprotein and fractionation of the fragments liberated. The Carlson condition of alkaline-β-elimination and radioactive ^{3}H-borotritide reduction was used to release carbohydrate chains from the protein core. To avoid liberation of large amounts of tritium gas in removing the excess radioactive sodium borohydride, Dr. Herbert Tabor of NIH made the excellent suggestion of adding acetone to destroy the borotritide by forming [^{3}H] isopropanol. Then 60 ml of acetone was added dropwise, while stirring, and the solution was kept at room temperature for 1 h and [^{3}H] isopropanol was removed by rotary evaporation *in vacuo*. This procedure was repeated three times. The isopropanol was disposed of through the Radiation Safety Office. This reaction was repeated three times. The isopropanol-free reaction mixture (175 ml) was transferred to a dialysis bag, and dialyzed against 20 vol. of distilled water at 0°C, changing the water every 30 min. After four changes, the dialyzable fraction (Dial I) was pooled (total volume about 10 liters). The reaction mixture was further dialyzed for 96 h (Dial II) with frequent changes of distilled water. The pooled dialysates were concentrated *in vacuo*. The nondialyzable fraction was centrifuged and the supernatant was lyophylized. Both Dial I and Dial II were neutralized with 0.1 N HCl to pH 7.0, concentrated by rotary vacuum evaporation below 39°C. Borate was removed by repeated evaporation with methanol. Dial I and Dial II were desalted by two passages through a retardion column 2.6x90 cm, then combined and fractionated on Bio-Gel P-2 and Bio-Gel P-6 columns. These fractionated materials were further purified on a charcoal-celite column, and by preparative paper and high-pressure liquid chromatography (See Fig. 8). (Reproduced from Wu *et al.*, Arch. Biochem. Biophys., 215:390-404, Fig. 3, 1982).

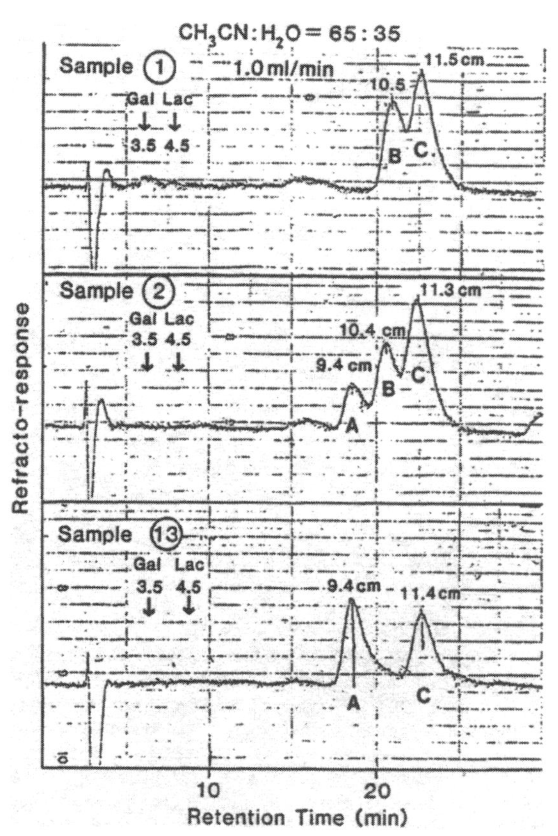

Fig. 7 Analytical HPLC profiles of three fractions of reduced hexasaccharides separated by Bio-Gel P-2, charcoal-celite column and preparative paper chromatography. HPLC was performed at room temperature on a Waters Associates Model ALC/1GPC 202/R-201 liquid chromatograph. A normal phase of 1/4 x 12 inch Bondapak/NH_2 column was used. The mobile phase was acetonitrile:H_2O in 65:35 proportions. Peaks were pooled based on continuous monitoring of the effluent by refractive index. Flow rate was from 1.0 ml/min corresponding to pressures of 780 psi. Samples, 150 µg in 600 µl CH_3CN/H_2O = 65/35 solution were injected. A, B, and C are hexa-isomers. They were identified by compositional analysis, GLC-MS (22), NMR (23, and Fig. 8 of this article) and FAB (34). [Reproduced from Wu A., *et al.*, J. Biol. Chem., 259:7178-7186, 1984, Fig. 1].

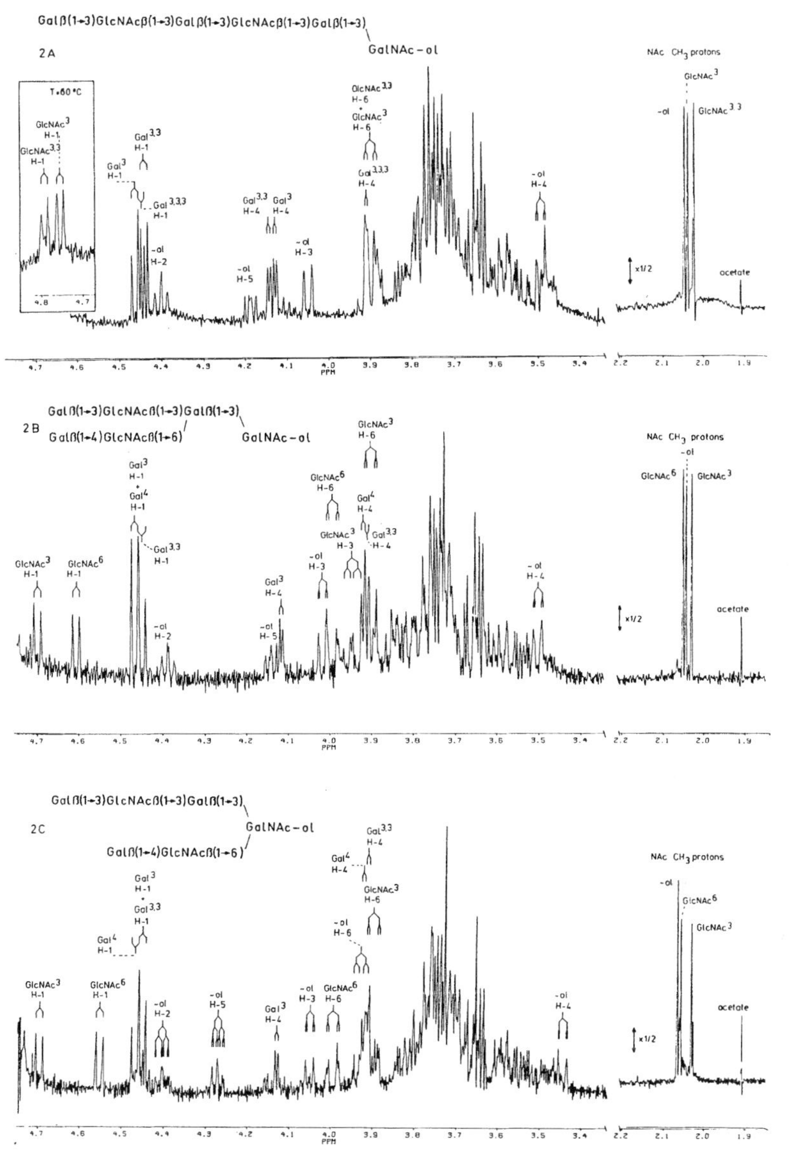

Fig. 8. 500-MHz ^{1}H-NMR spectrum of isomers of three hexa-saccharides, compounds 2A, 2B, and 2C (Fig. 7). The relative intensity scale of the *N*-acetyl proton region deviates from that of the other part of the respective spectra, as indicated. Samples were repeatedly exchanged in $^{2}H_2O$ (99.96 atom% ^{2}H), with intermediate lyophilization. 500-MHz ^{1}H-NMR spectra were recorded on a Bruker WM 500 instrument operating in the pulsed Fourier transform mode. The probe temperature was kept at 22.0 (± 0.1)°C, in order to ensure complete visualization of the H-1 signals in the spectral region 4.4 < δ < 4.8 ppm (δ $HO^{2}H$ = 4.81 ppm). The inset for compound 2A shows the GlcNAc H-1 signal at 60°C. Resolution enhancement of the spectra was achieved by Lorentzian-to-Gaussian transformation from quadrature phase detection. Chemical shifts (δ) are expressed at 22°C downfield from internal sodium 4,4-dimethyl-4-silapentane-1-sulfonate, but were actually measured by reference to internal acetone (δ = 2.225 ppm) with accuracy of 0.002 ppm.[reproduced from Mutsaers *et al.* Eur.J. Biochem., 157:139-146, 1986, Figs. 2 and 3].

The first method introduced is normal-phase chromatography on amino-(NH_2) bonded silica using a mobile phase of high polarity, which in principle is similar to paper chromatography. A second useful method is reverse-phase HPLC, which depends on hydrophobic bonding between the sample and the stationary phase. Hydrophobic binding of highly polar molecules such as carbohydrates to the alkyl groups of the C-18 stationary phase is expected to be weak. But for the weakest possible mobile phase, water, considerable binding occurs making the fractionation of oligosaccharides containing the *N*-acetylamino group possible. Recently, a combination of normal-phase and reverse-phase HPLC has been used to fractionate the oligosaccharide alditols from ovarian cyst glycoproteins (25-27). However, progress in the purification of cyst carbohydrate chains as well as other oligosaccharides is still being hindered by the unavailability of methods capable of separating closely related oligosaccharides larger than 7-9 sugars (3,17,22,25-27).

II-D. Structure Analysis of Carbohydrate Chains

To establish the primary structure of carbohydrate chains, it is necessary to analyze the composition and configuration of their sugar residues, and the sequence and linkage sites of their oligosaccharide chains. This information has traditionally been obtained by application of a combination of procedures such as compositional analysis by gas-liquid charomatography, mass spectrometry, permethylation studies, Smith degradation, partial acid or enzymatic hydrolysis, optical rotation measurements, etc. (32,34). During the past decade, the methods of structure determination have changed considerably. Most significantly, the introduction of high-resolution ^{1}H-NMR spectroscopy to provide the primary and secondary structure of a complex carbohydrate chain in solution has shown to be extremely useful because this method is rapid, quantitative, sensitive, and nondestructive (25-27,34). A set of isomers of three hexasaccharides analyzed by ^{1}H-NMR spectroscopy is shown in Fig. 8. The resulting structures are in good agreement with the result of other methods such as GLC-MS and FAB (22,35). Recently, the nuclear Overhauser enhancement (NOE) between anomeric protons and aglyconic protons has also been applied to establish the linkage of cyst glycoprotein oligosaccharide residues(25). NOE always identifies which residues are linked but not always the exact linkage position. Since NOE depends on the proximity of protons, it depends on conformational energy, NOE can be to protons adjacent to the aglycon sugar as in the case of Blood group A GalNAcα1→3Gal linkage, in which NOE is at H4 of Gal(27A).

III. Current Structural Concepts of Carbohydrate Side Chains

Regardless of their A, B, H, Le^a, Le^b, I and i activities, purified water soluble blood group active glycoproteins have a similar overall structure. They are polydisperse macromolecules (Mr 2.0×10^5 to several million) and exist in solution as random coil structures with high degrees of flexibility. Their peptide portions lack significant amounts of ordered tertiary structures. All of the glycoprotein preparations have a similar composition of 75% to 85% carbohydrate and 15% to 20% protein. As shown in Fig. 9, these glycoproteins consist of multiple heterosaccharide chains attached by an *O*-glycosidic linkage at their internal reducing ends to a serine or threonine residue of the polypeptide backbone.

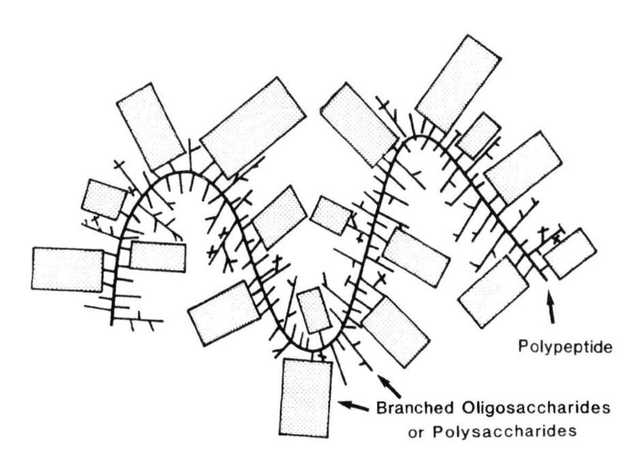

Fig. 9 Schematic structure of blood group active glycoproteins, purified from human ovarian cyst fluid. An overall structure of the branched complex is shown in Fig. 11. The structure shown above illustrates the branched heterosaccharides projecting from the polypeptide backbone. The continuous heavy line represents the protein core. The short lines, branched lines, or rectangles represent carbohydrate chains consisting of from one to about 120 sugar units ; ▭ , represents from 12-24 sugar units, and ▭, from 25-120 sugar units.

Table III

Size Distribution of Oligosaccharides Liberated by Carlson Degradation from First and Second Smith Degradations of MSS Blood Group A Active Glycoprotein (17)

Sources of oligosaccharides	Line	Fraction	Weight %	Mole %
1st stage Smith degradation	1	Nondialyzable[a]	37.7	9.6
	2	≥Heptasaccharides[b]	21.6	13.3
	3	<Hexasaccharides[b]	40.7[c]	77.1
2nd stage Smith degradation	4	Nondialyzable[a]	19.0	4.3
	5	≥Heptasaccharides[b]	35.6	13.7
	6	<Hexasaccharide[b]	45.4[c]	81.9

[a]The fraction of oligosaccharides retained in dialysis tubing with a molecular weight cut off of > 12,000.
[b]From Bio-Gel P-2 column. ≥Hepta is the excluded fraction. ≤Hexa is the included fraction.
[c]Chromogen was included.

About 90% of these carbohydrate side chains range from one to more than twenty-four sugars. Four-fifths of these side chain residues are smaller than twelve sugars, which can result from incomplete biosynthesis or by degradation in the cyst cavity. With a few exceptions, the carbohydrate binding determinants of glycoproteins are located at the nonreducing ends of the carbohydrate side chains. This Section is divided into three parts to construct a current structural concept of cyst carbohydrate chains. Heterogeneity of carbohydrate chains is described in Part A; a generalized overall structure, representative of the majority of carbohydrate chains, is mapped in Part B; and a group of short blood group active chains is defined in Part C.

III-A. Heterogeneity of the Carbohydrate Chains of Human Ovarian Cyst Glycoprotein

The carbohydrate chains of human ovarian cyst fluid glycoproteins are extremely heterogeneous with respect to both size and structure. Estimation of the size distribution and heterogeneity of carbohydrate side chains of blood group active glycoproteins can provide important insights into their interaction with compounds such as lectins or blood group specific antisera (36). Such data, however, are not easy to

Size and Mole %[a]	Structure of Carbohydrate Chains	Lectin Determinants
Monosaccharide[b] (21%)	1. GalNAcα1→Ser(Thr) of the protein core	Tn
Disaccharide (15.4%)	2. Galβ1→3GalNAcα1→Ser(Thr)	T
Trisaccharides (24.9%)	3. GlcNAcβ1,6↓ (major) GalNAcα1→Ser(Thr)	
	4. GlcNAcβ1,6↓ Galβ1→3GalNAcα1→Ser(Thr)	T(?)
	5. Galβ1→4GlcNAcβ1,6↓ GalNAcα1→Ser(Thr)	II
	6. GlcNAcβ1→3Galβ1→3GalNAcα1→Ser(Thr)	I
	7. Galβ1→3GlcNAcβ1,6(3)↓ GalNAcα1→Ser(Thr)	I
	8. Galβ1→3GlcNAcβ1,6↓ R[c]→GalNAcα1→Ser(Thr)	I
Tetrasaccharide (18.0%)	9. Galβ1→3GlcNAcβ1→3Galβ1→3GalNAcα1→Ser(Thr) (major)	I
	10. Galβ1→4GlcNAcβ1,6↓ Galβ1→3GalNAcα1→Ser(Thr)	I and T (?)
	11. GlcNAβ1→3Galβ1→4GlcNAcβ1,6↓ GalNAcα1→Ser(Thr)	
Pentasaccharides (4.2 mole%)	12. Galβ1→3GlcNAcβ1→3Galβ1→4GlcNAcβ1,6↓ GalNAcα1→Ser(Thr)	I
Hexasaccharides (6.1%)	13. Galβ1→3GlcNAcβ1→3Galβ1→3GlcNAcβ1→3Galβ1→3GalNAcα1→Ser(Thr)	
	14. Galβ1→3GlcNAcβ1→3Galβ1→3GalNAcα1→Ser(Thr) Galβ1→4GlcNAcβ1,6↑	I&II
	15. Galβ1→4GlcNAcβ1,6↓ Galβ1→3GlcNAcβ1→3Galβ1→3GalNAcα1→Ser(Thr)	I&II

Fig. 10. Heterogeneity and structure of major carbohydrate side chains of the MSS 1st Smith degraded blood group A active glycoprotein, purified from human ovarian cyst fluid. This fraction represents 77.1% of total carbohydrate residues and its chain size ranges from monosaccharide (Tn sequence) up to hexasaccharides (17,22,23). Three kinds of lectin receptors are available - Tn, T, and Types I/II [Galβ1→3(4)GlcNAcβ1→ linked residues], most of which are masked before Smith degradation. Human blood group Ii active glycoproteins (24) and mild acid hydrolyzed (pH 1.5, 100°C for 2 hours) blood group A and B active glycoproteins prepared from human ovarian cyst fluids (37,38) give a similar heterogeneity profile.

[a]About 10% is chromogen. [b]Chromogen was excluded.
[c]R, carbohydrate residues, cleaved by peeling reaction.

obtain, because considerable amounts of such chains (about 60%; Table III, sum of lines 1 and 2 in weight %) liberated from the first Smith degradation are large oligosaccharides or polysaccharides (17), which are recovered in the excluded fraction from Bio-Gel P-2 (> hexasaccharide). Moreover, combination of incomplete β-elimination (21) and peeling (19) make for further difficulty. However, size distribution of the side chains of the blood group glycoproteins can be estimated in several ways. Removal of GalNAc and other sugars at the nonreducing ends decreased by about 50% the size of carbohydrate side chains and simplified structural studies. Total number of side chains may be estimated from the number of GalNAc residues in the Smith degraded product and the short length carbohydrate chain residues in each fraction can be calculated from the number of GalNAc labeled with [^{3}H] by alkaline β-elimination and radioactive [^{3}H] borohydride reduction(17). The number of side chains in the nondialyzable fraction cannot be estimated by its radioactivity since addition of tritium to the unsaturated amino acids formed by alkaline β-elimination contributes additional radioactivity. Rather, it can be obtained by subtraction from the total side-chain units in the initial material of these in the dialyzable fractions. A small error in *N*-acetylgalactosaminitol in the dialyzable fraction could result in a larger error in estimating the nondialyzable fraction, but it can be calculated directly from the GalN residues in the nondialyzable fraction.

The nondialyzable fraction of the first Smith and Carlson degraded material represents about one third of the total carbohydrate (Table III line 1 and weight %). The main fraction (two-thirds of the total amount) shown on Bio-Gel P-60(50-100 mech) chromatography was near *Mr* 1.2 x 10^4 (unpublished data). The Gal/GalN is about 34.6/4.3 = 8/1(in reference 17, Table III). Therefore, it can be concluded that the major content of this nondialyzable fraction must be composed of polysaccharides of a size smaller than 60 sugar residues or glycopeptides containing about two or three carbohydrate side chains of 25 sugars each. This fraction is expected to be a highly branched polysaccharide and provides Type I/II determinants for binding GalNAc/Gal specific lectins and blood group I and i determinant reactive antibodies.

The first Smith and Carlson degraded material excluded from Bio-Gel P-2 is about one fifth of the total amount of carbohydrate. This fraction has an average sugar size ranging from six to twelve sugars, and each subfraction is composed of closely related isomers. This fraction also plays an important role in binding GalNAc/Gal I(II) specific lectins, chitin oligosaccharide specific lectins, and blood group I and i antibodies.

The structure and the size distribution of the small carbohydrate side chains (< hexasaccharide) from the 1st Smith degraded product of MSS blood group A active glycoproteins have been well established. As shown in Fig. 10, more than 15 different kinds of carbohydrate chains have been found. Thus, it is expected that the native cyst is more complicated. About 70% (90% from Fig. 10 x 0.77 from Table III, line 3 and Mole %) of the total carbohydrate residues are smaller than tetrasaccharide indicating that more than two-thirds of the total chain lengths are less than eight sugars in length (Smith degradation reduced about 50% of its size, therefore, it is assumed 4/0.5 = 8 sugar residues) and about four-fifths of the carbohydrate chains are less than 12 sugars.

III-B. Over-all Structure

III B-I. A Representive Structure for the Carbohydrate Chains

By the early sixties, Morgan, Watkins, Kabat, and their collaborators, by using mild acid hydrolysis, isolated two blood group active trisaccharides from both A and B blood group active glycoproteins (1-4,9,20).

Blood group

A	GalNAcα1→3Galβ1→3GlcNAc	Type I chain and
	GalNAcα1→3Galβ1→4GlcNAc	Type II chain
B	Galα1→3Galβ1→3GlcNAc	Type I chain and
	Galα1→3Galβ1→4GlcNAc	Type II chain

Two types of disaccharides, Galβ1→3GlcNAc (Type I chain) and Galβ1→4GlcNAc (Type II chain) were also isolated from blood group H active glycoproteins. Structurally, they are part of the A and B determinants, but they show neither A nor B activity. Thus, it can be assumed that these two disaccharides are the precursor disaccharides for A, B, and H determinants. In the late sixties, Lloyd *et al.*(3,15,16,19), using the alkaline β-elimination and peeling reactions, isolated two reduced branched oligosaccharides, this provided a primary idea for a core structure with three branches, which forms the basic structure for the current concept of the carbohydrate chains of the cyst glycoproteins (1-4, 15, 16, 19,20).

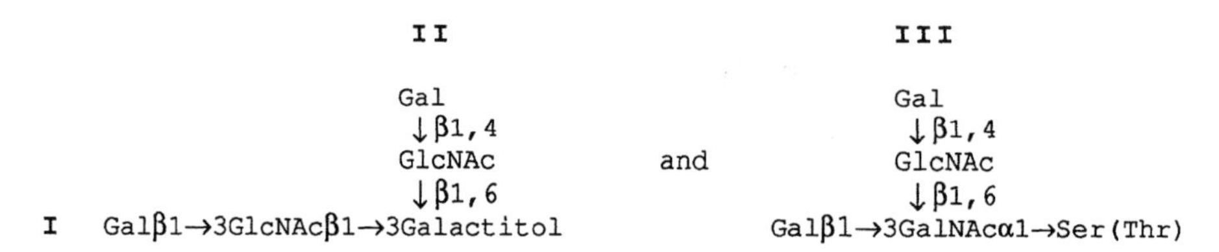

In 1973, an intact carbohydrate chain consisting of a complete core structure with two branches had been identified(5).

```
                     II

                     Gal
                      ↓ β1,4
            LFucα1→3GlcNAc
                      ↓ β1,6
I   Galβ1→3GlcNAcβ1→3Galβ1→3GlcNAcβ1→3Galβ1→3GalNAcα1→Ser(Thr)
          ↑α1,4
         LFuc
        Lewisa                              (Lewis  RIM8 0.78)
```

This oligosaccharide provides direct evidence to prove the proposed core structure (one unit of Type I chain and T structure) has two branches (**I** and **II**). As a result, a proposed composite core structure with three branches illustrated as follows has been cited for over a decade.

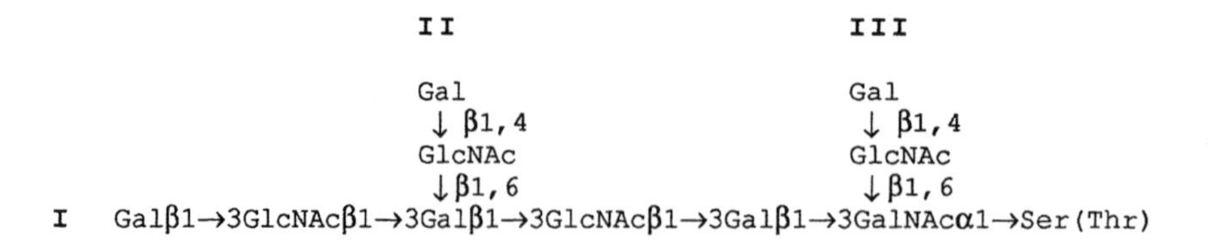

Branch **I** is a Type I chain and Branch **II** and **III** are Type II chains. In 1984, a revised core structure with four branches was constructed(22,23), as follows.

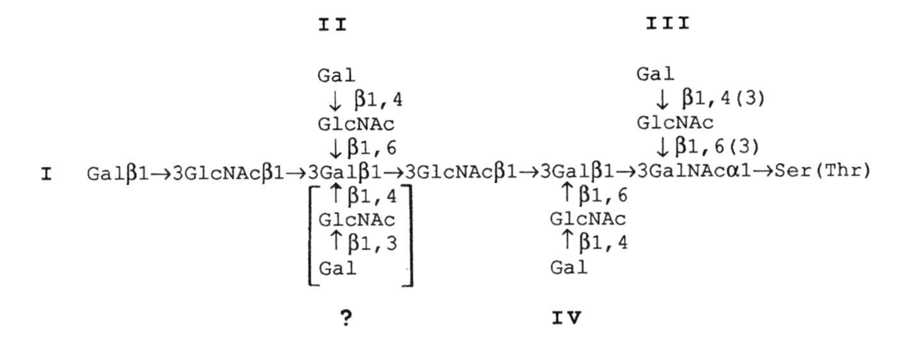

This structure is composed of a core chain with four branches. Branches **I** and **II** remain unchanged. Branch **III**, in most findings, is a Type II chain, but occasionally, it is found as a Type I chain [#8B, Galβ1→3GlcNAcβ1→3GalNAc-ol(23); and #R9, GalNAcα1→3[LFucα1→2]Galβ1→3GlcNAcβ1→3GalNAc-ol (25); and #L7, LFucα1→2Galβ1→3GlcNAcβ1→3GalNAc-ol(26)].

The chain size of Branch **III** can be lengthened by adding one or two sugars of the Type I sequence. This is supported by the following five oligosaccharides isolated.

	Oligosaccharide	Reference
1	Galβ1→3GlcNAcβ1→3Galβ1→4GlcNAcβ1→6GalNAc-ol (LFuc α1,4 on GlcNAc of Type I; Gal β1,3 on GalNAc-ol); Galβ1→3GlcNAcβ1 = Type I, Galβ1→4GlcNAcβ1 = Type II	Cb, Dua *et al.* 1985 (27) Lewis R_{IM5} 0.47 Rovis *et al.* 1973 (5)
2.	Galβ1→3GlcNAcβ1→3Galβ1→4GlcNAcβ1→6GalNAc-ol (Gal β1,3 on GalNAc-ol); Galβ1→3GlcNAcβ1 = Type I, Galβ1→4GlcNAcβ1 = Type II	Cd, Dua *et al.*, 1985 (27)
3.	GlcNAcβ1→3Galβ1→4GlcNAcβ1→6GalNAc-ol (Gal β1,3 on GalNAc-ol)	Cc, Dua *et al.*. 1985 (27)
4.	GlcNAcβ1→3Galβ1→4GlcNAcβ1→6GalNAc-ol (LFuc α1,4 on GlcNAc; Gal β1,3 on GalNAc-ol)	Ca, Dua *et al.*, 1985 (27)
5.	Galβ1→3GlcNAcβ1→3Galβ1→4GlcNAcβ1→6GalNAc-ol; Galβ1→3GlcNAcβ1 = Type I, Galβ1→4GlcNAcβ1 = Type II	3C, Wu *et al.*, 1986 (23)

In this revised structure, a new branch, defined as Branch **IV** of Type II chain, is added to the Gal nearest to the Ser or Thr of the protein moiety. It is mapped according to the following four oligosaccharides isolated.

```
1.  Galβ1→3GlcNAcβ1→3Galβ1→3GalNAc-ol              #2B ,Wu et al. 1986 (22,23)
                          ↑β1,6
                          GlcNAc
                          ↑β1.4
                          Gal
                          IV

                                       III
                                       Gal
                                       ↓β1,4
                LFuc                   GlcNAc
                ↓α1,4                  ↓β1,6
2.          Galβ1→3GlcNAcβ1→3Galβ1→3GalNAc-ol      Lewis R_IM8 1.28,
                              ↑β1,6                Rovis et al., 1973 (5)
                              GlcNAc
                              ↑β1,4
                              Gal
                              IV
        LFuc
        ↓α1,4
3.  Galβ1→3GlcNAcβ1→3Galβ1→3GalNAc-ol              Lewis R_IM8 3.5
                          ↑β1,6                    Rovis et al., 1973 (5)
                          GlcNAc
                          ↑β1,4
                          Gal
                          IV               III
                                           Gal
                                           ↓β1,4
                                           GlcNAc
                                           ↓β1,6
4.            Galβ1→3GlcNAcβ1→3Galβ1→3GalNAc-ol    Lewis R_IM8 1.95
                                ↑β1,6              Rovis et al. 1973 (5)
                                GlcNAc
                                ↑β1,4
                                Gal
                                IV
```

When the key sugars of the blood group determinants and/or the LFucα1→3 sequence are added to the appropriate position, a composite structure of the branched carbohydrate chain can be constructed and demonstrated as in Fig. 11. It is a megalosaccharide of twenty-four sugars. This polysaccharide has not been isolated. However, most of the cyst carbohydrate chains isolated are parts of this generalized structure. These structures may be a core structure with one or two branches or a complete or elongated Branch **III** with part of the core structure(5,22-26). A few irregular chains have also been reported as Compounds XI (GalNAcα1→3GalNAc-ol, Forssman disaccharide) and XII (Galβ1→3[GlcNAcα1→4Galβ1→6] GalNAc) by Maisonrouge-McAuliffe and Kabat(24) and the biological importance of Compound XII still is not known.

The blood group A, B, H, Le^a, and Le^b determinants attached on Branches **I** and **II** have been recognized for over two decades. However, such determinants attached on Branch

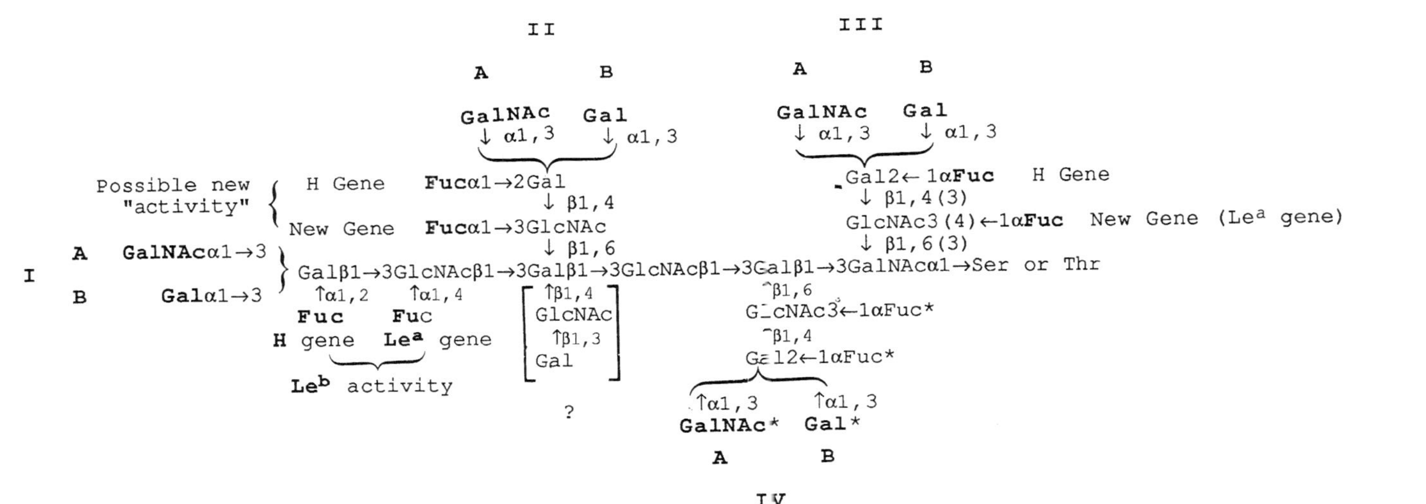

Fig.11 A generalized structure of the branched carbohydrate chains of blood group A, B, H, Le^a, Le^b and Ii active glycoproteins, prepared from human ovarian cyst fluid. The above proposed structure contains a core structure of Galβ1→3GlcNAcβ1→3Galβ1→3GalNAc with four branches - Branch **I** is type I (Galβ1→3GlcNAc) sequence, Branches **II** and **IV** are type II (Galβ1→4GlcNAc) sequence, and Branch **III** can be either type I or II. However, the type II sequence is more frequent than the type I sequence (linkage indicated in parentheses). Branch **III** can be longer than the one shown, such as Galβ1→3(Fucα1→4)GlcNAcβ1→3Galβ1→4GlcNAcβ1→6(Galβ1→3)GalNAc-ol (Lewis IM5 0.47) and Galβ1→3GlcNAcβ1→3Galβ1→4GlcNAcβ1→6GalNAc→Ser(Thr) (MSS 1st Smith degraded #3C, reference 23). Type II sequence is reactive with antisera to type XIV pneumococcal capsular polysaccharide and anti-I(Ma) cold agglutinin. The structure of the side chain in brackets has not been established but the existence of such a chain is indicated from the mechanism of alkaline borohydride degradation and the oligosaccharides isolated. "*" denotes the possible existence of these sugars, but they have not been isolated. The sugar residues responsible for A, B, H, Le^a, and Le^b activities are shown in **bold** type. They are added by glycosyltransferases to the branches. The majority of the carbohydrate chains are shorter than the one shown. In the core structure, Galβ1→3GlcNAcβ1→3Galβ1→3GlcNAc→Ser(Thr) or the incomplete core structure Galβ1→3GalNAc→Ser(Thr), GalNAcα1→3, Galα1→3, Fucα1→2 can also be attached to Gal at the nonreducing end, and/or Fucα1→3(4) to subterminal GlcNAc and show the corresponding blood group activities. Gal and GalNAc are in D form and Fuc is in L form. (Modified from Lloyd, K.O. and Kabat, E.A.; Proc. Natl. Acad Sci., U.S.A. 61:1470-1477, 1968; Rovis *et al.*, Biochemistry 12:5340-5354, 1973; Maisonrouge-McAuliffe, F., and Kabat, E.A. Arch. Biochem. Biophys. 179:90-113, 1976; and Wu *et al.*, J. Biol. Chem. 259:7178-7186, 1984; Dua *et al.*, J. Biol. Chem. 261:1599-1608, 1986).

III were not established until the following oligosaccharides were isolated.

In 1976, three Branch **III** oligosaccharides with blood group B determinant (key) sugar residue were demonstrated(24).

```
1.  III (Type II chain)                    2 (and 3)
        B                                            B
       Gal                                          Gal
        ↓α1,3                                        ↓α1,3
       Gal2←1αLFuc                                  Gal2←1αLFuc
        ↓β1,4                                        ↓β1,4
       GlcNAc                               LFuc    GlcNAc
        ↓β1,6                                ↓α1,2   ↓β1,6
  Galβ1→3GalNAcα1→Ser(Thr)       [Galα1→3]Galβ1→3GalNAcα1→Ser(Thr)

     Compound VII                 Compound VIII [and Compound IX*]
```

In 1986, five Branch **III** oligosaccharides with blood group A determinant (key) sugar residue were also isolated (25).

```
             A                                      A
1.R5       GalNAc                       2.R9      GalNAc
            ↓α1,3                                  ↓α1,3
           Gal2←1αLFuc                            Gal2←1αLFuc
            ↓β1,4                                  ↓β1,4
           GlcNAc                                 GlcNAc
            ↓β1,6                                  ↓β1,6
   Galβ1→3GalNAcα1→Ser(Thr)                       GalNAcα1→Ser(Thr)
```

Branch **III**, Type II chain with T structure

Branch **III**, Type II chain linked to Tn determinant

```
                            A                             A
3.R11                     GalNAc                 4.R14  GalNAc
                           ↓α1,3                          ↓α1,3
                          Gal2←1αLFuc  H                 Gal2←1αLFuc      H
                 H         ↓β1,4                          ↓β1,3
               LFuc       GlcNAc                 LFuc    GlcNAc
                ↓α1,2      ↓β1,6                  ↓α1,2   ↓β1,6
A   GalNAcα1→3Galβ1→3GalNAcα1→Ser(Thr)         Galβ1→3GalNAcα1→Ser(Thr)
```

Branch **III**, Type II chain with two sets of A and H determinants where the A determinant is active and H determinant is masked.

Branch **III** Type II sequence containing two sets of A and H key sugars

```
                                          A
                       5.R12            GalNAc
                                         ↓α1,3
                                        Gal2←1αLFuc    H
                                         ↓β1,3
                                        GlcNAc
                                         ↓β1,6
A   GalNAcα1→3Galβ1→3GlcNAcβ1→3Galβ1→3GalNAcα1→Ser(Thr)
```

Core structure with Branch **III** containing two sets of A and H key sugars.

These blood group A and B active oligosaccharides provide direct evidence for A, B, and H determinants on Branch **III** (Fig. 11).

The key sugars of the blood group A and B determinants that attach to Branch **IV** have not yet been reported. However, an intact Branch **IV** oligosaccharide, isolated (#2B in Scheme I ref.23) from Smith degraded blood group A active glycoprotein provides indirect evidence for the existence of some key sugar attached to Branch **IV**. Certainly, direct evidence is essential for proving this aspect.

The generalized structure of the branched megalo-saccharide of blood group A, B, H, Le^a, Le^b, I, and i active glycoproteins is shown in Fig. 11. It can be used to identify most of the carbohydrate chains isolated from human ovarian cyst fluids (16,21-27). Examples of cyst glycoproteins that have been isolated and their structural relationship to blood group antigenic determinants are shown in Fig. 12. These glycoproteins have been used as reagents for testing the binding properties of lectins for over a decade.

III B-2 Short Blood Group A, B, H, Le^a, and Le^b Active Chains

This group of carbohydrate chains is defined as the blood group key sugars that are attached to the core structure or T determinant. A list of short chain carbohydrates with blood group A activity is given in Fig. 13A. When the key sugars are attached to the T structure, some of them are identical to the blood group A active side chains isolated from porcine salivary glycoproteins (Section II-6 of this book). Examples of short chain carbohydrates with group B, H, and/or Le^a activities are illustrated in Fig. 13B and 13C. From these examples, it is concluded that the blood group active key sugars such as GalNAcα1→3, Galα1→3, and LFucα1→2 can also be attached to the terminal Gal of the core structure or T sequence of cyst chains.

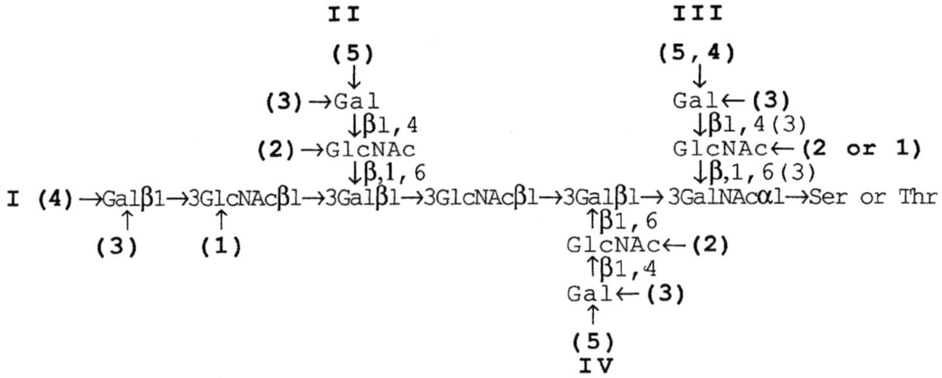

```
                    II                                   III
                   (5)                                  (5,4)
                    ↓                                     ↓
             (3)→Gal                                    Gal←(3)
                    ↓β1,4                                 ↓β1,4(3)
             (2)→GlcNAc                                 GlcNAc←(2 or 1)
                    ↓β,1,6                                ↓β,1,6(3)
I (4)→Galβ1→3GlcNAcβ1→3Galβ1→3GlcNAcβ1→3Galβ1→3GalNAcα1→Ser or Thr
        ↑        ↑                          ↑β1,6
       (3)      (1)                      GlcNAc←(2)
                                            ↑β1,4
                                          Gal←(3)
                                            ↑
                                           (5)
                                           IV
```

Proposed internal portion of carbohydrate side chains of blood group active glycoproteins.

Examples	Cyst blood group active glycoproteins	Human blood group determinant present	Sugar added	Site of addition
(I)	MSS 1st Smith degraded Beach phenol insoluble P1 Tij II phenol insoluble Tij II 20% 2X Tij II 10% 2X JS 1st Smith OG 10% from 20%.	Ii pneumococcus type XIV polysaccharide antigenic determinant	None	
(II)	N-1 phenol insoluble	Le^a unknown	Fucα1→4 Fucα1→3	(1) (2)
(III)	JS phenol insoluble Tighe phenol insoluble	H, Le^b	Fucα1→2 and as in example (II)	(1), (2), and (3)
(IV)	MSS, 10% 2X native MSM Cyst 9 Cyst 14	A	GalNAcα1→3 and as in examples (II) & (III)	(1), (2), (3) (4) and (5)
(V)	Beach phenol insoluble Tij II phenol insoluble Tij II 20% 2X Tij II 10% 2X	B	Galα1→3 as examples (II) & (III)	(1), (2), (3), (4) and (5)

Fig 12. Cyst glycoproteins and their structural relationship to blood group antigenic determinants. The structure shown is the proposed internal portion (core and four branches) of the carbohydrate moiety of blood group glycoproteins to which the residues responsible for A, B, H, Le^a, and Le^b activities are attached. This structure represents precursor blood group active glycoproteins (28) and can be prepared by Smith degradation of A, B, H active cyst glycoproteins (17). They are the sources of lectin A/Af, B, I/II, T, and Tn determinants (36). The residues responsible for A, B, H, Le^a, Le^b, and I activities are numbered in parentheses. The intact megalosaccharide of twenty-four sugars has not been isolated. However, most of the carbohydrate side chains isolated are part of this structure. (Modified from Wu *et al.*, Arch. Biochem. Biophys. 204: 622-639, Fig. 1, 1980).

Notative	Primary Structure	Structure Unit Analysis
1 R6	GalNAcα1→3Galβ1→3GalNAcα1→Ser(Thr) ↑α1,2 LFuc GlcNAc ↓β1,6	T Structure with A and H key sugars attached.
2 R7	GalNAcα1→3Galβ1→3GalNAcα1→Ser(Thr) ↑α1,2 LFuc Gal ↓β1,4 GlcNAc ↓β1,6	Same as R6 with incomplete Branch **III** of type II chain.
3 R8	GalNAcα1→3Galβ1→3GalNAcα1→Ser(Thr)	Same as R7 with complete Branch **III**, it shows blood group A & I activities.
4 R10	Gal ↓β1,4 GlcNAc ↓β1,6 GalNAcα1→3Galβ1→3GlcNAcβ1→3Galβ1→3GalNAcα1→Ser(Thr) ↑α1,2 LFuc	Complete core structure with A and H key sugars and Branch **III**, type II chain, showing both A & I Ma activities.
5 R11	GalNAc ↓α1,3 Gal2←1αFuc ↓β1,4 GlcNAc ↓β1,6 GalNAcα1→3Galβ1→3GalNAcα1→Ser(Thr) ↑α1,2 LFuc	Same as R6 with Branch **III** of type II chain where A and H key sugars attached. It has two A determinants.
6 R12	GalNAc ↓α1,3 Gal2←1αFuc ↓β1,4 GlcNAc ↓β1,6 GalNAcα1→3Galβ1→3GlcNAcβ1→3Galβ1→3GalNAcα1→Ser(Thr) ↑α1,2 LFuc	Complete core and Branch **III** of type II chain, with double A and H determinant sugars attached.

Fig. 13A. Structures of short chain blood group A active oligosaccharides, prepared from human ovarian cyst glycoprotein. In this group of carbohydrate chains, the key (determinant) sugar(s), GalNAcα1→3 for A activity and LFucα1→2, is(are) attached to the terminal Gal of the complete core structure, or T structure. (From Dua *et al.*, J. Biol. Chem., 261: 1599-1608, 1986).

Notation	Primary Structure and Structural Unit Analysis

1. Compound IV

```
                 H
               LFuc
                 ↓α1,2
     B  Galα1→3Galβ1→3GalNAcα1→Ser(Thr)
```

T structure with B and H key sugars attached.
It shows B activity.

2. Compound IX

```
                          III
                           B
                          Gal
                           ↓α1,3
                          Gal2←1αLFuc      H
                  H        ↓β1,4
                LFuc      GlcNAc
                  ↓α1,2    ↓β1,6
     B  Galα1→3Galβ1→3GalNAcα1→Ser(Thr)
```

T structure with Branch III, showing two
blood group B determinants

3. Compound X

```
           H
         LFuc
           ↓α1,2
     B  Galα1→3Galβ1→3GlcNAcβ1→3Galβ1→3GalNAcα1→Ser(Thr)
```

Core structure with one blood group B determinant

Fig. 13B Structures of short chain carbohydrates with blood group B activity, isolated from human ovarian cyst glycoprotein. In this group, the key determinant sugars Galα1→3 for B activity and LFucα1→2 for H activity are attached to the terminal Gal of the complete core structure, or T determinant. When Galα1→3 is attached, the H activity (LFucα1→2 linked) is masked (24).

Notation	Primary Structure	Structural Unit Analysis
1. Compound II, RL 0.32(24), L4(26), and JS RL 1.34(5)	LFuc ↓α1,2 Galβ1→3GalNAcα1→Ser(Thr)	T structure with H determinant
2. L6(26)	LFuc GlcNA ↓α1,2 ↓β1,6 Galβ1→3GalNAcα1→Ser(Thr)	Above structure with 1/2 Branch **III** of Type II chain
3. Compound VI, RL 0.35(24) and R8(26)	Gal ↓β1,4 LFuc GlcNAc ↓α1,2 ↓β1,6 Galβ1→3GalNAcα1→Ser(Thr)	Above structure with Branch **III** of Type II chain
4. JS RIM5 1.84(5) and L10(26)	**III** (type II) Gal2←1αLFuc ↓β1,4 LFuc GlcNAc ↓α1,2 ↓β1,6 Galβ1→3GalNAcα1→Ser(Thr)	T structure with Branch **III** of Type II chain and two H active key sugars
5. Compound VIII RIM5 0.92(24)	**B** Gal ↓α1,3 Gal2←1αLFuc **H** ↓β1,4 LFuc GlcNAc ↓α1,2 ↓β1,6 Galβ1→3GalNAcα1→Ser(Thr)	Above structure with B determinant on Branch **III**.
6. JS RIM5 0.92(5) and L12(26)	Gal2←1αLFuc **H** **H** ↓β1,4 LFuc GlcNAc3←1αLFuc **unknown** ↓α1,2 ↓β1,6 Galβ1→3GalNAcα1→Ser(Thr)	Above structure with LFucα1→3 to subterminal GlcNAc at Branch **III** (a possible new determinant)
7. LFuc ↓α1,2 Galβ1→3GlcNAcβ1→3Galβ1→3GalNAcα1→Ser(Thr)	JS RIM5 2.35(5)	Core structure with H determinant
Le^b^ (brace over H and Le^a^) **H** **Le^a^** 8. LFuc LFuc ↓α1,2 ↓α1,4 Galβ1→3GlcNAcβ1→3Galβ1→3GalNAcα1→Ser(Thr)	L11(26)	Core structure with H, Le^a^, Le^b^ determinants

Fig. 13C. Examples of short chain carbohydrates with blood group H determinant, prepared from human ovarian cyst glycoproteins. In this group, the key determinant sugar LFucα1→2 for H determinant is attached to the terminal Gal of the complete core structure, or T sequence (5,24,26).

IV. Biological Activities of Human Ovarian Cyst Glycoproteins

For decades, this group of glycoproteins has been considered one of the important sources for studying internal structures of glycoproteins as well as the structure of blood group determinants. Since the early seventies, the same methods (quantitative precipitin and precipitin inhibition assays) and the same materials (blood group active glycoproteins and blood group active oligosaccharides) have also been applied to investigate the binding properties of GalNAc/Gal specific lectins. Five out of six GalNAc/Gal specific lectin determinants (A/Af, B, I/II, T, and Tn) can be found in this group of glycoproteins (Table IV). Much of these studies has been contributed by Dr. E.A. Kabat[1] and his coworkers. A review is given in Section I-9 of this book. In this Section, biological properties of a blood group A_1 active cyst (MSS 10% 2X) glycoprotein before and after Smith degradation are compared. As expected, lectin A/A_f determinants or blood group A_1 activity are dominant in the native form and lectin I/II, T and Tn structure and blood group I Ma determinant are exposed only after the first Smith degradation.

IV-A. Blood Group I Ma Activity

When the MSS 10% 2X blood group A active glycoprotein is tested for blood group I Ma activity, as shown in Fig. 14-*1*, no precipitate was observed. However, its first Smith degraded product showed increased activity and precipitated all of the antibody N for the anti-I Ma serum used (30µl of 1/2 dilution) with 5µg glycoprotein required for 50% precipitation. The second Smith degraded product showed much lower activity, about 8X as much was required to precipitate 50% of the antibody N. These results indicate that all of the blood group I Ma determinants (Type II chain) are masked by A (GalNAcα1→3), H (LFuc α1→2), Le^a (LFuc α1→4), and/or unknown (LFuc α1→3) determinant sugars and are exposed after the first Smith degradation. Recently, three oligosaccharides with human blood group I Ma determinant were isolated from the first Smith degraded product.

Their structures are illustrated in Fig. 12 (oligosaccharides #10, #14 and #15).

IV-B. Reaction with Lectins

Human ovarian cyst glycoproteins, such as the MSS 10% 2X blood group A active substance and its first Smith degraded product, are some of the important sources for lectin A/A_f, I/II, T, and Tn determinants (Fig. 12). A list of GalNAc/Gal specific lectins and the determinants used to examine their binding properties is shown in Table IV. Four classes of eleven GalNAc/Gal specific lectins have been chosen. Based on precipitin-inhibition assays, DBL, HPL, and WFL are classified as Forssman specific lectins (GalNAcα1→3GalNAc)(36). Almost all of the glycoconjugates containing the disaccharide GalNAcα1→3GalNAc as the terminal sugar sequence at the nonreducing ends are found only in glycolipids. Therefore, quantitative precipitin data regarding **F**-specific lectins with Forssman active complex carbohydrates are not available. However, all of these lectins also react with the human blood group A structure. Consequently, the water soluble blood group A active glycoproteins are important probes for studying differential lectin binding affinities. When the above three lectins were tested, as shown in Fig. 14 and Table V, the original blood group A active substances (MSS 10% 2X) reacted strongly with DBL, precipitating over 90% of the lectin added. Its precipitating power was completely destroyed by the one stage of Smith degradation, implying that the DBL reacts strongly only with glycoconjugates possessing a terminal GalNAc residue. MSS 10% 2X glycoprotein reacted well with WFL and HPL, precipitating 85% of the lectins. Interaction of the first Smith degraded product with these two lectins still was substantial, but differed in the extent of reactivity. In the case of reaction with WFL, only 4 μg of the first Smith degraded product was required for 50% precipitation, while about 13 μg was required for the native glycoprotein, indicating that lectin I/II and Tn determinants play more important roles in the binding. Its precipitability after the second Smith degradation cycle was reduced as the lectin determinants of this product, in addition to solubility, is mainly depending on Tn only. When the Smith degraded products were tested with

Table IV

GalNAc/Gal specific lectins and their determinants tested for a blood group A active glycoprotein (MSS) before and after Smith degradation[1]

	Lectins tested	Lectin Determinants (Active carbohydrate sequence)
	F specific lectins	
a	*Dolichos biflorus* (DBL)	F> Af[2]>A
c	*Helix pomatia* (HPL)	F>A(>Af[3]) ≥ Tn
b	*Wistaria floribunda* (WFL)	A(>Af[3]), F > Tn, I(II)
	A specific lectins	
k	*Vicia villosa* (VVL) (A mixture of A4, A2B2, and B4)	A(>Af[2]) and Tn mainly
	T specific lectins	
g	Peanut (*arachis hypogaea*, PNA)	T >> I(II)
d	*Maclura pomifera* (MPL)	T> Tn
e	*Bauhinia purpurea alba* (BPL)	T > I(II) and Tn
f	*Sophora japonica* (SJL)[4]	T and I(II)
h	*Ricinus communis* toxin (Ricin, RCA2)	T>I(II) and Tn
	I(II) specific lectins	
i	*Ricinus communis* agglutinin (RCA1)	I(II)>T and B
m	*Geodia cydonium* (GLC)	I(II) and Tn

[1]Twenty Gal and/or GalNAc specific lectins have been divided into six classes according to their specificity for disaccharide sequence as all or part of the determinants and GalNAcα1→Ser(Thr) of the peptide chain(Section I-9 of this book). Abbreviation of the following six lectin determinants can also be used to classify these lectins. (1) **F** determinant (GalNAcα1→3GalNAc, Forssman specific disaccharide);(2) **A/Af** determinant (GalNAcα1→3Gal, Human blood group A specific disaccharide; Af, fucosylated A, GalNAcα1→3[LFucα1→2]Gal; (3) **Tn** determinant GalNAcα1→*O* to Ser (Thr) of the protein core, Tn antigen); (4) **T** determinant (T antigen, Galβ1→3GalNAcα1→Ser (Thr) of the protein core, the mucin type sugar sequence on the human erythrocyte membrane or Galβ1→3GalNAcβ1→ at the nonreducing end of ganglioside); (5) **I** and **II** determinants (human blood group type I and II carbohydrate sequences). Most of the lectins reactive to Galβ1→4GlcNAc(II) are also reactive to Galβ1→3GlcNAc (I). Lectin I/II determinants can be found at the nonreducing end of the carbohydrate chains derived from either *N*-glycosidic or *O*-glycosidic linkages; (6) **B** determinant (human blood group B specific disaccharide).

[2]Substitution of LFucα1→2 to subterminal Gal is an important factor for binding.

[3]Substitution of LFucα1→2 to subterminal Gal blocks binding.

[4]Some helping factor required for precipitation of the lectin.

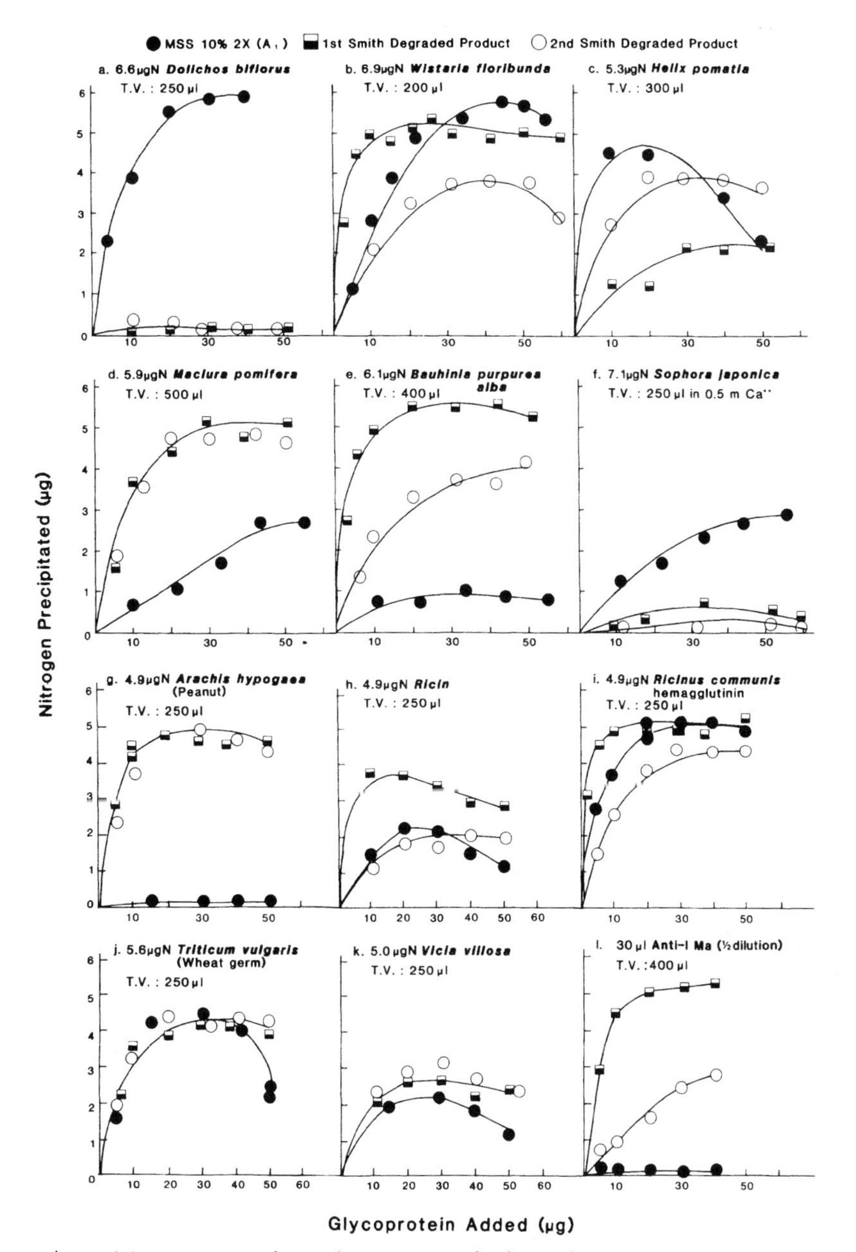

Fig. 14 Quantitative precipitation curves of blood group A active glycoprotein (MSS 10% 2X) and its Smith degraded products with anti-I Ma serum and with various lectins. Human blood group active glycoproteins and their chemical and/or enzyme treated products are important reagents to study binding properties of lectins, such as steric and/or charge effects on the lectin -glycoconjugate complex as well as the solubility of the complexes. The quantitative precipitation assay shown was performed by a microprecipitin technique using 5.0 to 7.1 μg N(Nitrogen) of various lectins for each determination. Total N in the washed precipitates was estimated by ninhydrin method. The native glycoprotein (MSS 10% 2X) shows blood group A1 activity, its active sites are equivalent to the A/Af determinant of GalNAc/Gal specific lectins. The first Smith degraded product provide T, Tn, I/II determinants, and the second Smith cycle yields a large number of Tn determinants (17,36). T.V., total volume of the reaction mixtures. (Revised from Wu *et al.*, Arch. Biophys. Biochem. 215, p. 395, 1982, Fig. 2, and Baker *et al.*, Biochemistry, 22:2741-2750, 1983, Figs 1-3).

Table V

Comparative Precipitating Activities of Blood Group A Active Glycoproteins(MSS 10% 2X) and Its Two Smith Degraded Products with Various Lectins[1].

			Percentage Maximum lectin N precipitated			Amount of glycoprotein required for 50% precipitation(µg)		
# in Fig 14	Lectin[2]	µg/N Lectin Added	Native	1stSmith degraded Product	2nd Smith degraded Product	Native	1st Smith degraded Product	2nd Smith degraded Product
a	DBL(39)	6.6	91	0	0	7.0	-	-
b	WFL(39)	6.9	85	78	56	13.0	4.0	38.0
c	HPL(39)	5.3	85	40	75	3.0	-	10.0
d	MPL(40)	5.9	44	92	90	-	7.5	7.5
e	BPL(41)	6.1	15	92	62	-	3.0	20.0
f	SJL(42)	7.1	42	7	3	-	-	-
g	PNA(17)	4.9	2	100	95	-	6.0	6.0
h	Ricin(17)	4.9	47	75	40	-	30.0	-
i	RCA_1(44)	4.9	100	100	92	5.0	1.0	13.0
j	WGA[3](17)	5.6	80	80	80	7.0	7.0	7.0
k	VVL(17)	5.0	46	54	60	-	20.0	20.0
m	GCL(45)	6.3	80	100	75	5.0	5.0	5.0
n	ConA[2](17)	5.0	0	0	0	-	-	-

[1]Reaction profiles are in Fig.14. [2]References are given in parenthesis. [3]WGA, Triticum vulgaris(wheat germ); ConA, from Jack bean(*Canavalia ensiformis*). (Reproduced with permission from Wu *et al.* Arch. Biochem. Biophys. 215:390-404, 1982, Table II).

HPL, the reactivity of these degraded compounds was substantially reduced. This is explained by the fact that interaction between HPL and the Smith degraded product is also dependent upon the Tn determinant.

The VVL used for this experiment is a mixture of isolectins A_4, B_2B_2, and B_4. It is considered to be an A specific lectin. MSS 10% 2X glycoprotein reacted intermediately with VVL and precipitated only about 45% of the lectin added. The first and second Smith degraded products increased only slightly the precipitability with this lectin.

PNA, MPL, BPL, SJL, and RCA_2 (ricin) are grouped together as T specific lectins. As shown in Fig. 14d to 14h, their reaction profiles varied; the MSS glycoprotein did not precipitate peanut lectin (Fig.14g), whereas its first and second Smith degraded products completely precipitated the lectin. MSS 10% 2X reacted weakly with MPL(40) and BPL(41) lectins but the first Smith degraded product showed a substantial increase in reactivity and precipitated over 90% of the lectins. The second Smith degraded products (Table II) precipitated 90% of the MPL, but only 62% of BPL. The first and second Smith degraded products showed identical precipitability with MPL, but four times more of the second Smith degraded product was needed for 50% precipitation. The resulting increments in reactivity of the first Smith degraded product with PNA(43), MPL, and BPL indicates that these lectin-binding determinants are covered or sterically hindered by long oligosaccharide chains and/or substituted by GalNAcα1→ linked and/or LFucα1→ linked, and were subsequently uncovered by Smith degradation.

MSS 10% 2X glycoprotein showed intermediate reactivity with ricin (Fig.14h) and *Sophora japonica* (Fig.14f); it precipitated about 40% to 45% of these lectins. The first Smith degraded product showed differences in reactivity with the above lectins; it precipitated 75% of ricin, with less than 3.0 μg of glycoprotein required for 50% precipitation, but reacted poorly with *Sophora japonica* lectin. The second Smith degraded glycoprotein reacted moderately with ricin but negligibly with *Sophora japonica* lectin.

RCA_1 and GCL are classified as blood group Type I and Type II chain specific lectins. Their reaction profiles with this glycoprotein before and after Smith degradation are also different. The MSS 10% 2X glycoprotein and its first Smith degraded product completely precipitated RCA_1 (Fig. 14i and Table IV). However, the amounts required for 50% precipitation were 1μg for the first Smith degraded product and 5μg for the original MSS, indicating that the number of lectin binding determinants had increased five times. The second Smith degraded product precipitated only 90% of the lectin with about 10μg of glycoprotein required for 50% precipitation.

The original MSS and its two Smith degradation products reacted well with *Geodia* lectin, the first Smith degraded product precipitating all of the lectin and the original MSS and second Smith degraded product, 80% and 75% respectively; 5μg of all three compounds yielded 50% precipitation (Table IV).

In addition to GalNAc/Gal specific lectins, two other sugar specific lectins (Wheat germ and Con A) were tested. Native MSS 10% 2X glycoprotein and its Smith degraded products gave the same reactivity with wheat germ lectin, precipitating about 80% of the lectin with approximately 6μg of glycoprotein required for 50% precipitation. The results of the interaction between MSS, and its sequential Smith degraded products and various GalNAc/Gal specific lectins, together with structural analysis of the carbohydrate side chains of the first Smith degraded product (Fig. 10), revealed that GlcNAc residues which alternate with Gal throughout the carbohydrate moiety, may be substituted by GalNAc, LFuc, and/or Gal and may have many GlcNAc residues close to the polypeptide backbone covered by long oligosaccharide chains. The fact that the activities of the native, first, and second Smith degraded glycoproteins with wheat germ lectin were the same indicates the ability of wheat germ lectin to react with GlcNAc residues both at the nonreducing ends and at positions within the interior of the chain. Con A, as expected, did not react with any of the three materials. From the above reaction profiles, it is believed that each lectin, even within the same class (Table III), has its own binding properties and that well defined glycoproteins, such as those given in this Section, are useful reagents for differential purposes.

SUMMARY

Regardless of the A, B, H, Le^a, Le^b, I and i activity, purified water-soluble blood group glycoproteins from human ovarian cyst fluid have a similar overall structure. They are polydisperse macromolecules (M_r 2.0 x 10^5 to several million) of similar composition (75 to 85% carbohydrate, 15 to 20% protein) and consist of multiple heterosaccharide side chains attached by an *O*-glycosidic linkage at their internal reducing ends to serine or threonine of the polypeptide backbone. About 90% of these carbohydrate side chains range in size from one to less than twenty-four sugar residues (twelve sugars in the internal structure and twelve key sugars specific as blood group determinants). Three-fourths of these side chains contain

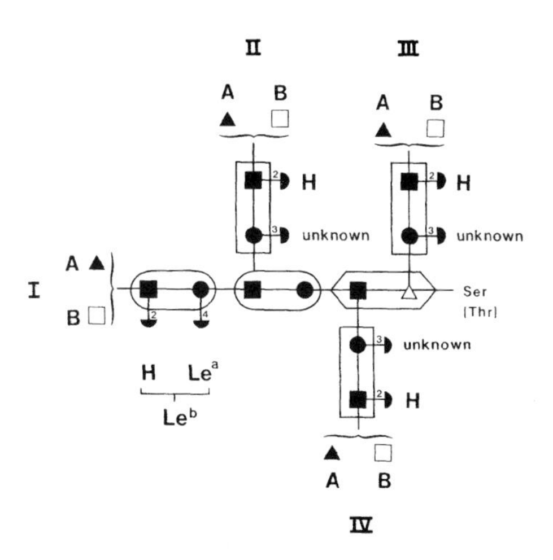

$\triangle$—Ser(Thr) =GalNAcα1→Ser(Thr), Tn determinant; ■— =Galβ1→3 in Type I chain and T structure,Galβ1→4 in Type II chain; □ , Galα1→3 as blood group B determinant; and ●— =GlcNAcβ1→3 in Type I chain, and GlcNAcβ1→6 in Type II chain; ▲ =GalNAcα1→3 as blood group A determinant; 2,3,or4,= LFuc α1→2,3,or 4 linkage. Three disaccharide units are: (■—●) ,Type I chain; [■—●] ,Type II chain; <■—△>—Ser(Thr) ,T structure.

fewer than twelve sugars. A generalized blood group active carbohydrate chain is shown above.

Three disaccharide units-Type I chain (Galβ1→3GlcNAcβ1→3), Type II chain (Galβ1→4GlcNAcβ1→6) and T determinant [Galβ1→3GalNAcα1→Ser(Thr)]-are used to elucidate the internal structure of the carbohydrate chains. The complete internal structure is considered to have a core structure with four branches, to which the blood group key sugars are attached at the appropriate locations. The core structure is a tetrasaccharide, composed of one unit of Type I chain at the nonreducing end and the T determinant at the other end, linked to Ser or Thr of the protein moiety. Branch **I** is Type I chain and Branch **II** is Type II chain. They are linked to Gal at the nonreducing end of the core structure. Branch **III** is usually a Type II chain, but may sometimes be a Type I chain, linked to the GalNAc of the reducing end. The length of Branch **III** can be increased by adding one or more monosaccharides of Type I chain sequence such as Galβ1→3GlcNAcβ1→3Galβ1→4GlcNAcβ1→6GalNAcα1→Ser(Thr), a combination of Type I and Type II chains. A new Branch **IV** is made up of Type II chain, which in turn is linked to the Gal end of the T determinant. The Type II chains react with the antibody to the type XIV pneumococcal capsular polysaccharide and with the anti-I(Ma) cold agglutinin. The sugar residues responsible for A, B, and H activities may be added by specific glycosyltransferases to any of the terminal Gal residues of the branches or the complete core chain Galβ1→3GlcNAcβ1→3Galβ1→3GalNAc→Ser(Thr), or to the incomplete core chain of T structure [Galβ1→3Gal NAcα1→Ser(Thr)]; the Le^a determinant, LFuc can be added in α1→3 linkage to a subterminal GlcNAc of Type I chains; and/or an unknown determinant (LFucα1→3) to the subterminal GlcNAc of Type II chains. The resulting structure exhibits the activities of the corresponding key sugar added. Oligosaccharides with the key sugars attached to the core chain or T structure are defined as **short blood group A, B, H, Le^a, and Le^b active chains.**

Acknowledgements

1. This work was aided by Grants from the Texas Agricultural Experiment Station (TAES H6194), USDA/SEA Formula Animal Health Funds (Project 6648) and the Robert J. and Helen C. Kleberg Foundation.

2. The author thanks Doretha A. Jones and Lee Park for their secretarial assistance and Drs. Byron Anderson and Anthony Herp for their valuable comments on this manuscript.

References

1. Watkins, W.M. (1980) Biochemistry and Genetics of the ABO, Lewis, and P Blood Group System Advan. Hum. Genet., 10:1-136.
2. Kabat, E.A. (1976) Antibody (and Lectin) Combining Sites for Elucidation of Structures of Antigenic Determinant. Structural Concepts *in* Immunology and Immunochemistry, 2nd ed., pp 167-200, Holt, Rinehart & Winston, New York.
3. Lloyd, K.O.(1976) Glycoproteins with blood group activity. International Review of Science Organic Chemistry Series Two, Vol. 7, Carbohydrates, (Aspinall, G.O., ed.) pp 251-281, London, Butterworths.

4. Watkins, W.M. (1972) Blood Group Specific Substance *In* Glycoproteins, (Gottschalk, A., ed.) 2nd ed., pp. 883-891, Elsevier, Amsterdam.
5. Rovis, L., Anderson, B., Kabat, E.A., Gruezo, F., and Liao, J. (1973) Structures of oligosaccharides produced by base-borohydride degradation of human ovarian cyst blood group H, Le^b, and Le^a active glycoproteins. Biochemistry, 12:5340-5354.
6. Schiffman, G., Kabat, E.A., and Leskowitz, S. (1960) Immunochemical studies on blood groups. XXIV. Some oligosaccharides isolated from dialysates after mild acid hydrolysis of human blood group B substances from ovarian cyst fluid. J. Amer. Chem., Soc., 82:1122-1127.
7. Schiffman, G., Kabat, E.A., and Leskowitz, S. (1962) Immunochemical studies on blood groups. XXVI. The isolation of oligosaccharides from human ovarian cyst blood group A substance including two disaccharides and a trisaccharide involved in the specificity of the blood group A antigenic determinant. J. Amer. Chem. Soc., 84:73-77.
8. Morgan, W.T.J. (1960) The Croonian lecture. A contribution to human biochemical genetics; the chemical basis of blood group specificity. Proc. Royal Soc., Ser B., 151:308-347.
9. Kabat, E.A. (1956) Blood Group Substances: Their Chemistry and Immunochemistry, pp 1-330, Academic Press, New York.
10. Watkins, W.M., Yates, A.D., and Greenwell, P. (1981) Blood group antigens and the enzymes involved in their synthesis: past and present. Biochem. Soc. Trans., 9:186-191.
11. Anderson, B., Hoffmann, P., and Meyer, K. (1964) A serine-linked peptide of chondroitin sulfate. Biochim. Biophys. Acta, 74:309-311.
12. Schiffman, G., Kabat, E.A., and Thompson, W. (1964) Immunochemical studies on blood groups. XXX. Cleavage of A, B, and H blood-group substances by alkali. Biochemistry, 3:113-120.
13. Rege, V.P., Painter, T.J., Watkins, W.M., and Morgan, W.T.J. (1964) Isolation of serologically active fucose-containing oligosaccahrides from human blood group H substance. Nature, 203:360-363.
14. Kabat, E.A., Bassett, E.W., Pryzwansky, K., Lloyd, K.O., Kaplan, M.E., and Layug, E.J. (1965) Immunochemical studies on blood groups. XXXIII. The effects of alkaline borohydride and of alkali on blood group A, B, and H substances. Biochemistry, 4:1632-1638.
15. Lloyd, K.O., Kabat, E.A., and Rosenfield, R.E. (1966) Immunochemical studies on blood groups. XXXV. The activity of fucose-containing oligosaccharides isolated from blood group A, B, and H substances by alkaline degradation. Biochemistry, 5:1502-1507.
16. Lloyd, K.O. and Kabat, E.A. (1968) Immunochemical studies on blood groups. XLI. Proposed structures for the carbohydrate portion of blood group A, B, H, $Lewis^a$ and $Lewis^b$ substances. Proc. Nat. Acad. Sci., 61:1470-1477.

17. Wu, A.M., Kabat, E.A., Pereira, M.E.A., Gruezo, F.G., and Liao, J. (1982) Immunochemical studies on blood group: the internal structure and immunological properties of water-soluble human blood group A substance studied by Smith degradation, liberation and fractionation of oligosaccharides and reaction with lectins. Arch. Biochem. Biophys., 215:390-404.
18. Vicari, G. Kabat, E.A. (1970) Structures and activities of oligosaccharides produced by alkaline degradation of a blood group substance lacking A, B, H, Le^a, and Le^b specificities. Biochemistry, 9:3414-3421.
19. Lloyd, K.O., Kabat, E.A., and Licerio, E. (1968) Immunochemical studies on blood groups. XXXVIII. Structures and activities of oligosaccharides produced by alkaline degradation of blood group $Lewis^a$ substance. Proposed structure of the carbohydrate chains of human blood group A, B, H, Le^a, and Le^b substances. Biochemistry, 7:2976-2990.
20. Kabat, E.A. (1982) Philip Levine Award Lecture. Contributions of quantitative immunochemistry to knowledge of blood group A, B, H, Le, I and i antigens. J. Amer. Clin. Pathol., 78:281-292.
21. Anderson, B., Rovis, L., and Kabat, E.A. (1972) Immunochemical Studies on Blood Groups. LIII. A Study of various conditions of alkaline borohydride degradation on human and hog blood group substances and on known oligosaccharides. Arch. Biochem. Biophys., 148:304-314.
21A Wood, E., Hounsell, E.F., and Feizi, T. (1981) Preparative, affinity chromatography of sheep gastric-mucins having blood-group I i activity, and release of antigenically active oligosaccharides by alkaline-borohydride degradation. Carbohydr. Res., 90:269-282.
22. Wu, A.M., Kabat, E.A., Nilsson, B., Zopf, D.A., Gruezo, F.G., and Liao, J. (1984) Immunochemical studies on blood groups. Purification and characterization of radioactive ^3H-reduced di- to hexasaccharides produced by alkaline β-elimination-borohydride ^3H reduction of Smith degraded blood group A active glycoproteins. J. Biol. Chem., 259:7178-7186.
23. Mutsaers, J.H.G.M., Van Halbeek, H., Vliegenthart, J.F.G., Wu, A.M., and Kabat, E.A. (1986) Typing of core and backbone domains of mucin-type oligosaccharides from human ovarian-cyst glycoproteins by 500-MHz ^1H-NMR spectroscopy. Eur. J. Biochem., 157:139-146.
24. Maisonrouge-McAuliffe, F., and Kabat, E.A. (1976) Immunochemical studies on blood groups. Structures and immunochemical properties of oligosaccharides from two fractions of blood group substance from human ovarian cyst fluid differing in B, I, and i activities and reactivity toward concanavalin A. Arch. Biochem. Biophys., 175:90-113.
25. Dua, V.K., Rao, B.N., Wu, S.S., Dube, V.E., and Bush, A.C. (1986) Characterization of the oligosaccharide alditols from ovarian cyst mucin glycoprotein of blood group A using high pressure liquid chromatography (HPLC) and high field ^1H NMR spectroscopy. J. Biol. Chem., 261:1599-1608.

26. Dua, V.K., Dube, V.E., and Bush, A.C. (1984) The combination of normal-phase and reverse-phase high-pressure liquid chromatography with NMR for the isolation and characterization of oligosaccharide alditols from ovarian cyst mucins. Biochim. Biophys. Acta, 802:29-40.
27. Dua, V.K., Dube, V.E., Li, Y.T., and Bush, C.A. (1985) Reverse phase HPLC fractionation of the oligosaccharide alditols isolated from an I-active ovarian cyst mucin glycoprotein. Glycoconjugate J., 2:17-30.
27A Bush, C.A., Yan, Z-Y., and Rao, B.N.N. (1986) Conformational energy calculations and proton nuclear Overhauser enhancements reveal a unique conformation for blood group A oligosaccharides. J. Amer. Chem. Soc., 108:6168-6173.
28. Maisonrouge-McAuliffe, F., Kabat, E.A. (1976) Immunochemical studies on blood groups LXIV. Fractionation, heterogeneity, and chemical and immunochemical properties of a blood group substance with B, I, and i activities purified from human ovarian cyst fluid. Arch. Biochem. Biophys., 175:71-80
29. Zinn, AB.,Plantner, J.J.,and Carlson, D.M. (1977) Nature of linkage between protein core and oligosaccharides *in* The Glycoconjuates. Vol.I (Horowitz, M.I. and Pigman, W., eds.) Academic Press, New York and London.
30. Feizi, T., Kabat, E.A., Vicari, G., Anderson B.,and Marsh, W.L. (1971) Immunochemical studies on blood groups XLVII. The I antigen complex-precursors in the A, B, H, Le^a, and Le^b blood group system-hemagglutination-inhibition studies. J. Exp. Med., 133:39-52.
31. Danishefsky,I., Whistler, R.L., and Bettelheim, (1970) Introduction to polysaccharide chemistry *in* The Carbohydrates, Chemistry and Biochemistry. 2nd ed. Vol. IIA, (Pigman, W. and Horton, D.,eds.) Academic Press, New York and London.
32. Marshall, R.D., and Neuberger, A. (1972) Structural Analysis of the Carbohydrate Groups of Glycoprotein *in* Glycoproteins, (Gottschalk, A., ed.) 2nd ed. Part A , pp.332-380, Elsevier, Amsterdam.
33. Wu, A.M., Wu, J.C., and Herp, A. (1978) Polypeptide linkages and resulting structural features as powerful chromogenic factors in the Lowry phenol reaction. Studies on a glycoprotein containing no Lowry phenol-reactive amino acids and on its desialylated and deglycosylated products. Biochem. J., 175:47-50.
34. Koerner, T.A.W., Yu, R.K., Scarsdale, J.N., Demou, P.C., and Prestegard, J.H. (1988) Analysis of complex carbohydrate primary and secondary structure via two-dimensional proton nuclear magnetic resonance spectroscopy, *in The Molecular Immunology of Complex Carbohydrates*, (Wu, A., ed.) IV-2, Plenum, New York and London.
35. Carr, S.A., Reinhold, V.N., Green, B.N., and Hass, J.R. (1985) Enhancement of structural information in FAB ionized carbohydrate samples by neutral gas collision. Biomed. Mass Spectrom., 12;288-295.

36. Wu, A.M., and Sugii, S. (1988) Differential binding properties of GalNAc and/or Gal specific lectins, *in The Molecular Immunology of Complex Carbohydrates*, (Wu, A. ed.) I-9, Plenum, New York and London.
37. Leskowitz, S. and Kabat, E.A. (1954) Immunochemical studies on blood groups XV. The effect of mild acid hydrolysis on the glucosamine and galactosamine in blood group substances. J. Amer. Chem. Soc., 76:5060-5065.
38. Allen, P.Z., and Kabat, E.A. (1959) Immunochemical studies on blood groups XXII. Immunochemical studies on the nondialyzable residue from partially hydrolyzed blood group A, B, and O(H) substances (Pl fractions). J. Immunol., 82:340-357.
39. Baker, D.A., Sugii, S., Kabat, E.A., Ratcliffe, R.M., Hermentin, P., and Lemieux, R.U. (1983) Immunochemical studies on the combining sites of Forssman hapten reactive hemagglutinins from *Dolichos biflorus, Helix pomatia* and *Wistaria floribunda*. Biochemistry, 22:2741-2750.
40. Sarkar, M., Wu, A.M., and Kabat, E.A. (1981) Immunochemical studies on the carbohydrate specificity of *Maclura pomifera* lectin. Arch. Biochem. Biophys., 209:204-218.
41. Wu, A.M., Kabat, E.A., Gruezo, F.G., and Allen, H.J. (1980) Immunochemical studies on the combining site of the D-galactopyranose and 2-acetamido-2-deoxy-D-galactopyranose specific lectin isolated from *Bauhinia purpurea alba* seeds. Arch. Biochem. Biophys., 204:622-639.
42. Wu, A.M., Kabat, E.A., Gruezo, F.G., and Poretz, R.D. (1981) Immunochemical studies on the reactivities and combining sites of the D-galactopyranose and 2-acetamido-2-deoxy-D-galactopyranose-specific lectin purified from *Sophora japonica* seeds. Arch. Biochem. Biophys., 209:191-203.
43. Pereira, M.E.A., Kabat, E.A., Lotan, R., and Sharon, N. (1976) Immunochemical studies on the specificity of the peanut (*Arachis hypogaea*) agglutinin. Carbohydr. Res., 51:107-118.
44. Wu, A.M., Sugii, S., Gruezo, F.G., and Kabat, E.A. (1987) Immunochemical studies on the Galβ1→4(3)GlcNAc specificity of *Ricinus communis* agglutinin. Carbohydr. Res., *(in* press).
45. Bretting, H., Phillips, S.G., Klumpart, H.J., and Kabat, E.A. (1981) A mitogenic lactose-binding lectin from the sponge *Geodia cydonium*. J. Immmunol., 127:1652-1658.

BIOCHEMISTRY AND LECTIN BINDING PROPERTIES OF MAMMALIAN SALIVARY MUCOUS GLYCOPROTEINS[1,2]

Anthony Herp, Carol Borelli, and Albert M. Wu[+]

Dept. of Biochemistry, New York Medical College, Valhalla, NY, 10595 USA
[+]*Dept. of Veterinary Pathology, Texas A&M University, College Station, TX,77843 USA*

Mucus is a complex exocrine secretion that covers the epithelial linings of higher animals. This secretion is derived from different types of epithelial glands which are composed of a variety of specialized cells. Consequently, tear, sputum, saliva, gastric juice, colonic and cervical mucus are all composed of a heterogeneous mixture of secretory products. Salivary mucus is produced by several glands (Table I). The bulk of it is derived from three main organ glands, namely the parotid which is serous in nature, the submandibular which contains both mucous and serous type acini, and the predominantly mucus secreting sublingual gland (1-5). In addition, numerous so-called minor salivary glands are dispersed throughout the oral soft tissue (6,7). These labial glands are composed mainly of mucous secreting cells, and consequently are a major source of the total mucins in saliva although they comprise only some 10% of the salivary volume produced daily. Mixed salivary secretions contain some 99.5% water; the remainder is made up of glycoconjugates, lipids, proteins, ions and small metabolites (8-12). Due to its great variety of components, saliva plays multiple

---

[1]The article is dedicated to **Dr. Ward Pigman**, one of the pioneers in the field of salivary mucin biochemistry, who died on September 30, 1977.
[2]Abbreviations are listed in Appendix I entitled "A Guide for Carbohydrate Specificities of Lectins".

TABLE I
DISTRIBUTION OF SECRETORY UNITS IN REPRESENTATIVE SPECIES OF MAMMALIAN SALIVARY GLANDS (2)

Animal	Submandibular Glands	Sublingual Glands	Parotid Glands	Other
Opposum	Mucous tubule; special serous	Mucous acini, serous demilunes	Serous acini	Mucous tubule (submaxillary)
Armadillo	Mucous tubule mucous demilunes	Gland not present	Seromucous acini	
Rabbit	Seromucous acini	Mucous acini	Serous acini	
Rat Mouse Hamster	Seromucous acini serous granular tubules	Mucous acini; Serous demilunes	Serous acini	
Guinea Pig	Serous acini	Mucous acini	Serous acini	
Dog Cat	Mucous acini; mucous demilunes	Special serous acini & mucous tubules	Seromucous acini	Mucous tubule (zygomatic & molar)
Pig	Mucous acini; seromucous demilunes	Mucous acini; seromucous demilunes	Serous acini	
Cow Sheep	Mucous acini; seromucous demilunes	Mucous acini; seromucous demilunes	Special serous (tubules)	
Horse	Mucous acini serous demilunes	Mucous acini; serous demilunes	Serous acini (3)	
Human Rhesus Monkey	Serous and mucous acini; serous demilunes	Mucous and serous acini; serous demilunes	Serous acini	

physiological roles. It moistens the mouth and coats and lubricates the food bolus from the oral cavity to the stomach. Salivary amylase, largely derived from the parotid (13), is helpful in the early digestion of starch containing nutrients. Salivary secretions possess a rich assortment of components, such as sIgA (14,15), lactoferrin (16), lysozyme (17), lactoperoxidase (18), and proline-rich proteins (19,20) which form an effective barrier against microorganisms (21), but perhaps the most efficient defense against toxins, viruses, and bacteria are the mucins. Being somewhat viscous and tacky, mucins provide not only a protective layer against mechanical injury of the oral epithelium beneath, but are also equipped to trap microorganisms, largely due to their lectin-like binding receptors (20-24).

Physicians are well aware of pathological conditions in which abnormal mucus can be a serious threat to health, hence the desire to study the nature of mucus, the cellular processes regulating its production, the functions mucus appears to perform and its interaction with surrounding structures.

Mucin constitutes about half of the total solid material of salivary secretions, and therefore affects the visco-elastic properties of saliva more than other components. The term mucin designates a family of glycoproteins which, while varying greatly in molecular weight (from 50,000 to several million), have a carbohydrate content that fluctuates between 50 and 80% by dry weight.

The importance of mucous glycoproteins was first recognized by the finding that, like the surface of red blood cells, salivary and other epithelial secretions contain blood-group active substances (25). Furthermore, it was observed (26) that bovine and ovine submandibular gland extracts inhibited heat-inactivated influenza virus induced haemagglutination, but lost this ability after treatment with the "receptor destroying enzyme" sialidase, identifying sialic acid as a peripheral sugar residue of mucins.

Fitting the various data together, Gottschalk (27) concluded that heamagglutination by heat-treated virus is mediated at the surface of red-blood cells by sialic acid, and confirmed the glycoconjugate nature of mucins, a concept that was first suggested by Hammarsten (28).

Mucin-type structural monosaccharide sequences are also found on the surface of erythrocytes. One such glycoprotein is glycophorin. Removal of sialic acid from glycophorin exposes an antigen, known as T-antigen, with the structure Galβ1→3GalNAc. This sequence (29), which is identical to the core region of several salivary mucins, binds to peanut lectin (30). The disaccharide is also the basic structure of the alkali-labile oligosaccharides that characterize the human MN blood-group active glycoproteins (31) and which possess sialic acid residues as non-reducing constituents (32,33). These studies emphasized the crucial function sialic acids play in biological processes.

The sialic acids which were first isolated from saliva, as indicated by the name, are a family of more than 20 derivatives of a 5-amino sugar acid containing nine carbon atoms, and have been identified in vertebrates, invertebrates and microorganisms as constituents of glycoproteins (34), glycolipids, and even proteoglycans (35-37). The most common form in humans is *N*-acetylneuraminic acid which is 5-acetamido-3,5-dideoxy-D-glycero-α-D-galactononulopyranosonic acid. In human saliva, additional forms that have so far been detected are the 9-*O*-acetyl-and 9-*O*-L-lactyl derivatives (38).

In the present article, we provide current information with respect to isolation and purification procedures of salivary mucins, the structural concepts of mucus glycoproteins and the significance and potential applications of their binding properties with lectins.

I. ISOLATION AND PURIFICATION OF SALIVARY MUCINS

Clearly, chemical and physical characterization of secretory mucins requires isolation procedures that are sufficiently mild to avoid or minimize irreversible alterations of the glycoproteins while insuring a reasonable yield.

A widely used method for the purification of mucins is that of Tettamanti and Pigman (39). Cetyltrimethylammonium-bromide (Cetavlon) serves to precipitate the mucin in the form of an ionic complex from the mildly acidified crude glandular extract; solubilization of the mucin-Cetavlon clot with 50% aqueous calcium chloride followed by fractional ethanol precipitation yields products that can usually be further resolved, based on their differential affinity for hydroxylapatite. The material that is recovered - between 50 and 75% ethanol - is dissolved in 0.01 M sodium phosphate buffer at pH 6.8, passed through a column of hydroxylapatite and eluted with the same solvent and then with 0.15 M sodium phosphate, pH 6.8. This treatment allows in most instances the recovery in the first stage of a "major" mucin and, at the higher ionic strength buffer, of a "minor" mucin fraction. This affinity for hydroxylapatite of the minor fraction has physiological importance and partly explains the adherence of saliva to the surface of the enamel in the form of a pellicle and its subsequent colonization by buccal bacterial flora (40). The proportion of the "minor" fraction may vary from none to as much as 50% of the native mucin. It contains less carbohydrate, but its most characteristic feature pertains to an amino acid composition which is enriched in phenylalanine and tyrosine as compared with the major mucous product (39). It is believed that the major mucin is derived from mucous acinar cells while the minor mucin originates from seromucous cells (35).

Initial precipitation of saline extracted mucosubstances by Cetavlon is especially advantageous when the starting material is available in relatively large amounts.

It should be noted that addition of proteinase inhibitors to the extraction solvent is recommended to avoid potential proteolysis; the same effect is obtained by briefly heating the solubilized mucoisolate at 100°C which also removes some of the extraneous proteins without loss of mucous material. Repeated preparative equilibrium density gradient centrifugation using caesium bromide or chloride (41), followed by gel-filtration chromatography, such as Sepharose 4B, preferentially with a buffer system containing 4 M guanidine HCl or 8 M urea are other methods in use (42). Tracheobronchial, gastrointestinal, and cervical fluids are often solubilized in the presence of reducing agents, preferably in 6M urea (43); further incubation with iodoacetate may yield an alkylated material (44) that can be resolved by gel-filtration and ion-exchange chromatographic techniques. However, in many instances, the viscous nature of the mucous extract precludes the use of a column; repeated application of small sized samples followed by elution with relatively high ionic strength buffer solutions at room temperature in order to decrease the viscosity of the material should facilitate the use of gel-filtration chromatography.

Snyder et al.(45) devised a method based on citraconylation to dissociate noncovalently bound basic proteins from mucins in human sputum. This procedure, which allows the recovery of the glycoprotein in high yield, should be applicable to other mucous secretions as well.

A feature inherent to mucins purified to homogeneity is their polydispersity with respect to composition, length, and charge of the carbohydrate side-chains; differences in the type of sialyl residues add yet another degree of variation to the complex nature of mucoisolates. These factors, together with the large molecular weight usually associated with mucins and the viscous nature of this material present problems in assessing purity.

Nevertheless, several criteria are useful as guides in verifying homogeneity - as opposed to polydispersity - of the isolated mucous glycoprotein. These include the absence of nucleic acids and, in most instances of mannose, a fairly constant ratio of carbohydrates, a characteristic amino acid composition, immunoelectrophoretic identification (46), and the lack of contaminating protein bands upon sodium dodecyl sulfate-low cross-linked poly(acrylamide) gel electrophoresis when performed in the presence of 0.2% β-mercaptoethanol in 6 M urea (47,48). This latter experimental condition also should reveal subunits, if any, of the purified mucin. Physical methods such as sedimentation velocity or equilibrium runs (49) in an analytical ultracentrifuge using Schlieren and interference optics in non-dissociating and dissociating conditions (50) are additional guides to ascertain the homogeneity of the isolated mucus product.

Also, care has to be taken when dissecting the glandular tissue. Rat and hamster sublingual glands are quite small, blend easily into the larger submandibular organ, so that in some instances, the data reported are actually for the mixed sublingual-submandibular glandular material (51). However, in the nine-banded armadillo, no anatomical distinction exists between the two glands (2) (Table I).

Many salivary secretions exhibit blood-group activity in the A, B, H, Le^a, and Le^b system, so that glands or saliva used for the purpose of structural analyses should be pooled accordingly (52,53).

II. STRUCTURAL ASPECTS OF SALIVARY MUCINS

A. *General Concepts*

Salivary mucins are characterized by a high carbohydrate to protein ratio; the saccharides account for 50 to 80% of the dry weight (Table II), the remainder being mainly proteins constituted by a single polypeptide chain, with clusters of seryl and threonyl residues dispersed throughout (Fig 1). The two hydroxyamino acids represent, for many purified salivary mucoisolates, as much as half of the total amino acids, a feature that seems to be unique to secretory mucous glycoproteins (Table III). The sugars found are L-fucose, D-galactose, 2-acetamido-2-deoxy-D-galactose, 2-acetamido-2-deoxy-D-glucose, and one or several kinds of sialic acid. In addition, mucins of some mammalian species contain significant amounts of D-mannose. Occasionally, *N*-acetylhexosamine carries a sulfate ester group. The notion that the oligosaccharide chains display, even within a single molecule, considerable heterogeneity has by now been accepted.

B. *Nature of the Carbohydrate Chains-Protein Linkage*

The linkage of the numerous oligosaccharide side chains attached to the polypeptide core is an *O*-glycosidic one, involving the 3-hydroxyl group of seryl and threonyl residues and the anomeric (C-1) carbon atom of *N*-acetylgalactosamine. This type of linkage, originally proposed for secretory mucous glycoproteins by Hashimoto and Pigman (64), is now commonly accepted as the characteristic covalent linkage by which the carbohydrate side chains are bound to the protein core in salivary and other exocrine mucins (Fig. 2). Traditionally, mild alkaline treatment (65-68), usually in the presence of sodium or potassium borohydride or boro-tritide has been used to cleave and separate such *O*-glycosidically bound sugar side chains, as more stringent conditions of

TABLE II
CHEMICAL COMPOSITION OF MAMMALIAN SALIVARY MUCINS
(in g/100 g dry weight)

Species[a]		Protein[b] LP	Protein[b] AAA	Sialic acid	Fuc	GalNAc	GlcNAc	Gal	Man	Sulfate	MWx10^5	Ref
Armadillo	SM	87.2	53.4	13.3	0.8	26.6	-	-	-	-	0.78	53
Bovine	SM	34.7	32.6	35.7	1.2	17.1	8.5	1.4	-	-	3.5	39
Canine	SM	34.5	36.5	8.6	10.4	15.2	12.5	14.6	-	3.8	ND	54
Goat	SM	48.9	48.9*	10.2	4.1	7.5	7.5	6.6	4.8	-	ND	55
Hamster	SL	-	33.0	39.2	1.6	16.2	-	5.3	-	-	ND	56
Hamster	SM	12.0	32.9	35.1	0.8	27.8	-	-	-	-	6.6	57
Human	SM	-	24.8-38.8	5.5-11.5	4.6-9.4	2.6-5.0	7.3-9.7	10.6-13.3	0.5-1.0	1.4-1.7	5-10	58
Murine	SL	-	34.1	25.9	0.8	15.4	3.4	16.2	3.9	0.4	ND	59
Murine	SM	-	18.9	30.7	0	11.9	15.8	16.5	6.1	-	1.4	60
Ovine	SM	39.6	33.4	30.7	0.7	22.7	-	0.7	-	-	4.0	39
Porcine	SM	36.2	35.8	15.6	7.3	26.2	-	10.1	-	-	6.3	61
Rat	SL	16.1	17.3	32.6	1.2	12.4	17.1	14.9	2.8	-	22.0	62
Rat	SM	-	35.0	10.6	5.8	37.8	-	11.0	-	-	1.0	63

[a]SL, sublingual; SM, submandibular; dash "-" indicates none or trace; ND = not determined.
[b]LP, determined by the Lowry et al. method, (J. Biol. Chem. *193*:265-275, 1951); AAA, by amino acid analysis

Table III
Amino Acid Composition of Mammalian Mucins (Major)
(in moles/100 moles)

Amino acid	ASM	BSM	CSM	GSM	HSL	HSM	Human SM	MSL	MSM	OSM	PSM	RSL	RSM
Lys	0.1	0.6	1.0	2.1	4.2	8.3	4.1	2.0	5.2	0.3	1.2	3.0	5.0
His	-	0.2	1.2	1.1	1.0	0.2	1.6	0.5	0.2	-	0.1	0.5	1.5
Arg	0.2	4.3	5.0	4.3	2.4	0.8	3.1	1.6	1.5	3.6	2.8	1.6	-
Asp	0.7	2.1	2.4	3.1	4.4	1.2	7.7	3.6	6.8	1.7	1.9	7.2	8.7
Thr	35.5	15.3	14.0	12.5	20.5	46.8	13.5	21.2	29.8	14.8	13.0	22.3	24.6
Ser	19.6	20.1	11.0	13.1	15.9	7.2	10.5	20.4	5.4	17.5	22.0	18.1	10.9
Glu	9.9	7.5	5.4	6.9	9.6	8.7	8.9	5.1	3.4	5.3	5.7	6.8	8.0
Pro	-	11.1	9.5	11.5	8.9	16.2	11.0	5.7	14.1	9.6	5.9	7.5	13.3
Gly	14.4	17.7	25.5	16.1	10.4	0.9	7.5	9.7	16.0	19.8	19.8	8.7	6.0
Ala	6.0	12.2	11.3	11.0	8.5	6.7	8.7	17.6	9.2	13.6	15.0	12.1	7.5
1/2 Cys	-	-	1.2	-	1.7	-	-	0.6	0.5*	-	0.1	0.8*	0.2
Val	13.2	6.7	4.9	7.3	4.6	0.8	6.4	3.7	1.7	6.0	7.3	5.9	2.7
Met	-	0.1	0.2	0.8	-	0.1	0.4	1.6	0.2	-	-	-	0.6
Ile	0.2	1.6	0.9	2.2	2.1	1.0	2.9	1.7	2.2	1.3	2.9	1.8	3.2
Leu	0.3	3.7	3.8	5.5	3.7	1.0	7.5	2.8	2.0	5.2	1.1	2.9	3.7
Tyr	-	-	0.6	1.0	1.0	0.2	2.1	0.8	0.6	-	0.4	0.2	1.0
Phe	-	0.4	2.2	1.2	1.0	0.1	3.2	1.4	1.5	1.6	0.4	0.7	1.4

SL, sublingual; SM, submaxillary, ASM, armadillo; BSM, bovine; CSM, canine; GSM, goat; HSL and HSM, hamster; MSL and MSM, murine; OSM, ovine; PSM, porcine; RSL and RSM, rat
For references, see Table II, except RSM (46); Dash "-" indicates none or trace
*as cysteic acid

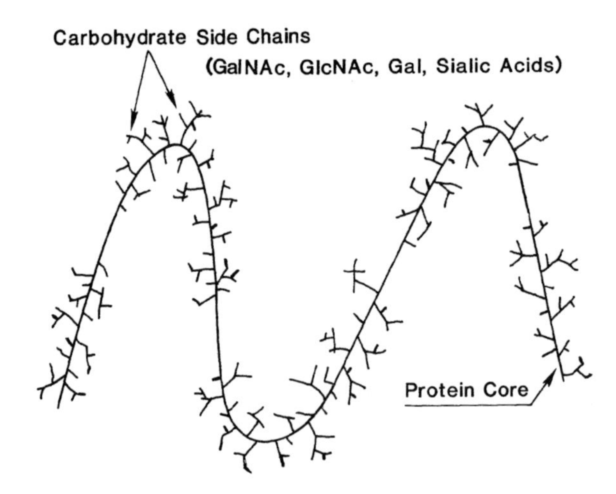

Fig. 1. Schematic Structure of Salivary Mucous Glycoproteins. The continuous heavy line represents the protein core. The short lines represent oligosaccharides or small polysaccharides. These side chains are asymmetrically distributed, and may be as small as 1 to 5 sugar residues of the type found in armadillo, ovine, bovine and porcine salivary mucus glycoproteins. Revised from Herp, A., Wu, A.M. and Moschera, J. Mol. Cell. Biochem. 23:27-44, Fig. 3, 1977).

alkali, temperature, and time of incubation are required for the removal of the *N*-glycosidic bond through *N*-acetyl-glucosamine to asparagine (69) or of the *O*-hydroxylysine linked galactose in collagen (70).

Reports have appeared (71,72) claiming that the mild alkaline reductive cleavage markedly affects also the stability of asparagine-linked glycoproteins and triggers the release of as much as 30% to 75% of these linkages, but such extensive cleavage of the *N*-glycosidic bond could not be confirmed (73,74). It seems then that the alkaline β-elimination reaction remains a useful distinguishing probe for *N*- and *O*-glycosidically linked sugar chains. When performed in

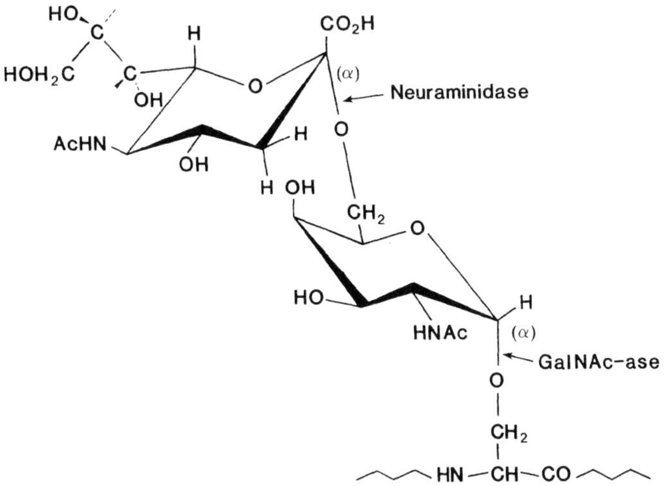

Fig.2. Structure of *N*-acetylneuraminyl ($\alpha 2 \rightarrow 6$)*N*-acetylgalactosamine in armadillo, bovine and ovine submandibular mucus glycoproteins. The disaccharide side chains are attached by an *O*-glycosidic bond to the hydroxyl group of seryl or threonyl residues which is alkaline labile and also susceptible to hydrolysis by *N*-acetylgalactosaminidase. Neuraminidase acts on the α-ketosidic bond as shown. (Revised from Herp, A., Wu. A.M., and Moschera, J. Mol. Cell. Biochem. *23*:27-44, Fig. 4, 1977).

the presence of $NaBH_4$, any *N*-acetylgalactosamine linked to the protein core is quantitatively converted into the corresponding alditol and should equal the combined disappearance of seryl and threonyl residues; as pointed out by Tanaka and Pigman (75), the reduction of these two amino acids can be made quantitative by the addition, at the end of the reaction, of $PdCl_2$ dihydrate. The conditions that are most suitable for a particular mucin-type glycoprotein as well as the mechanism of the reductive base catalysed β-elimination reaction have been reviewed (76).

Applying the alkaline reductive cleavage (0.6 M $NaBH_4$ in 0.1 M NaOH at 45°C, 5 hrs, and $PdCl_2$ as catalyst) to seven different mammalian mucins, Downs et al. (77) found that regardless of the molar amount of serine and threonine entering into the amino acid composition of a given mucin, which varies from 24 to 55 mol/100 moles amino acids for different mammalian species, the number of *O*-glycosidically linked hydroxyamino acid residues varied little among the species studied and represented approximately one fourth of the total amino acids of the protein core. In some instances, this suggests that all of the hydroxyl groups in the polypeptide backbone are substituted by *N*-acetylgalactos-amine.

Usually, treatment with alkaline sodium borohydride yields a mixture of neutral and acidic oligosaccharides which can be separated by ion-exchange chromatography. Each class is further resolved to homogeneity by thin-layer chromato-graphy.

Structural analyses have become facilitated by the availability of pure exoglycosidases which, together with permethylation and Smith degradation allow identification of individual sugars, their anomeric configuration, linkage point and precise sequence by gas-liquid chromatography-mass spectrometry. Recent advances in physical methods and instru-mentation, such as ^{1}H- and ^{13}C-nuclear magnetic resonance spectroscopies (78-80), because of their great resolving power, high sensitivity, and speed, should greatly facilitate structural and conformational characterization of carbohydrate chains in glycoproteins.

C. *Structural Characteristics of the Sugar Chains in Mucins*

Based on the compositional analysis and sequencing of sugar side chains of salivary mucous glycoproteins from a large number of mammalian species, a pattern begins to emerge.

The linkage of the oligosaccharide alditols to the core protein is 3-*O*-[2-acetamido-2-deoxy-α-D-galactopyranosyl]-L-serine (Fig. 2), or L-threonine. As first reported by Denny et al. (81) and since then confirmed by others (82), one notable exception is the mucin isolated from the submandibular glands of the mouse which contains, within the same glycoprotein molecule, both *N*- and *O*-glycosidic linkages, and may explain the unusually high proportion of mannose (Table II), a hexose generally associated with *N*-glycosylamine-type glycoproteins.

The oligosaccharide side chains may vary in length from one single *N*-acetylgalactosamine as in the case of the armadillo submandibular mucin (83) to as many as 15 sugar residues found to occur in rat sublingual glycoprotein (84), containing in addition, *N*-acetylglucosamine, fucose, galactose, and sialic acid. Occasionally, glucose and mannose are detected in small amounts, though these two hexoses are usually not involved in mucous oligosaccharide structures and, as first approximation, must therefore be considered as impurities. Hence, the occurrence of glucose as an integral monosaccharide constituent of the goat submandibular mucin (55) warrants reexamination.

Being the core sugar, *N*-acetylgalactosamine is shared by all mucin-type oligosaccharides. This amino sugar may be substituted at C-3 by galactose or *N*-acetylglucosamine, and at C-6 by a sialic acid.

Among the sugars constitutive of salivary mucins, sialic acid is perhaps the most characteristic monosaccharide residue, representing over one third by dry weight of a mucin containing no or only traces of fucose, and still about 15% in mucins for which the level of fucose exceeds 4.5% of the total dry material (Table II). Compilation of the available data reveals that the molar sum of sialic acids plus fucose approximately equals that of the *N*-acetylhexosamines.

Table IV
Sialic acid components of bovine submandibular glycoprotein*

Sialic acid	Amount (% of total)
N-Acetylneuraminic acid	~37
N-Acetyl-9-*O*-acetylneuraminic acid	~30
N-Glycolylneuraminic acid	~17
N-Acetyl-7-*O*-acetylneuraminic acid *N*-Acetyl-8-*O*-acetylneuraminic acid *N*-Acetyl-7,9-di-*O*-acetylneuraminic acid *N*-Acetyl-8,9-di-*O*-acetylneuraminic acid *N*-Acetyl-7,8,9-tri-*O*-acetylneuraminic acid *N*-Glycolyl-7-*O*-acetylneuraminic acid *N*-Glycolyl-9-*O*-acetylneuraminic acid *N*-Glycolyl-7,9-di-*O*-acetylneuraminic acid *N*-Glycolyl-8,9-di-*O*-acetylneuraminic acid *N*-Glycolyl-7,8,9-tri-*O*-acetylneuraminic acid	~16

**Data from Buscher et al. Eur. J. Biochem. 50:*71-82 (1974) *and Reuter et al. Eur. J. Biochem. 134:*139-143 (1983).

Fucose, which like sialic acid, occupies a peripheral position, is usually linked α1→2 to a galactosyl residue in salivary gland mucins. It is a critical sugar of many blood-group and antigenic determinants, whereas sialic acids usually mask recognition sites (36). The nature of the sialic acid varies with the animal species and notably with the secretory cell type. Often, a mixture of several sialic acid derivatives occurs within the same glandular mucin, with one or two kinds prevailing (85,86)(Table IV). Nevertheless, the most common form is *N*-acetylneuraminic acid, except for porcine submandibular gland mucins, where some 90% of the total sialic acid is in the form of *N*-glycolylneuraminic acid. Usually, one sialic acid residue exists per acidic oligosaccharide chain, most often linked α2→6 to the core *N*-acetylgalactosamine. This structure constitutes the principal side chain in many salivary mucins. However, the sialomucin of the rat sublingual gland contains as many as five *N*-acetylneuraminic acid residues in chains composed of 15 sugar units, attached to a terminal *N*-acetylglucosamine, to an internal galactose or to *N*-acetylgalactosamine, respectively (84). The mucin derived from hog glands has one fourth of its sialic acid in the form of *N*-glycolylneuraminyl α2→8*N*-glycolylneuraminyl residues(87), a structure also encountered in sialoglycoproteins isolated from brain tissue (88).

O-Sulfated oligosaccharide side chains, common in mucins of other exocrine secretions, such as the tracheobronchial (89,90) and gastric mucins (91) are rare but do occur in human (58) and canine (54) salivary mucous glycoproteins. It seems that the sulfate ester groups which are carried by *N*-acetylglucosaminyl residues in oligosaccharide chains devoid of sialic acid restore an anionic environment.

Among the mammalian species for which mucin structures have been established (Fig. 3), the armadillo submandibular glycoprotein A has the simplest side-chain structure, with 70% of the chains consisting of the core sugar GalNAcα1-*O*-Thr (Ser), and the remainder of NeuAcα2→6GalNAcα1-*O*-Thr(Ser)(83).

Some salivary mucins, such as the human (58), hog (66) and rat (63) submandibular gland secretions contain blood group active antigenic determinants. Based on differences in the oligosaccharide structures of the mucous glycoproteins isolated from individual porcine glands, three distinct phenotypes have been immunologically identified (98), A^+, H^+ and A^-H^-. Using the haemagglutination inhibition reaction on a total of 210 porcine glands, Payza et al. (100) found a frequency distribution for the A^+, H^+ and A^-H^- phenotypes of 3.5:3.5:3. As will be noticed (Fig. 3A) the various structures isolated from respectively A^+, H^+ and A^-H^- hog submandibular glands represent biosynthetically incomplete stages of the pentasaccharide of the mucous glycoprotein possessing blood group A activity (98).

The large variations in carbohydrate composition (Table II) observed for human submandibular mucins most likely reflect the secretor, non-secretor, and A, B, O status of the individual donors (101). A recent study indicates that saliva of individuals belonging to the Le(a^+b^-) blood group and, to a lesser degree of individuals having Le(a^-b^+) phenotype contains a mucin that binds to the monoclonal antibody 19-9. This antibody is produced by a hybridoma derived from the spleen cells of a mouse immunized with a human carcinoma cell line (102). The proposed saccharide chain of the sialylated salivary antigen is: NeuAcα2→3Galβ1→3[LFucα1→4]GlcNAcα1→3Gal. Purified rat submandibular gland mucin was shown to react with *Dolichos biflorus* agglutinin, confirming its blood

Armadillo submandibular glycoprotein A (83)

I. GalNAc (70%)

II. NeuAcα2→6GalNAc (30%)

Bovine submandibular (92,93)

I. NeuAcα2→6GalNAc (53%)

II. NeuAcα2→6[GalNAcβ1→3]GalNAc (22%)

III. GalNAc

IV. NeuAcα2→6[Gal β1→3] GalNAc

V. GlcNAcβ1→3GalNAc

VI. Galβ1→3GalNAc

VII. GlcNAcβ1→6[Gal β1→3] GalNAc

VIII. Fucα1→2Galβ1→3 [GlcNAc β1→6]GalNAc

and longer chains containing GalNAc, GlcNAc, Fuc, Gal

Canine submandibular (94)

I. Fucα1→2Gal β1→3[NeuAc α 2→6]GalNAc

II. Galβ1→3[NeuAc α2→6]GalNAc

III. NeuAcα2→6GalNAc

IV. GlcNAcβ1→6Galβ1→3GalNAc

V. Fucα1→2Galβ1→3(4 or 6)GlcNAc β1→6[Fuc α1→2]Gal β1→3GalNAc
↑3 or 4
OSO_3^-

VI. Fucα1→2Galβ1→3(4)GlcNAc β1→6Galβ1→3GalNAc
↑3 or 4
OSO_3^-

VII. GlcNAcβ1→6[Fuc α1→2]Gal β1→3GalNAc
↑3 or 4
OSO_3^-

Goat submandibular (95)

I. Man α1→6[Man α1→4]Glc α1→4GlcNAc α1→4[Fuc α1→2]Gal?1→ 3
[NeuAcα2→6]GalNAc

II. Man α1→6Glcα1→4GlcNAc β1→4[Fuc α1→2]Gal?1→3[NeuAcα2→6]GalNAc

III. Man α1→6[Manα1→4]Glcα1→4GlcNAc β1→4[Fuc α1→2]Gal?1 → 3Gal?1→3GalNAc

IV. Man α1→6Glcα1→4GlcNAcβ1→4[Fuc α1→2]Gal?1→3]GalNAc

V. Glc α1→4GlcNAc β1→4[Fuc α1→2]Gal?1 → 3GalNAc

VI. Glcα1→4GlcNAc β1→4Gal?1→3GalNAc

Human mixed salivary (96)

I. Gal β1→3GalNAc

II. Fucα1→2Galβ1→3GalNAc

III. NeuAcα2→3Galβ1→3GalNAc

IV. Fucα1→2Galβ1→4[Fuc α1→3]GlcNAcβ1→6[Gal β1→3]GalNAc or
Galβ1→4[Fuc α1→3]GlcNAcβ1→6[Fuc α1→2Galβ1→3]GalNAc

V. Galβ1→4[Fuc α1→3]GlcNAcβ1→6[NeuAc α2→3Galβ1→3]GalNAc

VI. Fuc α1→2Galβ1→4[Fuc α1→3]GlcNAcβ1→6[NeuAc α2→3Galβ1→3]GalNAc

continued on next page

Fig. 3A Carbohydrate side chains of some salivary mucous glycoproteins. (Part I)

Ovine submandibular (92,97)

I. NeuAcα2→6GalNAc (75%)

II. Galβ1→3GalNAc

III. Fucα1→2Galβ1→3GalNAc

IV. Galβ1→3[GlcNAcβ1→6]GalNAc

V. Galβ1→3[Gal β1→4GlcNAc β1→6]GalNAc

Porcine submandibular (66,98,99)

I. GalNAcα1→3[Fuc α1→2]Gal β1→3[NeuGly β2→6]GalNAc

II. Fucα1→2Gal β1→3[NeuGly α2→6]GalNAc

III. Galβ1→3[NeuGly α2→6]GalNAc

IV. NeuGly α2→6GalNAc

V. GalNAcα2→3[Fuc α1→2]Galβ1→3GalNAc

VI. Fucα1→2Galβ1→3GalNAc

VII. Galβ1→3GalNAc

VIII. GalNAc

IX. NeuAcα2→3Galβ1→3GalNAc

X. NeuGlyα2→3Galβ1→3GalNAc

XI. Galβ1→3[NeuAc α2→6]GalNAc

XII. Galβ1→3[NeuGly α2→6]GalNAc

Rat sublingual (84)

I.

GlcNAc β1→3Galβ1→4GlcNAc β1→3Galβ1→4GlcNAc β1→3Galβ1→4GlcNAc β1→3Galβ1→4GlcNAc β1→3GalNAc
↑ α 2,6 ↑ α 2,6 ↑ α 2,6 ↑ α 2,6 ↑ α 2,6
NeuAc NeuAc NeuAc NeuAc NeuAc

II. GlcNAc β1→3Galβ1→4GlcNAc β1→3Galβ1→4GlcNAc β1→3Galβ1→4GlcNAcβ1→3GalNAc
↑α2,6 ↑ α 2,6 ↑ α 2,6 ↑ α 2,6 ↑ α 2,6
NeuAc NeuAc NeuAc NeuAc NeuAc

III. GlcNAc β1→3Galβ1→4GlcNAc β1→3Galβ1→4GlcNAc β1→3Galβ1→4GlcNAc β1→3GalNAc
↑ α 2,6 ↑ α 2,6 ↑ α 2,6 ↑ α 2,6
NeuAc NeuAc NeuAc NeuAc

IV. GlcNAc β1→3Galβ1→4GlcNAc β1→3Galβ1→4GlcNAc β1→3GalNAc
↑ α 2, 6 ↑ α 2,6 ↑ α 2,6 ↑ α 2,6
NeuAc NeuAc NeuAc NeuAc

V. GlcNAcβ1→3Galβ1→4GlcNAc β1→3Gal β1→4GlcNAc β1→3GalNAc
↑ α 2,6 ↑ α 2,6 ↑ α 2,6
NeuAc NeuAc NeuAc

Fig. 3B. Carbohydrate side chains of some salivary mucous glycoproteins. (Part II)

group A reactivity, which was abolished upon partial deglycosylation by mild acid hydrolysis (63). Reactivity of the asialomucin with peanut lectin suggests the core sequence Galβ1→3GalNAc.

The mucins secreted by the salivary glands of the Chinese swiftlet (genus *Collocalia*) form the cementing substance of the bird's nest (102a). These mucous glycoproteins which are easily obtained in large quantities, have structures that are of general interest in the study of cell-cell interactions, as probes for determining the specificity of antibodies, and for characterizing the affinities of bacterial, mammalian, and plant lectins. Among the several oligosaccharides identified, some have the core structure Galβ1→3GalNAc-ol, specific for the Thomsen-Friedenreich antigen; another core structure is Galβ1→3[GlcNAcβ1→6]GalNAcα1→3GalNAc-ol. The most complex monosialyl oligosaccharide characterized has the novel sequence: NeuAcα2→3Galβ1→3[Galα1→4Galβ1→4Galβ1→4GlcNAcβ1→6] GalNAcα1→3GalNAc-ol; it contains the Galβ1→4GlcNAcβ1→6 disaccharide which reacts with the anti-I (Ma) blood-group specific antibody (102a).

For several of the mammalian mucins listed in Table II, no structural information is as yet available, but it is hoped that more data will be forthcoming, as newer methodologies become developed. It is also felt that reexamination of some of the proposed carbohydrate structures is highly desirable.

D. Core Protein and Glycoprotein Structures

For most salivary mucins, the core protein represents approximately one third by dry weight of the purified product and has a characteristic amino acid composition (Table III). Serine and threonine predominate, while other amino acids that occur in significant amounts are glycine, alanine and proline. The acidic amino acids, which account for about 10% of the total amino acids, always exceed the basic ones. Aromatic amino acids are found in higher than trace amounts only in the "minor" mucin fraction, and cysteine and methionine are usually absent (Table III). Only the pig salivary mucin seems to contain cysteine in amounts that may be of significance (61).

Based on the available data, the structural unit of a mucous salivary glycoprotein is thought to consist of a slightly hydrophobic protein backbone to which are attached *O*-glycosidically several hundred oligosaccharide side chains (103). The sugar chains provide a hydrophilic coat, impact a roughly spherically expanded and internally flexible molecule, and provide protection against proteolysis of the core protein. The presence of hydrophobic sites within the polypeptide chain has been confirmed by both C-13 nuclear magnetic resonance spectroscopy (104) and by fluorescent probe techniques (105). These hydrophobic binding sites may play a role in the formation of highly viscous complexes of saliva, resulting from the interactions between mucin, serum albumin, other serum transudates and lipids and which jointly influence the rheological properties of these secretions. Thus, alteration in the composition of secretions in patients with chronic obstructive pulmonary diseases may well modify the viscoelasticity and other characteristics of such secretions. Using gastric mucoisolates as the prototype, Allen (106) suggests that all mucins are composed of subunits; these subunits are thought to be joined together by disulfide bridges located in the nonglycosylated regions that terminate both ends of the protein core. These bare regions differ in amino acid composition from that found for the glycosylated domain of the polypeptide backbone by having a high proportion of aspartic and glutamic acids and by their susceptibility to easy proteolysis. This architecture would account for the solubilization of some mucous gels by thiol reducing reagents and hence for the loss of viscous and gel-forming properties of the native mucins. Similar models were proposed for bronchial (41,107) and cervical mucus (42). However, generalization of such a model may not be warranted as a universal scheme. Snyder et al. (45) could not confirm the existence of a polymeric structure in which the major effectors of physical properties are interchain disulfide bridges and cross-linking peptides that join subunits together. Using citraconylation or saline to solubilize bronchial mucus - which is a mixture of saliva and sputum (108) - no change was observed in the sedimentation coefficient or in the sodium dodecylsulfate polyacrylamide gel electrophoretic pattern by the addition of 2-mercaptoethanol in 6 M urea, in agreement with other studies (109-111). Houdret et al. (109) observed that thiol reducing agents activate mucolytic enzymes under non-dissociating conditions which would explain the discrepancy in the observed behavior of various mucin preparations; furthermore, Slayter et al. (89) cast doubt on the existence of bare peptide regions as an integral part of the protein core.

Studies on conformational aspects of the salivary mucins are relatively few. Gottschalk et al. (112) proposed for ovine submaxillary mucin a stiff rod-like shape imposed by the numerous peripheral sialic acid residues of the several hundred short oligosaccharide chains dispersed along the peptide core and which contribute to the viscous nature of the glycoproteins (Fig. 1). This model has since been considerably revised as more information becomes available concerning the nature of the core protein and the structural diversity of the oligosaccharide side chains. The polypeptide chain in ovine submaxillary mucin is composed of about 600 amino acids, one third of which are serine and threonine, each of which seems to be substituted by a sugar side-chain (113). Sequence analysis (114) indicates that the *O*-glycosylated hydroxyamino acids occur in clusters of three to nine residues interspersed by segments of four to seven amino acids devoid of seryl and threonyl residues, a feature that is similar to the hydrophilic amino-terminal region of glycophorin A which carries MN blood-group specificities. The presence of one or more prolyl residues in the vicinity of a glycosylated seryl or threonyl in the core protein seems to create a high probability of a β-turn structure (114a). Furthermore, ^{1}H-n.m.r. data for carbohydrate-peptide interactions of mucin type 2-acetamido-2-deoxy-D-galactopyranosyl-threonine containing tripeptides suggest the potential formation of an intramolecular hydrogen bond involving the glycosylically linked hydroxyamino acid (114b). Based on the calculated molecular weights for the protein core (0.6×10^5) and for the fully glycosylated ovine mucin (1.54×10^5), it has been suggested that ionic interactions between smaller subunits lead to a polymeric structure with a molecular weight of roughly 6×10^5. As the concentration in solution is increased, association leading to higher oligomers, and aggregation may ensue (45,49). Similar conclusions were reached concerning the polymeric structure of bovine salivary mucin. To gain information on the size of the core protein, Bhavanandan et al. (114c) examined the ability of mRNA isolated from bovine submandibular glands to direct protein synthesis *in vitro* using the rabbit reticulocyte lysate system. Specific antiserum raised against the chemically deglycosylated bovine mucin allowed the identification, among the translation products, of the nascent core protein. Its apparent molecular weight was estimated to be 60,000, which is the same size as that reported by Hill et al. (113) for the core protein of the ovine submandibular mucin. Based on a protein content of 35% (Table II), this would yield a monomer with a molecular weight of about 170,000. Again, self-association of such monomeric subunits results in the formation of higher oligomeric forms of the order of 10^6 daltons, as has been determined by physical methods (49). Electron microscopy of platinum unidirectionally shadowed preparations (114d) suggests that the highly hydrated ovine mucin has an elongated structure resembling extended filaments with average dimensions of 156 nm in length and 1.2 nm in diameter. The mucin assumes a globular conformation upon enzymatic removal of its saccharide units suggesting that the carbohydrate side chains limit the flexibility of the core protein (114f) and are essential to maintain the filamentous conformation of the fully glycosylated molecule. Obviously, such phenomena are of importance in respiratory disorders.

Sedimentation velocity runs of purified mucins usually yield a single, slow migrating peak which broadens with time reflecting their high degree of polydispersity. A similar pattern has been observed by Holden et al. (47) upon gel-electrophoresis using a highly porous gel medium under both dissociating and non-dissociating conditions. In the presence of disulfide bond cleavage reagents, porcine and canine submandibular mucins were converted into lower molecular weight species, but dithioerythritol had no such effect on bovine and ovine mucins (48). Shogren et al. (115), based on light scattering measurements, observed for hog salivary mucin a molecular weight of 7.4×10^6 in 0.1 M Nacl, 2×10^6 in 6 M guanidine HCl, interpreted as a result of dissociation between non-covalent bonds, and a further decrease to 9×10^5 in the presence of 2-mercaptoethanol, which confirms the data of Holden et al. (48) and suggests the occurrence in this salivary mucin of disulfide linked subunits. At a concentration of 2 mg/ml in 0.1 M NaCl, a sharp decrease in the translational diffusion coefficient, an increase in viscosity number and a decline in slope of the osmotic compressibility occurred. These phenomena have been interpreted as being due to molecular overlap, chain entanglements, and sugar-sugar interaction. The anomalous flow characteristics of mucus, the swelling of mucus as a function of salt concentration, and the regulation of hydration by the movement of electrolytes and water are factors that obviously influence the properties of salivary secretions.

The flexibility within sugar chains makes nuclear magnetic resonance spectroscopy a valuable tool in examining the spatial conformation of mucous glycoproteins in relation to saccharides, and to the binding of ions and other molecules, and should expand our understanding of mucin physiology.

III. MUCUS GLYCOPROTEIN LECTIN INTERACTIONS AND THEIR APPLICATION AND BIOLOGICAL SIGNIFICANCE

A. Lectin Binding Properties

Lectins are carbohydrate binding proteins that agglutinate erythrocytes and a variety of other cells (116-120). They widely occur in the seeds of plants but have also been detected in mammalian tissues, such as the integral membrane lectin of hepatocytes (118). This lectin represents a galactose-specific recognition system for circulating sialoglycoproteins having a subterminal galactosyl residue, which becomes exposed after removal of the peripheral sialic acid, leading subsequently to the internalized degradation of such glycoproteins. Several other vertebrate carbohydrate-binding lectins have been identified as being involved in the translocation of glycoproteins, with respectively, mannose 6-phosphate (121), fucose (122), *N*-acetylglucosamine, or mannose as the determining specific peripheral sugar residue (123).

Since the binding characteristics of lectins vary for specific sugar residues or sugar sequences (124,125), lectins are being widely explored as probes for the *in situ* localization of glycoconjugates, their isolation, and even for the separation of cell populations bearing distinctive carbohydrate profiles at their surface. Lectins have provided information of many biological phenomena, such as cell-cell recognition, cell-contact inhibition, and intercellular adhesive events as these events are at least in part mediated by cell surface glycoconjugates. Yet little is known about the forces, amino acids, sequence and spatial arrangements involved in these bindings except that lectins are endowed with bivalency or polyvalency (119).

Several techniques are used to test for and to identify lectin-binding components, such as fluorescein-(126), ferritin-(127) conjugated or ^{125}I-labeled lectin (128) affinity assays. Often, an appropriate lectin is coupled to Sepharose CL-4B, such as Concanavalin A-Sepharose (129) which is used to separate glycoconjugates that contain mannopyranosyl, glucopyranosyl, and sterically related carbohydrate moieties. In these instances, the binding sugar requires the C-3, C-4 and C-6 hydroxyl groups for reaction with this lectin (130,131).

It is thus clear, that a broad specificity exists between a given carbohydrate structure and its reactivity with various lectins (132). *Dolichos biflorus, Helix pomatia,* and *Wistaria floribunda* all share an affinity for α,β-D-GalNAc residues. This may be explained by the hydrophobic interactions that predominate in the binding reaction between saccharide and the haemagglutinins, although hydrogen-bonding and charge transfer forces have some bearing in determining the stability of the association. Lectin 1 of *Ulex Europaeus* binds Fucα1→2Galβ1→4GlcNAc and Fucα1→2Galβ1→GlcNAc. These two trisaccharides, although structurally different, apparently share similar topographical features that are recognized by this lectin.

This broad affinity of lectins for a variety of carbohydrate structures is thus comparable to that displayed by polyclonal antibodies and is most likely due to the occurrence of conformationally labile regions within the lectin protein. This flexibility in conformation enables the lectin to accommodate saccharide structures that are thermodynamically compatible (132).

For decades, mammalian mucous glycoproteins have been used as model compounds for studying glycoconjugate chemistry but knowledge concerning lectin binding properties is still somewhat limited. Recently, it has been found that mammalian mucous glycoproteins are among the best reagents for the sialic acid, Gal and GalNAc specific lectins. For example, *Wistaria floribunda, Dolichos biflorus, Helix pomatia, Bauhinia purpurea,* and *Vicia villosa* lectins react well with many desialized salivary glycoproteins such as ASM, BSM, RSL and OSM (for abbreviations, see Table III). Some reaction profiles, studied by a quantitative precipitin assay, are given in Fig.4. Ovine salivary mucous glycoprotein, in which 75% of the carbohydrate side chains are NeuAcα2→6GalNAcα1→*O* linked to the Ser or Thr of the protein core (Fig.3A), reacted only weakly with *Bauhinia purpurea alba* lectin, (133), however, its mild acid hydrolyzed product having GalNAcα1→*O* linked as the main carbohydrate side chain was a strongly active compound. It completely precipitated the lectin, requiring less than 2μg for 50% aggregation (Fig.4A). The desialized and defucosylated porcine submandibular mucous glycoprotein, in which the carbohydrate residues are mainly Galβ1→3GalNAcα1→*O* and GalNAcα1→3Galβ1→3GalNAcα1→*O* linked to Ser and Thr of the protein core, reacted strongly with *Dolichos biflorus* and precipitated over 75% of the lectin added (132), but the asialo ovine mucin which is left predominantly with the core sugar GalNAcα1→*O* showed little or no affinity for this lectin. Desialized PSM almost completely precipitated both *Wistaria* and *Helix* lectins. Desialized OSM also reacted well with *Wistaria floribunda* lectin, but it agglutinated only 2/3 of the *Helix pomatia* added (133). Some of the human salivary glycoproteins or their desialized products, which are probably built up in the same general pattern as the glycoproteins that occur in human ovarian cyst fluids (134-137), have a high affinity for Galβ1→3GalNAc specific lectins (138-142). As shown in Fig.4B. salivary glycoproteins with blood group B activity (PM phenol insoluble), and their mild acid hydrolyzed products, as well as blood group A_2 (WG phenol insoluble) specific glycoproteins precipitated over 75% of the *Bauhinia* lectin. Native PM glycoprotein aggregated over 85% of the *Sophora* lectin added; its mild acid hydrolyzed product completely precipitates the peanut lectin. From these and other binding properties (see below), it can be predicted that the mammalian salivary glycoproteins have the potential for being useful model compounds to study the receptor sites of lectins as well as for the purification of lectins.

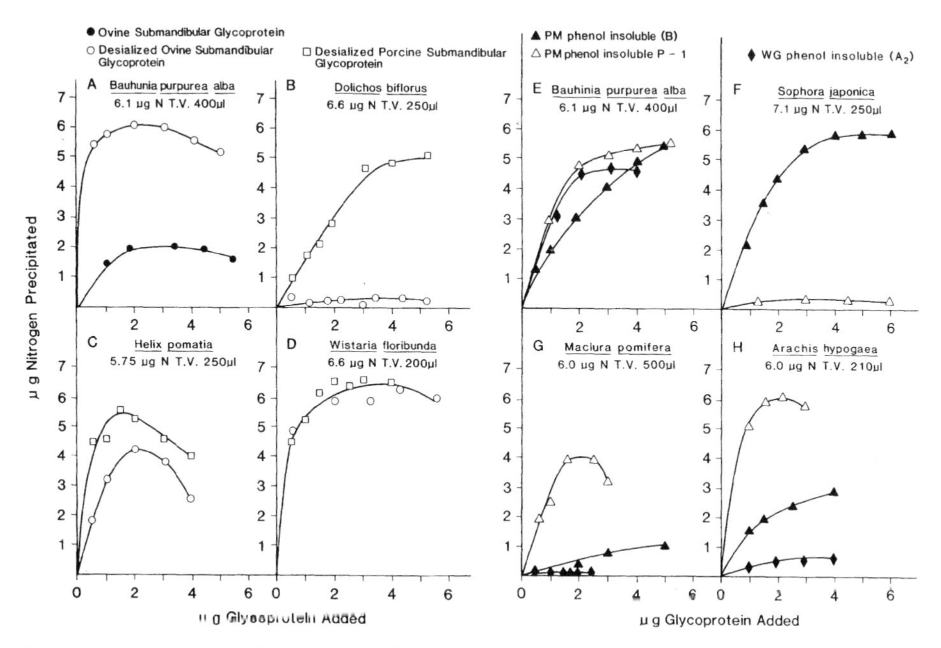

Fig. 4. Interaction of Salivary Mucous Glycoproteins with Gal and/or GalNAc Specific Lectins. The reactivities of the glycoproteins are analyzed by quantitative precipitin assay and expressed as the amount of lectin nitrogen precipitated (the amount of lectin added is shown at the right upper corner of each test; T.V., total volume of the reaction mixture). Quantitative precipitin curves of ovine submandibular glycoprotein (OSM), its desialized products, and desialized porcine submandibular glycoprotein with four GalNAc and/or Gal specific lectins are shown in Fig. 4A-4D (133,134). 20 μg of OSM precipitated completely 6.1 μg N *Bauhinia purpurea* lectin added (Fig. 4A). The precipitation profiles of human salivary blood group active glycoprotein with four Galβ1→3GalNAc specific lectins are illustrated in Fig. 4E-4H (132,140,142). PM phenol insoluble is a human blood group B active glycoprotein (insoluble in phenol). The B in parenthesis indicates the blood group activity. P-1 represents the nondialyzable portion of the blood group glycoprotein after mild acid hydrolysis at pH 1.5 to 2.0 at 100°C for 2 h. This treatment removed most of the αLFuc end group as well as some blood group B active oligosaccharides. WG phenol insoluble material is a fraction of human blood group A_2 active glycoprotein.

B. *Glycopeptides Prepared from Mucin Used as Inhibitory Reagents*

The carbohydrate binding affinities of several galactose specific lectins for pronase-digested ovine and porcine submandibular mucins and for an *N*-glycosidic peptide (B) derived from thyroglobulin were studied by the haemagglutination inhibition assay (142). Based on their respective reactivities, these phytohaemagglutinins were classified into three groups, according to the type of sugar sequence they recognize. This basic concept provides the core knowledge for the current classification of Gal and/or GalNAc specific lectins (see I-9, Differential Binding Properties of Gal and/or GalNAc specific lectins, and Appendix, A Guide for Carbohydrate Specificities of Lectins).

C. *Purification of Lectins or Mucins by Affinity Resin*

Salivary mucins covalently attached to Sepharose CL-4B are useful to isolate lectins. An application of this principle has been used for the purification of *Bauhinia purpurea* lectin from *B. purpurea alba* seeds (143). The seeds are extracted with phosphate buffered saline and the crude extract is subjected to ammonium sulfate fractionation. The precipitate (dissolved in the same buffer at pH 7) is dialyzed, applied to a column of desialized bovine submandibular mucin-Sepharose CL-4B, equilibrated with this buffer. The agglutinin is eluted with the above buffer containing 0.1 M lactose. This lectin is a Galβ1→3GalNAc specific binding protein for T antigen, but it is not specific regarding MN blood group antigens.

A bovine submandibular mucin substituted Sepharose 4B was also successfully used for the purification of *Limulus polyphemus* lectin (144). The active haemagglutinating limulin was resolved with 0.05 M Tris-HCl containing 1 M NaCl, pH 8.5. Finely tuned differences in structural features among several purified glandular mucins were readily determined. On a molar basis, NeuAcα2→6GalNAc was 5000 times as potent a haemagglutination inhibitor as free *N*-acetyl or *N*-glycolylneuraminic acids; comparable strengths were observed with bovine and porcine submandibular mucins, whereas those of ovine and equine mucins showed much weaker inhibitory activities (144).

In reverse, Tabak et al. (63) used *Dolichos biflorus*-agarose column chromatography to verify the homogeneity of purified rat submandibular gland mucin. The dissolved glycoprotein was first adsorbed to the affinity-gel equilibrated with 0.01 M *N*-acetylglucosamine and subsequently eluted with 0.01 M *N*-acetylgalactosamine.

Affinity chromatography using wheat germ agglutinin-Sepharose can simplify the isolation and purification of *O*-glycosyl linked mucin-type cell membrane sialoglycoproteins. The crude material, dissolved in a Tris-HCl buffer at pH 8 is loaded onto a small affinity column; impurities are eluted with this buffer. Changing to buffered (pH 8) 0.1 M *N*-acetylglucosamine as the eluant releases the purified glycoprotein from the column (145). Similarly, bovine and ovine submandibular mucins are retained on wheat-germ agglutinin-Sepharose columns, but the porcine mucin which contains mainly *N*-glycolylneuraminic acid as peripheral sialic residues, shows little affinity; it also fails to inhibit the agglutination of human blood-group (0)H erythrocytes by this lectin (145a). Bhavanandan et al. (145a) postulate that equatorial acetamido and hydroxyl groups at positions 5 and 4, respectively, of sialic acid are important for binding to the lectin.

D. *Lectins as Histochemical Tools, for the in situ Localizations of Mucous Glycoproteins*

Lectins have found wide-spread application as tools for histochemical techniques to study such phenomena as the subtle, gradual unfolding of premalignant changes of cells. Pretreatment of tissue sections with glycosidases (146) followed by application of lectins provides information concerning structural features of oligosaccharide chains, and insight into the organizational distribution of membrane components and of the extracellular matrix; such procedures identify *in situ* lectin receptors and define domains of interacting or connecting molecular species (147). Using a variety of horseradish-peroxidase conjugated lectins on paraffin sections of human salivary glands, Laden et al. (148)

confirmed a broad cellular heterogeneity in mucin production, based on the differences in affinity of the lectins applied. Variations also related to ABO blood groups as well as secretor status. Glands from nonsecretors revealed a staining pattern consistent with the presence of Le^a antigen. Mucous cells of nonsecretors stained with *Lotus tetragonolobus* (fucose) and *Pisum sativum* (mannose) but not with *Ulex europaeus* I (fucose). The study further indicated a higher proportion of fucose in human parotid and submandibular cells than in sublingual glands. The lectin binding characteristics of acinar cells are, as expected, affected by treatment with appropriate enzymes; thus, sialidase treated human submandibular glands showed enhanced binding by peanut and soybean agglutinins while mild digestion with trypsin induced their reactivity with *Ulex europaeus,* as compared to nontreated glands (148).

The same procedure, applied to murine (149) and rat salivary glands (150) revealed lectin binding patterns indicative of considerable heterogeneity in glycoconjugate structures even within a given cell population.

E. Effect of Salivary Glycoproteins on Oral Microbial Adherence

Bacterial lectins are of special relevance to oral physiology, as the attachment of bacteria to mucosal and tooth surfaces modulates the outcome of buccal flora and, in a broader sense, affects the pathophysiology of the respiratory, gastrointestinal, and cervical tracts.

The difference in affinity of various microbes in oral adherence and colonization seems to be attributable to both the type of microbial surface lectin and to the structural specificity of salivary glycoproteins. As already mentioned, salivary secretions contain glycoproteins that bind to a variety of bacteria acting as agglutinins. Murray et al. (151) isolated a lectin from *Streptococcus mitis* which seems to be specific for the sequence NeuAcα2→3Galβ1→3GalNAc. This sequence is identical to the core structure of the MN blood group type determinants of glycophorin on the surface of erythrocyte membranes. Purification of the bacterial lectin was achieved by affinity chromatography using CNBr activated

Sepharose 4B coupled to fetuin glycopeptides containing the above trisaccharide. The lectin was eluted with buffered 0.1 M *N*-acetylneuraminic acid. This specificity is at least similar, if not identical, to that of another lectin extracted from *Streptococcus sanguis* with lithium 3,5-diiodosalicylate (22). The binding specificity of this lectin was found to decrease in the order NeuAcα2→3Galβ1→3GalNac > Galβ1→3GalNAc, indicating that the *Streptococcus sanguis* strain KS32AR contains at least two separate carbohydrate components, with specificity towards *N*-acetylneuraminic acid and galactose, respectively. Recently, Wu found that the so-called "minor" glycoprotein fraction of bovine submandibular mucin not only contributes to the viscous nature of saliva but also acts as an excellent receptor for many sialic acid, galactose and/or *N*-acetylgalactosamine specific binding proteins (unpublished data). Gibbons et al. (24) observed that salivary mucous glycoproteins interact in a highly selective manner with microorganisms. Modification of the carbohydrate residues of these glycoproteins with periodate indicated that the sugar chains are essential for aggregation to occur (24). *Mycoplasma pneumoniae,* which is the most virulent infectious species of this organism, was shown to attach to airway epithelial cells through sialic acid receptors via *mycoplasma* binding sites thought to be lipoprotein in nature. Pretreatment of epithelial cells with neuraminidase destroyed the surface receptors and thus inhibited mycoplasma attachment to the cells (151). The importance of sialic acid as the terminal sugar residue in salivary glycoproteins to bind pathogenic bacteria was also demonstrated by McBride et al. (153). The saliva induced aggregation of *Streptococcus sanguis* was inhibited by treating saliva with neuraminidase; it was postulated that the salivary aggregating factors are of high-molecular weights. Similar conclusions were reached by Levine et al. (154) with respect to the ability of *Streptococcus sanguis* to bind to purified human salivary glycoproteins through sialic acid; however, desialylation of the glycoproteins did not affect the attachment of *Streptococcus mutans*. Saliva contains several glycoproteins (155,156) that act as agglutinins for *Streptococcus mutans* strains. It would be desirable to identify the binding sites of the components that are involved in modulating oral colonization by this microorganism since it is thought to be a major etiological agent in the development of dental caries.

A mannose containing glycoprotein, isolated from human saliva, was found to be a receptor for type I fimbriae of *E. coli* (mannose specific lectin) (157) in the oral cavity.

These few examples are evidence of the presence in saliva of a number of glycoproteins that possess bacterial agglutinating properties; some are beneficial in the clearance of microorganisms while others may favor colonization. The use of lectins is a valuable tool in clarifying the processes that are involved in the pathogenesis of oral infections by defining the *in situ* distribution of glycoconjugates in glandular tissues and by identifying the respective roles of salivary components in host-parasite interactions.

SUMMARY

The molecules responsible for the highly viscous properties of mucus are secretory glycoproteins referred to as mucins. Salivary mucins are characterized by a high sugar to protein ratio and are of a broad range of molecular weight from 7 X 10^4 to millions. With a few exceptions, they contain up to 30% of hexosamine (galactosamine and glucosamine), 8-33% of sialic acid, trace to 15% of galactose or fucose and little or no mannose. The size of carbohydrate side chains of these glycoproteins ranges from one to about fifteen units of sugar. These carbohydrate side chains are usually *O*-glycosidically linked through *N*-acetylgalactosamine to a peptidyl serine or threonine. In some instances, ester sulfate groups, mainly on *N*-acetylglucosamine, are also a structural feature. In many of these glycoproteins, the saccharide sequence is the same as that which determines the specificity of blood groups. Carbohydrate sequence analysis shows that salivary mucins exhibit considerable polydispersity, great diversity and remarkable structural flexibility not only among animal species but also within the same mucin molecule. Based on their lectin-binding ability, they can be used for purification of lectins, and lectins coupled to resin may be useful for the isolation of mucin-type glycoproteins. The epithelial mucous secretions modulate oral microbial flora; many secretory components serve as lectin-receptors for the attachment of microbes. The judicious use of lectins with widely differing binding characteristics has already been valuable in the *in situ* localization of salivary glycoproteins, in elucidating structural details, recording sugar density within a given tissue section, and defining host-parasite interactions.

It is hoped that their use, together with monoclonal antibody (158) and tissue culture techniques (159,160) will further clarify the roles of individual secretory mucous glycoproteins in health and disease.

Acknowledgement

1. This work was aided by grants from the Texas Agricultural Experiment Station (TAES H6194), USDA/SEA/ARS (Cooperative Agreement No. 58-6125-5-4), USDA/SEA Formula Animal Health Funds (Project 6648), and The Robert J. Kleberg and Helen C. Kleberg Foundation.

2. The authors thank Ms. Lee Park and Ms. Debbie Brantley for their secretarial assistance in preparing this manuscript.

REFERENCES

1. G. Quintarelli, Histochemical identification of salivary mucins. Ann. N.Y. Acad. Sci. *106*:339-363 (1963).
2. J.M. Shakleford and W.H. Wilborn, Structural and histochemical diversity in mammalian salivary glands, Alabama J. Med. Sci. *5*:180-203 (1968).
3. Banks, W.G., Applied Veterinary Histology. Williams Nilkins. Baltimore (1981).
4. Dellmanns, H.D., Veterinary Histology. Lea & Febiger, Philadelphia (1971).
5. G. Quintarelli, S. Tsuiki, Y. Hashimoto and W. Pigman, Studies of sialic acid-containing mucins in bovine submaxillary and rat sublingual glands. J. Histochem. Cytochem. *9*:176-183 (1961).
6. L.R. Eversole, The histochemistry of mucosubstances in human minor salivary glands. Arch. Oral Biol. *17*:1235-1239 (1972).
7. D.R. Green and G. Embery, Partial chemical characterization and biological activities of sulphated glycoproteins isolated from *in vivo* pilocarpine-stimulated secretions of rat minor salivary glands. Arch. Oral Biol. *29*:859-863 (1984).
8. R.C. Caldwell and W. Pigman, Disc electrophoresis of human saliva in polyacrylamide gel. Arch. Biochem. Biophys. *110*:91-96 (1965).
9. B.L. Slomiany, M. Aono, V.L.N. Murty, A. Slomiany, M.J. Levine and L.A. Tabak, Lipid composition of submandibular saliva from normal and cystic fibrosis individuals. J. Dent. Res. *61*:1163-1166 (1982).
10. A. Bennick, Salivary acidic proline-rich proteins. Mol. Cell. Biochem. *45*:83-99 (1982).

11. J.A. Young and C.A. Schneyer, Composition of saliva in mammalia. Australian J. Exp. Biol. Med. Sci.*59*:1-53 (1981).
12. M. Mogi, B.Y. Hiraoka, K. Fukasawa, M. Harada, T. Kage and T. Ching, Two-dimensional electrophoresis in the analysis of a mixture of human sublingual and submandibular salivary proteins. Arch. Oral Biol.*31*:119-125 (1986).
13. A.P. Vreugdenhil, A.V. Nieuw Amerongen, G.L. DeLange and P.A. Roukema, Localization of amylase and mucins in the major salivary glands of the mouse. Histochem. J. *14:* 767-780 (1982).
14. M.S. Finkelstein, M. Tanner and M.L. Freedman, Salivary and serum IgA levels in a geriatric outpatient population. J. Clin. Immunol. *4*:85-91 (1984).
15. M.R. Allansmith, J.L. Ebersole and C.A. Burns, IgA antibody levels in human tears, saliva and serum, Ann. N.Y. Acad. Sci. *409*:766-768 (1983).
16. R.R. Arnold, M.F. Cole and J.R. McGhee, A bactericidal effect of human lactoferrin. Science *197:*263-265 (1975).
17. J.D. Rudney, K.C. Kajander and Q.T. Smith, Correlation between human salivary levels of lysozyme, lactoferrin, salivary peroxidase and secretory immunoglobulin A with different stimulatory states and over time. Arch. Oral Biol. *30*:765-771, (1985).
18. B.L. Lamberts, K.M. Pruitt, E.D. Pederson and M.P.Golding, Comparison of salivary peroxidase system components in caries-free and caries-active naval recruits. Caries Res. *18*:488-494 (1984).
19. M.G. Humphreys-Behrer, Strain-specific differences in the proline-rich proteins and glycoproteins induced in rat salivary gland by chronic isoprenaline treatment. Biochem. J. *230*:369-378, (1985).
20. M.N. Hatton, R.E. Loomis, M.J. Levine and L.A. Taback, Masticatory lubrication. Biochem. J. *230*:817-820 (1985).
21. J.R. Clamp, The relationship between the immune system and mucus in the protection of mucous membranes. Biochem. Soc. Trans. *12*:754-756 (1984).
22. P.A. Murray, M.J. Levine, L.A. Tabak, and M.S. Reddy, Specificity of salivary-bacterial interactions: II. Evidence for a lectin on *Streptococcus sanguis* with specificity for a NeuAcα2$\rightarrow$3Galβ1$\rightarrow$3GalNac sequence. Biochem. Biophys. Res. Commun. *106:*390-396 (1982).
23. J. Parkkinen, J. Finne, M. Achtman, V. Väisänen and T.K. Korhonen, *Escherichia coli* strains binding neuraminylα2$\rightarrow$3 galactosides. Biochem. Biophys. Res. Commun. *111*:456-461 (1983).
24. R.J.Gibbons and J.V. Quershi, Selective binding of blood group-reactive salivary mucins by *Streptococcus mutans* and other oral organisms. Infect. Immun.*22*:665-671(1978).
25. K. Landsteiner and R.A. Harte, On group specific A substances. IV. The substance from hog stomach. J.Exp. Med. *71*:551-562 (1940).
26. F.M. Burnet, Mucins and mucoids in relation to influenza virus action. III. Inhibition of virus haemagglutination by glandular mucins. Australian J. Exp. Biol. Med. Sci. *26*:371-379 (1948).
27. A. Gottschalk, Carbohydrate residue of a urine mucoprotein inhibiting influenza virus haemagglutination. Nature *170*:662-663 (1952).

28. O. Hammarsten, Uber das Mucin der Submaxillardrüse. I. Darstellung, Zusammensetzung und Eigenschaften des Submaxillarismucins. Hoppe-Seyler's Z. *12*:163-195 (1888).
29. P. Vaith and G. Uhlenbruck, The Thomsen agglutination phenomenon: a discovery revisited 50 years later. Z. Immun. Forsch. *154*:1-14 (1978).
30. G.W.G. Bird, Anti-T in peanuts. Vox Sang. *9*:748-749 (1964).
31. D.B. Thomas and R.J. Winzler, Structural studies on human erythrocyte glycoproteins alkali-labile oligosaccharides. J. Biol. Chem. *244*:5943-5946 (1969).
32. E. Lisowska, Antigenic Properties of human erythrocyte glycophorins *in* Molecular Immunology of Complex Carbohydrates. Wu, A.M., Ed. Plenum Press. New York and London (1987).
33. G.F. Springer, P.R. Desai, M.S. Murthy, H.J. Yang and E.F. Scanlon, Precursors of the blood group MN antigens as human carcinoma-associated antigens. Transfusion *19*:223-247 (1979).
34. R. Schauer, Occurrence of Sialic Acids *in* Sialic Acids, Chemistry, Metabolism and Function. Springer Verlag, Wien, New York (1982) p. 5-27.
35. R. Schauer, Chemistry, metabolism and biological functions of sialic acids. Adv. Carbohydrate Chem. Biochem. *40*:131-234 (1982).
36. R. Schauer, Sialic acids and their role as biological masks. Trends Biochem. Sci. *10*:357-361 (1985).
37. D.C. Gowda, V.P. Bhavanandan and E.A. Davidson, Structures of *O*-linked oligosaccharides present in the proteoglycans secreted by human mammary epithelial cells. J. Biol. Chem. *261*:4935-4939 (1986).
38. J. Haverkamp, R. Schauer and M. Wember, Neuraminic acid derivatives newly discovered in Humans: *N*-acetyl-9-*O*-lactoyl-neuraminic acid, *N*-9-*O*-diacetylneuraminic acid and *N*-acetyl-2,3-dehydro-2-deoxyneuraminic acid. Hoppe-Seyler's Z. *357*:1699-1705 (1976).
39. G. Tettamanti and W. Pigman, Purification and characterization of bovine and ovine submaxillary mucins. Arch. Biochem. Biophys. *124*:41-50 (1968).
40. W.B. Clarke and R.J. Gibbons, Influence of salivary components and extracellular polysaccharide synthesis from sucrose on the attachment of *Streptococcus mutans* 6715 to hydroxyapatite surfaces. Infect. Immun. *18*:514-523 (1977).
41. J.M. Creeth, K.R. Bhasker, J.R. Horton, I. Das, M. Lopez-Vidriero and L. Reid, The separation and characterization of bronchial glycoproteins by density gradient methods. Biochem. J. *167*:557-569 (1977).
42. I. Carlstedt, H. Lindgren, J.K. Sheehan, U. Ulmsten and L. Wingerup, Isolation and characterization of human cervical-mucous glycoproteins. Biochem. J. *211*:13-22 (1983).
43. M. Mantle, D. Mantle and A. Allen, Polymeric structure of pig small-intestinal mucus glycoprotein. Dissociation by proteolysis or by reduction of disulfide bridges. Biochem. J. *195*:277-285 (1981).
44. G. Lamblin, M. Lhermitte, P. Degand, P. Roussel and H. Slayter, Chemical and physical properties of human bronchial mucus glycoproteins. Biochimie *61*:23-43 (1979).

45. C.E. Snyder, C.E. Nadziejko and A. Herp, Isolation of bronchial mucins from cystic fibrosis sputum by use of citraconic anhydride. Carbohydr. Res. *105*:87-93 (1982).
46. N. Fleming, M. Brent, R. Arellano and J.F. Forstner, Purification and immunofluorescent localization of rat submandibular mucin. Biochem. J. *205*:225-233 (1982).
47. K.G. Holden, N.C.F. Yim, L.J. Griggs and J.A. Weisbach, Gel electrophoresis of mucous glycoproteins. I. Effect of gel porosity. Biochemistry *10*:3105-3109 (1971).
48. K.G. Holden, N.C.F. Yim, L.J. Griggs and J.A. Weisbach, Gel electrophoresis of mucous glycoproteins. II. Effect of physical deaggregation and disulfide-bond cleavage. Biochemistry *10*:3110-3113 (1971).
49. N. Payza, M. Robert and A. Herp, The molecular weight of bovine and porcine submaxillary mucins. Int. J. Protein Res.*2*:109-115 (1970).
50. S.E. Harding, An analysis of the heterogeneity of mucins. Biochem. J. *219*:1061-1064 (1984).
51. W. Pigman, Submandibular and sublingual glycoproteins. *In* The Glycoconjugates (M. Horowitz and W. Pigman, Eds.). Vol. 1, 137-152 (1977), Academic Press, Inc. New York.
52. Nasir-Ud-Din, R.W. Jeanloz, G. Lamblin, P. Roussel, H. Van Halbeek, J.H.G. Mutsaers and J.F.G. Vliegenthart, Structure of sialyloligosaccharides isolated from bonnet monkey (*Macaca radiata*) cervical mucus glycoproteins exhibiting blood group activity. J. Biol. Chem. *261:* 1992-1997 (1986).
53. A.M. Wu and W. Pigman, Preparation and characterization of armadillo submandibular glycoproteins. Biochem. J. *161*:37-47 (1977).
54. C.G. Lombart and R.J. Winzler, Isolation and characterization of canine submaxillary mucin. Biochem. J. *128*:975-977 (1972).
55. B.B. Dutta, S. Ghosh, A. Das and C.V.N. Rao, Isolation and characterization of goat submaxillary-mucin. Carbohydr. Res.*101*:101-108 (1982).
56. F. Downs and A. Herp, Chemical studies on a hamster sublingual glycoprotein. Int. J. Peptide Protein Res. *10*:229-234 (1977).
57. F. Downs, M. Harris and A. Herp, The isolation and properties of a glycoprotein from hamster submaxillary gland. Arch. Oral Biol. *21:*307-311 (1976).
58. M.M. Baig, R.J. Winzler and O.M. Rennert, Isolation of mucin from human submaxillary secretions. J. Immunol. *111*:1826-1833 (1973).
59. P.A. Roukema, C.H. Oderkerk and M.S. Salkinoja-Salonen, The murine sublingual and submandibular mucins, their isolation and characterization. Biochim. Biophys. Acta *428:*432-440 (1976).
60. P.A. Denny and P.C. Denny, Purification and biochemical characterization of a mouse submandibular sialomucin. Carbohydr. Res. *87*:265-274 (1980).
61. N. Payza, S. Rizvi and W. Pigman, Studies of action of acids and bases on porcine submaxillary mucin. Arch. Biochem. Biophys. *129*:68-74 (1969).
62. J. Moschera and W. Pigman, The isolation and characterization of rat sublingual mucus-glycoprotein. Carbohydr. Res. *40*:53-67 (1975).
63. L.A. Tabak, L. Mirels, L.D. Monte, A.L. Ridall, M.J. Levine, R.E. Loomis, F. Lindauer, M.S. Reddy and B.J. Baum, Isolation and characterization of a mucin-

glycoprotein from rat submandibular glands. Arch. Biochem. Biophys. *242*:383-392 (1985).
64. Y. Hashimoto and W. Pigman, Action of proteolytic enzymes on purified bovine submaxillary mucins. N.Y. Acad. Sci. *106*:233-246. (1962).
65. B. Anderson, N. Seno, P. Sampson, J.G. Riley, P. Hoffman and K. Meyer, Threonine and serine linkage in mucopolysaccharides and glycoproteins. J. Biol. Chem. *239*:PC 2716-2717 (1964).
66. D.M. Carlson, Structures and immunochemical properties of oligosaccharides isolated from pig submaxillary mucins. J. Biol. Chem. *243*:616-626 (1968).
67. F. Downs, A. Herp, J. Moschera and W. Pigman, β-Elimination and reduction reactions and some applications of dimethylsulfoxide on submaxillary glycoproteins. Biochim. Biophys. Acta *328*:182-192 (1973).
68. C-C.W. Chao, J.P. Vergnes and S.I. Brown, *O*-Glycosidic linkage in glycoprotein isolates from human ocular mucus. Exp. Eye Res. *37*:533-541 (1983).
69. R.D. Marshall, Determination of the 4-*N*-2-acetamido-2-deoxy-β-D-glucopyranosyl-L-asparagine linkage in glycoproteins. Methods Carbohydr.Chem. *7*:212-220 (1976).
70. R.G. Spiro, Determination of the 5-*O*-β-D-galactopyranosylhydroxy-L-lysine linkage in glycoproteins. Methods Carbohydr. Chem.*7*:205-211 (1976).
71. S. Ogata and K.O. Lloyd, Mild alkaline borohydride treatment of glycoproteins - a method for liberating both *N*- and *O*-linked carbohydrate chains. Anal. Biochem. *119*:351-359 (1982).
72. J.R. Neeser, G.l.c. of methyloxime and alditol acetate derivatives of neutral sugars, hexosamines, and sialic acids: "one pot" quantitative determination of the carbohydrate constituents of glycoproteins and a study of the selectivity of alkaline borohydride reductions. Carbohydr. Res.*138*:189-198 (1985).
73. E.F. Hounsell, N.J. Pickering, M.S. Stoll, A.M. Lawson and T. Feizi, The effect of mild alkali and alkaline borohydride on the carbohydrate and peptide moieties of fetuin. Biochem. Soc. Trans. *12*:607-610 (1984).
74. H. Debray, G. Strecker and J. Montreuil, Effect of alkalis on *N*-glycosidic linkages of glycoproteins. Biochem. Soc. Trans. *12*:611-612 (1984).
75. K. Tanaka and W. Pigman, Improvements in hydrogenation procedure for demonstration of *O*-threonine glycosidic linkages in bovine submaxillary mucin. J. Biol. Chem. *240*:PC1487-1488 (1965).
76. A. Herp, A.M. Wu and J. Moschera, Current concepts of the structure and nature of mammalian salivary mucous glycoproteins. Mol. Cell. Biochem. *23*:27-44 (1979).
77. F. Downs, C. Peterson, V.L.N. Murty and W. Pigman, Quantitation of the β-elimination reaction as used on glycoproteins. Int. J. Peptide Protein Res. *10*:315-322 (1977).
78. J.F.G. Vliegenthart, L. Dorland and H. Van Halbeek, High resolution ^{1}H-nuclear magnetic resonance spectroscopy as a tool in the structural analysis of carbohydrates related to glycoproteins. Adv. Carbohydr. Chem. Biochem. *41*:209-374 (1983).
79. K. Dill, Natural-abundance, ^{13}C-nuclear magnetic resonance spectral studies of carbohydrates linked to amino acids and proteins. Adv. Carbohydr. Chem. Biochem. *43*:1-49 (1985).

80. J.H.G.M. Mutsaers, H. Van Halbeek, J.F.G. Vliegenthart, A.M. Wu, and E.A. Kabat, Typing of core and backbone domains of mucin-type oligosaccharides from human ovarian-cyst glycoproteins by 500-MHz ^{1}H-NMR spectroscopy. Eur. J. Biochem. *157*:139-146 (1986).
81. P.A. Denny and P.C. Denny, A mouse submandibular sialomucin containing both *N*- and *O*-glycosylic linkages. Carbohydr. Res.*110*:305-314 (1982).
82. A.V. Nieuw Amerongen, C.H. Oderkerk, P.A. Roukema, J.H. Wolf, J.J.W. Lisman and B. Overdijk, Murine submandibular mucin (MSM): a mucin carrying *N* - and *O*-glycosylically bound carbohydrate-chains. Carbohydr. Res. *115*:C1-C5 (1983).
83. A.M. Wu, A. Slomiany, A. Herp and B.L. Slomiany, Structural studies on the carbohydrate units of armadillo submandibular glycoprotein. Biochim. Biophys. Acta *578*: 297-304 (1979).
84. A. Slomiany and B.L. Slomiany, Structures of the acidic oligosaccharides isolated from rat sublingual glycoprotein. J. Biol. Chem. *253*:7301-7306 (1978).
85. H.P. Buscher, J.Casals-Stenzel and R. Schauer, Identification of *N*-glycoloyl-*O*-acetylneuraminic acids and *N*-acetyl-*O*-glycoloylneuraminic acids by improved methods for detection of *N*-acyl and *O*-acyl groups and by gas-liquid chromatography. Eur. J. Biochem. *50*:71-82 (1974).
85A. G. Reuter, R. Pfeil, S. Stoll, R. Schauer, Identification of new sialic acids derived from glycoprotein of bovine submandibular gland. Eur. J. Biochem. 134:139-143 (1983).
86. J.P. Kamerling, J.F.G. Vliegenthart, C.Versluis and R. Schauer, Identification of *O*-acetylated *N*-acylneuraminic acid by mass spectrometry. Carbohydr. Res. *41*:7-17 (1975).
87. B.L.Slomiany, A.Slomiany and A. Herp, Studies on the occurrence of disialosyl groups in glycoproteins of salivary glands. Eur. J. Biochem. *90*:255-266 (1978).
88. S. Ando and R.K. Yu, Isolation and characterization of a novel trisialoganglioside, G_{T1a} from human brain. J. Biol. Chem. *252*:6247-6250 (1977).
89. H.S. Slayter, G. Lamblin, A. LeTreut, C. Galabert, N. Houdret, P. Degand and P. Roussel, Complex structure of human bronchial mucus glycoprotein. Eur. J. Biochem. *142*:209-218 (1984).
90. M.W. Leigh, P-W. Cheng, J.L. Carson and T.F. Boat, Developmental changes in glycoconjugate secretion by ferret tracheas. Am. Rev. Respir. Dis. *134*:784-790 (1986).
91. B.L. Slomiany and K. Meyer, Isolation and structural studies of sulfated glycoproteins of hog gastric mucosa. J. Biol. Chem. *247*:5062-5070 (1972).
92. M. Bertolini and W. Pigman, The existence of oligosaccharides in bovine and ovine submaxillary mucins, Carbohydr. Res. *14*::53-63.(1970).
93. T. Tsuji and T. Osawa, Carbohydrate structures of bovine submaxillary mucin. Carbohyd. Res. *151*:391-402 (1986).
94. C.G. Lombart and R.J. Winzler, Isolation and characterization of oligosaccharides from canine submaxillary mucin. Eur. J. Biochem. *49*:77-86 (1974).
95. B. Dutta and C.V.N. Rao, Structures of carbohydrate chains of glycoprotein isolated from goat submaxillary mucin. Biochim. Biophys. Acta *701*:72-85 (1982).

96. M.S. Reddy, M.J. Levine and A. Prakobphol, Oligosaccharide structures of the low-molecular-weight salivary mucin from a normal individual and one with cystic fibrosis. J. Dent. Res. *64*:33-36 (1985).
97. D.H. Van Den Eijnden, W.E.C.M.Schiphorst and E.G. Berger, Specific detection of *N*-acetylglucosamine containing oligosaccharide chains on ovine submaxillary asialomucin. Biochim. Biophy. Acta *755*:32-39 (1983).
98. H. Van Halbeek, L. Dorland, J. Haverkamp, G.A. Veldink, J.F.G. Vliegenthart, B. Fournet, G. Ricart, J. Montreuil, W. Gathmann and D. Aminoff, Structure determination of oligosaccharides isolated from A^+, H^+ and A^-H^- hog-submaxillary gland mucin glycoproteins, by 360-MHz ^{1}H-nmr spectroscopy, permethylation analysis and mass spectrometry. Eur. J. Biochem. *118*:487-495 (1981).
99. A.V. Savage, P.L. Koppen, W.E.C.M. Schiphorst, L.A.W. Trippelitz, H. Van Halbeek, J.F.G. Vliegenthart and D.H. Van Den Eijnden, Porcine submaxillary mucin contains $\alpha2\rightarrow3$ and $\alpha2\rightarrow6$-linked *N*-acetyl- and *N*-glycolyl-neuraminic acid. Eur. J. Biochem. *160*:123-129 (1986).
100. N. Payza, L. Martinez and W. Pigman, Immunological and chemical studies on porcine submaxillary mucins. Anim. Blood Groups. Biochem. Genet. *1*:195-206 (1970).
101. R.C. Caldwell and W. Pigman, The carbohydrates of human submaxillary glycoproteins in secretors and non-secretors of blood group substances. Biochim. Biophys. Acta *101*:157-165 (1965).
102. M. Brockhaus, M. Wysocka, J.L. Magnani, Z. Steplewski, H. Koprowski and V. Ginsburg, Normal salivary mucin contains the gastrointestinal cancer-associated antigen detected by monoclonal antibody 19-9 in the serum mucin of patients. Vox Sang. *48*:34-38 (1985).
102a J.M. Wieruszeski, J.C. Michalski, J. Montreuil, G. Strecker, J.P. Katalinic, H. Egge, H. van Halbeek, J.H.G.M. Mutsaers, and J.F.G. Vliegenthart, Structure of the monosialyl oligosaccharides derived from salivary gland mucin glycoproteins of the Chinese swiftlet (genus *Collocalia*) Characterization of novel types of extended core structure, Galβ(1$\rightarrow$3)[GlcNAcβ(1$\rightarrow$6)] GalNAcα(1$\rightarrow$3) GalNAc(-ol), and of chain termination, [Galα(1$\rightarrow$4)]$_{0-1}$ [Galβ(1$\rightarrow$4)]$_2$ GlcNAcβ(1$\rightarrow$). J. Biol. Chem. *262*:6650-6657 (1987).
103. A. Gottschalk, The basic structure of glycoproteins and problems of their chemical and physicochemical analysis. N.Y. Acad. Sci.*106*:168-176 (1963).
104. K. Barrett-Bee, G. Bedford and P. Loftus, The use of high resolution carbon-13 NMR in the study of mucus. Adv. Exp. Med. Biol. *144*:109-111 (1982).
105. G.P. Sachdev, J.M. Zodrow and R. Carubelli, Hydrophobic interaction of fluorescent probes with fetuin, ovine submaxillary mucin and canine tracheal mucins. Biochim. Biophys. Acta *580*:85-90 (1979).
106. A. Allen, Mucus - a protective secretion of complexity. Trends Biochem. Sci. *8*:169-173 (1983).
107. G.P. Roberts, The role of disulfide bonds in maintaining the gel structure of bronchial mucus. Arch. Biochem. Biophys. *173*:528-537 (1976).

108. A.O. Jenssen and O. Smidsrød, Preparation of enzymatically active lysozyme from sputum and its distribution between the sol and gel phases. Eur. J. Respir. Dis. *63*:584-590 (1982).
109. N. Houdret, G. Lamblin, A. Scharfman, D.Humbert and P. Roussel, Activation of bronchial mucin proteolysis by 4-aminophenylmercuric acetate and disulfide reducing agents. Biochim. Biophys. Acta *758*:24-29 (1983).
110. I.P. Williams, R.L. Hall, R.J. Miller and P.S. Richardson, Analyses of human tracheobronchial mucus from healthy subjects. Eur. J. Respir. Dis. *63*:510-515 (1982).
111. P. Roussel, G. Lamblin, N. Houdret, M. Lhermitte and H.S. Slayter, Conformation of human mucus glycoproteins observed by electron microscopy. Biochem. Soc. Trans. *12*:617-618 (1984).
112. A. Gottschalk and H.A. McKenzie, Studies on mucoproteins. VIII. On the molecular size and shape of ovine submaxillary gland mucoprotein. Biochim. Biophys. Acta *54*:226-235 (1961).
113. H.D. Hill, Jr., J.A. Reynolds and R.L. Hill, Purification, composition, molecular weight, and subunit structure of ovine submaxillary mucin. J. Biol. Chem. *252*:3791-3793 (1977).
114. H.D. Hill, Jr., M. Schwyzer, H.M. Steiman and R.L. Hill, Ovine submaxillary mucin. Primary structure and peptide substrates of UDP-*N*-acetylgalactosamine:mucin transferase. J. Biol. Chem. *252*:3799-3804 (1977).
114a J.P. Aubert, G. Biserte, and M.H. Loucheux-Lefebvre, Carbohydrate-peptide linkage in glycoproteins. Arch. Biochem. Biophys. *175*:410-418 (1978).
114b N.J. Maeji, T. Inoue, and R. Chujo, The role of the *N*-acetyl group in determining the conformation of 2 acetamido-2-deoxy-D-galactopyranosyl-threonine-containing peptides. Carbohydr. Res. *162*:C4-C8 (1987).
114c V.P. Bhavanandan and J.D. Hegarty, Identification of the mucin core protein by cell-free translation of messenger RNA from bovine submaxillary glands. J. Biol. Chem. *262*:5913-5917 (1987).
114d M.C. Rose, W.A. Voter, H. Sage, C.F. Brown and B. Kaufman, Effects of deglycosylation of the architecture of ovine submaxillary mucin glycoprotein. J. Biol. Chem. *259*;3167-3172 (1984).
114e N. Jentoft, R.S. Shogren and T.A. Gerken, The conformation of mucins and *O*-glycosylated membrane proteins. Federation Proc. *46*:2150 (1987).
114f R.L. Shogren, N. Jentoft, T.A. Gerken, A.M. Jamieson, and J. Blackwell, Light-scattering studies of fractionated ovine submaxillary mucin. Carbohydr. Res. *160*:317-328 (1987).
115. R. Shogren, A.M. Jamieson and J. Blackwell, Solution properties of porcine submaxillary mucin. Biopolymers *22*:1657-1675 (1983).
116. W.T.J. Morgan and W.M. Watkins, The inhibition of the haemmagglutinins in plant seeds by human blood group substances and simple sugars. Brit. J. Exp. Pathol. *34*:94-103 (1953).
117. H. Lis and N. Sharon, Lectins as molecules and as tools. Ann. Rev. Biochem. *55*:35-67 (1986).
118. G. Ashwell and J. Harford, Carbohydrate-specific receptors of the liver. Ann. Rev. Biochem. *51*:531-554 (1982).

119. I.E. Liener, N. Sharon, and I.J. Goldstein, The Lectins. Properties, Functions, and Applications in Biology and Medicine. Academic Press, Orlando, FL, (1986).
120. I.J. Goldstein and I.E. Etzler, Chemical Toxonomy, Molecular Biology and Function of Plant Lectins. Alan R. Liss. New York, (1983).
121. G.G. Sahagian, The mannose 6-phosphate receptor: function, biosynthesis and translocation. Biol. Cell *51*:207-214 (1984).
122. M.A. Lehrman and R.L. Hill, The binding of fucose containing glycoproteins by hepatic lectins. Purification of a fucose-binding lectin from rat liver. J. Biol. Chem. *261*:7419-7425 (1986).
123. E.F. Neufeld and G. Ashwell, Carbohydrate recognition systems for receptor-mediated pinocytosis. *In* The Biochemistry of Glycoproteins and Proteoglycans (W.J. Lennarz, Ed.). Plenum Press, New York, pp 241-266 (1980).
124. A.M. Wu, Differential binding characteristics and applications of DGalβ1→3DGalNac specific lectins, Mol. Cell. Biochem. *61*:131-141 (1984).
125. A.M. Wu and A. Herp, A table of lectin carbohydrate specificities. *In* Lectins (T.C. Bøg-Hansen and J. Breborowicz, Eds.). W. de Gruyter & Co., New York, Vol. IV., pp 629-636, (1985).
126. P.J.A. Holt, J.H. Anglin and R.E. Nordquist, Localization of specific carbohydrate configurations in human skin using fluorescein-labeled lectins. Br. J. Dermatol. *100*:237-245 (1979).
127. G.L. Nicolson and S.J. Singer, The distribution and asymmetry of mammalian cell surface saccharides utilizing ferritin-conjugated plant agglutinins as specific saccharide stains. J. Cell Biol. *60*:236-248 (1974).
128. K. Burridge, Direct identification of specific glycoproteins and antigens in sodium dodecyl sulfate gels, Methods Enzymol. *50*:54-64 (1978).
129. E.V. Crean and E.F. Rosomando, Developmental changes in membrane-bound enzymes of *Dictyostelium discoideum* detected by concanavalin A-Sepharose affinity chromatography. Biochem. Biophys. Res. Commun. *75*:488-495 (1977).
130. I.J. Goldstein, C.E. Hollerman and E.E. Smith, Protein-carbohydrate interaction. II. Inhibition studies on the interaction of concanavalin A with polysaccharides. Biochemistry *4*:876-883 (1965).
131. Y.Ch. Sekharudu, M. Biswas and V.S.R. Rao, Complex Carbohydrates: 2. The modes of binding of complex carbohydrates to concanavalin A-a computer modelling approach. Int. J. Biol. Marcomol:*8*:9-19 (1986).
132. A.M. Wu, E.A. Kabat, F.G. Gruezo, and H.J. Allen, Immunochemical studies on the combining site of the D-galactopyranose and 2-acetamido-2-deoxy-D-galactopyranose specific lectin isolated from *Bauhinia purpurea alba* seeds. Arch. Biochem. Biophys *204*:622-639 (1980).
133. D.A. Baker, S. Sugii, E.A. Kabat, R.M. Ratcliffe, P. Hermentin and R.U. Lemieux, Immunochemical studies on the combining sites of Forssman hapten reactive hemagglutinins from *Dolichos biflorus*, *Helix pomatia*, and *Wistaria floribunda*. Biochemistry *22*:2741-2750 (1983).
134. E.A. Kabat, A. Bendich, A.E. Bezer, and S.M. Beiser, Immunochemical studies on blood groups. IV. Preparation

of blood group A substances from human sources and a comparison of their chemical and immunochemical properties with those of the blood group A substance from hog stomach. J. Exp. Med. *85*:685-699 (1947).

135. E.A. Kabat, A. Bendich, A.E. Bezer and V. Knaub, Immunochemical studies on blood groups. VI. The cross-reaction between type XIV antipneumococcal horse serum and purified blood group A, B, and O substances from hog and human sources, J. Exp. Med. *87*:295-300 (1948).
136. H.H. Baer, E.A. Kabat, and V. Knaub, Immunochemical studies on blood groups. X. The preparation of blood group A and B substances and an inactive substance from individual horse stomachs and of blood group B substance from human saliva. J. Exp. Med. *91*, 105-114 (1950).
137. W.M. Watkins, Blood group specific substances. *In* Glycoproteins, 2nd ed. (A. Gottschalk, ed.). Part B. pp 830-891.(1972) Elsevier Publ., New York.
138. A.M. Wu, E.A. Kabat, F.G. Gruezo and R.D. Poretz, Immunochemical studies on the reactivities and combining sites of the D-galactopyranose- and 2-acetamido-2-deoxy-D-galactopyranose-specific lectin purified from *Sophora japonica* seeds. Arch. Biochem. Biophys. *209*:191-203 (1981).
139. M.S. Sarkar, A.M. Wu and E.A. Kabat, Immunochemical studies on the carbohydrate specificity of *Maclura pomifera* lectin, Arch Biochem. Biophys. *209*:204-218 (1981).
140. Y. Takai, Y. Noda, S. Sumitono, S. Sagara and M. Mori, Different bindings to lectin in human submandibular gland after enzymatic digestion. Acta Histochem. *78*:111-121 (1986).
141. M.E.A. Pereira, E.A. Kabat, R. Lotan, and N. Sharon, Immunochemical studies on the specificity of the peanut (*Arachis hypogaea*) agglutinin, Carbohyd. Res. *51*:107-118 (1976).
142. T. Irimura and T. Kawaguchi,T. Terao and T. Osawa, Carbohydrate-binding specificities of the so-called galactose-specific phytohemagglutinins. Carbohyd. Res. *39*: 317-327 (1975).
143. T Osawa, T. Irimura and T. Kawaguchi, *Bauhinia purpurea* agglutinin. Methods Enzymol. *50*:367-372 (1978).
144. A.C. Roche, R. Schauer and M. Monsigny, Protein-sugar interactions. Purification by affinity chromatography of limulin, and *N*-acyl-neuraminidyl-binding protein, FEBS Let. *57*:245-249 (1975).
145. K. Furukawa, J.E. Minor, J.D. Hegarty and V.P. Bhavanandan, Interaction of sialoglycoproteins with wheat germ agglutinin-Sepharose of varying ratio of lectin to Sepharose. Use for the purification of mucin glycoproteins from membrane extracts. J. Biol. Chem. *261*:7755-7761 (1986).
145a V.P. Bhavanandan and A.W. Katlic, The interaction of wheat germ agglutinin with sialoglycoproteins. The role of sialic acid. J. Biol. Chem. *254*:4000-4008 (1979).
146. T. Menghi, A.M. Bondi, D. Accili, L. Fumagalli and G. Materazzi, Characterization *in situ* of the complex carbohydrates in rabbit oviduct using digestion with glycosidases followed by lectin binding, J. Anat. *140*:613-625 (1985).
147. T. Faraggiana, D. Villari, J. Jagirdar and J. Patil, Expression of sialic acid on the alveolar surface of adult and fetal human lungs. J. Histochem. Cytochem. *34*:811-816 (1986).

148. S.A. Laden, B.A. Schulte and S.S. Spicer, Histochemical evaluation of secretory glycoproteins in human salivary glands with lectin-horseradish peroxidase conjugates. J. Histochem. Cytochem. *32*:965-972 (1984).
149. B.A. Schulte and S.S. Spicer, Light microscopic detection of sugar residues in glycoconjugates of salivary glands and the pancreas with lectin-horseradish peroxidase conjugates. I. Mouse. Histochem. J. *15*:1217-1238 (1983).
150. B.A. Schulte and S.S. Spicer, Light microscopic detection of sugar residues in glycoconjugates of salivary glands and the pancreas with horseradish-peroxidase conjugates. II. Rat. Histochem. J. *16*:3-20 (1984).
151. P.A. Murray, M.J. Levine, L.A. Tabak and M.S. Reddy, Purification of a sialic acid binding lectin from *Streptococcus mitis*. Soc. Complex Carbohydr. Annual Meeting, *44* (1983).
152. O. Sobeslavsky, B. Prescott and R.M. Chanock, Adsorption of *Mycoplasma pneumoniae* to neuraminic acid receptors of various cells and possible role in virulence. J. Bacteriol. *96*:695-705 (1968).
153. B.C. McBride and M.T. Gisslow, Role of sialic acid in saliva-induced aggregation of *Streptococcus sanguis*. Infect. Immun. *18*:35-40 (1977).
154. M.J. Levine, M.C. Herzberg, M.S. Levine, S.A. Ellison, M.W. Stinson, H.C. Li and T. van Dyke, Specificity of salivary-bacterial interactions: role of terminal sialic acid residues in the interaction of salivary glycoproteins with *Streptococcus sanguis* and *Streptococcus mutans*. Infect. Immun. *19*:107-115 (1978).
155. T.Ericson and J. Rundegen, Characterization of a salivary agglutinin with a serotype *c* strain of *Streptococcus mutans*. Eur. J. Biochem. *133*:255-261 (1983).
156. C.W.I. Douglas and R.R.B. Russell, The adsorption of human salivary components to strains of the bacterium *Streptococcus mutans*. Arch. Oral Biol. *29*:751-757 (1984).
157. J.P. Babu, S.N. Abraham, M.K. Dabbous and E.H. Beachey, Interaction of a 60-kilodalton *D*-mannose-containing salivary glycoprotein with type i fimbriae of *Escherichia coli*. Infect. Immun. *54*:104-108 (1986).
158. T. Feizi, Demonstration by monoclonal antibodies that carbohydrate structures of glycoproteins and glycolipids are onco-developmental antigens. Nature *314*:53-57 (1985).
159. C.E. Snyder, C.E. Nadziejko and A. Herp, Human bronchial explants in long-term culture: establishing a baseline for secretion. In Vitro *20*:95-102 (1984).
160. D.A. Sens, D.S. Hintz, M.T. Rudisill, M.A. Sens and S.S. Spicer, Methods in laboratory investigation. Explant culture of human submandibular gland epithelial cells: evidence of ductal origin. Lab. Invest. *52*:557-567 (1985).

NEW TRENDS IN GANGLIOSIDE CHEMISTRY

Sandro Sonnino, Riccardo Ghidoni, Giuliano Gazzotti,
Domenico Acquotti, and Guido Tettamanti

Study Center for the Functional Biochemistry of Brain Lipids
Department of Biological Chemistry, The Medical School
University of Milan, Via Saldini 50, 20133 Milan, Italy

Gangliosides are normal components of the plasma membrane of vertebrate cells[1,2] and are particularly abundant in the nervous system.[1,3] Their concentration at the level of the cortex grey matter is about one-tenth that of total phospholipids[4]. They are asymmetrically located in the outer lipid layer of the membrane[5,6] and exhibit strong amphiphilic properties.[3,4] Gangliosides are glycosphingolipids constituted of a hydrophilic sialic acid-containing oligosaccharide and a hydrophobic ceramide portion, linked together by a glycosidic linkage.[1,3] The oligosaccharide portion protrudes from the outer membrane surface and the ceramide moiety is inserted into the lipid core of the membrane.[1,3] Gangliosides occur in nature in a large number of different species. The number and chemical features of sialic acid, and the number, type, and sequence of the individual sugars of the neutral oligosaccharide core are the basis for a wide variety of chemical structures.[1,3,7] Moreover, a ganglioside with a homogenous oligosaccharide portion can be a mixture of several molecular species differing in their ceramide composition (fatty acid of different hydrocarbon chain length, with or without double bonds or hydroxyl groups; long chain bases of different chain length, with or without double bonds).[1,3]

Glycosphingolipids are assumed to serve as recognition markers at the cell surface.[8] They can be involved in the various expressions of cell social behaviour. Increasing evidence implies that gangliosides serve as membrane receptors and in membrane-mediated transfer of information.[9] Their highly differentiated oligosaccharide chains provide a variety of sites for specific interactions with extracellular ligands. These interactions are followed by intramembrane events resulting in the activation of adenylate-cyclase and other systems controlling protein phosphorylation and dephosphorylation. A peculiar surface behaviour of gangliosides, promoted by their binding to external ligands, is the self association to form permanent "clusters" or "patches".[4] The formation of ganglioside-rich phases on the membrane may lead to dramatic changes in the local organization of membranes (determination of fusogenic conditions; formation of micellar structures and lipid channels).[10] These structural changes are able to induce conformational changes of membrane-bound proteins, and probably constitute the molecular basis for the cascade of reactions that transduce signals through the membrane.[10] Both the oligosaccharide and the lipid portions of the ganglioside molecule are involved in the above mentioned events. In fact, the formation of ganglioside clusters or patches rely mainly on the mutual interactions (hydrogen bonds; dipole-dipole interactions) within the polar head groups of gangliosides facilitated by the presence of cross-linking agents.[10] The ceramide portion of gangliosides, which cannot be excluded to influence cluster formation, will play a crucial role in the further changes of the membrane architecture that directly affect the inner hydrophobic core of the membrane and the interactions with membrane bound proteins.[10]

Gangliosides, as well as other glycosphingolipids, are considered "cryptic" components of the membrane, since under given experimental conditions they appear to be partially available to interact with specific ligands, such as enzymes, lectins, and antibodies.[8,11] This property is of great physiological importance. It may reflect the possiblity that the availability of gangliosides to interact with external ligands is regulated by "masking" or "unmasking" factors or conditions, which are part of the cellular machinery that controls membrane function. The cryptic properties of gangliosides are also dependent on the chemical reactivity of both the oligosaccharide and the ceramide portions of the molecule.

The maintenance of a constant composition in the sugar and lipid portions of glycosphingolipids appears to be essential for membrane function. This regularity is guaranteed by genetic control of the biosynthesis of the enzymes involved in glycosphingolipid metabolism, and by the fine regulation of their activities.[1,3] A dramatic demonstration of the importance of this phenomenon is given by malignant cells, where gene irregularities lead to different chemical expressions of cell surface glycolipids, intimately related to the abnormal social behaviour of cells.[14,15]

Ganglioside research has entered the stage of the functional implications of these natural compounds. What is important is to establish the specific role played by individual gangliosides in the complex architecture and functional performance of a membrane domain. This makes mandatory the use of absolutely pure and well characterized gangliosides. Hence, the necessity of sophisticated methodologies for the preparation of gangliosides homogenous in both their oligosaccharide and ceramide portions, and of isotopically labelled compounds in different sites of the molecule. The aim of this report is to present the most recent and advanced procedures for the preparation of (a) ganglioside species that are homogenous in their oligosaccharide, long chain base, and fatty acid moieties, and (b) gangliosides that are isotopically labelled with ^{3}H or ^{14}C in the oligosaccharide or ceramide portion.

ISOLATION OF INDIVIDUAL GANGLIOSIDES HOMOGENOUS IN THE OLIGOSACCHARIDE PORTION: A NEW APPROACH

Sialic acid is a characteristic component of gangliosides. Up to 7 residues of sialic acid can be present in a ganglioside molecule, with an average of 2-3 residues in brain gangliosides of most vertebrates.[3] Different forms of sialic acid can occur (see Table I). The most abundant form is N-acetyl-neuraminic acid. The sialic acid residue(s) is (are) glycosidically attached to a neutral oligosaccharide core, which may contain glucose, galactose, N-acetylgalactosamine or N-acetylglucosamine, and fucose. Conventionally, gangliosides are classified in several series on the basis of the neutral saccharide portions present in their molecules.[1] These are: galactose (Gal; Galβ1→); lactose (Lac; Galβ1→4Glcβ1→); gangliotriose ($GgOse_3$; GalNAcβ1→4Galβ1→4Glcβ1→); gangliotetraose ($GgOse_4$; Galβ1→3GalNAcβ1→4Galβ1→4Glcβ1→); lactotetraose ($LcOse_4$; Galβ1→3GlcNAcβ1→3Galβ1→4Glcβ1→); neolactotetraose ($LcOse_4$; Galβ1→4GlcNAcβ1→3Galβ1→4Glcβ1→). The schematic structures of the gangliosides that most commonly occur in mammalian nervous systems are shown in Table II. Gangliosides GM1, GD1a, GD1b, GT1b, and GQ1b are very often predominant and cover 70-90% of the total ganglioside content, the remainder being constituted by some 15-20 minor components.[19] In the nervous system of other animals, like amphibia, reptiles and fishes, the number of gangliosides is still higher (50-60 different structures).[2]

Gangliosides to be used in biological experiments must be prepared in relatively large amounts and with the highest degree of purity. Particularly they must be free of any contaminants of lipidic nature (including other gangliosides) and of amphilic polypeptides that might be extracted from tissues together with gangliosides.[20]

The first step in ganglioside preparation is the isolation of individual chemical entities having a homogenous oligosaccharide portion. This is carried out by means of chromatographic methods starting from crude tissue extracts. Owing to the great abundance of gangliosides in the nervous system, this tissue is generally employed as the source of gangliosides for preparative purposes. Procedures based on partition and absorption column chromatography, using silica gel of different characteristics and various solvent systems, are widely employed for the preparation of large amounts of highly purified gangliosides.[1,21,22] However, this approach suffers from a low resolution power. Therefore, several successive chromatographic runs are necessary for achieving an acceptable degree of purity, increasing both the cost and working time of the operation. Ion exchange chromatography is also used to separate gangliosides on the basis of their sialic acid content.[23] This fractionation procedure is necessarily followed by a silica gel chromatography in order to resolve the different ganglioside species within each class carrying the same sialic acid content. Ion exchange resins work at a high resin-ganglioside ratio, which limits their suitability for large-scale preparation of gangliosides.[23] A common disadvantage to all the above chromatographic procedures is the strong UV absorption of the solvent systems. Therefore, the elution profile of ganglioside cannot be followed by flow-through determination of UV absorption at the wavelengths that are typical for gangliosides (190-220 nm). Instead, the elution profile is monitored by thin layer chromatographic (TLC) analysis of the eluted fractions.

Table I. Names and abbreviations of natural sialic acids occurring in gangliosides

Name	Abbreviation
N-acetylneuraminic acid	Neu5Ac
N-glycolylneuraminic acid	Neu5Gc
4-*O*-acetyl-*N*-glycolylneuraminic acid	Neu4Ac5Gc
N-acetyl-9-*O*-acetylneuraminic acid	Neu5,9Ac_2
N-glycolyl-8-*O*-methylneuraminic acid	Neu5Gc8Me
N-acetyl-8-*O*-sulfoneuraminic acid	Neu5Ac8S
N-glycolyl-8-*O*-sulfoneuraminic acid	Neu5Gc8S

High performance liquid chromatography (HPLC), characterized by an extraordinarily high resolution power, has been recently introduced for the separation of individual gangliosides. The first attempts, which required prior derivatization of gangliosides with strong UV absorbing probes, were designed only for analytical purposes.[24,25] We developed a rapid and highly resolving normal phase HPLC

Table II. Structure of the gangliosides that most commonly occur in mammalian nervous system

IUPAC-IUB nomenclature[17]	Svennerholm nomenclature[18]	Schematic structure
$I^3\alpha$NeuAcGalCer	GM4	NeuAcα2→3Galβ1→1'Cer
$II^3\alpha$NeuAcLacCer	GM3	NeuAcα2→3Galβ1→4Glcβ1→1'Cer
$II^3\alpha$(NeuAc)$_2$LacCer	GD3	NeuAcα2→8NeuAcα2→3Galβ1→4Glcβ1→1'Cer
$II^3\alpha$NeuAcGgOse$_3$Cer	GM2	GalNAcβ1→4Gal(3←2αNeuAc)β1→4Glcβ1→1'Cer
$II^3\alpha$(NeuAc)$_2$GgOse$_3$Cer	GD2	GalNAcβ1→4Gal(3←2αNeuAc8←2αNeuAc)β1→4Glcβ1→1'Cer
$II^3\alpha$NeuAcGgOse$_4$Cer	GM1	Galβ1→3GalNAcβ1→4Gal(3←2αNeuAc)β1→4Glcβ1→1'Cer
$II^3\alpha$NeuAc$IV^3\alpha$NeuAcGgOse$_4$Cer	GD1a	NeuAcα2→3Galβ1→3GalNAcβ1→4Gal (3←2αNeuAc)β1→4Glcβ1→1'Cer
$II^3\alpha$(NeuAc)$_2$GgOse$_4$Cer	GD1b	Galβ1→3GalNAcβ1→4Gal (3←2αNeuAc8←2αNeuAc)β1→4Glcβ1→1'Cer
$II^3\alpha$NeuAc$IV^3\alpha$(NeuAc)$_2$GgOse$_4$Cer	GT1a	NeuAcα2→8NeuAcα2→3Galβ1→3GalNAcβ1→4Gal (3←2αNeuAc)β1→4Glcβ1→1'Cer
$II^3\alpha$(NeuAc)$_2IV^3\alpha$NeuAcGgOse$_4$Cer	GT1b	NeuAcα2→3Galβ1→3GalNAcβ1→4Gal (3←2αNeuAc8←2αNeuAc)β1→4Glcβ1→1'Cer
$II^3\alpha$(NeuAc)$_2IV^3\alpha$(NeuAc)$_2$GgOse$_4$Cer	GQ1b	NeuAcα2→8NeuAcα2→3Galβ1→3GalNAcβ1→4Gal (3←2αNeuAc8←2αNeuAc)β1→4Glcβ1→1'Cer
$II^3\alpha$NeuAc$IV^2\alpha$FucGgOse$_4$Cer	Fuc-GM1	Fucα1→2Galβ1→3GalNAcβ1→4Gal (3←2αNeuAc)β1→4Glcβ1→1'Ce
$II^3\alpha$(NeuAc)$_2IV^2\alpha$FucGgOse$_4$Cer	Guc-GD1b	Fucα1→2Galβ1→3GalNAcβ1→4Gal (3←2αNeuAc8←2αNeuAc)β1→4Glcβ1→1'Cer

procedure which is able to separate ganglioside mixtures into the individual entities without prior derivatization.[26] The adoption of a proper solvent system enabled us to follow the elution pattern by direct assay of UV absorption into the eluate. Analytical (μg of analyzed material) and semipreparative (mg of analyzed material) versions of the procedure were established with a high degree of reproducibility. The semipreparative method is briefly described here. Four-five mg (as bound sialic acid) of ganglioside mixture are dissolved in 100 μl of redistilled water in a microtube and introduced into a loading sample injection equipped with a 200 μl loop. The microtube is washed with 100 μl of redistilled water and the washing added to the previous sample. Chromatography is performed on a Lichrosorb-NH_2 column, 250 mm x 25 mm, (Merck, Darmstadt, FGR) at at flow rate of 39 ml/min, with a gradient of two solvents: solvent A, acetonitrile-5mM phosphate buffer, pH 5.6, 83:17 (v/v), and solvent B, acetonitrile, 20mM phosphate buffer, pH 5.6, 1:1 (v/v). The programmed elution is: 7 min with solvent A; 53 min with a linear gradient from solvent A to the mixture of solvent A and B, 66:34 (v/v); 20 min with a linear gradient from the mixture of solvent A and B, 66:34 (v/v); to the mixture of solvent A and B 36:64 (v/v). The elution profile is monitored by flow-through measurement of UV absorbance at 215 nm. The analytical method is a scaled down version of the semipreparative method, using much smaller columns and lower flow rates.

Figure 1 shows the fractionation of the total ganglioside extract obtained from calf brain after a single HPLC run. Each ganglioside is characterized by a single symmetric peak having a fixed and reproducible retention time. The identification of the gangliosides in the individual peaks is attained by comparison with the authentic standard gangliosides submitted to the same procedure. The purity of each ganglioside, as examined by analytical HPLC and by high performance thin layer chromatogrpahy (HPTLC)[22] showed a homogeneity exceeding 99%. For example, the HPLC analysis of purified GT1b is shown in Figure 2. Twenty nmoles of GT1b were injected and the elution profile was recorded at two different AUFS (absorbance units full scale) values in order to detect any minor contaminants. In this case, GT1b was over 99.5% pure.

ISOLATION OF GANGLIOSIDES WITH HOMOGENOUS LONG CHAIN BASE COMPOSITION

Ceramide, the lipid portion of gangliosides, contains a long chain base and a long chain fatty acid which are linked together by an amide bond. In mammalian nervous tissues, four different long chain bases are present in the ganglioside molecules (figure 3).[1,3] Two unsaturated bases at 18 and 20 carbon atoms (C_{18} and C_{20} sphingosines) account for about 90% of the total content. The remainder (10%) is the saturated compounds (C_{18} and C_{20} sphinganines). Conversely the fatty acid composition of the same gangliosides is close to being homogenous, since stearic acid (especially in purified gangliosides) accounts for 90% or more of the total fatty acid content. Therefore, the individual gangliosides obtained from mammalian brain can be considered practically homogenous in the fatty acid moiety.[1,3] The procedures based on silica gel chromatoqraphy or normal phase HPLC cannot fractionate a ganglioside with homogenous oligosaccharide portion into the molecular species having a single chain base moiety. In other words, the gangliosides obtained by the above methods are heterogenous in their lipid portion.

We recently developed a new fractionation procedure based on reversed phase HPLC.[27,28,29] This procedure has been successful for both preparative and analytical separations of mammalian gangliosides with homogenous oligosaccharide composition. An additional advantage is that the elution profile from the column can be monitored by direct measurement of the UV absorption of the eluate. The preparative procedure is briefly as follows: four-five mg (as sialic acid) of ganglioside are dissolved in redistilled water and introduced into the injector as described above (normal phase HPLC procedure). Chromatography is carried out on a Lichrosorb RP8 column (250mm x 25 mm, Merck, Darmstadt, F.R.G.) at a flow rate of 15 ml/min and with a solvent constituted by a combination of acetonitrile and 5 mM sodium

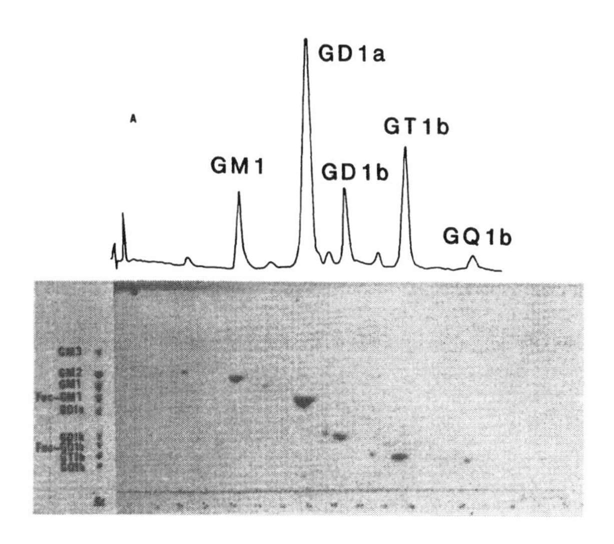

Fig.1. Application of the HPLC preparative method to the isolation of individual gangliosides from calf brain ganglioside mixture. **A**-HPLC elution profile, recorded by UV absorbance at 215 nm. **B**-HPLC elution profile monitored by HPTLC;20 ml fractions were collected and 0.1 ml aliquots were spotted after drying and redissolving in 20 µl of choloform/ methanol,2:1 (v/v).

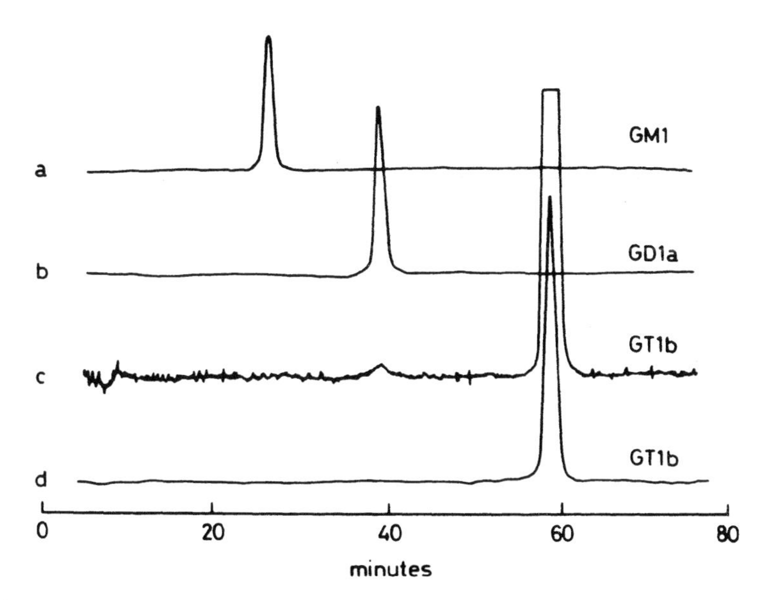

Fig. 2. Application of the HPLC analytical method for verification homogeneity of GT1b ganglioside purified by preparative HPLC from calf brain ganglioside mixture. **a**-10 nmoles of standard GM1, range 0.025 AUFS; **b**-10 nmoles of standard GD1a, range 0.025 AUFS; **c**-20 nmoles of purified GT1b, range of purified GT1b, range 0.005 AUFS; **d**-20 nmoles of purified GT1b, range 0.05 AUFS.

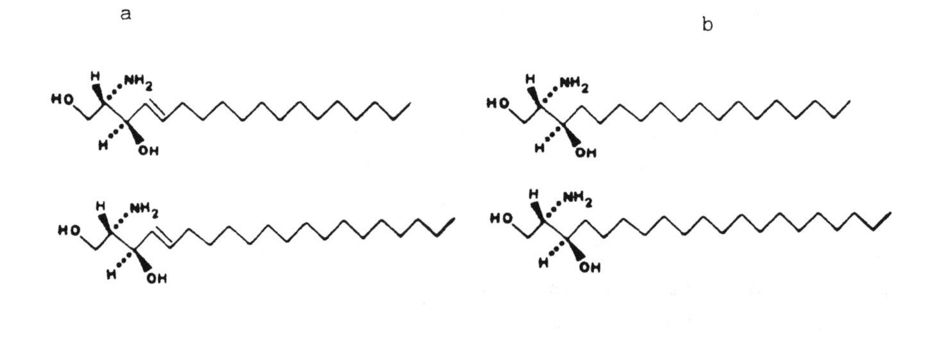

Fig.3. Structure of the major long chain bases present in the nervous systems' ganglioside molecules. a) C_{18} sphingosine; b) C_{18} sphinganine; c) C_{20} sphingosine; d) C_{20} sphinganine.

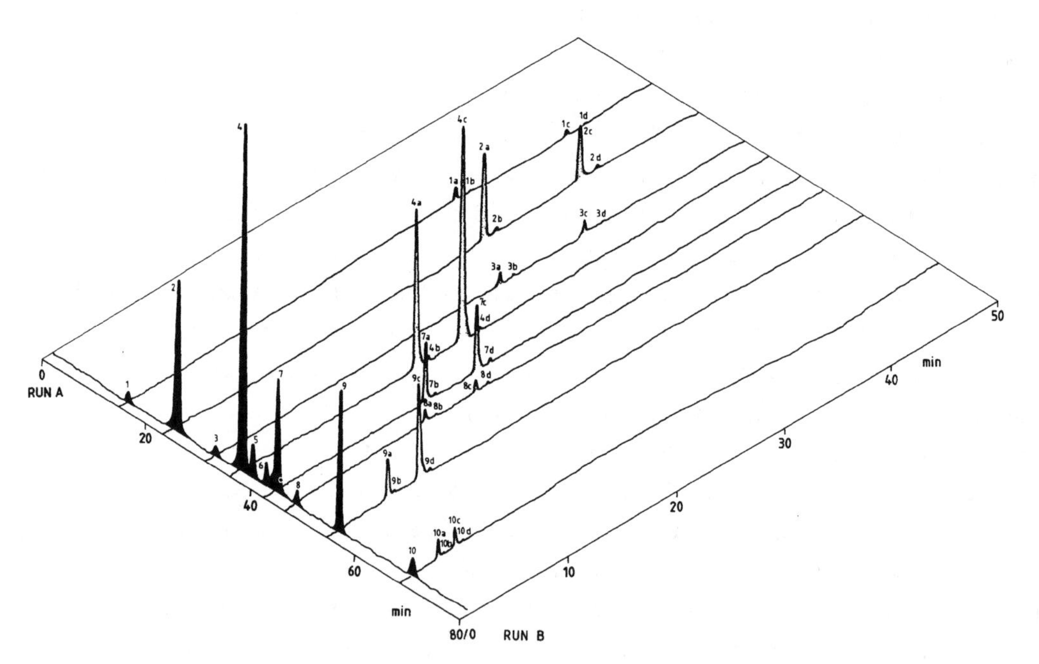

Fig. 4. Reversed phase HPLC fractionation (**Run B**) of each of the individual gangliosides from calf brain,prepared by normal phase HPLC (**Run A**-Fig. 1). **a,b,c,** and **d** correspond to each ganglioside molecular species containing C_{18} sphingosine, C_{18} sphinganine, C_{20} sphingosine and C_{20}sphinganine, respectively.

phosphate buffer, pH 7.0. The volume ratio between the two solvents is in the range 1:1 and 3:2, and the overall elution time varies from 7-45 min. The proportion of acetonitrile in the mixture and the run time are lower when the polarity of the ganglioside is higher. The elution profile is monitored by flow-through absorbance detection at 195 nm. The homogeneity of each ganglioside molecular species is determined by analytical reversed phase HPLC under the following conditions: a Lichosorb RP8 column of reduced dimension (150 mm x 4.6 mm), a flow rate of 1 ml/min.,the same above elution program with shorter run times, and 5-10 μg of starting ganglioside. Figure 4 shows the reversed phase HPLC fractionation of each of the individual gangliosides from calf brain, separated by normal phase HPLC (see also Figure 1). Each ganglioside is fractionated into four different peaks.[27] Two of them, always preponderant and accounting for 90-95% of the total injected material, proved[27] to correspond to the molecular species containing C_{18} sphingosine (lower retention time) and C_{20} sphingosine (higher retention time). The remainder (5-10%) of the injected material is represented by the two small peaks recorded immediately after the major peaks, and corresponding to the molecular species containing C_{18} sphinganine and C_{20} sphinganine, respectively.[27] By analytical reversed phase HPLC, each fraction was over 99% pure. A reversed phase TLC method has been also devised for the analysis of individual species of ganglioside having homogenous long chain base composition.[30] This method uses the reversed phase silica gel TLC plates (RP 18-HPTLC plates, Merck, Darmstadt, F.R.G.), and the solvent system methanol-acetonitrile-water, 19:5:1 (v/v). The reversed phase TLC method can be considered a useful alternative to the reversed phase HPLC method for analytical purposes. The reversed phase TLC separation of the GM1 molecular species is presented in figure 5.

PREPARATION OF THE GANGLIOSIDE SPECIES THAT CONTAIN SATURATED LONG CHAIN BASES

As already mentioned, the gangliosides from mammalian nervous system have less than 10% of the total long chain

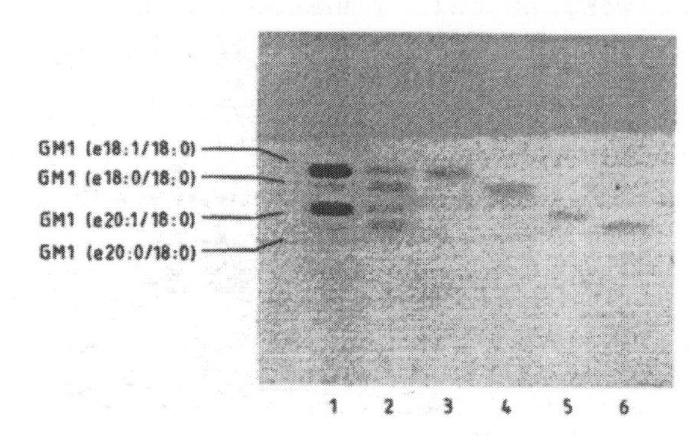

Fig. 5. Separation by reversed phase HPTLC of GM1 ganglioside molecular species. **1**-natural GM1; **2**-mixture of standard GM1 molecular species; **3,4,5,** and **6**-standard GM1 molecular species. Standards containing stearic acid (18:0) were used.

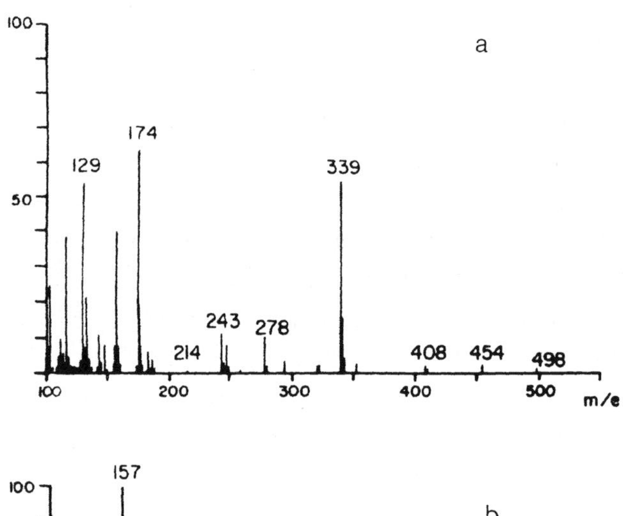

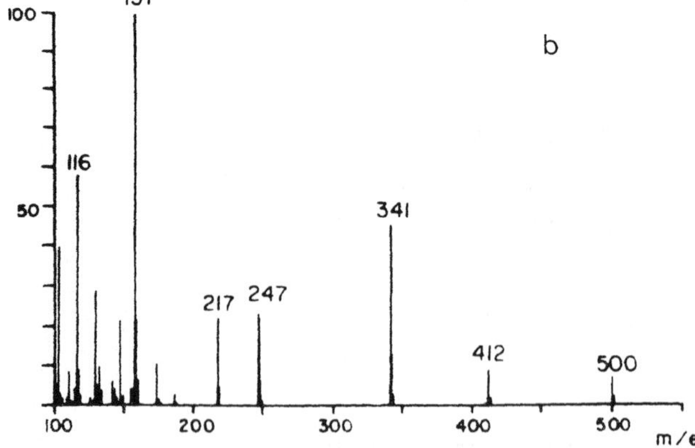

Fig. 6. MS spectra of the N-acetyl, trimethylsilyl long chain bases released from the GM1 molecular species containing C_{20} sphingosine, before (**a**) and after (**b**) hydrogenation treatment.

base content constituted by the saturated compounds, C_{18} and C_{20} sphinganine. Therefore, the large-scale preparation of ganglioside species containing these long chain bases, starting from natural gangliosides, is tedious and costly. An easy solution to the problem is selective chemical hydrogenation of the starting ganglioside followed by HPLC separation of the individual molecular species containing C_{18} and C_{20} sphinganine. A quantitative hydrogenation of gangliosides is obtained by the following procedure.[27] Ganglioside is dissolved in ethanol-water, 7:3 (v/v) (1 mg/ml) and submitted to catalytic hydrogenation in the presence of platinum dioxide (0.5 mg/mg of ganglioside). After standing at room temperature for 36 h under continuous stirring, the catalyst is removed by filtration, and the solution, evaporated to dryness, is suspended in redistilled water, then dialyzed and lyophilized. By this treatment, the double bond at C-4/C-5 positions of long chain base is quantitatively hydrogenated. Figure 6 shows the mass spectra of the C_{20} long chain base released from ganglioside GM1, before and after the process of hydrogenation. The fragmentation pattern is similar in both cases, but the m/e values of peaks in the two spectra differ by two units, indicating saturation of the double bond, in other words, transformation of C_{20} sphingosine into C_{20} sphinganine. The situation is identical for the C_{18} compounds. Of course, the hydrogenation procedure leads also to saturation, if present, of the fatty acids containing double bonds.

CHEMICAL SUBSTITUTION OF THE FATTY ACID MOIETY OF GANGLIOSIDES: PREPARATION OF GANGLIOSIDES WITH A GIVEN AND HOMOGENOUS FATTY ACID COMPOSITION

The fatty acid moiety of the gangliosides present in the nervous system is mainly stearic acid. Actually, the fatty acid composition of gangliosides of extraneuronal origin is much more heterogenous, with acyl chains differing in both hydrocarbon length and content of double bonds and hydroxyl groups. Intensive efforts were made to remove the fatty acid moiety from gangliosides and to substitute it with a new, and unique, fatty acid. The main difficulty was to find out

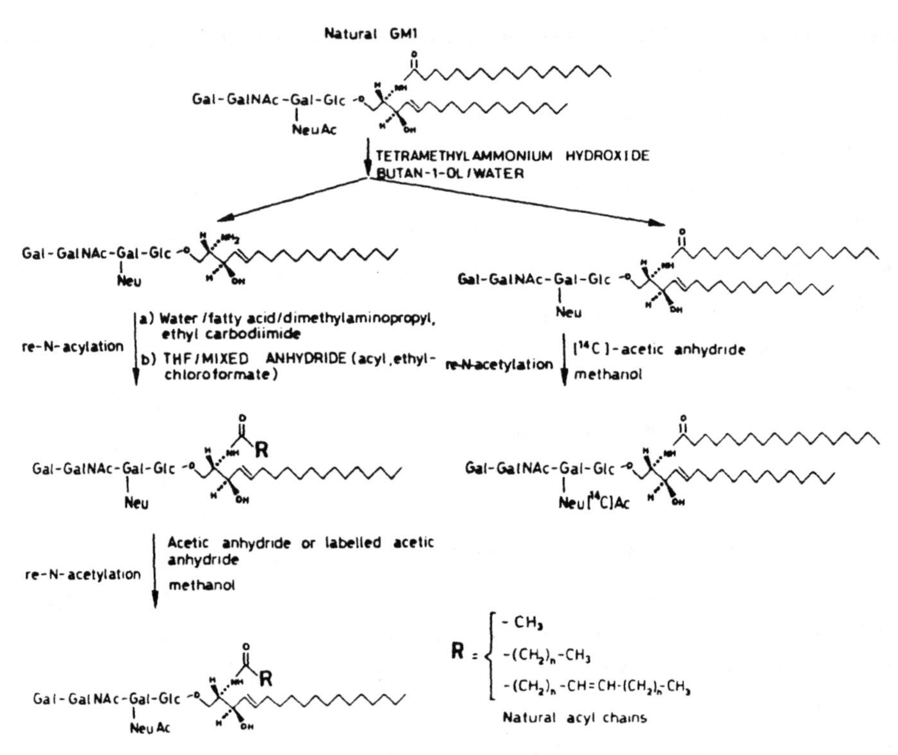

Fig. 7. Flow sheet of the procedural steps used for preparing GM1 with homogenous fatty acid moiety (left side) and GM1 with ^{14}C isotopically labelled sialic acid (right side).

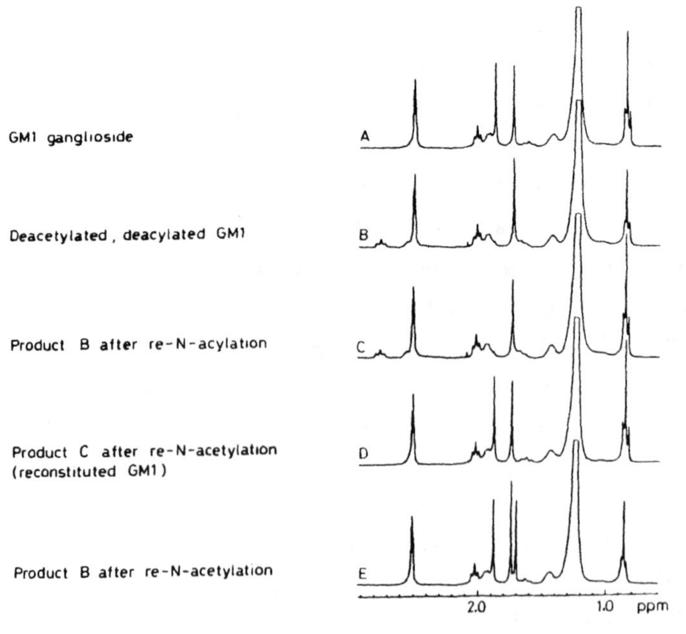

Fig. 8. Upfield proton NMR spectra (0.6-2.9 ppm) of GM1 and of some GM1 derivatives, obtained at 300 MHz and 25°C.

conditions suitable for achieving selective de-N-acylation and re-N-acylation without excessive degradation of the compound. Highly positive results have been reached in two laboratories.[31,32] The procedure developed in our laboratory,[32] schematically shown in Figure 7, is as follows. GM1 is dissolved in butan-1-ol (11 mg/ml) and mixed at 100°C with 9 volumes of an aqueous solution of tetramethylammonium hydroxide (final base concentration, 0.1M). The reaction mixture is refluxed at 100°C under continuous stirring for 13 h and then evaporated under vacuum at 50°C., dialyzed and lyophilized. The deacylated, and deacetlyated (at the level of sialic acid) GM1 is purified from the total mixture by chromatography on a silica gel 100 column (100 cm x 1 cm) equilibrated and eluted with methanol: butan-1-ol:water, 2:2:1 (v/v), and then submitted to re-N-acylation. [(Note: During alkaline treatment of ganglioside, besides the deacylated and deacetylated ganglioside, another product is obtained with a relatively high yield (30-35%). This is the derivative of ganglioside lacking only the acetyl group at the level of sialic acid (deacetylated ganglioside).] For this purpose, an aqueous solution containing fatty acid (final concentration, 4.5 mM) and 1-(3-dimethylaminopropyl)-3-ethylcarbodiimide hydrochloride (final concentration, 5 mM) is maintained at 50°C for 2 h under continuous stirring. Then the deacylated and deacetylated GM1 is added (molar ratio with fatty acid of 1:10), and the mixture is allowed to react for 5 additional hours under the same conditions. The re-N-acylated compound is purified from the mixture by chromatography on a silica gel 100 column (150 cm x 2 cm), equilibrated and eluted with the solvent system chloroform-methanol-water, 60:35:5, (v/v). In order to attain re-N-acetylation of the sialic acid moiety, the purified re-N-acylated compound is dissolved in magnesium dehydrated-methanol (1 mg/ml), added to an equal volume of sodium dehydrated-acetic anhydride and allowed to react overnight at room temperature. Alkaline hydrolysis of GM1 with tetramethylammonium hydroxide gives a yield of deacylated, deacetylated-GM1 of about 40%; re-N-acylation of deacylated, deacetylated GM1 has a final yield of 60%; final re-N-acylation undergoes with almost quantitative yield. The fatty acids which were used for acyl substitution in gangliosides were myristic, palmitic, stearic, arachidic, lignoceric, and oleic acid. However, any fatty acid can be successfully used for this purpose provided that the proper adaptations to the basic procedure are preliminarily ascertained. Verification of the main chemical events taking place during the above described processes has been performed by proton NMR measurements.[36] Particularly the right insertion of the acyl chain at the amino group of long chain base, and of the acetyl group at the sialic acid amino group has been checked. The NMR spectra of GM1 ganglioside and of the different derivatives obtained during treatment is presented in figure 8.

In the NMR spectrum of GM1, the acetyl protons of N-acetylneuraminic acid and of N-acetylgalactosamine resonate at 1.87 and 1.76 ppm, respectively; the methyl protons of the alkyl region, that is the terminal methyl groups of long chain base and fatty acid, resonate at 0.85 ppm. The NMR spectrum of the product of alkaline hydrolysis of GM1 completely lacks the peak centered at 1.87 ppm indicating that during alkaline hydrolysis the N-acetylneuraminic acid is transformed into neuraminic acid. In addition, peak intensity measurement shows that the ratio between methyl protons of the alkyl region and acetyl protons of N-acetylgalactosamine, which is about 2 in GM1, becomes 1 in the hydrolysis product, indicating the presence of a single alkyl chain, or, in other words, the successful removal of the fatty acid residue. In the NMR spectrum of the re-N-acylated compound the ratio between methyl portions of the alkyl region and acetyl protons of N-acetylgalactosamine goes back to the value of starting GM1. Finally, in the compound obtained after re-N-acetylation the peak at 1.87 ppm is recorded, indicating the insertion of the acetyl group. To exclude the possible final synthesis of an isomeric GM1, containing the acetyl group at the level of ceramide, and the acyl chain at the level of sialic acid, the product of alkaline hydrolysis of GM1 was also submitted to full re-N-acetylation in order to obtain a GM1 molecule with an acetyl group also at the level of ceramide.

The NMR spectrum of this compound, also presented in figure 8, shows, besides the peaks corresponding to the acetyl groups of sialic acid and N-acetylgalactosamine, a new peak resonating at 1.79 ppm and of the same intensity. This resonance derives from insertion of the acetyl group on the amino group of the long chain base. In parallel, the resonance due to the alkyl region has an equal intensity to that of the acetamide methyl protons, indicating a single alkyl chain.

All this evidence demonstrates that during the synthetic process the isomeric configuration of GM1 is maintained. The methodology for de-N-acylation and re-N-acylation of gangliosides is open to a number of applications. When combined to reversed phase HPLC, it allows one to prepare ganglioside molecular species which contain an individual long chain base and an individual and chosen fatty acid. Particularly, gangliosides can be selectively synthesized to carry short or long, mono- or poly unsaturated, branched or hydroxylated acyl chains similar to those occurring in extraneural tissues. The availability of these compounds can greatly help the understanding of the specific role played by the fatty acid residues in ganglioside functional performances. A further application, extremely useful for biological investigations, is the insertion into gangliosides of acyl moieties containing radioactive, or fluorescent, or spin-labelled probes.

PREPARATION OF RADIOLABELLED GANGLIOSIDES

An approach recently introduced to study the functional implications of gangliosides consists of the addition of exogeneous gangliosides to various *in vivo* and *in vitro* systems followed by determination of the exerted biological effects.[33,34,35,36] These effects are assumed to be derived from the insertion of gangliosides into the cell plasma membrane which results in the modification of the membrane properties and behaviour.[36,37] Moreover, experiments both *in vivo* and *in vitro* showed that exogenously administered gangliosides penetrate into cells, and undergo metabolic processing.[38,39] For this kind of study, as well as for inspecting the interactions of gangliosides with various ligands, it is required that gangliosides are available in radioactive form. The following methods have been designed for introducing radioactivity into gangliosides: a) the biosynthetic method, in which the labelled compound is biosynthesized starting from a radioactive precursor;[40,41] b) the galactose oxidase-sodium borohydride reduction method, in which the galactose (or N-acetyl-galactosamine) moiety in terminal position is first oxidized at carbon 6 by galactose oxidase, and then reduced back with tritiated sodium borohydride;[42,43,44,45,46] c) the periodate-sodium borohydride reduction method, in which sialic acid is oxidized at the level of the side chain, and then submitted to reduction with tritiated sodium borohydride;[47] d) the reduction method in which tritium is catalytically added to the double bond of sphingosines;[48,49] e) the dicyanodichlorobenzoquinone (DDQ) oxidation-sodium borohydride reduction method, in which the sphingosines are specifically oxidized by DDQ at carbon-3, and then reduced with tritiated sodium borohydride;[50] f) the de-N-acylation, re-N-acylation method in which the original fatty acid and/or sialic acid acetyl group are substituted with ^{14}C and/or ^{3}H labelled fatty or acetyl group, respectively. Compounds of relatively low specific radioactivity and with a relevant cost are obtained by the biosynthetic method. The periodate method is valuable in obtaining a high specific radioactivity and easy applicability. However, it cannot avoid marked structural modifications of ganglioside sialic acid residue(s) which may be important from the functional point of view. In fact the sialic acid side chain is known[51] to be involved in the formation of a network of weak linkages with cations such as Ca^{2+}, which can be seriously disturbed by the above chemical modifications. The reduction method produces completely saturated gangliosides that display physico-chemical properties quite different from those of the natural compounds. Methods b), e) and f) appear to have a number of advantages. They provide isotopic radiolabelling and allow introduction of radioactivity in the oligosaccharide or in the ceramide portion of the molecule, and to choose a labelling at the level of the long chain base or of the fatty acid moiety. Furthermore, double radiolabelling of the molecule, and ^{3}H or ^{14}C labelling is possible. All three methods are recommended to be available in laboratories aimed at studying ganglioside metabolism and functional involvement. Depending on the specific experimental purpose, the most adequate radiolabelling approach can be adopted. A brief description of these methods will be given here.

Preparation of gangliosides tritium labelled at the level of the terminal galactose and N-acetyl galactosamine residues

The procedure for tritium labelling of gangliosides at the terminal galactose or N-acetylgalactosamine residues by the galactose-oxidase-^{3}H $NaBH_4$ method, used in our laboratory, is briefly described. Twenty mg of ganglioside GM2, GM1 or GD1b, dissolved in 1 ml of chloroform-methanol, 2:1 (v/v) are mixed with 10 ml of the same solvent containing 180 mg of Triton-X 100. The mixture is dried under vacuum and the residue dissolved in 8 ml of 5mM EDTA-25 mM sodium phosphate buffer pH 7.0, containing 450 units of galactose-oxidase. The mixture is incubated at 37°C under constant stirring for 12 h, then 450 additional units of galactose-oxidase are added, and incubation prolonged for 6 h. The enzymatic reaction is stopped by the addition of 4 vol of tetrahydrofuran and vigorous shaking. The mixture is dried and chromatographed on a 110 cm x 2 cm silica gel 100 column, previously equilibrated with chloroform-methanol-water, 60:35:5 (v/v) and eluted with the same solvent. The purified oxidized ganglioside, over 98% pure, is dissolved in 5 ml of tetrahydrofuran-water-0.1 M NaOH, 4:2:1 (v/v) and treated at room temperature with 100 mCi of ^{3}H $NaBH_4$. After 5 h, 10 mg of cold $NaBH_4$ are added and the reaction mixture allowed to react further for 30 min.[4] The mixture is concentrated almost to dryness, diluted with distilled water and dialyzed. The radioactive ganglioside is finally purified from the mixture by chromatography on a silica gel 100 column (40 cm x 1 cm), equilibrated and eluted with chloroform-methanol-water, 60:35:5, (v/v).

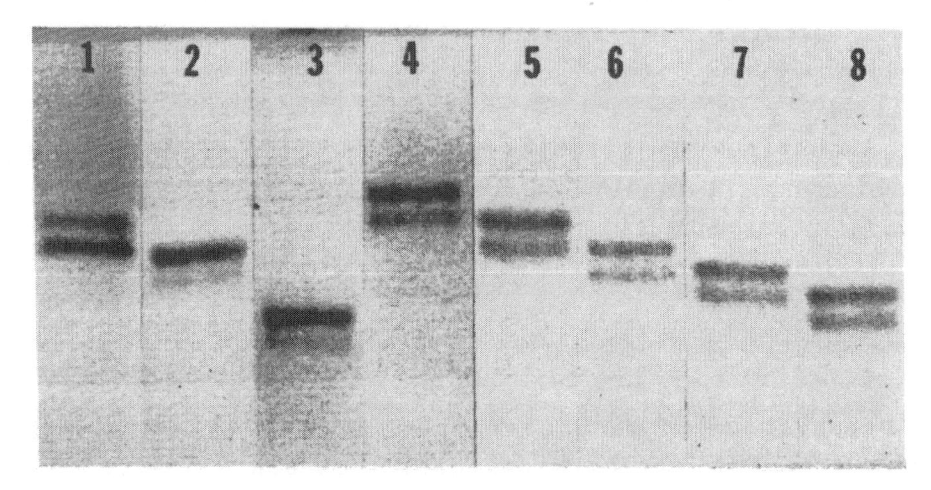

Fig. 9. TLC of the reaction products after ganglioside oxidation with galactose oxidase (1,2,3) and DDQ (4,5,6,7,8). **1**=GM2, **2**=GM1, **3**=GD1b, **4**=GM3, **5**=GM2, **6**=GM1, **7**=Fuc-GM1, **8**=GD1a. In all the chromatographs the upper spot corresponds to the oxidized ganglioside.

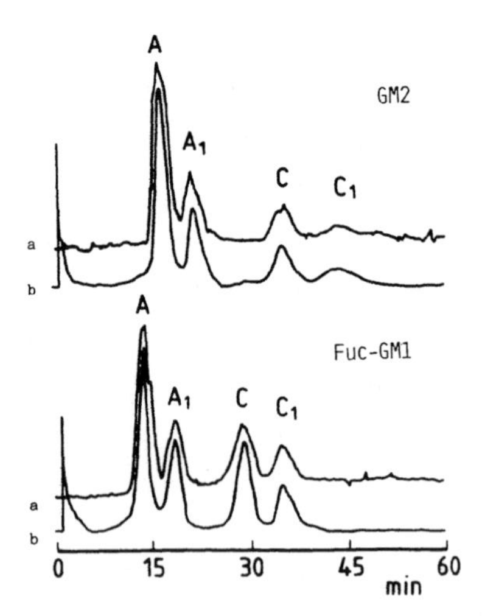

Fig. 10. HPLC fractionation of GM2 and Fuc-GM1, tritium labelled at the 3-position of long chain bases. A, A1, C and C1 correspond to the molecular species that contain erythro C_{18} sphingosine, threo C_{18} sphingosine, erythro C_{20} sphingosine and threo C_{20} sphingosine, respectively, a) eluted radioactivity; b) U.V. record.

Figure 9 shows the TLC pattern of the reaction products after 12 plus 6 h of incubation in the presence of galactose-oxidase. The percentage of the oxidized product in the mixture, as determined by quantitative densitometry,[52] is 30%, 80%, and 90% for ganglioside GM2, GD1b and GM1 respectively. After reduction with tritiated sodium borohydride some highly apolar radioactive compounds (10-20% of total radioactivity) appear to be present in the mixture. Their removal is achieved by the final column chromatography purification step. The purity of the final preparation of each ganglioside is over 98%. The specific radioactivity of the labelled gangliosides was constantly in a range of 1.3-1.5 Ci/mmol. The regularity of these values is due to the fact that the oxidized ganglioside is separated from the reaction mixture before reduction and that the reduction reaction occurs in an almost quantitative manner (90-95%). Therefore the final value of specific radioactivity depends on the starting specific activity of ^{3}H $NaBH_4$.

Preparation of gangliosides tritium labelled at the 3-position of sphingosines

Gangliosides can be tritium labelled at carbon-3 of unsaturated long chain bases by the dicyano-dichlorobenzoquinone (DDQ)/^{3}H $NaBH_4$ method.[50] In order to better understand this procedure, a few introductory comments are useful. DDQ is a specific oxidizing agent for alpha, beta-unsaturated alcohols, which are transformed in the corresponding ketones.[53] Gangliosides have a high content of sphingosines which contain a hydroxyl group in position 3 and a double bond in the vicinal 4-5 position. Therefore, they appear to be good substrates for this chemical oxidation, provided that their sphingosine moiety is made available to the oxidizing agents. This condition can be fulfilled in a toluene medium containing high quantities of Triton X-100, where gangliosides are present in the form of inverted micelles with the hydrophobic portion exposed on surface. When the product of oxidation in the case of gangliosides the 3-keto derivative is submitted to reduction with sodium borohydride, the starting compound is formed and this is the basis for radiolabelling. However, the reduction reaction is not stereo-specific and produces a mixture of the natural erythro and the unnatural threo form of sphingosine. Therefore the molecular species of gangliosides containing erythro sphingosines must be separated from the isomeric threo species after the final reduction step. The application of DDQ/^3H $NaBH_4$ method to tritium labelling of gangliosides developed in our laboratory,[50,52] is briefly described. Twenty mg of one of the following gangliosides, GM3, GM2, GM1, Fuc-GM1 and GD1a, are dissolved in 10 ml of chloroform-methanol, 2:1 (v/v), and mixed with 10 ml of a solution of Triton X-100 in the same solvent (60 mg/ml). The solvent is evaporated and the residue dissolved in 10 ml of a DDQ solution in sodium dehydrated toluene (60 mg/ml). The mixture is maintained at 37°C for 40 h under constant stirring in a screw capped tube. The solvent is then evaporated under vacuum at 37°C, and the dark brown residue suspended in 10 ml of cold acetone and sonicated in an ultrasonic water bath. After centrifugation at 12,000 x g, the supernatant, which contains Triton X-100 and DDQ, is discarded. The treatment is repeated until a white pellet is obtained (generally four times). The oxidized ganglioside is separated from the unreacted compound and from other contaminants by chromatography on a silica gel 100 column (100 cm x 2 cm), equilibrated and eluted with the mixture chloroform-methanol-water at the volume ratio of 60:35:3 for GM3 and GM2, and at the volume ratio of 60:35:5 for GM1, Fuc-GM1 and GD1a. The purified 3-keto ganglioside, dissolved in 5 ml of propan-1-ol-water, 7:3, v/v, is treated at room temperature for 30 min and under stirring with 100 mC of ^3H $NaBH_4$, dissolved in 0.5 ml of the same solvent. Then 2 mg of cold $NaBH_4$ are added and the reaction continued for 30 min. The mixture is concentrated almost to dryness, diluted with redistilled water, dialyzed and lyophilized. The residue, dissolved in redistilled water, is submitted to reversed phase HPLC, for the separation of the ganglioside molecular species containing erythro (C_{18} and C_{20}) sphingosine and threo (C_{18} and C_{20}) sphingosine. The reversed phase HPLC procedure is the same that has been described before. The threo form of ganglioside is eluted immediately after the erythro form without overlapping (figure 10). The elution profile is monitored by flow-through detection at 195 nm, or by direct counting of radioactivity in the eluate.

Figure 9 shows the TLC pattern of the reaction products of DDQ treatment of the various gangliosides, prior to purification of the corresponding 3-ketoderivatives. The yield in these oxidized compounds is, for all gangliosides, about 60-65%. The purity of the purified 3-keto derivatives before reduction is over 98%. Figure 10 shows the HPLC fractionation of tritium labelled GM2 and Fuc-GM1 into four well separated molecular species containing erythro C_{18} sphingosine, threo C_{18} sphingosine, erythro C_{20} sphingosine, and threo C_{20} sphingosine, respectively. All the final labelled compounds proved to be over 99% pure. The specific radioactivity of the labelled molecular species of the different gangliosides were in the range 0.85-1.3 Ci/mmol. The variability is higher and the values are lower than those obtained with the galactose-oxidase ^{3}H $NaBH_4$ method This is likely due to the starting neutral conditions used during ^{3}H $NaBH_4$ treatment which cause partial degradation of $NaBH_4$. On the other hand, neutral conditions during this treatment prevent degradation of the alpha, beta unsaturated ketones that are highly unstable under alkaline conditions.

Preparation of gangliosides ^{14}C labelled at the level of the fatty acid moiety

The de-N-acylation re-N-acylation procedure described above leads to produce radiolabelled gangliosides when radioactive fatty acids are employed. The application of this approach to labelling ganglioside GM1 in the fatty acid moiety is described here. One hundred µmoles of ^{14}C stearic acid (specific radioactivity, 56 mCi/mmol) are dissolved in 2 ml of dried tetrahydrofuran, cooled at -20°C by immersion in an ice-salt bath. Then 20 µl of triethylamine and 20 µl of ethylchloroformate are added and the mixture allowed to stand at -20°C for 5 min under continuous stirring. After addition of 100 µmoles of deacylated-deacetylated GM1 (prepared as described above), dissolved in 5 ml of tetrahydrofuran-water, 20:1, v/v, the reaction mixture is kept at room temperature for 10 min and under continuous stirring, and then evaporated under vacuum at 37°C. The wet residue is diluted with 2-3 ml of redistilled water, dialyzed and lyophilized. The re-N-acylated product is separated from the reaction mixture, submitted to re-N-acetylation (at the level of sialic acid) and further purified, as described above. The purity of the final compound is over 99%, and the specific radioactivity 55 mCi/mmol, reflecting that of the starting labelled fatty acid.

It should be noted that in the general procedure for re-N-acylation of de-N-acylation of gangliosides described previously, the molar ratio between deacylated, deacetylated ganglioside and fatty acid was 1:10 and the reaction was carried out in the presence of 1-(3-dimethylaminopropyl-3-ethylcarbodiimide hydrochloride. In the present description the mixed anhydride between the acyl chain and ethylchloroformate, with a ganglioside- fatty acid molar ratio of 1:1, is employed as condensing agents. This adaption was introduced in order to reduce the amount of the starting radioactive fatty acid (which is very expensive), and to maintain a relatively high yield (about 40%) of labelled ganglioside.

Preparation of gangliosides ^{14}C labelled at the acetyl group of sialic acid

Alkaline treatment of ganglioside under the conditions described in a previous section leads also to formation of a ganglioside derivative simply lacking the acetyl group of sialic acid (de-N-acetylated ganglioside) (see figure 7). Re-N-acetylation of this compound with ^{14}C acetic anhydride provides a radioactive ganglioside specifically and isotopically labelled at the level of sialic acid. The labelling procedure, worked out on ganglioside GM1 is as follows.[53] 7 mg of de-N-acetylated GM1, are dissolved in 1 ml of 5% triethylamine in magnesium-dehydrated methanol and 11.1 µl of 5% 1-^{14}C-acetic anhydride (111 mCi/mmol) in toluene are added. After stirring for 10 min at room temperature, 10 µl of cold acetic anhydride are added, and the mixture stirred for additional 30 min. The solution is then diluted with 100 ml of distilled water, dialyzed, lyophilized, and the residue (specific radioactivity: 53 mCi/mmole) solubilized in 1 ml of propan-1-ol-water, 7:3,v/v, is stored at 4°C.

CONCLUSION

The chemical technology of gangliosides is developing at a rapid rate. The application of HPLC procedures gave a powerful improvement to preparation of highly purified compounds, fulfilling the requirements for experimental approaches aimed at understanding the physiological role of these substances. The availability of a wide and well established strategy for radiochemical labelling of gangliosides, provides the opportunity to prepare the labelled compounds most suitable for specific research purposes. Finally, the success of the technology for manipulations of the ganglioside molecule, that produces either gangliosides with programmed chemical composition or ganglioside derivatives carrying special probes, greatly facilitates investigations on the specific role played by gangliosides in the functional behaviour of cell plasma membranes. On these premises it is not ambitious to expect in the near future a "renaissance" period in ganglioside research.

SUMMARY

New methods have been developed for the preparation of highly purified gangliosides, homogeneous in the saccharide, long chain base, and fatty acid moieties and gangliosides carrying different kinds of labelled probes. Gangliosides, homogeneous in the oligosaccharide portion, were prepared by preparative normal phase HPLC on a Lichrosorb-NH_2 column, using a gradient of acetonitrile-phosphate buffer, pH 5.6, as solvent system. Each class of ganglioside (from monosialo- to tetrasialogangliosides) was then submitted to reversed phase HPLC on a preparative RP-8 column, using acetonitrile-5mM phosphate buffer, pH 7, as solvent system, to obtain gangliosides homogeneous in the long chain base moiety. Gangliosides containing C_{18} and C_{20} sphinganine were prepared by catalytic hydrogenation of the corresponding unsaturated gangliosides. GM1 with homogeneous acyl chain was prepared by alkaline hydrolysis in the presence of tetramethylammonium hydroxide (which forms a GM1 deacetylated at the level of sialic acid, and a GM1 deacetylated at the level of sialic acid and deacylated at the level ceramide), followed by re-N-acylation, carried out in the presence of dimethylaminopropyl, ethylcarbodiimide and natural fatty acids, or of mixed anhydride of ethylchloroformate and ^{14}C-stearic acid, and re-N-acetylation performed with acetic anhydride or labelled acetic anhydride. The GM1 derivative, de-acetylated at the level of sialic acid, also produced by alkaline treatment of GM1, was submitted to re-N-acetylation with ^{14}C-acetic anhydride to produce specifically ^{14}C-labelled GM1. Re-N-acylation was carried out a) in the presence of dimethylaminopropyl, ethylcarbodiimide and natural fatty acids, b) with mixed anhydride of ethylchloroformate and ^{14}C-stearic acid. After re-N-acylations, re-N-acetylation was performed with acetic anhydride or labelled acetic anhydride. Gangliosides tritium labelled in the oligosaccharide moiety were prepared by the galactose oxidase/^{3}H $NaBH_4$ method, and gangliosides tritium labelled at carbon-3 of unsaturated long chain bases by the dicyano-dichlorobenzoquinone (DDQ)/^{3}H $NaBH_4$ method.

ACKNOWLEDGEMENTS

The experimental sections of this work were supported by grants from the Consiglio Nazionale delle Ricerche CNR (Grant n° 83.02855) Rome, Italy, and from the Ministero Pubblica Istruzione, Rome, Italy (Progetti di Ricerca di interesse nazionale).

REFERENCES

1. R.W. Ledeen and R.D. Yu, in: "Research Methods in Neurochemistry", N. Marks and R. Roonight, eds., Plenum Publishing Corp., N.Y., 371-410 (1978).

2. R. Ghidoni, S. Sonnino, V. Chigorno, A. Malesci and G. Tettamanti, in "Ganglioside structure, function and biomedical potential", R.W. Ledeen, R.K. Yu, M.M. Rapport and K. Suzuki, eds., Plenum Publishing Corp., N.Y., 307-318 (1984).
3. H. Wiegandt, Adv. Neurochem, 4:149-223 (1982).
4. G. Tettamanti, A. Preti, B. Cestaro, M. Masserini, S. Sonnino and R. Ghidoni, ACS Symp. Ser., 128:321-343 (1980).
5. C.G. Gahmberg and S.-I. Hakomori, J. Biol. Chem., 248:4311-4317 (1973).
6. T.L. Steck and G. Dawson, J. Biol. Chem., 249:2135-2142 (1974).
7. S. Ando, Neurochem. Int., 5:507-537 (1983).
8. S.-I. Hakomori, Ann. Rev. Biochem., 50:733-764 (1981).
9. R.O. Brady and P.H. Fishman, Adv. Enzymol., 50:303-323 (1979).
10. G. Tettamanti, S. Sonnino, R. Ghidoni, M. Masserini and B. Venerando, in: "Physics of Amphiphiles: Micelles, Vesicles and Microemulsion", M. Corti and V. Degiorgio, eds., XC Corso, Soc. Italiana di Fisica, Bologna, Italy, 607-636 (1985).
11. C.G. Gahmberg and S.-I. Hakomori, Biochem. Biophys. Res.Comm., 59:283-291 (1974).
12. C.G. Gahmberg and S.-I. Hakomori, J. Biol. Chem., 250:2438-2446 (1975).
13. Y. Kishimoto, S. Yahara and J. Podulso, in: "Biochemistry of Cell Surface Glycolipids", C.C. Sweely, ed., Am. Chem. Soc. Monogr., 10-23 (1980).
14. S.-I. Hakomori and R. Kannagi, JNCI, 71:231-251 (1973).
15. E. Nudelman, S.-I. Hakomori, R. Kannagi, S. Levery, M.Y. Yeh, K.E. Hellstrom and I. Hellstrom , J. Biol. Chem., 257:12752-12756 (1982).
16. R. Schauer, Adv. Carbohydr. Chem. Biochem., 40:131-234 (1982).
17. IUPAC-IUB Commission on Biochemical Nomenclature, Lipids, 12:455-475 (1977).
18. L. Svennerholm, J. Lipid Res., 5:145-156 (1964).
19. S. Sonnino, R. Ghidoni, V. Chigorno, M. Masserini and G. Tettamanti, Anal. Biochem., 128:104-114 (1983).
20. M.C. Byrne, R.W. Ledeen, F.J. Roisen, G. Yorke and J.R. Sclafani, J. Neurochem., 41:1214-1222 (1983).
21. G. Tettamanti, F. Bonali, S. Marchesini, V. Zambotti, Biochim. Biophys. Acta, 296:160-170 (1973).
22. R. Ghidoni, S. Sonnino, G. Tettamanti, N. Baumann, G. Reuter and R. Schauer, J. Biol. Chem., 255:6990-6995 (1980).
23. M. Iwamori and Y. Nagai, Biochim. Biophys. Acta, 528:257-267 (1978).
24. S.K. Kundu and D.D. Scott, J. Chromatog., 232:19-27 (1982).
25. R.H. McCluer and J.E. Evans, J. Lipid Res., 14:611-617 (1983).
26. G. Gazzotti, R. Ghidoni and S. Sonnino, Tenth Meeting of the International Society for Neurochemistry, Supplement to J. Neurochem, 44.
27. S. Sonnino, R. Ghidoni, G. Gazzotti, G. Kirschner, G. Galli and G. Tettamanti, J. Lipid Res., 25:620-629 (1984).
28. G. Gazzotti, S. Sonnino, R. Ghidoni, G. Kirschner and G. Tettamanti, J. Neurosc. Res., 12:179-197 (1984).

29. G. Gazzotti, S. Sonnino, R. Ghidoni, P. Orlando and G. Tettamanti, Glycoconjugate J., 1:111-121 (1984).
30. G. Gazzotti, S. Sonnino and R. Ghidoni, J. Chromat., 315:395-400 (1984).
31. S. Nevenhofer, G. Schwarzmann, H. Egge and K. Sandoff, Biochemistry, 24:525-532 (1985).
32. S. Sonnino, G. Kirschner, R. Ghidoni, D. Acquotti and G. Tettamanti, J. Lipid Res., 26:248-257 (1985).
33. K. Obata, M. Olde and S. Handa, Nature, 266:359-371 (1977).
34. J.I. Morgan and N. Seifert, J. Supramol. Struct., 10:1111-124 (1979).
35. F.G. Roisen, H. Bartfeld, R. Nagele and G. Yorke, Science, 214:577-578 (1981).
36. A. Leon, L. Facci, G. Toffano, S. Sonnino and G. Tettamanti, J. Neurochem., 37:350-357 (1981).
37. L. Facci, A. Leon, G. Toffano, S. Sonnino, R. Ghidoni and G. Tettamanti, J. Neurochem., 42:299-305 (1984).
38. R. Ghidoni, S. Sonnino, V. Chigorno, B. Venerando and G. Tettamanti, Biochem. J., 213:321-329 (1983).
39. S. Souderfeld, E. Conzelman, G. Schwarzmann, J. Burg, U. Himerichs and K. Sandhoff, Eur. J. Biochem., 149:247-255 (1985).
40. E.M. Kolodny, R.O. Brady, J.M. Quirk, J. Kanfer, J. Lipid Res., 11:144-149 (1970).
41. J.F. Tallman, P.M. Fishman and R.C.Menneberry, Arch. Biochem. Biophys., 182:556-562 (1977).
42. Y. Suzuki and K. Suzuki, J. Lipid Res., 13:687-690 (1972).
43. P. Orlando, G. Cocciante, G. Ippolito, P. Massari, S. Roberti and G. Tettamanti, Pharmacol. Res. Comm., 11:759-773 (1979).
44. R. Ghidoni, G. Tettamanti and V. Zambotti, Biochem. Exp. Biol., 13:61-69 (1977).
45. A. Novak, J.A. Lowden, Y.L. Gravel and L.S. Wolfe, J. Lipid Res.,20:678-681 (1979).
46. K.C. Leskawa, S. Dasgupta, J.L. Chien and E.L. Hogan, Anal. Biochem., 140:172-177 (1984).
47. R.W. Veh, A.P. Corfield, M. Sander, R. Schauer, Biochim. Biophys. Acta, 486:145-160 (1977).
48. G. Schwarzmann, Biochim. Biophys. Acta, 529:106-114 (1978).
49. J. Schraven, C. Cap, G. Nowoczek, K. Sandhoff, Anal. Biochem., 78:333-339 (1977).
50. R. Ghidoni, S. Sonnino, M. Masserini, P. Orlando and G. Tettamanti, J. Lipid Res., 22:1286-1295 (1981).
51. T.A.W. Koerner, J.H. Prestegard, P.C. Demou and R.K. Yu, Biochemistry, 22:2676-2687 (1983).
52. D. Burn, V. Petron and G.O. Westan, Tetrahedron Lett., 9:14-15 (1960).
53. V. Chigorno, M. Pitto, G. Cardace, D. Acquotti, G. Kirschner, S. Sonnino, R. Ghidoni and G. Tettamanti, Glycoconjugate J., 2:279-292 (1985).

STRUCTURAL ASPECTS OF BLOOD GROUP GLYCOSPHINGOLIPIDS IN THE GASTROINTESTINAL TRACT

Gunnar C. Hansson

Department of Medical Biochemistry
University of Göteborg
P.O. Box 33031, S-400 33 Göteborg, Sweden

The epithelial cell surface of the small intestine is very rich in carbohydrates. These consist of loosely bound mucous glycoproteins and of lipid and protein linked oligosaccharides bound to the brush border membrane. Although the majority is protein-linked, the large total amount makes the small intestine a rich source of glycosphingolipids. A large proportion of these glycolipids contains fucose and expresses blood group activities. The glycosphingolipids of the small intestine of the dog were the first to be studied more thoroughly[1] and a blood group-related variation between different individuals was found. A blood group individuality expressed on glycosphingolipids had earlier been observed on erythrocytes for the ABH, Lewis, P and Ii systems in man (for example see refs. 2-4). A species specificity of blood groups had been known[3] and a comparison of glycosphingolipids from the small intestine of several mammalian species showed a remarkable variation in pattern.[5] This variation was largely due to blood group type glycosphingolipids, but a variation in gangliosides was also found.[6] The small intestinal glycosphingolipids of man, dog[2,7] and rat[8] have been thoroughly studied, but the structures found in the stomach have only been described in detail for pigs[2,9] and in the large intestine of rats[10].

This review article will discuss the blood group type glycosphingolipids of the small intestine from several mammalian species. Information from other parts of the gastrointestinal tract will be included when available. A special focus will be made on the expression of blood group type glycosphingolipids in different parts of the rat gastro-intestinal tract as this is interesting and the first more complete study available within one animal species.

BLOOD GROUP ANTIGENS

The antigenic determinants of the human blood groups ABH, Lewis, X-Y, P and Ii systems and combinations of these are known to be due to specific carbohydrate sequences.[2,4,11] The structures of most of these determinants are gathered in Table 1. The nomenclature is based on the human blood group system and will also be used for other animals. Not all of the structures shown have been identified in human tissues. The X and Y system has a fucose linked $\alpha 1 \rightarrow 3$ to a type 2 saccharide chain (Galß1→4GlcNAc) instead of a Fuc$\alpha 1 \rightarrow 4$ to a type 1 chain (Galß1→3GlcNAc) as in the analogous Le^a and Le^b system. The Lewis and X-Y systems can also be combined with the A and B blood groups. The different determinants of Table 1 may reside on different core saccharides of glycosphingolipids. Relevant structures for this review are gathered in Table 2, where the chain type designations are according to Hakomori.[12]

COMPARISON OF DIFFERENT ANIMAL SPECIES

In the following section a description is made of the blood group active glycosphingolipid expressed in various parts of the normal gastrointestinal tract. The description is focused on the epithelial cells of small intestine as most is known of that part. It is meant to be a guide to what kind of different blood group antigens and core structures can be expected to be found as glycosphingolipids in mammalian species. Table 3 shows a summary of blood group antigens and core structures expressed as glycosphingolipids in the small intestine. The structures described vary between individuals

Table 1. Structure of blood group antigen determinants discussed

Blood group system	Structure
ABH system	
H	Fucα1→2Galβ1→
A	GalNAcα1→3Gal(2←1αFuc)β1→
B	Galα1→3Gal(2←1αFuc)β1→
Lewis system	
Le^a	Galβ1→3GlcNAc(4←1αFuc)β1→
Le^b	Fucα1→2Galβ1→3GlcNAc(4←1αFuc)β1→
ALe^b	GalNAcα1→3Gal(2←1αFuc)β1→3GlcNAc(4←1αFuc)β1→
BLe^b	Galα1→3Gal(2←1αFuc)β1→3GlcNAc(4←1αFuc)β1→
X-Y system	
X	Galβ1→4GlcNAc(3←1αFuc)β1→
Y	Fucα1→2Galβ1→4GlcNAc(3←1αFuc)β1→
A-Y	GalNAcα1→3Gal(2←1αFuc)β1→4GlcNAc(3←1αFuc)β1→
B-Y	Galα1→3Gal(2←1αFuc)β1→4GlcNAc(3←1αFuc)β1→
P system	
P^k antigen	Galα1→4Galβ1→4Glcβ1→Cer
P antigen	GalNAcβ1→3Galα1→4Galβ1→4Glcβ1→Cer
Forssman antigen	GalNAcα1→3GalNAcβ1→3Galα1→4Galβ1→4Glcβ1→Cer
Ii system	
i antigen	Galβ1→4GlcNAcβ1→3Galβ1→4GlcNAcβ1→3Galβ1→
I antigen	Galβ1→4GlcNAcβ1→6 Galβ1→4GlcNAcβ1→3 Galβ1→

Table 2. Core structures of glycosphingolipids discussed.

Designation	Structure
Type 1	Galβ1→3GlcNAcβ1→3Galβ1→4Glcβ1→Cer
Type 2	Galβ1→4GlcNAcβ1→3Galβ1→4Glcβ1→Cer
Type 4	Galβ1→3GalNAcβ1→3Galα1→4Galβ1→4Glcβ1→Cer
Ganglio series	Galβ1→3GalNAcβ1→4Galβ1→4Glcβ1→Cer

of the same species and no information is provided as to which structures to expect to be coexpressed in the same individual. The description does not contain the full structures of all glycolipids described so far. Many more blood group active structures in the gastrointestinal tract will certainly be described in the coming years.

HOMO

Epithelial cells of small intestine. The blood group active glycosphingolipids occupy a high proportion of the epithelial cell glycosphingolipids in man[13,14] and are the major structures besides monoglycosylceramide and sulfatide. The fucolipids of man's small intestine can express blood group H, A and B antigens with 5 and 6 sugars having a lacto-tetraosylceramide core sequence (type 1).[15-17] Le^a pentaglycosylceramide and Le^b hexaglycosylceramide is expressed dependent on the Lewis and Secretor status of the individual.[18,19] Structures due to the combined action of the ABH, Secretor and Lewis genes are found in the appropriate individuals as ALe^b- and BLe^b- heptaglycosylceramide.[14,20]

Human meconium, the first stool after birth, was shown by Larson and Karlsson[21] to contain large quantities of human blood group active glycosphingolipids. The neutral glycosphingolipids are probably mostly derived from extruded epithelial cells of the small intestine.[21] The structures found in adult epithelial cells discussed above were also found in meconia[21-23] of individuals with appropriate blood groups. The blood group determinants found in small intestine of man are gathered in Table 3.

Table 3. Blood group type antigens and core structures found in the glycosphingolipids of the epithelial cells and nonepithelial portion of the small intestine of 7 species

Animal	Epithelial Cells		Non-epithelial portion		References
	Blood group antigen	Core structure	Blood group antigen	Core structure	
Homo	H, A, B,	Type 1	P		13-20
	Le^a, Le^b, ALe^b, BLe^b	Type 1			
Rat	H, A	Lactose	P, isogloboside		8,48,52-56
	H, A	Type 1	H, B	Ganglio	
	H, A	Type 2, branched			
Mouse	H	Ganglio	Forssman		5,30-32
Dog	H, A, X, Y, A-Y	Type 2	Forssman		1,7,16,17,19,33,34
Rabbit	H, A, B, Y, B-Y	Type 2			5,35,36
Cat	A	Type 2	P, Forssman		5,37
Pig	H	Type 1			39-41

So far, no glycosphingolipid with 1→6 linked branches has been described in the human small intestine. Instead, longer straight-chain glycolipids have been found in both meconia and epithelial cells.[14,24] These glycolipids have both type 1 and type 2 determinants at their outer ends as revealed by the detection of Le^a, Le^b, X, and Y epitopes on immunostained thin-layer plates.[14] The type of chain closest to the ceramide in glycolipids with 6-sugar cores is not finally settled, but preliminary results from glycolipids in meconia propose that it can be a type 1 chain.[24] This suggests that type 1 chains can be repetitively elongated in the intestine. However, Kannagi et al. have claimed that type 1 chains cannot be repetitively elongated in contrast to type 2 chains.[25] Whether type 2 chains exist in longer glycosphingolipids of the human small intestine as found in the rat is not known.

Stomach and large intestine. Little information is available on these two parts, except for some very preliminary studies on the stomach.[26] Several studies have been devoted to adenocarcinoma cell lines derived from the large intestine.[2,27-29] A common finding is X and Y structures having repeated type 2 sequences with up to 3 fucoses.[27,28] Some of these structures have been found in normal epithelial cells.

MOUSE

Epithelial cells of small intestine. The glycolipids expressed in the mouse small intestine belong mainly to the ganglioseries of glycolipids. A major glycolipid in the epithelial cells is gangliotetraosylceramide[5,30-32] which is also fucosylated to give a blood group H similar compound.[30]

DOG

Epithelial cells of small intestine. The blood group active glycosphingolipids of the dog's small intestine have been thoroughly studied by McKibbin et al.,[1,7] and shown to be based only on type 2 sequences. Blood group H, A, X and Y structures have been described[16,17,19,33] as well as hybrids of the A and Y antigen.[34] Different patterns of these structures are found in different individuals.[7,19]

RABBIT

Epithelial cells of small intestine. The blood group active glycosphingolipids from the rabbit's small intestine have been deduced from analyses by mass spectrometry and immunological techniques on mixtures of glycolipids.[5,35] Type 2 chains are core saccharides as in the dog, which have also been confirmed by preparing and analysing the blood group H pentaglycosylceramide and B-Y heptaglycosylceramide (unpublished). Rabbits are known to be either blood group A-positive or A-negative[36] and the individual analysed by us[35] contained a blood group A active hexaglycosylceramide.

CAT

Small intestine. An individual cat small intestine was shown by mass spectrometry and immunological techniques to contain blood group A active components, probably based on type 2 sequences.[5,37] Indications were obtained for a 6-sugar structure without an internal hexosamine and with an A-type terminus, earlier described in hog gastric mucose.[38]

PIG

Epithelial cells of small intestine. The blood group active glycosphingolipids of porcine intestine have not been thoroughly described, but have been shown to contain a blood group H pentaglycosylceramide with a type 1 chain.[39,40] Other fucolipids, probably also being blood group active, have been indicated[41] and different blood group patterns of different individuals have been shown (B.E. Samuelsson, personal communication).

Gastric mucosa. In a series of papers, Slomiany et al. have shown both blood group H and A structures based on type 1 and type 2 chains.[38-44] These structures have been reviewed,[2,9] of which some are large, and proposed to contain up to 4 branches.[45-47]

Nonepithelial supporting stroma of small intestine

Most of the studies on the small intestine have been done on the whole tissue. After separation of epithelial cells from supportive stroma to give a nonepithelial residue, it was evident that there was a specific distribution of glycosphingolipids between these two compartments.[8,13,14,31,32] Globoside (blood group P and the basis for the type 4 series) was found in the nonepithelial compartment of man and rat (Table 3) and the Forssman antigen (Table 1) was nonepithelial in mouse, dog and cat.[5,31,32,37] The distribution in cat was implied by the ceramide composition which is usually less hydroxylated, i.e. containing less of hydroxy fatty acids and phytosphingosine in the nonepithelial tissue compared to the epithelial cells.

DISTRIBUTION OF BLOOD GROUP ANTIGENS ALONG THE RAT GASTROINTESTINAL TRACT

Nonepithelial supporting stroma of stomach, and small and large intestine

The nonepithelial residue after removal of epithelial cells in the rat small intestine is characterized by the presence of a series of structures having a terminal N-acetyl-galactosamine linked ß1→3 (Table 4B). The major compounds were globoside and isogloboside,[8,48] but a whole series of structures with up to 9 sugars were also found. Figure 1 shows an approximate conformation of the 9-sugar compound with the repeated Galα1→3Gal sequences lying in a rather straight chain, with every sugar turned 180°.

The nonepithelial residue contained a blood group B active hexaglycosylceramide based on a gangliotetraosylceramide core as preliminarily reported.[8] This structure is identical to that recently found in rat bone marrow cells.[49] Indications of a fucosylated gangliotetraosylceramide, that could be a H-type precursor, were also obtained during preparation of the B-active component.

Table 4A. Glycosphingolipids in Epithelial Cells of the White Rat Small Intestine.

Glycolipid Structures	Estimated amount (μg/animal)
Oligohexosylceramides	
Glcβ1→1Cer	880
Galβ1→4Glcβ1→1Cer	5
Galα1→4Galβ1→4Glcβ1→1Cer	410
Galα1→3Galα1→4Galβ1→4Glcβ1→1Cer	50
Galα1→3Galα1→3Galα1→4Galβ1→	
4Glcβ1→1Cer	< 2
Galα1→3Galα1→3Galα1→3Galα1→	
4Galβ1→4Glcβ1→1Cer	< 3
Fucolipids	
Fucα1→2Galβ1→4Glcβ1→1Cer	25
Fucα1→2Galα1→3Galα1→4Galβ1→	
4Glcβ1→1Cer	15
Fucα1→Hex→Hex→Hex→Hex→Hex→Cer	< 2
Fucα1→Hex→Hex→Hex→Hex→Hex→Hex→Cer	< 3
Fucα1→2Galβ1→3GlcNAcβ1→3Galβ1→	
4Glcβ1→1Cer	115
Fuc→Hex→HexN→Hex→Hex→Hex→Cer	< 3
Fuc→Hex→(HexN-Hex)→Hex→Hex→	
Hex→Cer?	< 4
Fucα1→2Galβ1→3GlcNAc	
↓ β1,6	
Galβ1→3GlcNAcβ1→3Galβ1→4Glcβ1→1Cer	50
↑ β1,3	
Fucα1→2Galβ1→3GlcNAc	
Fucα1→2Galβ1→4GlcNAc	
↓ β1,6	
Galβ1→3GlcNAcβ1→3Galβ1→4Glcβ1→1Cer	30
↑ β1,3	
Fucα1→2Galβ1→3GlcNAc	
Miscellaneous	
GlcNAc→Hex→Hex→Hex→Cer	< 3

Table 4B. Glycosphingolipids in Nonepithelial Supporting Stroma of the White Rat Small Intestine.

Glycolipid Structures	Estimated amount (μg/animal)
Oligohexosylceramides	
Glcß1→1Cer	170
Galß1→4Glcß1→1Cer	70
Galα1→4Galß1→4Glcß1→1Cer	50
Galα1→3Galß1→4Glcß1→1Cer	10
Galα1→3Galα1→4Galß1→4Glcß1→1Cer	80
Galα1→3Galα1→3Galα1→4Galß1→4Glcß1→1Cer	40
Galα1→3Galα1→3Galα1→3Galα1→ 4Galß1→4Glcß1→1Cer	25
Galα1→3Galα1→3Galα1→3Galα1→ 3Galα1→4Galß1→4Glcß1→1Cer	20
Galα1→3Galα1→3Galα1→3Galα1→ 3Galα1→3Galα1→4Galß1→4Glcß1→1Cer	< 3
Fucolipids	
Gal ↓ α1,3 Galß1→3GalNAcß1→4Galß1→4Glcß1→1Cer ↑ α1,2 Fuc	5
With Terminal GalNAc	
GalNAcß1→3Galα1→4Galß1→4Glcß1→1Cer	110
GalNAcß1→3Galα1→3Galß1→4Glcß1→1Cer	210
GalNAcß1→3Galα1→3Galα1→4Galß1→ 4Glcß1→1Cer	20
GalNAcß1→3Galα1→3Galα1→3Galα1→ 4Galß1→4Glcß1→1Cer	20
GalNAcß1→3Galα1→3Galα1→3Galα1→ 3Galα1→4Galß1→4Glcß1→1Cer	20
GalNAcß1→3Galα1→3Galα1→3Galα1→ 3Galα1→3Galα1→4Galß1→4Glcß1→1Cer	< 2
GalNAcß1→3Galα1→3Galα1→3Galα1→ 3Galα1→3Galα1→3Galα1→4Galß1→4Glcß1→1Cer	< 2
Miscellaneous	
Galß1→3GalNAcß1→4Galß1→4Glcß1→1Cer	10

Table 4C. Sialic Acid Containing Glycosphingolipids in Epithelial and Nonepithelial Supporting Stroma of the White Rat Small Intestine.

Glycolipid Structures	Estimated amount (μg/animal)
In Epithelial Cells	
NeuGc2→3Galß1→4Glcß1→1Cer	640
In Nonepithelial Cells	
NeuAc2→3Galß1→4Glcß1→1Cer	180
NeuGc2→3Galß1→4Glcß1→1Cer	25
NeuAc2→3Galß1→3GalNAcß1→4Galß1→4Glcß1→1Cer	15
NeuAc2→3Galß1→3GalNAcß1→ 3Galα1→3Galß1→4Glcß1→1Cer	30
Galß1→3GalNAcß1→4Galß1→4Glcß1→1Cer ↑ 2,3 NeuGc	15

Table 4D. Additional Glycosphingolipids Found in the Epithelial Cells of the Black-White Rat and Not in the White Rat Small Intestine

Glycolipid Structures	Estimated amount (μg/animal)
GalNAcα1→3[Fucα1→2]Galß1→4Glcß1→1Cer	90
GalNAcα1→3[Fucα1→2]Galß1→3GlcNAcß1→3Galß1→ 4Glcß1→1Cer	290
GalNAcα1→3[Fucα1→2]Galß1→3GlcNAc ↓ ß1,6 Galß1→3GlcNAcß1→ ↑ ß1,3 GalNAcα1→3[Fucα1→2]Galß1→3GlcNAc 3Galß1→4Glcß1→1Cer	60
GalNAcα1→3[Fucα1→2]Galß1→4GlcNAc ↓ ß1,6 Galß1→3GlcNAcß1→ ↑ ß1→3 GalNAcα1→3[Fucα1→2]Galß1→3GlcNAc 3Galß1→4Glcß1→1Cer	40

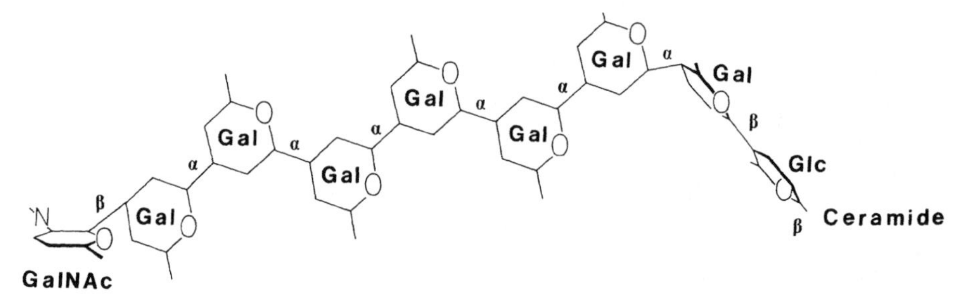

Fig. 1. Schematic formula showing a possible conformation of the 9-sugar glycosphingolipid with a globoside like terminus from the nonepithelial portion of the rat small intestine.[48]

Fig. 2 shows a thin-layer chromatogram of the neutral glycosphingolipids from the residue of stomach, small and large intestine. The epithelial cells of small and large intestine were removed by sequential EDTA-buffer incubations[8,50] and the residue was therefore relatively pure from epithelial cells. The epithelial cells of the glandular portion of the stomach were scraped off and some residual epithelial cells remained. When a monoclonal blood group B antibody was used to detect B-active components on the thin-layer plate, it labelled a faint band migrating as the purified blood group B-gangliohexaosylceramide, as indicated by the arrow in Fig. 2B. The other labelled bands were due to glycolipids from contaminating epithelial cells.

The patterns of neutral glycolipids and gangliosides of the nonepithelial residue of the different parts of the gastrointestinal tract were almost identical. Most of the structures shown in Table 4B and 4C for the small intestine are probably also present in the other parts. Part of the complexity of the glycolipids in the nonepithelial residue is probably due to the different kinds of cells in this tissue, e.g. smooth muscle, connective tissue, blood vessels, nerves, and lymphoid cells. Using specific antibodies we have, for example, shown that the blood group B glycolipid of the residue is found in nerve cells (C. Ernst and G.C. Hansson, unpublished).

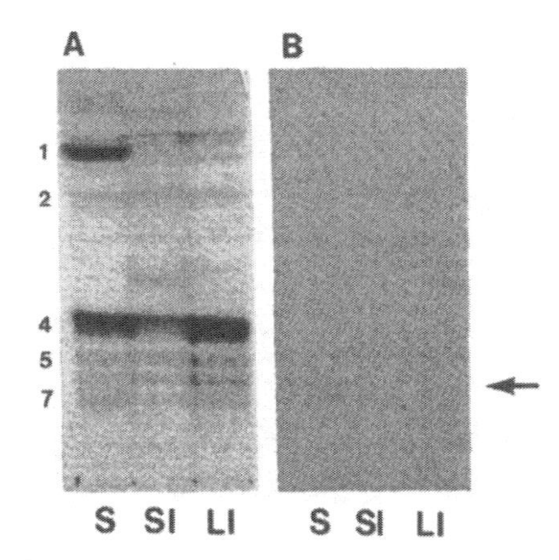

Fig. 2 Thin-layer chromatogram of the total nonepithelial glycosphingolipids from the stomach (glandular portion) marked S, small intestine (SI) and large intestine (LI) of the white rat strain. The chromatogram to the left (A) was chemically stained(the anisaldehyde reagent) and that to the right (B) is the autoradiogram obtained after labelling the separated glycolipids on the thin-layer plate by a monoclonal anti B antibody (DAKOpatts) followed by an ^{125}I-labelled anti mouse antibody.[73,74] Chloroform/methanol/water, 60:35:8 (by volume) was used for the chromatography. Numbers of sugars in the glycolipids are shown to the left and the arrow shows the migration of the blood group B-gangliohexaosyleramide.

Epithelial cells of small intestine

Forstner and Wherrett got indication of blood group type glycosphingolipids of rat small intestine mucosa by showing the presence of fucolipids.[51] Later a series of studies of the glycosphingolipids in the rat small intestine have been published.[8,48,52-56] The structures found are gathered in Table 4A. Thirteen of the structures described express blood group A and H activities. However, we have preliminary evidence of almost as many additional A and H structures.

In one of the rat strains (white) several types of blood group H active glycosphingolipids were expressed. The simplest contained only 3 sugars and had fucose added in an $\alpha1\rightarrow2$ linkage to lactosylceramide.[52] Other blood group H active structures without internal hexosamine were found to be based on the series of oligohexosylceramides having repeated Gal$\alpha1\rightarrow3$Gal sequences.[8,55] The structure of a pentaglycosylceramide was determined, whereas the structures of the hexa- and heptaglycosylceramides were proposed from the structures of the probable precursor penta- and hexaglycosylceramides[8,55] (see Table 4). Most of the lactoseries glycosphingolipids had type 1 chains as, for example, the major blood group H structure which was fucosyllactotetraosylceramide. Two different branched decaglycosylceramides[53] also had a type 1 chain in the common core and on its 3-linked branch. One of these also had a type 1 chain on the 6-linked branch, but the minor of these had a type 2 chain on the 6-linked branch. Two additional fucosylated glycosphingolipids with internal N-acetylhexosamine had a sequence deduced by mass spectrometry, as shown in Table 4A. These may also express blood group H activity.

DISTRIBUTION OF BLOOD GROUP B ANTIGENS IN THE EPITHELIAL CELLS

Fig. 3A shows the pattern of neutral and acid glycosphingolipids from the epithelial cells of stomach, small and large intestine from the white rat strain. Already the patterns revealed by chemical staining show remarkable differences. Immunostaining of the glycolipids on the thin-layer plate (Fig.3, B to D) further strengthens this impression. The distribution of A and B antigens are shown in Table 5.

Table 5. Blood group A and B reactivities of the glycosphingolipids in the epithelial cells of the stomach, small and large intestine of two different rat strains.

Rat strains[a]	Fore stomach	Stomach (Glandular portion)	Small intestine	Large intestine
White Strain	-	-	-	A
(Sprague Dawley)	-	B	-	B
Black-White	-	-	A	A
Strain	-	B	-	B

[a]These two rat strains have been inbred at the University of Gothenburg for more than 20 years.

Stomach. At least 3 blood group B active glycosphingolipids were detected in the glandular portion of the stomach (Fig. 3B). A blood group B active ganglioside was the major B active component. The structure of this ganglioside was determined by Drs. Bouhours and the author to be the ganglioside GM1 with a B-terminal.[57] This structure is identical to the B active ganglioside described by Holmes et al.[58] The core structures of the different blood group antigens are gathered in Table 6.

The neutral B-components have been characterized together with Dr. J.-F. Bouhours. The faster migrating compound in Fig. 3B was shown to be a blood group B terminal on gangliotetraosylceramide.[59] This structure was identical to the one found in the nonepithelial residue. An identical structure from rat bone marrow has been described.[49] The slower migrating band is a recently characterized heptaglycosylceramide, with a blood group B terminal on an isoglobotetraosylceramide core as shown in Table 6.[59] This is a novel type of blood group active glycosphingolipid.

Small intestine. No blood group B active glycosphingolipids have been found.

Large intestine. The blood group B active components of the large intestine have been described.[10] Fig. 3B shows the labelling of the blood group B active tetraglycosylceramide based on lactosylceramide.[60] The six sugar glycolipid stained is the B active hexaglycosylceramide based on a type 1 chain.[10,61] Recently the 7 sugar glycolipid has been purified and characterized.[61] This is only weakly stained with the monoclonal antibody used in Fig. 3B, but intensively stained with other anti blood group B monoclonals.[10] The structure was interestingly shown to be a difucosylated compound having a neolactotetraosylceramide core, i.e. a type 2 chain (Table 6). Thus the epithelial cells of the large intestine has a mono-fucosylated blood group B determinant on a type 1 structure and a difucosylated B determinant on a type 2 structure, i.e. with an extra fucose added to the N-acetylglucosamine.

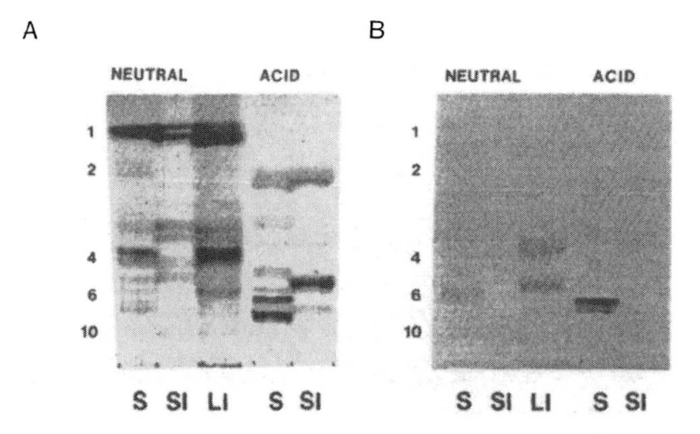

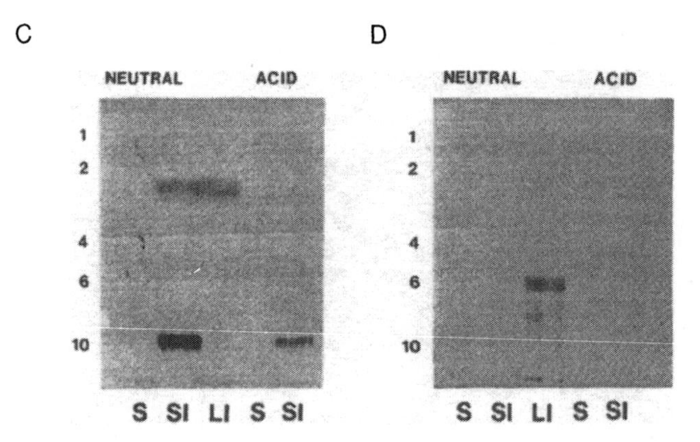

Fig. 3. Thin-layer chromatogram of the total neutral (left) and acid (right) glycosphingolipids of stomach (glandular portion) marked S, small intestine (SI)and large intestine (LI) of the white rat strain. The acid glycosphingolipids of the large intestine were omitted due to very small amounts present.[10] A was labelled with the anisaldehyde reagent. B-D are the autoradiograms obtained after labelling the separated glycosphingolipids on the thin-layer plate by a monoclonal anti-B antibody (DAKOpatts) in B, a monoclonal anti-H type 2 antibody (DAKOpatts) in C and by a monoclonal anti-A antibody (DAKOpatts) in D, followed by an ^{125}I-labelled anti-mouse antibody.[73,74] Chloroform/methanol/water, 60:35:8 (by volume) was used for the chromatography. Numbers of sugars are shown to the left.

DISTRIBUTION OF BLOOD GROUP H ANTIGENS IN THE EPITHELIAL CELLS

Figure 3C shows the immunostaining of the glycolipids with an anti blood group H type 2 antibody. This shows the H-type tri- and decaglycosylceramides in the small intestine but, for example, the abundant fucosylated lactotetraosylceramide was not labelled with this antibody (Table 4). The large intestine also contained fucosyllactosylceramide, but the fucosylated lactotetraosylceramide has not been found. The stomach lacked all lactoseries-based blood group H structures, but contained fucosylated GM1.[57]

DISTRIBUTION OF BLOOD GROUP A ANTIGENS IN THE EPITHELIAL CELLS

Stomach and small intestine. Both these parts of the gastrointestinal tract lacked blood group A active glycosphingolipids in the white rat strain.

Large intestine. This part contained several blood group A active components as shown in Fig. 3D. The monoclonal antibody used stains the blood group A active tetraglycosylceramide (Table 6) only faintly, as this antibody reacts only weakly with this structure (unpublished). The labelled hexaglycosylceramide has tentatively been identified as a blood group A structure with a type 1 chain.[10,61] These cells also contained a difucosyl 7-sugar glycolipid as shown earlier.[10] Difucosyl blood group A structures are not labelled with this antibody. The chemical characterization of this component has not yet been completed, as it exists in very small quantities. It is proposed to be analogous to the B-active difucosylheptaglycosylceramide with a type 2 core saccharide. The immunostaining of Fig. 3D indicates additional blood group A active structures in the large intestine.

A SECOND RAT STRAIN WITH BLOOD GROUP A ACTIVE GLYCOSPHINGOLIPIDS IN THE SMALL INTESTINE

Fig. 4 shows the chemically stained glycosphingolipids prepared from the epithelial cells of the small intestine from

Table 6. Core structures of the blood group A, B and H glycosphingolipids found in the epithelial cells of stomach, small and large intestine in the rat. A, B and H refer to the determinants as shown in Table 1 except that the common Gal is included also in the core structure. Lac means lactose (Galβ1→ 4Glc)

Part	Activity	Core structure
Stomach (Glandular portion)	B→	Galβ1→3GalNAcβ1→4Lacβ1→Cer (Gangliotetraosylceramide)
	B→	Galβ1→3GalNAcβ1→3Galα1→3Lacβ1→Cer (Isoglobotetraosylceramide)
	B→	Galβ1→3GalNAcβ1→4Lacβ1→Cer NeuAcα2→3 (to Lac) (GM_1)
Small Intestine	H→ A[a]→	Lacβ1→Cer (Lactosylceramide)
	H→ A[a]→	Galβ1→3GlcNAcβ1→3Lacβ1→Cer (Lactotetraosylceramide, Type 1)
	H[b]→ A[a]→	Galβ1→3GlcNAcβ1→6 Galβ1→3GlcNAcβ1→3Lacβ1→Cer Galβ1→3GlcNAcβ1→3 (Type 1, 1 and 1)
	H[b]→ A[a]→	Galβ1→4GlcNAcβ1→6 Galβ1→3GlcNAcβ1→3Lacβ1→Cer Galβ1→3GlcNAcβ1→3 (Type 2, 1 and 1)
Large Intestine	H→ A→ B→	Lacβ1→Cer (Lactosylceramide)
	A→ B→	Galβ1→3GlcNAcβ1→3Lacβ1→Cer (Lactotetraosylceramide, Type 1)
	A→ B→	Galβ1→4GlcNAcβ1→3Lacβ1→Cer Fucα2→3 (to GlcNAc) (X-pentaglycosylceramide, Type 2)

[a]Only found in the black-white rat strain.

[b]Only found in the white rat strain.

white rat strain has already been discussed. The second, "black-white" strain, differs from the white strain due to blood group A glycosphingolipids, which is lacking in the white strain. The additional four major components have been characterized[54,56] and are shown in Table 4D. These are the expected products from an α1→3N-acetylgalactosaminyl-transferase catalyzing the extension of the blood group H structures discussed for the white rat. The smallest was a tetraglycosylceramide with a lactosylceramide core. The hexa-glycosylceramide has a type 1 saccharide core.[54] There were two dodecaglycosylceramides[56] and their structures, together with molecular models are shown in Fig. 5. The difference between these two was identical to the one found between the blood group H decaglycosylceramides of the white rat strain previously discussed. The precursor blood group H structures with 3 and 5 sugars were also found in this strain, but the 10 sugar structures were lacking.[8]

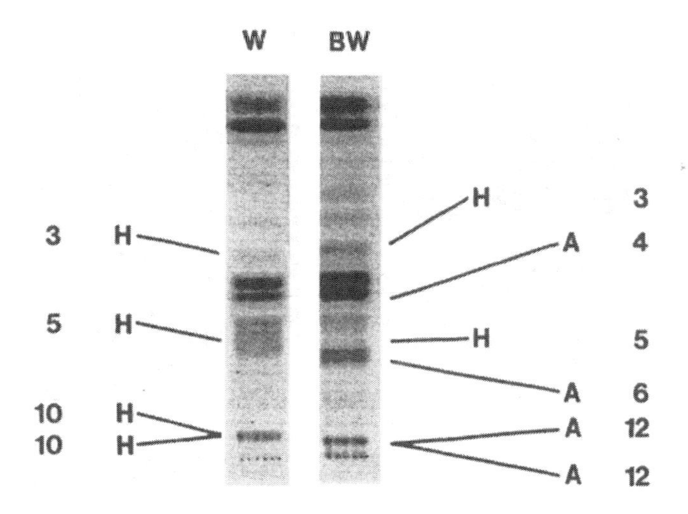

Fig. 4. Thin-layer chromatogram of the glycosphingolipids from the epithelial cells of the small intestine of the white (W) and black-white (BW) strain. The major H and A active glycosphingolipids in the two strains are indicated to the sides. The number of sugars in the bands are shown to the left and right. The plate was chemically stained after development in chloroform/methanol/water, 60:35:8 (by volume).

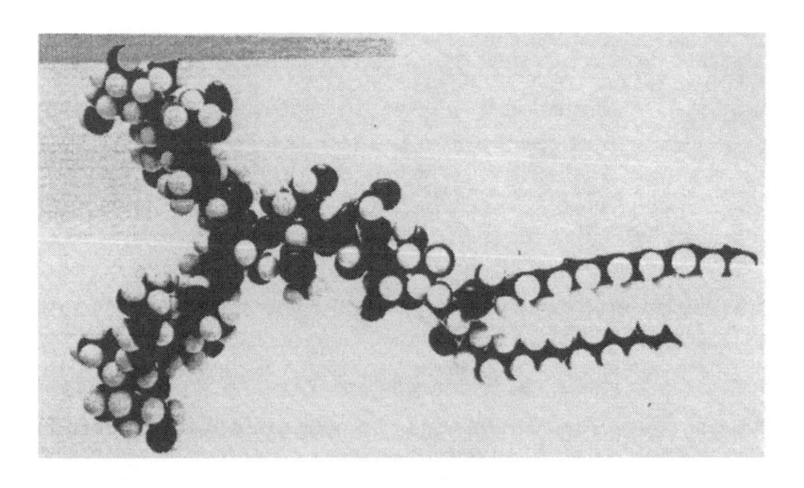

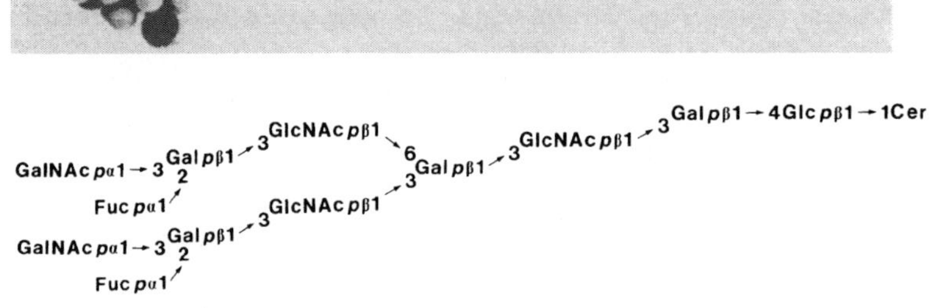

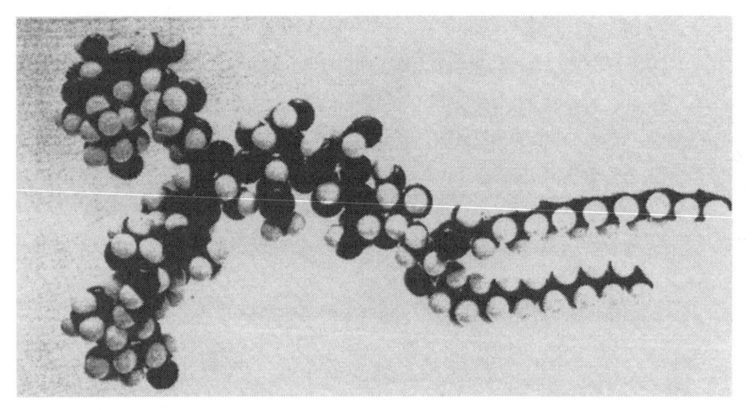

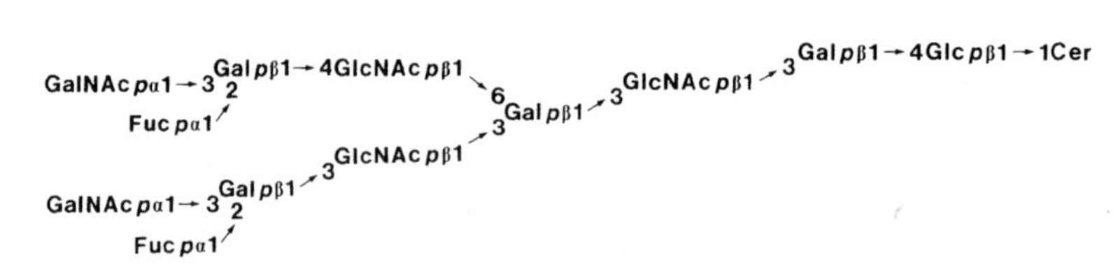

Fig. 5. Saccharide formulas and molecular models of the two blood group A active dodecaglycosylceramides from the epithelial cells of the small intestine in the black-white rat.[56] 60% was as the top structure with type 1 chains at all three positions and 40% as the bottom structure with a type 2 chain on the 6-linked branch.

The black-white strain lacked blood group A glycosphingolipids in the stomach(unpublished). The large intestine expressed identical blood group A structures to the white rat strain.

REGULATION OF GLYCOSPHINGOLIPID EXPRESSION IN THE GASTROINTESTINAL TRACT OF RAT

The ABH blood group antigens expressed on glycosphingolipids show a remarkable variation along the gastrointestinal tract of the rat. The expression of these antigens with different types of core saccharides is due to the combined result of a series of glycosyltransferases. From the result shown it is evident that several of these are independently regulated. The stomach is programmed for the ganglioseries structures, and the presence of both an α-Fuc1→2-transferase and an α-Gal1→3-transferase can convert this core sequence to blood group B active components with or without sialic acid. The large intestine is also expressing transferases for making blood group B components. The epithelial cells do not express any ganglioseries based structures, but instead B-active components based on lactosylceramide and lactotetraosylceramide (type 1). Thus, these epithelial cells also must contain an α-fucosyltransferase and an α-galactosyltransferase. We could early show the presence of an additional blood group B active component with 7 sugars and two fucoses, of which one was linked to the N-acetylhexosamine. Recently, we have been able to show that this component has a type 2 chain. Thus, this is not compatible with the B-hexaglycosylceramide (type 1) being an immediate precursor. Blood group A and B type 1 difucosyl compounds have been proposed to be synthesized from blood group A and B monofucosyl structures by the Lewis enzyme.[11] The type 2 isomers are probably similarly made by the X-enzyme, as this sequence has been shown for the biosynthesis of a blood group A heptaglycosylceramide with a type 2 chain.[62] The blood group B hexaglycosylceramide (type 1) in rat large intestine cannot be converted to a difucosyl compound, which is consistent with the lack of Lewis-type structures in animals other than primates. The biosynthetic pathway for the blood group B heptaglycosylceramide with a type 2 core should, according to proposed sequences, be the addition of a Fucα1→3 to a blood group B type 2 hexaglycosylceramide. This structure and any type 2 sequence has not been found so far in the rat large intestine. At the moment it is not known if a probable N-acetylglucosamine α1→3-fucosyltransferase is so effective that no precursor compounds are found or whether the B-heptaglycosylceramide is made by another as yet unresolved pathway.

The nonepithelial residue after removal of the epithelial cells also expressed a blood group B active glycosphingolipid with a ganglioseries core, identical to the structure found in the glandular portion of the stomach. Preliminary observations show that this is localized to nerve cells, which is known to be rich in gangliosides belonging to the ganglioseries.[63]

The small intestine is the only tissue in the gastrointestinal tract to have a galactose ß1→6-N-acetylglucosaminyltransferase working on glycosphingolipids to give branched structures. The small intestine must also contain galactosyltransferases for making both type 1 and type 2 sequences. However, the ß1→4-galactosyltransferase does only seem to work on GlcNAcß1→6Gal structures, whereas the ß1→3-galactosyltransferase works on both GlcNAcß1→3Gal and GlcNAcß1→6Gal sequences.

All rat strains studied so far express blood group A glycosphingolipids in the large intestine. However, only some express it in the epithelial cells of the small intestine. To further understand the expression of blood group A antigens in the small intestine, controlled breedings were performed with the blood group A (black-white) and non-A (white) strains. The results of these experiments are presented in Table 7 and show that the first generation (F1) of breeding black-white rats with white rats gave only blood group A offsprings. When these (F1) were further bred with the white strain, 3 out of 7 were expressing blood group A in the small intestine and 4 out of 7 were were not expressing. These results propose that the blood group A activity and thus the expression of a α1→3-galactosaminyltransferase were inherited as a Mendelian

dominant trait. The blood group A activity was independent of sex and colour of the animals.

With the presence of blood group A antigens in the large intestine of all strains and in the small intestine of some strains, the relation between the α1→3-N-acetylgalactosaminyltransferase in the epithelial cells of these two compartments is not understood at the moment. Several possibilities exist: One is that the rat has two enzymes independently regulated and separately expressed in the small and large intestine. Another possibility is that the same gene is under different control in the two compartments in such a way that its expression can be inherited independently. A third and more speculative possibility is the presence or absence of modulatory proteins, independently inherited in the small and large intestine, needed for activation or inactivation of the same α1→3-N-acetylgalactosaminyltransferase. The rat is an interesting model system for the regulation of the expression of glycosyltransferases in general and how the different enzymes are regulated in the gastrointestinal tract. Experiments to further explore this model system are under way and a clarification should help in understanding some of the mysteries of the tissue specific expression of glycosphingolipids.

Other systems for studying the regulation of glycosphingolipid expression in the gastrointestinal tract are the epithelial cell differentiation during migration from crypt bottom to villus tip and the development during the embryo and fetal periods. The former system of epithelial differentiation did not show any major differences in blood group type determinants of glycosphingolipids in crypt versus villus cells.[64,65,50] However, the studies by Bouhours and Bouhours on the development of the blood group H antigens of the epithelial cells showed impressive alterations.[66,67] A further insight in the regulation of the induction of different glycosyltransferases during the first week of life of the rat should be very interesting. Umesaki et al. showed that the conventionalization of germ-free mice induced or only increased the amount of α-fucosyltransferase.[30,68] This gave a transient increase in the amount of fucosylgangliotetraosylceramide.[30] Whether this is a specific process or only the result of an altered turnover of the epithelial cells by conventionalization is not known.

The biochemistry of glycosphingolipids has been studied for many years, but more precise physiological functions are still mysteries. An interesting aspect on glycolipids in tissues, such as the gastrointestinal tract, is their potential action as receptors for toxins and for indigenous and pathogenic bacteria,[69-72] (also article by Bock *et al.* in this book).

Table 7.

Breeding of white (W) and black-white (BW) rat and expression of blood group A glycosphingolipids in the small intestine.

Rat type	Blood group A
W	none
BW	all
F_1 (BW x W)	7/7
F_1 x W	3/7

Other approaches to glycolipid function and also the function of other types of glycoconjugates in the gastrointestinal tract, is to better know the distribution of saccharide sequences and antigenic determinants in both lipid and protein-linked forms of the brush border membrane together with the mucin-linked oligosaccharides from the same tissue compartment. This could also give additional information on how the biosynthetic machinery of the cells are handling different types of glycoconjugates. The techniques of molecular biology will give more final answers to the molecular basis of the regulation of the variable expression of glycosphingolipids in different cells of the gastrointestinal tract.

SUMMARY

The epithelial cells of the gastrointestinal tract from different species show a very variable expression of blood group active glycosphingolipids. The core saccharide sequences are typical for the species as, for example, type 1 chains (Galß1→3GlcNAc) are found in the small intestine of man, rat, and pig and type 2 chains (Galß1→4GlcNAc) are found in the small intestine of dog, rabbit, and cat. The mouse is atypical with the ganglioseries as the major core saccharide of the small intestine. Blood group A determinants can be found in the small intestine of man, rat, dog, rabbit, and cat, and the blood group B determinant in man and rabbit.

Studies on the blood group active glycosphingolipids along the gastrointestinal tract of rat have revealed a complex distribution. The glandular cells of the stomach and epithelial cells of the large intestine express blood group B active glycosphingolipids. The cores of these are the ganglioseries, and the isogloboseries in the stomach and the lacto- (type 1) and neolactoseries (type 2) in the large intestine. The type 2 component is only expressed as a difucosyl and the type 1 as a monofucosyl compound. The epithelial cells of the small intestine are devoid of blood group B glycolipids, but express blood group H structures of which some has a branched core saccharide. One rat strain is lacking blood group A structures in the small intestine, but another is converting the H precursors to blood group A compounds. Both these strains always express blood group A structures in the large intestine. The expression of blood group A glycosphingolipids in the small intestine is inherited as a dominant trait.

ACKNOWLEDGEMENTS

Parts of the experiments have been done together with Drs. J. Ångström, J.-F. Bouhours, M.E. Breimer, P. Falk, J. Holgersson, K.-A. Karlsson, H. Leffler, N. Strömberg, and J. Thurin and were supported by grants No. 3967 and 7461 from the Swedish Medical Research Council. Excellent secretarial assistance was provided by I. Loubier and B. Thylén. This review was supported by the Swedish Medical Research Council (No. 7461) and by the Swedish Society of Medicine.

REFERENCES

1. J.M. McKibbin. (1969) The composition of the glycolipids in dog intestine, Biochemistry 8:679-685.
2. J.N. Kanfer, and S.-i. Hakomori. (1983) Sphingolipid Biochemistry, Plenum Press, New York.
3. T. Yamakawa, and Y. Nagai. (1978) Glycolipids at the cell surface and their biological functions. Trends in Biol. Sci. 3:128-131.
4. R.R. Race, and R. Sanger. (1975) Blood Groups in Man, Blackwell, Oxford.
5. M.E. Breimer, G.C. Hansson, K.-A. Karlsson, and H. Leffler. (1981) Blood group type glycosphingolipids from the small intestine of different animals analysed by mass spectrometry and thin-layer chromatography. A note on species diversity. J. Biochem. 90:589-609.
6. M.E. Breimer, G.C. Hansson, K.-A. Karlsson, and H. Leffler. (1983). The preparative separation of sialic acid containing lipids from sulphate group-containing lipids from small intestine of different animals, analysis by thin-layer chromatography and detection of novel species. J. Biochem 93:1471-1485.
7. J.M. McKibbin, Fucolipids. (1978) J. Lipid Res. 19:131-147.
8. M.E. Breimer, G.C. Hansson, K.-A. Karlsson, and H. Leffler. (1982) Glycosphingolipids of rat tissues. Different composition of epithelial and nonepithelial cells of small intestine. J. Biol. Chem. 257:557-568.
9. B.L. Slomiany, and A. Slomiany. Glycosphingolipids and their role in the gastrointestinal tract. (1977) Prog. Gastroent. 3:349-371.
10. G.C. Hansson, K.-A. Karlsson, and J. Thurin. (1984) Glycosphingolipid patterns of the epithelial and non-epithelial compartments of rat large intestine. Biochim. Biophys. Acta 792:281-292.
11. W.M. Watkins. (1978). Genetics and biochemistry of some human blood groups. Proc. R. Soc. Lond. B. 202:31-53.
12. H. Clausen, S.B. Levery, E. Nudelman, S. Tsuchiya, and S.-i. Hakmomori. (1985). Repetitive A epitope (type 3 chain A) defined by blood group A_1-specific monoclonal antibody TH-1: Chemical basis of qualitative A_1 and A_2 distinction. Proc. Natl. Acad. Sci. USA 82:1199-1203.
13. K.-E. Falk, K.-A. Karlsson, H. Leffler, and B.E. Samuelsson. (1979) Specific pattern of glycosphingolipids enriched in a mucosa scraping of human small intestine. FEBS Lett. 101:273-276.
14. S. Bjork, M.E. Breimer, G.C. Hansson, K.-A. Karlsson, and H. Leffler. (1987) Structures of blood group glycosphingolipids of human small intestine. A relation between the expression of fucolipids of epithelial cells and the ABO, Le and Se phenotype of the donor. J. Biol. Chem. 262:6758-6765.
15. M.E. Breimer, K.-A. Karlsson, G. Larson, and J.M. McKibbin. (1983). Chemical characterization of a blood group H type pentaglycosylceramide of human small intestine, Chem. Phys. Lipids 33:135-144.
16. E.L. Smith, J.M. McKibbin, K.-A. Karlsson, I. Pascher, and B.E. Samuelsson. (1975). Characterization by mass spectrometry of blood group A active glycolipids from human and dog small intestine. Biochemistry 14:2120-2124.

17. J.M. McKibbin, E.L. Smith, J.-E. Månsson, and Y.-T. Li. (1977). Characterization of dog small intestinal fucolipids with human blood group A activity. Differences in dog and human A-active fucolipids. Biochemistry 16:1223-1228.
18. E.L. Smith, J.M. McKibbin, K.-A. Karlsson, I. Pascher, B.E. Samuelsson, Y.-T. Li, and S.-C. Li. (1975) Characterization of a human intestinal fucolipid with blood group Le^a activity. J. Biol. Chem. 250:6059-6064.
19. J.M. McKibbin, W.A. Spencer, E.L. Smith, J.-E. Månsson, K.-A. Karlsson, B.E. Samuelsson, Y.-T. Li, and S.-C. Li. (1982). Lewis blood group fucolipids and their isomers from human and canine intestine. J. Biol. Chem. 257:755-760.
20. M.E. Breimer. (1984) Tissue specificity of glycosphingolipids as expressed in pancreas and small intestine of blood group A and B human individuals. Arch. Biochem. Biophys. 228:71-85.
21. K.-A. Karlsson, and G. Larson. (1978) Molecular characterization of cell surface antigens of human fetal tissue. Meconium, a rich source of epithelial blood-group glycolipids. FEBS Lett. 87:283-287.
22. K.-A. Karlsson, and G. Larson. (1981). Molecular characterization of cell surface antigens of fetal tissue. Detailed analysis of glycosphingolipids of meconium of a human 0 Le(a-b+) Secretor. J. Biol. Chem.256:3512-3524.
23. J. Ångström, K.-E. Falk, K.-A. Karlsson and G. Larson. (1982). Chemical fingerprinting of non-acid glycosphingolipids in meconium of a human individual of blood group B Le(a-b+) and Secretor. Biochim. Biophys. Acta 712:274-282.
24. G. Larson. (1983) Trifucosyl glycosphingolipids of human meconium. In: Glycoconjugates, M.A. Chester, D. Heinegard, A. Lundblad, and S. Svensson, eds., p. 423, Rahms, Lund.
25. R. Kannagi, S.B. Levery, and S.-i. Hakomori. (1985). Le^a-active heptaglycosylceramide, a hybrid of type 1 and type 2 chain, and the pattern of glycolipids with Le^a, Le^b, X (Le^x), and Y (Le^y) determinants in human blood cell membranes (ghosts). J. Biol. Chem. 260:6410-6415.
26. M.E. Breimer, B. Cedergren, K.-A. Karlsson, K. Nilson, and B.E. Samuelsson. (1980). Glycolipid pattern of stomach tissue of a human with the rare blood group A,p. FEBS Lett. 118:209-211.
27. S.-i. Hakomori, E. Nudelman, S.B. Levery, and R. Kannagi. (1984). Novel fucolipids accumulating in human adenocarcinoma. J. Biol. Chem.259:4672-4680.
28. S.B. Levery, E. Nudelman, and S.-i. Hakomori. (1985) ^1H-NMR studies of polyfucosylated type 2 chain glycosphingolipids: Characterization of two new extended-Y structures isolated from human adenocarcinoma. In: Glycoconjugates, E.A. Davidson, J.C. Williams, and N.M. DiFerrante, eds., p. 620, Praeger, New York.
29. J. L. Magnani, B. Nilsson, M. Brockhaus, D. Zopf, Z. Steplewski, H. Koprowski, and V. Ginsburg. (1982) A monoclonal antibody-defined antigen associated with gastrointestinal cancer is a ganglioside containing sialylated lacto-N-fucopentaose II. J. Biol. Chem. 257:14365-14369.
30. Y. Umesaki, A. Suzuki, T. Kasama, K. Tohyama, M. Mutai, and T. Yamakawa. (1981). Presence of asialo GM_1 and glucosylceramide in the intestinal mucosa of mice and

induction of fucosyl asialo GM1 by conventionalization of germ-free mice. J. Biochem. 90:1731-1738.

31. A. Suzuki, and T. Yamakawa. (1981). The different distribution of asialo GM1 and Forssman antigen in the small intestine of mouse demonstrated by immunofluorescent staining. J. Biochem. 90:1541-1544.
32. G. C. Hansson, K.-A. Karlsson, H. Leffler, and N. Stromberg. (1982). Gangliotetraosylceramide is a major glycolipid of epithelial cells of mouse small intestine. FEBS Lett. 139:291-294.
33. E.L. Smith, J.M. McKibbin. K.-A. Karlsson, I. Pascher, and B.E. Samuelsson. (1975). Characterization of dog small intestinal fucolipids with human blood group H activity. Biochemistry 14:3370-3376.
34. E.L. Smith, J.M. McKibbin, M.E. Breimer, K.-A. Karlsson, I. Pascher, and B.E. Samuelsson. (1975). Identification of a novel heptaglycosylceramide with two fucose residues and a terminal hexosamine. Biochim. Biophys. Acta 398:84-91.
35. M.E. Breimer, G.C. Hansson, K.-A. Karlsson, H. Leffler, W. Pimlott, and B.E. Samuelsson. (1979) Selected ion monitoring of glycosphingolipid mixtures. Identification of several blood group type glycolipids in the small intestine of an individual rabbit. Biomed. Mass Spectrom. 6:231-241.
36. R. Oriol, and A.M. Dalix. (1977) Differences in the maturation of the immune response of A^- and A^+ rabbits. Immunology 33:91-99.
37. M.E. Breimer, G.C. Hansson, K.-A. Karlsson, H. Leffler, W. Pimlott, and B.E. Samuelsson. (1978). Structure determination of blood group type glycolipids of cat small intestine by mass fragmentography. FEBS Lett. 89:42-46.
38. B.L. Slomiany, and A. Slomiany. (1973). Blood group A active ceramide hexasaccharide lacking N-acetylglycosamine isolated from hog stomach mucosa. Biochim. Biophys. Acta 326:224-231.
39. G. Nilsson, S. Svensson, and A. Lindberg. (1983). The role of the carbohydrate portion of glycolipids for the adherence of *E. coli* K88 to pig intestine. In: Glycoconjugates, M.A. Chester, D. Heinegard, A. Lundblad, and S. Svensson, eds., p. 637, Rahms, Lund.
40. B.E. Samuelsson. (1984) Solid phase biosynthesis on high performance thin-layer plates of blood group glycosphingolipids II. FEBS Lett. 167:47-57.
41. C. Suzuki, A. Makita, and Z. Yosizawa. (1968) Glycolipids isolated from porcine intestine. Arch. Biochem. Biophys. 127:140-149.
42. B.L. Slomiany, A. Slomiany, and M.I. Horowitz. (1974). Characterization of blood group H active ceramide tetrasaccharide from hog stomach mucosa. Eur. J. Biochem. 43:161-165.
43. B.L. Slomiany, and A. Slomiany. (1977) Structural studies on branched fucosphingolipids of hog gastric mucosa. Chem. Phys. Lipids 20:57-69.
44. B.L. Slomiany, and A. Slomiany. (1977). Branched blood group A-active fucolipids of hog gastric mucosa. Biochim. Biophys. Acta 486:531-540.
45. A. Slomiany, and B.L.Slomiany. (1978). Blood-group (A+H) fucolipids - A new group of complex glycosphingolipids. FEBS Lett. 90:293-296.

46. B.L. Slomiany, and A. Slomiany. (1977). Complex glycosphingolipids with blood-group A specificity. FEBS Lett. 73:175-180.
47. A. Slomiany, and B.L. Slomiany. (1980) Structure of the ceramide octadekahexoside isolated from gastric mucosa. Biochem. Biophys. Res. Comm. 93:770-775.
48. J. Ångström, M.E. Breimer, K.-E. Falk, G.C. Hansson, and K.-A. Karlsson. (1982). Chemical characterization of penta-, hexa-, hepta-, octa-, and nonaglycosylceramides of rat small intestine having a globoside-like terminal. J. Biol. Chem. 257:682-688.
49. T. Taki, H. Kimura, S. Gasa, M. Nakamura, and M. Matsumoto. (1985). A new type of blood group B active glycosphingolipid in rat bone marrow cells. J. Biol. Chem. 260:6219-6225.
50. M.E. Breimer, G.C. Hansson, K.-A. Karlsson, and H. Leffler. (1981) Glycosphingolipids and the differentiation of intestinal epithelium, Exp. Cell Res. 135:1-13.
51. G.G. Forstner, and J.R. Wherrett. (1973) Plasma membrane and mucosal glycosphingolipids in the rat intestine. Biochim. Biophys. Acta 306:446-459.
52. M.E. Breimer, G.C. Hansson, K.-A. Karlsson, and H. Leffler. (1980). Glycolipids of rat small intestine. Characterization of a novel blood group H-active triglycosylceramide. Biochim. Biophys. Acta 617:85-96.
53. M.E. Breimer, K.-E. Falk, G.C. Hansson, and K.-A. Karlsson. (1982). Structural identification of two 10-sugar branched-chain glycosphingolipids of blood group H type present in epithelial cells of rat small intestine. J. Biol. Chem. 257:50-59.
54. M.E. Breimer, K.-E. Falk, G.C. Hansson, and K.-A. Karlsson. (1982). Structural identification of two 10-sugar branched-chain glycosphingolipids of blood group H type present in epithelial cells of rat small intestine. J. Biol. Chem 257:906-912.
55. J. Ångström, M.E. Breimer, K.-E. Falk, G.C. Hansson, K.-A. Karlsson, H. Leffler, and I. Pascher. (1982). Structural characterization of glycolipids of rat small intestine having one to eight hexoses in a linear sequence. Arch. Biochem. Biophys. 213:708-725.
56. G.C. Hansson. (1983). The structure of two blood group A-active glycosphingolipids with 12 sugars and a branched chain present in the epithelial cells of rat small intestine. J. Biol. Chem.258:9612-9615.
57. J.-F. Bouhours, D. Bouhours, and G.C. Hansson. (1987), submitted.
58. E.H. Holmes, and S.-i. Hakomori. (1982). Isolation and characterization of a new fucoganglioside accumulated in precancerous rat liver and in rat hepatoma induced by N-2-acetylaminofluorene. J. Biol. Chem.257:7698-7703.
59. G.C. Hansson, J.-F. Bouhours, and J. Ångström. (1987), Characterization of neutral blood group B-active glycosphingolipids of rat gastric mucosa. A novel type of blood group active glycosphingolipid based on isogloboside. J. Biol. Chem, in press.
60. G.C. Hansson, K.-A. Karlsson, and J. Thurin. (1980). Glycolipids of rat large intestine. Characterization of a novel blood group B-active tetraglycosylceramide absent from small intestine. Biochim. Biophys. Acta 620:270-280.
61. J. Ångström, P. Falk, G.C. Hansson, J. Holgersson, K.-A. Karlsson, N. Strömberg, and J. Thurin. (1987) The mono-

and difucosyl blood group B glycosphingolipids of rat large intestine differ in type of core saccharide. Biochim. Biophys. Acta, in press.

62. B.E. Samuelsson. (1983). Enzymatic synthesis of a blood group A related difucosyl heptaglycosylceramide with a type 2 carbohydrate chain. FEBS lett. 152:305-310.
63. H. Wiegandt. (1985). Gangliosides. In: Glycolipids, H. Wiegandt, ed., p. 199, Elsevier, Amsterdam.
64. J.-F. Bouhours, and R.M. Glickman. (1977). Rat intestinal glycolipids. II. Distribution and biosynthesis of glycolipids and ceramide in villus and crypt cells. Biochim. Biophys. Acta, 441:123-133.
65. J.-F. Bouhours, and R.M. Glickman. (1977) Rat intestinal glycolipids. III. Fatty acids and long chain bases of glycolipids from villus and crypt cells. Biochim. Biophys. Acta 487:51-60.
66. D. Bouhours, and J.-F. Bouhours. (1981). Developmental changes of rat intestinal glycolipids. Biochem. Biophys. Res. Comm.99:1384-1389.
67. D. Bouhours, and J.-F. Bouhours. (1985) Developmental control of the expression of two ten-sugar branched chain fucolipids in rat small intestine. Glycoconjugate J. 2:79-86.
68. Y. Umesaki, T. Sakata, and T. Yajima. (1982). Abrupt induction of GDP-fucose: asialo GM1 fucosyltransferase in the small intestine after conventionalization of germ-free mice. Biochem. Biophys. Res. Comm. 105:439-443.
69. J. Holmgren. (1981). Actions of cholera toxin and the prevention and treatment of cholera. Nature 292:413-417.
70. H. Leffler, and C.Svanborg Edén. (1980). Chemical identifiation of a glycosphingolipid receptor for *Escherichia coli* attaching to human urinary tract epithelial cells and agglutinating human erythrocytes. FEMS Microbiol. Lett. 8:127-134.
71. K. Bock, M.E. Breimer, A. Brignole, G.C. Hansson, K.-A. Karlsson, G. Larson, H. Leffler, B.E. Samuelsson, N. Strömberg, C. Svanborg Edén, and J. Thurin. (1985). Specificity of binding of uropathogenic *Escherichia coli* to Galα1→4Gal-containing glycosphingolipids. J. Biol. Chem. 260:8545-8551.
72. G.C. Hansson, K.-A. Karlsson, G. Larson, A. Lindberg, N. Stromberg, and J. Thurin. (1983). Lactosylceramide is the probable adhesion site for major indigenous bacteria of the gastrointestinal tract. In: Glycoconjugates, M.A. Chester, D. Heinegard, A. Lundblad, and S. Svensson, eds., p. 631, Rahms, Lund.
73. J.L. Magnani, D.F. Smith, and V. Ginsburg. (1980). Detection of gangliosides that bind cholera toxin: direct binding of ^{125}I-labeled toxin to thin-layer chromatograms. Anal. Biochem. 109:399-402.
74. G.C. Hansson, K.-A. Karlsson, G. Larson, J. McKibbin, M. Blaszczyk, M. Herlyn, Z. Steplewski, and H. Koprowski. (1983). Mouse monoclonal antibodies against human cancer cell lined with specificities for blood group and related antigens. Characterization by antibody binding to glycosphingolipids in a chromatogram binding assay. J. Biol. Chem. 258:4091-4097.

CHEMICALLY MODIFIED CAPSULAR POLYSACCHARIDES AS VACCINES

H.J.Jennings

Division of Biological Sciences, National Research Council of Canada
Ottawa, Ontario K1A OR6, Canada

Vaccination is one of the most useful scientific developments in the control and eradication of human disease. Early vaccines were based on whole organism preparations which, although efficacious in some cases (smallpox), had disadvantages which severely limited their general use. Because of the complexity of bacteria, it is difficult to maintain consistency of potency in terms of immunological protection, and, more importantly, to avoid severe toxic or other deleterious effects caused by indigenous component molecules (lipopolysaccharide or peptidoglycan) of some pathogenic bacteria. The discoveries of a "specific soluble substance" secreted by *Pneumococci* during growth[1] and the immunogenicity of these substances (capsular polysaccharides)[2] were new and important developments in vaccine technology. It was demonstrated[3] that this substance was in fact a type-specific polysaccharide that was able to quantitatively precipitate[4] antibodies produced in animals by injection of the homologous, whole organisms. In subsequent pioneering work[5] it was demonstrated that when used as human vaccines these purified polysaccharides provide type-specific protection against the development of pneumococcal infection. However, at this critical stage of development the phenomenal success of antibiotic therapy for treating bacterial infections overshadowed the early promise of polysaccharide vaccines.

Recently, interest in the prophylaxis of bacterial disease has been the subject of renewed intensified research[6,7] due in large part to the expanding incidence of antibiotic resistant bacterial strains.[8] In addition, recent important exacting clinical studies have also identified weaknesses in antibiotic therapy.[6,7] Thus "cured" *H. influenzae* type b meningitis is the leading cause of acquired mental retardation,[9] and epidemiological statistics indicate that deaths due to pneumococcal pneumonia occur at the same rate as in the pre-antibiotic era.[10] Current interest in the capsular polysaccharides has evolved simultaneously with the resurgence of interest in the prophylaxis of human bacterial disease because of their potential as good immunogens in providing protection against bacterial infections.

The success of capsular polysaccharides as vaccines against disease caused by encapsulated bacteria is based on the importance of the polysaccharide in the immune response to the organism. This is due to the location of polysaccharides on the surface of the bacteria, where they are situated at the interface of many host-bacteria interactions and thus constitute the principal antigens in most pathogenic gram-negative and gram-positive organisms.[7,11] Other antigens such as lipopolysaccharides, peptidoglycans, and other surface proteins also play a subsidiary role in the immunological response to encapsulated bacteria. In addition to being important antigens for the production of protective antibodies, capsular polysaccharides are also involved in the pathogenesis of bacteria following penetration of the bacteria into body tissue. Capsular polysaccharides are important virulence factors in disease caused by bacteria, and it is likely that the property of capsular polysaccharides which enhances the virulence of certain bacteria is the ability of the polysaccharide to mediate the hosts' immune response.[7]

Capsular polysaccharides can be readily isolated in their immunogenic high molecular weight form and the concept of using a pure capsular polysaccharide, devoid of its accompanying complex bacterial mass, is technically elegant. Polysaccharides are, with few execeptions, immunogenic in man, non-toxic, and free of other deleterious effects associated with whole organism vaccines.[6,7] Another important feature of polysaccharide immunogens is that they can be chemically and physically defined; criteria which add a greater measure of control over their efficacy as immunogens. In fact the groups A and C meningococcal polysaccharides were the first vaccines where the standards for licensure and release depended entirely on physico-chemical criteria. As a measure of the success of these vaccines, millions of people have been immunized with capsular polysaccharides resulting in a high degree of protection, with no fatalities or significant adverse effects. Currently[6,7] polysaccharide vaccines have been licensed for use in the immunoprophylaxis of pneumonococcal pneumonia and meningitis caused by *N. meningitidis* and *H. influenzae*. Obviously the concept of using capsular polysaccharides as human vaccines is capable of extension to diseases caused by other pathogenic bacteria, e.g. group B *Streptococcus*, *S. aureus*, *Klebsiella*, and *Pseudomonas*; however, their use will depend on their clinical importance and the presence of meaningful epidemiological studies.

Despite the success obtained with the use of polysaccharides as human vaccines, a number of problems have been identified which severely restrict the further development of this concept. Two of the most important problems are the poor immunogenicity of a few of the purified polysaccharides in all humans, of which the group B meningococcal polysaccharide is the salient example,[12] and the poor immunogenicity of all polysaccharides in infants.[13] This latter problem is particularly disturbing because this section of the population is the most susceptible to bacterial infection, especially bacterial meningitis. The above problems have stimulated much recent research, and focusing on the principal encapsulated bacteria responsible for human bacterial meningitis, this review is concerned with the delineation of these problems, and with attempts to solve them by the chemical manipulation of polysaccharides. This development has shown great promise in preserving the concept of using polysaccharide vaccines, because it has been established that chemical modification of polysaccharides can dramatically and favorably mediate the immune response to them. In order to accomplish the chemical modification of polysaccharides, it is essential to understand the interrelation between their structures and immunospecificities, and therefore this review also peripherally covers this aspect of research on polysaccharide antigens. It is concerned with the methodology associated with the basic structural elucidation of polysaccharides with emphasis on the most recent NMR spectroscopic techniques. These techniques are also invaluable in monitoring subsequent structural modifications made on the polysaccharides.

The review is also concerned with defining epitopes that are responsible for the production of polysaccharide-specific antibodies. In this regard emphasis is placed on the polysaccharides containing sialic acid residues because they are frequently associated with bacteria causing meningitis in humans, and often form very unconventional conformational determinants.[14] Although the scope of this review is restricted to encapsulated bacteria associated with meningitis in humans, it is supplemented by a more in-depth review on the use of polysaccharides as human vaccines.[7] Pertinent reviews have also been published on vaccines for the prevention of encapsulated bacterial diseases[6] and the application of NMR spectroscopy to polysaccharides of human pathogenic bacteria.[15,16,17] More general reviews on the structure[18] and immunological response[11] of polysaccharides are also available.

POLYSACCHARIDE STRUCTURE

Structural Determination

The immunological specificity of polysaccharide antigens resides in their structures. Therefore, two important steps toward understanding the immunology of polysaccharides are first, elucidating their basic structures and, secondly, ascertaining which part of the total structure (determinant or epitope) is responsible for its immunological specificity. On the basis of a wealth of chemical degradation, NMR spectroscopic and biosynthetic evidence, it is probably safe to assume that most bacterial polysaccharides, except for occasional minor substituent irregularities, are composed of long chains of fairly small identical repeating units. These repeating units consist of from 1 to 7 monosaccharide units, although the upper value cannot be considered to be the maximum possible. This is exemplified in the structures of the capsular polysaccharides shown in Table 1. Although the polysaccharides listed in Table 1 constitute only a fraction of those associated with human pathogenesis, they were selected in order to focus on the capsular polysaccharides of bacteria which cause the majority (approximately 90%) of cases of meningitis in humans. These polysaccharides are also of particular interest because they are prime vaccine candidates with which problems have been identified with their overall immunological performance in humans.[7]

Although the structures of the capsular polysaccharides listed in Table 1 are broadly representative of most bacterial capsules, they are unique in that they are composed of fairly simple repeating units and contain complex 3-deoxy-2-glyculosonic acid (sialic acid and KDO) components and phosphoric diester linkages which render them extremely fragile. This has important implications in limiting the kind of chemical techniques that can be used on them for the purposes of structural elucidation or for their chemical modification. This situation prompted a search for new techniques with which to tackle their structural elucidation. One that has been used with great success is nuclear magnetic resonance (NMR) spectroscopy.[7,15,16,17] In fact, except for the group A polysaccharide, on which some structural information was already available, all of the other structures were deduced almost entirely by NMR spectroscopy. The first NMR spectroscopic technique to be exploited was ^{13}C NMR which yielded valuable information on the structure and conformation of polysaccharides. Reviews pertinent to the application of

Table 1. Structures of capsular polysaccharides associated with human bacterial meningitis infections.

Group	Structure	References
Neisseria meningitidis		
A	→6)-αD-ManNAc-1-O-P(=O)(OH)-O- (3-OAc)	19
B	→8)αD-NeupAc(2→	20
C	→9)αD-NeupAc(2→ (7/8 OAc)	20
W-135	→6)-αD-Galp(1→4)αD-NeupAc(2→	21
Y	→6)-αD-Glcp(1→4)αD-NeupAc(2→(contains OAc groups)	21
Haemophilus influenzae		
b	→3)βD-Ribf(1→1)D-ribitol(5-O-P(=O)(OH)-O-	22,23
Streptococcus pneumoniae		
6A	→2)αD-Galp(1→3)αD-Glcp-αL-Rhap(1→3)-ribitol-(5-O-P(=O)(OH)-O-	24

Table 1 (Cont'd).

Group	Structure	References
Group B Streptococcus		
Ia	→4)-βD-Glcp-(1→4)-βD-Galp-(1→ 3 ↑ 1 βD-GlcpNAc 4 ↑ 1 αD-NeupNAc-(2→3)-βD-Galp	25
Ib	→4)-βD-Glcp-(1→4)-βD-Galp-(1→ 3 ↑ 1 βD-GlcpNAc 3 ↑ 1 αD-NeupNAc-(2→3)-βD-Galp	25
II	→4)-βD-GlcpNAc-(1→3)-βD-Galp(1→4)-βD-Glcp-(1→3)-βD-Glcp-(1→2)-βD-Galp-(1→ 6 ↑ 1 βD-Galp 3 ↑ 2 αD-NeupNAc	26
III	→4)-βD-Glcp-(1→6)-βD-GlcpNAc-(1→3)-βD-Galp-(1→ 4 ↑ 1 αD-NeupNAc-(2→6)-βD-Galp	27

this technique to polysaccharides of human pathogenic bacteria have been published.[7,15,16]

^{13}C-NMR spectroscopy of bacterial polysaccharides gives spectra containing signals associated with all the individual carbon atoms of their basic skeleton. Despite the large number of carbons in these molecules, the pattern of these signals is considerably simplified by the coincidence of the carbon signals of their individual identical repeating units. This is illustrated in the ^{13}C-NMR spectrum of the group B meningococcal polysaccharide[20] (Fig. 1) which, despite the large molecular size of the polysaccharide, is a simple eleven-resonance spectrum, containing one signal for each carbon in its α-(2→8)-linked sialic acid repeating unit. Thus that simple pattern of signals provides good evidence for the group B meningococcal polysaccharide being composed of the above simple repeating unit. In addition, further structural information can be obtained by comparing the chemical shifts of the carbon atoms of methyl α- and ß-ketosides of sialic acid with those of the sialic acid residues in the polysaccharide. Large chemical shift differences at C-8 and smaller ones at C-7 and C-9 confirm that the sialic acid residues are linked at _O_-8, and similarities in the chemical shifts of C-1, C-4, and C-6 of the methyl α-ketoside (signals

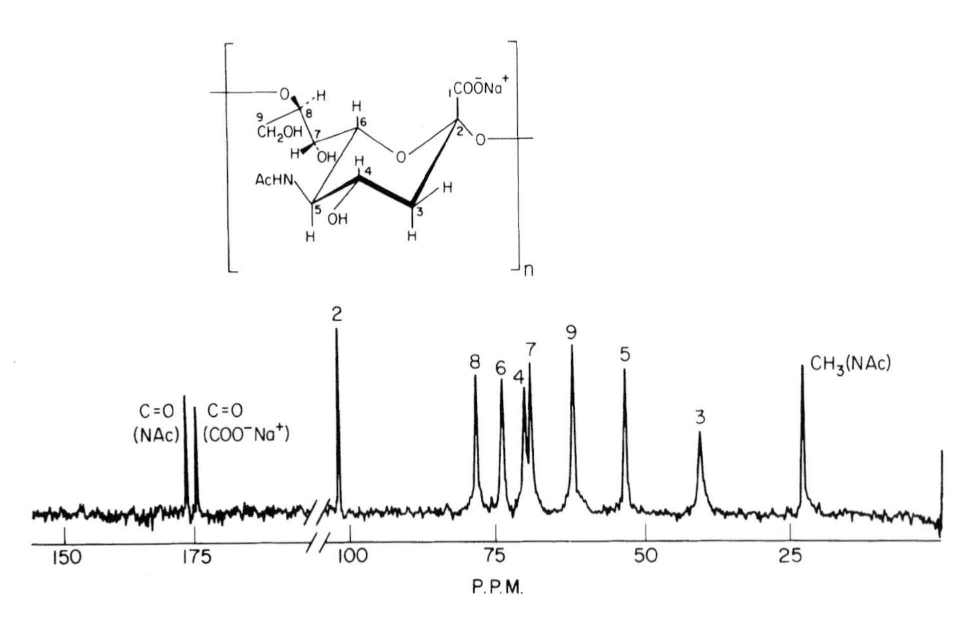

Fig. 1. Assignment of signals in the 100MHz-NMR spectrum of the group B meningococcal polysaccharide.

sensitive to the configuration of the interglycosidic linkage) with those of the sialic acid residues in the polysaccharide enable the α-configuration to be assigned to these latter residues. This type of analysis can also be used to locate the position of substituents (e.g. O-acetyl) on the polysaccharide[19] and chemical shifts have also been demonstrated to be conformationally sensitive.[14] Overall, ^{13}C-NMR spectroscopy has proved to be a powerful technique in the structural determination of bacterial polysaccharides but the analysis suffers the disadvantage of being empirical in nature.

With the advent of high resolution spectrometers, ^{1}H-NMR spectroscopy has also become a powerful tool in the structural analysis of complex carbohydrates, where a number of one dimensional, two dimensional and nuclear Overhauser enhancement (nOe) techniques are employed. The one dimensional techniques were largely utilized in the structural elucidation of glycoproteins;[28] however, despite the fact that polysaccharides generally yield more poorly resolved spectra due to their large molecular size and higher viscosity, the technique has been routinely used in elucidating polysaccharide structures. As in the case of ^{13}C-NMR spectra, the ^{1}H-NMR spectra of polysaccharides are considerably simplified by expressing the single repeating unit. Thus the chemical shifts of the proton signals can be used to make proton assignments, and changes in chemical shift following specific chemical modifications of polysaccharides can be used to provide valuable conformational information. The added advantage of ^{1}H-NMR spectroscopy is due to the phenomenon of proton-proton coupling ($^{3}J_{H,H}$), from which the relative orientation of vicinal protons can be established. However, because of the complexity of most one-dimensional proton spectra, this latter technique is usually only useful in making assignments on anomeric proton signals which resonate in a characteristic low field part of the spectrum.

The usefulness of NMR spectroscopy follows from the assignment of the individual proton and carbon resonances to particular hydrogen and carbon atoms. The ability to make these assignments in complex spectra has been enormously facilitated by the introduction of two dimensional (2-D) techniques.[17] Although these techniques have been used frequently on oligosaccharides, examples of the direct application of these techniques to polysaccharides are only now beginning to be reported. For example, the use of three of these techniques has recently been demonstrated in the structural elucidation of the groups I[29] and K[30] meningococcal polysaccharides. The K polysaccharide[30] is composed of the repeating unit shown in Fig. 2 and the 2-D experiments were carried out on the de-*O*-acetylated polysaccharide (Fig. 3). The ^{1}H-NMR spectrum of the de-*O*-acetylated K polysaccharide shown in Fig. 3 was difficult to assign directly because of its complexity. Therefore, a proton homonuclear shift correlated 2-D NMR (^{1}H-^{1}H COSY) experiment was performed and the spectrum is also shown in Fig. 3. As a result of this experiment it was possible, by tracing couplings between vicinal protons starting from the anomeric protons, to make unambiguous assignments of all the protons of both 2-acetamido-2-deoxy-*D*-mannopyranosyluronic acid residues (A and B) of the repeating unit (Fig. 2). Using the data, a heteronuclear shift correlated 2-D NMR (^{13}C-^{1}H COSY) experiment was performed and the ^{13}C-^{1}H correlation contour map is shown in Fig. 4. Those experiments enabled the assignment of all the carbons of the repeating unit (Fig. 2) to be made without reference to model compounds and all the above assignments, both ^{1}H and ^{13}C, were crucial to the eventual structural elucidation of the K polysaccharide.

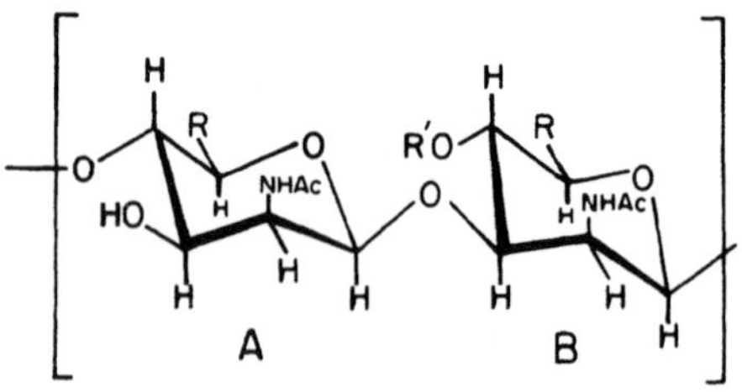

a) R = COO^{-}, R′ = $COCH_3$
b) R = COO^{-}, R′ = H
c) R = CH_2OH, R′ = H

Fig. 2. The repeating units of the native (a) de-*O*-acetylated, (b) carboxyl-reduced and (c) group K meningococcal polysaccharide.

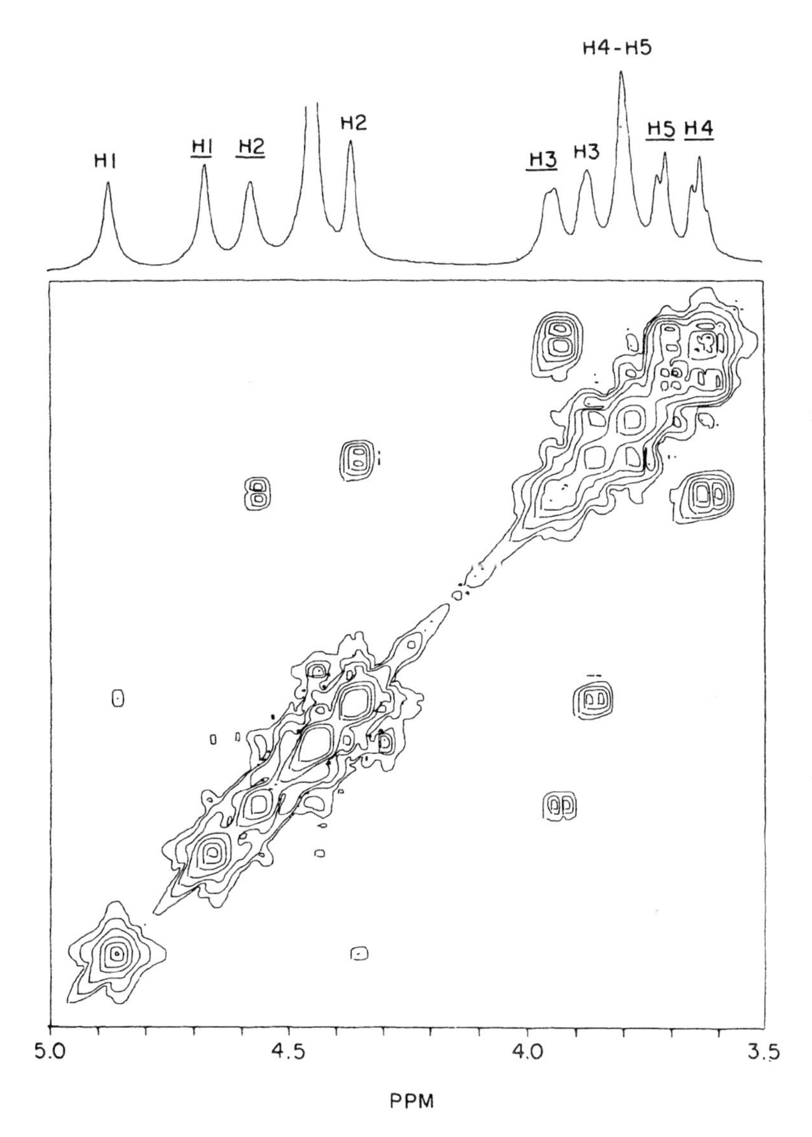

Fig. 3. Homonuclear 2-D shift correlated ($^1H-^1H$ COSY) spectrum of the de-*O*-acetylated K polysaccharide in D_2O (330°K) with the 1-D spectrum above. The corresponding assignments for residue A and residue B (underlined) of the repeating unit (Fig. 2) are also shown.

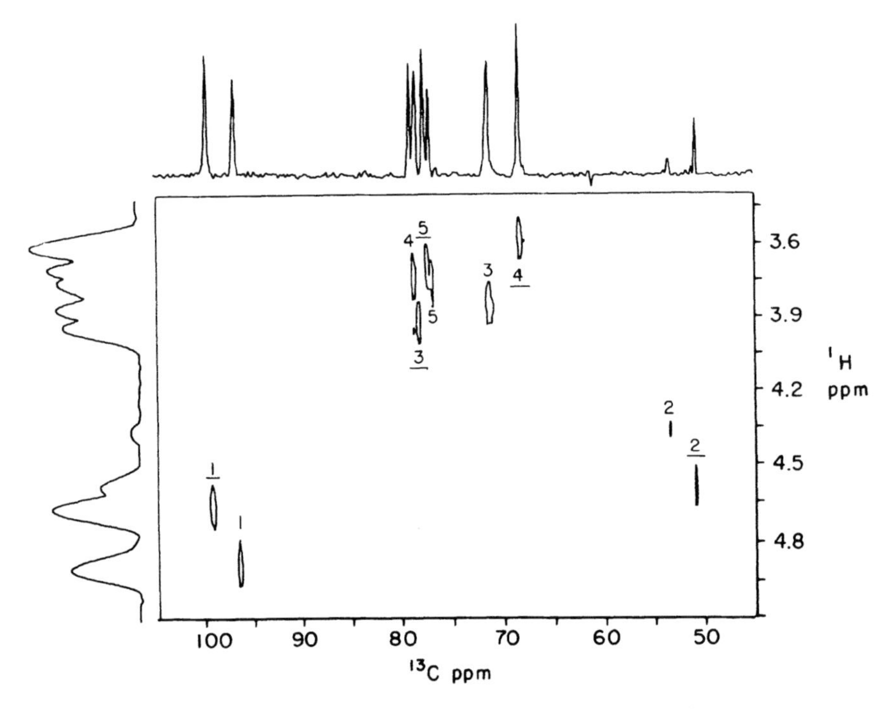

Fig. 4. Heteronuclear 2-D shift correlated (^{13}C-^{1}H COSY) spectrum of the de-O-acetylated K polysaccharide in D_2O (310°K) along with the f and f_2 projections. The resulting assignments of the C-H correlations for residue A and residue B (underlined) of the repeating unit (Fig. 2) are shown.

In addition to the Fig. 3 experiments, assignments of anomeric configuration (ß-D), conformation ($^4C_1$(D)) and linkage sites of residues A and B were obtained by 2-D nuclear Overhauser experiments shown in Fig. 5. A number of nuclear Overhauser enhancements were observed on the resonances of protons in close proximity to the anomeric protons of residues A and B (Fig. 2). Critical to their anomeric configurational assignments are the intro-residue enhancements observed between H1-H3, H1-H5 of ring A and H1-H3, H1-H5 of ring B, which indicated that H3, H5 and H3, H5 were both *cis* and 1,3-diaxially disposed to their respective anomeric protons. Inter-residue enhancements between H1 and H4-H5 (strongly coupled) were consistent with a ß-(1→4)-linkage between residues B and A which would place H4 in close proximity to H1. Similarly, inter-residue enhancements between H1 and H3

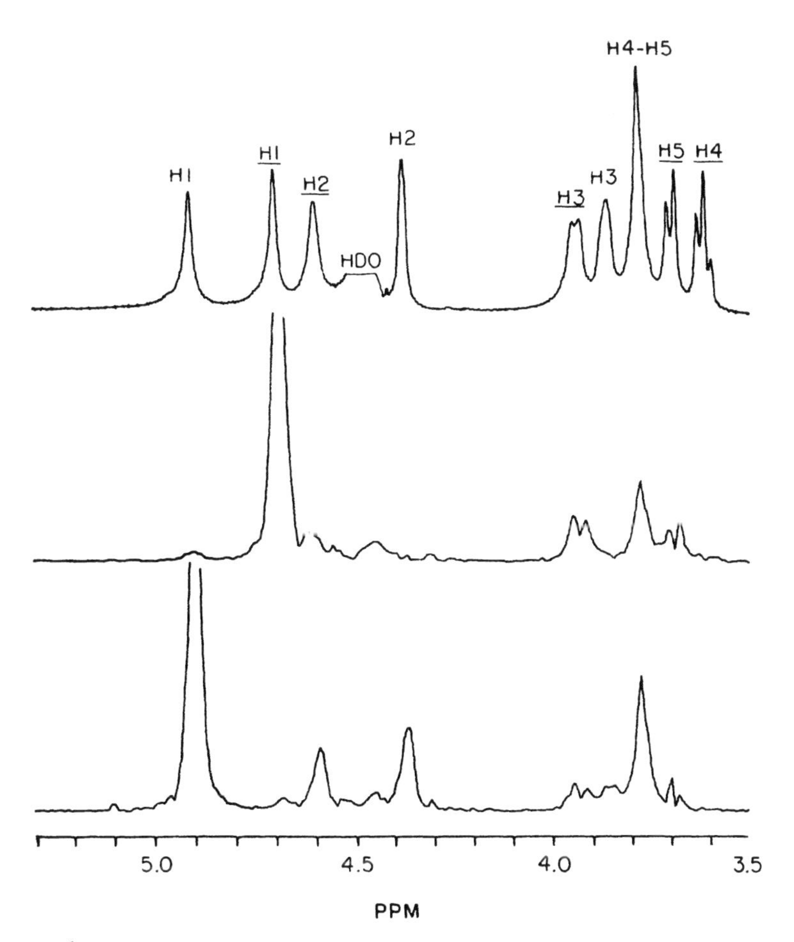

Fig. 5. ^{1}H-NMR spectrum (top) of the de-O-acetylated K polysaccharide in D_2O (330°K) with the assignments for residue A and residue B (underlined) of the repeating unit. (Fig. 2). Cross-sections of the 2D-NOESY contour map show the NOE's for H1 of residue B (middle) and H1 of residue A (bottom).

and between H1 and <u>H2</u> are a consequence of a ß-(1→3)-linkage between residues A and B. Since inter-residue nuclear Overhauser enhancements are dependent on the linkage conformation, the magnitude of these enhancements reflect the relative orientation of these residues with respect to each other. Similar 2-D experiments were used in the structural determination of the group I[29] meningococcal polysaccharide.

Structural Features of Biological and Immunological Significance

The polysaccharides, the structures of which are listed in Table 1, are located on the surface of bacteria responsible for the majority of cases of meningitis in humans. All of these polysaccharides are important virulence factors in disease and, although no single structural feature can be identified to account for the virulence of these bacteria in humans, it is probably more than circumstantial that eight of the eleven polysaccharides contain sialic acid residues. It is also noteworthy that of all the encapsulated bacteria, only a few are virulent in man, and two of these different species (group B *N. meningitidis* and *E. coli* K1) have a structurally and immunologically identical capsular polysaccharide.

One dominant structural feature among the sialic acid-containing polysaccharides is the occurrence of terminal sialic acid residues. The polysaccharides having this structural feature are present on the surface of groups B and C *N. meningitidis*, *E. coli* K1 and all the group B streptococcal polysaccharides. It is also possible that this is also true for the capsular polysaccharides of *N. meningitidis* groups W-135 and Y, although so far this has not been substantiated. The significance of terminal sialic acid in promoting virulence of the aforementioned bacteria is its known ability to inhibit the activation of complement by way of the alternative pathway.[7] This has been well documented for erythrocyte membrane surfaces[31] and also for the surface of type III group B streptococcal organisms.[32] In an attempt to delineate the structure-function relationship of terminal sialic acid residues, a number of chemical modifications of these residues was carried out while they were still present on the surface of erythrocytes[31] and bacteria.[32] All these chemical modifications resulted in converting these cells to activators of the alternative pathway. This evidence is consistent with the hypothesis that any change in the integrity of the sialic acid molecule will alter its capacity to modulate the complement system.

Another mechanism whereby these bacteria are able to evade the immune system and thus increase their potential as human pathogens is molecular mimicry.[7] If a bacterium is able to coat itself with molecules having a structure similar to that of those found in the hosts' tissue, the production of antibodies having a specificity for these structures would be suppressed, as they would be recognized as self-antigens. An examination of the sialic acid-containing capsular polysaccharides in Table 1 reveals extensive structural homology with human glycoprotein. Not only do they contain terminal sialic acid, which is not normally immunogenic because of its ubiquitous occurrence in human and animal tissues, but they also contain more extensive regions of homology. For example, α-Neu*p*Ac-(2→3)-β-D-Gal*p* and α-Neu*p*Ac-(2→6)-ß-D-Gal*p* found in the branches of the group B streptococcal polysaccharides have been respectively identified as end-group in the important human M and N blood group substances[33] and in human serotransferrin.[34] Further, both α-Neu*p*Ac-(2→3)-ß-D-Gal*p*-(1→4) ß-D-Glc*p*NAc and α-Neu*p*Ac-(2→6)-ß-D-Gal*p*-(1→4)-β-D-Glc*p*NAc have been identified as trisaccharide branches of complex glycose side chains in a variety of human and animal glycoproteins.[35] Oligosaccharides consisting of linear α-(2→8)-linked sialic acid, a structural feature of the group B meningococcal and *E. coli* K1 polysaccharides have also been identified in the glycopeptides of human and animal brain tissue.[36] Therefore, on the basis of the above evidence, it is not unreasonable to suppose that the incorporation of these structural features, which mimic human tissue antigens, was an evolutionary acquisition by bacteria.

Another structural feature, not referred to in Table 1, that has significance in terms of the biological function of some, and possibly all, of the polysaccharides, is the presence of small amounts of lipoidal residues.[37] This lipoidal material has been detected in the capsular polysaccharides of *N. meningitidis* groups A, B and C, *H. influenzae* type b, and various *E. coli* strains. Although this lipid component constitutes only a small fraction of the total mass of the polysaccharide, its effect on the physical and immunological characteristics of these polysaccharides is dramatic. Gotschlich and co-workers[38] were able to obtain the meningococcal capsular polysaccharides of groups A and C in their high-molecular weight form by their precipitation directly from the liquid culture using Cetavalon. These polysaccharides proved to be highly immunogenic in man. In contrast, the group A polysaccharide isolated from cultures concentrated by rotary evaporation had a much lower molecular weight (less than 50,000 daltons) and proved to be completely non-immunogenic. It was hypothesized that this depolymerization was the result of enzymatic degradation which had occurred in the culture medium during the evaporation process. The observation was then made that estimations of the molecular size of the group A polysaccharide by reducing end-group analysis were considerably lower than those obtained by gel filtration. Lui, et al.[37] hypothesized that this was due to aggregation of the polysaccharide chains, and that the lipid component was responsible for this aggregation. This type of aggregation was also identified in the groups B and C polysaccharides of *N. meningitidis* and the *E. coli* K92 polysaccharide. Gotschlich and co-workers[39] then demonstrated, as previously postulated, that a small proportion of the lipoidal material was attached to the above polysaccharides. Both di-O-palmitoyl glycerol and di-O-stearoyl glycerol were detected, the former constituting the major component. Evidence was submitted which indicated that these di-O-acylglycerols were glycosidically attached to the reducing end of the polysaccharides by phosphoric diester bonds. These results could be significant in our perceptions of capsular polysaccharides in that an apparently minor component can have such a profound effect on their physical (molecular size) and immunological (immunogenicity) properties. The function of these minor lipoidal components could be that they were originally involved in anchoring the capsular polysaccharides to the outer membrane of the bacterium.

Determinants

With the knowledge that polysaccharides are composed of long chains of fairly small identical repeating units, it is necessary to establish which part of the polysaccharide interacts with antibody. This part of the polysaccharide is called a determinant or epitope and from early studies by Goebel[40] it became apparent that determinants constitute only a small part of the large polysaccharide molecule. Therefore, because of their large molecular size, polysaccharides, with the possible exception of their non-reducing end-group determinants, are multivalent antigens. Antibodies made to the type 3 pneumococcal polysaccharide are strongly inhibited by cellobiouronic acid, the disaccharide repeating unit of the polysaccharide. Conversely, antibodies made to a cellobiouronic acid-protein conjugate cross-react with the type 3 pneumococcal polysaccharide. This procedure of using low molecular weight fragments of the polysaccharide (oligosaccharides) in order to inhibit the antigen-antibody precipitin reaction was used extensively by Kabat in the now classical studies on the linear dextran-antidextran reaction.[41] This model system is probably representative of all linear polysaccharides with the exeception of those terminating in nonreducing sialic acid groups.

These definitive studies[41] yield valuable information on the location and size of the determinants, and on the heterogeneous nature of antibodies in terms of their specificities. By using a series of oligosaccharides of the isomaltose series with human antidextran sera, it was established that the inhibitory power of the oligosaccharides increased with molecular size until it became more or less constant at the hexasaccharide. This was interpreted as being the optimum size of the combining site of the antibody molecule and this concept is still consistent with the wealth of physical and immunochemical data generated since that time. From the relative inhibitory power of these oligosaccharides it was also possible to calculate the contribution to the binding energy of each successive D-glucose residue. Those studies established that while the terminal D-glucose residue contributed to this binding energy, each succeeding D-glucose unit contributed incrementally smaller contributions. The non-reducing terminal D-glucose residues were called immunodominant. Another important finding in these studies was the heterogeneous nature of the antibodies in regard to their serological specificities.[42] Following absorption onto Sephadex, the antibodies were fractionated by elution with isomalto-oligosaccharides of different molecular sizes. Inhibition studies on the different fractions established that antibodies having specificities both to small (disaccharide) and large (hexasaccharide) determinants were present in the serum.

Branched polysaccharides in contrast to linear polysaccharides have many more terminal glycosyl residues. These residues are more exposed and therefore more accessible to the immune mechanism and in consequence are usually immunodominant, although not exclusively so, as smaller populations of antibodies with specificities for the backbone of the polysaccharide can usually be detected. The immunodominance of terminal glycosyl residues has now been firmly established and the phenomenon was first definitively resolved[43] in the classical studies on the serological determinants of the lipopolysaccharides of *Salmonella*. Similar to capsular polysaccharides, the O-chains of the lipopolysaccharides consist of a linear arrangement of oligosaccharide repeating units, some of which contain unique, terminal 3,6-dideoxyhexosyl residues in each repeating unit. These terminal saccharides are to a large extent responsible for the specificity of antibodies made to *Salmonella* organisms.

All of the phenomena elucidated above are probably applicable to the serological properties of the polysaccharides listed in Table 1 except for those containing sialic acid residues.[14] The presence of sialic acid in these polysaccharides tends to generate unconventional or non-classical determinants. The determinants are generated by a more selective response of the immune mechanism, which, except in some rare pathological conditions, avoids making antibody which might cross-react with structures associated with human tissue (self antigens). These structural features include terminal sialic residues alone, and as a part of larger sialooligosaccharides. However, sialic acid can be accommodated by the immune system where it is a part of an inner chain composite structure not cross-reactive with human tissue.[14] It is interesting to note that the human immune response to the type III group B streptococcal polysaccharide has revealed another important function of terminal sialic acid in that it is essential to the formation of the determinant responsible for the production of human protective antibodies,[14] despite the fact that it is non-immunogenic. The hypothesis proposed[14,44] to explain the above phenomenon is that sialic acid is able to exert conformational control over determinants remote from itself, by through-space interactions with other glycosyl residues in the polysaccharide. Another conformational determinant is also associated with interchain sialic acid residues when they exist in the form of α-(2→8)-linked sialic acid residues.[14,45] Using immunological and NMR spectroscopic techniques on the native and specifically modified polysaccharides, a number of these sialic acid-controlled determinants have been identified and located.

Determinants Involving Terminal Sialic Acid

The group B streptococcal polysaccharides serve as useful models to study the conformational function of terminal sialic acid when it is situated in a number of different molecular environments.[14] These polysaccharides are immunogenic, and are surprisingly immunospecific, despite the fact that they all contain sialic acid and even more extensive structural similarities. The hypothesis that terminal sialic acid was involved in the control of serological determinants originated as a result of studies on the type III group B streptococcal polysaccharide.[44] It was demonstrated[14,44] that anti-type III group B streptococcal rabbit sera contained two distinct populations of antibody with a specificity for determinants on the native type III polysaccharide antigen. That the specificity of the determinant corresponding to the major and protective population of antibodies is dependent on the presence of sialic acid was demonstrated in quantitative precipitin experiments with this antiserum using the native (Table 1), carboxyl-reduced native, desialylated (core), and degalactosylated core (backbone) type III antigens. The quantitative precipitin experiments shown in Fig. 6 indicate that only the native type III antigen is the complete antigen, the chemically modified antigens being capable of only partial precipitation and absorption of the total antibody. It is of interest to note that while the above studies indicate that sialic acid is crucial to the formation of the major native determinant, it cannot be immunodominant in the classical sense because of the following evidence. Both serotransferrin and the type III core antigen, molecules having all the additive and overlapping structural features of the native type III antigen, proved to be poor inhibitors of the homologous serological reaction. In addition, the removal of the side chain (C8 and C9) from all the sialic residues of the native type III antigen by controlled periodate oxidation did not alter its serological properties. On the basis of the above evidence, a conformational determinant was postulated, the formation of which was dependent on the carboxylate groups of the terminal sialic acid residues.

In addition to the above serological evidence, compelling evidence[14,44] for conformational change in the determinant caused by the loss or carboxylate reduction of the sialic acid residues was also obtained by comparing the chemical shifts of the anomeric and linkage carbons in the ^{13}C-NMR spectra of the native carboxylate reduced native and core type III antigens (Fig. 7). Chemical shift displacement involving these particular linkage carbons can be indicative of changes in interglycose orientation and the C4 signal of the 2-acetamido-2-deoxy-ß-D-glucopyranosyl residue (C) underwent significant displacements on carboxylate reduction or complete removal of sialic acid from the native type III antigen. Because of the remoteness of this linkage carbon from the sialic acid residue, this chemical shift was interpreted as being due to a change in conformation, involving the branches of the type III antigen rather than more extensive conformational changes in

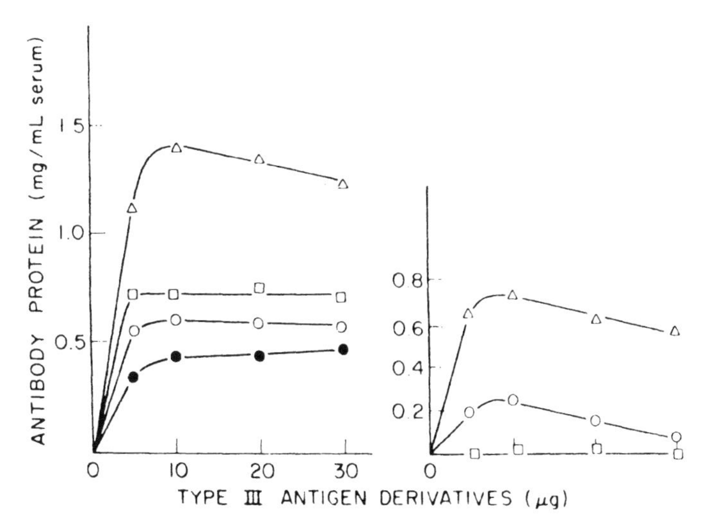

Fig. 6. Quantitative precipitin analysis of the native (Δ), reduced native (□), core (o), backbone (•) type III antigens with type III group B streptococcal anti-serum (left), and with the same antiserum previously absorbed with the reduced native antigen (right).

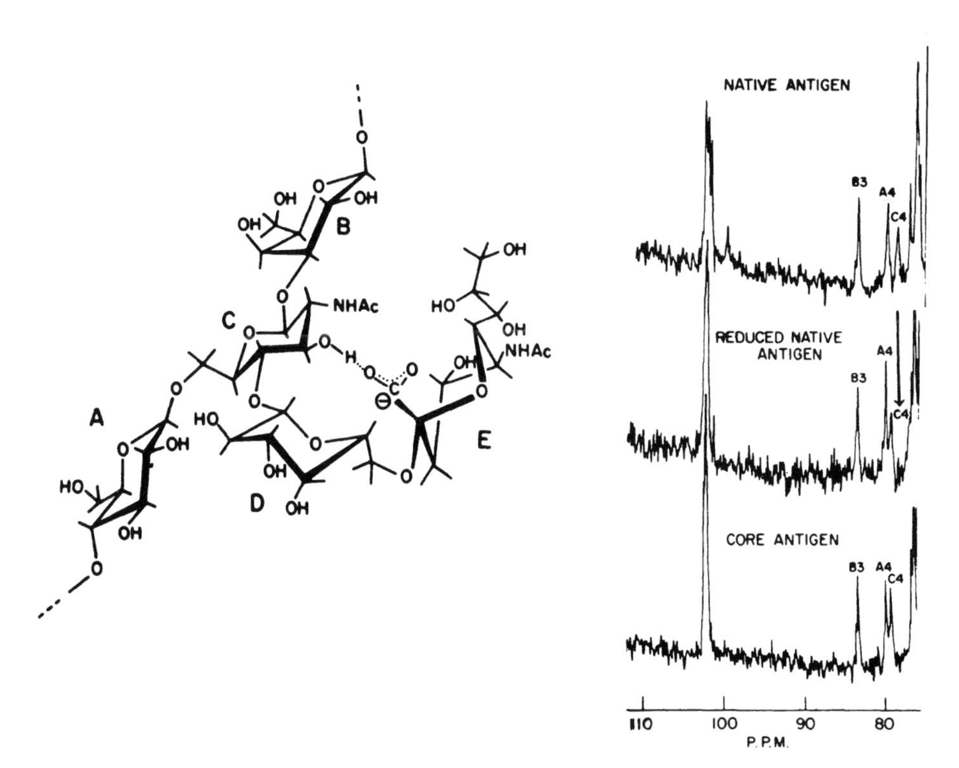

Fig. 7. Proposed conformation of the repeating unit of the type III group B streptococcal polysaccharide (left), and a spectral window on the 90 MHz ^{13}C-NMR spectra the type III antigens (right).

its backbone. The proposed conformation is shown in Fig. 7 and it was hypothesized that its formation probably involves interactions between the terminal sialic acid residues and the penultimate 2-acetamido-2-deoxy-ß-D-glucopyranosyl residues. The exact nature of this interaction has not yet been established but, because the carboxylate group is crucial to the formation of this determinant, it was hypothesized that it could possibly be involved in hydrogen bonding to the 2-acetamido-2-deoxy-ß-D-glucopyranosyl residue (C). Similar ^{13}C-NMR chemical shift displacements have also been observed on removal of the sialic acid residues from the identical trisaccharide, α-NeupNAc-(2→6)-ß-D-Gal*p*-(1→4)-ß-D-Glc*p*NAc, isolated from human urine.[46] The formation of the conformational determinant involving the sialotrisaccharide, however, cannot be solely responsible for the specificity of the native type III determinant because the identical trisaccharide failed to inhibit the homologous serological reaction of the native type III antigen.[14,44] Obviously this serological specificity must also depend on additional backbone residues, although the extent of their involvement has not yet been definitively established.

The above postulates are consistent with interpreting the immunological properties of sialic acid-dependent determinants in other group B streptococcal polysaccharides.[14] For example, the type II polysaccharide has a similar sialic acid-dependent determinant which produces highly specific antibodies.[47] Again using a combination of serological and NMR spectroscopic techniques[14] on the native and chemically modified native antigens, it was also possible to locate the critical structural feature involved in this determinant. The structure of the repeating unit of the type II antigen is shown in Fig. 8 and the conformation determinant is probably generated by interactions between sialic acid and other domains of the polysaccharide. These domains must be localized to regions of the backbone of the polysaccharide in the vicinity of the sialic acid residues because these latter residues have little lateral freedom as they are attached directly to the backbone of the polysaccharide. As in the case of the type III antigen, carboxyl reduction or removal of the sialic acid residues from the type II antigen destroyed one of the type-specific determinants and also produced a characteristic signal displacement in the ^{13}C-NMR spectra of the modified antigens. The displaced signal was assigned to C3 of residue B indicating an interaction between sialic acid and residues A and possibly B of the native type II antigen (Fig. 8). This evidence was also confirmed by chemical shift displacements observed in the anomeric proton signals of residues A and B on carboxylate reduction or removal of the sialic acid residues from the native type II polysaccharide. This latter evidence not only confirms the existence of the proposed interaction but, because the carboxylate group is responsible for the deshielding of both protons, uniquely defines with some precision the geometry of this part of the conformational determinant. The full dimensions of this determinant have not yet been established but it is highly probable that other backbone sugars are involved.

The types Ia and Ib polysaccharide antigens of group B *Streptococcus* are isomers, having trisaccharide branches terminating in sialic acid residues (Fig. 9). They differ only in the linkages between their branch ß-D-galactopyranosyl (C) and 2-acetamido-2-deoxy-β-D-glucopyranosyl (D) residues. In the former these residues are 1→3-linked, and this small difference has a profound effect on their individual immunological properties. As in the case of the types II and III group B streptococcal organisms, those of type Ia produce a distinct population of sialic acid-dependent antibodies.[14,48] This was demonstrated as in the former cases by serological experiments using an anti-type Ia group B streptococcal rabbit serum and a series of selectively degraded type Ia polysaccharide antigens shown in Fig. 9. The equivalent modifications of the type Ib polysaccharide are also shown in Fig. 9, the degalactosylated antigen being common to both the types Ia and Ib polysaccharides. In quantitative precipitin experiments the native Ia polysaccharide was able to precipitate further significant quantities of antibody following absorption of the antiserum with the desialylated and carboxyl-reduced native antigen. The serological specificity of the native type Ia antigen would mitigate against sialic acid being a part of its determinant; therefore we can again postulate the function of sialic acid as being

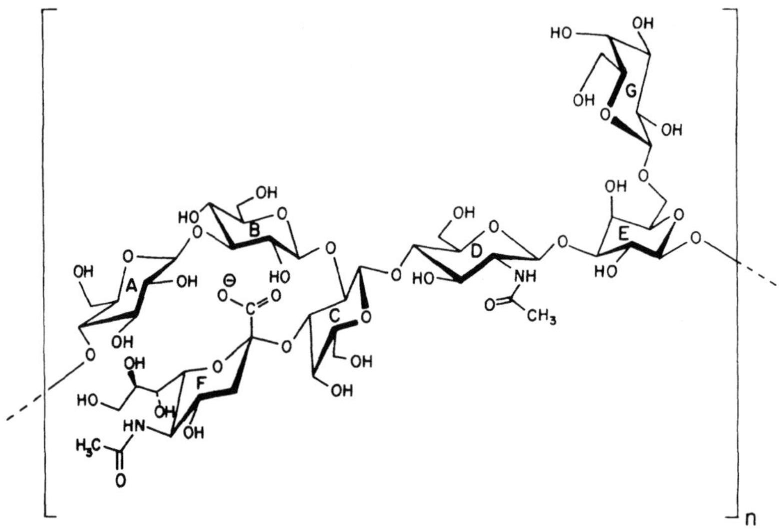

Fig. 8. The repeating unit of the type II Group B streptococcal polysaccharide.

one of conformational control over the determinant.[14] However, because of the complexity of this structure and the inability to locate the determinant from either ^{1}H- or ^{13}C-NMR chemical shift displacements in the spectra of the modified type Ia antigens, the part of the native type Ia polysaccharide with which the terminal sialic acid residue interacts has not been identified. In contrast to types Ia, II and III group B streptococcal organisms, when injected in rabbits, those of type Ib produce no sialic acid-dependent antibody to the native type Ib polysaccharide. This was demonstrated in serological studies when the native desialylated (core) (Fig. 9) and carboxyl-reduced native Ib antigens were all able to precipitate the same amount of antibody.[14,48,49] Thus the type Ib polysaccharide is a unique structure in which sialic acid is not immunodominant in either a direct or indirect sense. This can probably be attributed to its spatial location, terminal sialic acid being incapable of interacting with other branch components or other remote regions of the native type Ib polysaccharide, thus ensuring the antigenic equivalence of the native and core type Ib polysaccharides.[14]

The above arguments explain the unique function of sialic acid in the formation of the types Ia, II and III group B

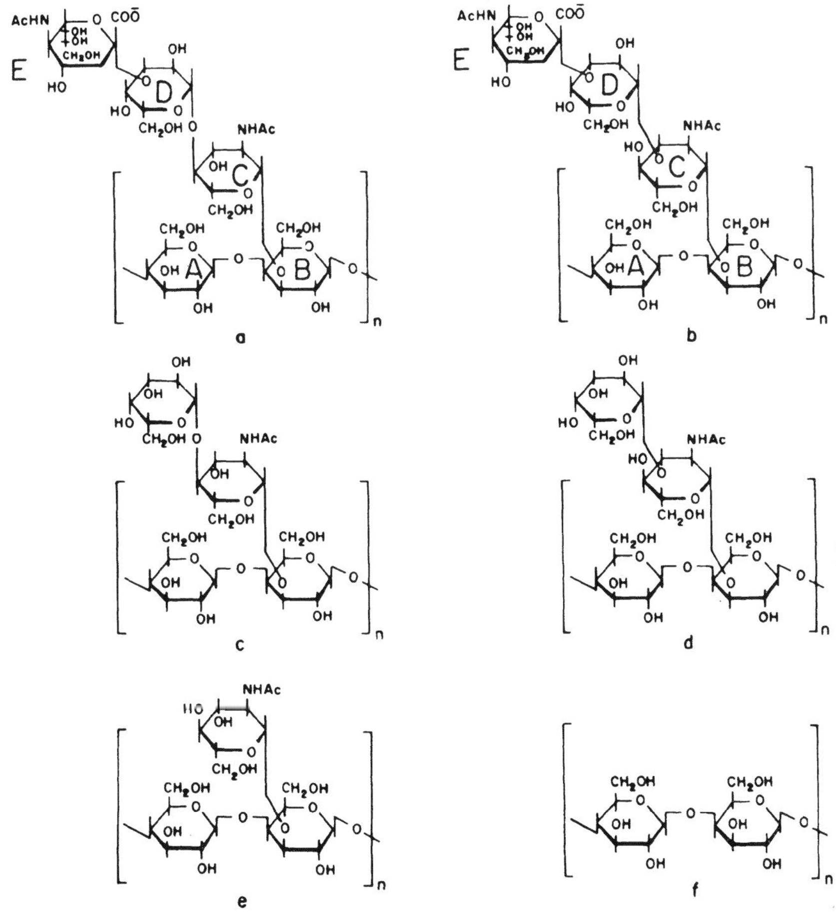

Fig. 9. Structures of the native (a), core (c), degalacto-sylated (e), and deglucosaminylated (f) type Ib Group B streptococcal polysaccharide. The structures of the equivalent type Ib antigens are (b), (d), (e), and (f) respectively.

streptococcal determinants. However, the evidence submitted is insufficient to ignore the possibility of other important factors,[14] not considered above. Certainly, in addition to hydrogen bonding, the contribution of other means of bonding must be considered in these through-space interactions. Also, because of the limitations of conventional inhibition techniques, the serological evidence supporting the conformational determinant is based on the non-immunogenicity of sialic acid as a whole and does not consider the possibility that a part of sialic acid could be included in the determinant. This might be important, for example, should the carboxylate group of sialic acid function as a small but highly critical part of the determinant.

Determinants Involving Terminal and Interchain Sialic Acid

The polysaccharides in Table 1 included in this category are those of groups B and C *Neisseria meningitidis.* The groups W-135 and Y meningococcal polysaccharides are probably also included, although there is no experimental evidence to indicate whether these polysaccharides do in fact have terminal sialic acid residues in addition to interchain sialic acid residues. The groups B and C polysaccharides of *N. meningitidis* are homopolymers of sialic acid, the sialic acid residues being linked α-(2→8) in the former and α-(2→9) in the latter (Table 1). Again, because of this linkage difference, the immunological properties of these two polysaccharides differ profoundly. While the C polysaccharide is highly immunogenic, and is currently used as a constituent of a vaccine against meningococcal meningitis in humans, the B polysaccharide is only poorly immunogenic.[12] It is also interesting to note that even in the case of the *Escherichia coli* K92 polysaccharide, which contains alternating α-(2→8)- and α-(2→9)-linked sialic acid residues, only antibodies with a specificity for the latter linkage could be detected in antisera to the homologous organism.[50]

The reason for the poor immunogenicity of the B polysaccharide is probably structure-related because α-(2→8)-linked sialooligosaccharides identical to those found in the basic structure of the B polysaccharide[20] have been identified in the glycopeptides of human and animal fetal brain tissue.[36] The sialooligosaccharides from human and rat fetal brain were demonstrated[36] to bind to B polysaccharide-specific horse antibodies and to inhibit the homologous serological reaction of these antibodies. The sialooligosaccharides from the glycopeptides of fetal brain are large and can consist of up to twelve linear α-(2→8)-linked sialic acid residues having in addition a terminal sialic acid residue.[51] As a consequence of this very extensive structural homology with such self antigens the immune system is confronted with an extremely complex problem in trying to avoid the production of tissue cross-reactive antibody. This probably accounts for the detection of extremely low levels of polysaccharide-specific antibodies produced in animals and humans by group B meningococcal organisms.[52] These antibodies are usually exclusively of the IgM isotype and are of low affinity. In addition, this structural homology with fetal brain glycopeptide antigens probably also accounts for the involvement of at least two and possibly three different complex determinants associated with the immune response to the B polysaccharide despite its apparent structural simplicity.[45] Two distinct populations of antibody to the B polysaccharide were detected in a horse antiserum to whole group B meningococcal organisms. They were differentiated on the basis of the ability of only one of them to be absorbed by the structurally related but lower molecular size (estimated at 30 sialic acid residues) colominic acid. The average size of the B polysaccharide is estimated to be approximately 50 sialic acid residues.

Experiments to define the determinant specificities associated with the two populations of group B polysaccharide-specific horse antibodies were carried out[45] by a radioimmunoassay inhibition technique using a series of linear α-(2→8)-linked oligomers of sialic acid to inhibit the binding of both extrinsically tritium-labeled colominic acid and the larger molecular size group B polysaccharide to the horse IgM antibodies. Oligosaccharides of from one to eighteen sialic acid residues were obtained by graded hydrolysis of colominic acid and separated on an ion-exchange column. The ability of the oligosaccharides to inhibit the binding of colominic acid to the group B polysaccharide-specific horse antibodies is shown in Fig. 10 and reveals that the colominic acid determinant is based on a linear arrangement of α-(2→8)-linked sialic acid residues, compatible with the proposed structure of colominic acid. However, the results indicate that the determinant is not conventional. Not only was sialic acid non-inhibitory, as expected from the previous results obtained with the group B streptococcal polysaccharides, but in addition a series of oligomers of from two to six sialic acid residues also proved to have poor inhibitory properties. Even when progressing to larger oligomers of up to seventeen sialic acid residues, of which the latter constitutes more than one-half of the colomic acid molecule, the inhibitory properties had not maximized. This behaviour provides strong evidence for the conformational dependence of the colominic acid-specific antibodies, because these latter oligosaccharides are considerably larger than the maximum estimated size of an antibody site.[41] The minimum size of oligomer which had the closest conformational resemblance to colominic acid was estimated by plotting the amount of the oligomers required to give 50% inhibition against the number of sialic acid residues in the oligomers. The curve is shown in Fig. 11 and extrapolation of the curve to the number of sialic acid residues' axis, indicated that the minimum size required to generate the conformational determinant was of the order of ten sialic acid residues. These results were confirmed by Finne and Mäkelä[51] who studied the direct binding of the same oligomers, obtained by endosialidase treatment of colominic acid, to the identical horse antibody. The oligomers were radiolabeled at the reducing end using sodium borotritide and it was again demonstrated that an oligosaccharide of at least ten sialic acid residues was required for binding. In similar inhibition studies[45] using the extrinsically labeled group B polysaccharide, all the oligosaccharides, including the larger ones (from twelve to seventeen sialic acid residues), proved to be extremely poor inhibitors. This sugggested that, if the basic structure of the B polysaccharide is in fact identical to colominic acid, as was proposed on the basis of ^{13}C-NMR spectroscopic data, then one would need to propose another determinant, having a different conformational dependence, but still being associated with the linear structure beyond the size of colominic acid. It is also conceivable, however, that this second determinant could be based on small but serologically important structural differences between the B polysaccharide and colominic acid which were not detected by ^{13}C-NMR spectroscopy.

In contrast to the B polysaccharide, the inhibition of the homologous serologic reaction of the group C polysaccharide followed a more conventional pattern,[45] with the exception of the identical phenomenon of the non-inhibitory properties of the terminal sialic acid residue. Experiments to inhibit the binding of the native C polysaccharide to its homologous rabbit antibodies using a series of α-(2$\rightarrow$9)-linked

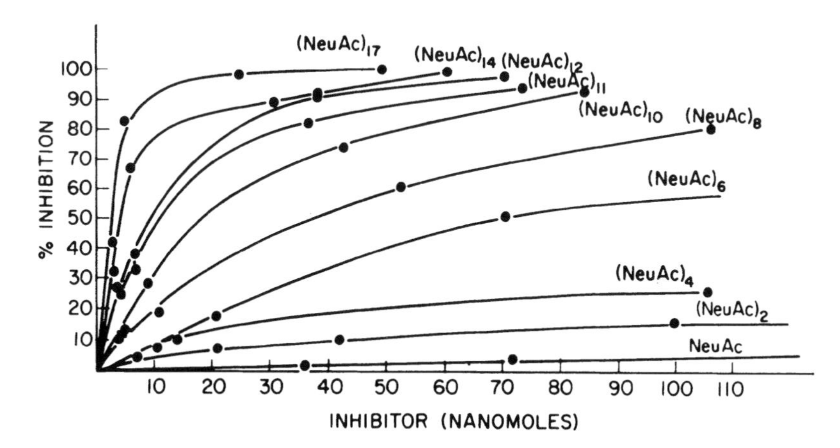

Fig.10. Quantitative precipitation inhibition of the reaction between group B meningococcal horse antiserum and colominic acid, using a series of linear α-(2→8)-linked sialic acid oligosaccharides (NeuAc)n.

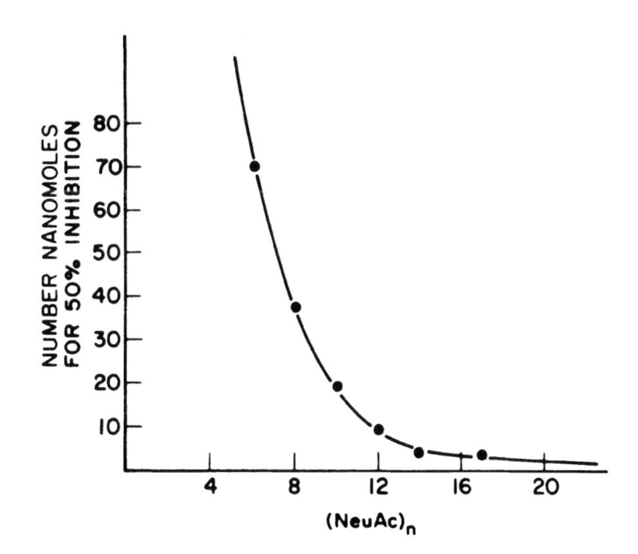

Fig.11. Quantity of linear α-(2→8)-linked sialic acid oligosaccharides of defined lengths required to cause 50% inhibition of the reaction between colominic acid and a group B meningococcal horse antiserum.

sialic acid oligomers were carried out and the inhibition curves are shown in Fig. 12. These curves indicated that the inhibitory properties of the oligomers increased rapidly with each successive addition of sialic acid and maximized with the pentamer. Thus the human immune system can respond to sialic acid residues when they are involved in interchain structures not associated with human tissue antigens. This situation probably also accounts for the good immunogenicity of the groups W-135 and Y meningococcal polysaccharides which both contain 4-linked sialic acid residues alternating with 6-linked galactopyranosyl and glucopyranosyl residues, respectively (Table 1).

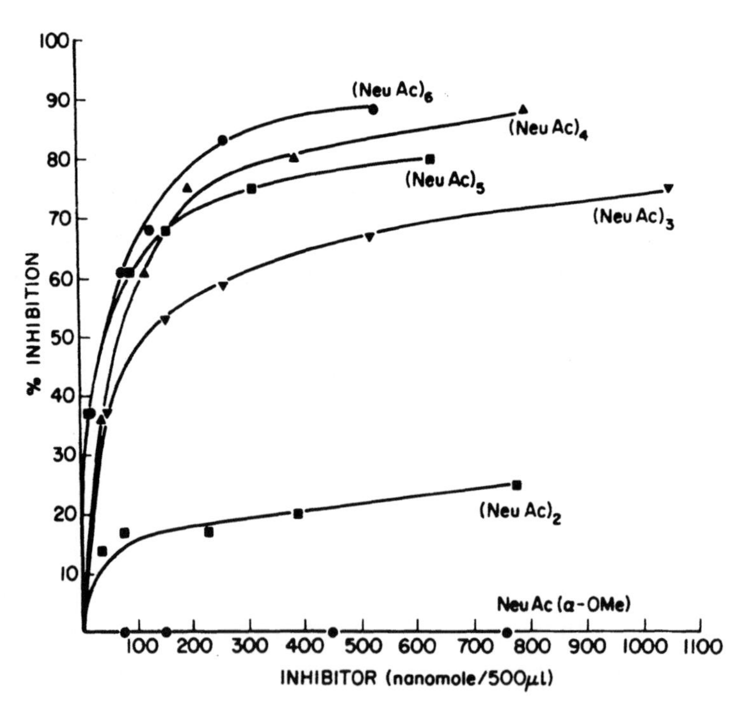

Fig.12. .Quantitative precipitation inhibition of the reaction between group C meningococcal rabbit anti-serum and the homologous group C polysaccharide using a series of α-(2→9)-linked sialic acid oligosaccharides (NeuAc)n.

MODULATION OF THE IMMUNE RESPONSE BY CHEMICAL MANIPULATION OF POLYSACCHARIDES

Human Immune Response to Polysaccharide Antigens

Before discussing the human immune response to modified polysaccharide antigens, it is necessary to understand some rudimentary facts concerning the human immune response to polysaccharide antigens. An important distinction must be made between the humoral response to a pure capsular polysaccharide and the same polysaccharide when it is an integral part of the bacterium.[7] Thus, the immunity received from infection by encapsulated bacteria, in terms of the polysaccharide antigen, differs from that generated by purposeful immunization with purified capsular polysaccharide vaccines. Fortunately, with the exception of infants, the polysaccharide vaccines are still able to stimulate protective antibody levels in humans, despite these differences.

For most antigens, the production of antibody (immunoglobulin) is based on the cooperative interaction of two types of lymphocytes, called T-cells (thymus-derived) and B-cells (bone marrow derived). The T-cells, preprimed with macrophage-presented antigen, stimulate the B-cells to secrete copious quantities of antibody. This type of response is associated with its ability to be boosted to higher levels on further exposure to the antigen and with the production of antibodies of the IgG isotype. However, on the basis of studies in mice, such polysaccharide antigens have been considered to be T-cell independent, in that they are only capable of inducing IgM antibodies; exhibiting no amnestic effect.[7] Although the above observations are basically correct, polysaccharides are not in fact T-cell independent in that the response to them is affected by modulator T-cells.[53] This type of T-cell has been shown to amplify or suppress the B-cell response to polysaccharides in mice, and, unlike the situation for whole bacteria, polysaccharides cause the suppressor mechanisms to be dominant.

If one projects the results obtained from experiments with mice to the human situation, the use of polysaccharides as efficacious human vaccines would in fact show little promise. However, the immunological response to polysaccharides in humans, with the exception of infants,[6,7,13] is different, and involves the production of a fuller range of antibody types. Thus polysaccharides are able to stimulate the production of IgG antibodies in humans,[54] together with those of IgM and IgA, but, as in the mouse experiments, they fail to exhibit a significant amnestic response to subsequent booster injections of the polysaccharide.[6,7] Polysaccharides maintain their use in human immunoprophylaxis because, despite the absence of a significant amnestic response, the antibody levels induced by polysaccharides in humans are generally high enough to provide effective protection against infection and remain efficacious for fairly long periods of time.[55] By comparison, the infant immune response to polysaccharide antigens is immature and follows the same pattern as that observed in mice.[13] Therefore, by extrapolation this immaturity can be conveniently related to T-cell suppressor function. However, because of the lack of evidence, it must be remembered that there are inherent risks in making such extrapolations.

Natural Immunity and Polysaccharide Serological Cross-Reactions

Serological cross-reactions among polysaccharides are a well documented phenomenon, due in large part to the extensive work conducted in this area by Heidelberger and co-workers.[56,57] This phenomenon can be attributed to the unique property of polysaccharides to retain domains of structural and conformational similarity, despite having structural differences. Heidelberger[56,57] and co-workers have used this phenomenon to advantage in probing polysaccharide structures, where an antiserum to a polysaccharide of known structure is used as a reagent to identify common structural features in other polysaccharides.

However, in addition to its analytical value, this phenomenon is also of profound significance to the human immune mechanism. This is because the most satisfactory explanation for the age-related maturation of the immune response to polysaccharides in humans is that it is due to exposure of infants to non-pathogenic microorganisms containing cross-reactive antigens.[6] It can be shown, for instance, that there is an age-related increase in natural antibodies to the group A meningococcal polysaccharide in children even though group A organisms are rarely isolated in North America.[58] Frequently found in normal human flora are cross-reacting organisms that could be responsible for natural immunity to group A, B and C *N. meningitidis* and type b *H. influenzae.* Robbins and co-workers have identified some of these important organisms and they are listed together with structural detail on their capsular polysaccharides in Table 2. Serological studies indicate that the capsular polysaccharides of these organisms are responsible for the cross-reactions and this is confirmed by the recognition of structural similarities between them (Table 2) and the polysaccharides from *N. meningitidis* and *H. influenzae* (Table 1). These structural similarities could form the basis of cross-reacting determinants. The exception to the rule is the polysaccharide of *Escherichia coli* K93 (Table 2) which is unconventional in that, although highly cross-reactive with the group A meningococcal polysaccharide (Table 1), does not share with it one single common glycose residue or linkage.[59] In this case it is impossible to recognize a common determinant from the structural information described in Tables 1 and 2. The probable explanation is that the common determinant can only be recognized by comparing the two dimensional structures of these two polysaccharides. The feasibility of the above phenomenon being involved in the human immune mechanism to pathogenic bacteria has been clearly demonstrated by Schneerson and Robbins[61] by deliberately feeding non-pathogenic *E. coli* possessing the K100 capsule to human-adult volunteers. Colonization readily occurred, and antibodies specific for the *H. influenzae* type b polysaccharide were induced.

Table 2. Polysaccharides of bacteria frequently found in human flora, that cross-react with the polysaccharide capsules of human pathogenic bacteria.

Pathogen	Cross-reacting organism	Structure	References
Neisseria meningitidis			
Group A	*Escherichia coli* K93	→4)βD-GlcpA(1→3)βD-[5,6-OAc]-Galf(1→	59
	Escherichia coli K51	→3)[6-OAc]αD-GlcpNAc(1-P(=O)(OH)-O-	59
	Bacillus pumilis	(1→6)-linked 2-acetamido -2-deoxymannosyl phosphate residues	6
Group B	*Escherichia coli* K1	→8)αD-NeupAc(2→ and its OAc+ variant	60
Group C	*Escherichia coli* K92	→8)αD-NeupAc(2→9)αD-NeupAc(2→	50
Haemophilus influenzae			16
Type b	*Escherichia coli* K100	→3)βD-Ribf(1→2)D-ribitol(5-O-P(OH)(=O)-O-	

Polysaccharide-Protein Conjugate Vaccines

The use of capsular polysaccharides as immunoprophylactic agents in human disease caused by bacteria is now firmly established. However, despite the many advantages of using capsular polysaccharides as human vaccines, a number of limitations to their general applicability have been identified. One of these limitations is associated with the poor immune response of these purified polysaccharides exhibited by infants.[6,7] This poor immune response is due to the immaturity of the infants' immune system, the basis of which has been previously discussed. This is a serious problem because young children are especially susceptible to bacterial meningitis and immunity to bacteria causing this disease has been clearly demonstrated to be due to the presence of circulating capsular polysaccharide-specific antibodies. Infant bacterial meningitis can be divided into that which occurs during the first weeks of life (neonatal meningitis) and the disease which occurs after this time. Susceptibility to neonatal meningitis has also been demonstrated to be due to a lack of polysaccharide-specific maternal antibodies.[13] It is also interesting to note that the two periods of susceptibility for the infant can also be differentiated by the type of bacteria involved. *Escherichia coli* K1 and group B *Streptococci* are important etiological agents in neonatal meningitis, whereas *Haemophilus influenzae* and *Neisseria meningitidis* are principally associated with disease in older infants together with a few types of *Streptococcus pneumonia*.[7]

A promising area of research to overcome the deficiencies of polysaccharide vaccines in infants is the design of a new generation of semi-synthetic vaccines based on the conjugation (covalent coupling) of the polysaccharides to protein carriers. Because immunization of the neonate is impractical, and the alternate vaccination strategy would involve the currently less acceptable immunization of pregnant women, most of the current research has evolved on the theme of developing a comprehensive infant vaccine against post neonatal meningitis. All of the capsular polysaccharides of the bacteria primarily responsible for post neonatal meningitis have been conjugated to protein carriers; these include type b *H. influenzae*,[62,63] groups A,[64,66] B[64] and C[64,65] *N. meningitidis*, and type 6A *S. pneumoniae*.[67] The choice of carrier proteins used in making the above conjugates was also influenced by the eventual goal of the vaccine and investigators have concentrated on using tetanus toxoid and diphtheria toxoid. Both proteins have already been extensively used for many years as components of infant vaccines.

The feasibility of the above approach is well established. Fifty years ago, Goebel and Avery[68] coupled the type 3 pneumoccocal polysaccharide to horse serum-globulin by the diazotization of p-aminobenzyl ether substitutuents on the polysaccharide. They demonstrated that this polysaccharide conjugate and a similar conjugate made with the oligosaccharide repeating unit (cellobiouronic acid) of the type III pneumococcal polysaccharide, were able to induce polysaccharide-specific antibody in rabbits unresponsive to the pure polysaccharide.[69] Goebel[70] also established that the cellobiouronic acid-conjugate was also able to confer immunity to challenge by pneumococci in mice and all these results have been substantiated by others. The coupling procedures employed in this early work, however, were far too drastic to be used on some of the highly sensitive polysaccharides currently used as human vaccines, and resulted in complete random coupling of the two molecules with the incorporation of highly undesirable structural features into the conjugates. Recently, more comprehensive studies on polysaccharide-protein conjugates, specifically directed to their use as human vaccines, have now been reported.

Two approaches to the synthesis of these conjugates have been described which involve either the use of random or selective activation sites in the polysaccharide as potential linkage sites.[71] In effect, the choice of the method employed is largely dictated by the molecular size of the polysaccharide; the larger molecule size polysaccharide of necessity requiring random activation in contrast to the facile end-group activation made possible by the use of oligosaccharides or polysaccharides of smaller size. The use of large molecular size polysaccharides in conjugates probably originated because of the deliberate development of polysaccharides in the form for use as human vaccines.[7] This strategy was based on the knowledge that high molecular size polysaccharides function as superior immunogens. Whether this remains true when polysaccharides are coupled to large immunogenic proteins is doubtful and this is substantiated in evidence from some recent comparative immunological studies. Although these were not extensive, they do indicate strongly that both oligosaccharide-protein conjugates and small molecular size polysaccharide-protein conjugates [63,64,72,73] are able to perform as immunogens, either better than, or at least as good as, their larger molecular size polysaccharide-protein conjugates. This result could be important in the development of polysaccharide-protein conjugate vaccines, because it is obviously advantageous to use as simple and easily definable immunogen as possible. While this principle is compatible with the end-group activation of polysaccharides, random activation interferes with their basic structures, promotes cross-linking and results in complex ill-defined conjugates.

Some examples of conjugation using the random activation technique are as follows. The large molecular size *H. influenzae* type b polysaccharide was conjugated to a number of proteins by Schneerson and co-workers[62] by activating the polysaccharide with cyanogen bromide and functionalizing the protein with an adipic dihydrazide spacer. The coupling occurs mainly through the formation of N-substituted isourea bonds but both the polysaccharide and protein retain unnecessary substituents, i.e. carbonate and underivatized spacer. The large molecular size group C polysaccharide from *N. meningitidis* was also successfully conjugated directly to tetanus toxoid by Beuvery and co-workers[65] using 1-(3-dimethylaminopropyl)-3-ethylcarbodiimidehydrochloride (DEC). This method introduces active O-acylisourea groups at many of the polysaccharides carboxyl groups, some of which, in the presence of protein, form the required amide bonds of the conjugate. However, the introduction of O-acylisourea groups can cause drastic structural changes in the polysaccharide as a result of internal lactonization,[74] and the quenching of the remainder with ethanolamine introduces permanent unwanted substituents into the polysaccharide.

To develop a more specific approach to the coupling of meningococcal polysaccharides to tetanus toxoid, they were monofunctionalized prior to conjugation.[64] Controlled periodate oxidation introduced unique terminal free aldehyde groups into the groups B and C polysaccharides and in the group A polysaccharide following the reduction of its terminal 2-acetamido-2-deoxy-D-mannose residue (Fig. 13). Those monovalent polysaccharides were then coupled to tetanus toxoid by reductive amination. The advantages of this method over those previously described are that it minimizes the possibility of polysaccharide or protein modification, and eliminates cross-linking. An even more effective method of conjugation would be the direct coupling of the unoxidized meningococcal polysaccharides to protein through their end-group reducing residues. However, this proved to be ineffective. This lack of activity could be associated with the molecular size of these polysaccharides, but it has also been established that 2-keto-3-deoxy-glyculosonic acid residues, which include the sialic end-group residues of the group B and C meningococcal polysaccharides, are extremely ineffective residues through which to carry reductively aminated coupling because of their preferential reduction with the reagent (sodium cyanoborohydride).[75] However, using the above method, Anderson[63] was able to effectively couple oligosaccharides of different sizes obtained by the deploymerization of the type b *H. influenzae* polysaccharide to a non-toxic version of diphtheria toxin (CRM 197).

A requirement for using the monofunctional approach to couple saccharides to proteins is that the saccharides have first to be obtained in an appropriate molecular size. This can be achieved by depolymerizing the native polysaccharide and subsequently fractionating the component saccharides by gel-filtration. Depolymerization methods employed have included autohydrolysis,[64] acid hydrolysis,[63] and controlled periodate oxidation[72] which in contrast to the first two methods results in the production of difunctional saccharides. Recently, ultrasonic irradiation[76] has been used to

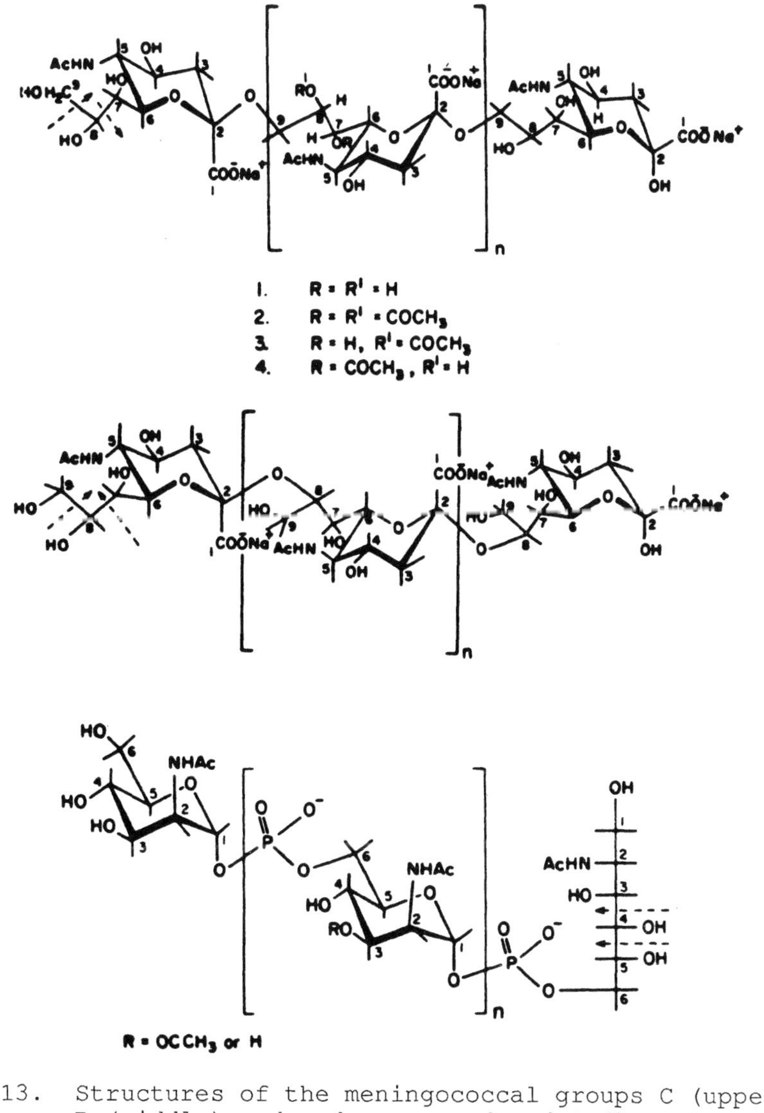

1. R = R' = H
2. R = R' = $COCH_3$
3. R = H, R' = $COCH_3$
4. R = $COCH_3$, R' = H

R = $OCCH_3$ or H

Fig. 13. Structures of the meningococcal groups C (upper), B (middle) and end group reduced A (lower) capsular polysaccharides depicting their periodate-sensitive terminal residues.

depolymerize a number of representative bacterial polysaccharides. This method is independent of structure and, unlike the other methods of depolymerization, has the distinct advantage of producing fragments of a finite and similar molecular size (ca. 50,000 daltons).

The groups A and C meningococcal polysaccharide-conjugates were able to stimulate in mice the production of polysaccharide-specific IgG antibodies indicative of the conversion of the polysaccharides to thymus dependent antigens.[64] The failure of the group B polysaccharide-conjugate to behave in the same way is associated with the inherent poor immunogenicity of the B polysaccharide; a problem discussed in more detail later. The thymus dependency of the group C polysaccharide-conjugate is demonstrated in Fig. 14 which depicts ELISA assays of the antisera from mice previously injected with the group C polysaccharide alone and in the form of its tetanus toxoid conjugate. In contrast to the pure group C polysaccharide, which elicited a very weak immune response in mice, the conjugate was able to induce in them high levels of polysaccharide-specific IgG antibodies. This phenomenon had been previously reported by Schneerson and co-workers[62] following the immunization of mice with the type b *H. influenzae* polysaccharide-tetanus toxoid conjugate and appears to be quite general as it has also been observed in similar animal experiments using other conjugates produced by many diverse procedures.[63,65,66,73]

From the immunological properties of these conjugates in animals one could predict that conjugate vaccines would have obvious potential in the formulation of future infant vaccines against bacterial meningitis and this was amply verified in recent highly successful human studies.[77,78] Two month old infants were immunized with protein-coupled oligosaccharides derived from the capsule of *H. influenzae* type b and observed an amnestic (IgG) polysaccharide-specific response following a second immunization with the conjugate.[77] Enhancements of polysaccharide-specific response in infants have also been obtained using an *H. influenzae* type b-diphtheria toxoid conjugate.[78] All the above studies indicate that polysaccharide-protein conjugate vaccines will be valuable

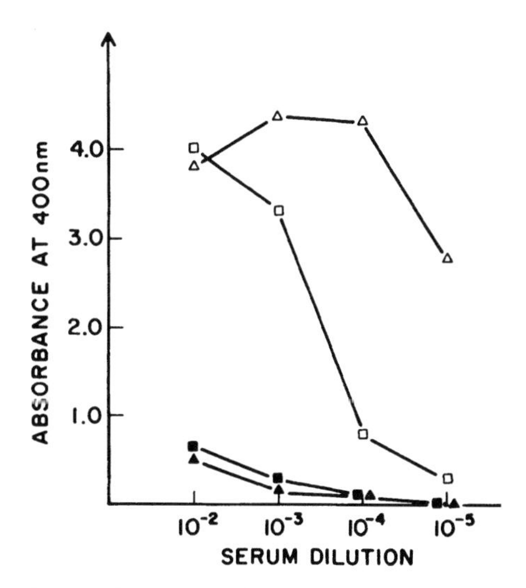

Fig. 14. Titration of an anti-group C polysaccharide-tetanus toxoid conjugate mouse serum in ELISA against different antigens. The wells were coated with the homologous conjugate (Δ) and the homologous group C meningocccal polysaccharide (□). Titration of the preimmune sera with the same antigens are marked with identical but solid symbols (▲, ■).

agents in the prevention of infant meningitis and that this technology is obviously capable of extension into other areas of immunoprophylaxis. However, the final optimal configuration of these conjugate vaccines will depend on criteria of their acceptability as human biologicals, as yet undefined, and on maximizing their immunological performance by varying the many structural parameters involved in their synthesis.

Chemical Modification of the Group B Meningococcal Polysaccharide

The poor immunogenicity of the group B meningococcal polysaccharide and the structurally identical *E.coli* K1 capsular polysaccharide preclude their use as human vaccines against meningitis caused by group B meningococcal and *E.coli* K1 organisms.[12] Although group B meningococcal organisms are able to produce low levels of group B polysaccharide-specific antibodies in animals and humans, these antibodies are almost exclusively IgM and of relatively low affinity.[52] The most probable explanation for the poor immunogenicity of these polysaccharides is that they induce in humans and animals a state of tolerance due to the presence of cross-reactive tissue components. This hypothesis is well supported by the identification of structurally similar α-(2→8)-linked oligomers of sialic acid in the glycopeptides of human and animal brain tissue.[36,51] Group B meningitis remains a major health problem in the world, and the lack of immunogenicity of the group B meningococcal polysaccharide prevents the formulation of a comprehensive polysaccharide vaccine against meningococcal meningitis. This has stimulated interest in alternative approaches to making a group B meningococcal vaccine based on carbohydrate antigens. One of these strategies involves chemical manipulation of the group B meningococcal polysaccharide. While the direct covalent coupling of the group B polysaccharide yielded a conjugate which failed to induce a polysaccharide-specific antibody response in rabbits, as determined by immunodiffusion analysis,[64] a measurable enhancement of its immune response in mice was detected using more sensitive assays (ELISA and radioimmunoassay).[79] The enhancement was only meager, however, and this fact, together with the inability to detect bactericidal activity in the antisera, suggests that the direct coupling of the B polysaccharide to protein will probably be of little importance in the production of a future human vaccine against group B meningococcal meningitis. One positive aspect of the above failure was that it prompted interest in a more esoteric approach, namely the direct chemical modification of the B polysaccharide itself. This was done with the idea of creating synthetic epitopes capable of modulating the immune system in such a way as to produce enhanced levels of cross-reactive B polysaccharide-specific antibodies. There is only one previous report [60] of a similar approach in the literature, but these authors did not use a synthetic antigen and the enhancement of immunogenicity they observed in rabbits was only marginal. The *E.coli* K1 polysaccharide is structurally and serologically identical to the group B meningococcal polysaccharide (Table 2), but a variant of the *E.coli* K1 organism produces a capsular polysaccharide randomly O-acetylated at O-7 and O-9 of its sialic acid residues. This variant (OAc^+) of the *E.coli* K1 organisms proved to be relatively more immunogenic in rabbits than the (OAc^-) variant, producing antibodies having specificities for both variants. This experiment in itself was not particularly encouraging but serendipity played a role when it was demonstrated that a synthetic structurally modified B polysacccharide antigen could be so constructed as to produce in mice high-titers of cross-reactive B polysaccharide-specific IgG antibodies.

In selecting possible chemical modifications of the group B polysaccharide two major requirements had to be met. First, the chemical modification had to be accomplished with facility and with the minimum degradation of the polysaccharide, and, secondly, the antigenicity of the modified polysaccharide to native B polysaccharide-specific antibodies had to be preserved. The most successful modification which satisfied the above criteria was that in which the N-acetyl groups of the sialic acid residues of the B polysaccharide were removed by strong base and replaced by N-propionyl groups (Fig. 15). Serological studies [79,80] on a number of different modifications of the B polysaccharide indicated that the retention of both the carboxylate and N-carbonyl groups of the sialic acid residues were essential to the formation of a cross-reactive determinant; factors which could be related to the preservation of its conformation. Due to the poor immunogenicity of the N-propionylated B polysaccharide itself in mice, it was subsequently conjugated to tetanus toxoid using procedures previously described [64] for the formation of the B polysaccharide-tetanus toxoid conjugate, thus yielding an artificial and virtually synthetic antigen.

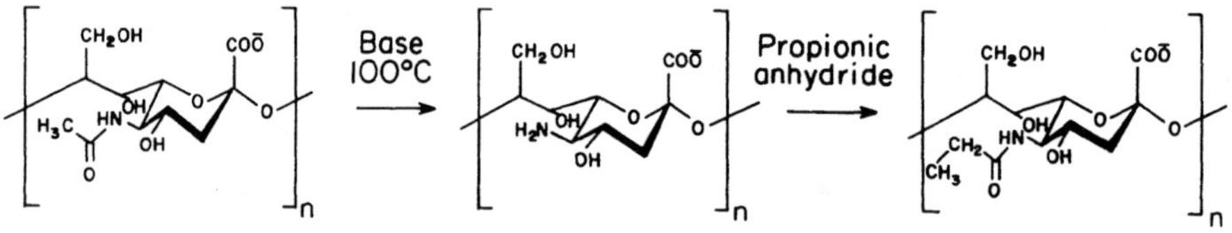

Fig. 15. Reaction sequence leading to the formation of the N-propionylated group B meningococcal polysaccharide from the native group B polysaccharide.

The potential of the N-propionylated B polysaccharide to enhance the induction of group B polysaccharide-specific antibodies was demonstrated by comparison of the immune response of the N-propionylated B polysaccharide-tetanus toxoid conjugate with that of the native N-acetylated B polysaccharide-tetanus toxoid conjugate in rabbits.[80] Precipitin analysis using the homologous polysaccharides indicated that the former conjugate not only elicited very high levels of antibody with a specificity for its homologous N-propionylated B polysaccharide, but also raised higher levels of native B polysaccharide-specific antibodies than the native N-acetylated B polysaccharide-tetanus toxoid conjugate itself. Studies on the binding of the B polysaccharide to the above conjugate antisera were also unusual in that antibodies raised to the N-propionylated B polysaccharide-conjugate had a much higher affinity for the native N-acetylated B polysaccharide than antibodies raised to the homologous B polysaccharide-conjugate.

The observations made in the above preliminary experiments were amply verified[79] by subsequently more comprehensive and statistically accurate experiments in which a comparison was made of the B polysaccharide-specific antibody response induced in mice by the N-propionylated and N-acetylated B polysaccharides and their respective tetanus toxoid conjugates (Fig. 16). While antisera from mice immunized repeatedly with the polysaccharides alone showed no significant binding to the group B polysaccharide, binding was detected in antisera obtained from mice immunized with each respective tetanus toxoid conjugate. However, as in the case of rabbits, the N-propionylated B polysaccharide conjugate produced much higher levels of B polysaccharide-specific antibody than the homologous N-acetylated B polysaccharide conjugate; a pronounced booster effect being particularly noticeable following three injections (1°, 2° and 3°). Interestingly, while the N-acetylated B polysaccharide-tetanus toxoid conjugate gave only a small booster effect in terms of B polysaccharide-specific antibody, it was able to significantly boost this same response in mice which had been previously primed with two previous injections of the N-propionylated B polysaccharide conjugate. This booster effect is indicative of a memory effect based on the participation of T-cells, and this fact is also substantiated by the large proportion of B polysaccharide-specific antibodies of the IgG isotype produced in this immune response (Fig. 16).

Because the N-propionylated B polysaccharide-tetanus toxoid conjugate is capable of inducing B polysaccharide-specific IgG antibodies in mice,[79] it must be considered as a prototype vaccine against meningitis caused by the bacteria group B *N. meningitidis* and *E.coli* K1. Experience would indicate that an immunogenic form of this capsular polysaccharide would be the ideal vaccine candidate, although, on the basis of structural homology between the B polysaccharide and the oligosaccharides of glycopeptides associated with fetal brain,[36,51] it might be inferred that its success as a vaccine could only be achieved at the risk of breaking tolerance. Certainly the production of B polysaccharide-specific IgG antibodies in mice indicates that the immunogen is capable of breaking tolerance. What the consequences of breaking tolerance to this antigen would be and whether in fact it could be avoided are unknown at present. However, because of the importance of developing a group B meningococcal vaccine, the exploration of chemically modified group B meningococcal polysaccharide-conjugate vaccines needs to be actively pursued.

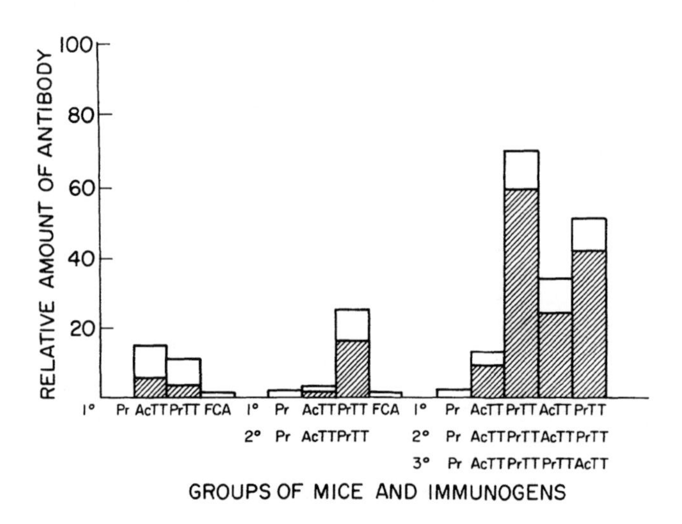

Fig. 16. Mean group B polysaccharide-specific antibody response in groups of mice immunized thrice (1°,2° and 3°) with different antigens as determined by radioactive antigen binding assay. The blank area of the histogram indicates the IgM response; the hatched area, the IgG response. The mice are grouped according to the immunogen in immunogens used, and the immunogens are referred to by standard abbreviated nomenclature, e.g. AcTT and PrTT = the tetanus toxoid conjugates of the native and N-propionylated group B polysaccharide, respectively. F.C.A. = Freunds' complete adjuvent.

Lipopolysaccharide Core-Protein Conjugates

An alternative approach to making a group B meningococcal vaccine, based on carbohydrate antigens, would be the use of subcapsular lipopolysaccharides, of which at least eight serologically different types (serotypes) are associated with group B organisms.[81] Obviously, because of their highly toxic nature, the intact lipopolysaccharides cannot be used as vaccines, but mild hydrolysis of the lipopolysaccharides yielded core oligosaccharides of molecular size 1200 to 1600 daltons which were the major serological determinants of the lipopolysaccharides.[82] The structure [83] of one of the major serotype (L3) oligosaccharides associated with meningococcal meningitis, minus its O-ethanolamine phosphate residues, is shown in Fig. 17 and, except for minor structural differences, this basic structure is representative of all the serotype R-type oligosaccharides studied to date.

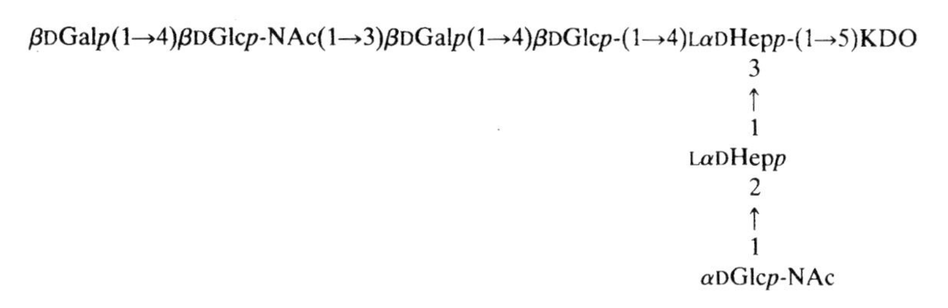

Fig. 17. Structure of the dephosphorylated oligosaccharide from the meningococcal serotype L3 lipopolysaccharide.

In order to form effective immunogens from these haptenic structures, they were conjugated to carrier proteins.[84] The simple direct coupling of the oligosaccharides to protein by reductive amination was not possible due to the reluctance of the 2-ketose group of the terminal 3-deoxy-ß-D-manno-2-octuloslylonic acid (KDO) residues to participate in this coupling reaction. Instead, the 2-ketose groups were preferentially reduced by sodium cyanoborohydride, thus inactivating the oligosaccharides completely in terms of their potential to couple to protein under these reaction conditions.[75] However, reductive amination of the KDO residues could be achieved using a large molar excess of a bifunctional small molecular size spacer containing at least one amino group.[84] Once attached to the oligosaccharide, the other functional group of the spacer could then be used to couple the oligosaccharide to protein. For the sake of convenience, these conjugates were carried out in preliminary studies using a previously published procedure,[85] involving the use of a 2-(4-isothiocyanatophenyl)-ethylamine spacer. This procedure resulted in the incorporation of from eighteen to thirty-eight oligosaccharides per molecule of tetanus toxoid.

When injected in rabbits, the tetanus toxoid conjugates of the dephosphorylated oligosaccharides obtained from the L2, L3, L5 and L10 lipopolysaccharides induced in them oligosaccharide-specific antibodies.[84] The specificity of these antibodies resembled very closely the organism-induced lipopolysaccharide-specific antibodies used to serotype meningococcal organisms. In bacterial assays listed in Table 3, all the conjugate-induced antisera had bactericidal activity against meningococcal organisms. This activity was predominantly associated with their respective homologous serotypes, although some minor cross-bactericidal activity was exhibited. From the perspective of using these oligosaccharide-conjugates as a human vaccine against group B meningococcal meningitis, the procedure warrants further study. However, two potential problems have been identified

Table 3. Bactericidal activities of oligosaccharide-conjugate antisera against homologous and heterologous meningococcal organisms.

Serotype conjugate antiserum	Bactericidal activity against strain			
	L2	L3	L5	L10
L2	128	0	32	0
L3	32	4	0	0
L5	0	0	512	8
L10	64	0	32	1024

in the above studies. One is that, on the basis of the number of serotypes associated with group B meningococcal organisms, it is likely that a complex multivalent oligosaccharide-conjugate would be required to provide complete protection from disease caused by these organisms, and the other concerns the poor immunogenicity and reduced bactericidal activity (Table 3) of the L3 oligosaccharide-conjugate. As in the case of the group B polysaccharide, this could again be due to immune tolerance because lacto-N-neotetraose, a structure associated with animal and human tissue antigens, e.g. glycosphing lipids of human plasma,[86] has also been identified as the longer of the two branches of the L3 oligosaccharide (Fig.17).

Summary

Capsular polysaccharides have assumed an important role as vaccines against disease caused by bacteria in humans. The concept of using pure definable polysaccharides devoid of their accompanying complex bacterial mass is technically elegant and is obviously capable of extension into other areas of immunoprophylaxis. However, problems have been identified which will need to be solved in order that the concept may be more widely adopted. Focusing on the meningococcal polysaccharides, possible solutions to two of these important problems, namely, the poor immunogenicity of the A and C polysaccharides in infants, and the poor immunogenicity of the B polysaccharide in all humans, are proposed. These solutions involve the use of a new generation of artificial synthetic antigens for modulating the immune response. For instance, conjugation of the A and C polysaccharides to tetanus toxoid converted them to T-cell dependent antigens in mice, thus making these conjugates potential infant vaccine candidates. Although a similar conjugation of the B polysaccharide failed to substantially enhance its immunogenicity in mice, this could be achieved by further chemical manipulation of the basic structure of the B polysaccharide. N-propionylation of the B polysaccharide, followed by its conjugation to tetanus toxoid, yielded an antigen, which when injected in mice, induced in them high titers of cross-reactive B polysaccharide-specific IgG antibodies.

The chemical modification of polysaccharides requires an understanding of the interrelation between their structures and immunospecificities, and the structural elucidation of polysaccharides and the resultant monitoring of their structural modifications, can be conveniently accomplished using a wide range of NMR spectroscopic techniques. The capsular polysaccharides of many of the bacteria which cause meningitis in humans contain sialic acid and have extensive structural homology with human tissue. As a result of this homology the immunospecificities of these polysaccharides are complex, being based on unconventional conformational determinants.

REFERENCES

1. Dochez, A. R. and Avery, O.T. (1917). The elaboration of specific soluble substance by Pneumococcus during growth. J. Exp. Med. 26:477-493.
2. Francis, T. Jr. and Tillet, W.S. (1930). Cutaneous reactions in pneumonia. The development of antibodies following the intradermal infection of type-specific polysaccharide. J. Exp. Med. 52:573-585.
3. Heidelberger, M. and Avery O.T. (1923). Soluble specific substance of Pneumococcus. J. Exp. Med. 38:73-79.
4. Heidelberger, M. and Kendall, F.E. (1935). Soluble specific substance of Pneumococcus III. A quantitative study and a theory of the reaction mechanism. J. Exp. Med. 61:563-591.
5. McLeod, C.M., Hodges, R.G., Heidelberger, M. and Bernhard, W.G. (1945). Prevention of pneumococcal pneumonia by immunization with specific capsular polysaccharides. J. Exp. Med. 82:445-465.
6. Robbins, J.B. (1978). Vaccines for the prevention of encapsulated bacterial diseases: Current status, problems and prospects for the future. Immunochemistry 15:839-854.
7. Jennings, H.J. (1983). Capsular polysaccharides as human vaccines. Adv. Carbohydr. Chem. Biochem. 41:155-208.
8. Finland, M. (1979). Emergence of antibiotic resistance in hospitals. Rev. Infect. Dis. 1:4-21.
9. Sell, H.W., Merrill, R.E., Dayne, O.E. and Zimsky, E.P. (1972). Long term sequelae of *Haemophilus influenzae* meningitis. Pediatrics 49:206-211.
10. Austrian, R. (1976). Vaccines of pneumococcal capsular polysaccharides and the prevention of pneumococcal pneumonia. In The Role of Immunological Factors in Infectious, Allergic and Autoimmune Processes. (Edited by Beers, R.F. Jr. and Bassett, E.G.). pp. 79-89. Raven Press, New York.
11. Bishop, C.T. and Jennings, H.J. (1982). Immunology of polysaccharides. In The Polysaccharides. (Edited by Aspinall, G.O.). pp. 291-330. Academic Press, New York.

12. Wyle, F.A., Artenstein, M.S., Brandt, D.L., Tramont, D.L., Kasper, D.L., Altieri, P., Berman, S.L. and Lowenthal, J.P. (1972). Immunologic response of man to group B meningococcal polysaccharide vaccines. J.Infect. Dis. 126:514-522.
13. Gotschlich, E.C., Goldschneider, I., Lepow, M.L. and Gold, R. (1977). The immune responses to bacterial polysaccharides in man. In Antibodies in Human Diagnosis and Therapy. (Edited by Haber, E. and Krause, R.M.). pp. 391-402. Raven Press, New York.
14. Jennings, H.J., Katzenellenbogen, E., Lugowski, C., Michon, F., Roy, R. and Kasper, D.L. (1984). Structure, conformation and immunology of sialic acid-containing polysaccharides of human pathogenic bacteria. Pure Appl. Chem. 56:893-905.
15. Jennings, H.J., Bhattacharjee, A.K., Bundle, D.R., Kenny, C.P., Martin, A. and Smith, I.C.P. (1977). Structures of the capsular polysaccharides of *Neisseria meningitidis* as determined by carbon-13 NMR spectroscopy. J. Infect. Dis. Suppl. 136:S78-S83.
16. Egan, W. (1980). Structure of the capsular polysaccharide antigens from *Haemophilus influenzae* and *Neisseria meningitidis* by ^{13}C NMR spectroscopy. In Magnetic Resonance in Biology. (Edited by Cohen, J.S.). pp.197-258. Wiley, New York.
17. Bax, A., Egan, W. and Kovac, P. (1984). New NMR techniques for structure determination and resonance assignments of complex carbohydrates. J. Carbohydr. Chem. 3:593-611.
18. Kenne, L. and Lindberg, B. (1983). Bacterial polysaccharides. In The Polysaccharides. (Edited by Aspinall, G.O.). pp. 277-363. Academic Press, New York.
19. Bundle, D.R., Smith, I.C.P. and Jennings, H.J. (1974). Determination of the structure and conformation of bacterial polysaccharides by carbon-13 nuclear magnetic resonance. J. Biol. Chem. 249:2275-2281.
20. Bhattacharjee, A.K., Jennings, H.J., Kenny, C.P., Martin, A. and Smith, I.C.P. (1975). Structural determination of the sialic acid polysaccharide antigens of *Neisseria meningitidis* serogroup B and serogroup C with carbon-13 NMR. J. Biol. Chem. 250:1926-1932.
21. Bhattacharjee, A.K., Jennings, H.J., Kenny, C.P., Martin, A. and Smith, I.C.P. (1976). Structural determination of the polysaccharide antigens of *Neisseria meningitidis* serogroups Y, W-135 and BO. Can. J. Biochem. 54:1-8.
22. Crisel, R.M., Baker, R.S. and Dorman, D.E. (1975). Capsular polymer of *Haemophilus influenzae* type b. Part. 1. Structural characterization of the capsular polymer of strain Eagan. J. Biol. Chem. 250:4926-4930.
23. Branefors-Helander, P., Erbing, C., Kenne, L. and Lindberg, B. (1976). Structural studies of the capsular antigen form *Haemophilus influenzae* type b. Acta. Chem. Scand. Ser. B. 30:276-277.
24. Rebers, P.A. and Heidelberger, M. (1961). The specific polysaccharide of type VI Pneumococcus. II. The repeating unit. J. Am. Chem. Soc.83:3056-3059.
25. Jennings, H. J., Katzenellenbogen, E., Lugowski, C. and Kasper, D.L. (1983). Structure of native

polysaccharide antigens of types Ia and Ib groups B *Streptococcus*. Biochemistry 22:1258-1264.
26. Jennings, H. J.,Rosell, K. -G., Katzenellenbogen, E. and Kasper, D.L. (1983). Structural determination of the capsular polysaccharide antigen of type II group B *Streptococcus*. J. Biol. Chem. 258:1793-1798.
27. Jennings, H.J., Rosell, K.-G. and Kasper, D.L. (1980). Structural determination and serology of the native polysaccharide antigen of type III group B *Streptococcus*. Can. J. Biochem. 58:112-120.
28. Vliegenthart, J.F.G., Dorland, L. and van Halbeek, H. (1983). High resolution ^{1}H-nuclear magnetic resonance spectroscopy as a tool in the structural analysis of carbohydrates related to glycoproteins. Adv. Carbohy. Chem. Biochem. 41:209-374.
29. Michon, F., Brisson, J.R., Roy, R., Ashton, F.E. and Jennings, H.J. (1985). Structural determination of the capsular polysaccharide of *Neisseria meningitidis* group I: A 2D-NMR analysis. Biochemistry 24:5592-5598.
30. Michon, F., Brisson, J.R., Roy, R., Ashton, F.E. and Jennings, H.J. (1985). Structural determination of the capsular polysaccharide of *Neisseria meningitidis* group K: A 2D-NMR analysis. Can. J. Chem. 63:2781-2786.
31. Fearon, D.T. (1978). Regulation by membrane sialic acid of ß-1H dependent decay dissociation of amplification complement C-3 convertase of the alternative complement pathway. Proc. Natl. Acad. Sci. 75:1971-1975.
32. Edwards, M.S., Kasper, D.L., Jennings, H.J. Baker, C.J. and Nicholson-Weller, A. (1982). Capsular sialic acid prevents activation of the alternative complement pathway by type III group B *Streptococci*. J. Immunol. 128:1278-1283.
33. Sadler, J.E., Paulson, J.C. and Hill, R.L. (1979). The role of sialic acid in the expression of human MN blood group antigens. J. Biol. Chem. 254:2112-2119.
34. Spik, G., Bayard, B., Fournet, B., Streker, G., Bouquelet, S. and Montreuil, J. Studies on glycoconjugates. Part 64. Complete structure of two carbohydrate units of human serotransferrin. FEBS Lett. 50:296-299.
35. Kornfield, R. and Kornfield, S. (1980). Structure of glycoproteins and their oligosaccharide units. In The Biochemistry of Glycoproteins and Proteoglycans. pp. 1-34. (Edited by Lennarz, W.J.). Plenum, New York.
36. Finne, J., Leinoren, M. and Mäkelä, P.H. (1983). Antigenic similarity between brain components and bacteria causing meningitis. Lancet 2:(8346)355-357.
37. Lui, T.-Y., Gotschlich, E.C., Egan, W. and Robbins, J.B. (1977). Sialic acid-containing polysaccharides of *Neisseria meningitidis* and *Escherichia coli* strain BOS-12 structure and immunology. J. Infect. Dis. Suppl. 136:S71-S77.
38. Gotschlich, E.C., Lui, T.-Y. and Artenstein, M.S. (1969). Human immunity to the meningococcus. Part 3. Preparation and immunochemical properties of the groups A,B and C meningococcal polysaccharides. J. Exp. Med. 129:1349-1365.
39. Gotschlich, E.C., Fraser, B.A., Nashimura, O., Robbins, J.B. and Lui, T.-Y. (1981). Lipid on capsular polysaccharides of gram negative bacteria. J. Biol. Chem. 256:8915-8921.

40. Goebel, W.F., (1938). Chemo-immunological studies on conjugated carbohydrate-proteins. XII. The immunological properties of an artificial antigen containing cellobiuronic acid. J. Exp. Med. 68:469-484.
41. Kabat, E.A. (1961). Kabat and Mayer's Experimental Immunochemistry. 2nd edition, Charles C. Thomas. Springfield, IL.
42. Gelzner, J. and Kabat, E.A. (1964). Specific fractionation of human antidextran antibodies. III. Fractionation of antidextran by sequential extraction with oligosaccharides of increasing chain lengths and attempts at subfractionation. Immunochemistry 1:303-316.
43. Jann, K. and Westphal, O. (1975). Microbial polysaccharides. In The Antigens. (Edited by Sela, M.). pp.1-125. Academic Press, New York.
44. Jennings, H.J., Lugowski, C. and Kasper, D.L. (1981). Conformational aspects critical to the immuno-specificity of the type III group B streptococcal polysaccharide. Biochemistry 20:4511-4518.
45. Jennings, H.J., Roy, R. and Michon, F. (1985). Determinant specificities of the groups B and C polysaccharides of *Neisseria meningitidis*. J.Immunol.134:2651-2657.
46. Berman, E. (1983). Determination of the structure of three oligosaccharides from normal human urine by using 60 MHz carbon-13 NMR spectroscopy. Carbohydr. Res. 118:9-20.
47. Kasper, D.L., Baker, C.J., Galdes, B., Katzenellenbogen, E. and Jennings, H.J. (1983). Immunochemical analysis and immunogenicity of the type II Group B streptococcal capsular polysaccharide. J. Clin. Invest. 72:260-269.
48. Schifferle, R.E., Jennings, H.J., Wessels, M.R., Katzenellenbogen, E., Roy, R. and Kasper, D.L. (1985). Immunochemical analysis of the type Ia and Ib Group B streptococcal polysaccharides. J. Immunol. In press.
49. Tai, J.Y., Gotschlich, E.C. and Lancefield, R.C. (1979). Isolation of type specific polysaccharide antigen from Group B type Ib *Streptococci*. J. Exp. Med. 149:58-66.
50. Egan, W., Liu, T.-Y. Dorow, D., Cohen, J.S., Robbins, J.D., Gotschlich, E.C. and Robbins, J.B. (1977). Structural studies on the sialic acid polysaccharide antigen of *Escherichia coli* strain BOS-12. Biochemistry 16:3687-3692.
51. Finne, J. and Mäkelä, P.H. (1985). Cleavage of the polysialosyl units of brain glycoproteins by a bacteriophage endosilidase. J. Biol. Chem. 260:1265-1270.
52. Mandrell, R.E. and Zollinger, W.D. (1982). Measurement of antibodies to meningococcal group B polysaccharide: low avidity-binding and equilibrium constants. J. Immunol. 129:2172-2178.
53. Baker, P.J., Amsbaugh, D.F., Stashak, P.W., Caldes, G. and Prescott, B. (1981). Regulation of the antibody response to pneumococcal *Streptococcus pneumoniae* polysaccharide by T derived cells. Rev. Infect. Dis. 3:332-341.
54. Yount, W.J., Dorner, M.M., Kunkel, H.J. and Kabat, E.A. (1968). Studies on human antibodies. IV. Selective

variations in subgroup composition and genetic markers. J. Exp. Med. 127:633-646.
55. Heidelberger, M., Dilapi, M.M., Siegll, M. and Walter, A.W. (1950). Persistence of antibodies in human subjects injected with pneumococcal polysaccharides. J. Immunol. 65:535-541.
56. Heidelberger, M. (1973). Immunochemistry of bacterial polysaccharides. In Research in Immunochemistry and Immunobiology. (Edited by Kwapinski, J.B.G.). pp. 1-40. University Park Press, Baltimore.
57. Heidelberger, M. and Nimmich, W. (1973). Immunochemical relationships between bacteria belonging to two separate families: *Pneumococci* and *Klebsiella*. Immunochemistry 13:67-80.
58. Goldschneider, I.M., Lepow, M.L., Gotschlich, E.C., Mauck, F.T., Bache, F. and Randolph, M. (1973). Immunogenicity of group A and Group C meningococcal polysaccharides in human infants. J. Infect. Dis. 128:769-776.
59. Guirguis, N., Schneerson, R., Bax, A., Egan, W., Robbins, J.B., Orskov, I., Orskov, E. and El Kholy, A. (1986). Studies on the immunochemistry, epidemiology and structure of K51 and K93 capsular polysaccharides of *Escherichia coli* cross-reactive with group A meningococcal polysaccharide. J. Exp. Med. In press.
60. Orskov, F., Orskov, I., Sutton, A., Schneerson, R., Wenlii, L., Egan, W., Moff, G.E. and Robbins, J.B. (1979). Form variation in *Escherichia coli* K1: determined by O-acetylation of the capsular polysaccharide. J. Exp. Med. 149:669-685.
61. Schneerson, R.S. and Robbins, J.B. (1975). Induction of serum *Haemophilus influenzae* type b capsular antibodies in adult volunteers fed cross reacting *Escherichia coli* 0-75-K-100-H-5. New Engl. J. Med. 292:1093-1096.
62. Schneerson, R., Barrera, O., Sutton, A. and Robbins, J.B. (1980). Preparation, characterization, and immunogenicity of *Haemophilus influenzae* type b polysaccharide-protein conjugates. J. Exp. Med. 152:361-376.
63. Anderson, P. (1983). Antibody responses to *Haemophilus influenzae* type b and diphtheria toxin induced by conjugates of oligosaccharides of the type b capsule with the non-toxic protein CRM 197. Infect. Immun. 39:233-238.
64. Jennings, H.J. and Lugowski, C. (1981). Immunochemistry of groups A, B and C meningoccocal polysaccharide-tetanus toxoid conjugates. J. Immunol. 127:1011-1018.
65. Beuvery, E.C., Miedema, F., Van Delft, F. and Haverkamp, J. (1983). Preparation and immunochemical characterization of meningococcal group C polysaccharide-tetanus toxoid conjugates as a new generation of vaccines. Infect. Immun. 40:39-45.
66. Beuvery, E.C., Kaaden, A..v.d., Kunhai, V. and Leussink, A.B. (1983). Physiochemical and immunological characterization of meningococcal group A polysaccharide-tetanus toxoid conjugates prepared by two methods. Vaccine 1:31-36.
67. Chu, C., Schneerson, R., Robbins, J.B. and Rastogi, S.C. (1983). Further studies on the immunogencity of *Haemophilus influenzae* type b and pneumococcal type 6A

polysaccharide-protein conjugates. Infect. Immun. 40:245-256.

68. Goebel, W.F. and Avery, O.T. (1931). Chemo-immunological studies on conjugated carbohydrate proteins. IV. Synthesis of the p-aminobenzyl ether of the soluble specific substance of type III pneumococcus and its coupling with protein. J. Exp. Med. 54:431-436.
69. Avery, O.T. and Goebel, W.F. (1931). Chemo-immunological studies on conjugated carbohydrate proteins. V. Immunological specificity of an antigen prepared by combining the capsular polysaccharide of type III *Pneumococcus* with foreign protein. J. Exp. Med. 54:437-447.
70. Goebel, W.F. (1940). Antibacterial immunity induced by artificial antigens. II. Immunity to experimental pneumococcal infection with antigens containing saccharides of synthetic origin. J. Exp. Med. 72:33-48.
71. Jennings, H.J. (1985). Polysaccharides and conjugated polysaccharides as human vaccines. In New Developments in Industrial Polysaccharides. (Edited by Crescenzi, V., Dea, I.C.M. and Stivala, S.S.). pp.325-344. Gordon and Breach Science Publishers, New York.
72. Beuvery, E.C., Jennings, H.J., Roy, R., Kanhai, V., Nagel, J. and Leussink, A.B. (1986). Vaccine potential of meningococcal group C O acetylated and non-O-acetylated polysaccharide-tetanus toxoid conjugates prepared by two procedures. In Ontogeny of Immune Function and Pathogenic Mechanisms Involved in Vaccine Development. (International Symposium on Bacterial Vaccines. Sept. 17-20, 1984. National Institutes of Health, Washington, DC.). (Edited by Robbins, J.B., Sadoff, G., Schneerson, R.S.and Klein, D.) Preger Press, Philadelphia.
73. Mäkelä, O., Peterfy, F., Outschoorn, I.G., Richter, A.W. and Seppälä, I. (1984). Immunogenic properties of α(1-6) dextran, its protein conjugates and conjugates of its breakdown products in mice. Scand. J. Immunol. 19:541-550.
74. Lifely, M.R., Gilbert, A.S. and Moreno, C. (1981). Sialic acid polysaccharide antigens of *Neisseria meningitidis* and *Escherichia coli*: Esterification between adjacent residues. Carbohydr. Res. 94:193-203.
75. Roy, R., Katzenellenbogen, E. and Jennings, H.J. (1984). Improved procedures for the conjugation of oliogosaccharides to protein by reductive amination. Can. J. Biochem. 62:270-275.
76. Szu, S.C., Zon, G., Schneerson, R. and Robbins, J.B. (1986). Ultrasonic irradiation of bacterial polysaccharides: Characterization of the depolymerized products and some applications of the process. Carbohydr. Res. In press.
77. Anderson, P., Pichichero, M.E. and Insel, R. (1985). Immunization of two-month-old infants with protein-coupled oligosaccharides derived from the capsule of *Haemophilus influenzae* type b. J. Pediatr. 107:346-351.
78. Eskola, J., Peltola, H., Mäkelä, P.H., Kayhty, H., Karanko, V., Samuelson, J. and Gordon, L.K. (1985).

Antibody levels achieved in infants by course of *Haemophilus influenzae* type b polysaccharide-diphtheria toxoid conjugate vaccine. Lancet 1:(8439)1184-1186.

79. Jennings, H.J., Roy, R. and Gamin, A. (1986). Induction of meningococcal group B polysaccharide-specific IgG antibodies in mice using an N-propionylated B polysaccharide-tetanus toxoid conjugate vaccine. J. Immunol. In press.
80. Jennings, H.J., Roy, R. (1985). Enhancement of the immune response to the group B polysaccharride of *Neisseria meningitidis* by means of its chemical modification. In The Pathogenic Neisseria: Proceedings of the Fourth International Symposium. (Edited by Schoolnik, G.K.). American Microbiological Society.
81. Zollinger, W.D. and Mandrell, R.E. (1977). Outer-membrane protein and lipopolysaccharide serotyping of *Neisseria meningitidis* by inhibition of a solid-phase radioimmunoassay. Infect. Immun. 18:424-433.
82. Jennings, H.J., Bhattacharjee, A.K., Kenne, L., Kenny, C.P. and Calver, G. (1980). The R-type lipopoly-saccharides of *Neisseria meningitidis*. Can. J. Biochem. 58:128-136.
83. Jennings, H.J., Johnson, K.G. and Kenne, L. (1983). The structure of an R-type oligosaccharide core obtained from some lipopolysaccharides of *Neisseria meningitidis*. Carbohydr. Res. 121:233-242.
84. Jennings, H.J., Lugowski, C. and Ashton, F.E. (1984). Conjugation of meningococcal lipopolysaccharide R-type oligosaccharides to tetanus toxoid as a route to a potential vaccine against group B *Neisseria meningitidis*. Infect. Immun. 43:407-412.
85. Svensson, S.B. and Lindberg, A.A. (1979). Coupling of acid labile *Salmonella* specific oligosaccharides to macromolecular carriers. J. Immunol. Methods 25:323-335.
86. Kundu, S.R., Diego, I., Osovitz, S. and Marcus, D.M. (1985). Glycosphingolipids of human plasma. Arch. Biochem. Biophys. 238:388-400.

STRUCTURAL AND IMMUNOCHEMICAL ASPECTS OF *BRUCELLA ABORTUS* ENDOTOXINS[3,4]

Albert M. Wu*[1], Neil E. MacKenzie+[2], L. Garry Adams*, and Roberta Pugh+

*Department of Veterinary Pathology and +Department of Veterinary Microbiology and Parasitology
College of Veterinary Medicine, Texas A&M University
College Station, Texas 77843 USA

The smooth lipopolysaccharide (sLPS)-protein complex of *Brucella abortus* has been identified as the immuno-dominant component, which reacts with antibody in the serological tests generally used for diagnosis of brucellosis (1). *B. abortus* sLPS carries the major A (*abortus*) antigen, the species-specific epitope located within, and the minor M (*melitensis*) determinant, which cross reacts with *B. melitensis* antigen. O-chain prepared from *B. abortus* LPS was reported to have protective antigen properties in mice (2). Unlike the LPS of other enterobacteriaceae, sLPS separates primarily into the phenol phase when *B. abortus* cells are extracted with hot phenol-water by the Westphal method (3-7). The crude fraction of sLPS, which was used to study the properties of smooth *B. abortus* in the early eighties, is associated with various proteins ranging from 10-30% of the weight composition,

---

[1]The corresponding author
[2]Present address: Department of Pharmaceutic Science, College of Pharmacy, University of Arizona, Tucson, Arizona 85720
[3]Most of the material in this article is from references 9 and 10. We thank Martinus Nijhoff/Dr. W. Junk Publishers for giving such permission. We use O-chain instead of O-hapten to express the O-antigen moiety of LPS. The terminology of O-antigen is not suitable in this case because it reacts as a hapten rather than as an antigen. See references 7,9, and 10.

most of which are non-covalently bound (3). These non-covalently linked proteins or nucleic acids can be removed by treatment with chaotropic agents and by gel filtration in the presence of chaotropic agents and detergents. Some of the enriched LPS fractions (f5p) contain approximately 6% residual protein, which is thought to be covalently linked to the lipid A moiety of LPS (1,3,5). High protein content fractions of smooth LPS have also been reported (1,8,9), but, they are not well characterized. In recent years, it was found that the O-antigen (O-chain) of the major fraction sLPS, prepared from strains 2308, 19, and 1119.3, is comprised of a repeat unit of *N*-formyl perosamine ($\alpha 1 \rightarrow 2$ linked 4,6-dideoxy-4-formamido-D-mannopyranose) (6,10). This O-antigen (O-chain) is identical to that of *Yersinia enterocolitica* serotype 0:9 (11) and similar to the O-antigen of the LPS of *Vibrio cholerae* 569B (1NABA) (12).

Despite the number of studies on *B. abortus* sLPS (3-7), several questions remain, they are (a) do the acetone-dried bacteria or lightly washed bacteria contain a mixture of sLPS from both membrane-bound and non-membrane-bound fractions of bacteria?, and what is the difference between membrane bound and nonmembrane bound LPS? (b) since the yield of LPS reported is low, does the sLPS prepared by the Baker and Wilson procedure (4) or modifications of their methods as described by Moreno *et al.*(3), represent all or part of sLPS?, (c) what is the definition of sLPS and its protein complexes, and (d) what are the criteria of their purity?, can its homogeneity be evaluated according to protein content, antigenicity, toxicity, or limulus lysate reactivity, or not? (e) what is the chain length distribution of the *B. abortus* O-chain (antigen)? In our current research (9,10), we have provided some information related to the above questions, (a) bacterial

---

[4]Abbreviations used: LPS, lipopolysaccharides; sLPS, smooth lipopolysaccharides; cLPS, crude lipopolysaccharides; AH, acid hapten; KDO, 3-deoxyoctulosonic acid; ELISA, enzyme-linked immunosorbent assay; LAL, limulus amoebocyte lysate test; HexN, Hexosamine; PS, phenol sulfuric acid method; O., orcinol method; Glc, DGlucopyranose; Gal, DGalactopyranose; Man, DMannopyranose; MeαMan, Methyl DMannopyranoside; MWCO, Molecular weight cut off.

antigens were separated into membrane bound and non-membrane bound fractions based on freeze-thaw cycles and ultrasonic technique; (b) the yield of LPS was improved by repeated phenol extractions; and (c) the membrane bound sLPS was subfractionated into groups A to C or more, according to the extent of dialysis and centrifugation. Characteristics of these subfractions are: less than 80 ng is required for positive ELISA; and about 0.2 ng for the Limulus test; the individual fractions react well with precipitating antibodies in the serum of a strain 2308 infected cow, demonstrating differences in precipitin curves; the protein content of these fractions varies from 16 to 42% as determined by dye binding test and from 17% to 60% by the Lowry phenol method using bovine serum albumin as standard; group f5A, which shows a standard bell shaped curve in the precipitin assay, is one of the major fractions in all three strains prepared (S1119.3, S19 and S2308); the amount of other subfractions obtained varies among batches and strains of bacteria. These findings give a different view on the immunochemical properties of *Brucella abortus* endotoxin. However, the function of the proteins associated with LPS still remain unresolved. It is highly possible that the associated protein is one of the major factors involved in the solubility of LPS-protein complex, used to subfractionate smooth LPS.

Recently, we reported our studies concerning the structural and immunochemical analyses of O-chains of membrane bound f5A, prepared from these three strains (10). These studies showed that the O-chains of *B. abortus* LPS can be grouped into several fractions according to a molecular weight cut-off (MWCO). Each fraction was characterized by compositional analysis; NMR spectroscopy and U.V. absorbance; furthermore their immunochemical reactivities were quantitated by a precipitin-inhibition assay.

The purpose of this review article is to present our current perspective of *B. abortus* endotoxins from three aspects - I. Chemical and immunochemical characterization of membrane bound endotoxins. II. Structural concepts of O-chain of smooth LPS f5A. III. Current concepts and future prospects.

I. Membrane Bound Smooth Lipopolysaccharides

I-A. Separation of Membrane-bound and Non-membrane Bound Fractions of Bacteria

The method of endotoxin preparations described in this review article is designed to study the location and the properties of *B. abortus* endotoxin. The procedure described is more complicated than most of the methods reported previously (3-6). It is not recommended as a standard method for all purposes. A scheme for the separation of membrane-bound and non-membrane bound fractions of bacteria is shown in Fig. 1. The inactivated bacteria can be centrifuged at 2.4 x 10^4g for 40 min. The supernatant, pre-washed fraction can be saved for antigen analysis. The pellet of bacteria is subjected to three sequential washes and recovered by centrifugation at 2.4 x 10^4g for 40 min. The supernatants are designated as Wash I, Wash II, and Wash III, according to the order of washes. After three washes, the bacterial pellet is sequentially subjected to 21 freeze-thaw cycles (-20°C to 50°C) and ultra-sonication at room temperature (RT) to separate soluble materials (Benchtop ultrasonic cleaners, Fisher, Springfield, NJ). The soluble materials are removed from the mixture after each 7 cycles by centrifugation at 2.4 x 10^4g for 1 hour and designated as FS-I, FS-II, and FS-III. The various soluble fractions of bacteria (i.e., prewash, Wash I to III, and FS-1 to III), are designated as non-membrane-bound fractions, and the ultrasonicated bacteria as the membrane-bound fraction.

Of the non-membrane bound fractions prepared from three strains (19, 2308 and 1119.3) of *B. abortus*, the prewashed fraction contains the major portion of the total material (60%) while Wash I and FS-I (23%) contain the major amounts of non-membrane-bound antigens, which can be detected by quantitative precipitin assay and/or ELISA. Low yields are found in Wash III; thus, it is assumed that most of the soluble material of the bacterial cell-wall surface has been removed by these treatments. Significant amounts of material can be obtained in FS-I and FS-II (19% of total non-membrane-bound fraction); the yield is decreased after 21 freeze-thaw cycles and ultra-sonication. About 15% of total bacteria (by

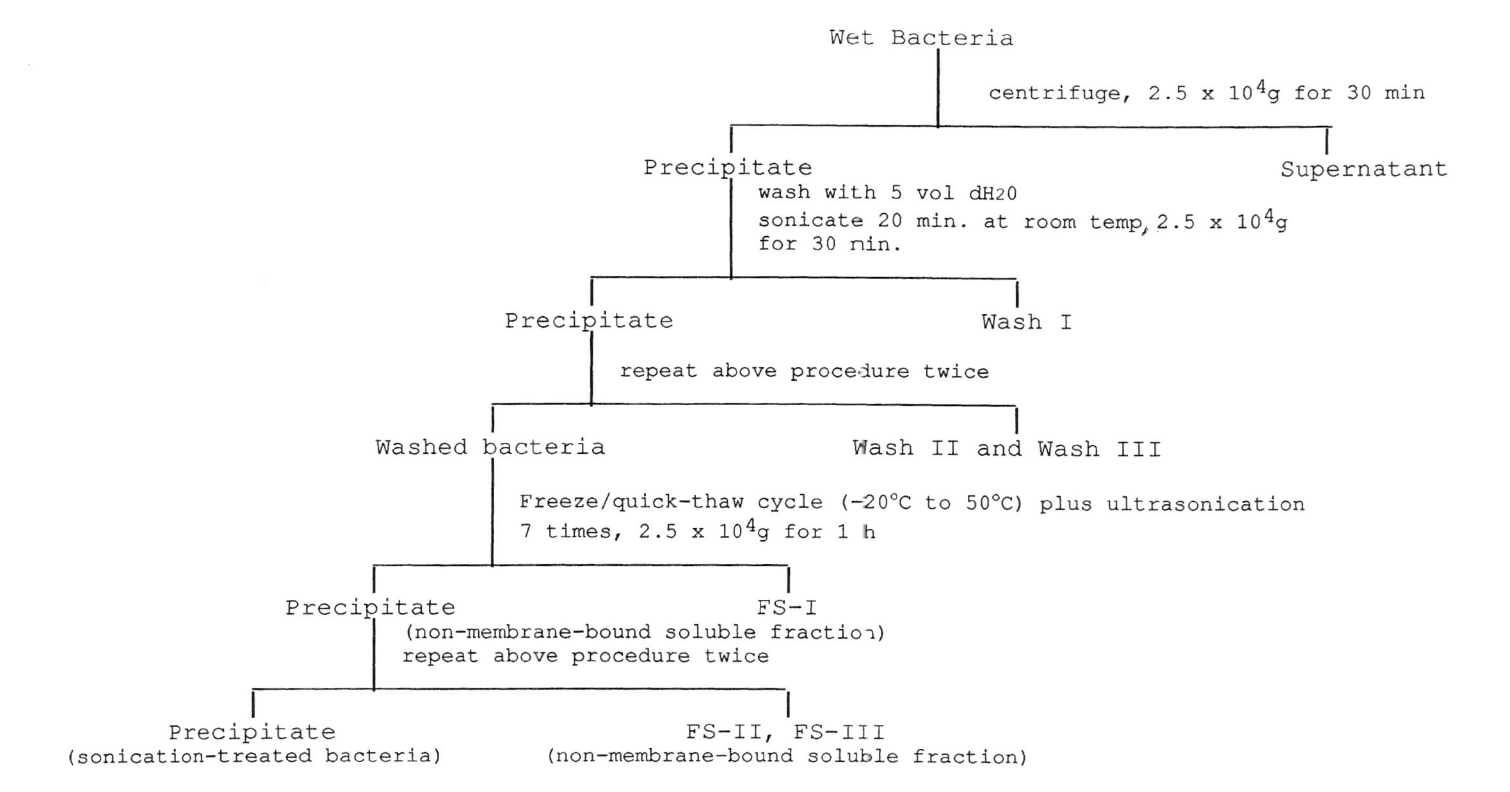

Fig. 1. Separation of membrane-bound and non-membrane bound bacterial antigens. In rough strains of *B. abortus*, such as strain 45/20 and rough strains of 2308, most of the immunoactive components are found in several of these fractions, which are LPS, but have not been purified and characterized.

dried weight) is recovered in the soluble fractions (i.e. equivalent to non-membrane-bound fraction). In the case of rough strains of *B. abortus*, strain 45/20 and the rough strain of 2308, most of the immunoreactive components are found in Pre-wash through Wash III. These immunoreactive components are mainly LPS. The partially purified material is about 1/5 to 1/10 as active as smooth LPS. The structural relationship between membrane bound and non-membrane bound LPS has not been established. Their differences will be investigated by SDS-PAGE electrophoresis.

I-B. Preparation and Subfractionation of Membrane Bound Smooth LPS

As shown in Fig. 2, the bacterial cell wall (envelope) remains almost intact after freeze-thaw and ultrasonication (FTU) treatment. At this stage, it is assumed that most of the non-membrane bound LPS has been removed. When the procedure of the modified method of the hot phenol extraction described in our recent report (9) is used to prepare smooth LPS, the membrane bound sLPS, based on repeated hot phenol extraction, differential centrifugation, and degree of dialysis (possibly due to size and charge effects), can be grouped into five fractions (Fig. 3).

In our modified method, phenol extraction is repeated three times and the extraction time is extended to 45 min.; thus, the yield of sLPS is significantly increased, ranging from 3.6 to 7.7% of dried bacteria, while the yields in previous works are less than 1.5% (3,6). The pooled phenol extract is washed 10 times, and significant amounts of chromogen, nucleic acids and polysaccharides are removed (unpublished data), suggesting that these washes can be used for purification. Analysis of these subfractions disclosed that: (a) the amount of each subfraction obtained varied with the strain and/or batches of bacteria (Table I and unpublished data); (b) less than 0.2 ng was required for a positive LAL result and less than 80 ng for a positive ELISA result (Table I); (c) each group reacted well with precipitating antibodies in the serum of a strain 2308-infected cow and exhibited unique precipitin curves (Fig. 4); (d) among the subfractions, f5A, which has a standard bell-shaped curve in the precipitin

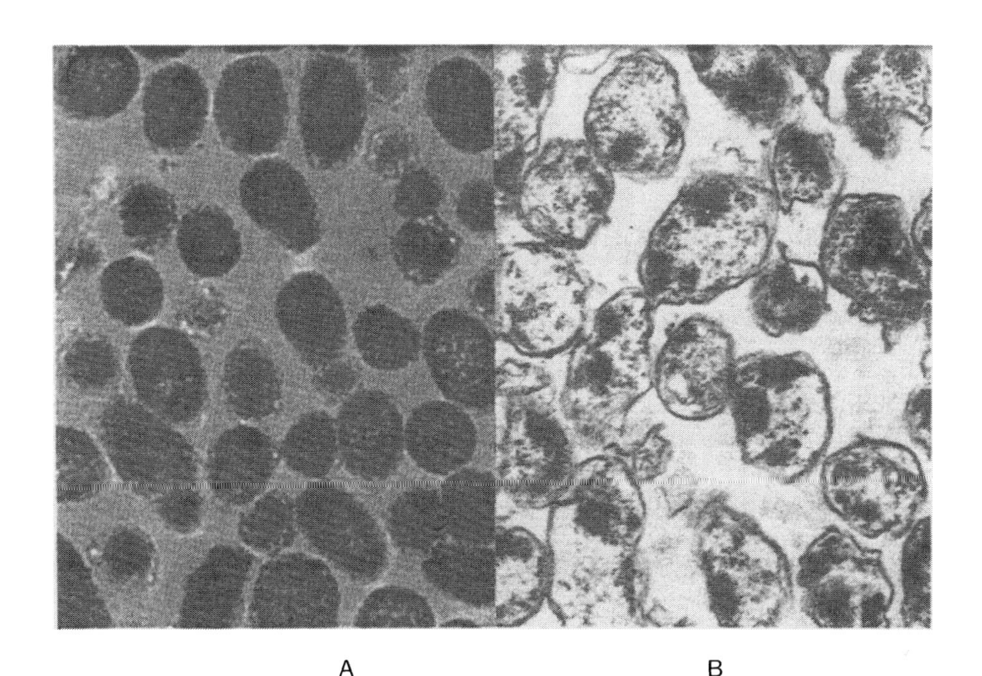

Fig. 2. Electron micrograph of *Brucella abortus* Strain 2308 before and after freeze-ultrasonication treatment. The bacteria, which were suspended in 2% agar and were then prepared for transmission electron microscopy following standard procedures, were stained with uranyl acetate and lead citrate, and were observed on an Hitachi-HS8 transmission electron microscope. A. Washed bacteria before ultrasonication. Compact and high density was shown on the cell surface. B. After freeze-ultrasonication treatment for 21 cycles, one hour each. The bacteria appear swollen and the shape changed, but the major cell wall still remains intact and most smooth lipopolysaccharide (85%) is still tightly bound to the membrane. The LPS isolated from these membranes is defined as membrane bound (smooth) LPS. (x 21,500)

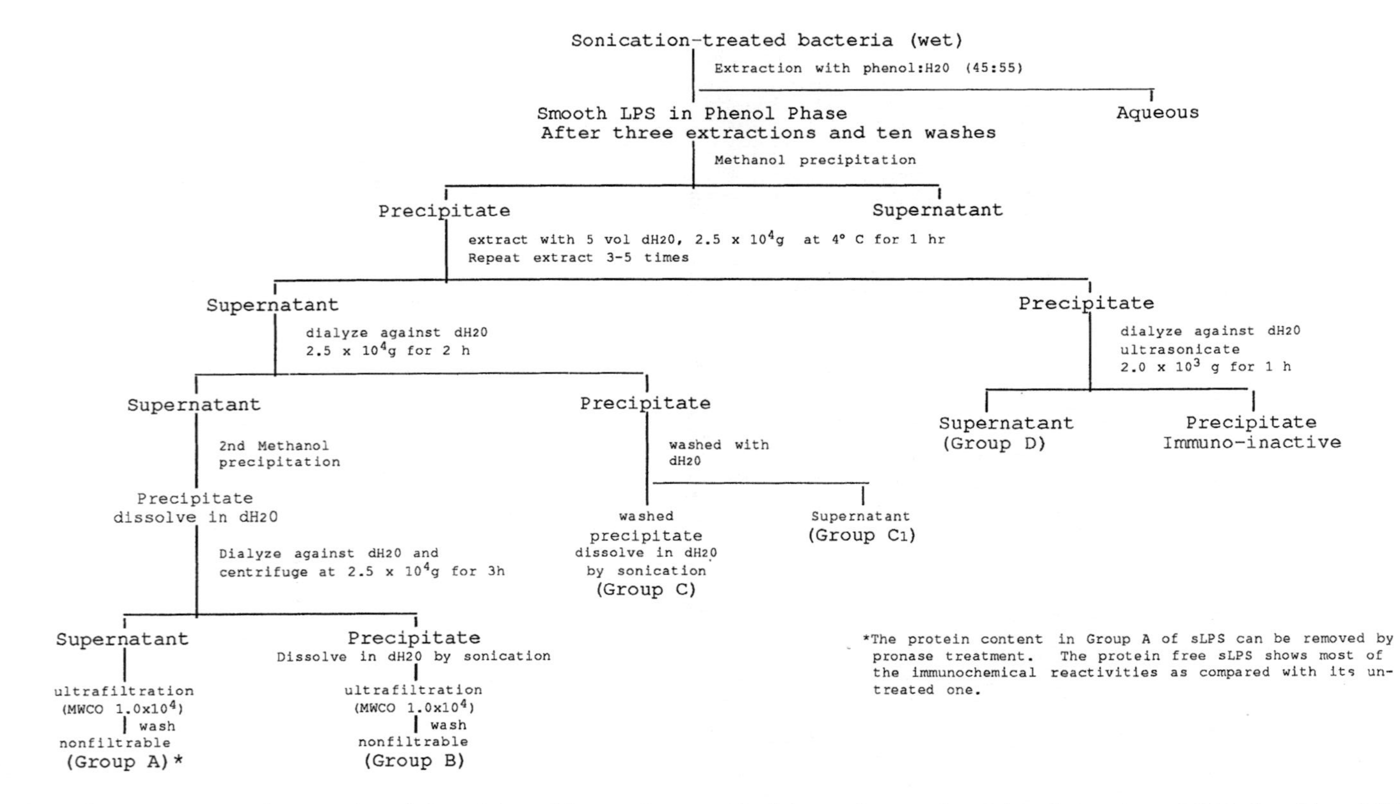

Fig. 3. Preparation and subfractionation of smooth lipopolysaccharide (strains 1119.3, 19, 2308). Based on differential centrifugation and degree of dialysis, the smooth lipopolysaccharide can be divided into more than three fractions.

Table I
Chemical and Immunochemical Analysis of Fractions of *Brucella abortus* Membrane-Bound Smooth Lipopolysaccharides[a]

Strain	g% dried Bacteria[b]	sLPS	g% of total	ELISA ng for end point	QPA µg N precip-itated at Max	Nitrogen	Protein[d] Lowry phenol	Protein[d] Dye binding	Phenol Sulfuric acid (PS)[e]	Orcinol[e] (O)	PS/O	HexN[f]
S1119.3	4.7	**A**	59	40	3.7	6.1	15.5[g]	22.1	19.5	5.2	3.8	4.0
		B	6	80	4.7	6.2	27.0	27.5	14.5	4.4	3.3	4.3
		C	10	80	5.3	7.5	35.0	32.0	13.3	4.5	3.0	4.9
		D	25	80	4.4	9.0	60.0	36.8	5.1	2.1	2.5	2.8
S19	3.6	**A**	54	40	4.0	5.8	17.3[g]	19.7	22.0	5.1	4.3	5.2
		B	4	40	4.0	8.4	26.0	32.6	24.9	7.0	3.6	5.7
		C	42	80	4.5	8.5	46.3	37.9	14.0	4.8	2.9	3.5
S2308	7.7	**A**	44	80	4.5	5.7	15.0[g]	16.2	18.8	5.6	3.4	3.4
		B	9.6	20	5.2	6.1	17.0	21.6	24.0	5.4	4.4	3.7
		C	37	40	4.9	8.4	47.3	32.0	13.4	3.1	4.3	2.8
		C_1	3.5	80	4.0	6.0	24.1	23.2	13.4	4.3	3.1	2.7
		D	5.8	80	2.4	7.7	45.7	41.6	8.9	2.7	3.3	1.8

a. Expressed as g% dried weight; All fractions of LPS were positive to the LAL test, <0.2 ng was the end point KDO, trace in F5D of S1119.3 and less than 1.0% in others; b. 22% of wet bacterial wt. is equivalent to dried wt. bacteria; c. QPA, quantitative precipitin assay; 10µl of cow 204 serum added; the N content from LPS corrected; d. Bovine serum albumin (Sigma S4503) as standard; e. Gal as standard; f. Analyzed by the Elson-Morgan method; g. The protein content in f5A can be removed by pronase digestion. The protein free product, which contains less than 0.8% of Lowry phenol positive material, retains about the same immunochemical reactivity.

assays, is one of the major fractions in all three strains (1119.3, 19, and 2308) (Fig. 4), and (e) the protein content varied from 17% to 60% by the Lowry phenol method and 16%-42% by the dye binding test (Table I).

When the method we used and the procedure described by Moreno *et al.* are compared, it is assumed that f5 of sLPS reported by Moreno *et al.* is equivalent to part of our f5A, f5B, and f5C (7). Analysis of *B. abortus* S2308 LPS by SDS-PAGE (12%) revealed that the protein content in f5A consists mainly of 3 bands near 7K, 8K and 14K (Wu and Tsai's unpublished data). The protein content in f5A can be removed by pronase digestion. The protein free product, which contains less than 0.8% of Lowry phenol positive material, retains about the same immunochemical reactivity. Among the sLPS fractions prepared, the yield and quality of f5A was much higher than the f5 (equivalent to parts of our f5A, f5B and f5C) or sLPS reported previously (3,6). Ultrogel (A-2 and A-4) and BioGel P-300 column chromatography, with PBS pH 7.0 as eluant, was used in an attempt to purify these crude endotoxins; but purification was hampered by (a) variable recovery of the sLPS applied, and (b) poor solubility of some fractions (especially f5D). Selection of a strong ionic strength solvent and improvement of solubilization and antigenicity of the protein rich fractions of sLPS (f5D) need to be investigated. Group f5A of strains 2308, 19, and 1119.3 were used to prepare O-chains and their immunochemical reactivities and size distribution have been determined (reference 10 and unpublished data). These will be described in the following section (**II**).

Our current studies support the concept of variation in protein content of the sLPS-protein complex, as has been previously reported (1,8). These complexes are linked covalently and/or noncovalently, and play an important role in the solubility and heterogeneity of sLPS. The results of these fraction studies provide a useful guide for further investigation. Future studies should include; (a) more precise criteria for grouping the sLPS and the procedure for improved purification; (b) further subdivision of some groups, especially f5A of S2308; (c) a better definition concerning criteria of purified sLPS or sLPS-protein complex; (d) elucidation of the structure and function of the protein moiety of LPS-protein complex, especially for the fraction of

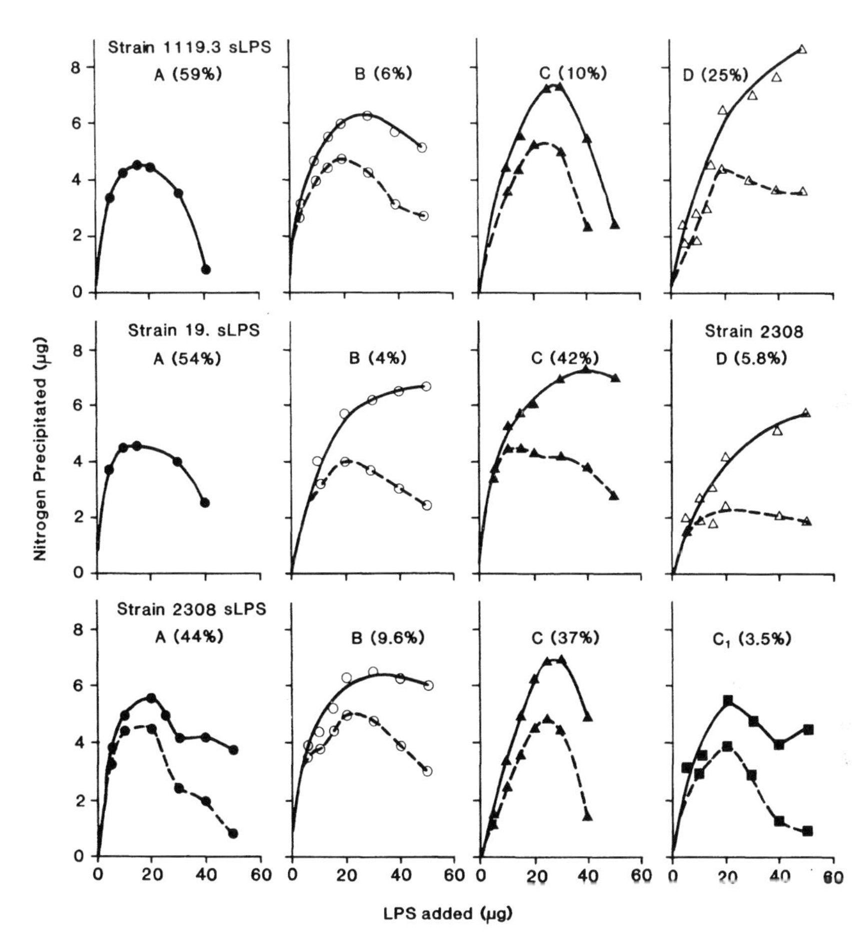

Fig. 4. Quantitative precipitin curves of subfractions of smooth LPS prepared froom *Brucella abortus* strains 1119.3, 19 and 2308 with *B. abortus*-infected cow serum (total volume 300 µl, 10 µl of serum added). Nitrogen content in precipitate of antigen-antibody complex expressed as solid curves (—) and nitrogen content in antibody precipitated (assumes that all LPS added is precipitated with antibodies and estimated by total nitrogen in precipitate minus the nitrogen of LPS added) expressed as dashed curves (···). Reproduced with permission from Wu et al., Mol. Cell. Biochem., 75:93-102, 1987, Fig. 3.

high protein content (f5D); and (e) a comparison of the LPS-protein complex before and after pronase treatment.

II. O Chains of *B. abortus* Endotoxins

II-A. Preparation and Fractionation

In early reports, the conditions used for the preparations of *B. abortus* O-chains (antigens) varied from 2 hours to 5 hours of hydrolysis of the LPS-protein complex in 1% acetic acid at 100°C. The carbohydrate moiety of the hydrolyzed product was designated as acid haptens, AH (13,14). In our studies, it was found that a 2 hour hydrolysis is sufficient to separate most of the Lipid A and protein from the LPS-protein complex.

As fraction 5A is the major membrane bound sLPS (9), the O-chains (antigens) prepared from this source must also represent the major fraction of the carbohydrate moiety of LPS. Furthermore, this fraction is relatively pure and therefore was chosen for our studies. The LPS f5A was hydrolyzed in 1% acetic acid (2 mg/ml) at 100°C for 2 hours. As shown in Fig. 5, the hydrolysate was centrifuged to remove protein and lipid A, and the supernatant was ultrafiltered using a MWCO membrane of 1.0×10^3 (YM2, Amicon, Danvers, MA). The non-filtrable fraction was washed three times and dialyzed against 20 volumes of dH_2O by using a MWCO dialysis tubing of 2.0×10^3 (Spectra/Por 6 membrane, Los Angeles, CA). The dialyzable fraction was concentrated and reultrafiltered on a MWCO 1.0×10^3 membrane. The filtrable fraction was concentrated and designated as fraction E (MWCO $< 1.0 \times 10^3$) and the non-filtrable fraction as fraction D (MWCO $1.0\text{-}2.0 \times 10^3$).

The nondialyzable fraction of MWCO 2.0×10^3 was subjected to ultrafiltration by using a MWCO 5.0×10^3 membrane (Amicon YM5). The filtrable fraction was concentrated and dialyzed against dH_2O using a MWCO 3.5×10^3 dialysis tubing. The dialyzable fraction was concentrated and designated as fraction C (MWCO $2.0\text{-}3.5 \times 10^3$), and the nondialyzable fraction as fraction B (MWCO $3.5\text{-}5.0 \times 10^3$). The latter is the major fraction in most preparations. The non-filtrable fraction of MWCO 5.0×10^3 was subjected to

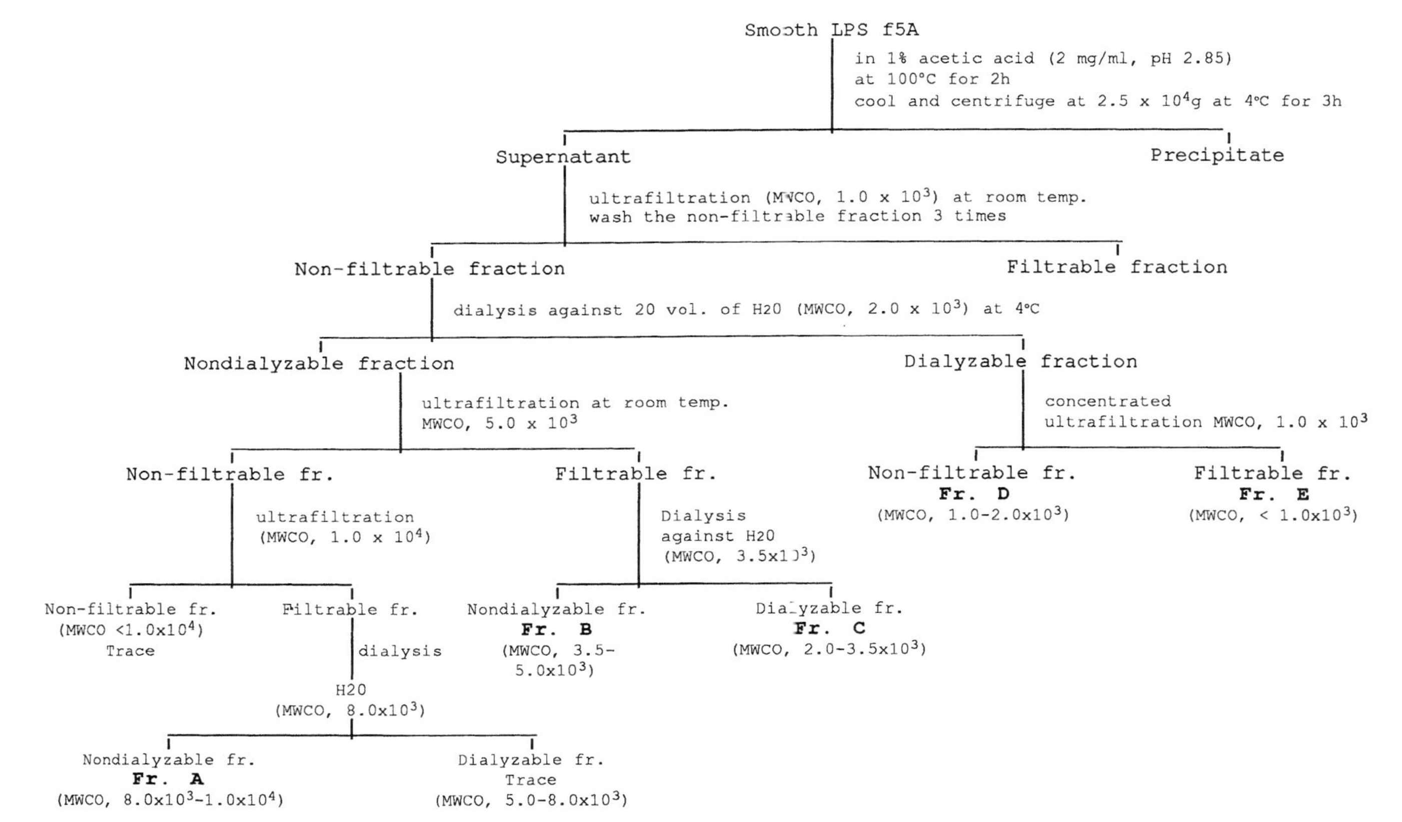

Fig. 5. Fractionation of O-chains of *Brucella abortus* smooth LPS f5A. Ultrafiltration at room temperature and centrifugation at 4°C.

ultrafiltration by using a MWCO membrane of 1.0×10^4 (YM10, Amicon). The filtrable fraction was concentrated and dialyzed against dH_2O using dialysis tubing of MWCO 8.0×10^3. Most of the hapten was not dialyzable and was designated as fraction A (MWCO 8.0×10^3-1.0×10^4).

The yields of O-chains obtained from fraction 5A of strains 19 and 2308 lipopolysaccharides were 45% and 52% respectively. When these haptens were further fractionated by both membrane ultra-filtration and dialysis (Fig. 5), the size distribution of O-chains of these endotoxins ranged from several oligosaccharides up to 1.0×10^4 MWCO (Table II). Fractions corresponding to MWCO 8.0-10.0 $\times 10^3$ (fraction A), 3.5-5.0 $\times 10^3$ (fraction B) and MWCO $<1.0 \times 10^3$ (fraction E) for both strains 19 and 2308 were the dominant fractions, and represented more than 85% of total immunoactive materials. Small O-chains of MWCO $<1.0 \times 10^3$ were ultrafiltered twice (Fig. I). Most of the non-immunoactive fragments were removed by the first ultrafiltration.

The sizes of O-chains obtained varied in molecular size from less than 1.0×10^3 to approximately 9.0×10^3 MWCO, thus, the specification of the MWCO filter is used to estimate the molecular weight of O-chains, the O-chains prepared must range from several up to 50 glycosyl units. However, the relationship between MWCO and the size of O-chains has not been determined. Moreover, the fraction of O-chains representing the native carbohydrate moiety (O-antigen) of LPS or its mild acid degraded product also remains to be elucidated. When *B. abortus* sLPS is analyzed by SDS-PAGE, two or perhaps three diffuse broad bands are seen at the middle and upper parts of the gel. Their sizes are equivalent to 11 to 15 and 21 to 30 repeating units of *S. typhimurium* O-chain, each composed of 4 to 5 monosaccharide residues (15). Thus the sizes of the two major bands, correspond to about 120 (84-150) and 60 (44-75) sugar residues, respectively (Wu and Tsai's unpublished data).

The average length of the *B. abortus* O antigen from S1119.3 was reported to range from 96 to 100 glycosyl units (6). The chain length of its O-antigen seems to be more uniform than that of other bacteria (16-18). From the size of the O-chains we prepared and based on the results from our SDS-PAGE analysis, the O-chains of *B. abortus* seem to be more heterogeneous than previously reported (6). The difference in O-chain length distributions require further investigation. It must be determined whether the uniform chain length (96 to 100 units) of the O-chain prepared from S1119.3 by Caroff (6) represents all of the O-chain in *B. abortus*, or only part of it.

II-B. Compositional Analysis and U.V. Absorbance

The nitrogen content of the O-chains ranged from 6.0 to 7.0% (Table II). These results are 13 to 25% lower than the calculated amount of nitrogen in terms of anhydroperosamine. No Lowry reactive components were found in fraction A and B (57.6%) of strain 2308 hapten and only 2% in fraction E, but small amounts of such components were detected in the major fractions (A, B, and D equivalent to 92.5%) of S19 O-haptens. These results indicate that most of the proteins (Lowry reactive components) were removed by centrifugation and ultrafiltration. As expected, O-chain prepared from protein free LPS contains none or only trace amounts of Lowry positive components (unpublished results).

The neutral sugar content in the O-chains of strain 2308 increased from 19% to 30% (about a 58% increase) based on the Phenol Sulfuric acid (PS) method and 5.6% to 9% (also about a 60% increase) by the Orcinol (O) method. Concomitant increases were found in the hexosamine content. These results imply that PS and O reactive carbohydrates were not degraded during hydrolysis with 0.1 M acetic acid and/or most of them stayed at the O-chain moiety. In the case of O-chains of S19, no significant increases were found in either the PS assay or the hexosamine determination, but some increase was found in the orcinol method. These data suggest that part of the PS and O reactive components in LPS of S19 were degraded during acid hydrolysis, or remained with the Lipid A moiety.

The analytical data of the fraction C chains (MWCO 2.0-3.0x10^3) are not given. The content was less than 2% of total O-chains and not pure enough (about 15% of Lowry protein) but, it represents a significant fraction in other preparations (unpublished data). As shown in Table II, some Lowry reactive components were present in fractions C and D. These Lowry reactive components are not found when protein free LPS is

TABLE II
Chemical and Immunochemical Analysis of *Brucella abortus* O-chains[a]

Strains	O-Haptens	Molecular Size[b] (MWCO x 10^3)	Distribution[c] (%)	PIA[d]	Nitrogen	Lowry Phenol[e]	Phenol Sulfuric Acid[f]	Orcinol[g] (O)	PS/ / / O	HexN[g]
	A	8.0 -10.0	35.7	0.29	6.1	0.	31.1	9.2	3.4	3.8
	B	3.5-5.0	21.9	0.29	6.7	0.	36.4	9.3	3.9	4.3
S2308	**D**	1.0-2.0	3.2	0.29	6.6	5.0	30.1	10.2	3.0	3.8
	E	<1.0	35.7	0.29	7.0	2.0	28.5	8.0	3.6	5.1
	A	8.0-1.0	11.5	0.29	6.1	0.5	23.0	7.8	3.0	2.8
	B	3.5-5.0	70.0	0.29	6.6	0.1	25.4	9.1	2.8	4.6
S19	**D**	1.0-2.0	11.0	0.29	6.1	2.7	17.2	6.4	2.7	2.6
	E	<1.0	4.6	0.29	6.0	4.3	20.4	8.3	2.5	2.9

a. expressed as g% dried weight. About or less than 1.0% of KDO were detected. Only trace amounts of nucleic acid were detected by the absorbance 280/260 in A and B fractions (10); b. determined by membrane filtration or dialysis. MWCO, Molecular weight cut-off, the relationship between molecular weight and MWCO has not been established; c. expressed as g% dried weight sample, about 3% in the fractions of 2.0-3.5 and 5.0-8.0 x 10^3 MWCO; d. PIA, precipitin inhibition assay, expressed as μmoles of anhydro N-formyl perosamine in O-chains (anhydrous M.W. 173) for 50% inhibition, the assay mixture contained 10 μl of cow 204 serum + 10μl of sLPS f5A prepared from S2308. (See Fig 8); e. Bovine serum albumin (Sigma; S4503) as standard.

used as a source for O-chain preparation. The theoretical nitrogen content in terms of anhydroperosamine is 8.1%; however, it was found to be 6.1% to 6.6% in most of our fractions, in agreement with in previously reported values (6), i.e. the actual nitrogen content is about 13% to 25% less than that predicted from structural considerations. Furthermore, the O-chains exhibited about 20-30% of PS reaction (with Gal as the standard), but contained less than 10% and 3% (2% Man and 1% glucose) of neutral sugar as determined by the orcinol method and gas liquid chromatography, respectively.

When U.V. absorbance was used to estimate the approximate amount of nucleic acids, less than 0.2% of nucleic acid was found in the major fractions of the O-chains. In particular, only 0.03% of nucleic acid equivalent compound was detected in the O-chain fraction A of strain 2308. Higher readings were found in fractions D and E of strain 19 O-chains, but these were minor fractions (total 15.6%; Table II).

II-C. NMR Spectroscopy

All major O-chain fractions of strains 2308, 19 and 1119.3 (see Table II) have been examined by ^{1}H and ^{13}C NMR spectroscopy. All fractions gave identical spectra, typified by the ^{13}C NMR spectrum shown in Fig. 6. Fraction A of Strain 2308 O-chains, (MWCO 8.0-10.0 x 10^{3}, 18 mgs, in 0.5 ml D_2O) gave a ^{13}C NMR spectfum with the most intense signals resonating at 166.16, 101.79, 78.27, 69.52, 68.86, 53.15, and 18.02 p.p.m.. This and all subsequent data are in complete agreement with the published assignment of the carbon atoms of α1,2-linked 4,6-dideoxy-4-formamido-α-D-mannopyranosyl units (6). The monomeric units are more commonly referred to as *N*-formylperosamine and a schematic of the polymer structure is shown as an insert in Fig. 6. The ^{1}H coupled ^{13}C NMR spectrum showed signals at 166.16 (d, J_{C-H}=197.3 Hz, -NH-CHO), 101.79 (d, $^{1}J_{C-H}$=172.5 Hz, C-1), 78.27 (d, $^{1}J_{C-H}$=151.7 Hz, C-2), 69.52 (d, $^{1}J_{C-H}$=145.9 Hz, C-3), 68.86 (d, $^{1}J_{C-H}$=146.2 Hz,C-5), 53.15 (d,$^{1}J_{C-H}$=143.0 Hz, C-4) and 18.02 (q, $^{1}J_{C-H}$=127.7 Hz, C-6) p.p.m. for which the indicated assignments were made. The partial ^{1}H NMR spectrum showed resonances at 8.18 (s,1H,

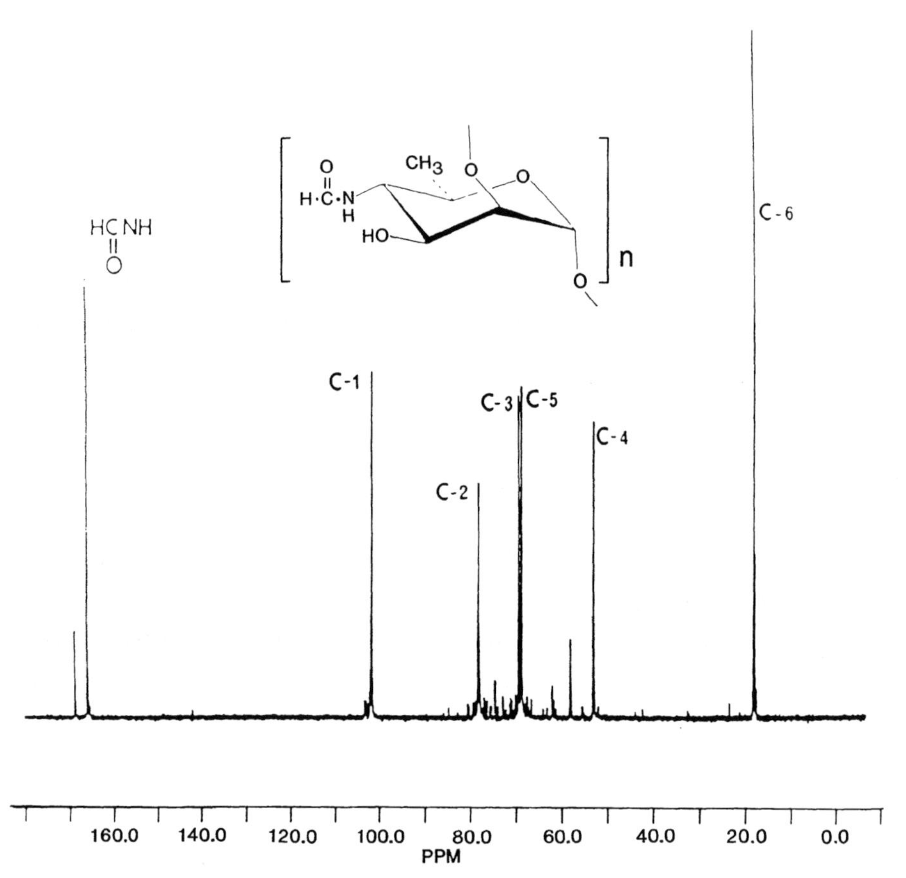

Fig. 6. ^{13}C NMR (125 MHz, 25°) of the major fraction of O-chains (Fr. A 18.6 mg, MWCO 8 x 10^3 to 1.0 x 10^4) of membrane bound sLPS (f5A), prepared from *B. abortus* S2308. Number of acquisitions and rate were 500,000 and 0.59 seconds, respectively. Lines were broadened by 2 Hz exponential multiplication of the free induction decay. An identical profile has been obtained from the O-chain of S19 and S1119.3. Reproduced with permission from Wu and MacKenzie, Mol. Cell. Biochem., 75:103-111, 1987, Fig. 2.

HNC$\underline{H}$O), 5.15 (1H, H-C1) and 1.19 (bd, 3H, $^{3}J_{H-H}$=9 Hz, H_3C-6), p.p.m., again with indicated assignments.

The major heterogeneity shown in the ^{13}C NMR spectrum is thought to arise from the presence of both E and Z conformations of the *N*-formyl group (6). The discernable ^{13}C-resonances for the other isomer occur at 169.06, 78.90, 58.28, and 18.00 p.p.m.. The ^{1}H coupled ^{13}C NMR spectrum (not shown) showed signals at 169.06 (d, $^{1}J_{C-H}$=191.2 Hz, -NH-CHO), 78.09 (d, $^{1}J_{C-H}$=151.7, C-2), 58.28 (d, $^{1}J_{C-H}$=143.0 Hz, C-4), and 18.00 (q, $^{1}J_{C-H}$=127.7 Hz, C-6) p.p.m. for which the indicated assignments were made. Again the partial ^{1}H NMR spectrum showed resonances at 8.02 (s,-HNC$\underline{H}$O) and 1.23 (bd, $^{3}J_{N-H}$=9 Hz, H_3-C6) p.p.m. The ratio of these isomers at 25° was ~5:1 as calculated from the relative ratio of the integrals of the formyl proton resonances.

Other resonances observed in NMR spectra have, in part, been assigned to de-formyl-polysaccharide, i.e. approximately 5% of the units have a perosamine structure. This is particularly apparent in the ^{1}NMR spectrum where the β-C4 proton resonates at 3.054 ppm ($^{1}J_{H-H}$=10.3 Hz) (H. Bretting, N. MacKenzie and A. Wu unpublished results).

As shown in Fig. 7, a hexamer consisting of repeating 1,2-linked 4,6-dideoxy-4-formamido-D-mannopyranosyl units was modeled by a minimum energy routine using MM2 parameters. The arrows in the picture indicate the location of the *N*-formyl substituents which has been considered to be the most important epitope in this antigen.

From the ^{13}C NMR spectroscopic profile it has been determined that the major fraction of these O-chains is composed of repeating units of *N*-formylperosamine and is identical to the O-chains of *B. abortus* S1119.3 and *Yersinia entercolitica* serotype 0:9. This is in good agreement with their immunochemical cross reactivity.

Bundle, *et al.* reported (19,19A,20) that the O-polysaccharide in the cell wall lipopolysaccharide of *Brucella melitensis* contained 4,6-dideoxy-4-formamido-D-mannopyranose as the only carbohydrate component, but unlike the O-antigen of *Brucella abortus* (6) which was shown to be a linear α1→2 linked homo-polymer of this sugar, the *B. melitensis* polysaccharide gave complex ^{1}H and ^{13}C NMR spectra. By one and two dimensional NMR analysis, the O-polysaccharide of *B. melitensis* was shown to

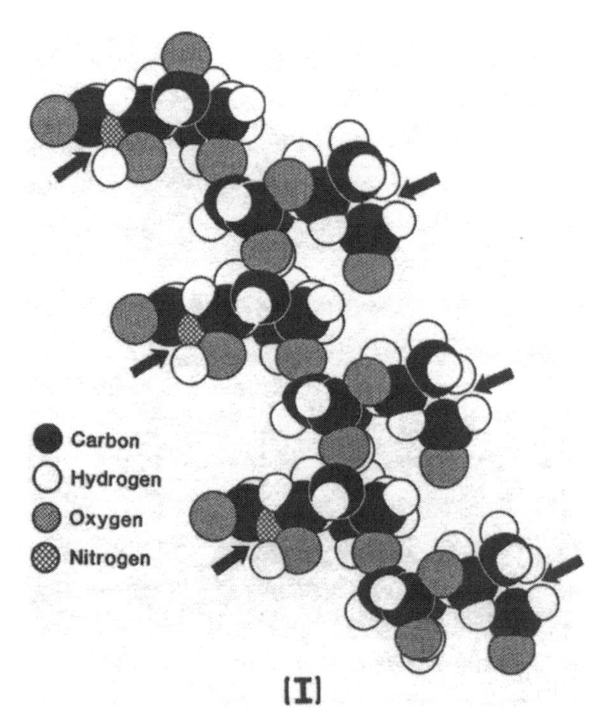

Fig. 7. A hexamer consisting of repeating 1,2-linked 4,6-dideoxy-4-formamido-D-mannopyranosyl units was modeled using the Chem Graf computer modeling program on a Lundy S5600 linked to a Vax 780. The configuration of the hexasaccharide chain shown (I) above was obtained by a minimum energy routine using MM2 parameters. The arrows in the figure above indicate the location of the *N*-formyl substituents (unpublished data).

possess a linear structure with a repeating pentasaccharide unit containing one 1→3 and four 1→2 linkages of *N*-formyl perosamine (4,6-dideoxy-4-formamido α-D-mannopyranosyl) residues. This is an important finding in the field of A and M antigens, however, more information seems to be required to establish the epitopes of these antigens.

II-D. O-Chain Reactivity

The immunochemical reactivities of the O-chains were analyzed by precipitin-inhibition assay. As shown in Fig. 8, when μmoles of inhibitors were expressed as monosaccharide of anhydro-*N*-formyl perosamine of M.W. 173, all chain sizes tested demonstrated similar inhibition profiles. Only about 0.29 μmoles of *N*-formyl perosamine in O-chains were required to inhibit 50% of the precipitin reaction. These O-chains were about 480 times as active as MeαMan or Man.

The reactivities of the O-chains (Fig. 8) seem to be independent of size. We are not sure whether it is due to the groove type of antibody combining sites, or to the heterogeneity of these antibodies in the serum used. This question shall be resolved by using monoclonal antibodies in the antigen-antibody inhibition reaction.

III. Current Concepts and Future Prospects

The smooth lipopolysaccharide (sLPS) of *Brucella abortus* is the most immunodominant component amoung the antigens of *B. abortus* isolated. Thus, one of the most important goals in our future research is to map structural and immunochemical relationships among the purified water extractable endotoxins (lipopolysaccharide, LPS), membrane bound smooth LPS (major in phenol phase) and rough LPS (minor in aqueous phase) by hot phenol extraction of the same strains and among the smooth LPS of various strains.

In most naturally infected cattle almost all anti-brucella antibodies (over 99.5%) are absorbed by protein free lipopolysaccharide (LPS). These antigen-antibody interactions can also be inhibited by the carbohydrate moiety of LPS (based on the results of HIG, ELISA and QPA-inhibition assays). The possibility of using exclusively proteins as the protective

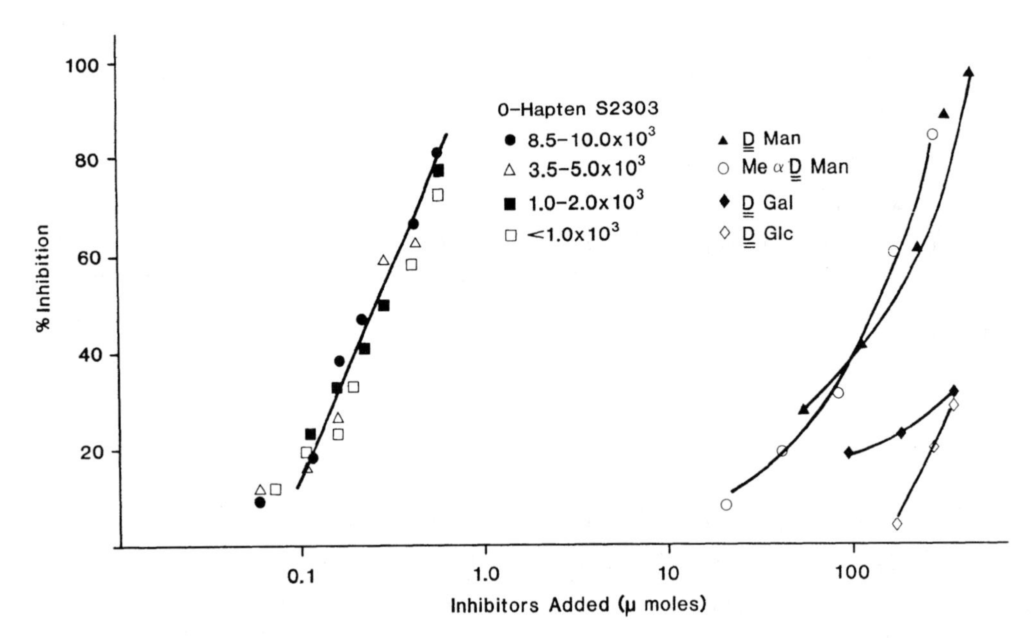

Fig. 8. Inhibition by monosaccharides and *B. abortus* O-chains of the precipitation of convalescent *B. abortus* S2308 infected bovine serum with smooth LPS-f5A prepared from *B. abortus* S2308. Total volume-350 μl. 10 μl of serum and 10 μg of LPS-f5A were in the reaction mixture. About 0.29 μmoles of *N*-formyl perosamine in O-chain and 140 μmoles of Man or MethylαMan are required for 50% inhibition. Reproduced with permission from Wu and MacKenzie, Mol. Cell. Biochem., 75:103-111, 1987, Fig. 3.

antigens (2) as well as for differential diagnosis in cattle is facing many biotechnology problems, especially antigenicity enhancement. Therefore, the area of research for serodiagnosis and vaccines should also focus equally on endotoxins. From our recent findings (9), it is obvious that the LPS-protein complex of S2308 is more complicated than that of strain 19. Several possible epitopes of S2308 lipopolysaccharides have been detected by using several cow sera infected with strain 2308. It is possible that detoxified lipopolysaccharides and/or O-chains coupled to protein carriers have potential application for both vaccine and serodiagnosis (2,21-26). Alkaline treatment of the LPS removes most of the ester linked fatty acids and results in a reduction of some toxicity (26). Finally, O-chain-protein conjugates have several advantages such as reduction of toxicity, enhancement of antigenicity , and the availability of protein carriers as vaccinated markers (27).

SUMMARY

Smooth lipopolysaccharide (sLPS) of *Brucella abortus*, which is the most immunodominant component among the antigens of *B. abortus* isolated, has been used for diagnosis for decades. High yields of sLPS can be prepared by a modification of the procedures of Moreno *et al.* (J. Bacteriol. 138:361-369, 1979). Washed *B. abortus* cells can be disrupted by 21 freeze-quick thaw cycles and ultrasonication to separate non-membrane-bound material; then phenol extraction is performed 3 times and the phenol fraction is washed with H_2O intensively. The membrane-bound sLPS can be fractionated into 3 to 5 groups according to the extent of dialysis and centrifugation. These membrane bound sLPS fractions show marked individual differences in their precipitin profile and chemical composition. Their protein content varies from 16% to 42% as determined by dye binding test and 17 to 60% by Lowry phenol method using bovine serum albumin as the standard, which indicates that these proteins associated with LPS may play important roles in the immunochemical interactions, solubility, and the heterogeneity of *B. abortus* lipopolysaccharides. Compared to previously published methods, a higher yield of sLPS, ranging from 3.6% to 7.7% of dried bacteria, is obtained. Group f5A, which has a standard bell shaped curve in the precipitin assay, is one of the major fractions in all three strains (1119.3, 19, 2308). The protein free sLPS (less than 1% of Lowry reactive component) can be prepared by pronase digestion. The immunochemical reactivity remains about the same before and after this treatment.

The O-chains of the major fraction (f5A) of *B. abortus* (Strains 2308 and 19) membrane bound smooth lipopolysaccharide (sLPS) are obtained by hydrolysis of f5A native sLPS in 1% acetic acid at 100°C for 2 hours. After hydrolysis, the O-chains are separated from the lipid A protein complex by centrifugation, and from small fragments by ultrafiltration of a molecular weight cut-off (MWCO) of 1.0×10^3. These carbohydrate haptens can be identified by precipitin-inhibition assay and further fractionated by both membrane filtration and dialysis. The size distributions of carbohydrate haptens of the endotoxins (f5A) ranged from several oligosaccharides up to 1.0×10^4 MWCO. Three major fractions of MWCO $8.0\text{-}10.0 \times 10^3$, $3.5\text{-}5.0 \times 10^3$, and $< 1.0 \times 10^3$ for both strains 2308 and 19 contain more than 85% of the total immunoreactive materials. These O-chain fractions are, by composition, ^1H, and ^{13}C NMR analysis, found to be homopolymers of $\alpha 1 \rightarrow 2$ linked 4,6-dideoxy-4-formamido-D-mannose (*N*-formylperosamine), which is identical to the O-chains of *B. abortus* strain 1119.3 and *Yersinia enterocolitica* serotype 0:9 and similar to *Vibrio cholera* 569B (INABA).

Fractions of O-chains have similar inhibitory reactivities in the precipitin-inhibition assay as expressed as µmoles of monosaccharide of anhydro-*N*-formylperosamine and are about 480 times as active as MeαMan.

The most immunodominant antigen isolated from *Brucella abortus* is the smooth lipopolysaccharide. Thus, our future research goal will be to map structural and immunochemical relationships among the purified water extractable endotoxins (lipopolysaccharide, LPS), membrane bound smooth LPS (major in phenol phase) and rough LPS (minor in aqueous phase?) by hot phenol extraction of the same strain and among the smooth LPS of various strains.

ACKNOWLEDGEMENT

1. Aided by grants from the Texas Agricultural Experiment Station (TAES H6194), and USDA/SEA Formula Animal Health Funds (Project 6648) and the Robert J. and Helen C. Kleberg Foundation.

2. The authors would like to thank Ms. Doretha A. Jones for her secretarial assistance in preparing this manuscript, and Mr. Miles S. Frey for technical assistance and provision of electron micrographs.

REFERENCES

1. Berman, D.T. (1981) The importance of being earnest with your antigens. *In:* The Ruminant Immune System (Edited by Butler, J.E.), pp. 217-224. Plenum Press, New York and London.
2. Montaraz, J.A., Winter, A.J., Hunter, D.M., Sowa, B.A., Wu, A.M. and Adams, L.G.(1986) Protection against *Brucella abortus* in mice and O-polysaccharide specific monoclonal antibodies. Infect. Immun., 51:961-963.
3. Moreno, E., Pitt, M.W., Jones, L.M., Schurig, G.G. and Berman, D.T. (1979). Purification and characterization of smooth and rough lipopolysaccharides from *Brucella abortus*. J. Bacteriol., 138:361 369.
4. Baker, P. and Wilson, J.B. (1965). Hypoferremia in mice and its application of the bioassay of endotoxin. J. Bacteriol., 90:903-910.
5. Marx, A., Jonescu, J. and Pop, A. (1983). Immunochemical studies on *Brucella abortus* lipopolysaccharide. Zbl. Bakt. Hyg. I Abt. Orig. A 253:544-553.
6. Caroff, M., Bundle, D.R., Perry, M.B., Cherwonogrodzky, J.W. and Duncan J.R. (1984).Antigenic S-type lipopolysaccharide of *Brucella abortus* S1119-3. Infect. and Immun., 46:384-388.
7. Wu, A.M., Heck, F.C., Adams, L.G. and Jones, K. (1984) Immunochemical studies on the binding properties of *Brucella abortus* lipopolysaccharides to bovine precipitating antibodies. Mol. Immun., 21:1123-1129.
8. Perera, V.Y., Winter, A.J., and Ganem, B. (1984) Evidence for covalent bondings of native hapten protein complexes to smooth lipopolysaccharide of *Brucella abortus*. FEMS Lett., 211:263-266.
9. Wu, A.M., Adams L.G. and Pugh, R. (1987) Immunochemical and partial chemical characterization of fractions of membrane-bound smooth lipopolysaccharide-protein complexes from *Brucella abortus*. Mol. Cell. Biochem., 75:93-102.
10. Wu, A.M. and MacKenzie, N.E. (1987) Structural and immunochemical characterization of the O-haptens of *Brucella abortus* lipopolysaccharides from strains 19 and 2308. Mol. Cell. Biochem., 75:103-111.
11. Caroff, M., Bundle, D.R. and Perry, M.B. (1984) Structure of the O-chain of the phenol-phase soluble cellular lipopolysaccharide of *Yersinia enterocolitica* serotype 0:9. Eur. J. Biochem. 139:195-200.
12. Redmond, J.W. (1978) The 4-amino sugars present in the lipopolysaccharides of *Vibrio cholerae* and related Vibrios. Biochim. Biophys. Acta., 542:378-384.

13. Moreno, E., Berman, D.T., and Boettcher, L.A. (1981) Biological activities of *Brucella abortus* lipopolysaccharides. Infec. Immun., 31:362-370.
14. Moreno, E., Speth, S.L., Jones, L.J., and Berman, D.T. (1981) Immunochemical characterization of *Brucella* lipopolysaccharides and polysaccharides. Infec. Immun., 31: 214-222.
15. Tsai, C.M. and Frasch, C.E. (1982) A sensitive silver stain for detecting lipopolysaccharides in polyacrylamide gels. Analyt. Biochem. 119:115-119.
16. Jann, B., Reske, K. and Jann, K., (1975) Heterogeneity of Lipopolysaccharides. Analysis of polysaccharide chain lengths by sodium dodecylsulfate-polyacrylamide gel electrophoresis. Eur. J. Biochem. 60:239-246.
17. Goldman, R.C., and Leive, L. (1980) Heterogeneity of antigenic side chain length in lipopolysaccharide from *Escherichia* 0111 and *Salmonella typhimurium* LT2. Eur. J. Biochem. 107:145-153.
18. Palva, E.T. and Mäkelä (1980) Lipopolysaccharide heterogeneity in *Salmonella typhimurium* analyzed by sodium dodecyl sulfate polyacrylamide gel electrophoresis. Eur. J. Biochem. 107:137-143.
19. Bundle, D.R., Perry, M.B., and Cherwonogrodzky, J.W. (1987) Structural elucidation of *Brucella melitensis* M antigen by high-resolution NMR at 500MHz, Biochemistry 26:8717-8726.
19A Moreno, E., Mayer, H., and Moriyon, I.(1987) Characterization of a native polysaccharide hapten from *Brucella melitensis*. Infec. Immun. 55:2850-2853.
20. Bundle, D.R., Cherwonogrodzky, J.W., Caroff, M. and Perry, M.B. (1987) The Lipopolysaccharide of *Brucella abortus* and *B. Melitensis*. Ann. Inst. Pasteur./Microbiol. 138:92-98.
21. MacIntyre, S., Lucken, R., and Own, P. (1986) Smooth lipopolysaccharide is the major protective antigen for mice in the surface extract from IATS serotype 6 contributing to the polyvalent *Pseudomonas aeruginosa* vaccine PEV. Infec. Immun., 52:76-84.
22. Cryz, S.J., Jr., Furer, E., and Germanier, R. (1984) Protection against fatal *Pseudomonas aeruginosa* burn wound sepsis by immunization with lipopolysaccharide and high-molecular weight polysaccharide. Infect. Immun., 43:795-799.
23. Pier, G.B.(1983) Immunochemistry of *Pseudomonas aeruginosa* lipopolysaccharides and high-molecular-weight polysaccharides. Rev. Infect. Dis. 5 (Suppl.):S950-S956.
24. Pollach, M., Pier, GG., and Prescott, R.K. (1984) Immunization with *Pseudomonas aeruginosa* high-molecular-weight polysaccharides prevent death from Pseudomonas burn infections in mice. Infect. Immun., 43:759-760.
25. Cryz, S.J., Furer, E., Sadoff, J.C., and Germanier, R. (1986) *Pseudomonas aeruginosa* immunotypes polysaccharide-Toxin A conjugate vaccine. Infect. Immun., 161-165.
26. Seid. R.C., and Sadoff, J.C. (1981) Preparation and characterization of detoxified lipopolysaccharide-protein conjugates. J. Biol. Chem., 256:7305-7310.
27. Enlanger, B.F., The preparation of antigenic hapten-carrier conjugates: A survey. (1980) Meth. Enzymol., 70:85-104.

CHEMICAL AND IMMUNOCHEMICAL STUDIES ON LIPOPOLYSACCHARIDES OF *COXIELLA BURNETII* PHASE I AND PHASE II

Hubert Mayer[1], Joanna Radziejewska-Lebrecht[2], and Stefan Schramek[3]

Max-Planck-Institut für Immunbiologie, Freiburg i.Br.,FRG[1]
Institute of Microbiology, University of Lódz´, Poland[2], and
Institute of Virology, Slovak Academy of Sciences, Bratislava, CSSR[3]

Coxiella burnetii, etiological agent of both acute and chronic Q-fever, is an obligate intracellular bacterial parasite. It is the only species of the genus Coxiella of the Rickettsiaceae family, and all strains so far studied belong to the same serotype (Bergey's Manual, 1984).

Coxiella burnetii stands unique amongst rickettsiae in showing a distinct phase variation: Virulent phase I cells occuring in nature change to avirulent phase II cells upon serial passage in embryonated chicken eggs. This phase variation, detected by serological means, is accompanied by differences on the surface of the cells but also by a change of many biological and physico-chemical properties (Kazár et al., 1974; Visacka et al., 1984).

Strains of *C. burnetii* differing in the number of yolk sac passages may also differ from each other and, therefore, a more distinct differentiation of the phase stage had to be introduced. The term "pure phase II" was introduced to characterize those strains having a long passage history in chicken embryo yolk sacs and can no longer convert to phase I by inoculation into laboratory mice. This conversion can be obtained, however, with cells being in the "transition phase II" stage (Brezina et al., 1978).

Isolation and analysis of lipopolysaccharides (LPS's) from cells of both phases provided evidence that the phase variation is a complex phenomenon resembling the well-studied S→R mutation occuring with many gram-negative (enteric) bacteria. LPS present in phase I cells was considered to be a major determinant of virulence expression and infection of *C.burnetii* (Baca and Paretsky, 1974; Kazár et al., 1974). Considering these aspects, the study of the composition and structure of LPS's from both phases was undertaken and the results were compared with those obtained with strains or species from other bacterial families.

Preparation of Lipopolysaccharides

Coxiella burnetii, mostly strain Nine Mile (9 mi.) in phase I (3rd–5th yolk sac passage; EP 3-5), in "transition phase II" (EP 21), or in "pure phase II" (EP 163) was propagated in chick embryo yolk sacs. Cells of phase I were in some cases also propagated in mouse spleen. Cells were killed with formalin or phenol and were then purified by differential centrifugation and ether treatment as previously described (Schramek et al., 1978). The purified cells were then treated with chloroform-methanol (Schramek and Mayer, 1982) and LPS's were extracted by using the hot phenol-water procedure. The LPS's were further purified by repeated high speed centrifugations (105,000 x g; 4 h each) (Schramek and Galanos, 1981). For the isolation of LPS from phase II cells, the phenol-chloroform-petroleum ether method of Galanos was additionally used (Schramek and Brezina, 1979), but because of higher yields of LPS, the phenol-water procedure was preferentially used. LPS extracted from phase I cells was designated as LPS I and that from "pure phase II" as LPS II.

Chemical Composition of Isolated LPS I and LPS II

A comparative analysis of the sugar composition of LPS's isolated from both phases was published by Baca et al. (1980). Most sugars found in LPS I were also detected in the LPS of phase II, although with quantitative differences. Our studies of LPS I and II, however, showed clear-cut differences between the respective sugar compositions (Schramek and Mayer, 1982). These results could be fully corroborated later on by Amano and Williams (1984). In Fig. 1 the sugar composition of both LPS-types is given, showing the sugars common to both phase types as full circles.

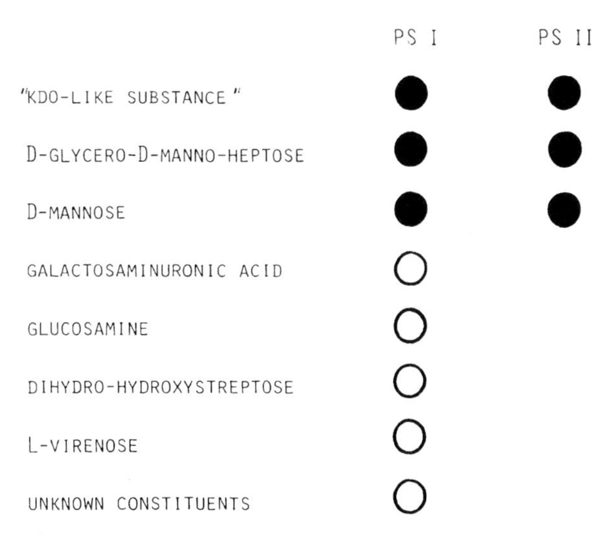

Fig. 1. Chemical compositions of the sugar parts (PS I and PS II) of *Coxiella burnetii* LPS I and II (the full circles indicate the sugars common to LPS I and II).

LPS II shows a rather simple composition: In addition to lipid A-derived glucosamine, it contains only mannose, the unusual D-glycero-D-manno-heptose and a thiobarbituric acid-positive KDO-like substance (KDO, 3-deoxy-D-manno-octulosonic acid). The latter can be differentiated from authentic KDO by high voltage electrophoresis (Schramek and Mayer, 1982).

In addition to sugars of phase II LPS, a number of (unusual) neutral and amino sugars could be detected in LPS I (Fig. 1 and 2). No differences in sugar composition were observable with cells propagated in yolk sacs of chick embryos, or grown in mouse spleen cells (unpublished). Two of the major neutral sugar constituents (designated as 1 and 2 in Fig. 2) could recently be identified as C-3-branched aldoses (Schramek et al., 1985). These sugars could easily be released from LPS I even by mild acetic acid hydrolysis (1%-10%) and be isolated in a pure state by thin-layer chromatography. Analysis of their deuterium-reduced alditol acetate derivatives (for details see Schramek _et al._,1985) showed that sugar 1 was a 6-deoxy-3-_C_-methyl-hexose and sugar 2 was a 3-_C_-(hydroxymethyl)-pentose (Fig.2).

Using a range of authentic naturally occuring 6-deoxy-3-_C_-methyl-hexoses (Table 1) and newly synthesized 3-_C_-(hydroxymethyl)-pentoses (Table 2) (Dahlman _et al._, 1986), sugar I was identified as a 6-deoxy-3-_C_-methyl-gulose, very probably the _L_-form of this sugar, and sugar 2 as 3-_C_-(hydroxymethyl)-_L_-lyxose (Schramek _et al._, 1985; Mayer _et al._, 1985). These two sugars have so far not been encountered in LPS and only rarely in other natural products. Related sugars have been described from antibiotics, e.g. _D_-virenose from the antitumour antibiotic virenomycin (Kulyaeva _et al._, 1978), _L_-virenose from chrysomycin A and B (Weiss _et al._, 1982) and 3-_C_-formyl-_L_-lyxose (hydroxy-streptose) from the antibiotics hydroxy-streptomycin and mannosido-hydroxystreptomycin (Stodola _et al._,1951). The unidentified alditol acetate, shown as U_1 in Fig. 2, is very probably the 1,5-anhydro-derivative of dihydro-hydroxystreptose [3-_C_-(hydroxymethyl)-_L_-lyxose], as indicated by characteristic fragments obtained in e.i.-mass spectrometry. The additional sugars or components designated as U in Fig. 2 are probably not related to LPS but nevertheless seem to be of rickettsial origin since they are equally present in LPS I obtained from yolk sacs and from mouse spleen cells (see below).

The hydrolysis kinetics of the two labile-linked branched sugars was recently investigated (Schramek _et al._, in preparation) and is depicted in Fig. 3. Hydrolysis of LPS I under stronger conditions (10% HOAc, 100°C, 90 min) resulted likewise only in the selective release of the branched sugars virenose and dihydro-hydroxystreptose. The two unusual sugars are depicted in Fig. 4-I and II. It was found by methylation analysis that virenose is present in pyranose form, whereas the furanose form was proven for the dihydro-hydroxystreptose. The terminal linkage (see below), the axial OH-groups, the bulky substituent on C-3 and the furanose form of the L-lyxoderivate are responsible for the acid-labile linkages of the branched sugars.

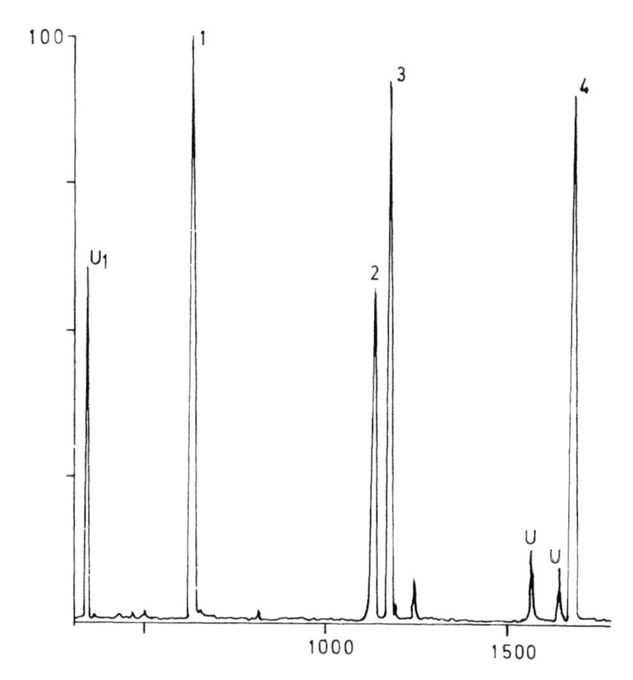

Fig. 2. Gas-chromatographaphic analysis of alditol acetates* of LPS I (SE-54 capillary column; 140-200°C; 2°/min), showing the position of (1) virenose, (2) dihydro-hydroxystreptose, (3) D-mannose and (4) D-glycero-D-manno-heptose (taken from Schramek et al., 1985). *Alditol acetates of both branched sugars have a free OH-group at C-3, which cannot be acetylated under normal conditions.

LPS isolated from "transition phase II" cells, obtained after pasage of *C.burnetii* phase I cells in chick embryo yolk sacs (EP 21) which resulted in partial unmasking of the phase II determinants, showed a composition which differed from both LPS I and LPS II. It contained strongly reduced amounts of the branched sugars and amino sugars. This result can mean that LPS from EP 21 cells is a mixture of LPS I and II but, since the ratio of the two branched sugars is differing from that of LPS I, it may also indicate that a distinct transferase or synthetase of O-chain sugars had been affected primarily by the mutation.

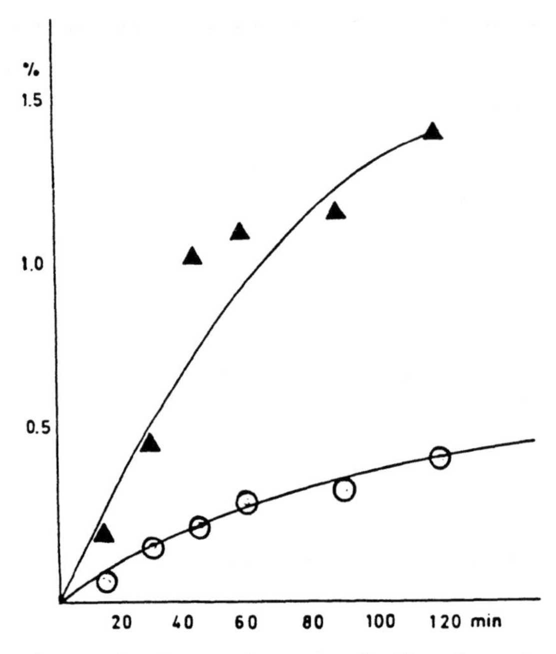

Fig. 3. Kinetics of the release of the two branched sugars with 1% acetic acid at 100°C. The liberated sugars were quantified by glc-analysis of their alditol acetates. ▲ indicates 3-C-(hydroxymethyl)-L-lyxofuranose; ○ indicates L-virenopyranose.

Table 1. Gas-liquid chromatography of alditol acetates* of authentic naturally occuring 6-deoxy-C-3-methylhexoses and of sugar 1 (L-virenose) from LPS-I of *Coxiella burnetii*. *See legend to Fig. 2.

Alditol acetate derivative of	t_R on		
	ECNSS-M 170°C	SE-30 170°C	Carbowax 20 M/220°C
Xylose	1	1	1
Coxiella phase I sugar (1)	0.88	0.92	1.02
6-Deoxy-3-C-methyl-D-mannose (evalose)	0.78	0.84	0.95
6-Deoxy-3-C-methyl-D-gulose (virenose)	0.89	0.92	1.00
6-Deoxy-3-C-methyl-L-talose (demethyl-vinelose)	1.19	0.96	1.19
6-Deoxy-3-C-methyl-2-0-methyl-L-talose (vinelose)	0.45	0.73	-

The unusual amino-hexoses reported to occur in LPS I (Schramek and Mayer, 1982) could recently be identified by Amano and Williams (1984) and Amano et al. (1985) as galactosaminuronic acid and the disaccharide D-galactosaminuronyl-α-(1,6)-D-glucosamine (see Fig. 4-III). The high extent of this disaccharide obtained by acid hydrolysis is explained by the aldobiuronic acid-type of the disaccharide and by the reported (Amano et al., 1985) partial occurence of unsubstituted amino groups (about 50% unsubstituted) in LPS I.

Table 2. Gas-liquid chromatography of alditol acetates* of synthetic 3-C-(hydroxymethyl)-pentoses and of sugar 2 [3-C-(hydroxymethyl)-L-lyxose] from LPS-I of *Coxiella burnetii*. *See legend to Fig.2 (Dahlman et al., 1986)

3-C-branched sugar		synthetic route	GLC of alditol acetate t (min)*
3-C-(hydroxymethyl)-		reaction of 3-uloses:	
"	D-ribose	with nitromethane	16.80
"	D-xylose	"	16.64
"	L-lyxose	with dithiane	16.90
"	sugar of PS I (2)		16.90

In a very recent paper, Hackstadt et al. (1985) described an additional mutant of *Coxiella burnetii* designated as 9 mi/Cr. It shows a structural complexity intermediate to phase I and phase II LPS and a changed serological specificity. It is still reacting with phase I antiserum, but to a much lesser extent than the wild-type, and shows no reactivity at all with phase II antisera. A very recent analysis of LPS 9 mi/Cr has shown that all sugars of LPS I do occur in this mutant in the usual amounts with the sole exception of L-virenose, which is totally absent, and some of the unidentified late peaks (U in Fig. 2) (Mayer, Hackstadt and Schramek, unpublished). This mutant will allow investigation of the localization of the acid-labile linked sugars by methylation analysis. The mutant was recognized already in 1965 by Marius G. Peacock and was isolated from placental tissue of a guinea-pig experimentally infected 11.5 months earlier with the wild-type strain 9 mi/I (Hackstadt et al., 1985). The fact that virenose can be lost without losing additional components substantiates that the biosynthesis cannot follow the way described for Enterobacteriaceae, where lipid-linked oligosaccharide repeating units are transferred to a core in polymerized form. It further indicates the terminal linkage of virenose which agrees with preliminary methylation studies showing that both branched sugars are in terminal positions.

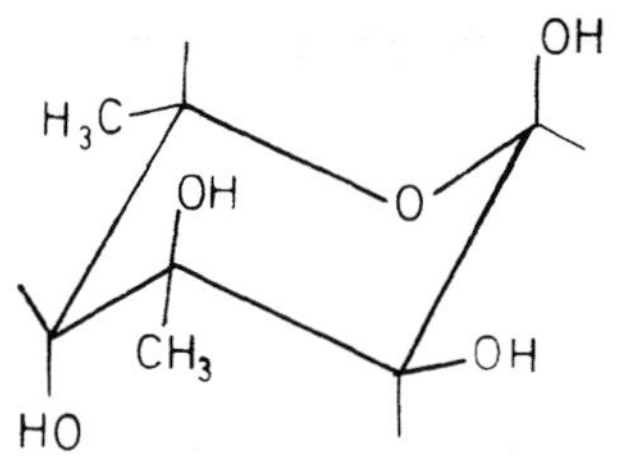

Fig. 4. Chemical structures of (I) 6-deoxy-3-C-methyl-L-gulo-pyranose (L-vireno-pyranose), (II) 3-C-(hydroxy-methyl)-L-lyxo-furanose (dihydro-hydroxy-L-strepto-furanose), and the amino sugar-disaccharide reported by Amano et al. (1985) (III) α-D-galactosaminuronyl-(1,6)-D-glucosamine.

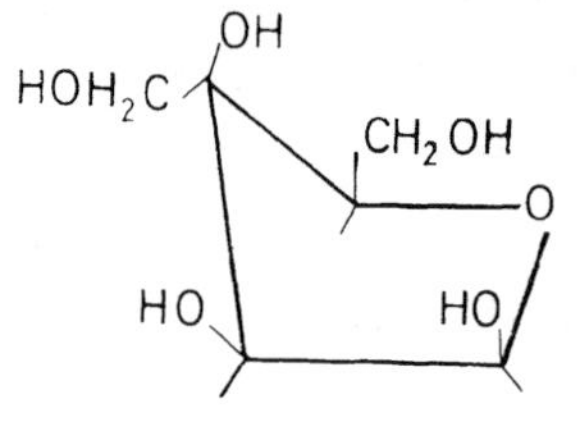

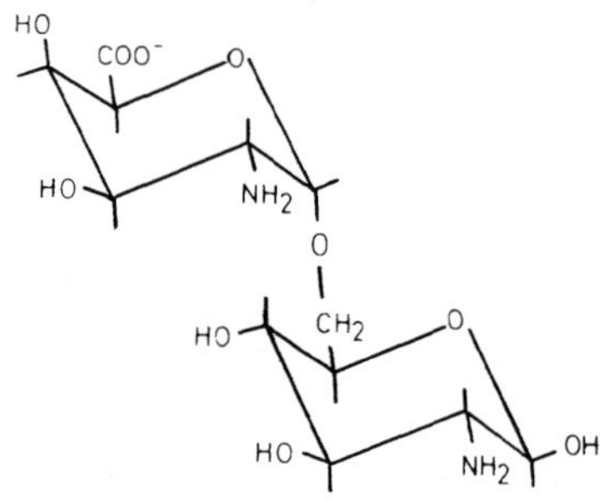

Until now we were not sure about the chemical nature of the components which were recognized by GLC-MS analysis of LPS I (designated as U in Fig. 2), which are present in much smaller quantities in phase II LPS. These components can withstand periodate oxidation and show a mass shift of one unit (or, in one case, two units) indicating the presence of reducing group(s). These components could be further degraded (or decomposed) using stronger hydrolytic conditions (2.5 N HCL, 100°C, 8h).

Serological Properties of *C.burnetii* LPS

Baca et al. (1980) demonstrated the serological identity of LPS I and LPS isolated from his phase II cells, using complement fixation and immunodiffusion tests. Our results

(Schramek *et al.*, 1983) differed in the finding that both LPS I and pure phase II LPS behaved as distinct antigens showing no serological cross reaction with each other, either in passive hemolysis, complement fixation, or in immunoprecipitation. For these tests, antisera obtained by immunizing rabbits with whole cells of *C.burnetii* of the corresponding phase were used. Amano and Williams (1984) also demonstrated that LPS I in immunodiffusion tests reacted only with hyperimmune sera to phase I cells but not with those against phase II. Conversely, no reactivity of LPS II was found with hyperimmune sera towards phase I.

It was observed that mild acid hydrolysis of LPS I led to a drastic decrease of its serological activity in the passive hemolysis test. This decrease occured earlier than the splitting of LPS in lipid A and "degraded polysaccharide" as indicated by the serological identification of unsubstituted lipid A by lipid A antisera (Schramek *et al.*, 1985). Under these conditions, only the two branched sugars were released from LPS I, indicating (in addition to the serological properties of 9 mi/Cr mutant) that both sugars are acting in LPS I as immunodominant constituents. The kinetics of the sugar release (Fig. 3) shows that dihydro-hydroxystreptose is released considerably faster than virenose (Schramek *et al.*, in preparation).

The Lipid A Component of *C.burnetii* LPS

C.burnetii cells, irrespective of their phase stage, show endotoxic activity (Schramek, 1978) which is attributable to their LPS content (Baca and Paretsky, 1974; Schramek and Brezina, 1976 and 1979). Again, as found with numerous gram-negative bacteria, it is the lipid A component which represents the endotoxic principle of LPS.

Free lipid A isolated from either phase of *C.burnetii* exhibited complete serological cross-reactivity with Salmonella lipid A (Schramek and Galanos, 1981). The most reduced lipid A structure known to still express complete lipid A antigenicity is represented by a ß-(1,6)-linked glucosamine disaccharide with at least one amide-linked fatty acid (Galanos et al., 1984). The complete serological cross-reaction with Salmonella lipid A indicates the presence of such a structure also in *C.burnetii* lipid A.

The endotoxic activity of *C.burnetii* LPS I and II and its lipid A component is considerably lower than that of Salmonella LPS and lipid A (Baca and Paretsky, 1974; Schramek and Brezina, 1976; Schramek and Galanos, 1981). This suggests that the lipid A component of *C.burnetii* may, nevertheless, differ structurally from that of enterobacterial LPS. For full expression of endotoxic activity, substitution of the central glucosamine disaccharide with phosphate and with ester-and amide-linked fatty acids is important (Galanos et al., 1977; Rietschel et al., 1984; Shiba and Kusomoto, 1984). Preliminary experiments (Wollenweber et al., in preparation) have indicated that the *C.burnetii* lipid A is lacking the phosphate substituent at C-4', which is present in the so-called "lipid A backbone" occuring in all enterobacterial lipid A and consisting of a ß-1,6-linked glucosamine disaccharide with phosphate substituents at C-1 and C-4' (Rietschel et al., 1982).

Common constituents of biologically active LPS's of various bacterial families are ester-bound (R)-3-hydroxy fatty acids or (R)-3-acylated hydroxy fatty acids (Rietschel et al., 1984). These were so far not encountered in *C.burnetii* lipid A (Wollenweber et al., 1985). Instead, a number of non-hydroxylated fatty acids (normal, iso- and anteiso-branched fatty acids) were found. Hydroxylated fatty acids were present, however, as amide-linked substituents of GlcN. A characteristic feature of *C.burnetii* lipid A is the presence of branched fatty acids and hydroxylated fatty acids (65% of the total amount), including the very rare (R)-3-hydroxylated iso- and anteiso-branched fatty acids. The finding of more than 50 different amide-linked 3-acyloxyacyl residues reflects the broad microheterogeneity of *C.burnetii* lipid A (Wollenweber et al., 1985). This complex fatty acid spectrum differs considerably from that of LPS of most (or all) other gram-negative bacteria.

Chemical Studies on the Sugar Moieties of LPS I and II

The detailed chemical structure of LPS II was recently investigated (Mayer *et al*., 1985) using methylation analysis, and NMR-techniques, as well as CrO_3-oxidation. LPS II consists of D-mannose and D-glycero-D-mannoheptose (in a molar ratio of 1:1) and a KDO-like substance, which could not be accurately quantified. GLC-MS analysis of hydrolysates of permethylated LPS II afforded the following methylated sugars: 2,3,4,6-tetra-O-methyl-D-mannose; 3,4,6,7-tetra-O-methyl-D-glycero-D-manno-heptose and 2,6,7-tri-O-methyl-D-glycero-D-manno-heptose in a molar ratio of 2:1:1. Using gated decoupled ^{13}C-NMR 4(-5) anomeric C-atoms could be detected showing coupling constants [$^{J}C1$-H1] of 158, 162, 174, and 174 Hz, indicating the presence of two α- and two ß-linkages in PS II. Oxidation of PS II with CrO_3 led to a complete destruction of mannose, whereas D-glycero-D-manno-heptose was only slightly affected. The two β-linkages have therefore been assigned to the mannose units, whereas the α-linkages are assigned to heptose.

Information on the attachment sites of phase I-specific sugars onto the phase II core was available by methylation analysis of LPS I. Preliminary analyses of permethylated LPS I afforded the following methylated sugars (as alditol acetates): 2,3,4,6-tetra-O-methyl-D-mannose and 2,3,6-tri-O-methyl-D-mannose (small amount); 4,6,7- and 2,6,7-tri-O-methyl-D-mannoheptose; 4,6-di-O-methyl-N-(methyl)acetyl-glucosamine and permethylated virenose and dihydro-hydroxystreptose. Mass spectra of the two latter components are shown in Schramek *et al*. (1985). These results, which are not yet complete and so far do not include methylated galactosaminuronic acid, may be used to indicate the two following points: (*i*) the attachment sites of phase I-specific sugars are the 3-position of the 1,2-linked D-glycero-D-manno-heptose and, to a lesser extent, the 4-position of one of the terminal mannoses; and (*ii*) the terminal location of virenopyranose, dihydro-hydroxystreptofuranose and D-mannose in LPS I has been established.

We do not yet know to which penultimate sugar(s) the branched aldoses are linked. Methylation of LPS of the mutant

9 mi/Cr may answer this question. Work on the detailed structure is presently going on in collaboration with Dr. Hackstadt, Hamilton, Montana. The preliminary structure showing phase II LPS and the attachment sites of the phase I-specific sugars is depicted in Fig. 5.

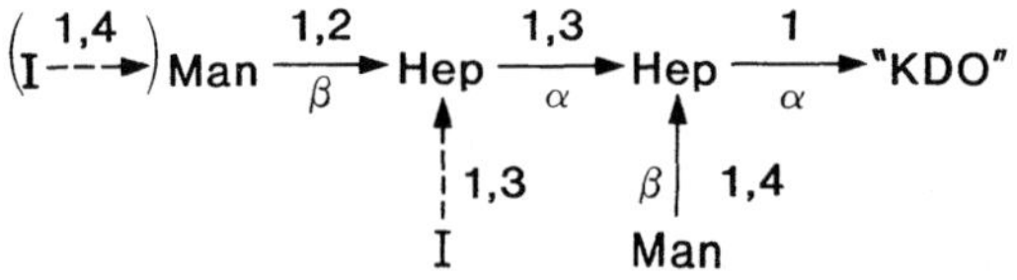

$$\left(I \xrightarrow{1,4} \right) Man \xrightarrow[\beta]{1,2} Hep \xrightarrow[\alpha]{1,3} Hep \xrightarrow[\alpha]{1} \text{"KDO"}$$

$$Hep \xleftarrow{1,3} I \quad (\text{first Hep}); \qquad Hep \xleftarrow[\beta]{1,4} Man \quad (\text{second Hep})$$

Fig. 5. Structure of LPS II with the attachment sites of phase I-specific sugars (I-→). Hep, D-glycero-D-manno-heptose; "KDO", KDO-like substance.

SUMMARY

Lipopolysaccharides of *Coxiella burnetii* phase I and II were comparatively investigated by chemical and immunochemical methods. LPS of phase I (LPS I) and phase II cells (LPS II) show no serological cross reaction, indicating that the serological determinants of LPS II are masked in LPS I. Chemical analysis of LPS I and II show that phase I and II cells can be considered as S and R forms of *Coxiella burnetii.*

The structure of LPS II has recently been elucidated and shows a dimannosylated core of an α(1,3)-linked heptose-disaccharide which is attached to a "KDO-like" substance. In enterobacterial core-types, α(1,3)-linked heptose-disaccharide is also part of the inner core structure, although the heptose occuring in enterobacterial R cores is the L-glycero-D-manno-heptose. In *Coxiella burnetii* we have only the rare D-glycero-D-manno-heptose which is the biosynthetic precursor of the former and is in many enteric LPS, present only in addition to L-glycero-D-mannoheptose. In these R-cores, it is occupying mostly terminal positions (Radziejewska-Lebrecht et al., 1981) and is absent from the main chain.

The complete structure of LPS I is not yet available, but some important points could recently be clarified. The immunodominant sugars in LPS I are C-3-branched sugars, 6-deoxy-3-C-methyl-L-gulose (L-virenose) and 3-C-(hydroxymethyl)-L-lyxose (dihydro-hydroxy-L-streptose). These two sugars have not been found so far in other lipopolysaccharides and the latter one not previously in any other natural product. Their identification is based on GLC-MS comparison with authentic and synthetic compounds. Both branched sugars (and in addition part of the mannose) are the terminal sugars in LPS I. Sites of attachment of phase I-specific sugars to the LPS II-core are: the 3-position of a branched heptose and, presumably, the 4-position of a terminal D-mannose.

The extreme acid-lability of the linkages of both branched sugars was investigated in detail and is caused by the nature of the branched sugars (deoxyhexose with bulky axial substituents; pentofuranose with axial OH-groups).

No information is so far available on the (penultimate) sugars to which the branched sugars are linked, but methylation analyses with LPS I, and with the recently described I/Cr mutant, which is selectively lacking the virenopyranose, are presently performed.

REFERENCES

1. Amano, K.I., and Williams, J.C. (1984) Chemical and immunological characterization of lipopolysaccharides from phase I and phase II *Coxiella burnetii*. J. Bacteriol. 160:994-1002.
2. Amano, K.I., Fukushi, K., and Williams, J. (1985). Chemical and ultrastructural properties of lipopolysaccharides from phase I and phase II. Abstr. 5th Nat. Conf. Am. Soc. Rickettsiology and Rickettsial Diseases. Laguna Beach, CA. March 1-3, 1985.
3. Baca, O.G., Martinez, I.L., Aragón, A.S., and Klassen, D. (1980 Isolation and partial characterization of a lipopolysaccharide from phase II *Coxiella burnetii*. Can. J. Microbiol., 26:819-826.
4. Baca, O.G., and Paretsky, D. (1974) Some physiological and biochemical effects of a *Coxiella burnetii* lipopolysaccharide preparation on guinea pigs. Infect. Immun. 9:939-945.

5. Bergey's Manual of Systematic Bacteriology. (1984). Vol. 1, pp. 701-704. "Coxiella" by Weiss, E. and Moulder, J.W., Williams and Wilkins, Baltimore, USA.

6. Brezina, R. (1978). Phase variation phenomenon in *Coxiella burnetii*, in: "Rickettsiae and Rickettsial Diseases" (eds), J. Kazár, R.A. Ormsbee, and I.N. Tarasevich, pp. 221-235, Veda, Bratislava.

7. Dahlman, O., Garegg, P., Mayer, H., and Schramek, S. (1986). Synthesis of three 3-C-hydroxymethylpentoses with the D-ribo, D-xylo and L-lyxo-configurations. Identification of the latter with a monosaccharide isolated from phase I *Coxiella burnetii* lipopolysaccharide. Acta Chem. Scand. Ser. B 40, 15-20.

8. Galanos, C., Lüderitz, O., Rietschel, E.Th., and Westphal, O. (1977) Newer aspects of the chemistry and biology of bacterial polysaccharides, with special reference to their lipid A component, in: "International Review of Biochemistry: Biochemistry of lipids II", vol 14, T.W. Goodswin (eds.), pp. 239-335, University Press, Baltimore.

9. Galanos, C., Freudenberg, M.A., Jay, F., Nerka, D., Veleva, K., Brade, H., and Strittmatter, W. (1984) Immunogenic properties of lipid A. Rev. Infect. Dis. 6:546-552.

10. Hackstadt, T., Peacock, M.G., Hitchcock, P.J., and Cole, R. (1985). Lipopolysaccharide variation in *Coxiella burnetii*: Intrastrain heterogeneity in structure and antigenicity. Infect. Immun. 48:359-365.

11. Kazár, J., Brezina, R., Schramek, S., Urvölgyi, J., Pospisil, V., and Kovácová, E. (1974). Virulence, antigenic properties and physicochemical characteristic of *Coxiella burnetii* strains with different chick embryo yolk sac passage history. Acta virol., 18:434-442.

12. Kulyaeva, V.V., Kudinova, M.K., Potanova, N.P., Rubasheva, L.M., Brazhnikov, M.G., Rozynov, B.V., and Bekker, A.R. (1978). Structure of the carbohydrate moiety of the antibiotic verenomycin. Bioorg. Khim. 1978:1087-1092.

13. Mayer, H., Dahlman, O., Garegg, P.J., Radziejewsk-Lebrecht, J., and Schramek, S. (1985) Lipopolysaccharides of *Coxiella burnetii*: Identification of virenose and dihydrohydroxystreptose in phase I and the structure of phase II lipopolysaccharide. Glycoconjugates, Proc. VIIIth Intern. Symposium, Houston, Texas, Sep. 8-13, Vol. I, 77-78.

14. Radziejewska-Lebrecht, J., Feige, U., Mayer, H., and Weckesser, J. (1981). Structure of the heptose region of lipopolysaccharides from *Rhodospirillum tenue*. J. Bact. 145:138-144.

15. Rietschel, E.Th., Galanos, C., Lüderitz, O., and Westphal, O. (1982). Chemical structure, physiological function and biological activity of lipopolysaccharides and lipid A, in: "Immunopharmacology and the regulation of leucocyte function", E.R. Webb (eds), pp. 183-229, M. Dekker, New York.

16. Rietschel, E.Th., Wollenweber, H.W., Russa, R., Brade, H., and Zahringer, U. (1984) Concepts of the chemical structure of lipid A. Rev. Infect. Dis. 6:432-438.

17. Schramek, S. (1978). Rickettsial endotoxic lipopolysaccharides, in "Rickettsiae and Rickettsial Diseases", J. Kazár, R.A. Ormsbee, I.N. Tarasevich (eds.), pp. 79-87, Veda, Bratislava.

18. Schramek, S., and Brezina, R. (1976) Characterization of an endotoxic lipopolysaccharide form *Coxiella burnetii*, Acta virol. 20:152-158.
19. Schramek, S., Brezina, R. (1979). Isolation of endotoxic lipopolysaccharide from Phase II *Coxiella burnetii*. Acta virol. 23:349.
20. Schramek, S., Brezina, R., and Kazár, J. (1978) Influence of mild acid hydrolysis on the antigenic properties of phase I *Coxiella burnetii*. Acta virol. 22:302-308.
21. Schramek, S., Brezina, R., and Visacká, E. (1983) Different antigenic properties of lipopolysaccharides isolated from *Coxiella burnetii* in phase I and pure phase II. Zbl. Bakt. Hyg., I Abt. Orig. A, 255:356-360.
22. Schramek, S., and Galanos, C. (1981). Lipid A component of lipopolysaccharides from *Coxiella burnetii*. Acta Virol 25:230-234.
23. Schramek, S., and Mayer, H. (1982). Different sugar composition of lipopolysaccharides isolated from phase I and pure phase II cells of *Coxiella burnetii*. Infect. Immun. 38:53-57.
24. Schramek, S., Radziejewska-Lebrecht, J., and Mayer, H. (1985). 3-C-branched aldoses in lipopolysaccharide of phase I *Coxiella burnetii* and their role as immunodominant factors, Eur. J. Biochem., 148:445-461.
25. Shiba, T. and Kusumoto, S. (1984). "Chemical Synthesis and biological activity of lipid A analogs". In: Handbook of Endotoxin, Vol. I: Chemistry of Endotoxin, E. Rietschel (ed.), Elsevier, Amsterdam and New York.
26. Stodola, F.H., Shotwell, O.L., Borud, A.M., Benedict, R.G., and Riley, A.C. Jr. (1951) Hydroxystreptomycin, a new antibiotic from *Streptomyces griseocarneus*. J. Am. Chem. Soc. 73:2290-2293.
27. Visacká, E., Schramek, S., Kazár, J., Kovácová, E., and Brezina, R. (1984) Changing of antigenic, immunogenic and chemical properties of a *Coxiella burnetii* strain during chick embryo yolk sac passaging. Acta virol. 28:300-308.
28. Weiss, U., Yoshihira, K., Highet, R.G., White, R.J., and Wei, T.Z. (1982) The chemistry of the antibiotics chrysomycin A and B. Antitumor activity of chrysomycin. A.J. Antibiot. (Tokyo) 35:1194-1201.
29. Wollenweber, H.W., Schramek, S., Moll, H., and Rietschel, E.T. (1985) Nature and linkage type of fatty acids present in lipopolysaccharides of phase I and phase II *Coxiella burnetii*. Arch. Microbiol., 142:6-11.

ABSTRACT

STRUCTURAL COMPARISONS OF *STREPTOCOCCUS PNEUMONIAE* SPECIFIC POLYSACCHARIDES OF GROUP 9 (9N, 9V, 9L, 9A) RELATED TO THE CHOICE OF VACCINE COMPONENTS

James C. Richards and Malcolm B. Perry

Division of Biological Sciences, National Research Council of Canada Ottawa, Canada K1A OR6

Within the group 9 *Streptococcus pneumoniae*, four serologically cross reacting types are recognized (9A, 9L, 9N and 9V) which together are responsible for 3-4% of pneumococcal disease in North America[1]. Microchemical methods of glycose analysis, gas-liquid chromatography-mass spectrometry and ^{13}C and ^{1}H n.m.r. has permitted the complete structural elucidation.of the capsular polysaccharides of the four group 9 serotypes[2-6]. The capsular antigens were found to be polymers of a repeating pentasaccharide unit sharing several similarities but differing in certain component glycoses and the presence of *O*-acetyl groups (Table).

TABLE Structures of the repeating pentasaccharide units of the 9N, 9A, 9L, and 9V pneumococcal specific polysaccharides

Type	Pentasaccharide unit	Ref.
9N	—4)-α-D-GlcA*p*-(1—3)-α-D-Glc*p*-(1—3)-β-D-ManNAc*p*-(1—4)-β-D-Glc*p*-(1—4)-α-D-GlcNAc*p*-(1—	4
9L	—4)-α-D-GlcA*p*-(1—3)-α-D-Gal*p*-(1—3)-β-D-ManNAc*p*-(1—4)-β-D-Glc*p*-(1—4)-α-D-GlcNAc*p*-(1—	3
9A, 9V	—4)-α-D-GlcA*p*-(1—3)-α-D-Gal*p*-(1—3)-β-D-ManNAc*p*-(1—4)-β-D-Glc*p*-(1—4)-α-D-Glc*p*-(1—	2,6,5
	OAc (on GlcA*p*) ... OAc (on β-D-Glc*p*)	

A detailed examination of the 500 MHz ^{1}H n.m.r. spectrum of the 9A polysaccharide[6] has indicated that the anomeric configurations of the two 1-4 linked glucopyranosyl residues were not as previously proposed,[2] but were the same as those observed for the 9V polysaccharide [5]. The only difference between the 9A and 9V antigens is the mode of *O*-acetyl substitution.
Hard sphere Exo-Anomeric effect calculations (cf. Ref. 7) in conjunction with ^{1}H-^{1}H n.O.e. measurements were used to estimate the minimum energy conformation molecular models of

these pneumococcal antigens can be of assistance in the choice of a sereotype as a candidate vaccine or in explaining immunochemically the protective spectrum of a particular structure.

References. 1) Szu, S.C., Lee, C.-J., Parke, C.J., Jr., Schiffman, G., Henrichsen, J., Austrian, R., Rastogi, S.C., and Robbins, J.B.b, Infect. Immun. 35, 777-782 (1982). 2) Bennett, L.G. and Bishop, C.T., Can. J. Chem. 58, 2724-2727, (1980). 3) Richards, J.C., Perry, M.B. and Kniskern, P.J., Can. J. Biochem. Cell Biol. 62, 1309-1320, (1984).
4) Rosell, K.-G. and Jennings, H.J., Can. J. Biochem. Çell Biol. 61, 1102-1107, (1983); Jones, C., personal communication.
5) Perry, M.B., Daoust, V. and Carlo, D.J., Can. J. Biochem. 59, 524-533, (1981). 6) Richards, J.C. and Perry, M.B., Unpublished result. 7) Bock, K., Pure Appl. Chem. 55, 605-622 (1983)

ABSTRACT

ELUCIDATION AND COMPARISON OF THE CHEMICAL STRUCTURES OF THE SPECIFIC CAPSULAR POLYSACCHARIDES OF *STREPTOCOCCUS PNEUMONIAE* GROUPS 11 (11F, 11B, 11C, AND 11A)

James C. Richards, Malcolm B. Perry, and M. Moreau

Division of Biological Sciences, National Research Council of Canada, Ottawa, Canada K1A OR6

The Group 11 *S. pneumoniae* is composed of four serologically cross-reacting types: 11F, 11C and 11A. Structural analysis of the four Group 11 capsular polysaccharides was achieved using a combination of one dimensional and two dimensional homo and heteronuclear n.m.r. experiments at high field in conjunction with microanalytical methods and specific chemical degradations. The capsular antigens of the *S. pneumoniae* serotypes 11F, 11B and 11C were found to be linear polymers with a common repeating tetrasaccharide unit (1), but differing in the nature of a phosphodiester substituent (compare 2 and 3) and the mode of *O*-acetyl substitution.[1] Substitution by *O*-acetyl groups was observed at C-3 of the -6)-α-D-GlcNAc*p*-(1- residues, and at C-2 and/or C-3 of the -4)-α-D-Gal*p*-(1-units.

The structure of the type 11A polysaccharide was found to differ from that of the other Group 11 polysaccharides in that the phospodiester substituted α-D-GlcNAc*p* units were replaced by α-D-Glc*p* residues (4)[2]. The proposed structure of the 11A antigen differs from that suggested previously.[3]

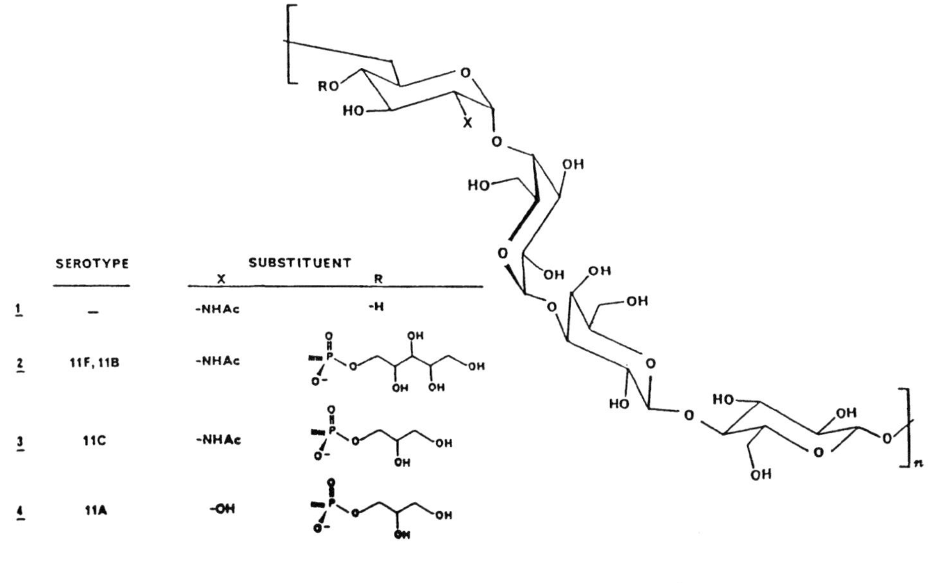

	SEROTYPE	SUBSTITUENT X	SUBSTITUENT R
1	–	-NHAc	-H
2	11F, 11B	-NHAc	
3	11C	-NHAc	
4	11A	-OH	

Differences in the immunogenic specificity of these antigens may be related to the nature of the phosphodiester substituent and to conformational variations imposed by the *O*-acetyl substituents.

References. 1) Richards, J.C., Perry, M.B., and Kniskern, P.J., Can. J. Biochem. Cell Biol. 63, 953-968 (1985). 2) Moreau, M., Richards, J.C., and Perry, M.B., Unpublished Results. 3) Kennedy, D.A. Buchanan, J.G., and Baddiley, J., Biochem. J. 115, 37-45 (1969).

ABSTRACT

APPLICATION OF TWO DIMENSIONAL NMR METHODS TO THE STRUCTURAL ELUCIDATION OF COMPLEX POLYSACCHARIDE ANTIGENS. THE STRUCTURE OF THE CAPSULAR POLYSACCHARIDE OF *STREPTOCOCCUS PNEUMONIAE* TYPE 22 F

James C. Richards and Malcolm B. Perry

Division of Biological Sciences, National Research Council of Canada, , Ottawa, Canada K1A OR6

The chemical structure and immunological properties of the cell-surface capsular polysaccharides of *Streptococcus pneumoniae* have received considerable attention because of their successful use as vaccines for the prevention of pneumococcal infections. Recently the capsular polysaccharide of *S. Pneumoniae* type 22F was included as a component of a 23 valent human pneumoccocal vaccine[1]. The purified antigen was subjected to structural analysis by high resolution ^{1}H and ^{13}C n.m.r. spectroscopy at 500 MHz and 125 MHz, respectively. Conventional one dimesional ^{1}H and ^{13}C n.m.r. data showed the polysaccharide to possesss a repeating hexasaccharide unit and a single *O*-acetyl substituent.(0.8 mole equivalent) Following removal of the *O*-acetyl groups by treatment with dilute alkali, the application of two dimensional ^{1}H homonuclear chemical shift correlation (COSY) and nuclear Overhauser enhancement (NOESY) experiments together with a two dimensional $^{1}H/^{13}C$ heteronuclear chemical shift correlation experiment permitted unambiguous assignment of the ^{1}H and ^{13}C resonances and provided the necessary sequence information for its complete structural elucidation. The location of the *O*-acetyl substituent was then determined by comparison of the ^{13}C n.m.r. spectrum of the *O*-deacetylated polymer with that of the native polysaccharide.

The proposed structure is shown below:

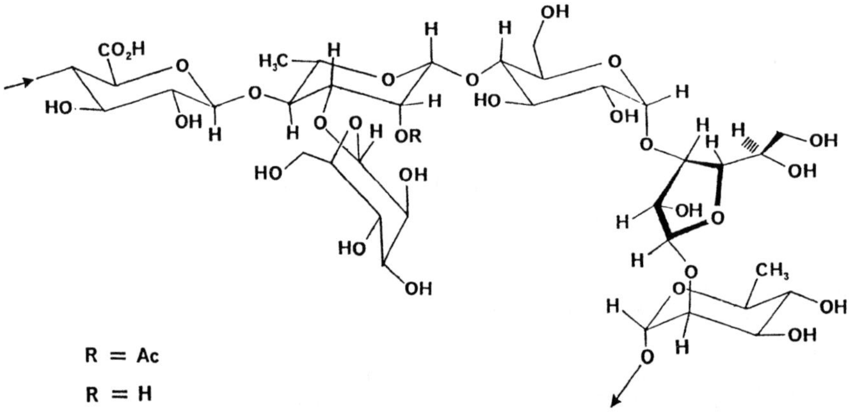

R = Ac

R = H

References. Broome, C.V., Rev. Infect. Dis. 3, S82-S96 (1982).

STRUCTURAL AND IMMUNOCHEMICAL INVESTIGATIONS ON SNAIL GALACTANS

Hagen Bretting, Gunter Jacobs, Ulrike Knels[1], Jochen Thiem[2] and Wilfried Konig[3]

[1]*Zoologisches Institut and Zoologisches Museum Hamburg*
[2]*Organisch-Chemisches Institut Munster*
[3]*Organisch-Chemisches Institut Hamburg*
Federal Republic of Germany

Snail galactans are highly branched polysaccharides which are synthesized in the albumen glands of pulmonate and prosobranchiate snails and added to the perivitelline fluid of their eggs as nutritive material for the growing embryos. (1) Snail galactans are predominantly composed of DGalactose linked glycosidically 1→3 and 1→6 (2,3). Depending on the species additional constituents were described, such as Lgalactose in *Helix pomatia, Arianta arbustorum* or *Capaea nemoralis* (14%), Lfucose in *Ampullarius spec.* (2%) (4) or phosphate groups as in *H. pomatia, C. nemoralis* and in *Biompholario glabrata* (2%) (3).
Permethylation of the galactans with subsequent analysis of the different methylation products obtained after hydrolysis revealed all to give 2,3,4,6-tetra- and 2,4-di-*O*-methyl-Gal derivatives in equal proportions. In the various species different quantitities of 2,3,4- and 2,4,6-tri-*O*-methyl Gal were found. In *H. pomatia, A. arbustorum* and *C. nemoralis* 3,4,6-tri-*O*-methyl Gal was also detected (2,3,4,5,6,7). Two consecutive Smith degradations disintegrated the galactan of *Lymnaea stagnalis* into small fragments, whereas a main portion remained as a polysaccharide in *H. pomatia* and *Strophocheilus oblongus* (3,6). This correlates with the systematic classification into fresh-water and land-bound snails respectively. Immunochemical studies with antisera raised in rabbits against certain snail galactans show a high degree of species-specificity which is probably due to the different length of side branches and varying position of branching points (3).

References: 1) Goudsmit, M.E., in: Chemical Zoology VII Mollusca M. Florkin and B.T. Scheer, Eds. (1972) Acad. Press, New York. 2) Baldwin, E. and Bell, D.J., Chem. Soc. 1461-1465 (1938). 3) Bretting, H., Whittaker, N.F., Kabat, E.A., Konigsmann-Lange, K. and Thiem, J., Carbohydr. Res. 98:213-235 (1981). 4) Feijo, M.A.L. and Duarte, J.H., Carbohydr. Res. 44:241-249 (1975). 5) Weinland, H., Z. Physiol. Chem., 305:87-96 (1956). (6) Segura, E.A.D. and Duarte, J.H., Carbohydr. Res. 52:159-167 (1976). (7) Duarte, J.H. and Jones, J.K.N., Carbohydr. Res. 16:327-355 (1971).

TUMOR-ASSOCIATED BLOOD GROUP ANTIGEN EXPRESSIONS AND IMMUNOGLOBULINS ASSOCIATED WITH TUMORS

Byron Anderson, Lyman E. Davis and Mario Venegas

Department of Molecular Biology
Northwestern University Medical School
Chicago, Illinois

There are very many glycolipid and glycoprotein carbohydrate sequences which may be expressed in varying quantities on mammalian cells from different tissues. The types and quantities of the carbohydrate sequences may also vary during the cell cycle, throughout the developmental stages of tissues, during cellular differentiation and in response to extracellular signals. The complexity of the carbohydrate sequences is evident and is further emphasized with reports of new sequences in the recent literature. Presumably, all such carbohydrate sequences are available for expression on transformed cells of benign and malignant tissue growths. Transformed cells may make use of this vast repertoire of sequences to maintain their viability in the various environments experienced in their metastatic ventures, to gain a growth advantage, and, in some cases, to maintain a de-differentiated state. Alterations in carbohydrate sequence expression by transformed cells may also allow such cells to escape the host defense mechanisms which would otherwise recognize and react with tumor-specific or -associated carbohydrate (as well as non-carbohydrate) antigenic differences.

A majority of the tumor-associated antigenic determinants have been characterized as carbohydrate sequences, some of which are related to the oligosaccharide determinants of the blood group antigens. The objectives of this review are to focus on the current status of certain blood group antigenic expressions of tumor tissues (particularly those of epithelial origin) and to summarize the nature of the human humoral response to tumor cells and immunoglobulins associated with tumor cells *in situ* (i.e., tumor-associated immunoglobulins). Relevant research in our laboratory is discussed.

TUMOR-ASSOCIATED BLOOD GROUP ANTIGENS

The blood group A, B and H oligosaccharide structures occur at the non-reducing termini of several different carbohydrate chains which are present as part of either glycolipids or glycoproteins, and comprised of both soluble and cell bound antigenic structures (1-4). Branched or unbranched type 1 $[(Gal\beta 1 \rightarrow 3GlcNAc)_n \beta 1 \rightarrow 3Gal \rightarrow R]$ and type 2 $[(Gal\beta 1 \rightarrow 4GlcNAc)_n \beta 1 \rightarrow 3Gal \rightarrow R]$ as well as the recently described (5,6) type 3 (present on type 1 chains) and type 4 $(\beta Gal1 \rightarrow 3\beta GalNAc1 \rightarrow 3R)$ carbohydrate chains serve to carry the A, B, H epitopes which are added by specific glycosyltransferases in a genetically dictated sequential manner (3). The A, B and H determinants are represented by $\alpha GalNAc1 \rightarrow 3(\alpha Fuc1 \rightarrow 2)$, $\alpha Gal1 \rightarrow 3(\alpha Fuc1 \rightarrow 2)$ and $\alpha Fuc1 \rightarrow 2$ structures, respectively, linked to the terminal Gal residues of the type 1, 2 or globoside sequences (Figs. 1 and 2). While glycolipids bear ABH determinants predominantly as adducts of type 2 chains, glycoproteins may express the antigens on either or both type 1 and 2 chains (1,4). Also, the blood group determinants, including those of the Lewis system, may be found on both mucin type and N-linked type glycoproteins. The relative contribution of glycoproteins or glycolipids as molecular carriers of A, B, and H carbohydrate structures, however, remains controversial even for the well studied erythrocyte membrane system. Some investigators suggest glycoplipids are the exclusive carriers of blood group antigens (7-9) while others favor glycoprotein substitution (10-14), still others indicate the involvement of both (15). The nature of the carrier molecules present on epithelial cells and their cancers are largely unknown (1,16) with a few exceptions (17-19). Due largely to the differential solubility of protein and lipid carriers, ABO blood group antigens can be classified as water and alcohol-soluble (20). The alcohol-extractable material represents mostly glycolipid with type 2 chain bound ABH determinants, while the alcohol-insoluble fraction represents predominantly type 1 and type 2 bearing glycoproteins (21) soluble in water. Classification based on solubility does not guarantee the nature of the carrier substance, however, due to the existence of water-soluble polyglycosylceramides (22) and protein associated blood group glycolipids (23).

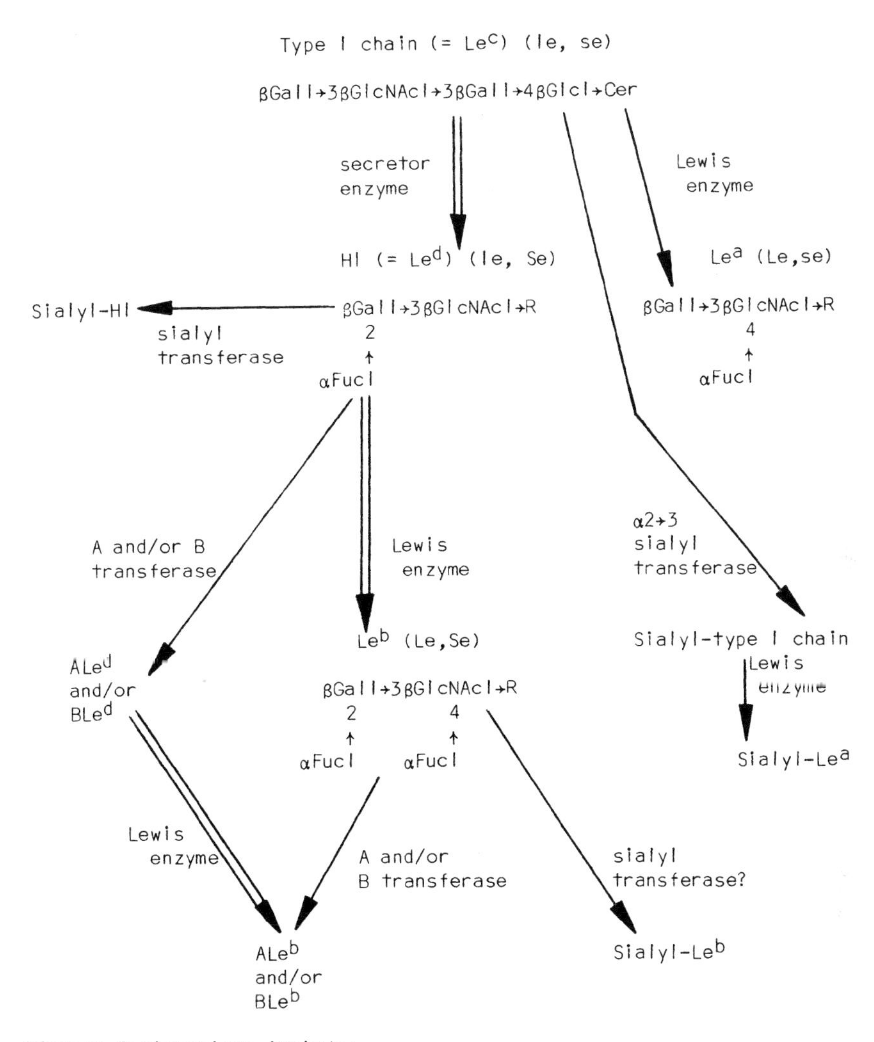

Diagram designations include:

(= abbreviation)	for commonly used structural abbreviation
(Le or le, Se or se)	for abbreviation of Lewis and secretor genotype required for expression of the structure where Le and Se represent expression of the gene product glycosyl-transferase and le and se represent non-expression.
⟹	for prefered pathway at branch points (requires expression of "favored" enzyme)

Gal, D-galactose; GlcNAc, N-acetyl-D-glucosamine; Glc, D-glucose; Fuc, L-fucose; Cer, ceramide; R = lactosyl-ceramide or oligosaccharide sequences plus the protein portions of N- or O-linked glycoproteins.

Fig. 1. ABH and Lewis Biosynthetic Pathways Based on Type 1 Chain Precursors

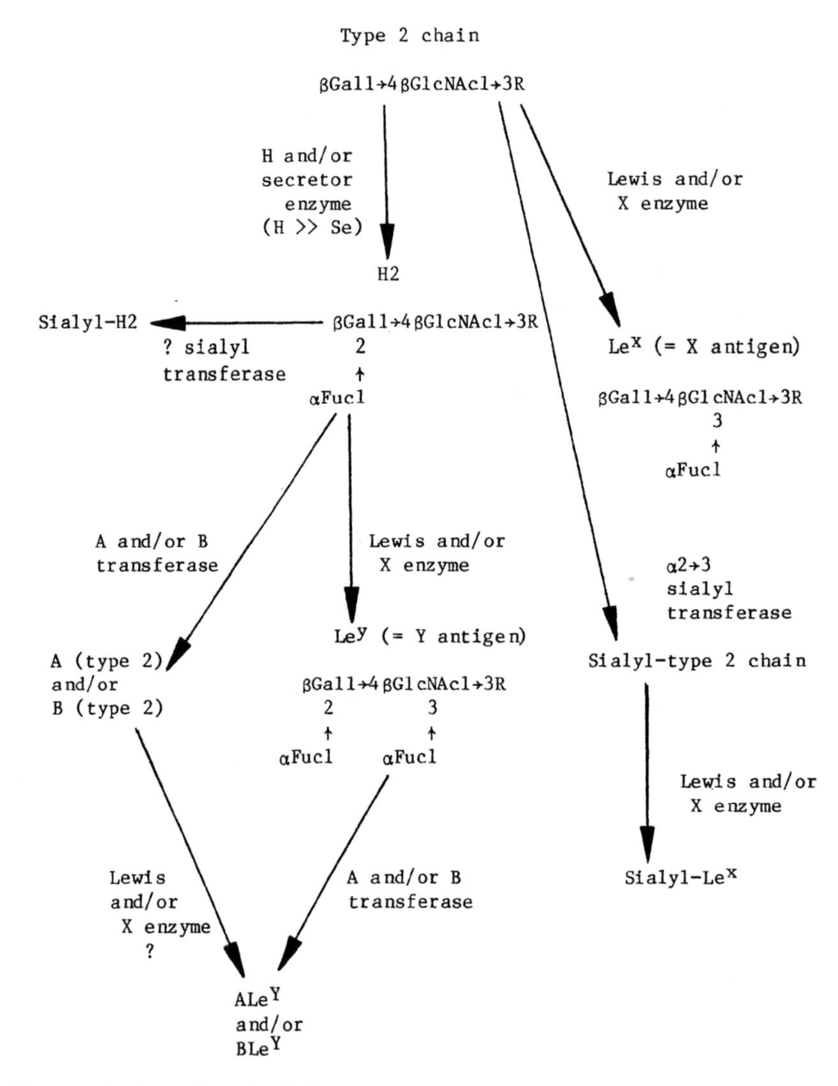

Diagram designations include:

(= abbreviation)	for commonly used structural abbreviation
?	for unknown pathways, enzymes and/or products
(H >> Se)	H gene encoded enzyme much more active than Se enzyme on substrate (type 2 chain)

Fig 2. ABH and Lewis Biosynthetic Pathways Based on Type 2 Chain Precursors

Tissue Distribution Of Blood Group Antigens

The tissue distributions of blood group antigens A, B and H have been recently summarized by Coon and Weinstein (16) and include erythrocytes, the plasma and the cell-membranes of endothelial cells and many normal epithelial tissues. Certain tissues, such as those of the distal colon, central nervous system, muscle and connective tissue cells apparently lack ABH expression (24-29).

The distribution of ABH antigens within normal epithelial tissues has generally been characterized using immunocytochemical staining techniques. The best studied model for such ABH carbohydrate sequence expressions has been in oral epithelium (30-34) where a sequential stepwise addition of carbohydrates with blood group specificities by glycosyltransferases apparently occurs as basal cells divide and differentiate to form the outer stratified epithelial cell layers. The deeper basal and parabasal cells demonstrate membrane precursor lacto-N-triosyl residues while H antigen is observed on the membranes of lower spinous and parabasal cells. Spinous cells express A and B blood group carbohydrates on their membranes (dependent upon the blood type of the individual). H type 2 chain structures, as well as type 2 chain precursor substances, are observed as intracellular cytoplasmic components of the upper 2/3 of the epithelium. Using *Ulex europaeus* I lectin, which has specificity for H type 2 chains and Y antigen (35), staining is localized to the granular epidermal layer (36) similar to that seen for A and B blood group structures. Blood group antigens in normal urothelium appear confined to the cell membranes of basal (evenly stained) and intermediary urothelial (unevenly stained) cells, as well as the luminal (but not the lateral) surface of superficial urothelial cells (37). Antigen expressions of cervical epithelium appear on the membranes of squamous cells (38), while those of normal colon are localized to luminal surfaces as well as within the cytoplasm where they may be heavily concentrated within the supra-nuclear region of columnar cells (28). Normal breast cells express ABH antigens on their plasma membranes (39).

Genetic And Biosynthetic Considerations Of Blood Group Antigen Expression

The amount of ABH substance which is expressed on an individual's epithelium appears to be dependent upon secretor status, ABO blood type, Lewis type and A subtype. The influence of secretor status is apparent in Figs. 1 and 2 outlining the formation of ABH antigens. Both the secretor gene (Se) and the H gene products are distinct $\alpha 1 \rightarrow 2$ fucosyltransferases (40) which synthesize blood group H from type 1 chain (Le^c) or type 2 chain precursors. The Se transferase acts on type 1, type 2 and type 3 chain precursors (preferentially type 1) while an H encoded product is responsible for blood group antigens based on type 2 precursors (and independent of the Se gene) such as those on erythrocytes, vascular endothelium (41) and in serum (42). It has been postulated by Oriol (42,43) that the Se structural genes act in (endodermally derived) epithelial locations preferentially on type 1 substrates while the H gene product(s) acts in mesodermally derived tissues (e.g., vascular endothelium and erythrocytes) and preferentially on type 2 acceptors. Recent biochemical evidence by Pendu and co-workers (44) suggests type 1 and type 3 structures are good substrates and type 2 chains are poor substrates for the Se gene encoded α-2-L-fucosyltransferase while the H gene enzyme demonstrates the exact opposite in substrate preferences.

Betteridge and Watkins (45) have additionally demonstrated that the rate of fucose transfer (relative Vmax) was markedly higher to type 1 than to type 2 acceptors for the secretor α-2-L-fucosyl transferase while the binding affinity (apparent Km) of Se enzyme for these substrates was similar. In contrast, these investigators found that the non-secretor (H encoded ?) α-2-L-fucosyltransferase had higher relative Vmax values and lower apparent Km (i.e., "tighter" binding) values with type 2 than with type 1 chain acceptors. Their finding that non-secretor tissues contained only ~ 5% of the α-2-L-fucosyltransferase activity present in secretor tissues, may explain the secretor status dependent quantitative differences in the expression of ABH structures reported for normal urinary (46-49), oral (30,50), duodenal and pyloric epithelia (51).

Additional explanations for observed differences in ABH antigen expressions between secretors and non-secretors are:

1. There is a differential availability of type 1, 2, 3 or 4 chain blood group H precursor substances which is organ-dependent (1) and are formed (at least in the case of type 1 and 2 chains) by different galactosyltransferases (52).

The availability of precursor chains may also be dependent upon the competition of precursor elongating glycosyltransferases for available precursors. In this regard, Kannagi, Levery and Hakomori (53) have recently presented evidence that type 2 chains can elongate repetitively [i.e., (type 2 chain)$_n$] while type 1 chains lack this ability and their addition to the non-reducing end of type 2 chains apparently terminates the chain elongation pathway with the formation of a type 1-(type 2)$_n$ hybrid.

2. There is competition between any of the various glycosyltransferases at biosynthetic pathway branch points (Fig. 1) as has previously been demonstrated for Lewis gene specified α-3/4 fucosyltransferase and the blood group A gene glycosyltransferase (54). Competition between the α-2-L-fucosyltransferase (Se or H gene encoded) with Lewis α-3/4-L-fucosyltransferase (or X gene product α-3-L-fucosyltransferase) and sialyltransferase for precursor (e.g., type 1 and 2 chains) should determine the amount of H_1 and H_2 available for formation of A and B blood group substances by A and B gene glycosyltransferases. Since the presence or absence of a given competing enzyme is genetically determined, the predominant operative biosynthetic pathways (and their products) are dependent on whether a given tissue (or sublocation within that tissue) potentially expresses a given gene (a function of the normal tissue differentiation pattern) as well as the presence or absence of the genes encoding the enzymes (the individual's genotype). This suggests that altered antigenic expressions as are observed in neoplasia can best be understood in the context of normal antigen differentiation patterns and a knowledge of genotype (e.g., Se or se, Le or le). An example of the modulation of antigenic expressions by the Se gene is the pattern of type 2 chain (X,Y,H) based structures in normal pyloric and duodenal surface epithelium of secretors and non-secretors (51). Individuals which were non-secretors (Le^a phenotype: Le,se) demonstrated X antigen with no Y or H type 2 structures. Individuals which were secretors (Le^b phenotype: Le, Se) demonstrated Y and H type 2 antigens with no X antigen. These results demonstrate that in epithelial tissue under control of the Se gene, the Se encoded α-2-L-fucosyltransferase of secretors competes favorably with the Lewis α-3/4-L-fucosyltransferase (and/or X gene α-3-L-fucosyltransferase) for type 2 chain precursor yielding H_2 and Y product, while without the Se gene product (non-se-

cretors), the Lewis (and/or X) fucosyltransferase predominates. Because of the correlation between Lewis type and secretor status (see Fig. 2) the co-expression of Le^b with Y and Le^a with X suggests Se gene control is operative. In Le^{a-b-} individuals (6% of the general population, 22% of the Black population and 6-14% of the Oriental population), one might expect X expression by secretors (the majority of Le^{a-b+} individuals) and Y expression by non-secretors by way of the type 2 (and H_2) specific α-3-L-fucosyltransferase (X enzyme) in the absence of the type 1 and type 2 reactive Lewis-enzyme (55). Recent examination of normal human urothelium by Juhl (48) reveals the absence of H_2 antigen and Y antigen [as detected by monoclonal anti-H_2 antibody and *Ulex europaeus* lectin (H_2 and Y reactive)] in non-secretor (Le^{a+b-}) individuals of all ABO blood types suggesting similar Se gene control within urothelial epithelium. This same investigator has demonstrated greater blood group antigen A expression in the urothelium of Le^{a-b+} (secretors) and Le^{a-b-} (nearly all secretors) compared to Le^{a-b-}(non-secretors) individuals. The limited availability of A precursors H_2 and Y antigen (Fig. 1) in Se gene controlled tissue may explain the secretor/non-secretor difference observed. The secretor fucosyltransferase may also compete with an α-2→3 sialyltransferase as recently outlined by Hansson and Zopf (56) in their biosynthetic scheme for the formation of the cancer-associated sialyl-Le^a antigen. The sialyltransferase is active on both normal secretor (Le^{a-b+}) and non-secretor (Le^{a+b-}) mucins, however, non-secretors generally express more of the sialyl-Le^a than do secretors which, as suggested by Brockhaus *et al*. (57), is probably the result of glycosyltransferase competition (Se and sialyltransferases). In addition to Se gene effects, the Lewis gene may also influence ABH expression in epithelial tissues. This is not surprising due to the relationship between secretor status and Lewis antigen expression. Individuals which lack the Lewis enzyme (Le^{a-b-}) have more precursor available for A and B antigen formation (Fig. 1) and in fact demonstrate larger amounts of A (58) and B (59) antigens than their Lewis enzyme positive counterparts. These individuals, although unable to form sialyl-Le^a may be able to form X and Y antigen by way of type 2 chain reaction with the X glycosyltransferase. The conversion of Le^y to ALe^y or BLe^y however would not be expected to be extensive since Y (like Le^b) is not a good acceptor for the A enzyme (41). A summary of the relationship between Lewis phenotype, secretor status and ABH expression is presented in Table I.

Table I. Relation between Secretor type, Lewis blood type and expected ABH expression.

Lewis phenotype	Secretor status	Relative ABH expression
Le^{a+b-}	non-secretor	low
Le^{a-b+}	secretor	high
Le^{a-b-}	secretor >> non-secretor	high

ABO blood type also affects the quantity of ABH antigens expressed in epithelial tissues. The amount of H antigen detected on erythrocytes varies dependent on blood type in the following order from high to low: O, A_2, A_2B, B, A_1 and A_1B (60,61) suggesting differential efficiencies for the A_1, A_2 and B glycosyltransferases. A similar differential expression of H antigen by normal human urothelium (48) suggests a similar phenomenon occurs in epithelial cells since the amount of H antigen in Le^{a-b+} and Le^{a-b-} individuals was correlated to ABO blood types when ranked from high to low for blood groups O, B, A, and AB.

Blood group A subtype may also influence ABH antigen expression. The A_1 and A_2 enzymes differ in Michaelis constants (62), catalytic efficiencies (63), pH optima (64), isoelectric points (65), and substrate acceptor specificities (41). Since A_2 enzyme is ~ 5-10 fold less efficient than its A_1 counterpart (63), it has been suggested (54) that the A_2 enzyme is a poorer competitor with Lewis or X gene fucosyltransferases for H acceptors than the A_1 enzyme (Fig. 1) explaining the significantly lower amounts of A determinant in saliva of A_2Le^b compared to A_1Le^b, A_1Le^d and A_2Le^d individuals. The difference in A_1 and A_2 expression on red blood cells has recently been attributed to the ability of A_1 enzyme, and the inability of A_2 enzyme, to act upon type 3 H and globo-H (type 4) chains (5). Whether the same substrate specificities and enzymatic efficiencies are operative in epithelial tissues has yet to be determined.

Methodologic Considerations

The detection of tissue blood group antigens depends not only on factors intrinsic to the source host (e.g., host genotype and tissue differentiation stage) as outlined above, but also is a function of extrinsic factors, particularly the limitations of the detection method used. Tissue expressed blood group structures are generally detected by either biochemical or immunohistochemical based methodologies. The advantages and disadvantages of these approaches have previously been outlined by Coon and Weinstein (16). The advantages of being able to process large numbers of tissues while requiring only small amounts of a given tissue and the ability to localize antigen distributions (and thus assess tumor heterogeniety) make immunohistochemical detection the more popular approach. Immunocytochemical methods of detection of tissue blood group antigens include the specific red cell adherence test as originally described by Davidsohn (66), immunofluorescence (37,67-69) and immunoenzyme (27,70-72) staining. Of the three, the immunoenzyme techniques offer the greatest advantages including providing a permanent staining record, a high sensitivity, as well as simultaneous histopathological and morphological evaluation along with antigen localization. There are many assay constructs employed in immunoenzyme staining the basic variations of which have been recently outlined (73-75). A frequently used variation is that employing avidin-biotin complex (ABC) using peroxidase as a label as originally described by Hsu et al. (76) and more recently using alkaline phosphatase label (77). A major advantage of using the ABC technique (or similar hapten sandwich labeling) is the ability to avoid background straining caused by cross-reaction with endogenous tissue immunoglobulin. The technique is not without its drawbacks, however (78). Irrespective of the immunohistochemical method chosen, one should be aware of inherent potential limitations. These include tissue processing, i.e., fixation or embedding induced alteration of tissue antigens and possible antibody crossreactions.

All conventional fixatives appear to cause a loss of antigenic activities (79) particularly surface-related antigens (80), including blood group antigens (36,46,47,69,81-84). While the general intent of fixation is to maintain tissue morphology and integrity, the fixative-induced molecular modifications may be the cause for antigenic derivatization, the solubilization of antigens (85), as well as potentially compromising immune recognition by limiting the accessibility of an antibody to its antigen (86). Fixative-induced masking of antigenic determinants may also result from the inclusion of extraneous antigens from serum (87) or in the immunobilization of endogenous host antibody associated with the antigen(s) of interest, preventing antigen recognition by exogenously added reagent antibodies. The extent of antigen preservation appears to depend both on the nature of the antigen examined and the type of fixative employed (88). Thus, urothelium blood group B antigens and (to a lesser extent) A antigens were shown to be more sensitive to formalin fixation/paraffin embedding than were H antigens (83) while the H antigen appeared more sensitive to alcohol than did the A or B antigen (47). The most denaturing step in conventional tissue processing appears to be antigen exposure to aldehyde or alcohol containing fixative, not the embedding in hot paraffin wax (79). However, both aldehyde and alcohol based fixation methods as well as paraffin embedding have previously been shown by Limas and Lange (47) to destroy epithelial blood group antigens; they demonstrated improved antigen detection in fresh frozen sections. These researchers also found that the extent of fixative related group antigen loss was dependent upon both tumor bearing status as well as secretor status of individuals. Tumor and non-secretor derived tissue exhibited significantly greater fixative-induced ABH antigen loss than their respective normal and secretor counterparts. The extent of ABH isoantigen destruction in formalin fixed paraffin embedded tissue appears to be a function of the length of tissue block storage time since no ABH isoantigen can be observed in urinary epithelial sections which were older than 10 years (82). Another consideration in the use of fixed tissue for immunocytochemical staining is the choice of monoclonal or polyclonal primary antibody reagents. In this regard, Loeffel _et al_. (89) and Taylor (80) have recently shown that routinely processed (fixed and embedded) tissue is more likely to have destroyed the single antigenic determinant detected by a monoclonal antibody, an important consideration with the increased use of monoclonal antibody reagents in immunohistochemical staining.

Recently, techniques have been developed to avoid the many limitations of the currently employed tissue fixation methods. These include the use of freeze-dried paraffin-embedded sections (79,90), the delaying of fixation until after incubation with primary reagents by the use of a dextran containing buffer (91), and the use of antibodies which recognize fixative modified antigens (92). While the problems of tissue processing can be avoided or overcome, those involving antibody cross-reactivity are more insidious. Cross-reactions due to the presence of multiple antibodies of different specificities, titers and affinities (as one might find in an antiserum) may be frequently eliminated by the use of absorption, affinity purification, the addition of excess competing liquid-phase cross-reacting antigen, checkerboard (cross-hatch) titrations and monoclonal antibody reagents (92-94). Cross-reactions due to shared similar or common antigenic determinants between different antigens require that the specificity of immune reagents used in staining be very well characterized. While monoclonal antibody technology may allow for selection of antibodies with exquisite discriminatory capabilities, the sole fact that an antibody is monoclonal in nature is no guarantee that the antibody will not recognize structures which although overall are apparantly different, share similar portions in three dimensions. Thus, although type 1 chain structures Le^a, Le^b and Le^d demonstrate expected structural differences compared to their type 2 chain counterparts (Le^x, Le^y and H type 2, respectively), they also share some structural similarities (95). Such similarities may explain the reactivity of polyclonal (96,97) and monoclonal (98) anti-Le^b with Le^y and the inhibition of Le^a reaction with Le^x (99). In the same manner, monoclonal antibody EGR/G49 directed to epidermal growth factor receptor has recently been shown to recognize both blood group ALe^b and ALe^y structures (100). Determination of the specificity of monoclonal antibodies raised to complex antigenic mixtures is further complicated by the fact that minor impurities in inhibitor preparations employed for specificity determinations may result in misassignments of specificity. A recent example is the initial proposal that epidermal growth factor reactive monoclonal antibody 101 was directed to the blood group H type I sequence (101) which on closer inspection revealed Le^y antigen specificity (102).

Blood Group Antigen Expressions of Tumors

The ABH blood group antigens represent a major group of carbohydrate structures altered in epithelial derived cancers, which comprise the majority of human cancers. The important tumor-associated changes of blood group structures have been extensively reviewed (103-108) and those involving the ABH antigens include: (a) the loss of normal expressed A and B antigen products with (or without) the accumulation of the corresponding normal masked precursor antigens (i.e., I and type I chain structures); (b) the inappropriate expression of A or B blood group antigens by individuals normally lacking those structures; and (c) the sialylation of ABH precursors (i.e., I and i sequences).

The most frequently observed alteration in human cancers (particularly carcinomas) is the complete or partial loss of blood group A and B antigens (109). Studies demonstrating the deletion of these determinants are numerous and include a variety of human neoplasms originating from the gastrointestinal tract (67,110-113), urinary bladder (114,115), uterine cervix (116,117), prostate (118), pancreas (119), lung (120), breast (121), oral cavity (31,122), and the larynx and head and neck (123-124) regions. The loss of ABH isoantigens is not generally an all-or-none phenomenon with approximately 60-80% of the carcinoma tissues examined exhibiting ABH deletions (37,82,124-126). Due to the apparent correlation between ABH antigen loss and tumorigenic potential, it has been suggested that ABH deletion could be of prognostic use in a variety of carcinomas as an indicator of invasive or malignant capability. Best examined in this regard are epithelial cancers of the bladder transitional cell carcinomas where the absence of the expected ABH antigens is indicative of probable malignant change (126). There are a number of studies using the transitional cell carcinoma model which suggest that ABH antigen loss could be used to predict low grade and tumor stage propensity to metastasize or invade (82,126-132). Similar correlations between ABH loss and tumor invasiveness has been reported for other cancers including those of the breast (121,133) and larynx (123) although such findings have not been demonstrated in certain laboratories (37,39, 134,135).

In studies conducted in this laboratory, decreases in expression of the ABH antigen reactivities have also been observed for squamous carcinomas of the head and neck regions. The immunocytochemical technique used is described in detail in the section below on immunoglobulins associated with these tumors. Briefly, 6 μ cryostat-cut tissue sections, after washing with a dextran-containing buffer, were reacted with either biotinylated antibody reagents to the A, B or H determinants or bound primary antibody was detected using a biotinylated second anti-immunoglobulin followed by avidin-biotinylated-peroxidase reagent and the diaminobenzidine (DAB) substrate.

The types of results obtained are shown in Table 2 comparing the A, B or H antigen expressions of 3 normal and 11 tumor tissue specimens. It is first noted that at best a partial deletion of A, B, H antigens may be concluded from the results. Two of the 4 A blood group positive individual tumors exhibited no A antigen reactivity but each, including the AB positive individual tumor specimen showed a strong H antigen expression. The one B blood group positive individual was negative for B antigen expression and the one AB blood group positive individual exhibited reduced tumor expression of both A and B antigens. The 5 O blood group positive individuals showed H anti-

Table 2. Expressions of A, B and H Antigenic Determinants of Normal Epithelial and Squamous Carcinoma Tissues[1].

Normal	Blood Type	Determinant H	Determinant A	Determinant B
1	0	+[2]	+/-	-
2	0	++	+/-	-
3	AB	-	+++	++
Tumor				
1	A	++	+	-
2	A	++	-	-
3	A	++	++	-
4	A	+++	+/-	-
5	AB	++	+	+
6	0	++	-	-
7	0	+	-	-
8	0	+++	-	-
9	0	+	-	-
10	0	+	-	-
11	B	-	-	-

1. The immunocytochemical technique on cryostat-cut tissue sections is described below in the section on tumor-associated immunoglobulins. The A, B and H (type 2) determinants were determined using monoclonal antibody reagents (Dako) and a biotinylated anti-mouse Ig reagent to detect bound antibody. The same results were obtained using anti-A and/or anti-B antibodies purified from human type O sera by adsorption and elution from type A or B erythrocytes, followed by biotinylation of the purified antibodies.

2. + indicates a definite positive reaction; +/-, a weak positive reaction is equivocal.

gen expression to varying degrees with no apparent complete absence in any tissue examined.

Those tumor tissues exhibiting decreases in A, B or H activities were pathologically defined as invasive tumors, however 3 of the tumor tissues showing little decrease in ABH expression were also described as invasive.

Several of the tumor tissues were also examined for the presence of the type 1 and type 2 precursor chains, the Lewis antigens, and for sialic acid terminated carbohydrate sequences (Table 3). The monoclonal antibodies designated H112 and H131 prepared to the N2 oligosaccharide as described in Figure 3 were used to detect the type I chain. An anti-I containing serum was used for the type 2 chain detection. Monoclonal antibodies H124 and H162 (Figure 3) and MCF-7 were used for detecting the Le^x determinant; the Limulus polyphemus and Limax flavus agglutinins (LPA and LFA) were used for sialic acid. Further details of the anitbody reagents and grading of results are given in the Table legend. The results of Table 3 are the compilation of two separate series of experiments, one which employed the first five tumor tissues and the second, tumor tissues 6 through 12.

The I and i determinant structures were present in the normal tissues and strongly reactive as well as in 3 of 5 tumor tissues. One of the tumor tissues was only weakly positive with the anti-I antibody reagent. The H112 and H131 monoclonal antibody to the type I determinant (βGal1→3βGlcNAc1→R) were only weakly reactive with 1 normal and weakly or moderately reactive with 2 tumor tissue specimens. For the monoclonal antibody reactivities to the four Lewis determinants, the Le^a and Le^b determinants were apparently only weakly expressed on one normal epithelial and one tumor tissue; an additional tumor tissue exhibited a focal staining with the anti-Le^b antibody reagent. The H124 and H162 anti-Le^x antibodies exhibited no reactivity to the 3 normal epithelial tissues tested but exhibited a strong reactivity with one tumor tissue specimen and weak, moderate or focal reactivity with 4 of the tumor tissues. The MCF-7 anti-Le^x antibody was reactive with 5 of 7 tumor tissuers tested, showed focal areas of reactivity in 2 tumor tissues where the H162 anti-Le^x antibody was negative and exhibited a more uniform reactivity with tumor tissue #12 whereas H162 was reactive with only areas of the same tissue. The anti-Le^y antibody reagent was reactive with 2 of the normal epithelial tissues and all but one of the 7 tumor tissues tested. The LPA lectin was reactive with all tissues tested (both

Table 3.

Type 1, Type 2, Lewis and Sialic Acid Determinants of Normal Epithelial and Squamous Carcinoma Tissues[1]

Tissue	Blood type	I	i	H112	H131	Le^a	Le^b	Le^x MCF-7	Le^x H124	Le^x H162	Le^y	LPA	LFA	sialyl-Le^a
Normal														
1	0	++	++	-	-	-	+		-	-		++		
2	0	+++	+++	+	++	-	-		-	-		+		
3	AB	+++	+++	-	-	-	-		-	-		++		
Tumor														
1	0	++	++	+/-	-	-	++		+++	++++				
2	0	+++	+++	++	+++	+/-	+/-		-	-		++		
3	0	+/-	+/-	-	-	-	-		-	-		+++		
4	A	+++	+++	+	++	-	-		+	++		+++		
5	B	+	-	-	-	-	-		-	-		++++		
6						+/-	-	1-3+[2]		1+	3-4+			-
7						-	+/-	+/-(2+)		-	1-2+		+/-	-(1+)
8						-	-	1+		-	2+		1+	-
9						-	-	-(2+)		-	3+		1-2+	-
10							-	+/-		-	+/-(1+)			+/-
11						-	-	-		+/-	1-2+		1-2+	-
12						+/-(2+)	-	1-3+		+/-(2+)	-			-(2+)

1. The various antibody reagents were as follows: Anti-I and anti-i antibodies were purified from human sera (provided by Dr. J. Moulds) by absorption and elution from adult and cord erythrocytes, respectively, and were biotinylated. The H112, H131, H124, and H162 monoclonal antibodies are described in Fig. 3. An anti-Le^a monoclonal was obtained from Dr. W. W. Young, and anti-Le^a, -Le^b and -Le^x (MCF-7) antibodies were kindly supplied by Dr. Z. Steplewski. The anti-Le^y was a gift from Dr. K. O. Lloyd and anti-sialyl-Le^a was supplied by Dr. B. Delvillano. All of the monoclonal antibody reactivities were detected using appropriate biotinylated anti-mouse Ig. The LPA and PFA lectins were biotinylated.

2. In the second series of tumor tissues examined (numbers 6-12) a different grading system was used. The 1 to 4+ designation means all tumor cells were stained whereas the ( ) indicates focal staining and the numbers in ( ) the intensity of the focal staining. A range of numbers, e.g., 1-3+ indicates that all tumor cells were stained but to different degrees in various portions of the specimen examined.

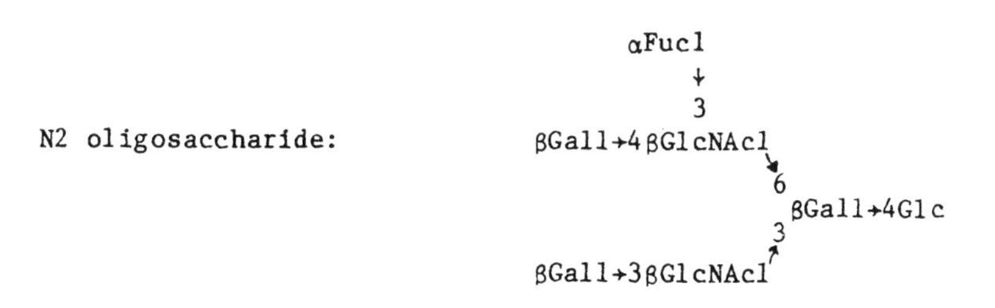

Antibody designation	Sequence specificity	Specificity
H124 C68	βGal1→4(αFuc1→3)βGlcNAc1→R	Le^x determinant; Fuc residue necessary for reactivity
H162 C30	βGal1→4(αFuc1→3)βGlcNAc1→R	Le^x determinant, lesser dependence on Fuc for reactivity
H112 C32	βGal1→3βGlcNAc1→3(→6)βGal1→R	type 1 chain plus some portion of the disubstituted Gal; Fuc residue on type 1 chain reduces reactivity
H131 C24 and H216 C45	βGal1→3βGlcNAc1→3(→6)βGal1→R	type 1 chain plus some portion of the disubstituted Gal; Fuc residue on the type 1 chain has no effect on reactivity

Fig. 3 Assignment of Specificities of Monoclonal Antibodies to the N2 Oligosaccharide

normal and tumor) and two of the tumor tissues exhibiting strong reactivities were essentially negative for reactivity with anti-type 1, -type 2 and -Lewis antibodies. Finally, the LFA lectin was weakly reactive with tumor tissues and the antibody reagent to the sialylated-Le^a determinant showed a limited and focal reactivity with 2 of the 7 tumor tissues tested.

Although the results in Table 3 are for a limited number of tissues, we can make some preliminary conclusions and speculations concerning these particular carbohydrate sequence expressions, relating these results to the discussion above on blood group antigen expression of tumors and the biosynthetic pathways. First, as with many other organ derived carcinomas, the Le^x determinant is expressed in head and neck squamous carcinomas, whereas the Le^y determinant may be expressed in both normal epithelium and carcinoma tissues. The type 1 precursor chain and Le^a and/ or Le^b do not appear to be expressed to any great extent whereas the type 2 precursor chain is clearly present in both the normal and cancer tissues. Furthermore, in a majority of tumor tissues there is a moderate to complete loss of A or B determinants, and a partial loss of H determinant in some tumor tissues. In consideration of the biosynthetic pathways diagrammed in Figs. 1 and 2, we speculate that the type 2 precursor sequences predominate in these epithelial tissues and in tumors are further fucosylated to form the Le^x and/or Le^y determinant. The type 2 chains may also be directly sialylated to yield the sialosyllacto-neo-tetraosyl-sequence or the Le^x and Le^y sequences may be further sialylated. These latter sialic acid substituted determinants were not examined in these studies. Furthermore, any apparent loss of A, B or H determinants in tumor tissues could be a consequence of this increased fucosylation of type 2 chain and H_2 precursors with a resultant less efficient transfer of A (αGalNAc1→3) or B (αGal1→3) determinant residues to the Le^y product structure, and a lack of transfer to the Le^x structure. Increased sialylation of the terminal Gal residues of the type 2 precursor chain would also result in decreased expression of A, B or H determinants. Thus, if the A or B and H determinants are expressed on squamous carcinoma cells, relative lesser amounts of Le^x, Le^y or sialylated determinant sequences may be expressed. Conversely, if Le^x and/or Le^y determinants are strongly expressed, H and A or B may be diminished. Likewise, the expression of the Le^x fucosyltransferase would also result in a decrease in the H determinant expression. For example, the tumor tissue specimen number 1 exhibited little H determinant expression and strong reactivity to the anti-Le^x monoclonals H124 and H162, and tumor tissue specimen 2 exhibited strong reactivity with anti-H antibody and no apparent reactivity with the anti-Le^x reagents. Furthermore, the degree of sialylated carbohydrate chain sequences could result in lesser apparent expressions of all the other determinants. Desialylation of the tissue sections by treatment with neuraminidase would then potentially uncover many determinants reactive with the corresponding antibody reagents.

These results suggest a sequence of study of the carcinoma tissues. If Le^x and/or Le^y, and/or sialic acid determinants are strongly reactive, such determinants (particularly the Le^x) may constitute a reasonable tumor marker. An intense staining with sialic acid reactive reagents, and weak reactivities with type 2 chain and the other Lewis determinants would suggest that further examination with monoclonals to sialylated determinants is warranted. These data also suggest that, in individuals with head and neck squamous carcinomas, an examination of sera for glycoproteins with Le^x, Le^y and various sialylated determinants would be merited.

TUMOR-ASSOCIATED IMMUNOGLOBULINS (TAIgs)

Associated with most malignant tumors are various types of molecules, or antigens if defined by immunochemical means, termed tumor-associated antigens (TAA). In this discussion, TAAs will be inclusive for those molecules which are not specific for tumors but may be found in other normal tissues, or in lesser amounts in the normal tissue progenitor of the tumor, and for tumor-specific antigens (TSA). TSAs are those tumor molecules which would be truely specific for tumor cells and not found on any normal tissue. It is difficult to assign any molecule or determinant as being a TSA because of possible experimental limits of detection of the same antigen or molecule in normal tissues.

The TAAs consist of many types, including those of viral, chemical, or spontaneous and unknown derivation. One or more TAAs manifested by a particular tumor are the putative elicitors for and the reactants with the many components of the immune system. Because it is very probable that the equilibrium between tumor growth, expression of TAAs, and the immune system's ability to destroy tumor cells, determines in part whether or not a tumor cell or cells will become established, grow, and metastasize, the factors that influence the equilibrium are important to define.

One part of this multiple factor equilibrium between the immune response and tumor cells is the interaction of antibody with tumor cell surface antigens. TSAs would be expected to elicit a humoral, antibody response in the individual with cancer in the same manner as any foreign antigen evokes an antibody response. In certain circumstances a TAA may also result in an antibody response even though the same antigen, or antigenic determinant(s), may be expressed to some extent in tissues other than that of the tumor origin or expressed in a developmental stage.

If antibodies are produced to TAAs it seems reasonable to expect that they would have access to the tumor cells if vascularization of the tumor growth occurs and/or if increased vascular permeability results from inflammation at the tumor site. The antibodies may bind to the TAAs at the tumor cell surfaces or to intracellular TAAs if the tumor cell is compromised and the membrane is disrupted. TAIgs would then be expected to be detectable. The many recent studies using monoclonal antibodies to TAAs for tumor imagining support the concept that circulating anti-TAA antibodies have access and bind to TAAs in tumor masses.

In this section the literature pertaining to immunoblobulin (Ig) presence in tumor tissues and antibodies reactive with human tumor tissues and/or cells will be reviewed. Unless directly related to TAIgs the following areas will not be discussed: aspects of tumor immunology dealing with types and numbers of TAAs and their operational definitions and characterizations, antibody reagents and their specificities to TAAs, the cell-mediated response to tumor antigens, autoantibodies and immune complexes associated with cancer, and other non-immunologic factors important to tumor growth, or spread, or vascularization.

Portions of the topics discussed below were reviewed by von Kleist et al. (136) and Ran et al. (137) focusing on the experimental evidence for tumor cell membrane bound antibodies and lymphocytotoxic antibodies. Also the extensive studies on circulating immune complexes in cancer have been reviewed recently (138,139), covering the topics of the relation of immune complexes to disease state, tumor burden and recurrence rate, problems in identification of antigens present in immune complexes, immunoglobulin characterization of immune complexes and their reactivities to tumor cells and soluble tumor antigens.

The following review is divided into several sections, the first discussing the evidence for Ig associated with tumor cells in vivo. The latter part of the section reviews studies that demonstrate possible antibody reactivities in cancer patient sera using cultured cells as antigen sources. These sections are followed by a further discussion of the experimental critera, controls and specificities for studies on TAIgs.

Immunoglobulins And Antibodies Associated Or Reactive With Tumor Cells

A careful review of the literature concerning evidence for Ig or antibody (Ab) reactive or associated with tumor cells shows that many of the findings are conflicting, and that it cannot be stated with a good degree of probability that autologous Ig is indeed associated or bound to cancer cells of various tissue origins. In a discussion of the studies, the terminology must first be considered. Tumor masses consist of the cancer cells themselves and a lymphoreticular cellular infiltrate, as well as normal cells constituting that part of a tissue in situ being invaded by cancer cells. In most of the earlier papers, "tumor cells" is an inclusive term for the malignant cells and lymphocytes, macrophages and other lymphoreticular cells. Because the membrane bound Ig on the latter cells is either Ig as an integral membrane constituent, or Ig as antibody-antigen (Ab-Ag) complexes bound through Fc or complement receptors, a differentiation of Ig or Ab associated with malignant cells is necessary. This differentiation is not made in earlier studies, making it difficult to assess the location of Ig in regards to malignant cells. Therefore, the review here of those studies will not attempt to decide which cells have Ig or Ab bound to them. In later reports, a differentiation is made of cell types, although there is still conflicting evidence concerning whether tumor cells bear Ig.

Witz (140,141) reviewed the earlier literature on tumor-associated Ig, primarily concentrating on studies in his own laboratory. He considered that antibodies should be reactive with TAAs on malignant cells if antibodies have access to the cells by suitable circulatory conditions (with vascularization at the tumor site and/or increased vascular permeability due to inflammatory components, the permeability change allowing larger molecular weight plasma components, and Ab, access to the extravascular regions). Antibodies should also be accessible to malignant cells where basement membranes are normally present but where the permeability barrier of the membrane has been disrupted. When certain tumors become invasive of surrounding tissue and when lymphoreticular cells have migrated to the tumor site, Ab and Ig would also be present and able to react or associate with malignant cells. In one of the systems studied by Witz and coworkers (mouse and hamster tumors) it was found that the main class of Ig eluted with low pH buffer was IgG of predominantly IgG2 subclass. In other studies by Cruse and Azar (142) and Takasugi and Hildemann (143), IgG2 had been identified as the subclass with Ab enhancing activities. The Ig in tumor eluates or bound to tumor cells may thus be a subfraction of the total Ig and may have varying activities in the immune reactivities to tumor cells as regards complement fixing abilities or cytotoxic potential. Also, the plasma Ig of tumor-bearing animals (and the Ig at tumor sites) may exhibit the blocking activity of lymphocyte-mediated cytotoxicity (144). Witz (141) also discussed the possibility that Ab binding to cell surface antigens may have the effect of eliminating the expression of certain antigens or causing their redistribution and subsequent loss of activity. Such antigenic modulation has been described by a number of laboratories.

The studies of Brandtzaeg (145) and Brandtzaeg _et al_. (146-148) are of importance in the understanding of the conditions under which Ig may be found in tissue sections, the application of the immunofluorescence technique, control of the reactions involved, and the interpretation of immunohistochemical results. Mucosal specimens were collected and divided into small pieces and either fixed immediately with cold ethanol, or washed with phosphate buffered saline (PBS) to remove non-cell associated Ig. The latter were then fixed with cold ethanol, dried with xylene, and paraffin-embedded. Tissue sections were cut from the blocks, dried at 37° and stored at 4°. The main purpose of Brandtzaeg's studies was to demonstrate that Ig containing immunocytes of the different classes of Ig were present in mucosal tissue of individuals with different levels of serum Ig and at sites of inflammation. Tissues of nasal and oral mucosa at the inflammatory sites which had been fixed without PBS washing showed intense immunofluorescent staining for Ig throughout the connective tissue and epithelium. Those tissues washed with PBS showed the removal of the Ig and the immunocytes became discernible and were brightly stained for IgG or IgA depending on the fluorescent-labeled anti-Ig used.

Thus, it seems that Ig is accessible to epithelium, and the Ig is removed readily by washing procedures if not cell-bound. It is interesting to note that, although antibody to epithelial cell antigens may be present in sera of individuals, such antibodies may not have access to the epithelium. For example, Quismorio et al. (149) showed that IgM deposits are found at the dermo-epidermal junction of individuals with lepromatous leprosy, similar to the types of deposits seen in certain arthritic diseases, with no staining of the epithelium. However, some of the sera of the individuals with leprosy contained antibodies of IgG class which bound to intracellular antigens of the epithelial cells. This was demonstrated by indirect immunofluorescence following incubation of the diluted sera with normal human or animal skin, or esophageal tissue sections. In those experiments, the normal tissue sections showed no fluorescence in the non-inflamed connective tissue layers.

Izsak et al. (150) attempted to correlate malignant potential of human tumors with amounts of cell bound Ig. The content of Ig was measured on dissociated tumor cells (23 different carcinomas and 2 sarcomas) using ^{125}I-labeled anti-human globulin, and ^{131}I-labeled globulin of a non-immunized rabbit to discern non-specific binding of labeled Ig to tumor tissues. A ratio of the two labels with an increment of greater than 0.15 was considered significant. Of the 25 human tumor tissues examined, 14 had Ig coating. Nine of the 14 were tumors from individuals with a high grade of malignancy, and 5 with a low malignancy index. One problem with this study is that it would be expected that the ^{125}I-antihuman globulin would react with the cell membrane Ig of lymphoid cells. Thus, the higher ^{125}I to ^{131}I ratios may reflect the lymphoid cell content of tumors rather than the content of Ig bound to tumor cells. In a later report from the same laboratory (151), evidence was presented that the Ig bound in tumor cells could be non-specific and was of lymphoid cell origin. They also showed that immune complexes could be bound to cells freshly prepared from the tumors (murine ascites tumor), but would not bind to tumor cells propagated in vitro. Thus, in that particular system the tumor cells themselves apparently do not contain Fc receptors, and perhaps the greater portion (depending on the particular tumor situation being tested) of Ig in a tumor mass may be derived directly from the infiltrating lymphoreticular cells in the tumor mass.

The studies of Lewis et al. (152,153) presented evidence that in vivo melanoma cells do not contain bound Ig. They made suspensions of the cells from tissue specimens and tested them for the presence of bound Ig by direct immunofluorescence. Small lymphocytes and plasma cells were differentiated by morphology; macrophages were not mentioned. None of the 37 cell preparations of different individuals were positive by their fluorescence criteria, even though most of those cell preparations could be rendered positive by first incubating the cells with diluted patient sera. Furthermore only 2 of the 26 specimens were positive and from individuals who had received autoimmunizations with irradiated tumor cells. However, 9 of 10 individuals given intravenous phytohemagglutinin were positive for bound Ig. The authors state that their and others' evidence support the concept that, except under special circumstances, Ig is not bound to malignant melanoma tumor cell surfaces. The authors left open the question of whether similar or different results would be obtained with other tumor tissues. The studies of Gerber et al. (154) also do not support the concept that Ig is bound to tumor cells in situ. They studied cryostat-cut sections of ovarian carcinoma tissues and, although the purpose of their study was to ascertain antibody reactivities in cancer patient sera, their immunofluorescence asays should have detected any Ig already bound to the tissue sections. No such reactions were seen.

In contrast to the above studies, Richman (155) reported the presence of Igs in human breast cancer tissues. He compared normal, benign, and malignant human mammary tissues by immunofluorescence using fluorescein-conjugated reagents specific for IgA, IgM, IgG, C3 and albumin. IgM and IgA were demonstrated in extravascular regions of normal and benign breast tissues as would be expected (Richman discusses extravascular Ig in relation to Brandtzaeg's studies); no IgG was seen. In the malignant human mammary tissues, IgA, IgM, and IgG were found associated with epithelial cell surfaces; no C3 was detected in the same location. All controls were appropriate and negative. Igs have also been delineated as associated with squamous cell carcinomas of the head and neck regions by this laboratory (91,156,157). Those studies are described in detail below.

Evidence For Tumor-Bound Ig In Animal Tumor Systems

There are several reports which show that in animal model systems antibody is associated with tumor cells which were virally-induced or transformed. Moav et al. (158-159) used a syngeneic polyomavirus-induced ascites tumor (SEYF-a) in mice, isolated IgG fractions (and from control mice for specificity studies), radiolabeled (^{125}I) the IgG, and tested the binding of the ^{125}I-IgG to live SEYF-a cells (or mouse spleen cells) in suspension. The results were analyzed in terms of the binding constants of the Ig to antigen of the target cells. The IgG eluted from the SEYF-a tumor cells by pH 3.5 citrate buffer washes consisted mainly of the IgG2a subclass. It is interesting that the authors further determined that the IgGs from 11-15 days after inoculation of tumor cells consisted only of high binding constant affinities whereas IgGs eluted from later times, as well as the IgGs from sera and ascites fluids from any time after inoculation of tumor bearing mice, consisted of both high and low binding constant IgGs.

Using the same tumor system, Braslawsky _et al_. (151) showed that the antibody-containing older tumor cell populations were not capable of complement activation. It is perhaps notable that these authors used goat anti-mouse IgG to quantitate Ig bound to the SEYF-a cells, a method which Brown _et al_. (160) stated detected Ig bound non-specifically to cell surfaces. The evidence presented by Braslawsky _et al_. (151) suggested that the SEYF-a cells have Fc-like sites which bind immune complexes spatially in close proximity to the SEYF-a antigens to which antibodies bind. An Fc receptor on these polyoma-induced cells is interesting in view of similar Fc-like receptors found on herpes simplex viral infected cells. Perhaps part of the antitumor response is to the virally-induced Fc receptor of SEYF-a cells, the Fc receptor in this system being a neoantigen to the murine system.

In a study similar to those above, Ting (161) presented evidence for possible anti-TAA antibody presence early in tumor growth in several virally-induced tumors in mice. The antibody presence seemed to parallel tumor growth. It is interesting to note that ^{51}Cr-release of the complement-dependent cytotoxicity assay detected antibody at much later times than did the isotopic antiglobulin technique (reaction of target cells first with serum from a tumor-bearing mouse followed by ^{125}I-labeled antimouse globulin). It should be noted that a portion of the antibody (Ig) bound to the target cells may be nonspecific, as noted by the author, and was not cytotoxic for that reason.

Pertinent to the studies of Moav _et al_. (158,159), Braslawsky _et_ al. (151), Izsak et al. (150) and Ting (161) are the details of the binding experiments reported by Brown et al. (160). These authors reported on the use of ^{125}I-labeled staphyloccocal protein A (SPA) for detection of Ig adherent to tumor cells. The tumor cells were mouse sarcomas; alloantisera were used as antibody sources. The assay system was highly reproducible and sensitive and showed less non-specific binding than with radiolabeled goat anti-mouse IgG. The advantages of the SPA reagent may be due to its preferentially binding to antibodies complexed to antigens. Thus, the SPA assay may not measure the low affinity associated Ig observed by Moav et al. (158, 159), giving a better approximation of Ig in antibody-antigen complexes. The SPA will bind to immune complexes, therefore, the method can not be used to differentiate immune complexes in Fc receptors from Ig bound as antibody to cell surface antigens.

Evidence For Antibodies In Cancer Patient Sera That Bind To Tumor Tissues, Dissociated Cells, Or Tissue Cultured Tumor Cell Lines

There have been many literature reports on attempts to demonstrate antibodies reactive to tumor cells in sera of cancer patients (and antibodies in circulating immune complexes, cf. references 138 and 139). The results have been variable and some of the reasons for the variability are discussed below. These studies have utilized as antigen targets either established tissue cultured cells or cells from tumor tissue (imprints, dissociated tissues, and cryostat-cut tissue sections).

Wood and Barth (162) presented a detailed study which used melanoma tissues to make touch imprints on coverslips as antigen targets to test a large number of sera of melanoma patients, normal patients, and patients with other types of cancer, for their reactivities. Indirect immunofluorescence was the assay system. All three groups of sera gave similar percentages of positive reactions, at dilutions of sera of 1:4, although the mean titer of melanoma patient sera (310) was greater than that for normal sera (18, $p<0.05$). Blocking experiments showed there were similar reactivities in sera of the three groups. The specificities of the reactions were not great, as the sera reacted with tumor tissues of other histologic types. The fluorescence staining obtained in their studies was cytoplasmic in nature. The findings that certain positive sera were not reactive with all cell preparations suggested that TAAs of cytoplasmic origin were being detected and not autoantibody reactivities. Although normal tissues were not tested, the authors stated that normal appearing cells in the imprints were not stained.

The studies of Lewis and Phillips (163) utilized melanoma cell suspensions obtained by mechanically dispersing fresh tumor specimens. The cells were incubated with 1:4 diluted patient sera and the indirect immunofluorescence technique was used to detect bound antibody. The authors' discussion of problems and controls for the immunofluorescence method and interpretation of results is important. Their results are of particular interest because they demonstrated patient-specific reactions by absorption means. Thus, absorption of several patient sera with the same melanoma cell reduced the fluorescence to autologous cells exclusively (results obtained using a cross-pattern of seven sera and seven cell preparations). These results suggest that the melanoma cells are expressing individual-specific antigens for which each patient has a specific antibody, specificity being different in each patient. The cross-reactions seen between all sera with each cell line would then represent an additional common antigen. These results are similar to those obtained from Old's laboratory discussed below. The advantage of this study may be in the use of several non-cultured melanoma specimens, the cells being a much closer approximation of the _in vivo_ situation than cultured cells. Not being in culture, the cells may not have had the opportunity to acquire such cross-reactive antigens as the FCS antigens (see below) nor to lose antigens (and TAAs) expressed _in vivo_.

Gerber _et al_. (154) used cryostat-cut sections of four ovarian carcinoma tissues and indirect and complement fixation fluorescence to show anti-tumor antibody reactivities in sera of patients with ovarian carcinoma. Only 4 of 91 carcinoma sera reacted (titer of 1:10 or higher) with carcinoma tissue sections, 3 of those were shown to be against a mitochondrial antigen, and the fourth against microsomal antigens. One third of the carcinoma sera showed evidence of non-organ specific autoantibodies (compared to 6% of control sera). The authors made a valid and important point that many reactivities observed in other studies may not be due to tumor-specific (or -associated) antigens, but rather due to autoantibodies to homologous antigens. The growing number of reports of such autoantibodies in sera of cancer patients is similar to that found in chronic inflammatory conditions (particularly well-characterized in the arthritic diseases). Such reactivities must be kept in mind in any similar type study.

Chapuis _et al_. (164) investigated possible antibody activity to autologous acute myelogenous leukemia (AML) cells of seven patients who were receiving immunotherapy with irradiated allogeneic cryopreserved AML cells. No activity to autologous cells was observed although activity to allogeneic leukemia cells was seen and interpreted to mean a response to histocompatability antigens.

Antibody reactivities to lung carcinoma tissues have been obtained from low pH (3.2, glycine buffer) eluates of solid tumors of the lung as reported by Paluch and Ioachim (165). Following PBS washes of minced tissues, the Igs were precipitated from the neutralized glycine buffer eluates with 50% ammonium sulfate. Indirect immunofluorescence was the assay system, using lung (and other tumor and control cells) carcinoma cells dissociated by trypsin from the tumor tissues, or cells recovered from effusions. IgG was the main class of Ig recovered. The data on the specificities of the 11 eluates tested were very high. Approximately 75% were positive for squamous carcinoma and adenocarcinoma cells. No positive reactions were seen against normal adult or fetal lung cells, or colon, breast and oat cell carcinoma cells. Similar results were obtained with Ig isolated from pleural effusions of squamous cell carcinoma and adenocarcinoma of the lung (only 6 positive reactions were seen in 140 tests of various control cells). Ig from the pleural effusions of oat cell carcinoma were unreactive against all cells, and pleural and peritoneal effusion Igs gave only a small number of positive reactions against the same panel of cells. The type of fluorescence observed was granular cytoplasmic; in the figures of this study, it appears as though some nuclear staining may also have occurred. Because the cells were acetone fixed before the indirect immunofluorescence technique, the high degree of cytoplasmic fluorescence obscured possible cell surface staining. The very high degree of specificity and percentage of positive reactions observed by the authors is in contrast to other reports. These authors employed Ig from tumor tissues or effusions, and cells derived from tumors or effusions which were kept for short time periods in culture. Perhaps these experimental details account for their positivity and specificity results. It would also be of interest if the authors had tested the Ig preparations against established cell lines of squamous cell origin and had tested sera of lung cancer patients against their panel of cells.

Perhaps different specificities (and less specificity?) may be obtained with cancer sera, or perhaps established cell lines lose or are unable to synthesize the TAAs observed in their study.

The sera of glial neoplasm patients were tested by Sheikh et al. (166) against individual patients' own and others' glial cells obtained by dissociation from excised tumors. Of 17 sera tested using indirect immunofluorescence, only 2 were positive for Ig binding to cell surfaces, whereas 5 showed reactivity for a cytoplasmic component and 10 sera were positive for a cross-reacting cytoplasmic component. Controls consisted of observing no reaction with normal brain tissue. Furthermore, adsorption by glial tumor tissue removed reactivity from the 3 sera tested to glial neoplastic cells.

Gupta et al. (167,168) used membranes isolated from malignant melanoma tissue to prepare affinity columns with cyanogen-bromide activated agarose. Various sera were processed through the columns, and bound components were eluted with 2.5 M $MgCl_2$. Antibodies of the IgG and IgM classes were eluted and were shown to react with melanoma, sarcoma, and carcinoma cells. Evidence was presented for both anti-oncofetal and other anti-TAA type reactivities. The methods used by these authors suggest techniques that could be useful for isolation of antibody reactivities to carcinoma cells and for their characterization as to class and antibody specificities.

There have been many studies from different laboratories where data has been acquired on antibody reactivities in patient sera to cultured established cell lines. As with the above cited studies, variable results in terms of percent positive reactions, and on the specificities of the reactions have been obtained.

A high degree of specificity toward cultured cell lines of breast carcinoma compared to a large number of other cell lines of non-breast origin using breast cancer patient sera was shown by Edynak et al. (169). By immunofluorescence, cytoplasmic fluorescence was obtained with 91% of the sera obtained 7-10 days after surgery, whereas only 20% of control sera were positive. The controls were matched in terms of age and parity of donors to the breast cancer individuals. The positive reactions were obtained with 2 cell lines of breast carcinoma origin, whereas 9 other lines of the same origin were negative, indicating, as in so many other studies, that the sources of antigen target, particularly when they are cell cultures, can influence results greatly. The work of these authors illustrates another problem in the use of primary cell cultures. The 2 carcinoma cell lines that expressed reactive antigen were short-lived, and further studies on the nature of the antigen were not possible.

The studies by Sofen and O'Toole (170) suggested that there is a common antibody specifity in the IgM class to squamous cell carcinomas. The authors used sera of nine patients with various organ type squamous carcinomas and 16 control sera (8 non-cancer and 8 with non-squamous cell carcinomas) and 3 established carcinoma cell lines (one skin and two bladder) as antigen targets; several other cell lines were controls for the antigen. Quantitation of the fluorescence obtained by the indirect technique against live cells was performed after formaldehyde fixation. The fluorescence intensity per cell was approximately 2 to 8 times higher for the sera of squamous carcinoma patients, as compared to all control sera. Likewise, no positive reactions were obtained with the various control cell lines. By separation of class of Ig in sera after gel filtration and by use of fluorescein-conjugated antibody specific to human IgM, the class of antibody reactive with the cells was shown to be IgM. The authors interpreted these results to indicate a common squamous antigen. There is no interpretation offered as to why these data did not detect IgG reactive antibodies as in several other studies.

Rosenberg _et al_. (171) have provided a careful study of different methods for detecting antibodies against surface antigens of tissue cultured cells. The fibroblastic cells were derived from two individuals with osteogenic sarcoma; control cells consisted of fibroblasts initiated in culture from tissues far from the tumor sites. Serum, with high allogeneic antigen specificity (cytotoxic to >99% of the lymphocyte population), obtained from a Waldenstrom's macroglobulinemia patient, was used as antibody source and apparently contained mainly IgG antibodies to an antigen of the fibroblastic cells. The most sensitive assay of three types used was the mixed hemadsorption technique. This technique used fibroblastic cells in microtiter wells as targets and Rh-D positive erythrocytes coated with anti-D followed by anti-IgG antibodies as indicator cells. A microcytotoxicity assay measuring cells that did not remain adherent to test plates after antibody and complement, and a ^{51}Cr release complement-dependent cytotoxicity assay were also used. One of several important aspects of this report is the use and discussion of optimization of the assay system, noting ways to improve accuracy in comparison of different antibody-containing sera and different cell lines, and to reduce numbers of false-negative results. The authors also discuss the various advantages and disadvantages of the different assay systems. They stress the importance of comparisons of tumor to normal cell lines, and showed that the titer of the allogeneic serum was greatly different against cell lines derived from the two individuals, but essentially the same when tumor and normal cell lines were compared from the same individual. Another problem that may be encountered in studies using cell lines is that certain antisera or populations of antibody may be ineffective in producing the end-point of the particular assay system utilized. This was shown by Boyle _et al._ (172), where a guinea pig hepatoma cell line was not killed by a complement dependent cytotoxicity assay unless the cells were pretreated with various proteases, lipases or neuraminidase. Thus, a false-negative could be due to resistance of a cell line to bind sufficient antibody or to allow complement lysis even though the antigens may be present. Such considerations are part of the general principle that the results obtained in any study are accurate to the extent of the sensitivity, specificity and the limitations of the particular assay systems used, and the nature of the antigen targets.

In two studies from the same laboratory Nordquist _et al._ (173) and Lerner _et al._ (174) used the BOT-2 cell line (breast tumor origin) and showed variable antibody reactivities to the cell line with sera of breast cancer patients as antibody source. When the antigen targets were acetone fixed cells, high percentages of reactivity were found: 70% of carcinoma patient sera, 91% of fibroadenoma, and 71% of cystic mastitis sera, compared to 0 of 10 positive reactions of normal controls. The results using live cells in suspension (scrape-harvested) were uniformly less. Most of the reactions were interpreted as cytoplasmic fluorescence, with a smaller percentage also exhibiting a nuclear location of the fluorescence. The high positivity in benign or possible pre-malignant breast disease patients' sera were interpreted to mean that an organ specific response was being obtained rather than one with tumor specificity. However, because cross-reactions and blocking experiments were not performed with patient sera from different groups, it cannot be ascertained if the same or different specificities were being observed with the antigen sources. It is of interest that one of the antigens of the BOT-2 cell line studied by this group was a glycoprotein similar to the M, N and T determinant containing blood group component. If, however, reactivity were being observed to the T antigenic determinant, it would have been expected that a high percentage of reactivity would be obtained in controls because all individuals apparently have anti-T antibody specificities (175).

The report by Ferrone and Pellegrino (176) in which melanoma patients' sera were tested for antibody reactivities to five cultured melanoma cell lines, takes into account many of the necessary controls realized by experience in previous work in this area. The authors tested for cytotoxic antibodies by the complement dependent microcytoxicity assay. The percent of positive reactions for cytotoxic serum antibodies of melanoma patients, controls, and patients with tumors other than melanoma, were 37%, 21%, and 33%, respectively. Although the differences are not great by inspection, the incidence of melanoma sera cytotoxic antibodies compared to the controls was significant at a P value of <0.02. Melanoma sera, compared to sera of patients with other types of tumor, were not significantly different. It is interesting to note that in this report, similar to findings discussed above in other studies using cell lines, some sera showed selective reactivities to one or more (and all five) cell lines. Thus, particular cell lines would result in differing percentages of positivity. This report is also notable for the technical detail, as the authors provide the experimental development for optimizing the assay system.

Zeltzer and Seeger (177) in a report similar to Brown _et al._ (160), developed the use of radiolabeled (^{131}I) staphylococcal protein A for determining Ig bound to tumor cells. The characteristics of this assay system were carefully described and the sensitivity was about 0.25 ng Ig. The authors used several cultured tumor cells lines in monolayers and heterologous antisera to test, develop, and determine specificities in the assay system. They also reported preliminary data on sera of children with neuroblastomas, and showed that part of their antibody reactivities were to fetal calf serum antigens. They, as others, point out the necessity of sequential adsorption experiments to ascertain such cross-reactivities and for the purpose of determining specificity and reactivities which would approach showing specificity to TAAs.

Hahn _et al._ (178) employed complement-dependent cytotoxicity as well as antibody-dependent cell-mediated cytotoxicity (ADCC) to evaluate antibody reactivities in colon cancer patient sera using as targets several colon cancer cell lines and ^{51}Cr release as the indicator system. It is notable that in none of the assays were normal control sera reactive. The authors found that 8 of the cancer sera failed to demonstrate anti-HLA and anti-CEA (which can mediate ADCC against cultured colon lines) activities. As in the previously discussed report, there were variable reactivities present in the colon cancer patient sera depending on the assay system and the target cells used. Almost 80% of the sera were positive against at least one target cell line and by one of the assays, whereas with either assay system the chance of detecting a positive reaction was half as great. The antibody reactivity resided in a DEAE-fractionated IgG fraction. Peripheral blood lymphocytes from both cancer patients and controls could equally well mediate ADCC. The authors did not state whether immune complexes in cancer patient sera may have inhibited some of the possible ADCC reactions. There were no absorptions done on the positive sera to ascertain if, for example, some of the reactivity was due to specificity to FCS antigen. It should be noted, however, that antibodies bound to FCS adsorbed antigens on cultured cells will not mediate either the complement dependent cytotoxicity or ADCC lysis reactions, although the FCS antigens can readily be detected by the indirect immunofluorescence assay (179-181). Those results have been interpreted to mean that such antigens may be in a peripheral cell surface location such that, although complement activation may occur, the geometry of complement-mediated or cell-mediated lysis is not properly obtained. One consequence of such findings is that conclusions on types of antibody specificities in cancer sera will in part be dependent on the assay system used. Thus, a conclusion that anti-FCS antigen activities are not present in cancer sera may not be valid for complement-dependent cytotoxicity or ADCC assays.

The studies of Gupta _et al_. (167,168) are important with regard to cancer sera antibody reactivity to FCS antigens acquired by tumor cells in culture. A high incidence (51%) of natural antibody to the FCS antigens was shown in both normal donor sera and sera of sarcoma patients. However, the authors showed that although the FCS antigen may be present on cultured cells, and antibodies exist to those antigens, their presence did not influence the reactivity differences noted with the complement fixation assay utilized. Thus, it was shown that the percent positive reactivities of the sarcoma-derived sera to human sarcoma cell extracts were much higher (86%) than in normal sera (32%). The authors gave an extensive and important discussion of their results compared to previous findings on cancer sera antibody reactivities to cultured cells. Again, it should be emphasized that the type of assay can greatly influence results in part because certain cell surface antigens, although reactive with antibody, may not produce the end-point indicator of the assay system.

A comprehensive study of possible human antibody reactivities in sera of individuals with cancer has been pursued by Oettgen et al. and Houghton et al. (182,183). A large number of individual cell lines derived from melanomas and other tumor types were established in culture. Four different immune adherence assays and serologic analyses were used to define types of circulating antibodies. Also, adsorptions of the serum specimens with autologous and allogeneic melanoma cell lines, other tissue type cancer cells and normal tissues and cells were used to define cross-reactivities. The results of these studies showed that the antibody reactivities could be divided into three classes: I, antigens specific for each patient's individual tumor; II, antigens shared by tumors of the same type; III, antigens found widely distributed on normal and malignant cells. The study included patients with melonoma, astrocytoma, renal cancer and acute leukemia. Five patients exhibited antibodies to the class I tumor antigens which were of the IgG class and one of the IgM class. Two of the class I melanoma antigens have been characterized as glycoproteins of approximately 40,000 and 90,000 daltons. Similar results have been discussed by Gupta and Morton (138) demonstrating that autologous anti-tumor antibody can be isolated from circulating complexes of patients with melanoma. Although the antibodies to class I antigens have been found in low frequency in melanoma patients, the antibodies may approximate tumor-specific reagents and are similar to animal model anti-TAAs.

TAIgs Of Squamous Carcinomas Of The Head And Neck

In studies from our own laboratory (91,156,157), we demonstrated that Igs could be detected in squamous carcinoma tissues of the head and neck regions. The studies by Popovic et al. (156) and Koneval et al (157) were initiated in order to examine the reactivity of anti-squamous carcinoma sera produced in the rabbit to possible TAAs of the carcinoma tissue. Binding of rabbit Igs (antibodies) to washed frozen sections of tumor tissues was to be detected using commerically obtained fluorescienated (FITC)-goat anti-rabbit Igs. In the control reactions, we consistently observed that the FITC-anti-rabbit Ig reagent control gave a weak but positive reaction with all tumor tissues. Attempts to reduce this "non--specific" reactivity were not successful until we showed that the anti-rabbit Ig reagent had a small but significant cross-reactivity to human Igs and the reactivity to tumor tissues could be blocked by the addition of human Igs to the FITC-anti-rabbit Ig incubation step.

With the indication that human Igs were present on the tumor tissues the foucs of the research was changed to investigate this finding. Using FITC-conjugated anti-human IgG, IgM, IgA or C3. All tumor tissues examined were positive for IgG, I each of 7 were positive for IgA or IgM, and 5/7 were positive for C3. Other plasma components were removed in the washing of the tissues as shown by negative reactivities for albumin and transferrin. Furthermore, we observed that tissue sections marginal to the tumor but not containing histologically identifiable tumor cells were also positive for IgG. All normal epithelia tissues were negative except for one each which were reactive with the FITC-anti human IgA and -albumin reagents. We assumed that the positive Ig presence of tumor cells was not due to some phenemonon associated with inflammation as normal epithelia of inflammed tonsilar tissues were negative for Igs.

The fluorescence was localized to intercellular spaces or at the cellular surfaces of the carcinoma cells; no intracellular fluorescence was observed. The Ig reactivity was not removed by prolonged incubation of the tissues with buffer at 37°C. Also, using the erythrocyte adherence test for Fc receptors (184), the carcinoma cells were negative and lymphoreticular cells in submucosal regions positive. Thus, the Ig present was not correlated to regions of Fc receptors. The Igs could be removed from tumor tissues by elution with low pH (2.1) glycine-HCl buffer and in one experiment, the addition of diluted human serum (or an IgG fraction of serum) to glycine buffer washed tissue sections resulted in Ig being added back to the tissues.

The finding of positive reactivity to IgG with the epithelial cells of marginal non-tumor tissue sections was of interest and suggested that tumor-associated changes were occurring adjacent to tumor cell areas. There are several reports on pre-malignant changes of marginal tissues of several types of cancer (185-189). These will not be discussed here.

In a subsequent study, Pankow _et al_. (91) confirmed the earlier results using the avidin-biotin-peroxidase complex (ABC) reagents and post-primary antibody fixation method. In order to maintain good morphology of the tissues, a number of washing buffers were studied to remove plasma proteins in tumor tissues resulting from the increased vascular permeability and inflammation associated with the tumor areas. A buffered-dextran wash was found to remove plasma components from tissue sections, while maintaining morphology and tissue integrity. Glutaraldehyde fixation was performed after primary reagent incubation (in dextran buffer). The tissue sections could also be washed with either low or high pH buffers to remove the TAIgs with maintenance of tissue and cellular integrity. The details of the procedures are outlined in Fig. 4.

A sandwich type of enzyme-linked immunosorbent assay was developed to quantitate the human Igs eluted from tumor tissue sections. The assay has also proved to be useful for the quantitation of mouse monoclonal antibodies, or other antibody reagents, reactive with antigens of tissue sections. Dynatech microFLUOR 96-well plates were coated with affinity purified anti-human or anti-mouse IgG (H and L chains reactive) and backcoated with BSA. Known amounts of human or mouse IgG as standards, or glycine-HCl, pH 2.1, eluates of tumor or normal tissues sections were added to the wells. For the human Igs from tumor tissue sections, the tissues were pre-washed with buffered dextran and eluted for 5, 10 or 15 minutes with the low pH glycine-HCl buffer. For the mouse IgG, a blood type O positive tissue was prewashed with buffered dextran and mouse monoclonal anti-H (Dako) antibody applied. Following incubation and washing, the mouse anti-H antibody was removed by low pH elution. The low pH eluates were neutralized and incubated with anti-human or anti-mouse IgG coated plate wells, washed and β-galactosidase conjugated anti-human or anti-mouse IgG added. The amount of the β-galactosidase conjugated bound second antibody was measured with 4-methyl-umbelliferyl-β-galactoside substrate and readings of relative fluorescence intensity units obtained on the Dynatech microFLUOR plate reader. This assay system could accurately quantitate between 0.5 and 10 ng of human Ig and 0.5 to 100 ng of mouse IgG. The amounts of human IgG removed from tumor tissue sections with the buffered dextran wash and low pH eluates are listed in Table 4.

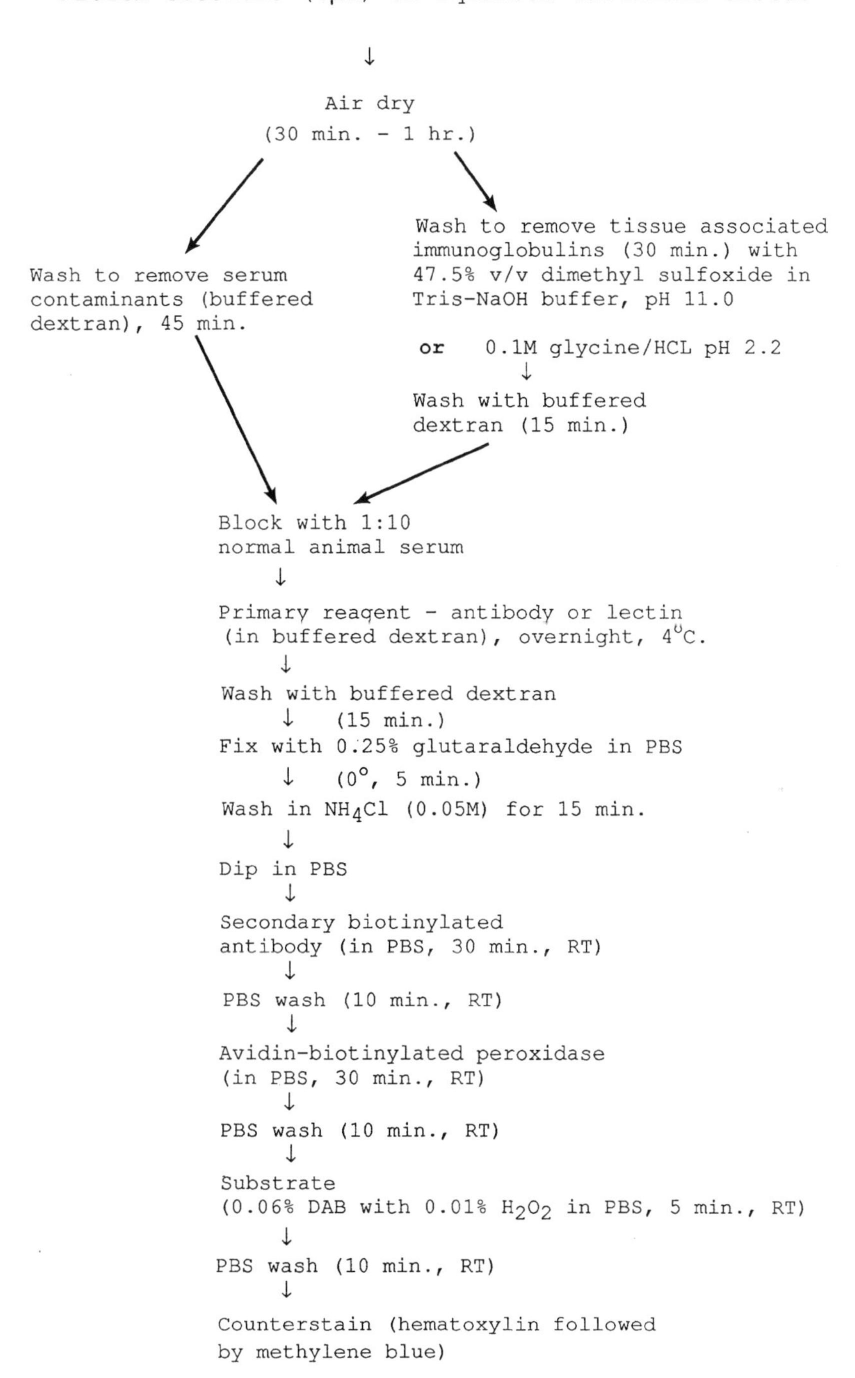

Fig. 4. IMMUNOHISTOCHEMICAL STAINING PROTOCOL

Table 4. Quantitation of Human IgG Eluted From Tumor Tissue Frozen Sections

Tumor Tissue Designation	ng Igs/Tissue section			
	Dextran Buffer Wash	Glycine-HCl, pH 2.1 Eluates 5 min.	15 min.	30 min.
Mu(56)	22	0.9	0.8	0.7
Sc(28)	200	14	10	9.0
Sch(38)	26	3.2	2.8	2.6

The amounts of IgG, not bound to tissue components and removed by the buffered dextran wash were greater than those removed by low pH glycine-HCl buffer. Thus, there is considerable non-bound human IgG present in tissue sections. Such IgG may be fixed by conventional (formalin) fixation techniques used for pathologic examination purposes.

The 5 minute low pH glycine-HCl treatment was sufficient to remove all detectable bound IgG from the tissue sections; the sections were negative for IgG by the immunostaining procedure after glycine-HCl treatment. The consistent small decrease in amounts of IgG quantitated with increasing time of glycine-HCl treatment is probably due to denaturation of the Igs at the low pH.

In the experiment where mouse monoclonal anti-H antibody was applied to a blood group O positive tissue section, 400 ng of mouse IgG antibody was removed with the 5 min. glycine-HCl elution buffer thus allowing quantitation of the amount of reactive antibody and H antigen present on the sections. The eluted tissue section was negative for mouse IgG showing that essentially all anti-H antibody was removed by the 5 min. elution procedure.

In other experiments we have shown that antibody reactivity is retained throughout these procedures. We incubated tissue sections of a blood group AB gum epithelial tissue with monoclonal anti-A antibody (Dako), removed the anti-A by the low pH glycine-HCl procedure and the pH neutralized eluate was reapplied to the treated tissue sections and to non-treated tissue sections. Positive cell surface staining for the blood group A antigen was achieved demonstrating effective preservation of antibody activity following low pH glycine buffer elution of anti-A from tissue sections.

The results of these studies suggest several important conclusions and considerations when studies of mouse and human monoclonal antibody reactivities to human tumor tissue sections are undertaken:

1. Igs, particularly IgG, are apparently associated with tumor tissues and localized to tumor cell surfaces and intercellular spaces. The Igs are in addition to those Igs or immune complexes that may be bound to Fc receptors of infiltrating lymphoreticular cells. The major question is whether the Igs have antibody reactivities to TAAs and, if so, to what type of TAAs and antigenic determinants. This is a major interest of this laboratory and such studies are being presently pursued.

2. The presence of TAIgs and possibly antibodies bound to TAAs raises the consideration that certain TAAs are covered by the TAIgs. This may contribute to the observed tumor tissue heterogeneity of TAA expressions if portions of the TAAs are blocked from reactivity with exogeneously added anti-TAA antibodies. Other factors such as the differential synthesis and expression of TAAs undoubtedly also contribute to tumor tissue TAA hetereogeneity.

3. For those studies that use radiolabeled monoclonal anti-TAA antibodies for radio-imaging of tumors _in vivo_, it should be kept in mind that TAIgs, and the differing quantities of TAIgs, may reduce the amounts of the exogeneous monoclonal anti-TAAs that can bind to tumor masses. Larger amounts of TAIgs may reduce the amounts of the exogeneous monoclonal anti-TAAs that can bind to tumor masses and thus the uptake of such anti-TAA antibody at tumor sites.

4. For studies using human antibody sources, or human monoclonal antibodies, to putative TAAs on tissue sections, the TAIgs are a major concern. The intrinsic TAIgs are removed by procedures that dissociate antibody-antigen complexes, e.g., low and high pH buffer elutions, and the exogenously added human antibody needs to be discerned from the intrinsic TAIg. Because the TAIg seems to be mainly of the IgG class, this may not present a problem with human IgM antibody reagents (such as anti-I antibodies), however, at least in one case, IgM was found as a TAIg.

As noted above, this laboratory's present study is on the possibility of discerning antibody reactivities of TAIgs eluted from tumor tissue sections by testing their reactivity to known TAA determinants.

Further experiments will possibly allow us to define the TAIg antibody specificities and to thus better delineate a humoral antibody response to tumors. Also, the methodology developed will allow for the study of possible anti-TAA antibodies in human serum because the intrinsic TAIgs can be removed by low or high pH buffer elutions while apparently maintaining the antigenicity of some components, and maintaining tissue and cellular morphology.

Summary Of Considerations Important To The Delineation Of TAIgs And Experimental Approaches Used

Several criteria should be applied in evaluating studies on Igs associated with tumor tissues and to evidence for host antibodies to TAAs. These criteria and considerations have been mentioned in the discussions above and are summarized as follows:

1. Ig may be found in small amounts in normal tissues although the amounts in tumor tissue are greater and found in higher frequency. As documented by Brandtzaeg et al. (145-148), Ig is found in large quantities at sites of inflammation other than tumor tissues. The latter type of Ig is apparently not bound in antigen-antibody complexes and can be removed by washing of the tissue slices before examination by particular assay systems. In the case of many other non-cancer diseases, antigen-antibody complexes can be discerned with locations characteristic of the disease.

2. Non-tissue component bound Igs must be removed by washing before TAIgs may be delineated. Fixation of tissues prior to sectioning will necessarily fix a portion of Igs from the plasma filtrate to tissue components.

3. Plasma proteins other than Igs may be found in eluates of tumors. As discussed by Witz (141), high concentrations of the larger molecular weight plasma components can gain access to the tumor site by reason of the increased vascular permeability. Access of these components to the tumor cells would result if the microcirculation is in close proximity to the tumor cells and, in the case of epithelial cell cancers, if the integrity of the basement membrane has been damaged. Disruption of the basement membrane with invasion of the tumor mass into submucosal connective tissue is a common finding for tumor masses which have been excised. In the case of early cancer development it would be expected that the basement membrane is intact and would limit large molecular weight plasma component access to the

tumor cells. The studies by Tickle *et al.* (190) on the morphological juxtapositions of carcinoma cells with underlying tissues in the early phases of invasiveness of epithelial tumors are important to assessing these factors.

4. The IgG found associated with tumors may be restricted to certain subclasses as discussed in some of the studies cited above. It is important to recognize this possible restriction of Ig subclass such that the reagent or assay system detecting the Ig does not exclude a particular subclass or react poorly with it.

5. In some tumor systems Ig which can coat tumor cells disappears with time. One explanation is that the antigen-antibody complexes formed on cell surfaces are internalized by endocytosis or extruded (shed) as immune complexes with consequent antigen (TAA) depletion of the cells. Thus, a negative result may reflect that phenomenon and not the lack of Ig with specificity to TAAs.

6. Tumor sites contain lymphoreticular cells of varying content and types, depending on the particular tumor analyzed. These cells can contain Ig as an integral portion of their cell surfaces and Ig and immune complexes bound to Fc receptors. Complement components, particularly C3 or its split products, may also be bound to complement receptors on immunocytes or macrophages at the tumor site. Thus, tumor-bound Ig may be associated with the lymphoreticular cells and not with the tumor cells themselves. It is sometimes difficult to differentiate the cell types in a tissue section. Also, the possibility exists that some types of tumor cells may contain Fc or Fc-like receptors with bound Ig or immune complexes. Most of the studies that have described Ig eluted from tumor tissues, or isolated from effusions, have not discerned the origin of the Ig obtained.

In the studies from our laboratory (91,157), we provided morphological evidence that suggested that the Ig was associated with tumor cells. In most tissue sections, the mucosal layer was clearly separated from the submucosal regions which was almost always heavily infiltrated by lymphoreticular cells. Both H & E and immunofluorescent stains exhibited this differentiation. In tissue sections where the carcinoma was extensively invasive, the basement membrane was disrupted or not apparent and the mixture of cell types was also extensive and not clearly differentiated. Even in those latter tissue sections, one could usually follow the invasive carcinoma cell mass. In all cases, however, the lymphoreticular cell infiltrate would stain brightly in the immunofluorescence procedure, presumably because of the high content of cell bound Ig and immune complexes. The fluorescence of tumor cell-associated Ig was always less in intensity. Futhermore, we performed assays to determine Fc-bearing cells using the Ig-coated sheep erythrocytes in an adherence assay on the tissue slices for which serial sections had been processed through the indirect immunofluorescence procedure. It was evident that the highly fluorescing regions were also those that contained Fc receptors, and that apparent tumor cells were devoid of Fc receptor activity. In the case of tumor tissue sections where the demarcartion of mucosal and submucosal layers were maintained, the Fc receptor bearing cells were limited to the submucosal layers. Thus, we felt it unlikely that tumor cells contain the Fc receptor or that immune complexes would be bound to those cells in that manner. We also processed the tissue sections by washing at 37° in PBS under conditions that should remove Ig from Fc receptors. This showed only a slight diminution of fluorescence intensity in the Fc receptor bearing cell regions. The 37° buffer washes, however, may not remove a portion of immune complexes bound to Fc receptors. The latter consideration is dependent on the solubility of immune complexes in nearly neutral pH buffer after dissociation from Fc receptors.

7. The methods for preparation of the antigen target tissue or cells are also important. For cell targets, the use of live or fixed cells is an important distinction since with fixed cells intracellular reactive antigens are more likely to be detected. Furthermore, in the case of fixed-cell immunofluorescence (or other detection methods) a strong positive cytoplasmic immunofluorescence can obscure a cell surface reactivity. The use of live cells as antigen targets necessarily excludes the possibility of detecting intracellular TAAs. The use of fixed cells is more likely to detect autoantibodies as most of those characterized have been shown to be of an intracellular specificity. Such autoantibody reactivities can be controlled for by use of normal cell or tissue targets and by adsorption with normal tissues or cells, or their extracts. Cells grown in tissue culture are more likely to contain FCS adsorbed antigens for which a high percentage of individuals apparently have antibody specificities (see above discussions, especially the papers by Gupta _et al_., 167,168). For studies with tissue, the possibility exists that any method of preparation of tumor tissue may elute or denature the antigenic determinants of TAAs. Thus, although washing of tissues obtained

from an inflammatory site may be necessary to remove non-cell-bound Ig, a TAA may likewise be removed. Fixation of tissue or cells with any of the substances commonly used, e.g., formaldehyde, ethanol, acetone or even just air drying, may denature antigenic determinants, as described by Kluchapeva _et al._ (191) and as discussed in detail in the section above on blood group antigen reactivities. Also, the use of organic solvents can solubilize lipids and glycolipids, some of which may satisfy the criteria of being TAAs.

8. Nature of the target cells: It has been emphasized in the many reports and discussions that positivity and cross-reactivities of cancer patient sera are greatly variable, at least partly depending on the particular cells and tissues used as antigen targets.

9. Methods of assay: Many methods have been utilized to ascertain type and extent of reactivities. The limitation and possible selectivities of those methods have been discussed above.

SUMMARY

As outlined in Figures 1 and 2, the biosynthetic pathways for the expression of the A, B and H, and the Lewis determinant carbohydrate sequence structures, as well as sialylated structures, involves both type 1 and type 2 precursor chains (which may be present as glycolipids and N- or O-linked glycoproteins), and many glycosyltransferases. For tumor cells, there appears to be increased expressions of fucosyl- and sialyltransferases yielding such structures as the Le^x, sialyl-Le^a, and many other similar determinants, which are not found on the normal cell progenitor of the tumor. The types of structures expressed on tumor cells is dependent on the particular fucosyl-, sialyl- and other glycosyltransferase genes activated in the transformation and tumor progression events, the availability of the substrates for the glycosyltransferases (both the precursor sequences and the nucleotide-sugar substrates) which is partly dependent on metabolites available to the tumor mass, and on the genotype of the individual regarding particular glycosyltransferases.

Both the loss of A, B and/or H blood group antigen expressions of tumor cells and the relative expressions of the Lewis and sialylated-oligosaccharide determinants may be a consequence of the competing biosynthetic pathways and the glycosyltransferases for common substrate sequences, as well as due to the loss of particular glycosyltransferases concommitant with transformation. All of these factors probably account for the variable expressions of the complex of carbohydrate sequence determinants when comparing tumor sections of different individuals as well as the heterogeneity of expression of particular determinants within a single tumor tissue section. As described above, the A, B and/or H determinants, and precursor sequences, are also expressed to differing extents on epithelial cells depending on the tissue type and cellular location in the tissue. Thus, the differentiation state of the particular epithelial cell also determines the quantity and types of carbohydrate sequences expressed. However, because of the complex nature of the competing biosynthetic pathways for the carbohydrate sequences of glycolipids and glycoproteins, and the relative activations of fucosyl- and sialyltransferases of tumor cells, it would seem that simple deductions as to the state of differentiation of particular tumors with A, B, H and precursor sequence expressions is not warranted.

A reasonable conclusion derived from the very large number of studies on carbohydrate sequence determinant expressions of tumors seems to be that each tumor needs to be characterized with respect to the many tumor-associated carbohydrate determinants, in particular those of the fucosylated and sialylated-oligosaccharides. Both the quantity and the heterogeneity (location) of the tumor-associated sequences are important to delineate in consideration of tumor imaging and other diagnostic and therapeutic purposes. Whether the relative expressions of the many tumor-associated carbohydrate sequences will correlate with the differentiation state of a particular tumor, with metastatic predictability and/or other tumor progression indices needs yet to be determined. Such studies will necessarily have to be of a rather comprehensive nature and will require a large number of well-defined monoclonal antibody reagents to the carbohydrate sequence determinants.

In terms of tumor-associated Ig (TAIg) it seems probable that at least certain and perhaps most tumors have host Ig associated with tumor cells _in situ_. There is recent and ample evidence that TAIg and immune complexes isolated from individuals with cancer contain antibodies to tumor-associated antigens (138,139).

It is to be emphasized that the demonstration of TAIgs involves careful attention to the experimental details and approaches described above. Methodologic considerations in these types of studies are of primary importance and we have described in detail one approach as outlined in Figure 4 (reference 91).

The presence of endogenous TAIg, which is not removed by washing of frozen tissue sections, needs to be considered when antibody reactivities to tumor tissues are investigated. For studies employing mouse or rat monoclonal anti-TAA antibodies, cross-reactivities of anti-mouse or anti-rat Ig reagents with the human Igs present may yield false negative results, or at least perceived background problems. In particular, the use of human-derived antibody reagents may be obscured by the presence of the endogenous TAIgs. Also, the Ig associated with the tumor cells may be antibody that is covering certain tumor-associated antigenic determinants thus blocking the reactivity of other mono- or polyclonal antibody or lectin reagent reactivities. In addition, such TAIgs may be responsible for the apparent heterogeneity of tumor-associated determinants in tissue sections if portions of the determinants are blocked by the TAIgs. The evidence for TAIgs is sufficiently substantial that consideration of their presence is required in studies of any tumor-associated determinant.

ACKNOWLEDGEMENTS

The studies in this laboratory have been made possible by a grant from the National Cancer Institute, CA 30070, by a postdoctoral fellowship to Dr. Davis from the Cancer Research Institute/Donald P. Moore Fellowship, and by the National Cancer Institute's support of the Cancer Center of Northwestern University (grant no. CA 15145). The expert secretarial assistance in typing this manuscript by Ms. Ileen Neal is gratefully acknowledged.

REFERENCES

1. Hakomori, S. (1981) Blood Group ABH and Ii Antigens of Human Erythrocytes: Chemistry, Polymorphism, and Their Developmental Change, Semin. Hematol. 18: 39-62.
2. Lemieux, R. U. (1978) Human Blood Groups and Carbohydrate Chemistry (Hayworth Memorial Lecture), Chem. Soc. Rev. 7: 423-427.
3. Watkins, W. M., Yates A. D. and Greenwell, P. (1981) Blood Group Antigens and the Enzymes Involved in Their Synthesis: Past and Present. Biochem. Soc. Trans. 9: 186-191.
4. Kabat, E. A. (1982) Contributions of Quantitative Immunochemistry to Knowledge of Blood Group A, B, H, Le, I and i Antigens, Am. J. Clin. Pathol. 78: 281-292.
5. Clausen, H., Levery, S. B., Nudelman, E., Tsuchiya, S. and Hakomori, S. (1985) Repetitive A Epitope (Type 3 Chain A) Defined by Blood Group A_1-Specific Monoclonal Antibody TH-1: Chemical Basis of Qualitative A_1 and A_2 Distinction, Proc. Natl. Acad. Sci. USA 82: 1199-1203.

6. Clausen, H., Levery, S. B., Kannagi, R. and Hakomori, S. (1986) Novel Blood Group H Glycolipid Antigens Exclusively Expressed in Blood Group A and AB Erythrocytes (Type 3 Chain H), J. Biol. Chem. 261: 1380-1387.
7. Gardas, A. and Koscielak, J. (1971) A, B and H Blood Group Specificities in Glycoprotein and Glycolipid Fractions of Human Erythrocyte Membrane. Absence of Blood Group Active Glycoproteins in Membranes of Non-secretors, Vox Sang. 20: 137-149.
8. Anstee D. J. and Tanner, M. J. A. (1974) The Distribution of Blood-Group Antigens on Butanol Extraction of Human Erythrocyte Ghosts, Biochem. J. 138: 381-386.
9. Dejter-Juszynski, M, Harpaz, N., Flowers, H. M. and Sharon, N. (1978) Blood-group ABH-Specific Macroglycolipids of Human Erythrocytes: Isolation in High Yield from a Crude Membrane Glycoprotein Fraction, Eur. J. Biochem. 83: 363-373.
10. Jarnefelt, J., Rush, J., Li, Y. and Laine, R.A. (1978) Erythroglycan, a High Molecular Weight Glycopeptide With The Repeating Structure [Galactosyl-(1→4)-2-deoxy-2-acetamido-glucosyl (1→3)] Comprising More Than One-third Of The Protein-bound Carbohydrate of Human Erythrocyte Stroma, J. Biol. Chem. 253: 8006-8009.
11. Whittemore, N.B., Trabold, N.C.H., Reed, C. F. and Weed, R. I. (1969) Solubilized Glycoprotein from Human Erythrocyte Membrane Possessing Blood Group A, B and H Activity, Vox Sang. 17: 289-299.
12. Karhi, K.K. and Gahmberg, C.G. (1980) Identification of Blood Group A-active Glycoproteins in the Human Erythrocyte Membrane, Biochim. Biophys. Acta 622: 344-354.
13. Schenkel-Brunner, H. (1980) Blood Group - ABH Antigens of Human Erythrocytes. Quantitative Studies on the Distribution of H Antigenic Sites Among Different Classes of Membrane Components, Eur. J. Biochem. 104: 529-534.
14. Wilczynska, Z., Miller-Podraza, H. and Koscielak, J. (1980) The Contribution of Different Glycoconjugates to the Total ABH Blood Group Activity of Human Erythrocytes, FEBS Lett. 112: 277-279.
15. Mehta, N. G. (1980) ABO(H) Blood Group Antigens Of The Human Erythrocyte Membrane: Contribution of Glycoprotein and Glycolipid, J. Membrane Biol. 52: 17-24.
16. Coon, J. S. and Weinstein, R. S. (1983) Blood Group Antigens in Tumor Cell Membranes, in Biomembranes - Pathological Membranes, Vol. 11, Nowotny A., ed., Plenum Press, New York, pp. 173-205.
17. Courtney, M., Jallat, S., Tessier, L., Benavente, A., Crystal, R.G. and Lecocq, J. (1985) Synthesis in E. coli of α-Antitrypsin Variants of Therapeutic Potential for Emphysema and Thrombosis, Nature. 313: 149-152.
18. Bathurst, I. C., Stenflo, J., Errington, D.M. and Carrell, R.W. Translation and Processing of Normal (Pi MM) and Abnormal (Pi ZZ) Human α_1-Antitrypsin, FEBS Lett. 153: 270-273.
19. Foreman, R.C., Judah, J.D. and Colman, A. (1984) Xenopus Oocytes can Synthesize but do not Secrete the Z Variant of Human α_1-Antitrypsin. FEBS Lett. 168: 84-88.
20. Lambert, R.M., Zelenski, S.K. and Kratzel, R.J. (1976) Blood Group Activity in Butanol Extracts of Human Tissues, in: Human Blood Groups 5th Int. Convocation on Immunology, Mohn, F., Plunkett, R.W., Cunningham, R.K., and Lambert, R.M., eds. Karger, Basel, pp. 410-416.
21. Sakamoto, J., Yin, B.W.T. and Lloyd, K.O. (1984) Analysis of the Expression Of H, Lewis, X, Y and Precursor Blood Group Determinants in Saliva and Red Cells Using a Panel of Mouse Monoclonal Antibodies, Mol. Immunol. 21: 1093-1098.

22. Koscielak, J., Miller-Podroza, H., Krauze, R. and Piasek, A. (1976) Isolation and Characterization Of Poly(glycosyl) Ceramides (Megaloglycolipids) with A, H, and I Blood-group Activities, Eur. J. Biochem. 71: 9-18.
23. Sharom, F. J. and Grant, C.W.M. (1978) A Model for Ganglioside Behaviour in Cell Membranes, Biochem. Biophys. Acta 507: 280-293.
24. Abdelfattah-Gad, M. and Denk, H. (1980) Epithelial Blood Group Antigens in Human Carcinomas of The Distal Colon: Further Studies on Their Pathologic Significance, J. Natl. Cancer Inst. 64: 1025-1028.
25. Cooper, H. and Haesler, W.E. (1978) Blood Group Substances as Tumor Antigens in the Distal Colon, Am. J. Clin. Pathol. 69: 594-598.
26. Denk, H.S., Tappeiner, G., Davidovits, A., Eckerstorfer, R. and Holzner, J.H. (1974) Carcinoembryonic Antigen and Blood Group Substances in Carcinomas of the Stomach and Colon, J. Natl. Cancer Inst. 53: 933-942.
27. Ernst, C., Thurin, J., Atkinson, B., Wurzel, H., Meenhard, H., Stomberg, N., Civin, C. and Koprowski, H. (1984) Monoclonal Antibody Localization of A and B Isoantigens in Normal and Malignant Fixed Human Tissues, Am. J. Pathol. 117: 451-461.
28. Wiley, E.L., Murphy, P., Mendelsohn, G. and Eggleston, J.C. (1981) Distribution of Blood Group Substances in Normal Human Colon. Use of the Unlabeled Antibody (PAP) Immunoperoxidase Technic to Identify A and B Blood Group Substances, Am. J. Clin. Pathol. 76: 806-809.
29. Holborow, E.J., Brown, P.C., Glynn, L., Gresham, G.A., O'Brien, T.F. and Coombs, R.R.A. (1960) The Distribution of the Blood-Group A Antigen in Human Tissue, Br. J. Exp. Pathol. 41: 430-437.
30. Dabelsteen, E., Vedtofte, P., Hakomori, S. and Young, W.W. (1982) Carbohydrate Chains Specific for Blood Group Antigens in Differentiation of Human Oral Epithelium. J. Invest. Derm. 79: 3-7.
31. Dabelsteen, E., Vedtofte, P., Hakomori, S. and Young, W.W. (1983) Accumulation of a Blood Group Antigen Precursor in Oral Premalignant Lesions, Cancer Res. 43: 1451-1454.
32. Vedtofte, P., Dabelsteen, E., Hakomori, S. and Young, W.W. (1984) Regional Variations of Cell Surface Carbohydrates in Human Oral Stratified Epithelium, Differentiation 25: 221-228.
33. Vedtofte, P. (1985) Distribution of Type 1 and 2 Blood Group Chains in Normal and Pathological Odontogenic Epithelium Defined by Monoclonal Antibodies Specific for Le, A and H Type 2, Acta. Path. Microbiol. Immunol. Scand. Sect. A, 93: 265-276.
34. Vedtofte, P., Pindborg, J.J. and Hakomori, S. (1985) Relation of Blood Group Carbohydrates to Differentiation Patterns of Normal and Pathological Odontogenic Epithelium, Acta. Path. Microbiol. Immunol. Scand. Sect. A, 93: 25-34.
35. Hindsgaul, O., Norberg, T., Le Pendu, J. and Lemieux, R.U. (1982) Synthesis of Type 2 Human Blood Group Antigenic Determinants. The H, X, and Y Haptens and Variations of the H Type 2 Determinant as Probes for the Combining Site of the Lectin I of Ulex europaeus, Carbohydr. Res. 109: 109-142.
36. Dabelsteen, E. and Fejerskov, O. (1974) Distribution of Blood Group Antigen A in Human Oral Epithelium, Scand. J. Dent. Res. 82: 206-211.
37. Coon, J.S. and Weinstein, R.S. (1981) Detection of ABH Tissue Isoantigens by Immunoperoxidase Methods in Normal and Neoplastic Urothelium, Am. J. Clin. Pathol. 76: 163-171.
38. Bonfiglio, T.A. and Feinberg, M.R. (1976) Isoantigen Loss in Cervical Neoplasia, Arch. Pathol. Lab. Med. 100: 307-310.

39. Lee, A.K., Delellis, R.A., Rosen, P.P., Saigo, P.E., Gangi, M.D., Bagin, R., Groshen, S. and Wolfe, H.J. (1985) ABH Blood Group Isoantigen Expression in Breast Carcinomas - An Immunohistochemical Evaluation Using Monoclonal Antibodies. Am. J. Clin. Path. 83: 308-319.
40. Kumazaki, T. and Yoshida, A. (1984) Chemical Evidence that Secretor Gene, Se, is a Structural Gene Encoding a Specific Fucosyltransferase, Proc. Natl. Acad. Sci. USA 81: 4193-4197.
41. Watkins, W. M. (1980) Biochemistry and Genetics of the ABO, Lewis and P Blood Group Systems, Adv. Hum. Genet 10: 1-136.
42. Le Pendu, J., Lemieux R. U., Lambert F., Dalix, A. M. and Oriol, R. (1982) Distribution of H Type 1 and Type 2 Antigenic Determinants in Human Sera and Saliva, Am. J. Hum. Genet. 34: 402-415.
43. Oriol, R., Danilovs, J. and Hawkins, B.R. (1981) A New Genetic Model Proposing that the Se Gene is a Structural Gene Closely Linked to the H Gene, Am. J. Hum. Genet. 33: 421-431.
44. Le Pendu, J., Cartron, J.P., Lemieux, R. U. and Oriol, R. (1985) The Presence of at Least Two Different H-Blood Group-Related β-D-Gal α-2-L-Fucosyltransferases in Human Serum and the Genetics of Blood Group H Substances, Am. J. Hum. Genet. 37: 749-760.
45. Betteridge, A. and Watkins, W. M. (1985) Variant Forms of α-2-L-Fucosyltransferase in Human Submaxillary Glands from Blood Group ABH "Secretor" and "Non-secretor" Individuals, Glycoconjugate J. 2: 61-78.
46. Limas, C. and Lange, P. (1980) Altered Reactivity for A, B, H Antigen in Transitional Cell Carcinomas of the Urinary Bladder. A Study of the Mechanisms Involved, Cancer 46: 1366-1373.
47. Limas, C. and Lange, P. (1982) A, B, H Antigen Detectability in Normal and Neoplastic Urothelium. Influence of Methodologic Factors. Cancer 49: 2476-2484.
48. Juhl, B. R. (1985) Semiquantitative Immunohistochemical Evaluation of H-Antigen Expression in Human Ureters of Different ABO- and Lewis Types, J. Histochem. Cytochem. 33: 867-874.
49. Juhl, B.R. (1985) Methodologic and Genetic Influence on Immunohistochemical Demonstration and Semiquantitation of Blood Group Antigen A in Human Ureter Urothelium, J. Histochem. Cytochem. 33: 21-26.
50. Vedtofte, P., Hansen, H.E. and Dabelsteen, E. (1981). Distribution of Blood Antigen H in Human Buccal Epithelium of Secretors and Nonsecretors, Scan. J. Dent. Res. 89: 188-195.
51. Mollicone, R., Bara, J., Le Pendu, J. and Oriol, R. (1985) Immunohistologic Pattern of Type 1 (Le^a, Le^b) and Type 2 (X, Y, H) Blood Group-Related Antigens in the Human Pyloric and Duodenal Mucosae, Lab. Invest. 53: 219-227.
52. Sheares, B. T. and Carlson, D.M. (1983) Characterization of UDP-Galactose: 2-Acetamido-2-Deoxy-D-Glucose 3-Beta-Galactosyltransferase from Pig Trachea, J. Biol. Chem. 258: 9893-9898.
53. Kannagi, R., Levery, S. B. and Hakomori, S. (1985) Le^a-active Heptaglycosylceramide, a Hybrid of Type 1 and Type 2 Chain, and the Pattern of Glycolipids with Le^a, Le^b, X (Le^x) and Y (Le^y) Determinants in Human Blood Cell Membranes (Ghosts), J. Biol. Chem. 260: 6410- 6415.
54. Le Pendu, J., Lemieux, R. U., Dalix, A.M., Lambert, F. and Oriol, R. (1983) Competition Between ABO and Le Gene Specified Enzymes, Vox Sang. 45: 349-358.
55. Johnson, P.H. and Watkins, W. M. (1985) Acceptor Substrate Specificities of Human α-3- and α3,4-L-Fucosyltransferases, in: Glycoconjugates; Proceedings of the VIII Scientific International Symposium, Davidson, E.A., Williams, J.L. and DiFerrante, N.M., eds., vol. 1, Praeger Scientific, New York, pp. 222-223.

56. Hansson, G. C. and Zopf, D. (1985) Biosynthesis of the Cancer-Associated Sialyl-Le^a Antigen, in: Glycoconjugates - Proceedings of the VIIIth International Symposium, Davidson E.A., Williams J.C. and DiFerrante N.M., eds., vol. 2, Praeger Scientific, New York, pp. 554-555.
57. Brockhaus, M., Wysocka, M., Magnani, J., Steplewski, Z., Koprowski, H. and Ginsburg, V. (1985) Normal Salivary Mucin Contains the Gastrointestinal Cancer-Associated Antigen Detected by Monoclonal Antibody 19-9, in the Serum Mucin of Patients, Vox Sang. 48: 34-38.
58. Crookston, M. C. (1980) Blood Group Antigens Acquired from the Plasma, in: Immunobiology of the Erythrocyte, Sandler, G.S., Nusbacher, J. and Schamfield, eds., Alan R. Liss, New York, pp. 99-114.
59. Oriol, R., Danilovs, J., Lemieux, R., Terasaki, P. and Bernoco, D. (1980) Lymphocytotoxic Definition of Combined ABH and Lewis Antigens and their Transfer from Sera to Lymphocytes, Human Immunol. 1: 195-205.
60. Race, R. R. and Sanger, R. (1975) in: Blood Groups in Man, 6th ed., Blackwell Scientific Publications, Oxford.
61. Matsukura, Y. (1976) Electron Microscopic Observations on the H Antigen Sites of Human Erythrocytes using Ferritin Antibody Conjugates, Vox Sang. 31: 321-331.
62. Schachter, H., Michaels, M.A., Tilley, C.A., Crookston, M.C. and Crookston, J.H. (1973) Qualitative Differences in the N-Acetyl-D-galactosaminyltransferases Produced by Human A^1 and A^2 Genes, Proc. Natl. Acad. Sci. USA 70: 220 224.
63. Schachter, H., Michaels, M.A., Crookston, M.C., Tilley, C.A. and Crookston, J.H. (1971) A Quantitative Difference in the Activity of Blood Group A-Specific N-acetylgalactosaminyltransferase in Serum from A_1 and A_2 Human Subjects, Biochem. Biophys. Res. Commun. 45: 1011-1018.
64. Carton, J.P., Ropars, C., Calkovska, Z. and Salmon, C. (1976) Detection of A_1 A_2 and A_1 A_m A^1 Heterozygotes Among Human A Blood Group Heterozygotes, J. Immunogenet. 3: 155-161.
65. Topping, M.D. and Watkins, W. M. (1975) Isoelectric Points of the Human Blood Group A^1, and A^2 and B Gene-Associated Glycosyltransferases in Ovarian Cyst Fluids and Serum, Biochem. Biophys. Res. Commun. 64:89-96.
66. Davidsohn, I. (1972) Early Immunologic Diagnosis and Prognosis of Carcinoma, Am. J. Clin. Path. 57, 715-730.
67. Davidsohn, I., Kovarik, S. and Lee, C.L. (1966) A, B, and O Substances in Gastrointestinal Carcinoma. Arch. Pathol. 81: 381-390.
68. Dabelsteen, E. (1972) Quantitative Determination of Blood Group Substance A of Oral Epithelial Cells by Immunofluorescence and Immunoperoxidase Methods. Acta. Pathol. Microbiol. Scand. Sect. A, 80: 847-853.
69. Thorpe, S.J., Abel, P., Slavin, G. and Feizi, T. (1983) Blood Group Antigens in the Normal and Neoplastic Bladder Epithelium, J. Clin. Pathol. 36: 873-882.
70. Farr, A.G. and Nakane, P.K. (1981) Immunohistochemistry with Enzyme Labeled Antibodies: A Brief Review, J. Immunol. Methods 47: 129-144.
71. Sesterhenn, I., Mostofi, F.K. and Davis. C.J. (1984) Immunopathology in Prostate and Bladder Tumors, in: Immunocytochemistry in Tumor Diagnosis, Russo, J., ed., Martinus Nijhoff, Boston, pp. 337-361.
72. Alroy, J., Ucci, A.A. and Pereira, M.E.A. (1984) Lectins: Histochemical Probes for Specific Carbohydrate Residues, in: Advances in Immunohistochemistry, Delellis, R.A., ed., Masson Publ., New York, pp. 67-88.

73. Iman, A. and Taylor, C.R. (1985) Application of Immunohistochemical Methods in the Diagnosis of Malignant Disease, Cancer Investigation 3: 339-359.
74. Taylor, C.R. (1983) Immunoenzyme Techniques and Their Application to Diagnostic Studies, Ann. N.Y. Acad. Sci. 420: 115-126.
75. Mason, D.Y., Abdulaziz, Z., Falini, B. and Stein, H. (1983) Single and Double Immunoenzymatic Techniques for Labeling Tissue Sections with Monoclonal Antibodies, Ann. Acad. Sci. 420: 127-133.
76. Hsu, S.M., Raine, L. and Fanger, H. (1981) Use of Avidin-Biotin-Peroxidase Complex (ABC) in Immunoperoxidase Techniques: A Comparison Between ABC and Unlabeled Antibody (PAP) Procedures. J. Histochem. Cytochem., 29: 577-580.
77. Ormanns, W. and Schaffer, R. (1985) An Alkaline-Phosphatase Staining Method in Avidin-biotin Immunohistochemistry. Histochem. 82: 421-424.
78. Bayer, E.A. and Wilchek, M. (1984). The Avidin-Biotin Complex in Immunology, Immunol. Today 5: 39-43.
79. Stein, H., Gatter, K., Asbahr, H. and Mason, D.Y. (1985) Methods in Laboratory Investigation: Use of Freeze-Dried Paraffin-Embedded Sections for Immunohistologic Staining with Monoclonal Antibodies, Lab. Invest. 52: 676-683.
80. Taylor, C.R. (1985) Monoclonal Antibodies and 'Routine' Paraffin Sections, Arch. Pathol. Lab. Med. 109: 115-116.
81. Weinstein, R.S., Coon, J., Alroy, J. and Davidsohn, I. (1981) Tissue-Associated Blood Group Antigens in Human Tumors, in: Diagnostic Immunohistochemistry, (DeLellis, R.A., ed.,) Masson Publ., New York, pp. 239-261.
82. Chapman, C.M., Allhoff, E.P., Proppe, K.H. and Prout, G.R., Jr. (1983) Use of Monoclonal Antibodies for the Localization of Tissue Isoantigens A and B in Transitional Cell Carcinoma of the Upper Urinary Tract, J. Histochem. Cytochem. 31: 557-561.
83. Seal, G.M., Rowland, R.G., Thomalla, J.V., Rudolph, R.A., Pfaff, D.S., Kamer, M. and Eble, J.N. (1985) A, B and H Antigens in Normal Urothelium: An Immunohistochemical Study using Monoclonal Antibodies with the Avidin-Biotin Complex Technique, J. Urol. 133: 513-516.
84. Dabelsteen, E., Buschard, K., Hakomori, S. and Young, W.W. (1984) Pattern of Distribution of Blood Group Antigens on Human Epidermal Cells During Maturation, J. Invest. Dermatol. 82: 13-17.
85. Horikawa, M., Chisaka, N., Yokoyama, S. and Onoe, T. (1976) Effect of Stirring During Fixation Upon Immunofluorescence. Results with Distribution of Albumin-Producing Cells in Liver, J. Histochem. Cytochem. 24: 926-932.
86. Larson, L.-I. (1981) Peptide Immunocytochemistry. Prog. Histochem. Cytochem. 13: 1-83.
87. Coruh, G. and Mason, D.Y. (1980) Serum Proteins in Human Squamous Epithelium, Br. J. Dermatol. 102: 497-505.
88. Brandtzaeg, P. and Rognum, T.O. (1984) Evaluation of Nine Different Fixatives. 2. Preservation of IgG, IgA and Secretory Component in an Artificial Immunohistochemical Test Substrate, Histochemistry 81: 213-219.
89. Loeffel, S.C., Gillespie, G. Y. and Mirmiran, S.A. (1985) Cellular Immunolocalization of S100 Protein within Fixed Tissue Sections by Monoclonal Antibodies, Arch. Path. Lab. Med. 109: 117-122.
90. Nemes, Z., Thomazy, V. and Szeifert, G. (1983) Demonstration Of Light Chain Monotypia in B Cell Non-Hodgkins Lymphomas Using Unfixed Freeze-Dried and Formalin-Fixed Trypsinized Paraffin Sections, J. Clin. Pathol. 36: 883-893.

91. Pankow, M.L., Davis, L.E., Becker, S.P., Ossoff, R.H., and Anderson, B.E. (1984) Use of Dextran and Post-Primary Antibody Fixation in Immunoperoxidase Staining of Fresh Frozen Tissue: Detection of Immunoglobulin Associated with Squamous Carcinomas of the Head and Neck, J. Histochem. Cytochem. 32: 771-777.
92. Landis, D.M.D. (1985) Promise and Pitfalls in Immunocytochemistry, Trends Neur. Sci. 8: 312-317.
93. Taylor, C.R. (1978) Immunoperoxidase Techniques - Practical and Theoretical Aspects, Arch. Pathol. Lab. Med. 102: 113-121.
94. Petrusz, P. (1983) Essential Requirements for the Validity of Immunocytochemical Staining Procedures, J. Histochem. Cytochem. 31: 177-179.
95. Thogersen, H., Lemieux, R.U., Bock, K. and Meyer, B. (1982) Further Justification of the Exo-Anomeric Effect: Conformational Analysis Based on Nuclear Magnetic Resonance Spectrosocpy of Oligosaccharides, Can. J. Chem. 60: 44-57.
96. Marcus, D. M. and Grollman, A. P. (1966) Studies of Blood Group Substances: I. Caprine Precipitating Antisera to Human Le^a and Le^b Blood Group Substances, J. Immunol. 97: 867-875.
97. Marr, A.M.S., Donald, A.S.R., Watkins, W. M. and Morgan, W.T.J. (1967) Molecular and Genetic Aspects of Human Blood-Group Le^b Specificity, Nature 215: 1345-1349.
98. Messeter, L., Brodin, T., Chester, M.A., Karlsson, K.A., Zopf, D. and Lundblad, A. (1984) Immunochemical Characterization of a Monoclonal Anti-Le^b Blood Grouping Reagent, Vox. Sang. 46: 66-74.
99. Young, W. W., Johncon, H.S., Tamura, Y., Karlsson, K.A., Larson, G., Parker, J.M.R., Khare, D.P., Spohr, U., Baker, D.A., Hindsgaul, O. and Lemieux R. U. (1983) Characterization of Monoclonal Antibodies Specific for the Lewis a Human Blood Group Determinant, J. Biol. Chem. 258, 4890-4894.
100. Gooi, H.C., Picard, J.K., Hounsell, E.F., Gregoriou, M., Rees, A.R. and Feizi, T. (1985) Monoclonal Antibody (EGR/649) Reactive with the Epidermal Growth Factor Receptor of A431 Cells Recognizes The Blood Group ALe^b and ALe^Y Structures, Mol. Immunol. 22: 689-693.
101. Fredman, P., Richert, N.D., Magnani, J.L., Willingham, M.C., Pastan, I. and Ginsburg, V. (1983) A Monoclonal Antibody that Precipitates the Glycoprotein Receptor for Epidermal Growth Factor is Directed Against the Human Blood Group H Type 1 Antigen, J. Biol. Chem. 258: 11206-11210.
102. Le Pendu, J., Fredman, P., Richter, N.D., Magnani, J.L., Willingham, M.C., Pastan, I., Oriol, R. and Ginsburg, V. (1985) Monoclonal Antibody 101 that Precipitates the Glycoprotein Receptor for Epidermal Growth Factor is Directed Against the Y Antigen, Not the H Type 1 Antigen, Carbohydr. Res. 141: 347-349.
103. Feizi, T. (1985) Demonstration by Monoclonal Antibodies that Carbohydrate Structures of Glycoproteins and Glycolipids are Oncodevelopmental Antigens, Nature 314: 53-57.
104. Hakomori, S. (1984) Tumor-associated Carbohydrate Antigens, Ann. Rev. Immunol. 2: 103-126.
105. Lloyd, K.O. (1983) Human Tumor Antigens: Detection and Characterization with Monoclonal Antibodies, in: Basic and Clinical Tumor Immunology, Herberman, R.B., ed., Martinus Nijhoff, Boston, MA, pp. 159-214.
106. Koprowski, H. (1982) Embryonic Precancerous and Cancerous Human Antigens Recognized by Monoclonal Antibodies, in: Fetal Antigens and Cancer (Ciba Foundation Symposium no. 96) Evered, D. and Whelan, J., eds., Pitman, London, pp. 204-229.

107. Damjanov, I. and Knowles, B.B. (1984) Monoclonal Antibodies and Tumor-Associated Antigens, in: Advances in the Biology of Disease, Rubin, E. and Danjanor, I., eds. Williams and Wilkins, Baltimore, MD, pp. 181-196.
108. Kuhns, W.J. and Primus, F.J. (1985) Alterations of Blood Groups and Blood Group Precursors in Cancer, in: Progress in Clinical Biochemistry and Medicine, vol. 2., Springer-Verlag, Berlin, pp. 49-95.
109. Hakomori, S. and Kannagi, R. (1983) Glycosphingolipids as Tumor-Associated and Differentiation Markers, J. Natl. Cancer Inst. 71: 231-251.
110. Cowan, W. K. (1962) Blood Group Antigens on Human Gastrointestinal Carcinoma Cells, Br. J. Cancer 16: 535-540.
111. Davidsohn, I., Ni, L. Y. and Stejskal, R. (1971) Tissue Isoantigens A, B, and H in Carcinoma of the Stomach, Arch. Pathol. 92: 456-464.
112. Eklund, A. E., Gullbring, B. and Langerlof, B. (1963) Blood Group Specific Substances in Human Gastric Carcinoma. A Study using the Fluorescent Antibody Technique, Acta. Pathol. Microbiol. Scand. 59: 447-455.
113. Sheahan, D. G., Horowitz, S. and Zamcheck, N. (1971) Deletion of Epithelial ABH Isoantigens in Primary Gastric Neoplasms and in Metastatic Cancer, Am. J. Dig. Dis. 16: 961-969.
114. Kay, H.E. M. (1957) A and B Antigens of Normal and Malignant Cells, Br. J. Cancer 11: 409-414.
115. Lange, P.H., Limas, C. and Fraley, E.E. (1978) Tissue Blood-group Antigens and Prognosis in Low Stage Transitional Cell Carcinoma of the Bladder, J. Urol. 119: 52-55.
116. Davidsohn, I., Kovarik, S. and Ni, L.Y. (1969) Isoantigens A, B and H in Benign and Malignant Lesions of the Cervix, Arch. Pathol. 87: 306-314.
117. Davidsohn, I. and Stejskal R. (1972) Tissue Antigens A, B and H in Health and Disease, Haemotologia (Budap) 6: 177-184.
118. Gupta, R. K., Schuster, R. and Christian, W. D. (1973). Loss of Isoantigens A, B and H in Prostate, Am J. Pathol. 70: 439-447.
119. Davidsohn, I., Ni, L.Y. and Stejskal, R. (1971) Tissue Isoantigens A, B and H in Carcinoma of the Pancreas. Cancer Res. 31: 1244-1250.
120. Davidsohn, I. and Ni, L.Y. (1969) Loss of Isoantigens A, B and H in Carcinoma of the Lung, Am. J. Pathol. 57: 307-334.
121. Gupta, R. K. and Schuster, R. (1973) Isoantigens A, B, and H in Benign and Malignant Lesions of Breast, Am. J. Pathol. 72: 253-260.
122. Liu, P. I., McGregor, D.H. and Liu, J.G. (1974) Carcinoma of the Oral Cavity Evaluated By Specific Red Cell Adherence Test, Oral Surg. 32: 56-64.
123. Lin, F., Liu, P.I. and McGregor, D.H. (1977) Isoantigens A, B, and H in Morphologically Normal Mucosa and in Carcinoma of the Larynx, Am. J. Clin. Pathol. 68: 1372-376.
124. Davis, L.E., Pankow, M.L., Sobut, R., Keesara, S. and Anderson, B. (1984) Blood Group (BG) Antigen (Ag) Expression by Normal and Squamous Carcinoma Epithelia, Fed. Proc. 43: 1751.
125. Fujita, J., Matsumoto, K., Kishi, K. and Ishiyama, I. (1981) Synthesis of ABH Blood Group Substances in Bladder Tumors, Br. J. Urol. 53: 448-452.
126. Hall, L., Faddoul, A., Saberi, A. and Edson, M. (1982) The Use of the Red Cell Surface Antigen to Predict the Malignant Potential of Transitional Cell Carcinoma of the Ureter and Renal Pelvis, J. Urol. 127: 23-25.

127. Emmott, R. C., Javadpour, N., Bergman, S. M. and Soares, T. (1979) Correlation of the Cell Surface Antigens with Stage and Grade in Cancer of the Bladder, J. Urol. 121: 37-39.
128. Richie, J.P., Blute, R.D. and Waisman, J. (1980) Immunologic Indicators of Prognosis in Bladder Cancer: The Importance of Cell Surface Antigens, J. Urol. 123: 22-24.
129. Decenzo, J. M., Howard, P. and Irish, C.E. (1975) Antigenic Deletion and Prognosis of Patients with Stage A Transitional Cell Bladder Carcinoma, J. Urol. 114: 874-878.
130. Newman, A. J., Carlton, C.E. and Johnson, S. (1980) Cell Surface A,B or O (H) Blood Group Antigens as an Indicator of Malignant Potential in Stage A Bladder Cancer, J. Urol. 124: 27-29
131. Bergman, S. and Javadpour, N. (1978) The Cell Surface Antigen A,B or O (H) as an Indicator of Malignant Potential in Stage A Bladder Carcinoma: Preliminary Report, J. Urol. 119: 49-51.
132. Cooper, H.S. and D'Elia, F. L. (1982) Comparison Between the Methods of Indirect Immunofluorescence and Specific Red Cell Adherence in Detecting ABH Iso-antigens in Bladder Carcinoma, Am. J. Clin. Pathol. 77: 548-554.
133. Strauchen, J.A., Bergman, S.M. and Hanson, T.A.S. (1980) Expression of A and B Tissue Isoantigens in Benign and Malignant Lesions of the Breast, Cancer, 45: 2149-2155.
134. King, C. T., Clark, T.D., Lovett, J., Cash, J.B., Primus, F.J., McRoberts, J.W. and Flanigan, R.C. (1983) A Comparison of Clinical Course with Blood Group Antigen Testing by Specific Red Cell Adherence and Immunoperoxidase in Ureteral and Renal Pelvic Tumors, J. Urol. 130. 871-873.
135. Flanigan, R. C., King, C.T., Clark, T.D., Cash, J.B., Greenfield, B. J., Sniecinski, I.J. and Primus, F.J. (1983) Immunohistochemical Demonstration of Blood Group Antigens in Neoplastic and Normal Human Urothelium, J. Urol. 130: 499-503.
136. Von Kleist, S., King, M. and Huet, C. (1980) Evidence for Membrane Bound Antibodies Directed Against Antigens Expressed on Tumors, in: Contempary Topics in Immunobiology, vol. 10, Witz, I.P. and Hanna, Jr., M.G., eds., Plenum Press, New York, pp. 177-189.
137. Ran, M., Yaakubowicz, M., Amitai, O. and Witz, I.P. (1980) Tumor-localizing Lymphocytoxic Antibodies, in: Contemporary Topics in Immunobiology, vol. 10, Witz, I.P. and Hanna, Jr. M.G., eds., Plenum Press, New York, pp. 191-211.
138. Gupta, R.K. and Morton, D.L. (1985) Clinical Significance and Nature of Circulating Immune Complexes in Melanoma Patients, in: Immune Complexes and Human Cancer, Contemporary Topics in Immunobiology, vol. 15, Salinas, F.A. and Hanna, Jr., M.G., eds., Plenum Press, New York, pp. 1-53.
139. Salinas, F.A., Wu, K.H. and Silver, H.K. (1985) Clinical Relevance of Immune Complexes, Associated Antigen and Antibody in Cancer, in: Immune Complexes and Human Cancer, Contemporary Topics in Immunobiology, vol. 15, Salinas, F.A. and Hanna, Jr., M.G., eds., Plenum Press, New York, pp. 55-109.
140. Witz, I.P. (1971) Tumor-Associated Immunoglobulins, Isr. J. Med. Sci. 7: 230-238.
141. Witz, I.P. (1973). The Biological Significance of Tumor-Bound Immunoglobulins, Curr. Top. Microbiol. Immunol. 61: 151-171.
142. Cruse, J.M. and Azar, M.M. (1967) Immunologic Enhancement Induced by 7S Antibody Globulin, Z. Immun.-Forsch. 132: 240-246.
143. Takasugi, M. and Hildemann, W. H. (1969) Regulation of Immunity Toward Allogeneic Tumors in Mice. I. Effect of Antiserum Fractions on Tumor Growth, J. Natl. Cancer Inst. 43: 843-856.

144. Sjogren, H.O., Hellstrom, I., Bansol, S.C. and Hellstrom, K.E. (1971) Suggestive Evidence that the "Blocking Antibodies" of Tumor-Bearing Individuals May be Antigen-Antibody Complexes, Proc. Natl. Acad. Sci. 68: 1372-1375.
145. Brandtzaeg, P. (1974) Mucosal and Glandular Distribution of Immunoglobulin Components. Immunohistochemistry With a Cold Ethanol-Fixation Technique, Immunology 26: 1101-1114.
146. Brandtzaeg, P. and Baklien, K. (1972) Extra- and Intracellular Distribution of Immunoglobulins and Secretory Component in Human Intestinal Mucosa as Revealed by Immunofluorescence, Scand. J. Gastroenterol., Suppl. 16: 12-24.
147. Brandtzaeg, P., Surjan, L., Jr., and Rerdal, P. (1978) Immunoglobulin Systems of Human Tonsils. I. Control Subjects of Various Ages: Quantification of Ig-producing Cells, Tonsillar Morphometry and Serum Ig concentrations. Clin. Exp. Immunol. 31: 367-381.
148. Brandtzaeg, P., Fjellanger, I. and Gjeruldsen, S.T. (1970) Human Secretory Immunoglobulins. I. Salivary Secretions from Individuals with Normal or Low Levels of Serum Immunoglobulins, Scand. J. Haemat., Suppl. 12: 3-83.
149. Quismorio, F.P., Rea, T. H., Levan, N.E. and Friou, G. J. (1975) Immunoglobulin Deposits in Lepromatous Leprosy Skin. Presence of Deposits in Apparently Uninvolved Skin and Occurrence of Serum Antiepithelial Antibodies, Arch. Dermatol. 111: 331-334.
150. Izsak, F.C., Brenner, H.J., Landes, E., Ran, M. and Witz, I.P. (1974) Correlation Between Clinicopathological Features of Malignant Tumors and Cell Surface Immunoglobulins, Isr. J. Med. Sci. 10: 642-646.
151. Braslawsky, G. R., Ran, M. and Witz, I.P. (1976) Tumor Bound Immunoglobulins: The Relationship Between the In Vivo Coating of Tumor Cells by Potentially Cytotoxic Anti-Tumor Antibodies, and the Expression of Immune Complex Receptors, Int. J. Cancer 18: 116-121.
152. Lewis, M.G., Avis, P.J.G., Phillips, T.M. and Sheikh, K.M.A. (1973) Tumor-Associated Antigens in Human Malignant Melanoma, Yale J. Biol. Med. 46: 661-668.
153. Lewis, M.G., Hartman, D. and Jerry, L.M. (1976) Antibodies and Anti-Antibodies in Human Malignancy: An Expression of Deranged Immune Regulation, Ann. N.Y. Acad. Sci. 276, 316-327.
154. Gerber, M.A., Koffler, D. and Cohen, C.J. (1977) Circulating Antibodies in Patients with Ovarian Carcinoma, Gynecol. Oncology 5: 228-232.
155. Richman, A.V. (1976) Immunofluorescence Studies of Benign and Malignant Human Mammary Tissue, J. Natl. Cancer Inst. 57: 263-267.
156. Popovic, D., Gill, L., Sisson, G., Applebaum, E. and Anderson, B. (1976). Detection and Localization of Tumor-Associated Immune Components of Head and Neck Squamous Cell Carcinomas, Tr. Am. Acad. Opth. Otol. 82: 119-130.
157. Koneval, T., Applebaum, E., Popovic, D., Gill, L., Sisson, G., Wood, G.W. and Anderson, B. (1977) Demonstration of Immunoglobulin in Tumor and Marginal Tissues of Squamous Cell Carcinomas of the Head and Neck, J. Natl. Cancer Inst. 59: 1089-1097.
158. Moav, N., Hochberg, Y., Cohen, G. and Witz, I.P. (1978) Binding Patterns of Immunoglobulins from Tumor-Bearing Mice to the Corresponding Tumor Cells, J. Immunol. Methods 22: 37-49.
159. Moav, N. and Witz, I.P. (1978) Characterization of Immunoglobulins Eluted from Murine Tumor Cells: Binding Patterns of Cytotoxic Anti-Tumor IgG, J. Immunol. Methods 22: 51-62.

160. Brown, J.P., Klitzman, J.M. and Hellstrom, K.E. (1977) A Microassay for Antibody Binding to Tumor Cell Surface Antigens Using ^{125}I-Labelled Protein A from Staphylococcus Aureus, J. Immunol. Methods 15: 57-66.
161. Ting, C. C. (1976) Detection of Anti-Tumor Antibody in Virally Induced Tumors and Its Relationship to Tumor Growth, Int. J. Cancer 18: 205-215.
162. Wood, G. W. and Barth, R.F. (1974) Immunofluorescent Studies of the Serologic Reactivity of Patients with Malignant Melanoma Against Tumor-Associated Cytoplasmic Antigens, J. Natl. Cancer Inst. 53: 309-316.
163. Lewis, M.G. and Phillips, T.M. (1972) The Specificity of Surface Membrane Immunofluorescence in Human Malignant Melanoma, Int. J. Cancer 10: 105-111.
164. Chapuis, B. J., Powles, R. and Alexander, P. (1978) Inability to Demonstrate Lytic Antibodies to Autologous Leukemia Cells in the Sera from Remission Patients with Acute Myelogenous Leukaemia Treated with Active Specific Immunotherapy, Clin. Exp. Immunol. 32: 253-258.
165. Paluch, E. and Ioachim, H.L. (1978) Lung Carcinoma-Reactive Antibodies Isolated from Tumor Tissues and Pleural Effusions of Lung Cancer Patients, J. Natl. Cancer Inst. 61: 319-325.
166. Sheikh, K.M.A., Apuzzo, M.L.J., Kochsiek, K.R. and Weiss, M.H. (1977) Malignant Glial Neoplasms: Definition of a Humoral Host Response to Tumor-Associated Antigen(s), Yale J. Biol. Med. 50: 397-403.
167. Gupta, R.K., Irie, R.F. and Morton, D.L. (1978) Antigens on Human Tumor Cells Assayed by Complement Fixation with Allogeneic Sera, Cancer Res. 38: 2573-2580.
168. Gupta, R.K., Silver, H.K.B., Reisfeld, R.A. and Morton, D.L. (1978) Isolation and Characterization of Antitumor Antibodies from Cancer Patient Sera by Affinity Chromatography. Proc. Am. Assoc. Cancer Res. 19: 133.
169. Edynak, E.M., Hirshaut, Y., Bernhard, M. and Trempe, G. (1972) Fluorescent Antibody Studies of Human Breast Cancer, J. Natl. Cancer Inst. 48: 1137-1143.
170. Sofen, H. and O'Toole, C. (1978) Anti-Squamous Tumor Antibodies in Patients with Squamous Cell Carcinoma, Cancer Res. 38: 199-203.
171. Rosenberg, S. A., Schwarz, S., Anding, H., Hyatt, C. and Williams, G.M. (1977) Comparison of Multiple Assays for Detecting Human Antibodies Directed Against Surface Antigens on Normal and Malignant Human Tissue Culture Cells. J. Immunol. Methods 17, 225-239.
172. Boyle, M.D.P., Ohanian, S.H. and Borsos, T. (1976) Lysis of Tumor Cells by Antibody and Complement. VI. Enhanced Killing of Enzyme-Pretreated Tumor Cells, J. Immunol. 116: 661-668.
173. Nordquist, R.E., Schafer, F.B., Manning, N.E., Ishmael, D.R. and Hoge, A.F. (1977) Antitumor Antibodies in Human Breast Cancer Sera as Detected by Fixed Cell Immunofluorescence and Living Cell Membrane Immunofluorescence Assays, J. Lab. Clin. Med. 89: 257-261.
174. Lerner, M.P., Anglin, J.H. and Nordquist, R.E. (1978) Cell-Surface Antigens From Human Breast Tumor Cells, J. Natl. Cancer Inst. 60: 39-44.
175. Springer, G.F., Desai, P.R. and Scanlon, E.F. (1976) Blood Group MN Precursors as Human Breast Carcinoma-Associated Antigens and "Naturally" Occurring Human Cytotoxins Against Them, Cancer 37: 169-176.

176. Ferrone, S. and Pellegrino, M.A. (1977) Cytotoxic Antibodies to Cultured Melanoma Cells in the Sera of Melanoma Patients, J. Natl. Cancer Inst. 58: 1201-1204.
177. Zeltzer, P.M. and Seeger, R.C. (1977) Microassay Using Radioiodinated Protein A from Staphylococcus Aureus for Antibodies Bound to Cell Surface Antigens of Adherent Tumor Cells, J. Immunol. Methods 17: 163-175.
178. Hahn, W.V., Kagnoff, M.F. and Hatlen, L.H. (1978) Immune Responses in Human Colon Cancer. II. Cytotoxic Antibody Detected in Patients' Sera, J. Natl. Cancer Inst. 60: 779-784.
179. Sloan, T.B., Martincic, R.R. and Anderson, B. (1981) Synovial Cell Antigens - Production of Heterologous Anti-Human Synovial Cell Sera and General Reactivities of the Antisera, Exp. Cell Biol. 49: 20-33.
180. Sloan, T.B., Martincic, R.R. and Anderson, B. (1981) Synovial Cell Antigens - Differences of Antigen Compositions Between Rheumatoid Arthritic (RA)- and Non-RA-Derived Synovial Cells Detected with Anti-Synovial Cell Sera, Exp. Cell Biol. 49: 195-206.
181. Sloan, T.B., Martincic, R.R. and Anderson, B. (1981) Differences of Antigen Composition of Rheumatoid Arthritic (RA) and Non-RA Derived Tissue Cultured Synovial Cells as Detected by Monkey Anti-RA Synovial Cell Sera, J. Rheumatol. 8: 204-213.
182. Oettgen, H.F., Shiku, H., Takahashi, T., Carey, T., Resnick, L., Ueda, R., Pfreundschuh, M. and Old, L.J. (1978) Human Cancer Immunology, in: Brain Metastasis, Weiss, L., Gilbert, H.A. and Posner, J.B., eds., vol. 2, G.K. Hall and Co., Boston, pp. 165-185.
183. Houghton, A.N., Taormina, M.C., Ikeda, H., Watanabe, T., Oettgen, H.F. and Old, L.J. (1980) Serological Survey of Normal Humans for Natural Antibody to Cell Surface Antigens of Melanoma, Proc. Natl. Acad. Sci. 77, 4260-4264.
184. Wood, G.W. and Gillespie, G.Y. (1975) Studies on the Role of Macrophages in Regulation of Growth and Metastasis of Murine Chemically Induced Fibrosarcomas, Int. J. Cancer 16: 1022-1029.
185. Bleiberg, H. and Galand, P. (1976) In Vitro Autoradiographic Determination of Cell Kinetic Parameters in Adenocarcinomas and Adjacent Healthy Mucosa of the Human Colon and Rectum, Cancer Res. 36: 325-328.
186. Filipe, M.I. (1975) Mucous Secretion in Rat Colonic Mucosa During Carcinogenesis Induced by Dimethylhydrazine. A Morphological and Histochemical Study, Br. J. Cancer 32: 60-77.
187. Dawson, P.A. and Filipe, M.I. (1976) An Ultrastructural and Histochemical Study of the Mucous Membrane Adjacent to and Remote From Carcinoma of the Colon, Cancer 37: 2388-2398.
188. Lipkin, M. (1975) Biology of Large Bowel Cancer: Present Status and Research Frontiers, Cancer 36: 2319-2324.
189. Koutcher, J.A., Goldsmith, M. and Damadion, R. (1978) NMR in Cancer. X. A Malignancy Index to Discriminate Normal and Cancerous Tissue, Cancer 41: 174-182.
190. Tickle, C., Crawley, A. and Goodman, M. (1978) Mechanisms of Invasiveness of Epithelial Tumors: Ultrastructure of the Interactions of Carcinoma Cells with Embryonic Mesenchyme and Epithelium, J. Cell Sci. 33: 133-155.
191. Kluchapeva, T.E., Matveeva, V.A. and Deichman, G.I. (1978) Sensitivity of TSTA and Species-Specific Cell Membrane Antigens of Tumor Cells to Glutaraldehyde Treatment, Neoplasma 25: 273-278.

THE CHEMICAL BASIS FOR EXPRESSION OF THE SIALYL-Le^a ANTIGEN

David Zopf[*] and Gunnar C. Hansson[#]

[*]*Laboratory of Pathology, National Cancer Institute National Institutes of Health, Bethesda, MD, 20892 and*
[#]*Department of Medical Biochemistry, University of Gothenburg, P.O. Box 33031, S-400 33 Gothenburg, Sweden*

Tumor antigens can be defined by antibodies that bind cancer cells and/or their secreted products, but not cells or secreted products of normal tissues from which the cancer cells derive (1,2). Tumor antigens are of interest for three major reasons: (i) they are strategic targets for immunodiagnosis and immunotherapy; (ii) they are potential receptors mediating physiological events that determine cellular organization in tumors; (iii) they are substances produced due to altered biosynthetic control, understanding of which might provide insights into mechanisms for malignant transformation. During the past few years, several laboratories have produced monoclonal antibodies that specifically bind human cancer cells. Many of these recognize complex carbohydrate antigens (3). Such antigens sometimes are present at specific stages of normal embryonic development but disappear completely during later development or persist only in isolated, specialized tissues in the mature organism (4). Carbohydrate tumor antigens probably result from aberrant expression of glycosyltransferases that normally appear and disappear in an orderly and selective fashion according to the genetic program for histologic development and cytodifferentiation (reviewed in 5-7). The sialyl-Le^a (SLe^a) antigen, originally defined by antibody 19-9 (8), occurs during early development in the gastrointestinal tract, but after birth is virtually absent from cells that line the stomach and intestines, persisting only in ducts and secretions of the digestive glands, respiratory tract, and reproductive system (9,10). The antigen is synthesized from type 1 blood group precursor chains by the sequential actions of two glycosyl transferases, an α2-3 Gal sialyltransferase, and the Lewis fucosyltransferase (11). The amount of SLe^a expressed is strongly influenced by ABO blood group secretor status due to competition for precursor between the Secretor fucosyl transferase and α2-3 Gal sialyltransferase (12). This review will discuss the known sites of occurrence of SLe^a in normal and malignant tissues, its chemical structure and biosynthesis, the relationship of SLe^a to other blood group antigens, and some implications for use of SLe^a as a marker for immunodiagnosis of cancer.

OCCURRENCE OF SLe^a ANTIGEN

The SLe^a antigen was first detected as the target of monoclonal antibody 19-9, an immunoglobulin that binds to the surface membranes of the SW1116 human colorectal carcinoma cell line but fails to bind normal colonic epithelium (8,13-15). The membrane antigen of SW1116 cells is a monosialoganglioside, the major ganglioside produced by these cells (15). The SLe^a antigen only occurs in individuals who express the Le^a and/or Le^b blood group antigens (12,16). Table 1 summarizes the known sites of occurrence of SLe^a in normal and malignant human tissues. During early development (12-17 weeks gestation) the SLe^a antigen is detectable in fetal intestinal epithelium by immunohistologic staining with monoclonal antibody 19-9 (9,10). Gangliosides with SLe^a activity, probably shed from these cells, are present in meconium, the first feces of the newborn (15). In healthy adults, the antigen is undetectable by immunohistologic staining in epithelial cells lining the small intestine and colon, but is detectable in single midzone goblet cells and in areas of intestinal metaplasia in the stomach (9,10). In contrast, SLe^a is strongly expressed in the adult in normal epithelial cells lining ducts of the salivary glands and pancreas, as well as in the gall bladder and hepatobiliary tree (9,10). It also occurs in the small bronchial glands of normal lung, in the lining cells of the uterine endocervix, and in ducts of the prostate gland (9,10). Slight reactivity of some Le(a+b-) erythrocytes is detectable by hemagglutination (17) and SLe^a-active free oligosaccharides are excreted in small amounts in human milk (18,19). In most histologic sections of normal SLe^a-positive glandular tissues, staining with 19-9 antibody is observed in intralumenal secretions as well as within goblet cells, suggesting that the SLe^a antigen may be carried on mucins secreted into the ducts (9,10). Association of SLe^a with mucins in normal tissue is confirmed by analysis of saliva using antibody 19-9 in ELISA (12) and by direct chemical analysis of mucin purified from seminal plasma (20). Under normal physiologic conditions, most SLe^a-containing mucins probably are excreted into compartments where they are degraded or excreted, and thus are excluded from blood.

Table 1. Occurrence of the Sialyl-Le^a Antigen

Tissue	Secretion	References	Comments
Fetal Intestine	meconium	9,10,15	12-17 weeks gestation
Adult normal tissues:			
pancreas	mucin	9,10,	duct epithelium
salivary gland	mucin	9,10,12	duct epithelium
biliary system	mucin	9,10	gall bladder, small intrahepatic ducts
lung	mucin	9,10	small bronchial glands
uterine cervix	mucin	10	endocervical lining cells
prostate	mucin	9,20	duct epithelium, seminal plasma
erythrocytes	none	17	trace amounts
lactating mammary gland	milk	18,19	free oligosaccharides
Carcinomas:			
Pancreatic	mucin	9,10,15,	≥ 80% serum positive
Gastric	mucin	9,10,15, 24-26	30-50% serum positive
Colorectal	mucin	9,10,15, 13,21,25	40-50% serum positive
Hepatobiliary	mucin	9,10,25	few reported
Endometrial and Ovarian	-	9,10	few reported
Colorectal Ca cells in culture	-	8,13-15	major ganglioside in SW1116 cells

Adenocarcinomas of pancreas commonly express the SLe^a antigen both as gangliosides in tumor cell membranes (15,21,22) and as mucins detectable in blood after release by tumor cells that invade across submucosal barriers (23). The frequency of elevated SLe^a detected in serum of patients with pancreatic carcinoma is about the same (≥80%) as the frequency of individuals who carry the Le gene in the same population (27). In contrast, only 30-50% of patients with gastric or colorectal cancer have elevated levels of SLe^a-containing mucin in tumor tissue (28) or serum (13,24,25-27). Thus, tumors that arise from cells lining the gastrointestinal tract, where SLe^a normally occurs only during prenatal development, appear to express the antigen less frequently or in smaller amounts than tumors arising in pancreatic ducts where SLe^a is normally expressed. As only a few cases of hepatobiliary, endometrial, and ovarian adenocarcinoma have been examined (9,10), the true frequency of SLe^a expression in these tumors is undetermined.

As noted above, SLe^a antigen is lacking in saliva of individuals who belong to the Le(a-b-) blood group. The amount of SLe^a in saliva of Le(a+b-) individuals is generally greater than in Le(a-b+) individuals (12). Feizi, et al (28) reported small amounts of SLe^a in normal gastric mucin from Le(a+b-) individuals but it is not clear that the samples analyzed were free of contaminating salivary mucin, known to be rich in SLe^a antigen.

STRUCTURE AND BIOSYNTHESIS OF SLe^a

The antigen recognized by antibody 19-9 was first purified from a chloroform/methanol extract of human SW 1116 colorectal carcinoma cells by Magnani, et al (14). The antigen molecule was observed to elute from DEAE-Sephadex as a monosialoganglioside and to migrate on thin-layer plates between standard gangliosides GM1 and GD1a. Antigenic activity was abolished by treatment with neuraminidase. Monosaccharide composition and methylation analysis of the native and asialo forms of the antigen (15) produced data consistent with either of two isomeric structures corresponding to the milk oligosaccharides lacto-N-fucopentaose III (LNF III) and lacto-N-fucopentaose II (LNF II) (see Table 2). Cleavage of the asialoganglioside by trifluoroacetolysis and conversion of the released oligosaccharide to a permethylated, N-trifluoroacetylated alditol gave a product with the same retention time in gas-liquid chromatography as authentic milk oligosaccharide lacto-N-fucopentaose II (15,29) (see Table

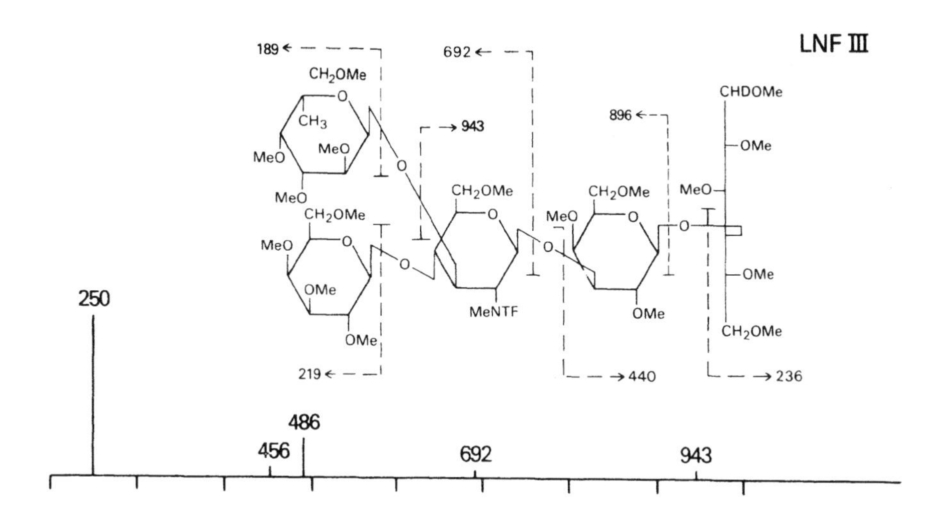

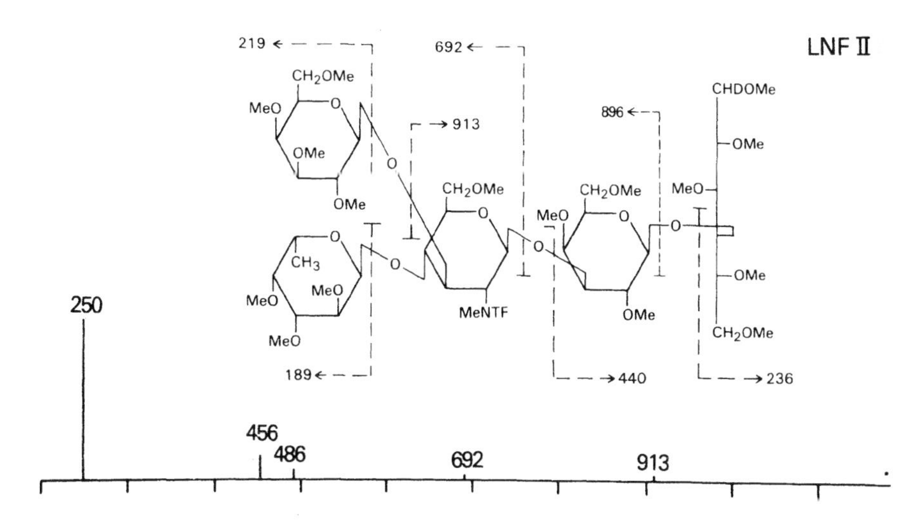

Fig. 1. Relative intensities of selected ions from electron impact mass spectra of N-trifluoroacetylated and permethylated oligosaccharide alditols of LNF III and LNF II (complete spectra are given in reference 15).

2 for oligosaccharide structures). By electron impact mass spectrometry the unknown oligosaccharide could be differentiated from lacto-N-fucopentaose III (Figure 1). Ions resulting from elimination or cleavage of substituents from carbon 3 of GlcNAc are preferentially stabilized by the highly electronegative N-trifluoroactyl group at carbon 2 of GlcNAc (15,29) and provide clear evidence for the relative positions of fucose and galactose (Figure 2).

Table 2. Structures of Oligosaccharides

Name	Structure
lacto-N-tetraose	Galβ1-3GlcNAcβ1-3Galβ1-4Glc
lacto-N-fucopentaose II	Galβ1-3GlcNAcβ1-3Galβ1-4Glc 4 | Fucα1
lacto-N-fucopentaose III	Galβ1-4GlcNAcβ1-3Galβ1-4Glc 3 | Fucα1
LSTa	NeuAcα2-3Galβ1-3GlcNAcβ1-3Galβ1-4Glc
LSTb	Galβ1-3GlcNAcβ1-3Galβ1-4Glc 6 | NeuAcα2
LSTc	NeuAcα2-6Galβ1-4GlcNAcβ1-3Galβ1-4Glc

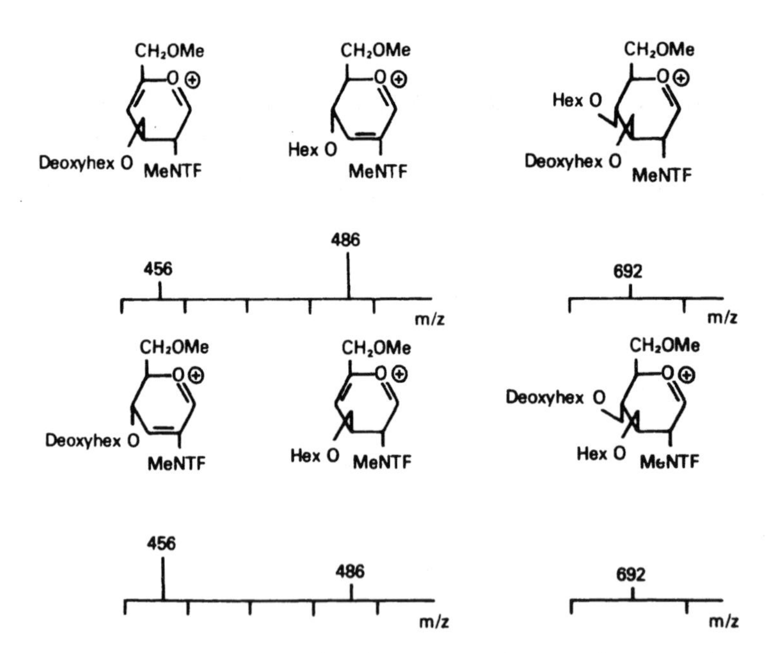

Fig. 2. Elimination of hexose (Hex) and deoxyhexose (Deoxyhex) from m/z 692, an ion derived from homolytic cleavage of hexosaminidic linkage of the permethylated N-trifluoroacetylated alditols of lacto-N-fucopentoase III (top) and lacto-N-fucopentaose II (bottom) (see figure 1). Elimination of hexose from the 3,4-di-O-substituted hexNTF residue gives m/z 456, whereas elimination of deoxyhexose gives m/z 486. Since elimination from position 3 is favored, the positions of hexose and deoxyhexose can be inferred from the relative abundances of the secondary fragments.

Thus, the antigen recognized by antibody 19-9 has the structure,

```
NeuAcα2-3Galβ1-3GlcNAcβ1-3Galβ1-4Glc-Cer (15).
                 4
                 |
              Fucα1
```

The asialo form of this carbohydrate chain is identical to the milk oligosaccharide lacto-N-fucopentaose II, a strong inhibitor of anti-Le^a sera (30). The presence of sialic acid masks Le^a activity and and creates a new specific antigen that requires both the sialyosyl and fucosyl residues for tight antibody binding (15). Thus, the sialylated structure is called the "sialyl-Le^a" (SLe^a) antigen. The SLe^a structure was confirmed in a 19-9-active ganglioside isolated from human pancreatic carcinoma tissue (21). The same carbohydrate structure has been identified and as a free oligosaccharide in human milk (18,19) and carbohydrate chains with the sugar sequence,

```
NeuAcα2-3Galβ1-3GlcNAcβ1-3Galβ1-3GalNAc...
                 4
                 |
              Fucα1
```

have been identified in mucin isolated from human seminal plasma (20).

Biosynthesis of SLe^a starting from the Type 1 precursor chain (Galβ1-3GlcNAcβ1-...) might proceed by either of two alternate pathways:

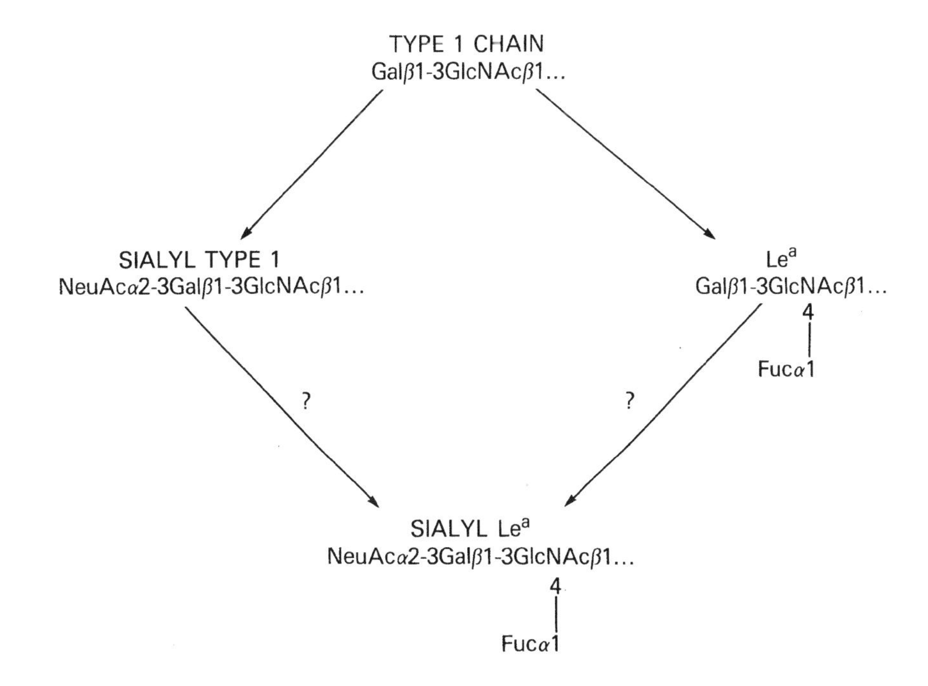

Carbohydrate chains containing the non-reducing trisaccharide sequences present in both putative intermediates are known in milk oligosaccharides (31,32), glycolipids (33-37), and glycoproteins (38-41), and glycosyltransferases that catalyze their synthesis have been purified and characterized (42-47). Thus, the major task in elucidating the biosynthesis of SLe^a has been to determine whether the final step involves addition of sialic acid to the Le^a-active precursor structure or, alternatively, addition of fucose to the sialylated type 1 precursor. Experiments using enzymes contained in microsomal membranes prepared from SW1116 human colorectal carcinoma cells have shown that SLe^a is formed by fucosylation of the sialylated type 1 precursor (11). The final step was demonstrated by enzymatic addition of $[^{14}C]$-fucose to the precursor LSTa, an oligosaccharide purified from human milk (Table 2). The product was isolated by affinity chromatography on a column containing monoclonal antibody 19-9 bound to sepharose beads and was observed to migrate on paper chromatography as a single component, well-behind LSTa (Figure 3). After treatment with neuraminidase the material migrated as a single component with the same mobility as authentic LNF II.

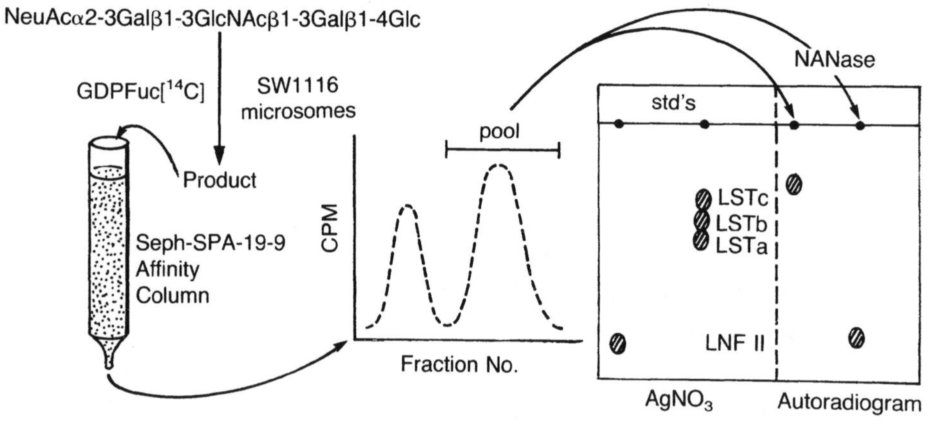

Fig. 3. Schematic depiction of experiments demonstrating biosynthesis of the SLe^a antigen. A microsomal enzyme preparation from SW1116 cells was incubated with $[^{14}C]$-GDPfucose and the milk oligosac-charide precursor LSTa (see table 2). After deionization of the reaction mixture, the product was recovered by affinity chromato-graphy on a column containing antibody 19-9 bound to Staphylococcal protein A-Sepharose. Material in the retarded peak, well separated from free $[^{14}C]$-fucose, was desalted and a portion treated with neuraminidase. On paper chromatography the untreated material migrated just behind LSTc, whereas the treated oligosaccharide comigrated with authentic LNF II.

The microsomal preparation could also convert an H-type 1 glycolipid to the difucosylated Le^b antigen, confirming the presence of the Lewis fucosyltransferase in SW1116 cells (11). The Le^a antigen is widely distributed in normal epithelial organs (48-51) and is a simpler structure than SLe^a, so it is natural to suspect that SLe^a also could be synthesized by sialylation of previously-existing Le^a-active structures (12,20,52). However, when the Le^a-active oligosaccharide LNF II was tested as an acceptor, no product was retarded by the affinity column under conditions that would detect less than 1% of the material formed when fucose was added to LSTa as described above. Sialyltransferase activity capable of converting the type-1 tetrasaccharide lacto-N-tetraose to LSTa was readily demonstrated. Thus, the pathway for biosynthesis of SLe^a in SW1116 cells is as follows:

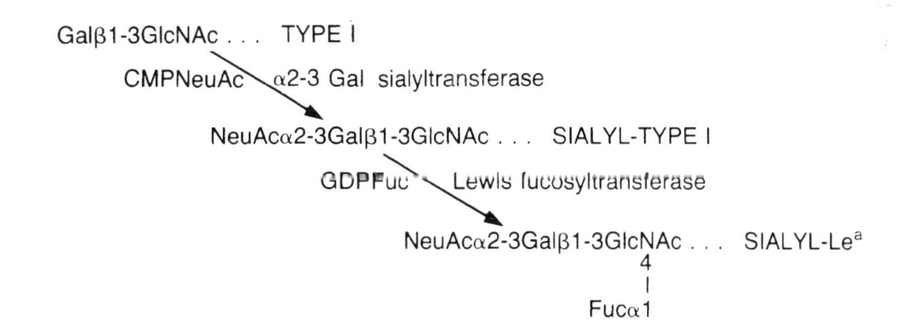

SLe^a AS A MARKER FOR IMMUNODIAGNOSIS

An ideal tumor marker should be absent from blood of individuals without cancer and released by all malignant tumors in detectable amounts at a very early stage of neoplastic proliferation. Like all other known tumor markers, SLe^a does not completely satisfy these criterea. Present in normal mucin-secreting organs of many individuals, SLe^a may be released into blood during inflammatory destruction of glandular tissues, and, in fact, has been proposed as a marker for cystic fibrosis (53), a disease characterized by impaired drainage of glandular mucins and chronic destruction of many mucin-producing organs. Increased levels of SLe^a also can occur in patients with rheumatoid arthritis (54). Similar release of SLe^a-active mucins into blood might accompany obstruction of pancreatic or biliary ducts, phenomena which may be associated with either malignant or benign conditions affecting these organs. As there is potential for false positive detection of SLe^a as a tumor marker in screening populations at risk for gastrointestinal cancer (e.g., patients with ulcerative colitis), the utility of measuring SLe^a as a cancer screening procedure remains to be established.

A more significant potential difficulty in routine interpretation of serum levels of SLe^a during followup evaluaton of patients with known gastrointestinal cancer is the substantial variability in levels of the antigen among patients with clinically similar disease (24,25). Wide quantitative variation in antigen expression among individuals probably is an inherent feature of the SLe^a antigen that can be at least partly explained the biosynthetic relationship of SLe^a to inherited blood group isoantigens.

Biosynthetic Relationships Between SLe^a and A, B, H, Le^a, and Le^b Antigens

Figure 4 summarizes the known pathways of biosynthesis for the A,B,H,Le^a, Le^b, and SLe^a antigens based on the type 1 precursor chain. Expression of SLe^a in normal and malignant cells is limited to individuals who also produce Le^a and/or Le^b antigens (12,16). The genetic and chemical basis for expression of Le^a and Le^b was proposed by Watkins and Morgan (55) and demonstrated by Ginsburg and collaborators (for review see ref. 56). Two independently-inherited Mendelian dominant genes, the Lewis gene (<u>Le</u>) and the secretor gene (<u>Se</u>), and their amorphic alleles, <u>le</u> and <u>se</u>, exist at two separate loci. <u>Le</u> is required for expression of both Le^a and Le^b antigens and codes for a fucosyltransferase that adds fucose to a type 1 precursor chain in α1-4 linkage to GlcNAc to form the Le^a antigen:

```
   Type 1                                      Lea
Galβ1-3GlcNAcβ1...  ------------------>  Galβ1-3GlcNAcβ1...
                     Lewis                          4
                     fucosyltransferase             |
                                                  Fucα1
```

<u>Se</u> is required for expression of A,B, and H blood group antigens in secreted mucins and codes for a fucosyltransferase that adds fucose in α1-2 linkage to terminal galactose on type 1, type 2, and type 3 precursor chains in secretory organs:

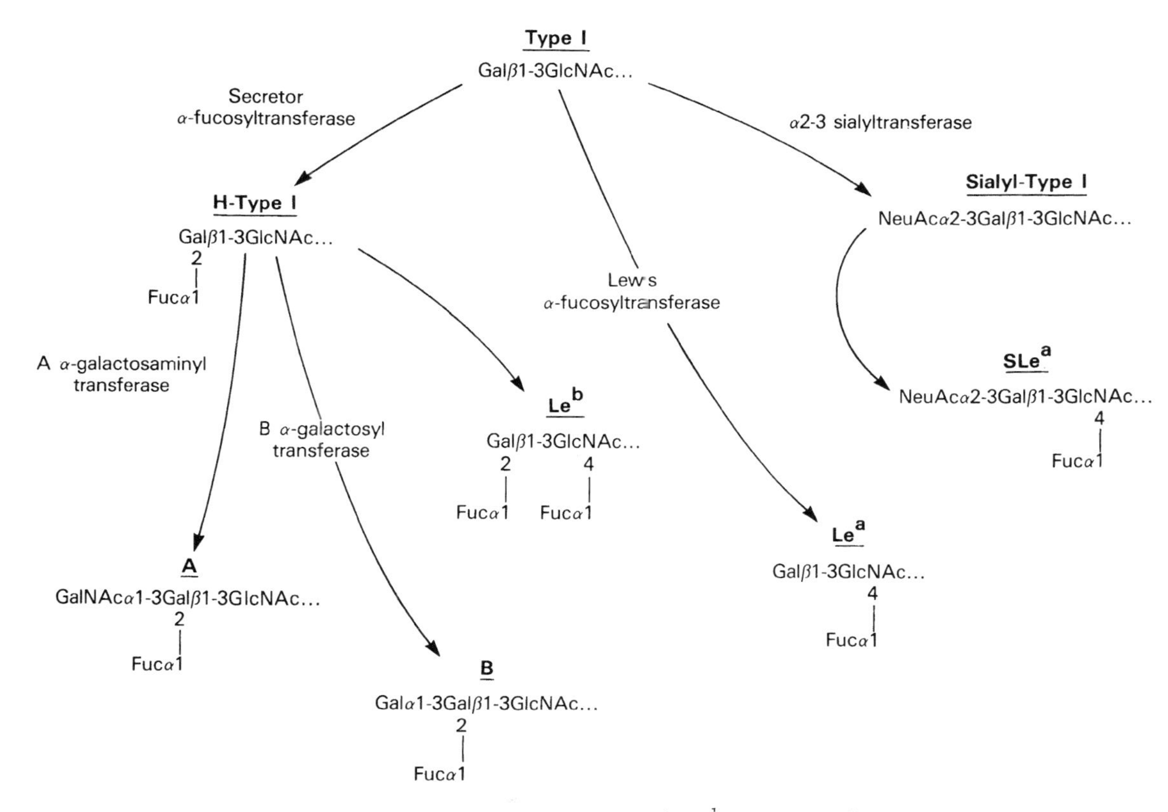

Figure 4. Biosynthesis of A,B,H,Le^a,Le^b, and SLe^a antigens.

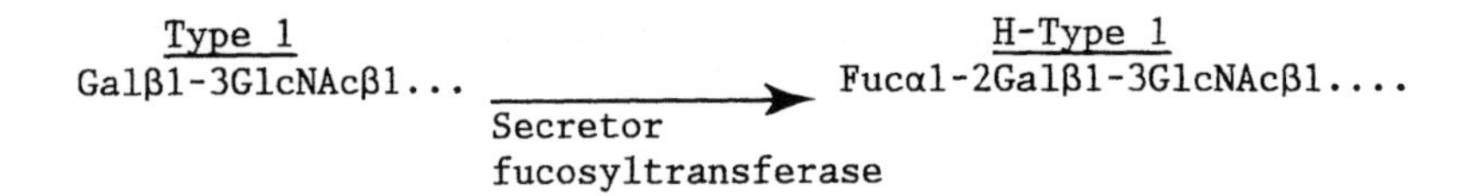

The non-reducing terminal structure, Fucα1-2Gal..., constitutes the H-antigen and is the obligate substrate for glycosyltransferases that form the A and B blood group antigens by addition of GalNAc or Gal, respectively (see Figure 4). When both Le and Se are present in the same individual, H-type 1 structures on growing carbohydrate chains attached to mucins can be modified by the Lewis fucosyltransferase to form the Le^b antigen:

H-Type 1 Fucα1-2Galβ1-3GlcNAcβ1... —(Lewis fucosyltransferase)→ Le^b Fucα1-2Galβ1-3GlcNAcβ1....
(4 | Fucα1)

Addition of fucose to the Le^a antigen to form Le^b cannot be demonstrated experimentally (56). Thus, the Lewis enzyme appears to have an unusually broad specificity for oligosaccharide precursors, including Galβ1-3GlcNAc..., Fucβ1-3Galβ1-3GlcNAc..., and NeuAcα2-3Galβ1-3GlcNAc. In fact, the activity of the Lewis fucosyltransferase may even include addition of fucose α1-3 linked to the type 2 precursor, as Prieels, et al (42) and Johnson, et al (43) observed copurification of α1-3 and α1-4 fucosyltransferase activities.

Possible Mechanisms for Regulation of SLe^a Expression and Appearance of NeuAcα2-3Galβ1-3GlcNAc... as a Tumor Marker

Several possible mechanisms could account for renewed synthesis of SLe^a in transformants of colorectal epithelial cells that normally express Le^a and/or Le^b but normally lack SLe^a during adult life. Increased synthesis or up-regulation of preexisting α2-3 Gal sialyltransferase could lead to increased numbers of chains terminating in NeuAcα2-3Galβ1-3GlcNAc... which can be converted to SLe^a by the Lewis fucosyltransferase. In Le(a-b-) individuals, increased sialyltransferase activity would lead to accumulation of sialylated type 1 chains. A ganglioside with the sialylated type 1 structure has been identified in cultured PA1 human embryonal carcinoma cells (37). If increased sialyltransferase activity is a consistent feature of malignant colorectal epithelial cells, a monoclonal antibody with specificity broad enough to recognize both the sialylated type 1 and SLe^a antigens may be useful in detecting increased sialylated type 1 chain as a general cancer marker for all individuals, regardless of Lewis blood group (36). Alternatively, a mixture of two antibodies such as 19-9 and K4 (37), which is directed against sialylated type 1 chain, might serve the same purpose.

Other mechanisms that might lead to increased expression of SLe^a as a consequence of malignant transformation include, (i) altered biosynthesis of the core regions of type 1 chains to produce new precursors for which sialyltransferases may have unusually high activity, (ii) expression of modifier proteins that may alter sialyl- or fucosyltransferase specificities, or (iii) structural translocation of enzymes in the golgi apparatus, altering the order in which enzymes may gain access to elongating sugar chains.

Influence of ABH Secretor Status on Expression of SLe^a

Decreased expression of SLe^a in Le^b-active mucins relative to Le^a-active mucins can be explained by competition for precursor between the enzymes that initiate the synthesis of these two antigens along mutually exclusive pathways. Ginsburg (56) has pointed out that glycosyltransferases that utilize the same precursor, such as the enzymes that form the A, B, and Le^b antigens, may coexist in one individual and compete for substrate. This competition has been proposed to explain, 1) variations in amounts of A- and B-active glycolipids in persons of different Lewis blood groups (57), 2) differing expression of Le^b-active glycolipids among individuals with blood groups A_1 and A_2 (58), and 3) variations in the level of Le^b-active oligosaccharides excreted in the urine of pregnant and lactating Le(a-b+) women belonging to different ABO blood groups (59). In cells with the secretor phenotype, modification of type 1 precursor chains by addition of fucose linked α1-2 to Gal commits the chain along the pathway to become an H-type 1 antigen or, by further modification, an A, B, or Le^b antigen if the corresponding glycosyltransferases are present in the cell. Likewise, modification of the type 1 precursor by the Lewis fucosyltransferase creates an Le^a-active structure which apparently cannot be sialylated to form SLe^a (11). Thus, the relative increase in SLe^a-activity in mucins from individuals of the Le(a+b-) over the Le(a-b+) blood group (12) probably reflects a relative increase in type 1 precursor chains available for sialylation in non-secretors. Synthesis of sialylated type 1 chains in Le(a-b-) individuals also is likely to vary significantly between secretors and non-secretors, limiting the utility of this antigen as a tumor marker (36). On the basis of known metabolic pathways (Figure 4), ABO blood group is expected to exert little or no metabolic effect on the level of expression of SLe^a antigen, as the enzymes that create the A and B antigens do not directly compete with α3-sialyltransferase for the type 1 precursor.

Sialylated Le^x, a Tumor-associated Antigen Related to SLe^a

An isomer of the SLe^a fucoganglioside, termed "sialyl-Le^x", with the structure,

```
NeuAcα2-3Galβ1-4GlcNAcβ1-3Galβ1-4Glc
                   3
                   |
                 Fucα1
```

was first described as a component of normal human kidney (60). Monoclonal antibodies recognizing SLe^x (54) detect the antigen frequently in sera of patients with adenocarcinoma of stomach (19%), lung (34%), colon (35%), and breast (53%) but rarely in normal sera (3%). The structure has been identified in α_1-acid glycoprotein purified from liver metastases of lung, colon, and breast tumors (61). Biosynthesis of SLe^x proceeds in a manner similar to that of SLe^a, i.e, sialic acid is first transferred to the type 2 blood group precursor chain followed by addition of fucose (62,63):

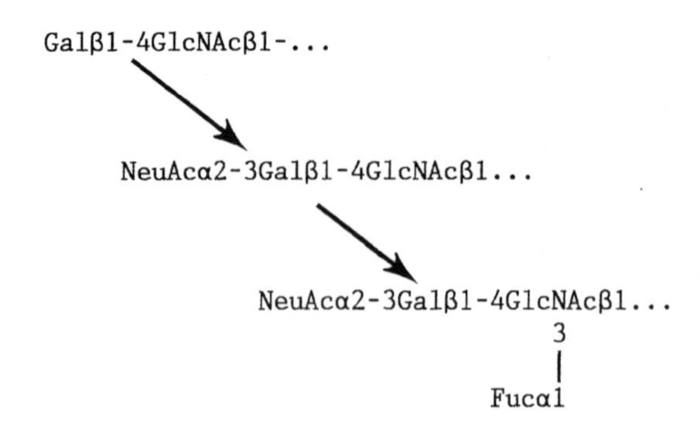

The incidence of this antigen in patients of different Lewis blood groups has not been reported. However, studies by Johnson and Watkins using low molecular weight acceptors suggest that fucosylation may be catalyzed by either the α-3-fucosyltransferase purified from human serum or the α-3/4-fucosyltransferase purified from human milk. Whether the amount of SLe^x expressed in different individuals varies with Lewis blood group and secretor status is a question that deserves further investigation. As the fucosyl- transferase controlled by the secretor gene can act on both type 1 and type 2 precursors, it seems unlikely that ABH secretor status would significantly influence expression of SLe^x.

Future Directions

Large differences in plasma SLe^a observed for patients with clinically similar disease and heritable inability to produce the antigen in Le(a-b-) individuals limit the utility of measuring SLe^a as a tumor marker. Different overall rates of mucin production and many other growth-related variables among gastrointestinal tumors undoubtedly influence the rate of release of mucins into the circulation. These variables are difficult to evaluate in individual tumors and their importance relative to heritable differences in glycosyltransferase levels has not been established. On the other hand, our current understanding of the structure and metabolic origins of the SLe^a antigen, together with previous knowledge of the chemical basis for expression of the A, B, H, Le^a and Le^b antigens, provides a rational framework for interpretation of the variable SLe^a levels observed in normal and malignant tissues of different individuals. The genetic propensity for an individual patient to express SLe^a in normal mucin can be assessed easily by determination of blood group isoantigens in saliva (11). In the future it may be profitable to compare plasma levels of SLe^a with expression of Le^a, Le^b, SLe^a, and the sialylated type 1 determinant in saliva in a large population of patients with different Lewis blood groups and secretor phenotypes. If a correlation can be established between salivary expression of SLe^a and the amount of SLe^a antigen in plasma at a given stage of malignant diesase, the value of this antigen as a predictive or follow-up test for human malignant disease will be greatly enhanced.

Summary

The SLe^a antigen, originally defined by monoclonal antibody 19-9, is a complex carbohydrate epitope that differs from the normal human blood group Le^a antigen only by the presence of an additional sialic acid residue. SLe^a-active oligosaccharides occur in both gangliosides and mucin-like glycoproteins in developing embryonic gut, as well as in many normal adult glandular tissues and secretions, but the antigen is virtually absent from normal adult gastrointestinal lumenal epithelial cells. Following malignant transformation of adult gastrointentinal lining epithelium and many other endodermally-derived glandular epithelia, SLe^a-active mucins released from the ensuing tumor appear in blood plasma. The level of circulating SLe^a antigen is currently being investigated as a means of following tumor recurrence, progression, and therapy. Recent studies on the biosynthesis of SLe^a explain the observations that, 1) the antigen does not occur in individuals of Le(a-b-) blood group , and 2) individuals that belong to the Le(a+b-) blood group express SLe^a more strongly than Le(a-b+) individuals. Further, the biosynthetic studies predict a new tumor antigen, NeuAcα2-3Galβ1-3GlcNAcβ1.... (the immediate precursor to SLe^a) that should be expressed in Le(a-b-) individuals in nearly the same tissue distribution as found for the SLe^a antigen in Le(a+b-) and Le(a-b+) individuals. Based upon studies of SLe^a expression in normal saliva and the pathway for biosynthesis of SLe^a, it seems likely that future clinical studies could be profitably directed towards improving the predictive value of the plasma SLe^a level by adjusting the quantitative results according to the Lewis blood group and ABH secretor phenotype of the individual patient.

REFERENCES

1. I. Damjanov and B. B. Knowles (1983) Biology of Disease: monoclonal antibodies and tumor-associated antigens, Lab. Invest. 48: 510-525.
2. D. J. Kennel, K. Flynn, L. Foote, and T. Lankford (1984) Monoclonal antibodies in cancer detection and therapy, BioScience 34: 150-156.

3. V. Ginsburg, J. L. Magnani, S. L. Spitalnik, P. F. Spitalnik, K. K. Roberts, and C. Dubois (1985) Carbohydrate antigens detected by monoclonal antibodies, Glycoconjugates: Proc. VIIIth Int. Symp. (Abst.), Vol 2, pp 550-551, Praeger, New York.
4. S. Hakomori and R. Kannagi (1983) Glycosphingolipids as tumor-associ ated and differentiation markers, J.N.C.I. 71: 231-251.
5. S. Hakomori (1984) Tumor-associated carbohydrate antigens, Ann. Rev. Immunol. 2: 103-126.
6. T. Feizi (1983) Carbohydrate differentiation antigens, Biochem. Soc. Trans. 11 :263-271.
7. M. Fukuda (1985) Cell surface glycoconjugates as onco-differentiation markers in hematopoietic cells, Biochim. Biophys. Acta 780: 119-150.
8. H. Koprowski, Z. Steplewski, K. Mitchell, M. Herlyn, D. Herlyn, and P. Fuhrer (1979) Colorectal Carcinoma antigens detected by hybridoma antibodies, Somat. Cell Genet. 5: 957-972.
9. B. F. Atkinson, C. S. Ernst, M. Herlyn, Z. Steplewski, H. F. Sears, and J. Koprowski (1982) Gastrointestinal cancer-associated antigen in immunoperoxidase assay, Canc. Res. 42; 4820-4823.
10. J. W. Arends, C. Verstynen, F. T. Bosman, J. Hilgers, and Z. Steplewski (1983) Distribution of monoclonal antibody-defined monosialoganglioside in normal and cancerous human tissues: an immunoperoxidase study, Hybridoma 2: 219-229.
11. G. C. Hansson and D. Zopf (1985) Biosynthesis of the cancer-associated sialyl-Le^{a} antigen, J. Biol. Chem. 260: 9388-9392.
12. M. Brockhaus, M. Wysocka, J. L. Magnani, Z. Steplewski, H. Koprowski, and V. Ginsburg (1985) Normal salivary mucin contains the gastrointestinal cancer-associated antigen detected by monoclonal antibody 19-9 in the serum mucin of patients, Vox Sang. 48: 34-38.
13. H. Koprowski, M. Herlyn, Z. Steplewski, and H. F. Sears (1981) Specific antigen in serum of patients with colon carcinoma, Science 212: 53-55.
14. J. L. Magnani, M. Brockhaus, D. F. Smith, V. Ginsburg, M. Blaszczyk, K. F. Mitchell, Z. Steplewski, and H. Koprowski (1981) A monosialoganglioside is a monoclonal antibody-defined antigen of colon carcinoma, Science 212: 55-56.
15. J. L. Magnani, B. Nilsson, M. Brockhaus, D. Zopf, Z. Steplewski, H. Koprowski, and V. Ginsburg (1982) A monoclonal antibody-defined antigen associated with gastrointestinal cancer is a ganglioside containing sialylated lacto-N-fucopentaose II, J. Biol. Chem. 257: 14365-14369.
16. H. Koprowski, M. Blaszczyk, Z. Steplewski, M. Brockhaus, J. Magnani, and V. Ginsburg (1982) Lewis blood-type may affect the incidence of gastrointestinal cancer, The Lancet, June 12, 1332-1333.
17. J. Picard, D. Loveday, and T. Feizi (1985) Evidence for sialylated type 1 blood group chains on human erythrocyte membranes revealed by agglutination of neuraminidase-treated erythrocytes with Waldenstrom's macroglobulin IgM^{WOO} and hybridoma antibody FC 10.2, Vox Sang. 48: 26-33.
18. J-M. Wieruszeski, A. Chekkor, S. Bouquelet, J. Montreuil, G. Strecker, J. Peter-Katalinic, and H. Egge (1985) Structure of two new oligosaccharides isolated from human milk: sialylated lacto-N-fucopentaoses I and II, Carb. Res. 137: 127-138.
19. D. F. Smith (1985) Sialylpentasaccharides of human milk: identification by specific anti-oligosaccharide antibodies, Fed. Proc (Abst.) 44:1087.

20. F-G. Hanisch, G. Uhlenbruck, and C. Dienst (1985) Structure of tumor-associated carbohydrate antigen Ca 19-9 on human seminal-plasma glycoproteins from healthy donors, Eur. J. Biochem. 144: 467-474.
21. K-E. Falk, K-A. Karlsson, G. Larson, J. Thurin, M. Blaszczyk, Z. Steplewski, and H. Koprowski (1983) Mass spectrometry of a human tumor glycolipid antigen being defined by mouse monoclonal antibody NS-19-9, Biochem. Biophys. Res. Comm. 110: 383-391.
22. J-E. Mansson, P. Fredman, O. Nilsson, L. Lindholm, J. Holmgren, and L. Svennerholm (1985) Chemical structure of carcinoma ganglioside antigens defined by monoclonal antibody C-50 and some allied gangliosides of human pancreatic adenocarcinoma, Biochim. Biophys. Acta 834: 110-117.
23. J. L. Magnani, Z. Steplewski, H. Koprowski, and V. Ginsburg (1983) Identification of the gastrointestinal and pancreatic cancer-associated antigen detected by monoclonal antibody 19-9 in the sera of patients as a mucin, Canc. Res. 43: 5489-5492.
24. B. C. Del Villano, S. Brennan, P. Brock, C. Bucher, V. Liu, M. McClure, B. Rake, S. Space, S. Westrick, H. Schoemaker, and V. R. Zurawski, Jr. (1983) Radioimmunometric assay for a monoclonal antibody-defined tumor marker, CA 19-9, Clin. Chem. 29: 549-552.
25. T. Yoshikawa, K. Nishida, M. Tanigawa, K. Fukumoto, and M. Kondo (1985) Carbohydrate antigenic determinant (CA 19-9) and other tumor markers in gastrointestinal malignancies, Digestion 31: 67-76.
26. S. Hirohashi, Y. Shimosato, Y. Ino, Y. Tome, M. Watanabe, T. Hirota, and M. Itabashi (1984) Distribution of blood group antigens and CA 19-9 in gastric cancers and non-neoplastic gastric mucosa, Gann 75: 540-547.
27 R. R. Race and R. S. Sanger (1975) Blood Groups in Man, Blackwell, Oxford pp323-349.
28. T. Feizi, H. C. Gooi, R. A. Childs, J. K. Picard, K. Uemura, L. M. Loomes, S. J. Thorpe, and E. F. Hounsell (1984) Mucin-type glycoproteins, tumour-associated and differentiation antigens on the carbohydrate moieties of mucin-type glycoproteins, Biochem. Soc. Trans. 12: 591-596.
29. B. Nilsson and D. Zopf (1983) Oligosaccharides released from glycolipids by trifluoroacetolysis can be analyzed by gas chromatography/mass spectrometry, Arch. Biochem. Biophys. 222: 628-648.
30. W. M. Watkins and W. T. J. Morgan (1957) Specific inhibition studies relating to the Lewis blood-group system, Nature 180: 1038-1040.
31. R. Kuhn, H. H. Baer, and A. Gauhe (1958) Die Konstitution der Lacto-N-fucopentaose II. Ein Beitrag zur Spezifität der Blutgruppensubstanz Le^a, Chem. Ber. 91: 364.
32. R. Kuhn and A. Gauhe (1965) Bestimmung der Bindungsstelle von Sialinsäureresten in Oligosacchariden mit Hilfe von Perjodat, Chem. Ber. 98: 365.
33. M. Blaszczyk, G. C. Hansson, K.-A. Karlsson, G. Larson, N. Stromberg, J. Thurin, M. Herlyn. Z. Steplewski, and H. Koprowski (1984) Lewis blood group antigens defined by monoclonal anti-colon carcinoma antibodies, Arch. Biochem. Biophys. 233: 161-168.
34 S. Hakomori and H. D. Andrews (1970) Sphingoglycolipids with Le^b activity and the copresence of Le^a-, Le^b-active glycolipids in human tumor tissue, Biochim. Biophys. Acta 202: 225-228.
35. E. L. Smith, J. M. McKibbin, K.-A. Karlsson, I. Pascher, B. E. Samuelsson, Y.-T. Li, and S.-C. Li (1975) Characterization of a human intestinal fucolipid with blood group Le^a activity, J. Biol. Chem. 250: 6059-6064.

36. O. Nilsson, J. E. Mansson, L. Lindholm, J. Holmgren, and L. Svennerholm (1985) Sialosyllactotetraosylceramide, a novel ganglioside antigen detected in human carcinomas by a monoclonal antibody, FEBS Lett. 182: 398-402.
37. M. N. Fukuda, B. Bothner, K. O. Lloyd, W. J. Rettig, P. R. Tiller, and A. Dell (1986) Structures of glycosphingolipids isolated from human embryonal carcinoma cells; the presence of mono- and di- sialosyl glycolipids with blood group type 1 sequence, J. Biol. Chem. (in press).
38. W. T. J. Morgan (1960) A contribution to human biochemical genetics; the chemical basis of blood group specificity, Proc. Royal Soc. B 151: 308-347.
39. E. A. Kabat (1973) Immunochemical studies on the carbohydrate moiety of water soluble blood group A, B, H, Le^a, and Le^b substances and their precursor I antigens, in: Carbohydrates in Solution (H. Isbell, ed.) Advances in Chemistry Series No. 117. pp 334-361, American Chemical Society, Washington, D. C.
40. W. M. Watkins (1974) Genetic regulation of the structure of blood group specific glycoproteins, Biochem. Soc. Symp. 40: 125-146.
41. T. Mizuochi, K. Yamashita, K. Fujikawa, W. Kisiel, and A. Kobata (1979) The carbohydrate of bovine prothrombin. Occurrence of Galβ1-3GlcNAc grouping in asparagine-linked sugar chains J. Biol. Chem. 254: 6419-6425.
42. J.-P. Prieels, K. Monnom, M. Dolmans, T. A. Beyer, and R. L. Hill (1981) Co-purification fo the Lewis blood group N-acetylglucosaminide α1-4fucosyltransferase and an N-acetylglucosaminide α1-3fucosyl-transferase from human milk, J. Biol. Chem. 256: 10456-10463.
43. P. H. Johnson, A. D. Yates, and W. M. Watkins (1981) Human salivary fucosyltransferases: evidence for two distinct α-3-L-fucosyltransferase activities one of which is associated with the Lewis blood group Le gene, Biochem. Biophys. Res. Commun. 100: 1611-1618.
44. J. E. Sadler, J. I. Rearick, J. C. Paulson, and R. L. Hill (1979) Purification to homogeneity of a β-galactoside α2-3 sialyltransferase and partial purification of an α-N-acetylgalactosaminide α2-3 sialyltransferase from porcine submaxillary glands, J. Biol. Chem. 254: 4434-4443.
45. J. I. Rearick, J. E. Sadler, J. C. Paulson, and R. L. Hill (1979) Enzymatic characterization of β-D-galactoside α2-3 sialyltransferase from porcine submaxillary gland, J. Biol. Chem. 254: 4444-4451.
46. J. Weinstein, U. de Souza-e-Silva, and J. C. Paulson (1982) Purification of a Galβ1-4GlcNAc α2-6 sialyltransferase and a Galβ1-3GlcNAc α2-3 sialyltransferase to homogeneity from rat liver. J. Biol. Chem. 257: 13835-13844.
47. J. Weinstein, U. de Souza-e-Silva, and J. C. Paulson (1982) Sialylation of glycoprotein oligosaccharides N-linked to asparagine, enzymatic characterization of a Galβ1-3GlcNAc α2-3 sialyltransferase and a Galβ1-4GlcNAc α2-6 sialyltransferase from rat liver, J. Biol. Chem. 257: 13845-13851.
48. A. E. Szulman and D. M. Marcus (1973) The histologic distribution of the blood group substances in man as disclosed by immunofluorescence VI. The Le^a and Le^b antigens during fetal development, Lab. Invest. 28: 565-574.
49. R. Oriol, J. P. Cartron, J. Cartron, C. Mulet (1980) Biosynthesis of ABH and Lewis antigens in normal and transplanted kidney, Transplantation 29: 184-188.

50. R. U. Lemieux, D. A. Baker, W. M. Weinstein, C. M. Switzer (1981) Artificial antigens: antibody preparations for the localization of Lewis determinants in tissues, Biochemistry 20: 199-205.
51. C. Ernst, B. Atkinson, M. Wysocka, M. Blaszczyk, M. Herlyn, H. Sears, Z. Steplewski, and H. Koprowski (1984) Monoclonal antibody localization of Lewis antigens in fixed tissue, Lab. Invest. 50: 394-400.
52. H. Clausen, S. B. Levery, J. M. McKibbin, and S. Hakomori (1985) Blood group A determinants with mono- and difucosyl type 1 chain in human erythrocyte membranes, Biochemistry 24: 3578-3586.
53. D. R. Roberts, D. L. Monsein, R. C. Frates, Jr., M. S. Chernick, and V. Ginsburg (1986) A serum test for cystic fibrosis using monoclonal antibody 19-9, Arch. Biochem. Biophys. 245: 292-294.
54. D. Chia, P. I. Terasaki, N. Suyama, J. Galton, M. Hirota, and D. Datz (1985) Use of monoclonal antibodies to sialylated Lewis^x and sialylated Lewis^a for serological tests of cancer, Cancer Res. 45: 435-437.
55. W. M. Watkins and W. T. J. Morgan (1959) Possible genetical pathways for the biosynthesis of blood group mucopolysaccharides, Vox. Sang. 4: 97-119.
56. V. Ginsburg (1972) Enzymatic basis for blood groups in man, Adv. Enzymol. 36: 131-149.
57. C. A. Tilley, M. C. Crookston, B. L. Brown, and J. R. Wherrett (1975) A and B and A_1Le^b substances in glycosphinglipid fractions of human serum, Vox Sang. 28: 25-28.
58. D. A. Zopf, V. Ginsburg, P. Hallgren, A.-C. Jonsson, B. S. Lindberg, and A. Lundblad (1979) Determination of Le^b-active oligosaccharides in urine of pregnant and lactating women by radioimmunoassay, Eur. J. Biochem. 93: 431-435.
59. T. E. Rohr, D. F. Smith, D. A. Zopf, and V. Ginsburg (1980) Le^b-active glycolipid in human plasma: measurement by radioimmunoassay, Arch. Biochem. Biophys 199: 265-269.
60. H. Rauvala (1976) Gangliosides of human Kidney, J. Biol. Chem. 251: 7517-7520.
61. E. V. Chandrasekaran, M. Davila, D. Nixon, and J. Mendicino (1984) Structures of the oligosaccharide chains of two forms of α_1-acid glycoprotein purified from liver metastases of lung, colon, and breast tumors, Cancer Res. 44: 1557-1567.
62. P. H. Johnson and W. M. Watkins (1985) Sialyl compounds as acceptor substrates for the human α3- and α3/4-L-fucosyltransferases, Biochem. Soc. Trans. 13: 1119-1120.
63. E. H. Holmes, G. K. Ostrander, and S. Hakomori (1986) Biosyntheisis of the sialyl-Le^x determiant carried by type 2 chain glycosphingolipids ($IV^3NeuAcIII^3FucnLc_4$, $VI^3NeuAcV^3FucnLc_6$, and $VI^3NeuAcIII^3V^3Fuc_2nLc_6$) in human lung carcinoma PC9 cells, J. Biol. Chem. 261: 3737-3743.

GLYCOCONJUGATES AND TUMOR METASTASIS

Tatsuro Irimura, Motowo Nakajima, Takao Yamori, David M. Ota*, Karen F. Cleary**
and Garth L. Nicolson

Departments of Tumor Biology, Surgery and Pathology***
The University of Texas M.D. Anderson Hospital and Tumor Institute
Houston, Texas USA

Metastasis is one of the most dangerous characteristics of malignant tumors, and it seems to be a biological property separate from rapid and uncontrolled growth. The clinical significance of metastasis as opposed to tumor growth was suggested when we studied the records of 391 colorectal cancer patients surgically treated at The University of Texas M. D. Anderson Hospital and Tumor Institute between 1959 and 1975 to see if there was any correlation between the size of a primary tumor and the incidence of local or distant metastasis. The mean size of primary tumors from patients having regional or distant metastasis was significantly smaller than the mean size of primary tumors from patients without metastasis (Miller et al., 1985), and the five-year survival rate of patients having metastasis at the time of surgery was much poorer than that of patients having no metastasis (Miller et al., 1985). These results indicated that metastatic capacity, not rapid proliferation, is the major cause of death by colorectal cancer. This hypothesis is presumably true for many other malignant tumors. The other important conclusion drawn from this study was that the tumor cells of each colorectal carcinoma may be unique in their metastatic potential. In addition, intratumoral heterogeneity may exist. These findings are consistent with the biological nature of metastatic tumors as determined in experimental tumor systems (Nicolson and Poste, 1982, 1983).

Successful metastasis requires a series of complex interactions between the tumor cells and their host environments (Nicolson, 1984). Cell-surface and extracellular molecules produced by malignant cells, as well as by host cells, play important roles in regulating these interactions. Glycoproteins, glycolipids, proteoglycans and carbohydrate-reactive substances such as glycosidases and lectins are cell surface and extracellular matrix molecules that have been implicated in various steps of the metastatic cascade, such as tumor cell dissemination from the primary site, escape from specific or nonspecific host immune defense mechanisms in the circulation, and implantation and colonization at specific distant organ sites. It has become more and more evident that highly metastatic tumor cells simultaneously express a wide variety of cellular properties that are necessary for their metastatic potential. Therefore, it must be emphasized that no single molecular difference is enough to explain the metastatic capacity of a particular tumor cell.

Data on the properties required for successful metastasis are mainly derived from studies on experimental animal tumor models. For example, mouse melanoma sublines have been used by many investigators as an experimental model for metastatic organ colonization. Fidler selected a variant cell line (B16-F10) from mouse B16 melanoma that is highly metastatic to the lung (Fidler, 1973). Subsequently, brain- (B16-B10b and B16-B15b) and ovary-colonizing (B16-O10) sublines were established (Brunson et al., 1978; Miner et al., 1982; Brunson and Nicolson, 1979). In the B16 melanoma system metastasis did not result from random survival of cells released from the primary tumor but did result from the selective growth of highly metastatic subpopulations of cells (Fidler and Kripke, 1977).

Various cellular properties are specifically associated with the highly metastatic B16 melanoma sublines. These include: expression and topographical distribution of cell-surface molecules (Brunson et al., 1978; Nicolson et al., 1978; Raz et al., 1980a,b; Miner et al., 1982; Reiber and Reiber, 1981; Yogeeswaran et al., 1978; Irimura and Nicolson, 1984), metabolism of intracellular mediators such as prostaglandins and cyclic nucleotides (Stringfellow and Fitzpatrick, 1979; Sheppard et al., 1984), and production and secretion of enzymes, and particularly degradative enzymes (Dobrossy et al., 1981; Liotta et al., 1980; Sloane et al., 1981; Nakajima et al., 1983; Wang et al., 1980). However, the exact function of these molecules in the metastatic process has not been elucidated.

One approach to determining the function of various properties in the metastatic process is to select so-called phenotypic variant cells. For example, a wheat germ agglutinin (WGA)-resistant B16 melanoma subcell line was found to be less metastatic than its poorly metastatic parental line and this variant subline possessed distinct cell surface changes in particular glycoconjugates (Finne et al., 1980). This result suggested that the presence of certain cell surface glycoconjugates is a requirement for the metastatic phenotype of tumor cells.

Similar studies have been conducted with other animal tumor models of metastasis. For example, the RAW117 lymphoma is a large cell lymphoma of pre-B cell lineage that was originally generated by viral transformation of BALB/c mouse splenocytes (Raschke et al., 1975). These tumor cells which form rare solid tumor nodules in the liver, lung, and spleen were subjected to sequential *in vivo* selection for enhanced liver colonization (Brunson and Nicolson, 1978). Biochemical and immunochemical analysis of such selected sublines revealed decreased expression of RNA tumor virus envelope glycoprotein gp70 in the cells of high liver-colonization potential (Reading et al., 1980a,b). This glycoprotein antigen was identified by its concanavalin A (ConA)-binding and could serve as a recognition site for host macrophage surveillance (Miner and Nicolson, 1983; Reading et al., 1983). Other properties of RAW117 cells, such as their net negative surface charge (Miner et al., 1981), and their expression of a oncofetal antigen that cross-reacts with fetal liver cells (Nicolson et al., 1982).

Based on the findings above, some fundamental questions concerning the role of glycoconjugates as determinants of metastasis can be summarized as follows:

(a) What are the biological functions of the molecules that are differentially expressed between poorly and highly metastatic tumor cells?

(b) Is it possible to block tumor metastasis by modulating the structure, display, or metabolism of these molecules?

(c) Do genetic or cellular mechanisms control the simultaneous expression of a variety of metastasis-related cellular phenotypes within a cell population?

(d) Can metastasis-associated cellular properties be used as markers for the diagnosis and prognosis of metastasis in humans?

In this contribution we will summarize and discuss some of our recent efforts to answer these questions using animal tumor models and clinical specimens.

DIFFERENTIAL EXPRESSION OF CELL SURFACE SIALOGLYCOPROTEINS OF B16 MELANOMA VARIANTS

Early attempts to discover differences in the cell-surface components of highly metastatic B16 melanoma cells were relatively unsuccessful. For example, Raz et al. (1980b) compared low and high lung-colonizing B16 melanomas, but they only found differences in the degree ^{3}H incorporation into a sialoglycoprotein component following periodate oxidation and reduction with ^{3}H-sodium borohydride. They proposed that negatively charged groups on highly metastatic B16-F10 cells had a more clustered distribution (Raz et al., 1980a). Other investigators have reported that vectorially radioiodinated cell-surface components of B16-F10 cells migrate differently upon $NaDodSO_4$ polyacrylamide gel electrophoresis (Reiber and Reiber, 1981). Differential expression of a cell-surface component capable of being radiolabeled with iodine was shown on brain-colonizing variants of B16 melanoma (Brunson et al., 1978; Miner et al., 1982). This component was expressed at higher levels on brain colonizing melanoma cell lines and clones, and appeared to be a glycoprotein of $M_r \sim 90{,}000$. This component has now been isolated, and it has properties similar or identical to the transferrin receptor (Nicolson et al., 1985).

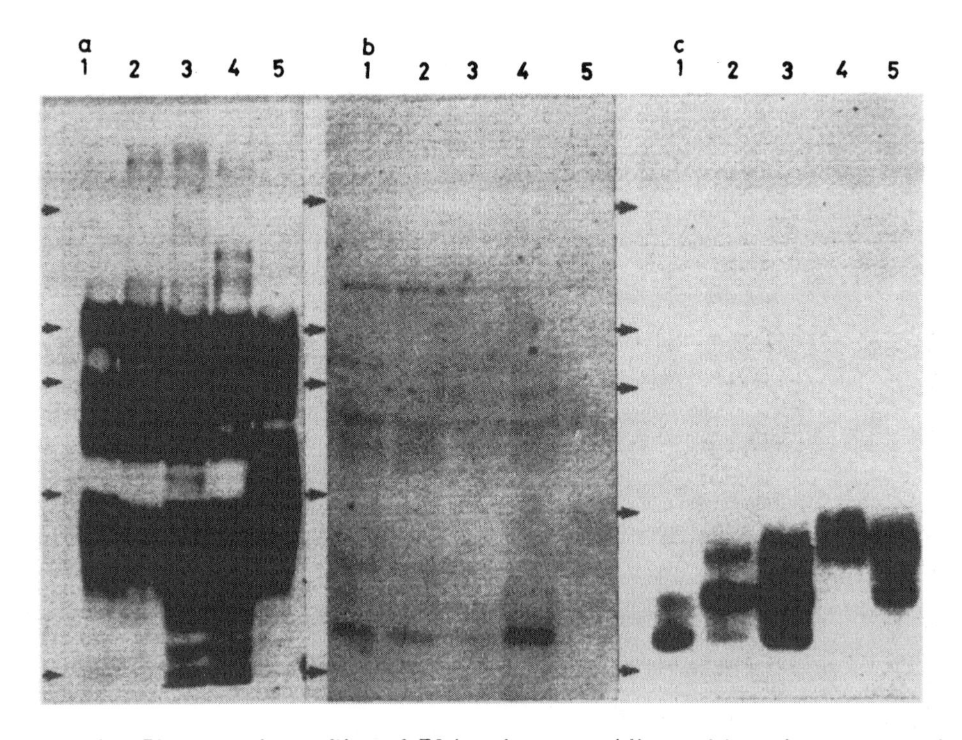

Figure 1. Glycoprotein profiles of B16 melanoma sublines with various metastatic potentials revealed by the binding of lectins after electrophoretic separation in the presence of $NaDodSO_4$. Panel a, stained with ^{125}I-WGA; panel b, stained with ^{125}I-LCA; panel c, stained with ^{125}I-PNA after the sialic acid in the glycoproteins was removed by mild acid hydrolysis *in situ*. Lane 1, a lysate from B16-F1 melanoma cells (selected once for lung colonization); lane 2, a lysate from B16-F10 melanoma cells (selected ten times for lung colonization); lane 3, a lysate from B16-BL6 melanoma cells (selected *in vivo* and *in vitro* for invasion of the basement membrane); lane 4, a lysate from B16-O13 melanoma cells (selected 13 times for ovary colonization); and lane 5, a lysate from B16-B15b melanoma cells (selected 15 times for brain colonization). Arrows indicate the positions of M_r markers from the top to the bottom myosin ($M_r \sim 200{,}000$), β-galactosidase ($M_r \sim 116{,}000$), phosphorylase b ($M_r \sim 93{,}000$), serum albumin ($M_r \sim 66{,}000$), and ovalbumin ($M_r \sim 43{,}000$).

Using lectins as probes differences in the cell surfaces of B16 melanoma sublines with different organ-colonization potentials have been revealed (Nicolson and Irimura, 1984; Irimura and Nicolson, 1984a). In these studies we estimated the class of carbohydrate chains in the glycoproteins of lung-, ovary-, and brain-colonizing B16 melanoma variants by direct binding of lectins to the glycoproteins separated by polyacrylamide gel electrophoresis in combination with in situ chemical modifications of the carbohydrate chains (Irimura and Nicolson, 1983a). The results indicated that the majority of glycoprotein components on the B16 sublines selected in vivo were similar. However, a few differences were revealed by the binding of Lens culinaris agglutinin (LCA), Arachis hypogaea (peanut) agglutinin (PNA) and WGA (Figure 1). A WGA-reactive sialoglycoprotein with an M_r of $\sim$70,000 was seen predominantly with brain-colonizing B16-B15b, whereas the expression of PNA-reactive (after removal of sialic acid) sialoglycoproteins of M_r 56,000 and 61,000 were more prominent on all metastatic sublines except B16-F1. These results indicated that lectins can reveal subtle structural differences in cell-surface glycoconjugates among tumor cells having different metastatic abilities. Because the melanoma cell lines used in this study were selected in vivo for their abilities to colonize certain organs, the differences found are probably related to the specific organ-colonizing capacities of the B16 cells.

FUNCTIONAL ROLE OF B16 MELAMOMA CELL SURFACE SIALOGLYCOPROTEINS IN METASTASIS

The functional significance of a class of sialoglycoproteins in blood-borne melanoma lung colonization has been determined by the use of tunicamycin as an inhibitor of glycoprotein formation (Irimura et al., 1981; Irimura and Nicolson, 1981a). After incubation with a nontoxic dose of tunicamycin in vitro, B16 melanoma cells were altered in their morphologies. Studies of Ricinus communis agglutinin-I (RCA_I) binding to the cells, as well as to the glycoproteins separated by polyacrylamide gel electrophoresis, indicated that cell-surface glycoproteins were deglycosylated or disappeared altogether. The cells that were treated lost their abilities to adhere to an endothelial cell monolayer and to endothelial extracellular matrix (Irimura et al., 1981; Irimura and Nicolson, 1981a). The tunicamycin-modified cells formed very few experimental lung tumor colonies after i.v. injection (Table 1) (Irimura et al., 1981; Irimura and Nicolson, 1981a). These results suggested that melanoma cell surface sialoglycoproteins play significant roles in blood-borne lung colonization. However, sialic acid residues of the glycoproteins did not seem to be involved in this process because the sialidase treatment of B16 melanoma cells did not affect their adhesion or lung colonization (Figure 2 and Table 2).

Table 1. Lung and Extrapulmonary Tumors Induced by Injection of B16 Melanoma Sublines

Cell line	Treatment[a]	Number of lung tumor colonies/animal	Median (range)	Number of animals with extrapulmonary tumor	Location and number of extrapulmonary tumor
B16-F1	-	0, 0, 0, 0, 0, 1, 2, 3, 4, 5, 10, 13, 28, 32, 35, 40	4 (0-40)	2/16	subcutaneous 2
	+	0, 0, 0, 0, 0, 0, 0, 0, 0, 0, 0, 0, 1, 3, 5	0 (0-5)	1/15	ovary 1
B16-F10	-	19, 27, 28, 32, 35, 44, 48, 57, 62, 66, 83, 91, 127	48 (19-127)	6/13	ovary 5 subcutaneous 1 mesentery 1
	+	0, 0, 0, 0, 0, 0, 0, 0, 1, 1, 1, 1, 1, 2, 2	0 (0-2)	2/15	ovary 2 mesentery 1

[a]Tunicamycin 0.5 µg/ml for 36 hr.

[b]Twenty days after injection of 4×10^4/0.1 ml B16 melanoma cells intravenously.

Additional evidence also supported the concept that cell-surface molecules play important roles in B16 melanoma blood-borne organ colonization (Figure 3). Fidler found that prolonged trypsin treatment of B16 melanoma cells reduced blood-borne implantation and lung colonization (Fidler et al., 1978), and Poste and Nicolson (1980) demonstrated that modification of cell surfaces of poorly metastatic B16-F1 cells by fusing them with spontaneously shed membrane vesicles from highly

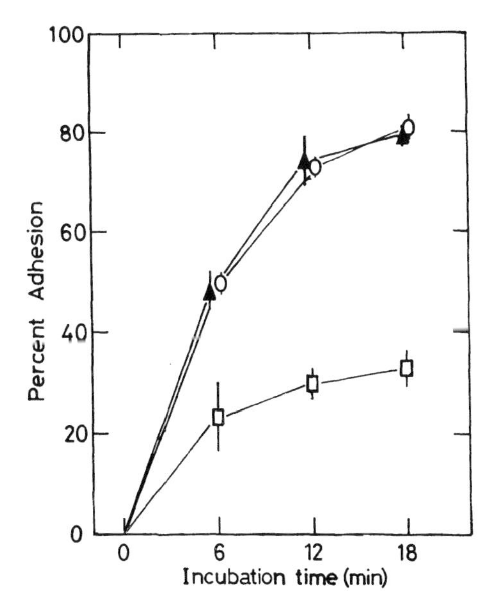

Figure 2. Effects of tunicamycin or neuraminidase treatment on the rate adhesion of B16-F10 melanoma cells to endothelial extracellular matrix. The endothelial matrix was isolated from a confluent monolayer of bovine aortic endothelial cells as previously described (Kramer et al., 1982). The B16 melanoma cells were pretreated with tunicamycin (0.5 μg/ml) for 36 hrs, or with neuraminidase (20 m unit/ml) for 1 hr and were labeled with Na $^{51}CrO_4$ 2 hr before the harvest. Adhesion assays were performed in Dulbecco's minimum essential medium containing 1% bovine serum albumin and 18 mM HEPES at 37°C. (O) Intact B16-F10 cells; (▲) neuraminidase-treated B16-F10 cells; (□) tunicamycin-treated B16-F10 cells.

metastatic B16-F10 cells elevated the metastatic potential. LeGrue (1982) found that certain cell membrane molecules could be selectively removed with a mild aqueous butanol extraction without affecting cell viablility, and the modified B16 melanoma cells expressed higher metastatic capabilities. These results clearly demonstrated the importance of cell-surface molecules in tumor metastasis.

Table 2. Effect of Neuraminidase Treatment on Lung Colonization of B16 Melanoma Sublines

Subline	Treatment[a]	Number of lung tumor colonies/animals[b]	Median (range)	Number of animals with extrapulmonary tumors	Location and number of extrapulmonary tumors
B16-F1	-	2, 2, 6, 9, 9, 12, 20, 22, 42, 44, 73	12 (2-73)	1/12	Ovary 1
B16-F1	+	5, 6, 11, 12, 19, 37, 50, 80, 104, 125, 125	37 (5-125)	0/12	-
B16-F10	-	95, 103, 112, 125, 132, 150 185, 190, 200, 250+, 250+	150 (95-250+)	6/12	mesentery 3 peritoneal cavity 5 ovary 4 lymph node 3
B16-F10	+	50, 114, 133, 135, 136, 146 156, 165, 185, 190, 250+	146 (50-250+)	4/12	mesentery 2 peritoneal cavity 2 ovary 3 lymph node 2

[a] *Vibrio cholerae* neuraminidase: 20 mU/ml, 37°C, 1 hr.

[b] Groups of C57BL/6 mice were injected i.v. with 10^5 B16 cells in 0.2 ml; lung tumor colonies were determined after 15 days.

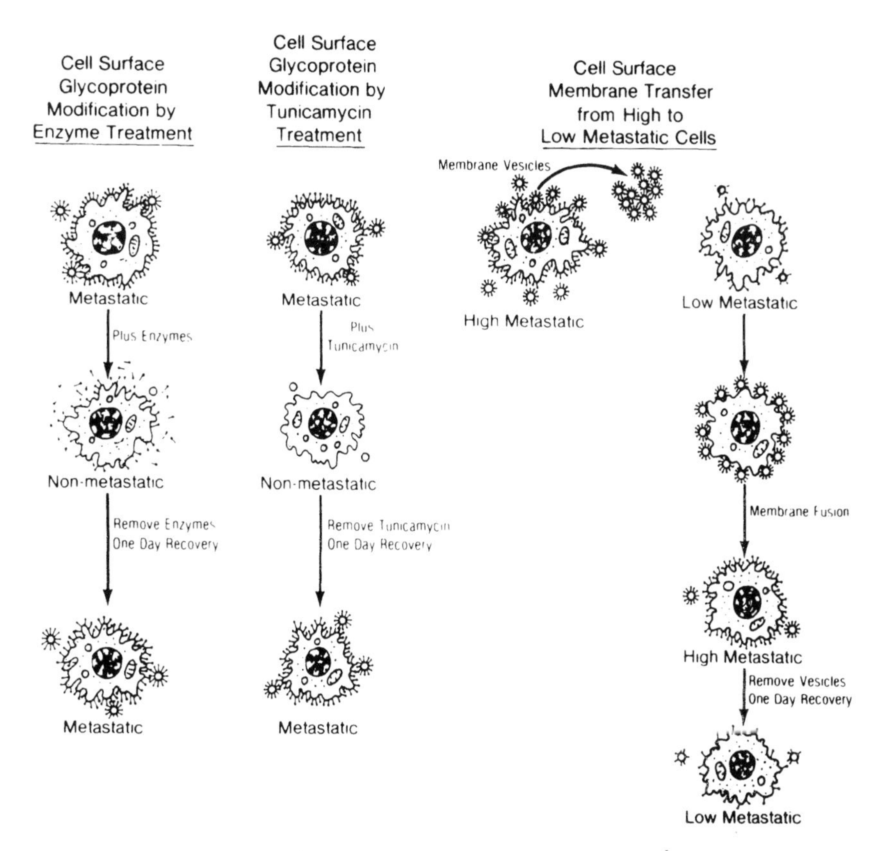

Figure 3. Three lines of evidence indicating that the cell surface has a role in blood-borne lung colonization of B16 melanoma cells. Left panel demonstrates that enzymatic modification of the cell-surface component by trypsin can change implantation properties. Center panel indicates that selective biosynthetic modifications of cell surface glycoprotein can modulate metastatic processes. Right panel shows that the fusion of plasma membrane-derived vesicles from highly metastatic tumor cells with poorly metastatic cells can increase the lung colonization potential of the latter (from Nicolson and Poste, 1982).

In order to determine which of the sialoglycoproteins were involved in adhesive interactions between B16 melanoma cells and the subendothelial matrix, we attempted to analyze adhesive molecules on these cells. The B16 melanoma surface sialoglycoproteins were first labeled by the periodate ^{3}H-sodium borohydride method. Solubilized sialoglycoproteins in NP40 were freed from other cellular components by affinity chromatography on WGA-agarose. The labeled sialoglycoprotein fractions from B16-F10 were incubated at 37°C with the isolated endothelial extracellular matrix (produced by bovine-aortic endothelial cells in tissue culture). A very small portion (approximately 1%) of the added radioactivity was found associated with the matrix. The matrix interaction was time and temperature dependent, and the dissociation constant was estimated as $\sim 10^{-7}$ M. The endothelial matrix together with bound tumor cell surface components was dissolved by heating at 100°C in sample buffer for $NaDodSO_4$-electrophoresis that contained 2%

$NaDodSO_4$ and 1% 2-mercaptoethanol. After conducting electrophoresis in 7.0% gels, the gels were cut into 2-mm pieces, dissolved in NCS (Amersham) and counted in a liquid scintillation counter. As shown in Figure 4, a sialoglycoprotein component with an M_r of $\backsim$67,000 was found associated with the subendothelial matrix (Irimura and Nicolson, 1981b). Terranova et al. later claimed that metastatic tumor cells utilize laminin for their adhesion. Liotta et al. (1984) have reported the isolation of laminin receptor molecules from mouse PMT-2 carcinoma, and they were glycoproteins with an M_r of $\backsim$67,000. Thus, one may speculate that tunicamycin somehow had an influence on the biosynthesis or recycling of laminin receptors on B16-melanoma cells.

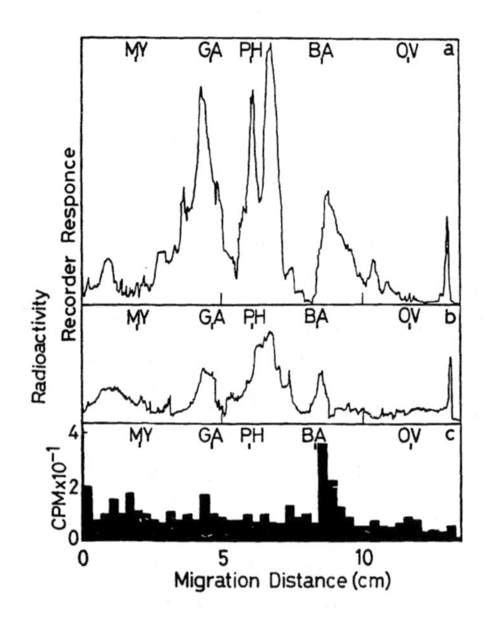

Figure 4. Association of WGA-reactive sialoglycoproteins of B16 melanoma cells with endothelial extracellular matrix. (a) Densitometric profile of WGA-binding glycoproteins of B16-F10 (revealed by the direct binding of ^{125}I-WGA) after electrophoresis on 7% polyacrylamide gels in the presence of $NaDodSO_4$. (b) Densitometric scan of a fluorogram of purified sialoglycoproteins previously labeled by periodate oxidation and reduction with ^{3}H-sodium borohydride. The surface-labeled cells were lysed in 0.5% NP-40 and chromatographed on WGA-agarose in the presence of 0.1% NP-40 in DPBS. The bound glycoprotein was eluted with 0.2 M N-acetylglucosamine. (c) Labeled sialoglycoproteins were layed over an endothelial extracellular matrix and incubated at 37°C for 1 hr. After rinsing in DPBS and 5 mM Tris-HCl, 0.25 M sucrose, 50 µM $CaCl_2$, and 10 µM phenylmethyl sulfonyl fluoride (pH 7.5), the matrix together with bound sialoglycoproteins were dissolved in 2% $NaDodSO_4$, 1% 2-mercaptoethanol, 0.5 mM EDTA, 10% glycerin, and 65.2 mM Tris-HCl buffer (pH 6.8) by heating at 50 - 60°C for 10 min. The lysates were separated by polyacrylamide gel electrophoresis. The gels were stained with Coomassie brilliant blue and cut into 3 mm pieces. The gel pieces were then dissolved in NCS and counted with a liquid scintillation cocktail.

Using the MDAY carcinoma Dennis et al. (1962) suggested that sialylation cell surface differences affect the adhesion of these cells to collagen and fibronectin. They found structural differences in carbohydrate chains between a nonmetastatic WGA-resistant variant line and a related metastatic line derived from MDAY cells (Dennis et al., 1984). The metastatic cells possessed a sialylated poly(N-acetyllactosamine)-type carbohydrate chain, whereas nonmetastatic cells only produced neutral short-chain oligosaccharides lacking sialic acid and galactose, despite the fact that nonmetastatic cells have a similar level of galactosyltransferase activity (Dennis et al., 1984). We have examined cells for the presence of poly(N-acetyllactosamine)-type carbohydrate chains by means of the pokeweed mitogen lectin that is specific for this type of carbohydrate structure (Irimura and Nicolson, 1983b). We found that the B16-melanoma and RAW117 lymphoma cell lines with various metastatic potentials did not express branched poly(N-acetyllactosamine)-type carbohydrate chains (Irimura and Nicolson, 1984b). Although one of the isolectins of pokeweed mitogen (Pa-4) labeled several glycoprotein components of rat 13762NF mammary adenocarcinoma cell clones there was no correlation between Pa-4 reactivity and metastatic potential.

Cell surface carbohydrates have been linked to diverse biological mechanisms in tumor cells. As described above, Finne et al. (1980) proposed that there are highly fucosylated asparagine-linked carbohydrate chains in poorly metastatic WGA-resistant B16 melanoma variants. By using these cells, Sheppard et al. (1983) suggested that modulation of cellular hormonal response can occur through structural changes in cell-surface carbohydrate chains. They are also involved in cell-cell interactions. For example, sialoglycoproteins may also be involved in metastasis by their ability to induce platelet aggregation as suggested by Pearlstein et al. (1980). Yogeeswaran and Salk (1981) attempted to associate metastasis with sialic acid by simply analyzing the amount of membrane sialic acid on B16 melanoma and other tissue culture cells. They concluded that the sialic acid content was positively correlated with metastatic potential. However, tumor metastasis is a complicated multistep process, and such a generalized argument seems meaningless, unless the concept serves as useful background information in the search for metastasis-associated markers.

SIALOGLYCOPROTEINS OF HUMAN MELANOMA CELLS

Human melanoma cell lines that show low- and high-metastatic potential when assayed in nude mice have been recently developed (Kozlowski et al., 1984a,b). We have used various plant lectins to compare the sialoglycoprotein of the human melanoma A375 parental line with the A375 MM line that shows a greater incidence of lung metastasis. Interestingly, the profiles of mannose-rich glycoproteins revealed by ConA labeling were similar between human and mouse melanoma cells, whereas sialoglycoprotein profiles of human A375 melanoma cells revealed by WGA, RCA_I (after removal of sialic acid) or PNA (after removal of sialic acid) were distinct from those of mice. For example, in A375 cells the major PNA-reactive glycoproteins possessed M_r of ~80,000 instead of 56,000 and 61,000 as described above for B16 melanoma cells (Figure 5). The most remarkable metastasis-associated difference between the A375 parental and MM cell lines was a glycoprotein of M_r ~240,000. This glycoprotein appeared to contain both complex- and mucin-type carbohydrate chains because of its reactivity with PNA and RCA_I after removal of sialic acid (Irimura et al., 1975). The asparagine-linked chains appeared to possess nonbranched α-mannosyl residues, as estimated from their LCA reactivity (Kornfeld et al., 1981). It would be extremely interesting to determine the function of these molecules with regard to tumor progression and lung metastasis.

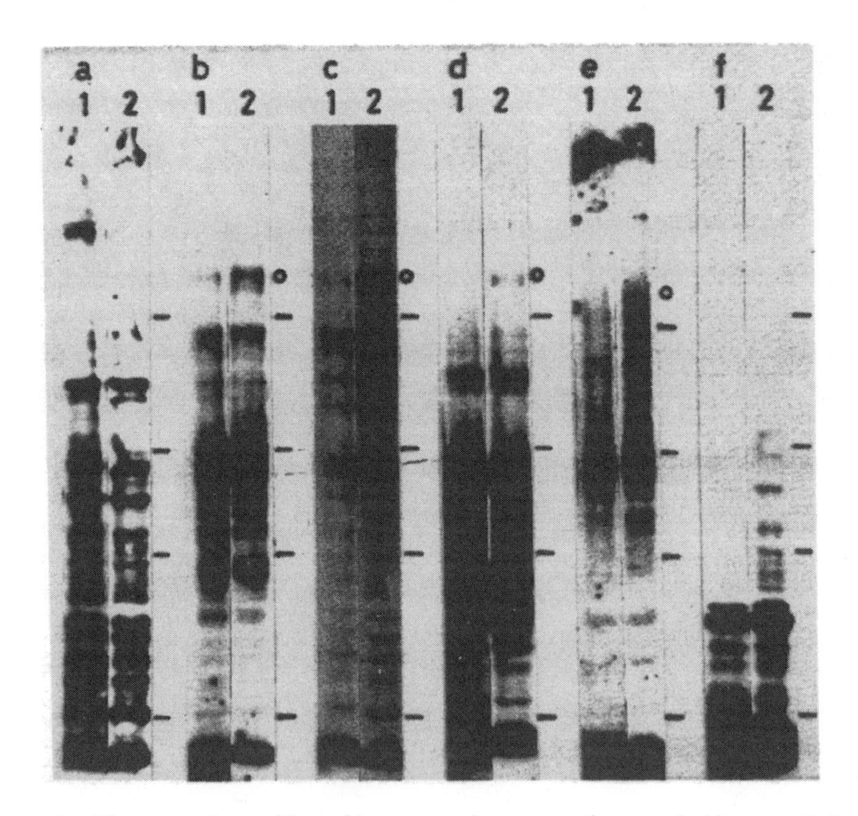

Figure 5. Glycoprotein profiles of human melanoma cells revealed by reactivity with lectins. Panels are the autoradiograms of lectin-stained nitrocellulose membranes on which electrophoretically separated melanoma components were transferred. Lane 1, A375 - Parental melanoma; Lane 2, highly metastatic A375-MM melanoma. Lectins used were: (a) ConA; (b) RCA_I (after removal of sialic acid); (c) LCA; (d) PNA (after removal of sialic acid); (e) WGA; (f) UEA_I. Bars indicate the positions of prestained molecular weight markers (BRL, Rockville, MD); myosin, phoshphorylase b, serum albumin, and ovalbumin. Circles indicate the position of a differentially expressed sialoglycoprotein.

SIALOGLYCOPROTEINS IN LIVER METASTASIS OF RAW117 LYMPHOMA

Glycoprotein analysis using lectins has provided interesting information on the involvement of cell surface molecules in determining liver metastasis of lymphoma cells in mice. As described above, ConA was used for identifying and quantitating oncogenic viral envelope glycoprotein gp70 expressed on these RAW117 cells (Reading et al., 1980a,b). Apparently, RAW117 cell lines consist of mixtures of heterogeneous cell populations having different metastatic and cell-surface characteristics. Immobilized-WGA has been successfully used for eliminating the highly metastatic cell subpopulation from the mixture (Reading et al., 1980a). Differences in the number of WGA-binding sites between the highly metastatic selected population and poorly metastatic parental cell line were demonstrated by direct binding of ^{125}I-WGA to intact cells (Irimura et al., 1985a,b).

Using $NaDodSO_4$ polyacrylamide gel electrophoresis the WGA-reactive cellular components were identified as glycoproteins with a wide range of M_r. WGA bound to the terminal sialic acid of the carbohydrate chains (Figure 6), because its binding was diminished after sialic acid removal by in situ mild acid hydrolysis. Differences between highly and poorly metastatic sublines cannot be explained by altered sialylation as demonstrated in other systems (Dennis et al., 1984; Schirrmacher et al., 1982; Altevogt et al., 1983); because the differential expression of this class of glycoproteins was also shown by RCA_I-reactivity after removal of sialic acid in situ. After in situ Smith degradation of desialyzed RAW117-glycoproteins, WGA reactivity restored to the same extent in poorly and highly metastatic sublines. Furthermore, the differential expression of WGA-reactive sialoglycoproteins was revealed by polypeptide portions of cell surface glycoproteins vectorially labeled with iodogen and separated by WGA-agarose affinity chromatography. These results indicated that the expression of the sialoglycoprotein was altered, not its carbohydrate structure.

The apparent heterogeneity of WGA-bound sialoglycoproteins of RAW117 cells after $NaDodSO_4$-polyacrylamide gel electrophoresis was not due to smearing or sialylation. Glycoprotein profiles of sialidase-treated RAW117 cells were almost identical to those derived after sialic acid was chemically removed from the glycoprotein in situ (Figure 6). In addition, the sialoglycoproteins isolated from RAW117 cells by WGA-agarose affinity chromatography can be subfractionated further according to their molecular size by gel filtration on Sepharose CL-6B in the presence of $NaDodSO_4$. These results indicate that highly metastatic RAW117 have greater number of sialoglycoproteins bearing similar carbohydrate chains, but these molecules are heterogeneous regarding their size.

Do the sialoglycoproteins play functional roles in liver-specific metastatic colonization of RAW117 lymphoma cells? This question is difficult to answer because increased expression of sialoglycoproteins and decreased expression of leukemia viral envelope antigen gp70 were observed in variants of high liver colonization potential, regardless of the selection method used to establish the variants (Reading et al., 1980a,b; Irimura et al., 1985a,b). Therefore, the relatively high liver-colonization potential of RAW117 cells may be due to either decreased expression of gp70, increased expression of sialoglycoproteins, or both. It is doubtful that the expression of these cell surface molecules is concomitantly regulated, and it seems more likely that each of these cell surface changes adds an increment to the overall metastatic properties of RAW117 cells.

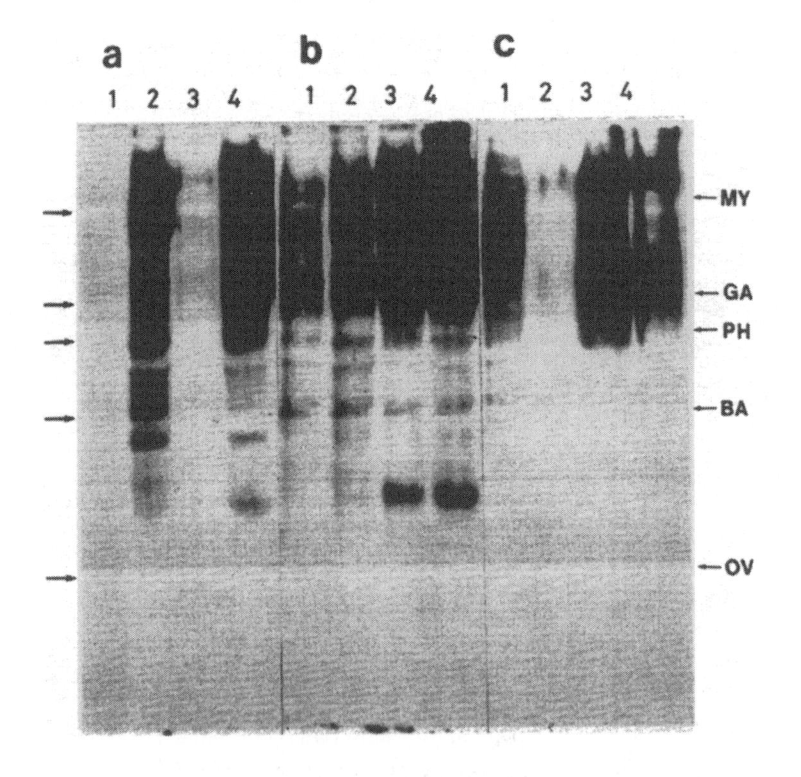

Figure 6. Effects of neuraminidase treatment of RAW117 parental and H10 cells on the electrophoretic mobility and lectin binding to sialoglycoproteins. (a) polyacrylamide slab gel stained with ^{125}I-RCA$_1$; (b) polyacrylamide gel treated under mild acid hydrolytic condition, and stained with ^{125}I-RCA$_1$; (c) polyacrylamide gel stained with ^{125}I-WGA. Lane 1, lysates of RAW117-P cells; lane 2, lysates of RAW117-P cells previously treated with neuraminidase 25 m unit/ml at 37°C for 60 min; lane 3, lysates of RAW117-H10 cells; lane 4, lysates of neuraminidase-treated RAW117-H10 cells.

The involvement of sialoglycoconjugates in the metastatic properties of another murine lymphoma system has been studied by Schirrmacher et al. (1982). Based on their studies of the Eb/ESb T-lymphoma system, they suggested that the sites of sialylation on oligosaccharide chains of poorly and highly metastatic cells were different (Altevogt et al., 1983). This may influence the adhesiveness of the lymphoma cells to hepatocytes (Schirrmacher et al., 1980), and thought to be an important determinant of liver colonization. The WGA-reactive sialoglycoproteins of liver-metastatic hamster lymphosarcoma cells were studied by Chan et al. (1984). In this study expression of this class of sialoglycoproteins was greatly influenced by the organ environment where the cells were grown; reversible elevation of the number of sialoglycoproteins was detectable in liver metastases (Chan et al., 1984). Finally, our studies on metastatic RAW117 lymphoma indicated that structural differences in carbohydrate chains did not always explain the increase in net negative charge, nor did the presence of more sialoglycoconjugates on the cell surface. Our studies also suggested that there are limitations to the use of lectin-resistant variant cells for proving the functional significance of a specific cell-surface molecule.

SULFATED GLYCOCONJUGATES PRODUCED BY LIVER METASTATIC RAW117 LYMPHOMA

Sulfated proteoglycans are another class of cell surface molecules expressed at higher levels on RAW117 highly metastatic to the liver. This finding was suggested earlier in studies where extensive sialidase treatment of RAW117-P and H10 cells failed to diminish the net surface negative charge of these cells as determined by countercurrent distribution in two-polymer aqueous phases (Miner et al., 1981). Analytical approaches to the characterization of sulfated materials were not undertaken until recently, when RAW117 sublines were grown in tissue culture in the presence of [^{35}S]sulfate. Polyacrylamide gel electrophoresis in $NaDodSO_4$ indicated that initially all of the sulfate label was incorporated into a high molecular-weight component that stayed at the top of a 7.5% running gel. Another aliquot of the labeled cells was suspended in Dulbecco's phosphate buffered saline (DPBS), precipitated and washed with 80% ethanol. The residue was digested with pronase (4 mg/ml) at $50^{\circ}C$ for 48 hr and sulfated glycosaminoglycans (GAGs) were isolated by gel filtration on Bio-Gel P-10. The amounts of [^{35}S]sulfate incorporated into the void volume fractions were compared based on the total amount of protein in the ethanol-precipitated pellet. The [^{35}S]label incorporated into GAGs by highly metastatic RAW117-H10 cells was two-fold greater than that incorporated by RAW117-P cells. The GAGs were a highly polydispersed component that had an average M_r of ~12,000 as determined by high-speed gel-permeation chromatography (Irimura et al., 1983). DEAE-cellulose ion-exchange chromatography and cellulose acetate sheet electrophoresis revealed that the sulfate-labeled GAG apparently consisted of dermatan sulfate-like and chondroitin sulfate-like substances (Figure 7). All the sulfate-labeled GAG was degraded by treatment with chondroitinase ABC.

The production and maintenance of the heparan sulfate proteoglycan at the surface of tumor cells seems to be associated with a nonproliferating stage of the cell cycle in tissue culture, whereas increased production of chondroitin sulfate and hyaluronic acid has been noted in malignant tumor cells (Cohen et al., 1976; Ninomiya et al., 1980; Angello et al., 1982a,b). Little is known of the influence of cell surface proteoglycans on the metastatic ability of tumor cells. In B16 melanoma cell lines, increased retention of sulfated GAGs was observed in highly metastatic

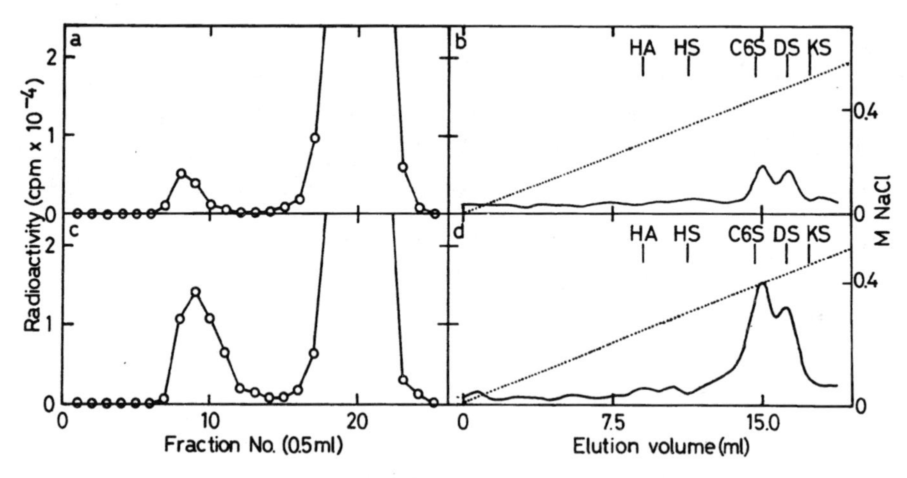

Figure 7. Analysis of cell-associated GAGs produced by poorly and highly metastatic RAW117 cells. ^{35}S-sulfate labeled GAGs were isolated by pronase digestion. Panel a and b, RAW117-P cells; panels c and d, RAW117-H10 cells. The ^{35}S-labeled materials derived from the equivalent number of cells were analyzed first by gel filtration on Bio-Gel P-10 (a,c); and then by ion exchange chromatography on minicolumns of DEAE cellulose (b, d). The eluent for the Bio-Gel P-10 column was water. The void column fractions were applied to 0.8 x 1.0 cm of DEAE-cellulose, previously equilibrated with 5 mM Tris-HCl buffer pH 8.0; and then eluted with a linear gradient of 0-0.6 M sodium chloride in the same buffer. The ^{3}H-labeled GAGs at their reducing terminal were used as standards, and the eluting positions were indicated at the top of each panel.

cell clones (Maniglia et al., 1985), whereas in FM-3A mammary adenocarcinoma, the more metastatic cells produced increased amounts of hyaluronic acid (Kimata et al., 1983). Because chondroitin and dermatan sulfate did not specifically interact

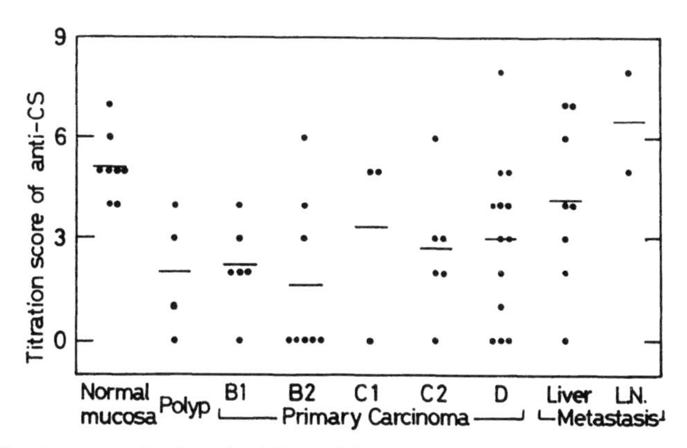

Figure 8. Contents of chondroitin-sulfate proteoglycans in human colorectal carcinoma tissues surgically resected from patients. The tissues were minced and extracted with 4 M guanidine hydrochloride and 4% Zwittergent 3-12 0.1 M sodium acetate buffer (pH 6.0) containing protease inhibitors at a tissue-buffer ratio of 10 mg/100 µl. After 18 hr, the mixture was centrifuged at 9000 g for 5 min, and the supernatant was collected and diluted to make the protein concentration 2 mg/ml in the same buffer. Serially diluted extracts were blotted onto nitrocellulose membrane (0.1 micropore size) and reacted with anti-chondroitin sulfate monoclonal antibody; and then with biotinylated anti-mouse IgM, followed by ABC-peroxidase reagent. The titration score was determined by positive reaction with 4-chloro-1-naphthol. Each dot in the figure represents the titration score of a primary tumor from a patient with colorectal cancer at various stages (according to Astler and Coller, 1954), or with a metastasis to the liver or a lymph node. The adjacent normal mucosa were analyzed. Relatively, a high score for normal mucosa maybe because of its lower protein on a per weight basis.

with other extracellular matrix components, the presence of these glycosaminoglycans on the RAW117 cell surface may result in less adherent cells and facilitate their detachment. In addition, the strong negative charge associated with these molecules could interfer with host immune recognition cell surface antigens. Increased expression of cell surface mucin-like glycoproteins is also known to partially conceal transplantation antigens (Codington et al., 1978), and this could presumably facilitate metastasis of mammary carcinoma cells (Steck and Nicolson, 1983).

CHONDROITIN SULFATE PROTEOGLYCAN PRODUCED IN COLORECTAL CARCINOMA TISSUES

Increased production of chondroitin sulfate proteoglycan has been found in colorectal tumor tissue (Iozzo et al., 1982). This appeared contradictory to the histochemical observation known as decreased expression of sulfo-mucin (Filipe, 1969), but Iozzo has recently shown that colorectal carcinoma cells produce factors that stimulate host-derived fibroblastic cells to produce chondroitin sulfate proteoglycan (Iozzo and Wight, 1982; Iozzo, 1985). This change, apparently a part of pathologic host response that is also called desmoplasia, may facilitate dissemination of tumor cells from primary organ sites. Thus, we compared the content of chondroitin sulfate proteoglycan in colorectal carcinoma tissues with that in liver metastases using a monoclonal antibody specific for chondroitin sulfate (Avnur and Geiger, 1984). Tumor tissues freshly resected from the patients were extracted with 4 M guanidine-HCl containing protease inhibitors and the extracts were serially diluted. The samples were then blotted onto nitrocellulose membranes, washed with Tris-buffered saline containing 0.05% Tween 20 and reacted with the monoclonal antibody followed by reaction with biotinylated rabbit anti-mouse IgM, ABC-peroxidase reagent (Vector Lab, Burlingame, CA), and 4-chloro-1-naphthol as peroxidase substrate. After analyzing 40 tumor specimens, we concluded that reactivity of tumor tissue with this antibody varied from one specimen to another, and that there was no statistically significant correlation between antibody reactivity and the stage of colorectal carcinomas (Figure 8).

GLYCOSIDASES IN MELANOMA METASTASIS

Since the early 1970's, when Bossman et al. (1973) reported differences in various glycosidase activities between poorly and highly metastatic tumor cells little attention has been paid to these phenomena. When we were studying tumor cell-endothelial cell interactions <u>in vitro</u> as a model for colonization of blood-borne metastatic tumor cells (Nicolson, 1982), and B16 melanoma cells were placed on an endothelial extracellular matrix previously labeled with radioactive precursors, products derived from heparan sulfate degradation were detected (Kramer et al., 1982). Heparan sulfate from an intact extracellular matrix of bovine aortic endothelial cells had an estimated M_r of $\sim$25,000, whereas the average M_r of the degraded products was $\sim$10,000. These results suggested that metastatic B16 melanoma cells produced an endo-glycosidase capable of degrading heparan sulfate. This enzyme has now been implicated in the vascular extravasation of malignant tumor cells (Kramer et al., 1982). Using purified heparan sulfate, we demonstrated that B16 melanoma sublines that had a higher potential for metastasis to the lung degraded heparan sulfate more rapidly than poorly metastatic sublines (Nakajima et al., 1983).

In order to facilitate analysis of the degradation products of GAGs, we developed high-speed gel-permeation chromatography for glycosaminoglycans and their fragments (Irimura et al., 1983). The size of the fragments obtained from heparan sulfate ranged from 5,000 to 10,000; apparently the size was dependent on the source of heparan sulfate (Irimura et al., 1983; Nakajima et al., 1984; Wang et al., 1985). The malignant cell degradative enzyme appeared to be very specific for a certain linkage that periodically occurs in heparan sulfate, judging from the size of the products and from the fact that the same melanoma cell lysates failed to degrade any other glycosaminoglycans. Interestingly, heparin, which is structurally and biosynthetically closely related to heparan sulfate, inhibited degradation of heparan sulfate by this enzyme.

The reducing terminal monosaccharide of the degradation products from bovine lung heparan sulfate was determined after reduction with $[^3H]NaBH_4$ and hydrolysis. More than 90% of the radioactivity was recovered as a reduction product of glucuronic acid, indicating that the enzyme was endo-β-glucuronidase. Because this is distinct from the well-known heparan sulfate degrading elimination enzyme heparitinase, which is derived from bacteria, we decided to call it heparanase. Although a similar enzyme was reported to occur in platelets (Oldberg et al., 1980; Oosta et al., 1982), mastocytomas (Ogren and Lindahl, 1975), and lymphocytes (Savion et al., 1985), melanoma heparanase probably possesses unique specificity for the penultimate monosaccharides in heparan sulfate chains. It was known that heparan sulfate proteoglycans in various organs are different and heterogeneous in their arrangement of the heparan sulfate chains, chain lengths, degrees of sulfation, and periodic structures within the chains. It will be interesting to determine if the heterogeneity of heparan sulfate is somehow related to the specific organ colonization properties of B16 melanoma cells selected <u>in vivo</u>.

In order to further characterize the metastasis-associated heparanase, we developed a solid-phase substrate. Using this assay system, we have analyzed 14 cultured human melanoma cell lines, including A375, SK-mel-23, SK-mel-93, M40, HS294, HS852, and HS939. All human malignant melanoma cells exhibited heparanase activity comparable to highly metastatic mouse B16 melanoma sublines, and the heparanase activity of A375 melanoma sublines correlated with the metastatic potentials of these cells in nude mice. The solid-phase assay has also been used in the purification and characterization of melanoma heparanase from Triton X-100 extracts of B16-BL6 cells. This B16 melanoma heparanase has been purified to a single molecular species on $NaDodSO_4$-polyacrylamide gel electrophoresis by a combination of chromatographies on concanavalin A-Sepharose, heparin-Sepharose, hydroxylapatite, and Sepharose CL-6B.

Heparanase is not the only enzyme responsible for the destruction of the vascular basement membrane. Collagenase degrading type IV collagen, cathepsin B, and plasminogen activator are enzymes known to be associated with metastatic tumor cells (Liotta et al., 1980; Sloane, 1981; Wang et al., 1980). These enzymes may act in concert, and the enzyme combination may determine the organ-specific nature of tissue invasion.

GLYCOCONJUGATE INHIBITORS OF METASTATIC MELANOMA HEPARANASE

Because metastasis-associated heparanase may play an important role in metastatic processes, inhibitors against this enzyme could be useful in confirming the biological significance of the enzyme, and also for developing antimetastatic substances. The argument has been made that blocking the process of metastasis will not be effective therapeutically, because micrometastatic foci have been established by the time the primary tumor is diagnosed. On the other hand, it is well known that certain therapeutic modalities such as chemotherapy, radiotherapy and surgery may enhance dissemination of tumor cells, and it is also known that metastases can metastasize. Therefore, antimetastasis treatments may be more useful in combination with other therapies than first thought.

Heparin is a potent inhibitor of metastatic melanoma heparanase, but it may not be useful for development as an antimetastasis agent. The reason is that heparin is involved in dozens of biological activities, including anticoagulation. In fact, it has been reported that heparin acts protectively against metastases in certain experimental systems (Hagmar and Norby, 1970; Tsubura et al., 1977; Hilgard, 1984), and in some systems heparin has been shown to enhance tumor cell dissemination, for example, to facilitate extrapulmonary metastasis (Maat, 1978; Hagmar and Boeryd, 1969; Chan et al., 1980). These effects of heparin have been considered a consequence of the inhibition of blood coagulation and platelet aggregation that facilitates the arrest of tumor cells in organ capillary systems. Furthermore, heparin is a mixture of polysaccharides that have intrinsic heterogeneity and molecular diversity. The complex nature of its biological activities has been attributed to its structural heterogeneity.

We have attempted to determine the structural requirement for heparanase inhibitory activity and to find a heparanase inhibitor without anti-coagulation activity by chemically modifying heparin. Modification reactions such as desulfation-deacetylation, sulfation-acetylation and carboxy-reduction were performed. The resultant modified heparins eluted as single peaks from high-speed gel-permeation chromatography (Irimura et al., 1983) and migrated as single spots after cellulose acetate sheet electrophoresis in pyridine-acetate buffer pH 5.5.

Among the 8 chemically modified heparins, we have tested 3 derivatives: N-acetylate N-desulfated heparin; N-resulfated N-,O-desulfated heparin; and carboxy-reduced heparin. These were shown to maintain their heparanase inhibitory activity using the solid-phase heparanase assay (Table 3), and these substances inhibited the release of ^{35}S-labeled materials from the endothelial matrix produced by mouse lung endothelial cells in tissue culture by B16 melanoma cells. When highly metastatic B16-BL6 melanoma cells were preincubated with N-acetylate N-desulfated heparin, N-resulfated N-heparin, and carboxy-reduced heparin, significant reduction of lung colonization was observed (Table 4; Irimura et al., submitted for publication).

CONCLUDING REMARKS

Tumor metastasis is a complex phenomenon involving both host mechanisms and tumor cell properties, yet it is only a part of the reason why cancer is such a destructive disease. Future studies on the genetic mechanisms of carcinogenesis may explain how malignant tumor cells arise, but the molecular mechanisms underlying the malignant behavior of tumor cells are not likely to be elucidated anytime soon. The observed phenotypic changes associated with highly metastatic tumor cells, such as alterations in cell surface sialoglycoproteins, proteoglycans, and glycosidases, are by no means specifically characteristic of malignant tumor cells. However, these properties would serve as prognostic markers for metastasis under specific circumstances. Also, by modulating cellular properties that are functionally important for the metastasis of a particular cancer, we may be able to control the outcome of each disease.

Table 3. Heparanase Inhibitory Activity of Heparin Derivatives

	Inhibition by Heparin Derivative (%)	
	1.25 mg/ml	0.25mg/ml
Heparin	100.7	9.5
N-desulfated heparin	16.9	4.7
N-acetylate N-desulfated heparin	88.5	63.7
N- and O-desulfated heparin	0.0	0.0
N-acetylate-N- and O-desulfated heparin	12.2	0.0
N-resulfate-N- and O-desulfated heparin	40.3	30.2
Carboxy-reduced	53.2	8.0

Table 4. The Effects of Chemically-Modified Heparins on Blood-Borne Lung Colonization of B16-BL6 Melanoma Cells in Mice

Treatment[a]	Median	Number of Colonies
None	75	8, 20, 28, 33, 75, 85, 106, 116, 193
Heparin	16	0, 1, 8, 9, 16, 18, 21, 82, 174
N-acetylate-N-de-sulfated heparin	43	11, 19, 22, 23, 43, 89, 109, 199
Carboxy-reduced heparin	29	0, 15, 25, 25, 29, 36, 45, 46, 53

[a]The cells were incubated with chemically modified heparins (500 μg/ml) at 4°C for 2 hr before injection.

ACKNOWLEDGEMENTS

Supported by U.S. Department of Health and Human Service Grants RO1-CA39319 and BRSG RR-5511-23 to T.I. and grant RO1-CA42346 to G.L.N. and a grant from the National Foundation for Cancer Research to G.L.N.

REFERENCES

Altevogt, P., Fogel, M., Cheingsong-Popov, R., Dennis, J., Robinson, P. and Schirrmacher, V. (1983). Related high and low metastatic tumor lines show different patterns of lectin binding and cell surface sialylation. Cancer Res., 43:5138-5144.

Angello, J. C., Danielson, K. G., Anderson, L. W. and Hosick, H. L. (1982a). Glycosaminoglycan synthesis by subpopulations of epithelial cells from a mammary adenocarcinoma. Cancer Res., 42:2207-2210.

Angello, J. C., Hosick, H. L. and Anderson, L. W. (1982b). Glycosaminoglycan synthesis by a cell line (C1-S1) established from a preneoplastic mouse mammary outgrowth. Cancer Res., 42:4975-4978.

Astler, V. B. and Coller, F. A. (1985). The prognostic significance of direct extension of carcinoma of the colon and rectum. Ann. Surg., 139:846-851.

Avnur, Z. and Geiger, B. (1984). Immunochemical localization of native chondroitin-sulfate in tissues and cultured cells using specific monoclonal antibody. Cell, 38:811-822.

Bossman, H. B., Bieber, G. F., Brown, A. E., Case, K. R., Gevsten, D. M., Kimmerer, T. W. and Leone, A. (1973). Biochemical parameters correlated with tumor cell implantation. Nature, 246:487-489.

Brunson, K. W. and Nicolson, G. L. (1978). Selection and biologic properties of malignant variants of murine lymphosarcoma. J. Natl. Cancer Isnt., 61:1499-1503.

Brunson, K. W. and Nicolson, G. L. (1979). Selection of malignant melanoma variant cell lines for ovary colonization. J. Supramol. Struct., 11:517-528.

Brunson, K. W., Beatti, G. and Nicolson, G. L. (1978). Selection and altered properties of brain-colonizing metastatic melanoma. Nature, 272:543-545.

Chan, S.-Y. and Pollard, M. (1980). Metastasis enhancing effect of heparin and its relationship to a lipoprotein factor. J. Natl. Cancer Inst., 64:1121-1125.

Chan, W. S., Jackson, A. and Turner, G. A. (1984). Differences in surface expression of WGA-binding protein of cells from a lymphosarcoma and its liver metastases. Br. J. Cancer, 89:181-191.

Codington, J. F., Klein, G., Cooper, A. G., Lee, N., Brown, M. C. and Jeanloz, R. W. (1978). Further studies on the relationship between large glycoprotein molecules and allotransplantability in the TA3 tumor of the mouse: Studies on segregating TA3-HA hybrid. J. Natl. Cancer Inst., 60:811-818.

Cohen, R., Cassiman, J. and Bermfield, M. (1976). Relationship of transformation, cell density, and growth control to the cellular distribution of newly synthesized glycosaminoglycans. J. Cell Biol., 71:280-294.

Dennis, J., Waller, C., Timple, R. and Schirrmacher, V. (1982). Surface sialic acid residues attachment of metastatic tumor cells to collagen type IV and fibronectin. Nature, 300:274-476.

Dennis, J. W., Carver, J. P. and Schachter, M. (1984). Asparagine-linked oligosaccharides in murine tumor cells: Comparison of a WGA-resistant (WGA^r) nonmetastatic mutant and a related WGA-sensitive (WGA^s) metastatic line. J. Cell Biol., 99:1034-1044.

Dobrossy, L., Pavelic, Z. P. and Bernacki, R. J. (1981). A correlation between cell surface sialyltransferase, sialic acid, and glycosidase activities and implantability of B16 murine melanoma. Cancer Res., 41:2262-2266.

Felipe, M. I. (1969). Value of histochemical reactions for mucosubstances in the diagnosis of certain pathological conditions of the colon and rectum. Gut, 10:577-586.

Fidler, I. J. (1973). Selection of successive tumor line for metastasis. Nature New Biol., 242:148-149.

Fidler, I. J. and Kripke, M. L. (1977). Metastasis results from preexisting variant cells within a malignant tumor. Science, 197:998-1003.

Fidler, I. J., Gersten, D. M. and Hart, I. R. (1978). The biology of cancer invasion and metastasis. Adv. Cancer Res., 28:199-250.

Finne, J., Tao, T.-W. and Burger, M. D. (1980). Carbohydrate changes in glycoproteins of a poorly metastasizing wheat germ agglutinin-resistant melanoma clone. Cancer Res., 40:2580-2587.

Hagmar, B. and Boeryd, B. (1969). Disseminating effect of heparin on experimental tumor metastasis. Pathol. Eur., 4:274-282.

Hagmar, B. and Norby, K. (1970). Evidence for effects of heparin cell surface influencing experimental metastasis. Int. J. Cancer 5:72-84.

Hilgard, P. (1984). Anticoagulant and tumor growth: Pharmacological considerations. In "Cancer Invasion and Metastasis: Biologic and Therapeutic Aspects (Nicolson, G.L. and Milas, L., eds.), Raven Press, NY, pp 353-360.

Iozzo, R. V. (1985). Neoplastic modulation of extracellular matrix: Colon carcinoma cells release polypeptides that alter proteoglycan metabolism in colon fibroblasts. J. Biol. Chem., 260:7464-7473.

Iozzo, R. V. and Wight, T. N. (1982). Isolation and characterization of proteoglycan synthesized by human colon and colon carcinoma. J. Biol. Chem., 257:11135-11194.

Iozzo, R. V., Bolender, R. P. and Wight, T. N. (1982). Proteoglycan changes in the intercellular matrix of human colon carcinoma. An integrated biochemical and stereologic analysis. Lab. Invest. 47:124-138.

Irimura, T. and Nicolson, G. L. (1981a). The role of glycoconjugates in metastatic melanoma blood-borne arrest and cell surface properties. J. Supramol. Struct. Cell., Biochem., 14:325-336.

Irimura, T. and Nicolson, G. L. (1981b). Affinity isolation of cell surface sialoglycoproteins from B16 melanoma and their interaction with endothelial cell basal lamina. J. Cell. Biol., 91:118a.

Irimura, T. and Nicolson, G. L. (1983a). Carbohydrate chain analysis using lectin binding to mixtures of glycoproteins separated by polyacrylamide slab gel electrophoresis with in situ chemical modifications. Carbohydr. Res., 115:209-220.

Irimura, T. and Nicolson, G. L. (1983b). The interaction of pokeweed mitogen with poly(N-acetyllactosamine)-type carbohydrate chains. Carbohydr. Res., 120:187-195.

Irimura, T. and Nicolson, G. L. (1984a). Carbohydrate chain analysis by lectin binding to electrophoretically separated glycoproteins from murine B16 melanoma sublines of various metastatic properties. Cancer Res., 44:791-798.

Irimura, T. and Nicolson, G. L. (1984b). Use of pokeweed mitogen for the study of the occurrence of poly(N-acetyllactosamine)-type carbohydrate chain in metastatic tumor cells. Fed. Proc., 43:1695.

Irimura, T., Kawaguchi, T., Terao, T. and Osawa, T. (1975). Carbohydrate binding specificity of so-called galactose-specific phytohemagglutinins. Carbohydr. Res., 39:317-327.

Irimura, T., Gonzalez, R. and Nicolson, G. L. (1981). Effects of tunicamycin on B16 metastatic melanoma cell surface glycoproteins and blood-borne arrest and survival properties. Cancer Res., 41:3411-3418.

Irimura, T., Nakajima, M., DiFerrante, N. and Nicolson, G. L. (1983). High-speed gel permeation chromatography of glycosaminoglycans: Its application to the analysis of heparan sulfate of embryonic carcinoma and its degradation products by tumor cell derived heparanase. Anal. Biochem., 130:461-468.

Irimura, T., Nakajima, M. and Nicolson, G. L. (1986). Chemically modified heparins as inhibitors of heparan sulfate specific endo-β-glucuronidase (heparanase) of metastatic melanoma cells. Biochemistry, (in press).

Irimura, T., Tressler, R. J. and Nicolson, G. L. (1985a). Cell surface glycoproteins associated with liver colonization of metastatic variants of RAW117 large cell lymphoma. J. Cell Biol., 101:4802a.

Irimura, T., Tressler, R. J. and Nicolson, G. L. (1985b). Sialoglycoproteins of murine RAW117 large cell lymphoma/lymphosarcoma sublines of various metastatic colonization properties. Exp. Cell Res., (in press).

Kimata, K., Homma, Y., Okayama, M., Oguri, K., Hozumi, M. and Suzuki, S. (1983). Increased synthesis of hyaluronic acid by mouse mammary carcinoma cell variants with high metastatic potential. Cancer Res., 43:1347-1354.

Kornfeld, K., Reitman, M. L. and Kornfeld, R. (1981). The carbohydrate binding specificity of pea and lentil lectins: Fucose is an important determinant. J. Biol. Chem., 256:6633-6640.

Kozlowski, J. M., Fidler, I. J., Campbell, D., Xu, Z., Kaighn, M. E. and Hart, J. R. (1984a). Metastatic behavior of human tumor cell lines grown in the nude mouse. Cancer Res., 44:3522-3529.

Kozlowski, J. M., Hart, I. R., Fidler, I. J. and Hanna, N. (1984b). A human melanoma line heterogeneous with respect to metastatic capacity in athymic nude mice. J. Natl. Cancer Inst., 72:913-917.

Kramer, R. H., Vogel, K. and Nicolson, G. L. (1982). Solubilization and degradation of subendothelial matrix glycoproteins and proteoglycans by metastatic tumor cells. J. Biol. Chem., 257:2678-2686.

LeGrue, S. J. (1982). 1-Butanol extraction and subsequent reconstitution of membrane components which mediate metastatic phenotype. Cancer Res., 42:2126-2134.

Liotta, L. A., Tryggvason, K., Garbisa, S., Hart, I. R., Fortz, C. M. and Shafie, S. (1980). Metastatic potential correlates with enzymatic degradation of basement membrane collagen. Nature, 284:67-68.

Liotta, L. A., Rao, N. C., Terranova, V. P., Barsky, S. and Thorgeirsson, U. (1984). Tumor cell attachment and degradation of basement membranes. In "Cancer Invasion and Metastasis: Biologic and Therapeutic Aspects" (Nicolson, G. L. and Milas, L., eds.). Raven Press, N.Y., pp. 169-176.

Maat, B. (1978). Extrapulmonary colony formation after intravenous injection of tumor cells into heparin-treated animals. Br. J. Cancer, 37:369-376.

Maniglia, C. A., Gomez, J. J., Luikart, S. D. and Sartorelli, A. C. (1985). Glycosaminoglycan production and distribution in cloned B16 murine melanoma cell lines exhibiting different lung colony-forming efficiencies. J. Natl. Cancer Inst., 75:111-120.

Miller, W., Ota, D., Giacco, G., Guinee, V., Irimura, T., Nicolson, G. L. and Cleary, K. (1985). Absence of a relationship of size of primary colon carcinoma with metastasis and survival. Clin. Expl. Metastasis, 3:189-196.

Miner, K. M. and Nicolson, G. L. (1983). Differences in the sensitivities of murine metastatic lymphoma/lymphosarcoma variants to macrophage-mediated cytolysis and/or cytostasis. Cancer Res., 43:2063-2067.

Miner, K. M., Walter, H. and Nicolson, G. L. (1981). Subfractionation of malignant variants of metastatic murine lymphosarcoma cells by countercurrent distribution in two-polymer aqueous phases. Biochemistry, 20:6244-6250.

Miner, K. M., Kawaguchi, T., Uba, G. W. and Nicolson, G. L. (1982). Clonal drift of cell surface, melanogenic, and experimental metastatic properties of in vivo-selected brain meninges-colonizing B16 melanoma. Cancer Res., 42:4631-4638.

Nakajima, M., Irimura, T., Di Ferrante, D., Di Ferrante, N. and Nicolson, G. L. (1983). Heparan sulfate degradation: Relation to tumor invasive and metastatic properties of mouse B16 melanoma sublines. Science, 220:611-613.

Nakajima, M., Irimura, T., Di Ferrante, N. and Nicolson, G. L. (1984). Metastatic melanoma cell heparanase. Characterization of heparan sulfate degradation fragments produced by B16 melanoma endoglucuronidase. J. Biol. Chem., 259:2283-2290.

Nicolson, G. L. (1982). Metastatic tumor cell attachment and invasion assay utilizing vascular endothelial cell monolayers. J. Histochem. Cytochem., 30:214-220.

Nicolson, G. L. (1984). Cell surface molecules and tumor metastasis. Regulation of metastatic diversity. Exp. Cell Res., 150:3-22.

Nicolson, G. L. and Irimura, T. (1984). Estimating glycoprotein carbohydrate chain structures by lectin reactivities in polyacrylamide gels. Biol. Cell, 51:157-164.

Nicolson, G. L. and Poste, G. (1982). Tumor cell diversity and host response in cancer metastasis. Part 1: Properties of metastatic cells. Curr. Probl. Cancer, 7(6):1-83.

Nicolson, G. L. and Poste, G. (1983). Tumor cell diversity and host responses in cancer metastasis. Part 2: Host immune responses and therapy of metastasis. Curr. Probl. Cancer, 7(7):1-42.

Nicolson, G. L., Mascali, J. J. and McGuire, E. J. (1982). Metastatic RAW117 lymphosarcoma as a model for malignant-normal cell interactions: Possible roles for cell surface antigens in determining the quantity and location of secondary tumors. Oncodev. Biol. Med., 4:149-159.

Nicolson, G. L., Van Pelt, C., Irimura, T. and Kawaguchi, T. (1985). Stabilities and characteristics of brain meinges-colonizing murine melanoma cells. Prog. Exp. Tumor Res. 29:17-35.

Ninomiya, Y., Hata, R. and Nagai, Y. (1980). Glycosaminoglycan synthesis by liver parenchymal cell clones in culture and its change with transformation. Biochim. Biophys. Acta, 629:349-358.

Ogren, S. and Lindahl, U. (1975). Cleavage of macromolecular heparin by an enzyme from mouse mastocytoma. J. Biol. Chem., 250:2690-2697.

Oldberg, A., Heldin, C.-H., Westson, Å., Busch, C. and Hoole, M. (1980). Characterization of a platelet endoglycosidase degrading heparin-like polysaccharides. Biochemistry, 19:5755-5762.

Oosta, G. M., Faureau, L. V., Beefer, D. L. and Rosenberg, R. D. (1982). Purification and properties of human platelet heparitinase. J. Biol. Chem., 257:11249-11255.

Pearlstein, E., Salk, P. L., Yogeeswaran, G. (1980). Correlation between spontaneous metastatic potential, platelet aggregating activity of cell surface extracts and cell surface sialylation in 10 metastatic variant derivatives of rat renal sarcoma cell line. Proc. Natl. Acad. Sci., U.S.A., 77:4336-4339.

Poste, G. and Nicolson, G. L. (1980). Blood-borne tumor cell arrest and metastasis modified by fusion of plasma membrane vesicles from highly metastatic cells. Proc. Natl. Acad. Sci. U.S.A., 77:399-403.

Raschke, W. D., Ralph, P., Watson, J., Sklar, M. and Coon, H. (1975). Oncogenic transformation of murine lymphoid cells by in vitro infection with Abelson leukemia virus. J. Natl. Cancer Inst., 54:1249-1253.

Raz, A., Bucana, C., McLellan, W. and Fidler, I. J. (1980a). Distribution of membrane anionic sites on B16 melanoma variants with differing lung colonization potential. Nature, 284:363-364.

Raz, A., McLellan, W. L., Hart, I. R., Bucana, C. D., Hoyer, L. C., Sela, B-A., Dragsten, P. and Fidler, I. J. (1980b). Cell surface properties of B16 melanoma variants with differing metastatic potential. Cancer Res., 40:1645-1651.

Reading, C. L., Belloni, P. N., and Nicolson, G. L. (1980a). Selection and in vivo properties of lectin attachment variants of malignant murine lymphosarcoma cell lines. J. Natl. Cancer Inst., 64:1241-1249.

Reading, C. L., Brunson, K. W., Torriani, M. and Nicolson, G. L. (1980b). Malignancies of metastatic murine lymphosarcoma cell lines and clones correlate with decreased cell surface display of RNA tumor virus envelope glycoprotein gp70. Proc. Natl. Acad. Sci. U.S.A., 77:5943-5947.

Reading, C. L., Kramer, P. M., Miner, K. M. and Nicolson, G. L. (1983). In vivo and in vitro properties of malignant variants of RAW117 metastatic murine lymphoma/lymphosarcoma. Clin. Exp. Metastasis, 1:135-151.

Reiber, M. and Reiber, M.S. (1981). Metastatic potential correlates with cell surface protein alterations in B16 melanoma variants. Nature, 293:74-76.

Savion, N., Vlodavsky, I. and Fuks, Z. (1985). Interaction of T-lymphocytes and macrophages with cultured vascular endothelial cells: Attachment, invasion, and subsequent degradation of the subendothelial extracellular matrix. J. Cell. Physiol., 118:169-178.

Schirrmacher, V., Cheingsong-Popov., R. and Arnheiter, H. (1980). Hepatocyte-tumor cell interaction in vitro. I. Conditions for rossette formation and inhibition by anti H-2 antibody. J. Exp. Med. 151:984-989.

Schirrmacher, V., Altevogt, P., Fogel, M., Dennis, J., Waller, C. A., Barz, D., Schwaltz, R., Cheingsong-Popov, R., Springer, G. F., Robinson, P.J., Nebe, T., Brossmer, W., Vlodavsky, I., Poweletz, N., Zimmerman, H.-P. and Uhlenbruck, G. (1982). Importance of cell surface carbohydrates in cancer cell adhesion, invasion and metastasis. Does sialic acid direct metastatic behavior? Invasion Metastasis, 2:313-360.

Sheppard, J. R., Kerr, S. T., Brown, D. R. and Burger, M. M. (1983). Lectin resistant B16 melanoma cell exhibits an altered response to MSH and cholera toxin. Exp. Cell Res. 149:577-581.

Sheppard, J. R., Koestlev, T. P., Corwin, S. P., Buscarino, C., Doll, J., Lester, B., Greig, R. G. and Poste, G. (1984). Experimental metastasis correlates with cyclic AMP accumulation in B16 melanoma clones. Nature, 308:544-507.

Sloane, B. F., Dunn J. R. and Honn, K. V. (1981). Lysosomal cathepsin B: Correlation with metastatic potential. Science, 212:1151-1153.

Steck, P. A. and Nicolson, G. L. (1983). Cell surface glycoproteins of 13762NF mammary adenocarcinoma clones of differing metastatic potentials. Exp. Cell Res., 147:255-267.

Stringfellow, D. A. and Fitzpatrick, F. A. (1979). Prostaglandin D2 controls pulmonary metastasis of malignant melanoma cells. Nature, 282:76-78.

Terranova, V. P., Liotta, L. A., Russo, R. and Martin, G. R. (1982). Role of laminin in the attachment and metastasis of tumor cells. Cancer Res., 42:2265-2269.

Tsubura, E., Yamashita, T., Kobayashi, M., Higuchi, Y. and Isobe, J. (1977). Inhibitory mechanism of blood-borne pulmonary metastasis by sulfated polysaccharides. Gann Monogr. Cancer Res., 20:147-161.

Wang, B. S., McLoughlin, G. A., Richie, J. P. and Mannick, J. A. (1980). Correlation of the production of plasminogen activator with tumor metastasis in B16 melanoma cell lines. Cancer Res., 41:4629-4636.

Wang, Z-W., Irimura, T., Nakajima, M., Belloni, P.N. and Nicolson, G. L. (1985). Characterization of the extracellular matrix-associated glycosaminoglycans produced by untransformed and transformed bovine corneal endothelial cells in culture. Eur. J. Biochem. (in press).

Yogeeswaran, G. and Salk, P.L. (1981). Metastatic potential is positively correlated with cell surface sialylation of cultured murine and tumor cell lines. Science, 212:1514-1516.

Yogeeswaran, G., Stein, B.S. and Sebastian, H. (1978). Altered cell surface organization of ganglisides and sialoglycoproteins of mouse metastatic melanoma variant lines selected in vivo for enhanced lung implantation. Cancer Res., 38:1336-1344.

GLYCOLIPIDS AND GLYCOPROTEINS IN PLASMA MEMBRANE OF HEPATOCELLULAR CANCER

Tian-Jue Gu

Department of Biochemistry
Shangai First Medical College
Shangai, Peoples's Republic of China

The recent literature contains numerous studies of glycolipids and glycoproteins of tumor cells. Diverse changes in chemical composition and/or organization have been found in tumor cells of various sources: chemical and viral transformation *in vitro* and tumor tissues *in vivo*.[1,2] They have incited wide interest although little is known about their exact biological significance. Theoretically, they may lead to an understanding of some essential problems, such as the malignant behavior of tumor cells and the role of gene mutation in the development of cancer. In addition, their practical potentialities as tumor markers has also been extensively explored.

China is a relatively high risk area of hepatoma with an incidence that is the third highest of all malignant tumors of the digestive system. In certain high risk maritime areas of southeast China, the occurence of the disease may be as frequent as 60-80 per 100,000.[3] Therefore, a special research group, including epidemiologists, immunologists, toxicologists, biochemists, and pathologists was established in 1973 in our college to deal with the problem. Over the past 12 years some progress has been made. For example, many incipient and subclinical cases were searched out by means of mass screening of serum alpha-fetoprotein (AFP) levels in a high-risk population, hence the operability and 5-year survial ratio were greatly promoted.[4] Within the purview of the present report, we will discuss research work done in our laboratory on glycolipids and glycoproteins on cell surface membrane of hepatoma. To our great encouragement, the study revealed tumor-related changes both in glycolipids (gangliosides) and glycoproteins (ß-galactoside specific hepatic lectin, receptors of plant lectins and transferrin receptors).

THE CHANGES OF GLYCOLIPIDS IN HEPATOMA

The earliest report on the changes of glycosphingolipids (GLS) in hepatoma was published at the end of the sixties. Brady[5] pointed out there was a decrease of gangliosides with low sialic acids accompanied by an increase of GM_3 and GM_1 in liver cancer cells. The study of Morris hepatoma by Hakomori[6] also found a decrease of complex GLS with two or more sialic acids and increase of neutral GLS.[2] These are supported by the further study of Dyatoviskaya (1974).[7] However, Dnistrian's work[8] on Morris hepatoma indicated an increase of disialic acid-containing gangliosides and no trisialic acid gangliosides. The reports since 1968 are summarized in Table 1.

Only two specimens of human hepatoma listed in Table 1 have ever been analyzed; the results are contradictory. Our findings[16,17] were not included in this table.

Dnistrian[8] (1977) reported that the composition of neutral glycolipids in Morris hepatoma was similar to that of normal liver tissue, predominantly ceramidemonohexoside (CMH) and ceramidedihexoside (CDH) (about 83-86%) and a little ceramidetrihexoside (CTH) and ceramidetetrahexoside (CQH), but the total neutral glycolipids increased. Walter's study (1980)[18] revealed that hepatoma tissue had 5-30 times more neutral glycolipids than normal liver tissue and in plasma the neutral glycolipids were 5-9 times higher than normal.

Since 1976, the progress in methodology[19,20] has made the isolation, purification and structural analysis of glycolipids more efficient and credible. It seemed worthwhile, therefore, to reevaluate those pioneering studies.[21] Since 1981, we have systematically investigated the gangliosides and neutral glycolipids of all types of hepatomas available: human surgical hepatoma samples, cultured human hepatoma cell strains, Hep A (methylcholanthrane-induced mice ascitic hepatoma), BERH-2 (dimethyl-nitrosamine-induced rat hepatoma) and 3'-methyl,4-(dimethyl)amino-diazobenzene (3'Me-DAB)-induced rat hepatoma. We have also examined human embryonic liver of different fetal ages.

Table 1. CHANGES OF GLYCOSPHINGOLIPIDS PATTERN IN HEPATOMA

Types of hepatoma	Glycosphingolipids	Ref.
Morris hepatoma	$GD_1\downarrow\downarrow$ [a], $GM_3\uparrow$[b]	(5)
Morris hepatoma	$GT\downarrow\downarrow$, $GD_{1a}\uparrow$, $GM_1\uparrow$	(6)
Morris hepatoma	$GT\downarrow\downarrow$, $GD_{1a}\uparrow$, $GM_1\uparrow$	(19)
Morris hepatoma (plasma membrane)	$GT\downarrow\downarrow$, $GD_{1a}\uparrow\uparrow$[c], $GM_1\uparrow\uparrow$	(8)
	tetraosylceramide	
Rat hepatoma 27	$GT\downarrow\downarrow$, $GD_{1b}\uparrow\uparrow$	(9)
Rat Novikoff hepatoma	$GT\downarrow\downarrow$, $GD\downarrow\downarrow$	(10)
Rat hepatocarcinogenesis by N-2-fluorenylacetamide	Inhibition of disialogang-side pathway, enhancement of monosialoganglioside pathway ($GD\downarrow$[d], $GM\uparrow$)	(11,12,13)
Human hepatoma	unchanged	(14)
Human hepatoma	$GM_3\downarrow$, $GM_2\uparrow$, $GD_3\uparrow$, $GM_1\uparrow$	(15)

a. "↓↓" means deletion of gangliosides.
b. "↑" means increase of gangliosides.
c. "↑↑" means accumulation of gangliosides.
d. "↓" means decrease of gangliosides.

GM_3 accounts for about 95% of the total gangliosides of normal adult liver and no difference was found in embryonic liver (4-10 months). But the ganglioside pattern of human surgical hepatoma specimens was strikingly different: GM_3 dropped to about 58.5% of the total, while the more complex gangliosides, especially GD_3, increased (Fig. 1.).[16,17] Further analysis of 5 surgical samples of hepatoma and four hepatoma cell strains, which were first proliferated to about

1 gram in nude mice for 6 weeks, gave similar ganglioside patterns (Fig.2,3). We have concluded from these experiments that the decreased GM_3 and increased GD_3 is a common phenomena in the hepatomas that we have studied (Table 2).

The gangliosides of normal rat liver consist of more multisialic acid gangliosides in comparison with those of human liver, However, rat hepatoma also contains highly elevated levels of GD_3 as well as reduced GM_3 (Table 2).[22,23,24,25] There are changes in other components of gangliosides, but they are not hepatoma specific (Fig. 4,5).

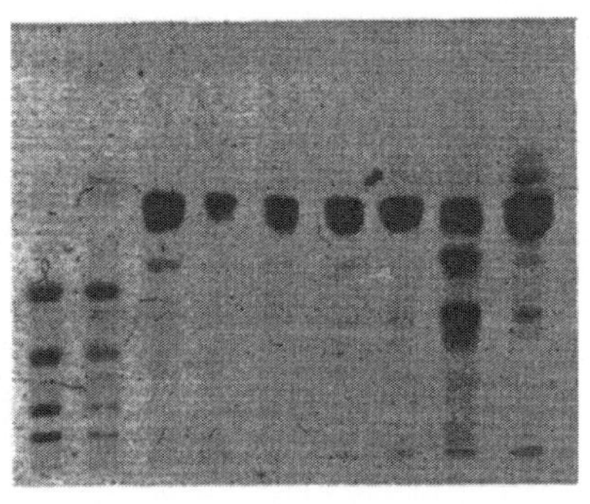

Fig. 1. Thin-layer chromatogram of gangliosides from human hepatoma and human embryonic liver (HEL). 1. Human brain gray matter; 2. Human brain white matter; 3. Four months HEL; 4. Six months HEL; 5. Eight months HEL; 6. New born; 7. Adult liver; 8. Human hepatoma; 9. Peripheral to tumor nodule.

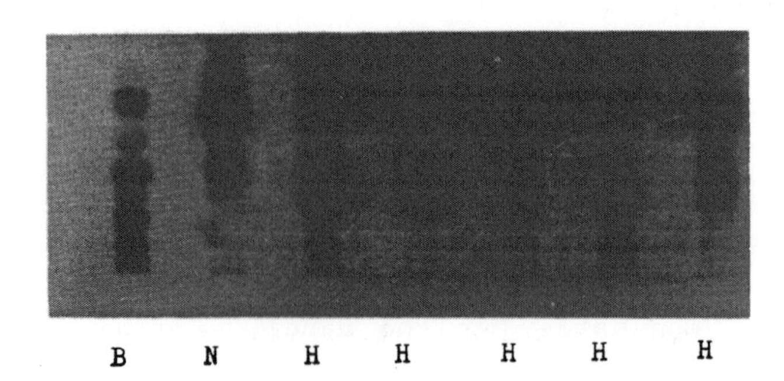

Fig. 2. Thin-layer chromatogram of gangliosides from normal liver and human hepatoma. **B** is human brain white matter, **N** is normal human liver, and **H** is human hepatoma.

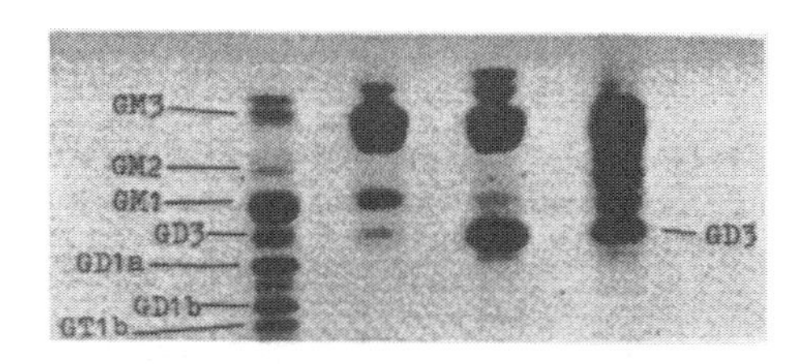

B N H H

Fig. 3. Thin-layer chromatogram of gangliosides from human liver and cultured human hepatoma cells. **B** is human brain white matter, **N** is normal liver, and **H** is cultured human hepatoma cell.

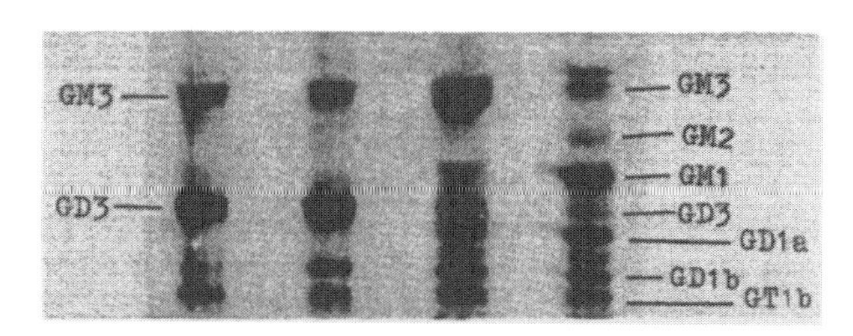

BERH-2 N B

Fig. 4. Thin-layer chromatogram from rat liver and rat BERH-2 hepatoma. **B** is human brain white matter, **N** is normal rat liver, and **BERH-2** is rat BERH-2 hepatoma.

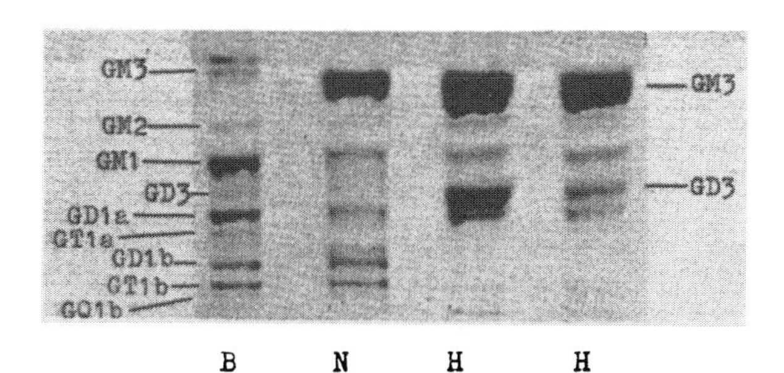

B N H H

Fig. 5. Thin-layer chromatogram of gangliosides form rat liver and rat hepatocarcinogenesis by 3'Me-DAB. **B** is human brain white matter, **N** is normal rat liver, and **H** is rat hepatocarcinogenesis by 3'Me-DAB.

Table 2. COMPARISON OF THE CONTENT PERCENTAGE OF GM_3 WITH GD_3 FROM VARIOUS LIVER TISSUES

Types	n sample	GM_3 %	GM_3 Ave.	GD_3 %	GD_3 Ave.
Normal human liver	3	95.9 95.0 93.0	94.6	1.2 0.7 3.0	1.6
Surgical human hepatoma	5	58.5 32.7 81.7 58.4 67.4	59.1	10.5 16.5 7.3 10.0 20.3	12.9
Cultured human hepatoma cell strain	2	68.0 47.8	57.9	10.5 5.5	8.0
Normal rat liver	3	55.4 41.2 32.7	43.1	4.3 7.6 9.0	7.0
BERH-2	2	17.5 20.8	19.2	49.2 43.1	46.2
Rat hepatocarcinogenesis by 3'Me-DAB	2	51.2 48.6 49.9	15.2	22.6	18.9

The possible role of cell proliferation on ganglioside pattern of hepatoma was ruled out by the study of regenerating liver. We closely monitored the ganglioside pattern for 3 days at an interval of 24 hours after the resection of three-fourths of rat liver. The ganglioside patterns had no appreciable changes from that of the normal as illustrated in Fig. 6,7.[26] Therefore, the increase of GD_3 is unique to hepatoma.

The above mentioned results clearly showed that there was considerable discrepancy between hepatoma cells and normal liver cells. Further studies of the properties of cancer cell membrane are important to an understanding of the behavior and abnormal gene expression of tumor cells, and will be helpful in developing new methods for cancer diagnosis and treatment.

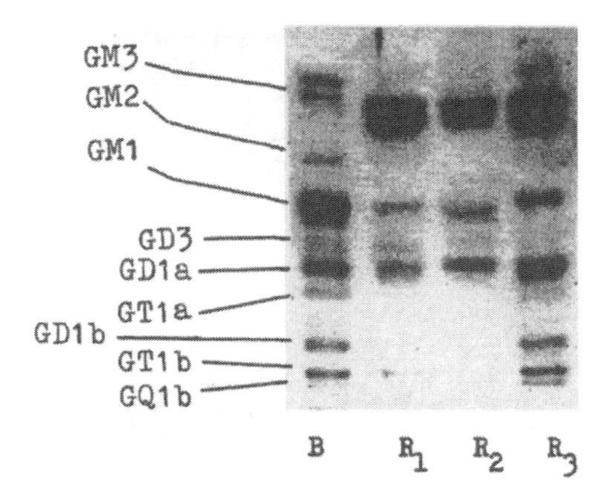

Fig. 6. Thin-layer chromatogram of ganglioside from regenerating liver of rat. **B** is brain white matter, **R1** is 24 hours after operation, **R2** is 48 hours after operation, and **R3** is 72 hours after operation.

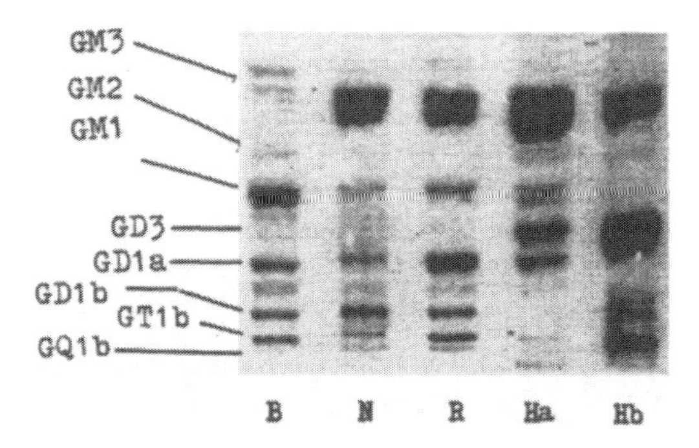

Fig. 7. Thin-layer chromatogram of gangliosides from different liver samples of the rat. **N** is normal liver, **B** is brain white matter, **R** is regenerating liver, **Ha** is hepatoma induced by 3'Me-DAB, **Hb** is hepatoma induced by nitrosamine.

LECTIN RECEPTORS ON THE SURFACE OF HEPATOMA CELLS

Another way to detect changes of oligosaccharide chains on the cell surface is to analyze the agglutinability of cells by lectins. Lectins are carbohydrate-binding proteins which are ubiquitous in plants and animals. Many papers have reported that, in contrast to normal cells, tumor and embryonic cells can be agglutinated by lectins in very low concentration.[26,27] This is also the case in hepatoma; we have examined human and animal experimental hepatoma cells of different sources: surgical samples, cell cultured lines and chemically induced tumors, all of which showed significant agglutination to concanavalin A and soy bean lectin at concentrations of 30 μg/ml. This is about one-twentieth the concentration that causes the agglutination of their normal counterparts.[28]

The proposed mechanisms of this simple phenomenon are different. Many factors have been investigated: the fluidity of plasma, the function of microtubules and microfilaments in cell agglutination and the changes of oligosaccharide chains on the cell surface (lectin receptors). According to our experiments on hepatoma cells, the presence or absence of intact microtubule and microfilament did not affect the cell agglutinability very much;[29] the determining factor was the affinity of lectin receptors of the cell surface.[30,31,32] We used the radioreceptor methods to examine the receptor number and the dissociation constant of lectin-receptor binding on hepatoma and normal liver cells. Lectin receptors (Con A and soybean lectin) were found to be almost equal in number on the two kinds of cells, but the dissociation constant (or its reciprocal association constant), as calculated from the Scatchard plots, revealed that there were two kinds of receptors on hepatoma cells. The Scatchard plot of the hepatoma cell is an upward concave curve instead of the straight line of the normal adult liver cell (Fig. 8). There are two ways to explain the Scatchard plot: the presence of two different kinds of receptors, or the presence of negative cooperation among receptors. These can be distinguished by certain experiments, such as heat inactivation or dilution inhibitory dissociation tests. Our results revealed that there are two kinds of receptors (low and high affinity) on hepatoma cells but only one type (low-affinity) on normal liver cells (Table 3). It is possible that the high agglutinability of tumor cells by lectins is related to the presence of high-affinity receptors on the cells.

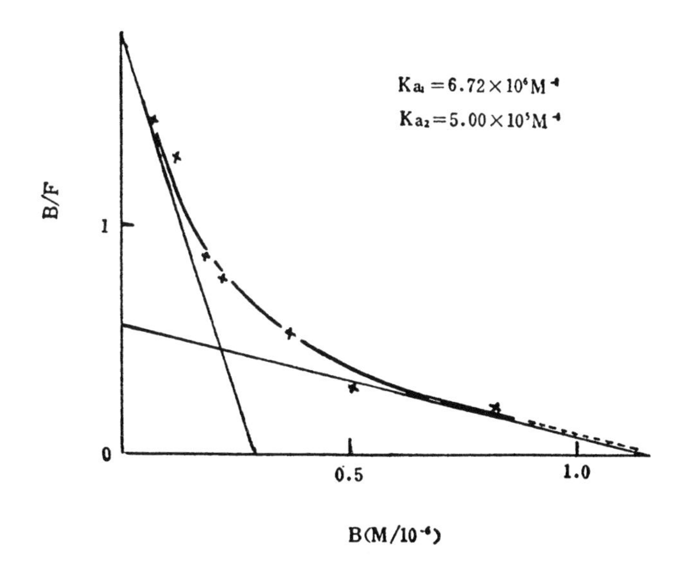

Fig. 8. The Scatchard plot of specific binding of Hep A with ^{125}I-Con A. Reaction system: ^{125}I-Con A (0.4μg) 10μl, unlabeled; Con A of different concentrations (20-100μg.) 10μl.; Hep A cells (5 x 10^6) 100μl.

Table 3. COMPARISON BETWEEN CON A RECEPTORS OF NORMAL LIVER CELLS AND THOSE OF HEP A CELLS

Cell membrane	Scatchard plot	Affinity constant (M^{-1})	No. of receptors per cell	No. of receptors per μm^2
Normal liver cell	linear	K=7.89x10^5	4.1x10^5	2.45x10^5
Hep A cell	upward concave curve	K_1=6.72x10^6 K_2=5.00x10^5	R_1=4.69x10^6 R_2=2.13x10^7 $R_1 + R_2$ = 2.60x10^7	5.68x10^4

HEPATIC LECTIN IN HEPATOMA

Hepatic lectin itself is a glycoprotein, which acts as a carbohydrate-binding protein.[33] Due to its high affinity to certain asialo-glycoproteins and cells in blood, it is assumed to be related to the degradation of these blood components.[34] Some authors also reported that this lectin might be responsible for the trapping of circulating metastatic tumor cells by the liver.[35] It is reasonable to infer that this molecule may play an important role in cell adhesion and recognition.[36] One of the characteristics of liver cancer is its ready metastasis, both within the liver and other organs. Therefore, it is interesting to know whether there is a difference in levels of the lectin between normal and liver cancer cells.

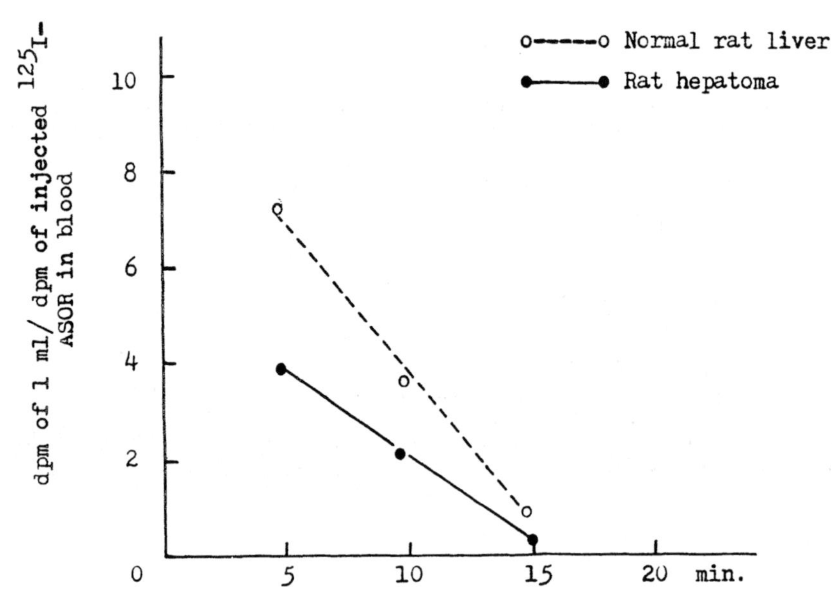

Fig. 9. Clearance of ^{125}I-ASOR in rat blood.

We first observed the clearance rate of ^{125}I-labelled asialoorosomucoid (^{125}I-ASOR) given intravenously from blood on normal rats and rats bearing hepatoma induced by 3'Me-DAB. Both had approximately the same clearance time of 14 minutes, as shown in Fig. 9. However, autoradiography demonstrated an uneven distribution in liver cancer, with tumor nodules having very low concentration of radioactivity (Fig. 10), which strongly suggests the loss of hepatic lectin activity in hepatoma cells. Further work was done to determine the relevance between this phenomenon and liver cancer. We investigated the lectin activity of the liver cancer from different sources: 3'Me-DAB-induced and nitrosamine-induced

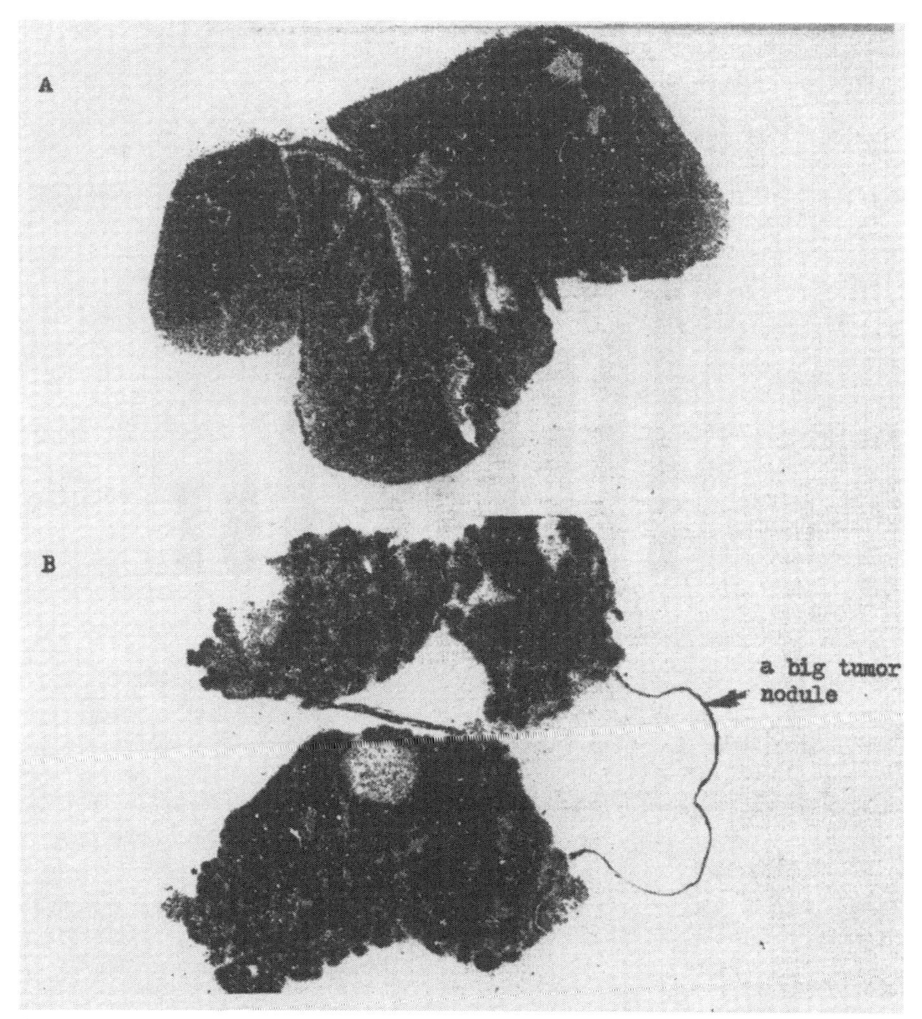

Fig. 10. Radio-autography of liver (14 minutes after injection of ^{125}I-ASOR). **A** is the normal rat liver, **B** is the 3'Me-DAB-induced rat hepatoma.

rat hepatoma (BERH-2), methylcholanthrane-induced mice ascitic hepatoma (Hep A), human hepatoma cell strain proliferated in nude rat and surgical samples of human hepatoma by radio-receptor method, as well as the dynamic changes of lectin activity during carcinogenesis. The results are shown in Fig. 11. Taking the lectin activity of a normal rat as 100%, BERH-2 has only 18.7% and 3'Me-DAB (fed 9 weeks) 32.9%. Lectin activity of human liver cancer is also decreased, as low as 36.2% of the normal human liver for surgical samples, and 29.5% for human liver cancers grown in nude rats. The results were checked with enzyme-linked immuno-chemical stain of liver slices and they matched well with that of the radio-receptor

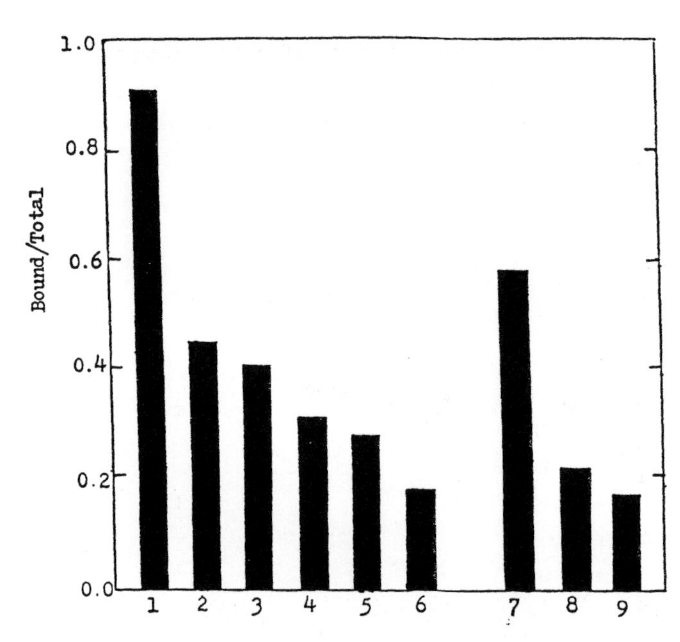

Fig. 11. Comparison of binding activity between normal liver lectin and liver cancer lectin. **1** is normal rat liver; Rat liver hepatoma induced by DAB: **2** is at 6 wks., **3** is at 8 wks., **4** is at 9 wks., **5** is BERH-2, **6** is Hep A, **7** is normal human liver, **8** is human liver cancer, **9** is human liver cell implanted in a nude rat.

method. The staining of cancer cells was much weaker than the normal.

There were two possibilities accounting for the reduction of hepatic lectin activity; one was the diminishing of molecular number and another, changes in microenvironment or molecular structure as Stockert demonstrated, that removal of sialic acid resulted in total loss of lectin activity.[46] Further study in our laboratory proved that the former was the predominant reason. We inhibited the lectin activity using specific polyclonal antibody and found that less amount of antibody was needed to produce same inhibitory effect in hepatoma cells compared with normal liver cells, approximately paralleling with the reduction of lectin activity (unpublished), which clearly indicated that the membrane of hepatoma cells had fewer lectin molecules. The observation of dynamic changes of lectin activity during the procedure of carcinogenesis by 3'Me-DAB revealed a very interesting phenomenon, e.g., the decrease of lectin activity occurred much earlier than the appearance of morphologically malignant cells. We think it is most likely due to the modified gene expression at the early stage, although the direct toxic effect of 3'Me-DAB may also play a role.

TRANSFERRIN RECEPTORS ON HEPATOMA CELL SURFACE

Transferrin receptor is a specific cell surface receptor for transferrin, an iron-transporting protein in plasma.[38] It is by nature a glycoprotein with a molecular weight of 100,000 daltons.[39] Recent studies indicate that the receptor correlates with cell proliferation and most of the tumor cells have an increased amount of transferrin receptors.[40,41,42,43] It is natural to assume that transferrin is an essential factor for cell growth, the supposed mechanism of which is that the iron transported by it may be required for key reactions of catalysis of energy metabolism and DNA synthesis.[44] However, it appears to be a more complicated process than simply a reflection of iron requirement of cells.[45,46]

Hepatoma cells, like other malignant cells, exhibit a high density of transferrin receptors on their surface.[47] However, we found neither transferrin nor antitransferrin receptor antibody demonstrated such excellent specificity to hepatoma cells *in vivo* as they did to hepatoma slices *in vitro*. ^{125}I-labelled transferrin or antireceptor IgG was injected intraperitoneally in mice bearing hepatoma or mammary cancer induced by nitrosamine.[48] The mice were sacrificed every 24 hours up to 120 hours after injection and the radioactivity of different organs was measured. As shown in Fig. 12, radioactivity in hepatoma tissue was only 3.49 times higher than that of normal liver tissue (hepatoma/ normal liver 3.49), when ^{125}I-transferrin was used as a tracer, and it was even lower (hepatoma/normal liver 2.18) if ^{125}I-antireceptor IgG was used. To our great interest, appreciable amounts of transferrin and anti-receptor IgG accumulated in the lung *in vivo*, but we could not detect transferrin receptor by immuno-peroxidase staining *in vitro*. This was also claimed by other authors using several kinds of monoclonal antibodies.[37] How does one explain the discrepancy found

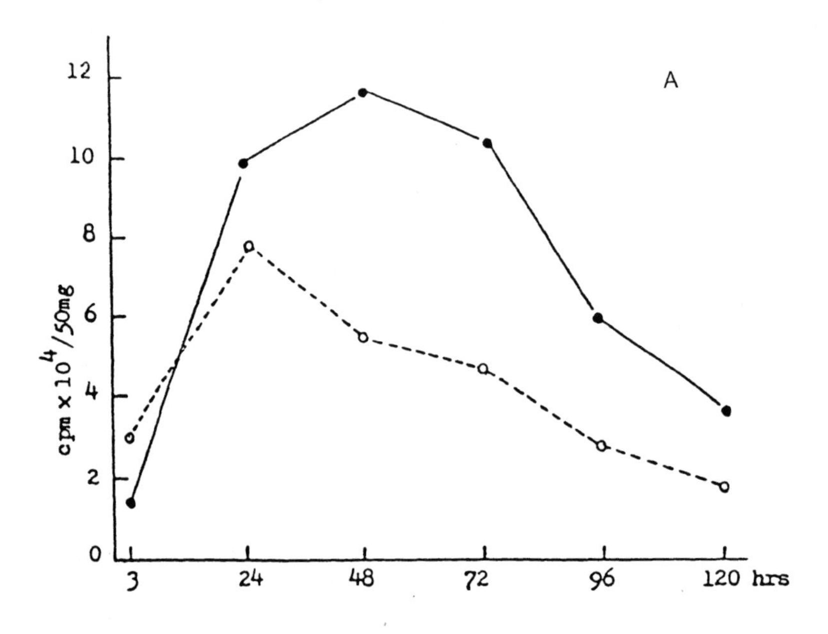

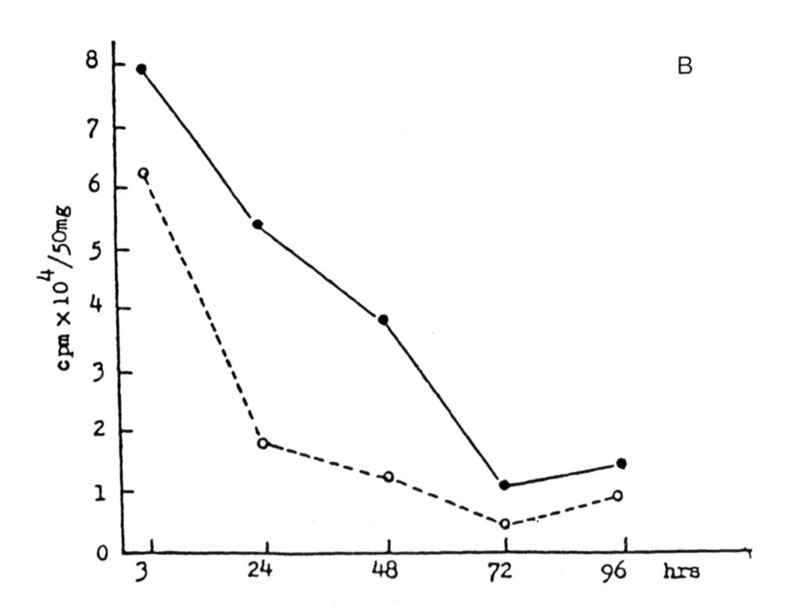

Fig.12. The distribution of ^{125}I-transferrin **(A)**, and ^{125}I-anti-transferrin receptor IgG **(B)**, in nude mice with transplanted human hepatoma.

o-----o Normal liver of nude mice.

•——• Transplanted human hepatoma.

between experiments *in vitro* and those *in vivo* ? It does not seem possible to us that the distribution pattern *in vivo* is caused by the effects of diffusion rate. The reason is that asialoorosomucoid, a protein with a molecular weight lower than transferrin, is taken up exclusively in liver within a short time.[37] We have tried a few other "tumor specific" antibodies such as anti-α-fetoprotein[49] and anti-acid ferritin antibodies.[50] They all have the same problem. It is, perhaps, the greatest challenge we must face if we will eventually apply radioimmune scanning and immunotoxin treatment to improve the diagnosis and therapy of cancer.

ACKNOWLEDGEMENTS

This work was supported in part by a grant from The Fund for Sciences from the Chinese Academy of Sciences.

REFERENCES

1. S. Hakomori. Monoclonal antibodies directed to cell-surface carbohydrates, in "Monoclonal antibodies and functional cell lines." pp. 67-100. R.H. Kennet, K.B.Bechtol, and T.J. McKaarn ed., Plenum Publishing, New York. (1984).
2. R.F. Irie, L.L. Sze, and R.E. Saxton. Human antibody to OFA-1, a tumor antigen, produced *in vitro* by Epstein-Barr virus-transformed human B-lymphoid cell lines. Proc. Natl. Acad. Sci. USA 79:5666-5670, (1982).
3. Shun-zhang Yu. Epidemiology of primary liver cancer, *in* "Subclinical Hepatocellular Carcinoma," Z.Y. Tang ed., pp.189-211. China Academic Publishers, Beijing, Spring-Verlag, Berlin. (1985).
4. Zhao-you Tang. A new concept of the natural history of AFP positive hepatocellular carcinoma, *in* "Subclinical Hepatocellular Carcinoma," Z.Y. Tang ed., pp. 171-177. China Academic Publishers, Beijing, Spring-Verlag, Berlin. (1985).
5. R.O. Brady, C. Borek, and R.M. Bradley. Composition and synthesis of ganglioside in rat hepatocyte and hepatoma cell lines. J. Biol. Chem. 244:6552-6554, (1969).
6. S. Hakomori. Change of glycolipid pattern in Morris hepatoma 5123 and 7800. Cancer Research, 30:2930-2936, (1970).
7. E.V. Dyatlovitskaya. Gangliosides of rat liver and hepatoma 27. Biochimya (Russ), 39:552-556, (1974).
8. A.M. Dnistiran, V.P.Skipski, M. Barclay, and C.C. Stock. Alterations in glycosphingolipids of plasma membranes

from Morris hepatoma 5132TC. Cancer Res., 37:2182-2187, (1977).

9. E.V. Dyalovitskaya, A.M. Novikov, N.P. Gorkova, and L.D. Bergelson. Gangliosides of hepatoma 27, normal and regenerating rat liver. Eur. J. Biochem., 67:357-364, (1976).

10. R.P. Van Hoeven and P. Emmelot. Plasma membrane lipids of normal and neoplastic tissues, in "Tumor lipids: Biochemistry and Metabolism", pp. 126-138. E. Wood ed., Amer. Oil Chemist's Soc. Press: Illinois, (1973).

11. W.D. Merritt, C.L. Richardson, T.W. Keenan, and D.J. Morre. Gangliosides of liver tumors induced by N-2-fluorenylacetamide I. Ganglioside alterations in liver tumorigenesis and normal development. J. Nat'l. Cancer Inst., 60:1313-1327, (1978).

12. W.D. Merritt, C.L. Richardson, T.W. Keenan, and D.J. Morre. Ganglioside of liver tumors induced by N-2-fluorenylacetamide II. Alterations in biosynthetic enzymes. J. Nat'l. Cancer Inst., 60:1329-1337, (1978)

13. D.J. Morre, T.M. Kloppel, W.D. Merritt, and T.W. Keenan. Glycolipids as indicators of tumorigenesis. J. Supramol. Struct., 9:157-177, (1978).

14. J. Kawanami. Lipids of cancer tissue III. Glycolipids of human hepatoma tissue. Jap. J. Exp. Med., 38:11-18, (1968).

15. Tanno Munehiko, et al. Copmparison of gangiosides pattern in cirrhotic liver and liver cell cancer. Liver, (Japan) 23:684, (1982).

16. Neng-hua Gho and Tian-Jue Gu. Ganglioside pattern in hepatoma. Acta. Acad. Med. Primae, Shangai, 12:149-151, (1985).

17. Zheng-mei Zhu, De-sheng Zhang,Ji-ding Yuan, and Tian-jue Gu. The distribution pattern of ganglioside in normal, embryonic, and cancerous human liver. Acta. Acad. Med. Primae., Shangai, 10:234-236, (1983).

18. V.P. Walter, T.M. Kloppel, I.G. Deimling, and D.J. Morre. Alterations in neutral glycosphingolipids from transplantable hepatomas and in sera of rats bearing transplantable hepatomas. Cancer Biochem. Biophys., 4:145-151, (1980).

19. T. Momoi, S. Ando, and Y. Magai. High resolution preparative column chromatographic system for ganglioside DEAE-Sephadex and a new porous silica, Iatrobeads. Biochim. Biophys. Acta, 441:448-497, (1976).

20. R.W. Ledeen and R.K. Yu. Gangliosides: structure, isolation, and analysis, in "Methods on Enzymology" V. Ginsberg ed., 83:139-191, (1982).

21. S. Hakomori. Tumor associated glycolipid antigens defined by monoclonal antibodies. Bull. Cancer, 70:118-126, (1983).

22. Neng-hua Guo and Tian-Jue Gu. Ganglioside pattern in hepatocellular hepatoma. Acta Biochim. Biophys. Sinica, 17:538-543, (1985).

23. Xia-juan Xia, Zhao-chun Cui, and Tian-Jue Gu. Isolation and purification of the hepatoma associated ganglioside. Chinese Biochem J. 1(2):13-17, (1985).

24. Xia-juan Xia, Zhao-chun Cui, and Tian-Jue Gu. Composition and carbohydrate sequence analysis of the hepatoma

associated ganglioside. Acta Biochim. Biophys. Sinica, 17:615-623, (1985).

25. Xia-juan Xia, De-sheng Zhang, Zhao-chun Cui, Lan-nian Zhang, and Tian-Jue Gu. Permethylation analysis of hepatoma associated ganglioside. Acta Biochim. Biophys. Sinica, 17:635-642, (1985).
26. M. Inbar and L. Sachs. Interaction of the carbohydrate-binding protein Concanavalin A with normal and transformed cells. Proc. Nat'l. Acad. Sci. USA, 63:1418-1425, (1969).
27. L.D. Berman. Lack of correlation between growth characteristics, agglutinability by plant lectins and the malignant phenotype. Int. J. Cancer, 15:973-979, (1975).
28. Zheng-Mei Zhu and Tian-Jue Gu. The agglutinative effect of Con A and PHA on normal, embryonic, and tumor cells. Chinese J. Cell Biol., 4(1):35-37, (1982).
29. Zheng-Mei Zhu and Tian-Jue Gu. The function of cell membrane in cell agglutination. Chinese J. Cell Biol. 4(2):16-19. (1982).
30. Zheng-Mei Zhu and Tian-Jue Gu. Con A receptors on normal liver and hepatoma cells. Chinese J. Cell Biol. 4(3):21-24, (1982).
31. Zheng-Mei Zhu and Tian-Jue Gu. Agglutinin and cell agglutination IV characterization of Concanavalin A receptors on hepatoma cells. Acta Biochim. Biophys. Sinica, 14:295 301, (1982).
32. Xie-lin Ren, Jun-Cai Lin, and Tian-jue Gu. Comparison of soybean agglutinin receptors on hepatoma and normal mouse liver cells. Tumor, 2(5):4-6, (1982).
33. J.U. Baenziger and Y. Maynard. Human hepatic lectin purification. J. Biol. Chem., 225:4607-4613, (1980).
34. A.G. Morell, G. Gregoriadis, and I.H. Scheinberg. The role of sialic acid in determining the survival of glycoproteins in the circulation. J. Biol. Chem., 246:1461-1467, (1971).
35. G. Uhlenbruck, J. Beuth, and V. Weidtman. Liver lectins: mediators for metastases? Experentia, 39:1314-1315, (1983).
36. S.H. Berondes. Lectins: their multiple endogenous cellular functions. Ann. Rev. Biochem., 50:207-231, (1981).
37. Hui-ren Zhou, Zheng-Mei Zhu and Tian-Jue Gu. ß-galactoside specific lectins in normal anmd cancer bearing rat liver cells. Acta Acad. Med. Primae, Shangai, 11:361-364, (1984).
38. J.H. Jandel and J.H.. Katz. The plasma-to-cell cyle of transferrin. J. Chin. Invest. 42:314-326, (1963).
39. I.S. Trowbridge, R.A. Newman, and D.L. Domingo. Transferrin Receptors: structure and function. Biochem. Pharmcol., 33:925-932, (1984).
40. W.P. Faulk and G.M.P. Galbraith. Trophoblast transferrin and transferrin receptor in the host parasite relationship of human pregnancy. Proc. Royal Soc. Lond.(Biol), 204:83-97, (1979).
41. G.M. Galbraith, R.H. Galbraith, and W.P. Faulk. Transferrin binding by human lymphoblastoid cell lines and other transformed cells. Cell Immun., 49:215-222, (1980).
42. R. Sutherland, D. Delia, C. Schneider, R. Newman, J. Kemshead, and M. Greaves. Ubiquitous cell surface

glycoproteins on tumor cells is proliferation associated receptor for transferrin. Proc. Nat'l. Acad. Sci. USA, 78:4515-4519, (1981).

43. D. Delia, M.F. Greaves, R. Newman. Modulation of T-leukemic cell phenotype with phorbol ester. Int. J. Cancer, Int. J. Cancer,29:23-31, (1982).

44. P. Reichard. From deoxynucleotides to DNA synthesis. Federation Proc., 37:9-14, (1978).

45. L. Vodinelich, R Sutherland, C. Schneider, R. Newman, and M. Greaves. Receptor for transferrin may be a target structure for natural killer cells. Proc. Natl. Acad Sci. USA, 80:835-839, (1983).

46. R. J. Stockert, A. G. Morell and I.H. Scheinberg. Hepatic binding protein: The protective role of its sialic acid residues. Science 197:66-68, (1977).

47. Hui-ren Zhou, Jin-yu Ma, and Tian-Jue Gu. Detection of transferrin receptor on cancer cell mebrane and the preparation of transferrin receptor and its antibody. Acta. Acad Med. Primae Shangii 11:165, (1984).

48. Hui-ren Zhou, Jian-nan Yeh, and Tian-Jue Gu. The uptake of ^{125}I-transferrin in N-nitrosurea induced rat mammary tumor. In press.

49. Yu-ding Kuo, Z.Y. Tang, Z.Y. Lin, Z.S. Ma, K.G. Cheng, and W.Y. Yhao. Observation of radioimmunodetection of ^{131}I-anti AFP in nude mice bearing human hepatocellular carcinoma. Tumor, 51:4-6, (1985).

50. D. Yu and Z.Y. Tang. Radioscanning of hepatocellular cancer using ^{131}I-anti-acid ferrin antibody. Thesis for M.S. degree, (1984).

ABSTRACT

CANCER-ASSOCIATED CARBOHYDRATE ANTIGENS IN MUCINS

John L. Magnani

Laboratory of Structural Biology, NIDDK, National Institutes of Health Bethesda, MD (USA)

Many monoclonal antibodies which detect differentiation or cancer-associated antigens are directed against carbohydrates. Of about 500 monoclonal antibodies that we have obtained from different laboratories, 124 bind carbohydrates. We have determined the carbohydrate structure of the epitope for 93 of these antibodies. Most tumor markers of gastrointestinal and pancreatic cancer are sialylated oligosaccharides found in mucins in patient's sera. Some of these oligosaccharides, such as sialylated Le^a (19-9), are found in both glycolipids (1) and mucins (2). Determining the structure of this particular antigen has been helpful in evaluating its genetic distribution in cancer patients. As predicted by the Lewis blood group association of this structure, Le(-) individuals lack this antigen, Le(b+) individuals synthesize low or undetectable levels, while Le(a+) individuals systhesize high levels of this antigen (3). This knowledge helps improve the reliability of current assays for 19-9 in patients' sera.

Other oligosaccharide epitopes on mucins are not found in glycolipids and must be released intact from glycoproteins to be analyzed. A new method has been developed to detect these oligosaccharides by chromatography on thin layer plates followed by immunostaining with monoclonal antibodies after covalent attachment of the oligosaccharides to the silica gel (4). Using this method 10 pmol of carbohydrate antigen can be detected. By analyzing the structure of these oligosaccharides in mucins it may be possible to predict better tumor markers for gastrointestinal and pancreatic cancer.

Many antibodies that recognize the same carbohydrate sequence react differently with tissues. Some antibodies bind slightly different epitopes, while others bind the same epitope with different affinities (5). The binding of these antibodies to cells depends on the density of antigens on the cell surface, each antibody requiring a different density. Thus, cells containing antigen below a certain threshold concentration may not bind low-affinity antibodies.

Reference

1) Magnani, J.L., Nilsson, B., Brockhaus, M., Zopf, D., Steplewski, Z., Koprowski, H., and Ginsburg, V. (1982) J. Biol. Chem. 257:14365-14369; 2) Magnani, J.L., Steplewski, Z., Koprowski, H., and Ginsburg, V. (1983) Cancer. Res. 43:5489-5492; 3) Brockhaus, M., Wysocka, M., Magnani, J.L., Steplewski, Z., Koprowski, H. and Ginsburg, V. (1985) Vox Sang. 48:34-38; 4) Magnani, J.L. (1985) Anal. Biochem. 150:13-17; 5) Magnani, J.L., Ball, E.D., Fanger, W.W., Hakomori, S., and Ginsburg, V. (1984) Arch. Biochem, Biophys. 233:501-506

N-GLYCOLYLNEURAMINIC ACID-CONTAINING GANGLIOSIDES AS A TUMOR ASSOCIATED ANTIGEN IN HUMAN: EXPRESSION OF HANGANUTZIU-DEICHER ANTIGEN ACTIVE GANGLIOSIDES ON HUMAN COLON CARCINOMA AND MELANOMA TISSUES

Yoshio Hirabayashi and Makoto Matsumoto[1]
Hideyoshi Higashi and Shiro Kato[2]

[1]*Department of Biochemistry, Shizuoka College of Pharmacy, Oshika, Shizuoka-shi 422, Japan*
[2]*Department of Pathology, Research Institute for Microbial Diseases, Osaka University, Suita, Osaka 565, Japan*

The function of glycosphingolipids on cell surface membranes as "cellular" antigen has recently received much attention. Hanganutziu-Deicher (HD) antigen is an antigen recognized by a human heterophile antibody, HD antibody. Glycosphingolipids containing *N*-glycolylneuraminic acid (NeuGc) were found to possess HD antigen activity (1). NeuGc is widely distributed in most animals except human and chickens. Recently, the antigen has been found in malignant tissues and sera from patients with various tumors (2,3). Here we described isolation and characterization of HD antigen-active gangliosides as a tumor associated antigen in human colon carcinoma and melanoma tissues.

In order to characterize HD-antigenic gangliosides as tumor associated antigen for human, we have developed a method to purify anti-HD antibody (4), and used the antibody for direct demonstration of HD-active gangliosides in the TLC-separated glycosphingolipids. The procedure consists of immuno-reaction among NeuGc-containing gangliosides, affinity-purified anti-GM3(NeuGc) and horseradish peroxidase-conjugated rabbit anti-chicken IgG, and the peroxidase reaction using 4-chloro-1-naphthol as a chromogenic substrate. Quantitative determination was performed by densitometric scanning of the enzyme-immunostained spots on the TLC plate. As little os 0.2 pmol of HD-active gangliosides such as GM3(NeuGc), NeuGc-nLcOse$_4$Cer and NeuGc-nLcOse$_6$Cer could be detected with a good S/N ratio (5). The chemical structures were further characterized by the method combining with 2d-TLC/enzyme-immunostaining and treatment with specific glycosidases or chemical reagent.

One to six species of the HD-antigenic gangliosides were isolated from 7 of 16 cases of colon cancers. Tissues from different patients showed different patterns of molecular species of the antigenic gangliosides. Four species of antigenic gangliosides could be identified as GM3(NeuGc), 4-*O*-Acetylated NeuGc-containing GM3, GM2(NeuGc) and NeuGc-nLcOse$_4$Cer by their mobilities on 2d-TLC and the effect of mild alkaline treatment, sialidase treatment, β-hexosaminidase/GM2-specific activator digestion (Fig.) and endo-β-galactosidase treatment. Densitometric determination revealed that HD-antigenic sialic acid accounted for only about 1% or less of the total lipid-bound sialic acids.

HD-active gangliosides were also detected in human meconium and fetal intestinal tissues. Three species of antigenic gangliosides in pooled meconium were identified as GM3(NeuGc), NeuGc-nLcOse4Cer and NeuGc-nLcOse6Cer by their migration on 2d-TLC and endo-β-galactosidase treatment. GM3(NeuGc) was a sole HD-active ganglioside in one fetal intestine from 3 individuals. Since HD-antigenic gangliosides can never be detected in normal human colorectal tissues, they are characterized as onco-fetal antigens.

Human melanoma tissues (5 of 6 patients) also expressed the *N*-glycolylneuraminic acid-containing gangliosides. Tissues from different patients gave different patterns of molecular species of the HD-active gangliosides. Four molecular species could be determined to be GM3(NeuGc), GM2(NeuGc), GD3(NeuGc-NeuAc) or GD3(NeuAc-NeuGc) and GD3(NeuGc-NeuGc) by their behavior on 2d-TLC.

It is very important to understand the precise tissue distribution of the tumor-associated antigens with HD antigen activity and mechanism of the expression in cancer tissues. For this purpose, the specific antibodies against those antigens will be very useful. Indeed, Kasai *et al.* could prepare the antibody specific for GM2(NeuGc) (6) and use the antibody for detection of the antigen in human colon cancers.

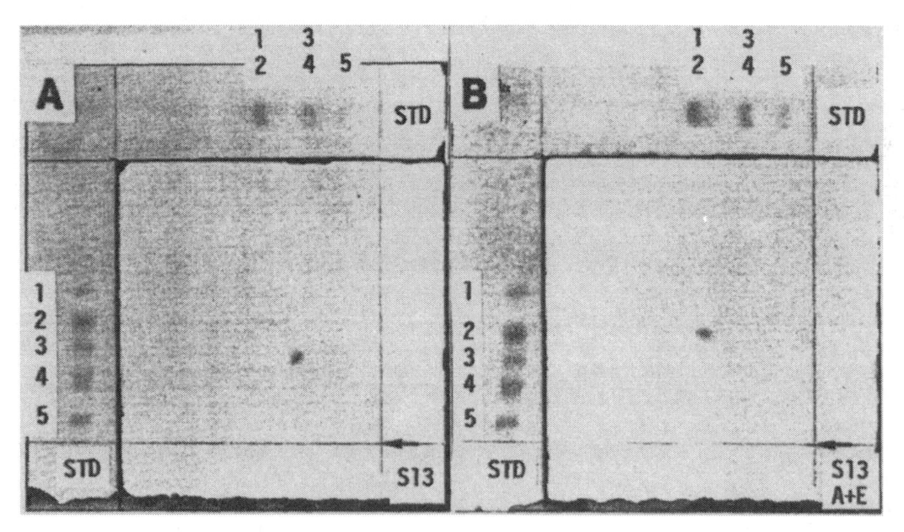

Fig. Detection of GM2(NeuGc) in human colon cancer by Two-dimensional TLC/enzyme-immunostaining analysis. Plate A, HD-active gangliosides isolated from colon tissues; plate B, the HD antigenic gangliosides after incubation with β-hexosaminidase A in the presence of GM2-specific activator isolated from guinea pig kidney.

References: 1) Higashi, H. *et al.*, Biochem. Biophys. Res. Commun. 79:388-395 (1977). 2) Ikuta, K. *et al.*, Biken J. 25:47-50 (1982). 3) Nishimaki, T. *et al.*, J. Immunol. 122:2314-2318 (1979). 4) Hirabayashi, Y. *et al.*, J. Biochem. 94:327-330 (1983). 5) Higashi, H. *et al.*, J. Biochem. 95:1517-1520. 6) Kasai, N. *et al.*, Biochem. Biophys. Res. Commun. 129:334-341 (1985).

STRUCTURAL ELUCIDATION OF COMPLEX CARBOHYDRATES

Vernon N. Reinhold and Sitthivet Santikarn

Department of Nutrition
Harvard School of Public Health
Boston, Massachusetts 02115

The ability to recognize the chemical markers that distinguish "self" from "foreign" molecules is the first and most important step in the initiation of an immune response. The recognition process essentially involves two components, the antibody/receptor and the antigen. The complete structures of these two components need to be established so that the highly specific process can be understood at the molecular level. In the past decade the application of recombinant DNA technology to dissect the immune system has provided a great deal of insight into the structure and function of antibodies and antigens.

Complex carbohydrates represent a significant class of antigenic determinants whose structures are not readily studied by the recombinant DNA techniques; the use of a combination of various enzymes, chemistry, and separation and analytical instruments have yielded some results. However, our ability to establish structures of complex carbohydrates is woefully inadequate when it is compared with our ability to determine structures of other biopolymers such as proteins and DNA's. The reason for this is partly due to the fact that oligosaccharide structures are compounded with a multiplicity of linkage types, chain branching, structural isomerism, and a great diversity of conjugate structures that are not generally observed in other biopolymers. These are the features that provide antigens with great molecular diversity and demand highly specific analytical approaches.

Mass spectrometry may well provide a solution to some of the above problems. The recent technological developments for the analysis of biological molecules (i.e., high molecular weights, polar and thermally unstable molecules) by mass spectrometry (MS) have yet to be fully exploited for the study of complex carbohydrates. In this chapter we will briefly describe some applications of state-of-the-art mass spectrometry instrumentation in the structure determination of complex carbohydrates. The use of gas chromatography coupled to a mass spectrometer (GC-MS) for the linkage and branching determination has recently been reviewed (Van Langenhove & Reinhold, 1985). Specifically, the results that can be obtained from, and the limitations of the direct chemical ionization (DCI), fast atom bombardment (FAB), and collisionally induced decomposition (CID) mass spectrometry will be discussed under appropriate headings.

Sequence information obtained by MS is dependent on a periodic polymer fragmentation that is characteristic of monomer array. For oligosaccharides, the major fragments from these processes can be explained by cleavage on either side of the glycosidic oxygen followed by a proton transfer. These cleavages have been observed under chemical and fast atom bombardment ionization, as well as for neutral gas collisions. The exceptional glycosidic lability can probably be related to two structural aspects: That of bond energetics; and the fact that intervening ring rupture requires the cleavage of two bonds for depolymerization. Factors that enhance glycosidic rupture relative to ring fragmentation would be expected to accentuate sequence information. These features are demonstrated below with direct chemical ionization.

DIRECT CHEMICAL IONIZATION

Chemical ionization (CI) mass spectrometry provides a technique to maximize molecular and high mass information by replacing the conventional highly exothermic process of electron ionization with a low energy, chemical process involving ion-molecule reactions. The essential feature of CI involves electron ionization of a reagent gas maintained at near atmospheric pressures which results in the generation of an ion plasma. Introduction of a sample into this plasma causes positive ion production by one of three processes: (a) charge exchange, by using a non-hydrogen containing gas; (b) proton transfer, where the reagent gas functions as a BrØnsted acid; or by (c) formation of collision-stabilized complexes or adducts, the reagent gas functions as a Lewis acid. Processes (b) and (c) are thermodynamic in nature and distribution of ion products is dependent on reactant structure and their proton affinities. The topic has been reviewed by one of the major contributors to this area (Munson, 1977) and an excellent book covering the topic has recently been published (Harrison, 1983).

Direct chemical ionization (DCI) involves the placement of sample on an electrically resistant wire (emitter) which can be inserted directly into an ion plasma. The materials are "distilled" from this emitter by a programmed heating current (Fig. 1). Samples analyzed by this direct method frequently provide mass spectra with enhanced high mass fragments. Considering the overall ease of operation, the high sensitivity, the ionization flexibility available by using different and isotopically labeled reagent gases, and the excellent opportunity this provides for the study of molecular fragmentation, it is surprising that DCI has not seen widespread application.

The essential features and limitation of the DCI technique can be directly related to the heat initiated desorptive process. Spectra acquired early during sample desorption (minimal heat) show molecular weight-related ions only. Subsequent spectra, acquired at higher emitter heating currents, show increase in low mass fragments with decreasing abundance of molecular weight-related ions. This increasing pyrolytic component of the DCI method can be advantageous in that it provides additional structural detail within a single sample analysis. However, as one proceeds to the analysis of higher molecular weight and more polar samples, the thermal

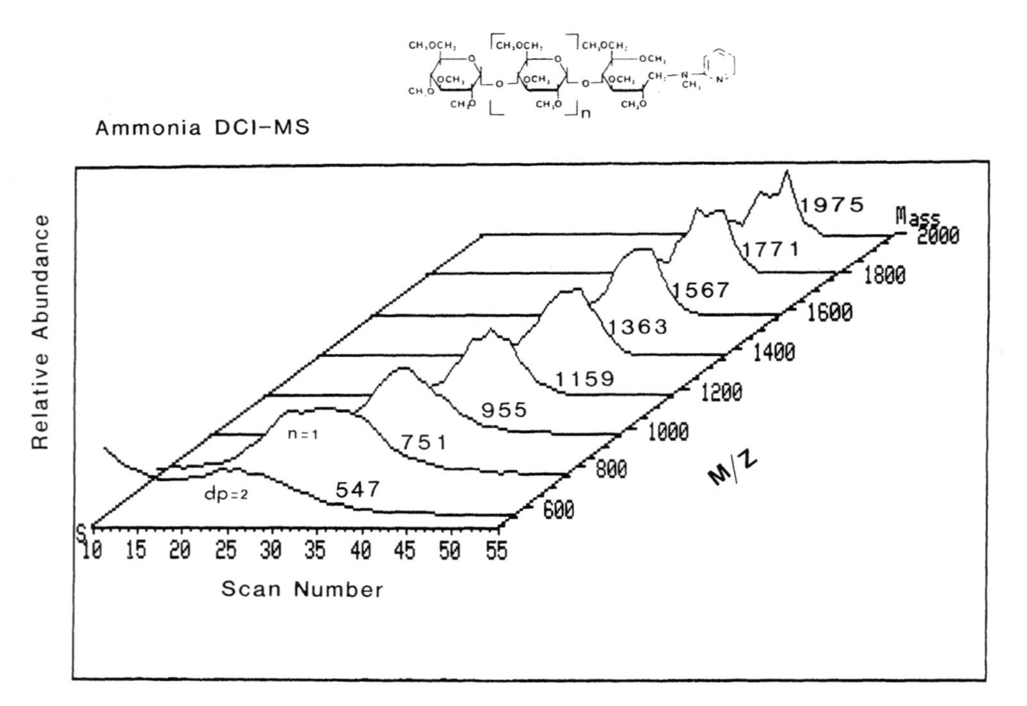

Fig. 1. Selected ion chromatograms for a series of permethylated pyridinylamino oligosaccharide derivatives prepared from corn syrup. The low molecular weight components appear in the early scans (low emitter temperature) while the high molecular weight components appear at high temperature.

energy needed to initiate desorption becomes excessive and the spectra exhibit a proponderance of pyrolytic fragments at the expense of the molecular weight-related ions. Therefore, the application of this technique is somewhat limited in terms of sample mass range (ca. 5000 daltons) and sample polarity. Various derivatization methods (e.g., acetylation, methylation, esterification) can be used to reduce sample polarity and increase its thermal stability.

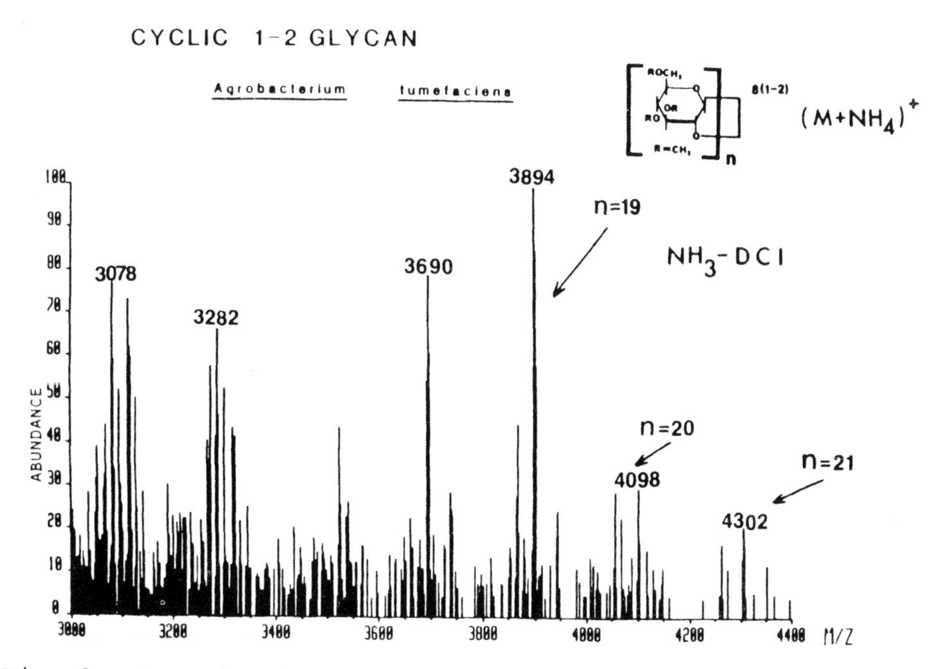

Fig. 2. Ammonia DCI mass spectrum of permethylated mixture of cyclic glycans from *Agrobacterium tumefaciens* (Miller,et al., 1986).

For carbohydrates, we have found DCI most successful on permethylated materials when using ammonia as the reagent gas (Fig. 2). Excellent sequence information can be obtained for samples with molecular weights of up to 3,000 daltons and the technique appears to be three to five times more sensitive than fast atom bombardment mass spectrometry. The principle features and advantages of ammonia DCI are illustrated in Figure 3 for the ammonia DCI of a permethylated heptasaccharide. The molecular weight can be determined from the ammonium adduct ion at m/z 1508 and sequence ions are observed for each saccharide moiety (Reinhold,et al.,1982). Some of these fragments are isomeric but can be differentiated by

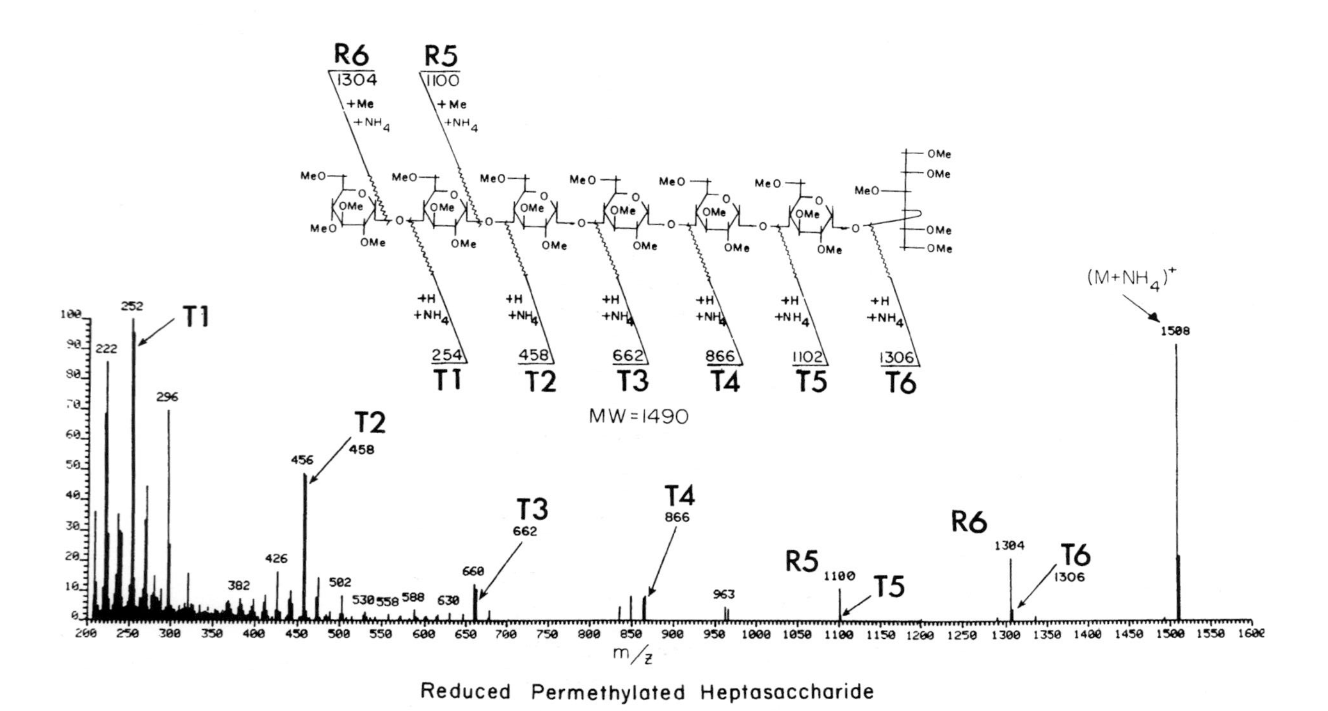

Fig. 3. Direct chemical ionization MS of a reduced and permethylated glucoheptulose a(1-4) glycan using ammonia as a reagent gas, (Reinhold and Carr, 1982, with permission).

reduction of the glycose end of the polysaccharide with $NaBH_4$. These studies, in combination with the use of isotopically labeled reagent gases ($^{15}N^1H_3$, $^{14}N^2H_3$) has provided a detailed understanding of oligosaccharide fragmentation. The $^{15}N^1H_3$ experiments have consistently shown that all ion fragments are incremented by one mass unit. The clustering of ammonia (ammonia/ammonium) with itself $\{(NH_3)_nNH_4\}^+$, parent ions $(M + NH_4)^+$ and fragments $(F + NH_3)^+$ as well as neutral species provides a situation in which a large portion of desorbed sample and its fragments are observed. This fact has been illustrated earlier (Reinhold and Carr, 1984) and is in contrast with many other MS techniques where neutral species are produced and remain undetected.

The origins of oligomer sequence information can be shown from the fragmentation of a simple glucoside (Scheme 1). From this scheme and the fragment ion abundance in Figure 3, it can be seen that all major fragments in ammonia DCI analysis are related to glycosidic cleavage; and important feature when studying sequence. Does ammonia serve to catalyse or react chemically at the glycosidic linkage to induce rupture (aminolysis)? There have been reports of ammonia reacting with aldehydes (Hunt, 1974) and ketones (Tabet and Fraisse, 1981; DeMark and Klein, 1981; Rudewicz and Munson, 1985) under CI conditions, thus, it may be that the nucleophilicity of ammonia enhances sequence ion information by gas phase chemistry (pathway b, Scheme 1).

The application of ammonia DCI to glycolipid materials has proven equally successful by providing molecular weight, sequence and lipid ester detail in a single analysis (Carr and Reinhold, 1984). This can be illustrated in Figure 4 for a permethylated and reduced ceramide trihexoside. Reduction of this amide fatty acid conjugate, previous to analysis (Karlsson, *et al.*, 1974), enhances DCI elimination of the methylated alkyl amine. These fragments provide a separate evaluation of the fatty ester composition (i.e., m/z 342, 368, 370, 398, 400) which is frequently the major cause of molecular ion heterogeneity, $(M + H)^+$. The opportunity to compare these two groups of ions is of considerable value especially for relating structural features to the sphingosine moiety. Carbohydrate sequence ions can be observed which are identical to those in Figure 3 (e.g., m/z 254, 458, 662) and represent the major fragments in the spectra. All of these fragments are ammonium adduct ions. Alterations in structure are readily observed by DCI as exemplified below with a related sample. Extension of the carbohydrate chain by an amino sugar moiety (globotetraglycosylceramide, GL-4) provides the same sequence information with the appropriate shift in mass(Fig. 5). Because of the basicity of this amino sugar residue, these sequence ions (m/z 266, 470, 674, 878) appear as protonated fragments and not as ammonium adduct ions as in Figure 3 and 4. However, the spectral information is not compromised and a complete structural assignment is still very easy . The spectra suggests that the terminal amino sugar residue is protonated and the sequence ions are derived by cleavage or pyrolysis at glycosidic linkages remote from this charged site (for further discussion on this point see Jensen,<u>et</u> <u>al</u>.,1985). This positional location of polar or charged groups and its relationship to sequence information is considered again in the next section.

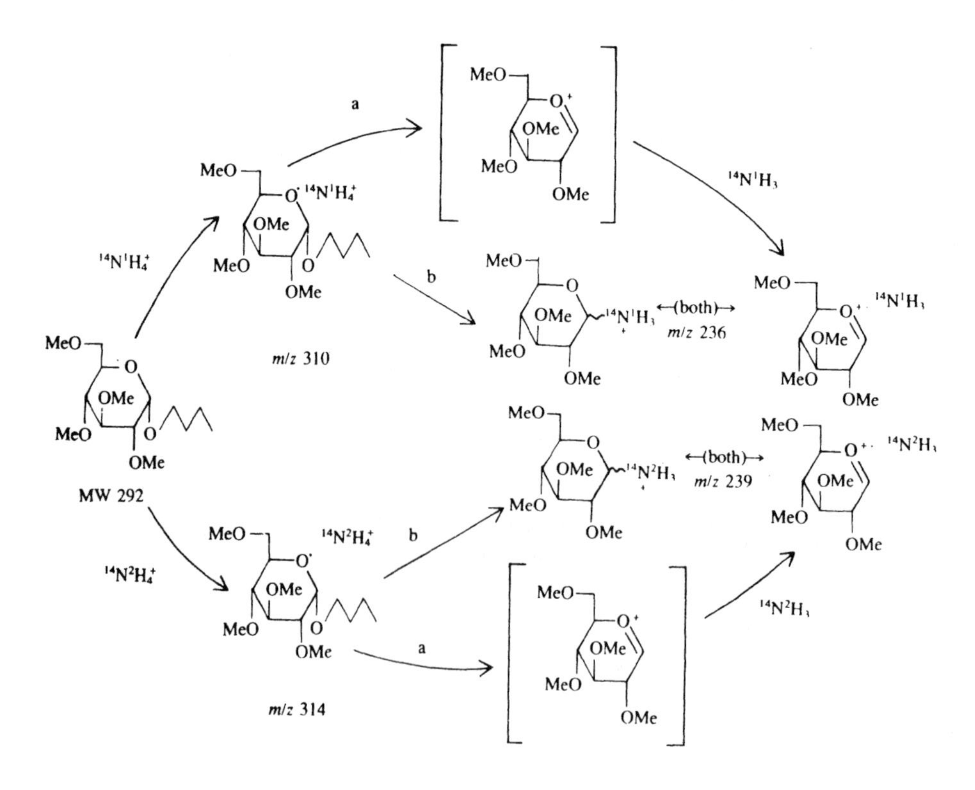

Scheme 1

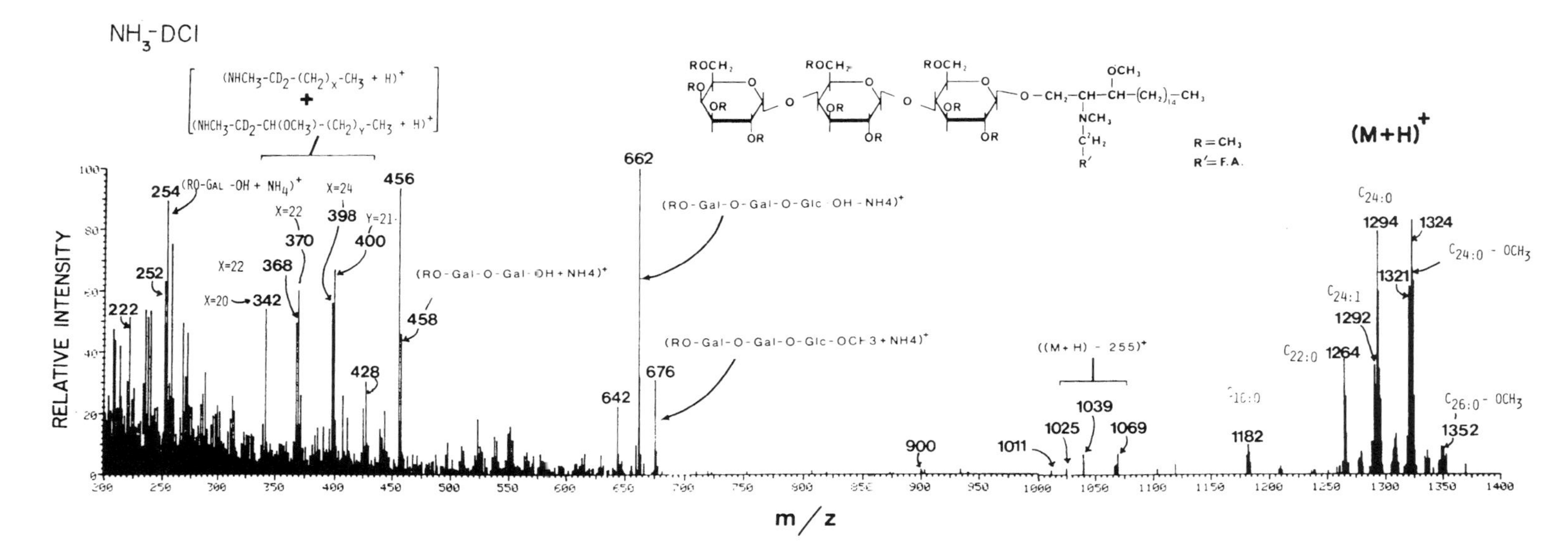

Fig. 4. Direct chemical ionization MS of a permethlyated and reduced cerebrotrihexoside. Ammonia was used as the reagent gas, (Carr and Reinhold, 1984, with permission).

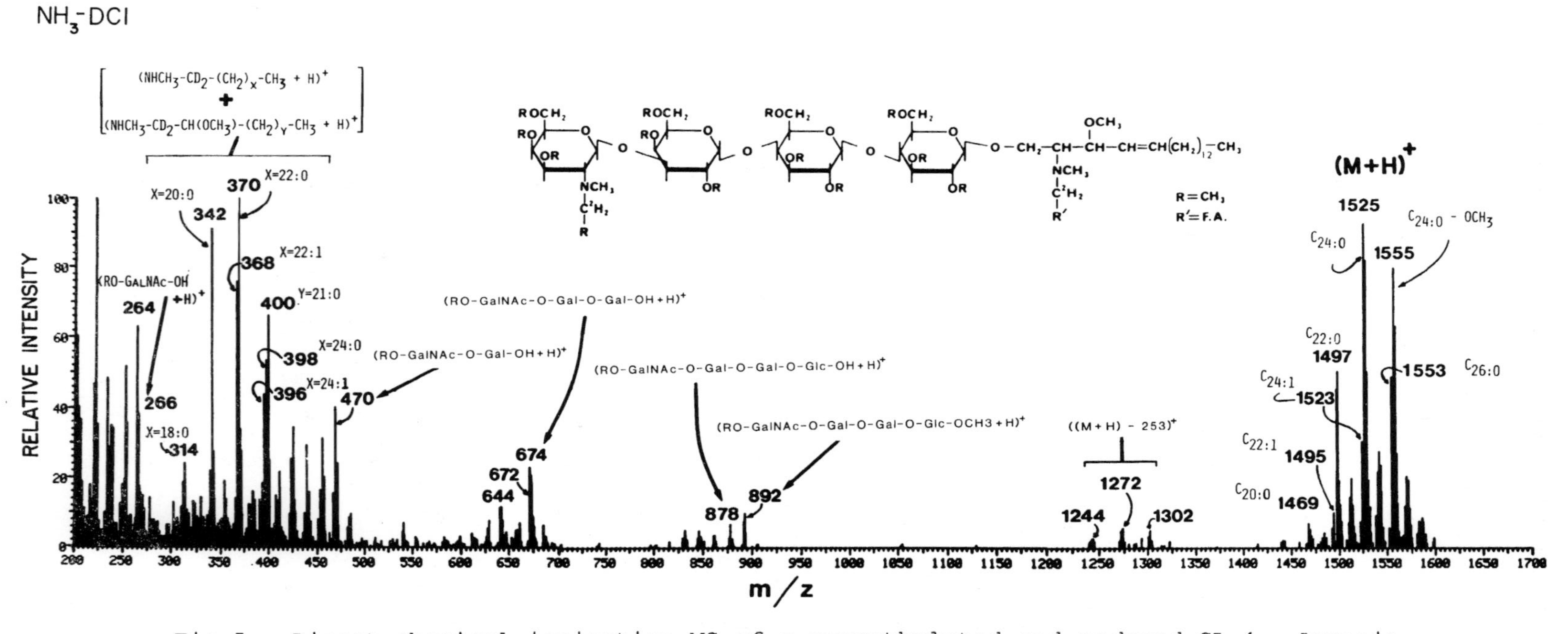

Fig 5. Direct chemical ionization MS of a permethylated and reduced GL-4. Ammonia was used as the reagent gas, (Carr and Reinhold, 1984, with permission).

FAST ATOM BOMBARDMENT (FAB) MASS SPECTROMETRY

Since sample heating is not necessary, beam assisted desorption techniques (such as laser desorption, LD; secondary ion mass spectrometry, SIMS; and FAB) do not suffer from the same drawbacks as the DCI techniques. The most successful, and therefore the most widely used technique in this category is FAB-MS (Barber, M., et al., 1981) wherein the sample of interest is deposited in a liquid matrix (e.g., glycerol, thioglycerol, triethanolamine) and bombarded with 4-8 keV particles (e.g., Xe, Ar, Cs^{+}). Charged species (usually protonated, deprotonated, or sodiated molecular ions) desorbed from the matrix generally have sufficiently low internal energies so that abundant intact molecular weight-related ions have been observed up to a mass range of 10,000 daltons or so. This presents a unique opportunity for accurate molecular weight analysis of large oligosaccharides (see Dell and Taylor, 1983).

An example of this important capability has been demonstrated for the molecular weight analysis of beta glycans from gram-negative bacteria (Dell,et al.,1984). We have also investigated these beta glycans for their possible role in osmotic adaption (Miller,et al., 1986). As anticipated, permethylation and reductive cleavage followed by GC-MS showed only one component, that of 2-0-acetyl-1,5-anhydro-3,4,6-tri-O-methyl glucitol. Thus, the glycan was a 1-2 linked cyclic polysaccharide consisting only of glucose. The molecular weight distribution of the cyclic glycans from *Agrobacterium tumefaciens* indicated $(Glucose)_{19}$ to be in highest concentration with an envelope of homopolymers related by one glucose residue. The dp range from 17 through 23 is presented in Figure 6. There appears to be no precipitous fall off in concentration of these cyclic glycans, weaker signals could be detected well below and above this dp range. It is difficult to determine the concentration of these materials because of the inability to find an appropriate internal standard. The relative distribution, however, may be of more inherent value for following the metabolic course of these periplasmic components.

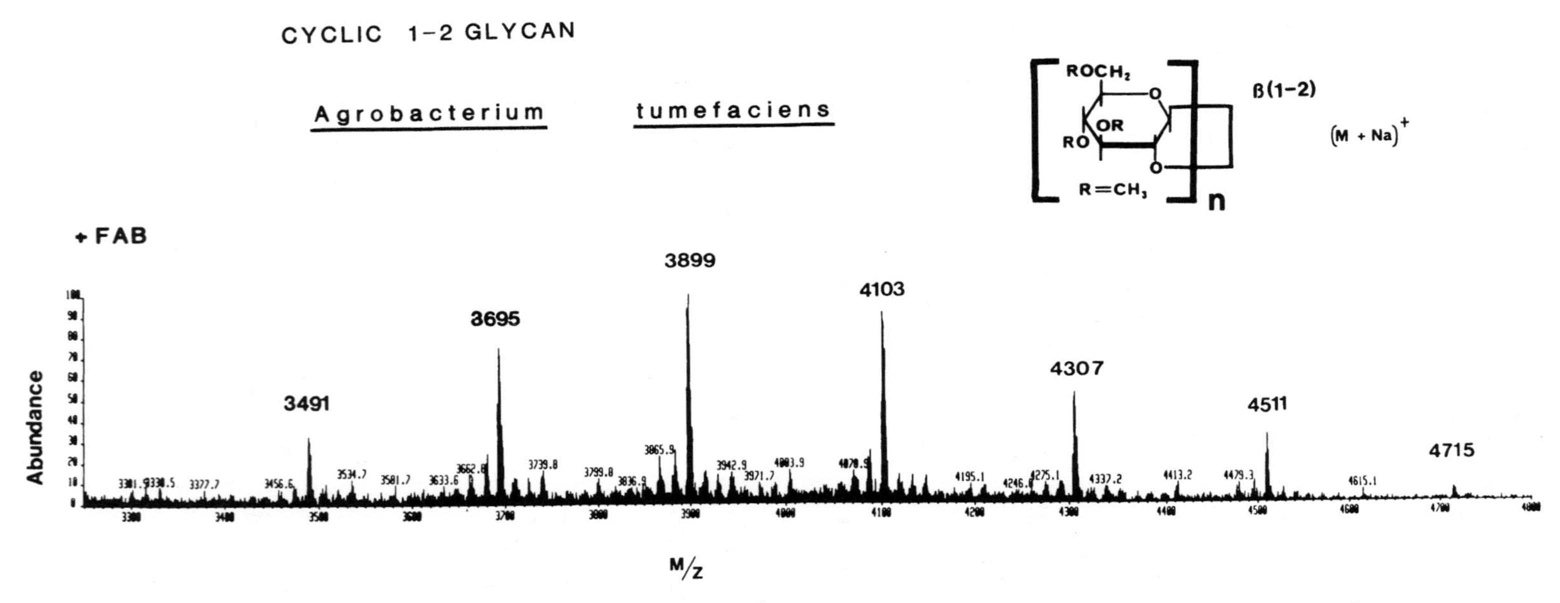

Fig. 6. Fast atom bombardment MS of β(1-2) glycan isolated from *Agrobacterium tumefaciens*, (Miller, et al., 1986, with permission).

In the simplest cases as in the case described above, the complete structure of the cyclic glycans can be established from the linkage analyses and molecular weight information alone. Unfortunately most of the carbohydrates of interest are composed of more than one saccharide component and linkage type. While the linkage information cannot be generally derived from FAB mass spectra, in some favorable cases the sequence information can be derived. An example of FAB mass spectrum that provide fragment ions indicative of oligomer sequence, in addition to molecular weight, information has been reported with a mycobacterial 6-O-methyl glucose polysaccharide (Dell and Ballou, 1983a, 1983b). Two structural factors may have enhanced these sequence ion fragments; first, the sample is partially methylated (naturally occurring), a feature which seems to direct fragmentation to glycosidic rupture (see discussion above); and the second, the oligomer is terminated with an acidic group. Under such circumstances, charged terminating groups provide a series of fragments with incrementing glycosidic residues, (note similar effects under DCI, Fig. 4 & 5).

Obtaining sequence information, however, is usually less predictable and more difficult than in the case described above. For oligosaccharides, the complications appear in the selection of a suitable matrix and instrumental operation rather than limitations imposed by the sample. This can be demonstrated with a pentasaccharide derived from an amino glycoside antibiotic, viridopentaose A, (Harada, *et al.*, 1983; The sample was kindly provided by Professor Suzuki, Meijo University, Nagoya, Japan). For greater sensitivity, positive ion extraction using a thioglycerol matrix provided abundant molecular weight-related ions (Fig. 7a, note that in this analysis the molecular weight-related ions are divided between the protonated and sodiated species), while the greatest structural detail was obtained with triethanolamine as the matrix and negative ion extraction (Fig 7b, Santikarn and Reinhold, 1985). It should be noted that the choice of the matrix, the presence or absence of impurities (such as salts, buffers, grease, acids), and the ion extraction mode can greatly modify the sequence ions and in some cases suppress them totally, (e.g., compare m/z 564 and 334, Fig. 7b). Beam

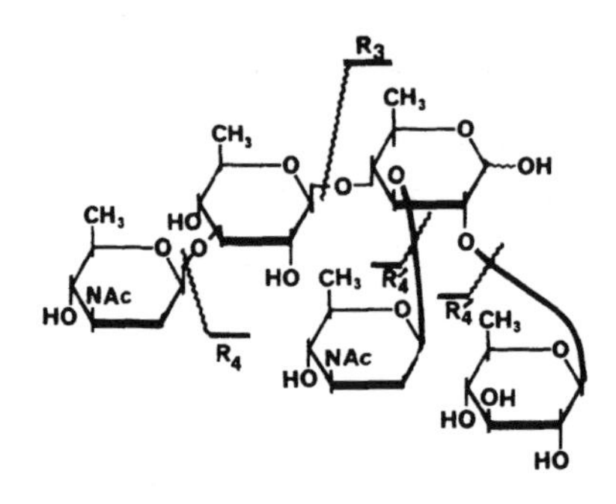

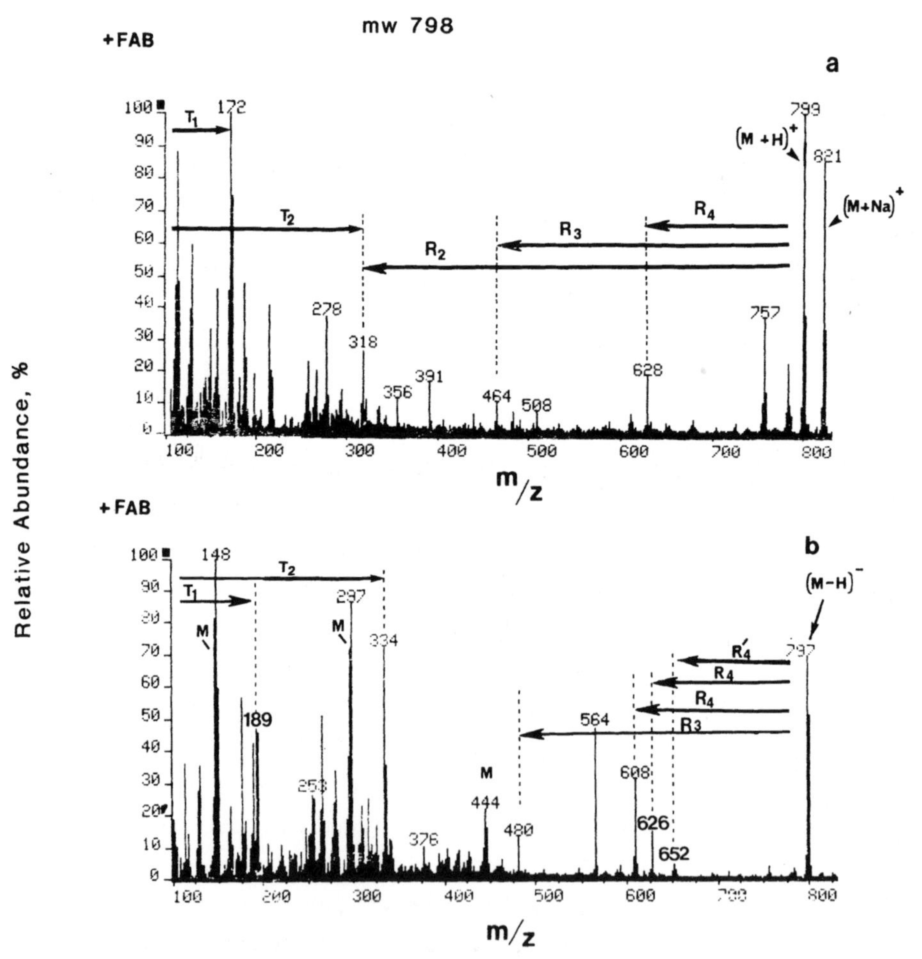

Fig. 7a. Fast atom bombardment MS of viridopentose A using thioglycerol as a matrix and positive ion extraction, (Santikarn and Reinhold, 1985).

7b. Fast atom bombardment MS of viridopentose A using triethanolamine as a matrix and negative ion extraction, (Santikarn and Reinhold, 1985).

longevity, molecular ion intensity and sequence related information may not maximize using the same liquid matrix, and the choice of matrix depends on the information desired and the sample to be analyzed. Thus, it may be inappropriate to state unequivocally that one matrix is superior for all carbohydrate samples. Moreover, the great diversity of carbohydrate structures will probably result in a diversity of suitable matrices. There has been much discussion on sample solubility and distribution with the matrix and these variables influence the resultant spectra greatly (Chan and Cook, 1983; Keough, 1985; for a discussion of matrices see Gower, 1985). Trace sample analysis offers the usual difficulties because of contaminants; with FAB these problems are further complicated due to their ability to cause suppression of sample ions and modify the fragmentation pattern. If adequate amounts of sample are available, three or four analyses using a cross section of matrices usually becomes most informative and is recommended, (see also the FAB-moving belt section).

As discussed above for viridopentaose A, our preliminary studies with N-linked oligosaccharides have also provided spectra with sequence information when recorded in the negative ion mode. Additionally, these materials provide an opportunity to evaluate the influence of a polar group located at the terminal end of the oligomer combined with the spectral variations caused by chain branching. The N-linked oligosaccharides are usually prepared chemically by hydrazinolysis, or enzymatically from glycoproteins. An example of the spectral detail provided by FAB-MS for a group of isomeric N-linked glycans ($Man_8GlcNAc$) is presented in Figure 8. These materials were prepared from the yeast glycoprotein invertase by enzyme hydrolysis (kindly provided by Dr. Trimble, State of New York Department of Health, Albany, NY). From the spectrum one can clearly determine the molecular weight and a series of fragments showing single losses of hexose units, R_8 through R_1. The R fragments are related to the reducing end of the structure and the number subtended represents the number of carbohydrate residues in the designated ion. The T fragments correspond to the terminal end of the polymer. The terminal location of the amino hexose residue appears to provide a

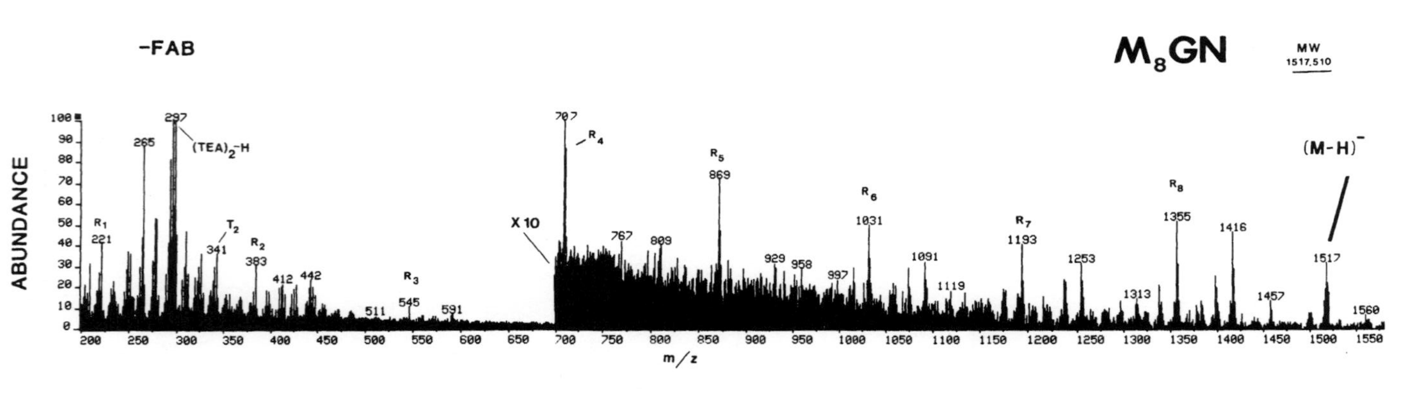

Fig. 8. Fast atom bombardment MS of $Man_8GlcNAc$ using triethanolamine as a matrix and negative ion extraction, (Santikarn, et al., 1985).

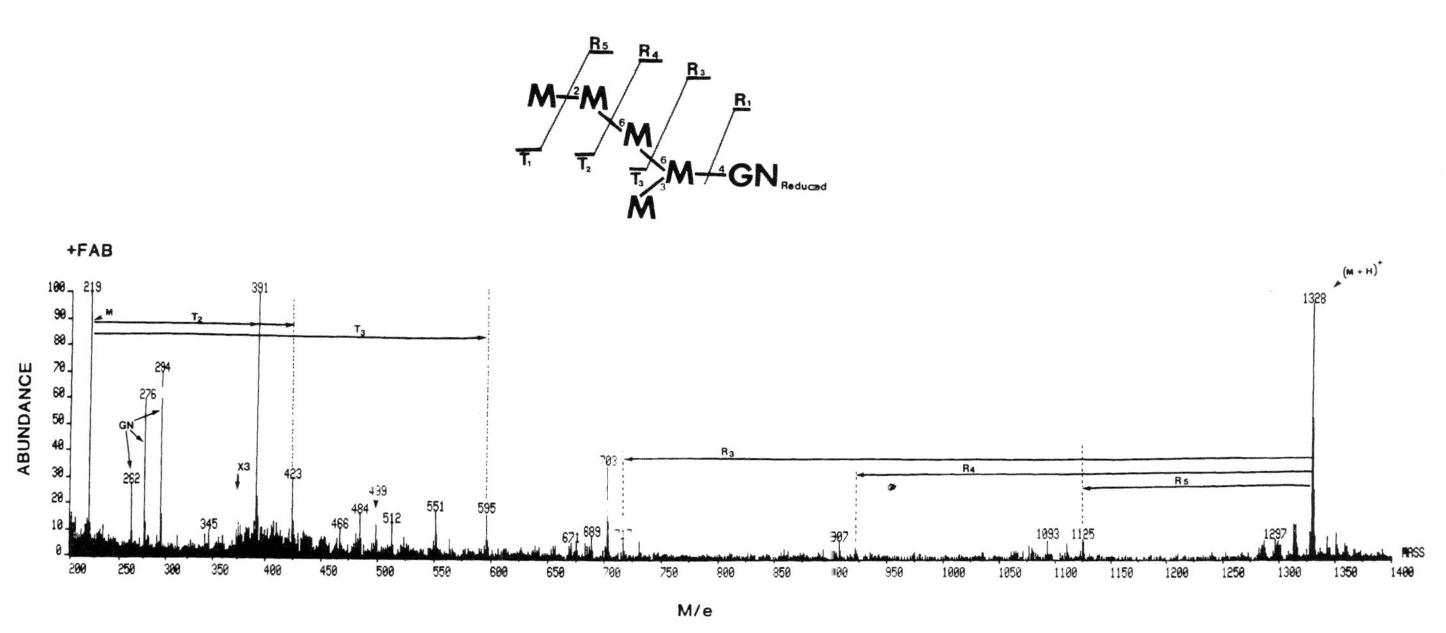

Fig. 9. Fast atom bombardment MS of permethylated doubly branched $Man_5GlcNAc$ using thioglycerol as the matrix and positive ion extraction, (Santikarn and Reinhold, 1985).

negatively charged "handle" for extraction of a series of oligomer fragments of shorter chain length. Closer examination of this material for specific branching information would not be fruitful because this sample is known to be contaminated with isomeric structures.

The determination of oligosaccharide branching poses a unique analytical problem for carbohydrate researchers. To ascertain the ability of FAB desorption to differentiate different branching patterns within an isomeric structure, an HPLC fraction of $Man_5GlcNAc$ was permethylated and analyzed (Fig. 9). This N-linked glycan has a branched residue at the penultimate mannose on the non-reducing terminus. The spectrum indicates a protonated molecular ion at m/z 1312 with facile loss of methanol to give a major fragment, m/z 1280. The next grouping of ions (m/z 1062, 1076, 1090; combination of hydrogen or methyl rearrangements) indicate a loss of one hexose unit. No single cleavage of this glycan can occur that would result in the loss of two hexose residues and, if detected, double cleavage or contamination with an isomeric structure must be considered. Double cleavages would be expected to have a diminished probability and hence ion abundance. The spectra of this doubly branched isomer indicates R_5, R_3, and R_1 fragments, but, no R_4 fragment. Unfortunately, some ion fragments appear in the spectrum which can be explained by isomer contamination (a singly branched isomer at a penultimate mannose residue) or by considering double cleavage, R_2 fragments (e.g., m/z 499, loss of a mannose residue from m/z 703, Fig. 9. However, the absence of specific residues is very suggestive that branching information may be determined by FAB-MS, but clearly, the area requires further study.

In summary, two questions are of considerable interest in the structural analysis of these FAB ionized materials: Can the terminal location of a polar or charged residue provide greater sequence information?; and, can chain branching interrupt the sequence fragmentation pattern in such a way as to provide oligomer branch points? The answer to both questions is yes, if the charged oligomer undergoes primary cleavages only, and randomly, at glycosidic linkages remote from the charged site. Under these conditions the spectral pattern should represent a primary sequence with characteristic deletions at branching point. Unfortunately, glycosidic cleavage during FAB desorption is not generally random (Egge, *et al.*, 1983; Hanfland, *et al.*, 1983a; 1983b; Kamerling, *et al.*, 1983); selectivity has also been observed under neutral gas collision and our preliminary data suggests some secondary cleavage. Considering the complications inherent in the desorptive processes, it may not be possible for this technique to accurately assign oligomer branching *ab inito.*

Certainly the most troublesome aspect of the FAB ionization when considering sequence information is the spectral variations observed with changes in target matrix. In practice, biological isolates frequently bring their own contaminants, thereby adulterating the matrix and skewing the resultant spectra. These matrix effects become most evident with the analysis of trace components. With these complications in mind, it may be asking too much to expect great reproducibility in FAB fragmentation. This important point will be considered again in the next sections.

DESORPTION FROM A MOVING BELT INTERFACE BY FAB

Since the presence of a liquid matrix or contaminants in the sample represent one of the major problems in the analysis of carbohydrates by FAB-MS, we have been studying an alternative FAB approach; using a moving polyimide belt instead of the liquid matrix. The moving belt was originally developed as a means of introducing liquid chromatographic eluents into the vacuum of a mass spectrometer for the direct online analysis (McFadden, *et al.*, 1976). It was observed that polar organic samples can be directly desorbed from the belt by the FAB technique without the need to add liquid matrix as is the case for the stationary target (Stroh, *et al.*, 1985). Our initial investigation of the moving belt as a substitute for the liquid matrix is very promising in three respects

(Santikarn, *et al.*, 1986). In comparison with liquid assisted FAB mass spectra, the spectra of samples desorbed from a moving belt show:

1) improved signal-to-noise ratios for sample related ions, particularly at low mass (i.e., below 400 daltons) where the matrix derived ions can sometimes dominate the spectra (Fig. 10). The detection sensitivity can be up to an order of magnitude better in some cases; the detection limit for viridopentose B, for example, is about 100 pmole at present (Santikarn, *et al.*, 1986).

2) less susceptibility to the presence of trace components and impurities which tend to plague most of the real life samples (i.e., non-synthetic samples). Therefore one can expect more reproducible mass spectra when they are recorded using a moving belt instead of liquid matrix.

3) some control of the amount of fragmentation. By adjusting the belt speed the abundances of the fragmentation ions can be partly adjusted. The spectra recorded at the slower belt speed tend to show a greater amount of fragment ions and less molecular related ions than the spectra recorded at higher belt speeds (Fig. 11). These observations can be rationalized in terms of the extent of surface charging and the resultant energies of the desorbed species (van der Peyl, *et al.*, 1985). Briefly, the lower belt speed results in an increase in surface charging which leads to an increase in internal energies in ions desorbed from the belt and, we believe, consequently greater abundances of fragment ions are produced.

With these advantages in hand we again ask, can carbohydrate sequence and branching be established by moving belt FAB-MS? If careful attention is paid to belt speed, the answer is tentatively yes. This is illustrated in the belt FAB mass spectrum of $Man_5GlcNAc_2$-ol which shows an abundant sodiated molecular ion at 1259 daltons; fragment ions corresponding to losses of Man, Man_3, and Man_5 can be observed, while those representing losses of Man_2 and Man_4 are absent (Fig. 12). The structure proposed shows that a single bond cleavage can result in a loss of 1,3 or 5 mannose units from the molecular ions, whereas in order to break off 2 or 4 mannose units at least two covalent bonds must be ruptured.

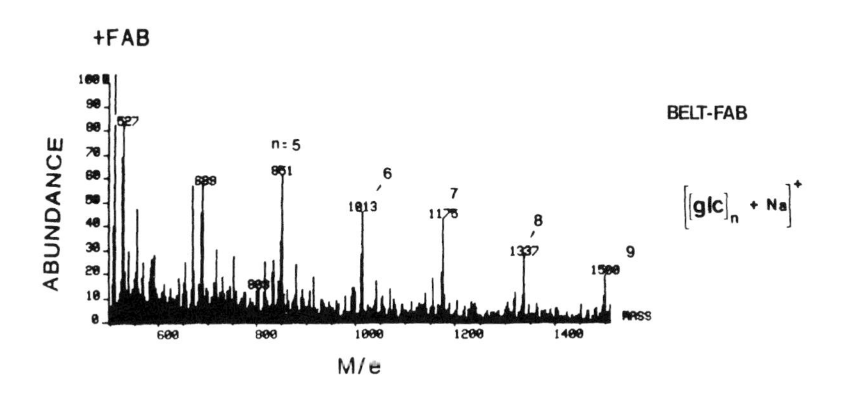

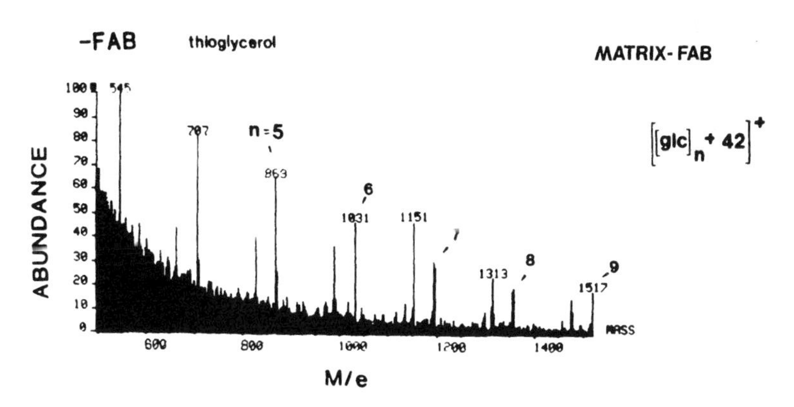

Fig. 10. Belt-FAB vs matrix FAB-MS of corn syrup oligomers.

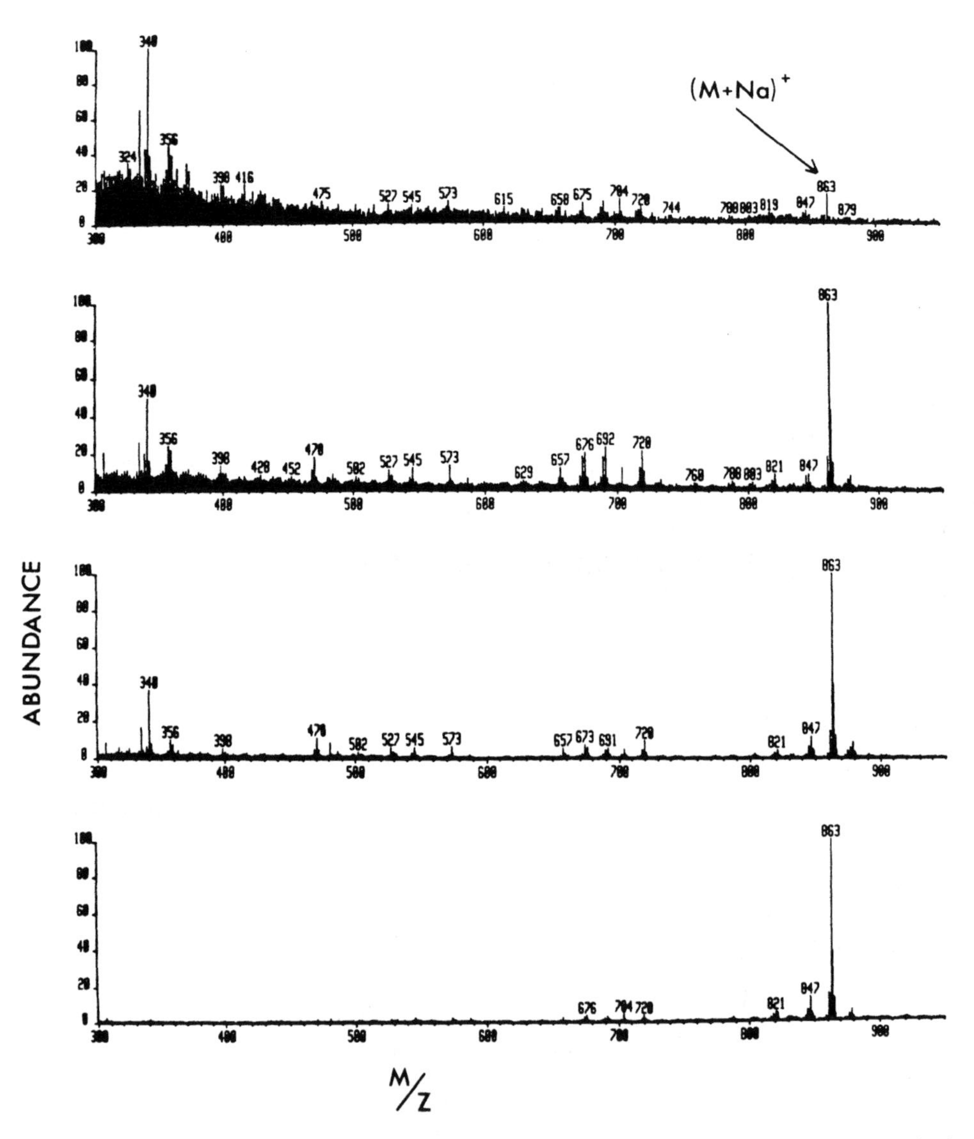

Fig. 11. Belt-FAB mass spectrum of viridopentaose B recorded at various belt high speeds: a) 0.5 cm/sec, b) 1.0 cm/sec, c) 2.0 cm/sec, d) 3.0 cm/sec. (Santikarn, *et al*., 1986).

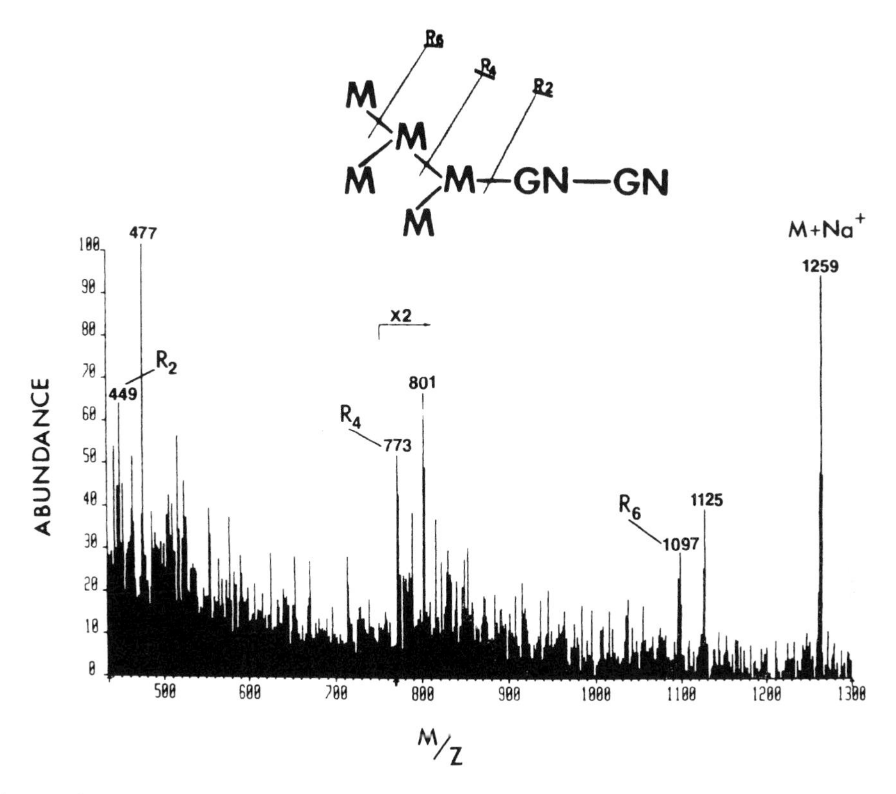

Fig. 12. Belt-FAB mass spectrum of $Man_5GlcNAc_2$-ol. Belt speed 2.0 cm/sec. (Santikarn,et al., 1986).

The observation that the single bond cleavages can be controlled with the proper belt speed was further tested by recording FAB mass spectra of a cyclic heptasaccharide, beta-cyclodextran. A comparison was made for sample deposited on a belt and dissolved in a liquid matrix. In order to produce fragment ions from beta-cyclodextrin (other than simple loss of OH group) at least two bonds must be broken. As expected, the belt FAB mass spectrum of beta-cyclodextrin shows no fragment ions while the matrix FAB mass spectrum show successive losses of hexose units (n x 162) from the protonated molecular ion (Santikarn,et al., 1986).

The major benefit of the moving belt interface lies in the function for which it was designed, that is, the coupling of a liquid chromatograph to a mass spectrometer. The advantages of direct chromatographic coupling have been well appreciated and applied (e.g., GC-MS); most of these advantages have a corollary with LC-MS interfacing. FAB ionization, however, adds a new dimension to this adjunct technique for biopolymer studies since it provides both molecular weight and sequence information. The application of this HPLC-interface to larger molecular weight materials and the problems of determining oligomer branching information using this belt as a substrate for FAB could prove most interesting.

There are several limitations to the belt FAB interface at present. The range over which the belt speeds can be varied is limited by several factors: below about 0.5 cm/sec the surface charging becomes excessive and no molecular related ions can be detected, above 2.5 cm/sec both the vacuum lock and wash bath do not function efficiently. However, the major problem with the belt interface is the instability of the ion flux issuing from the belt, which we believe is a consequence of uneven sample deposition on the belt.

SEQUENCE STUDIES FOLLOWING GAS PHASE COLLISIONS

A fourth MS technique that can provide oligosaccharide sequence information is that of ion collision with neutral gases. In the discussions above, oligomer sequence information was more or less a by-product of molecular ionization. In this section gas phase collisions will be considered only in their ability to produce fragmentation, separate from, and outside of the ion source.

The details of collision induced fragmentation are poorly understood for polyatomic systems but the fragmented products are known to vary with target gas and its pressure, scattering angles and ion kinetic energies. Collisions not only activate some of the ions for uni-molecular decomposition but also induce charge neutralization, inversion, and stripping as well as ion scattering (a recent text covering a full length discussion of collision spectroscopy has been written, Cooks, 1978). The loss of sample signal is considerable following collision but part of these losses can be "recovered" by signal amplification. The focusing of single ions (parent ions) for collision effectively filters out organic background, "chemical noise", therefore the daughter ions show good signal-to-noise ratios despite their low abundances.

The implementation of molecular ionization, enhancing fragmentation by collision (more frequently referred to as collision induced dissociation, CID), mass analysis, and detection can be performed with a wide assortment of instruments having multiple analyzers. Generically, the caption MS/MS has been given to these techniques and the instruments are characterized by the number and types of analyzers in final configuration. Multiple analyzers are not new to the field of mass spectrometry. They have long been used to study metastable ions and for obtaining higher mass resolution on fragment ions. Fast atom bombardment ionization has renewed interest for the use of multiple analyzers for two major reasons; first, in the "purification" of complex biological mixtures, and second, to resolve structural detail following gas phase collisions. Reviews have been written about MS/MS (Kondrat and Cooks, 1978; McLafferty and Bockhoff, 1978; Yost and Enke, 1979; Yost and Fetterolf, 1983), and an excellent book (McLafferty, 1983) has been published on this topic.

Our interest in CID experiments was motivated, in large part, by the need to generate abundant fragment ions for the determination of oligomer sequences. Considering the current state-of-the-art method for inducing glycosidic cleavage, gas-phase collisions may not be the foremost technique to consider. However, other aspects of this preliminary study may turn out to have greater relevance and these points will be considered next, (Carr, *et al.*, 1985). For these experiments, two mass spectrometers in tandem (BE-EB) were utilized interposed with a neutral gas collision cell (C) making the final configuration, BE-C-EB. Two features of MS/MS were evaluated; first, to assess the specificity of fragmentation by collision, isomeric hexasaccharides components were compared; and second, to study the separation and identification capability of these coupled (four) analyzers, protonated molecular ions in a complex mixture were isolated and characterized. The samples were ionized by FAB and selected by the double focusing combination BE. Daughter ions produced in the collision cell (C) between BE and EB were scanned simultaneously by the EB sectors such that the ratio E/B was maintained constant.

The structural specificity realized with molecular ion collision can be seen in the next figures. Presented in Figures 13a and 13b is a comparison between a linear and branched hexasaccharide. The linear and branched saccharides each provided a protonated molecular ion, $(M + H)^+$ at m/z 1116, which was selected for collision. Helium was used as the neutral gas. The spectrum of the linear hexasaccharide (Fig. 13a) shows a series of fragments incrementing from the terminal end, T, as well as a related series decreasing in mass for each carbohydrate residue from the reducing end, R. As noted above for FAB desorption, there is also an enhanced selectivity for rupture at amino sugar linkages, (e.g., compare relative abundances Fig 13a, m/z 366, 731; Fig. 13b, m/z 366, 751). A comparison of the two isomers shows a distinct ion abundance difference in the fragments m/z 731 and m/z 751. For the linear hexasaccharide these are accounted for as the T_4 and R_4 fragments, (Fig 13a). For the branched hexasaccharide the base ion in Figure 13a (m/z 731) now becomes a minor fragment ion while the abundance of the R_4 fragment has increased. The fact that these results are obtained by a combination of isobaric ion selection and collision under identical instrumental operating conditions, insures the reproducibility. Two fragments in Figure 13b are still difficult to account for on the basis of a branched structure, m/z 589 and m/z 731. These fragments are the R_3 and T_4 cleavages which can be accounted for only in the linear structure and their presence in the branched hexasaccharide is unexpected. An explanation may be that multiple cleavages occur or the sample is contaminated with the isomeric linear hexasaccharide. The samples were isolated from the same organism and separated by HPLC where incomplete separation could be a possibility (personal communication, Dr. A.M. Wu, Texas A&M). The differences between the two collision spectra indicate that sequence, including branching, information may

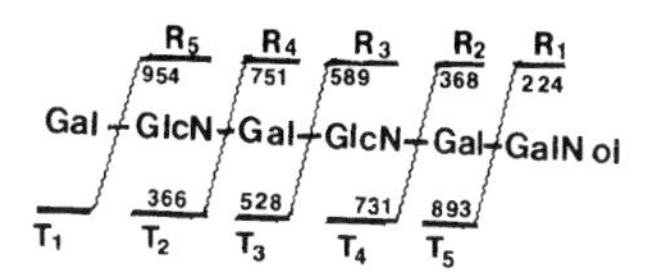

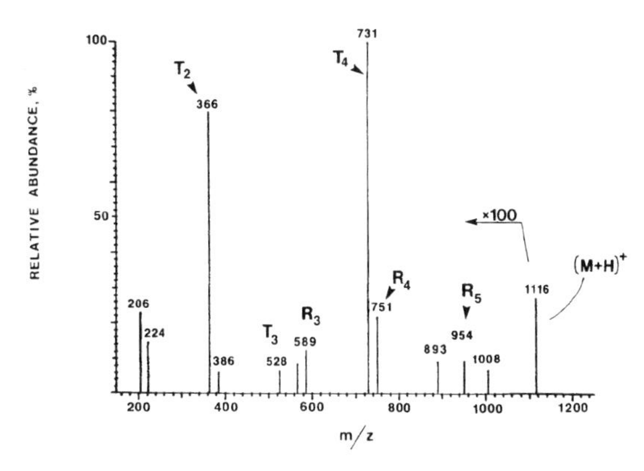

Fig. 13a. Spectrum of linear hexasaccharide obtained with MS/MS instrument. Fast atom bombardment desorption, BE analyzer isolation, m/z 1116, collision with helium gas and EB linked scan of daughter ion products, (Carr, et al., 1985, with permission).

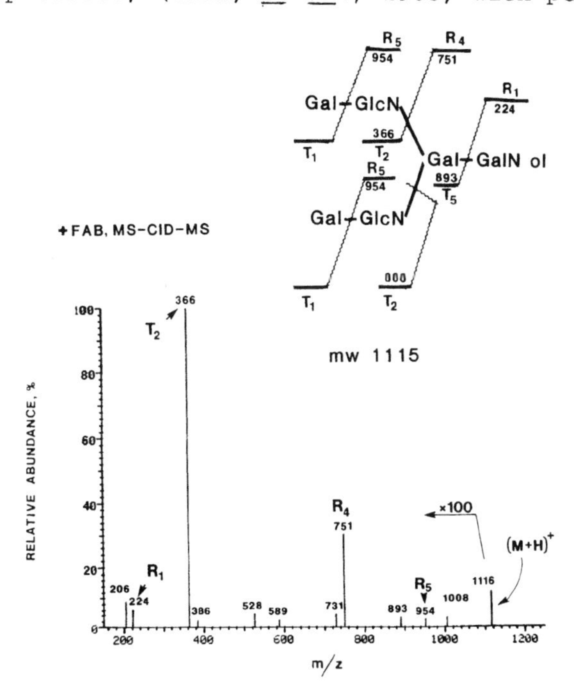

Fig. 13b. Spectrum of branched hexasaccharide obtained with MS/MS instrument. Fast atom bombardment desorption, BE analyzer isolation, collision with helium gas and EB linked scan of daughter ion products, (Carr, et al., 1985, with permission).

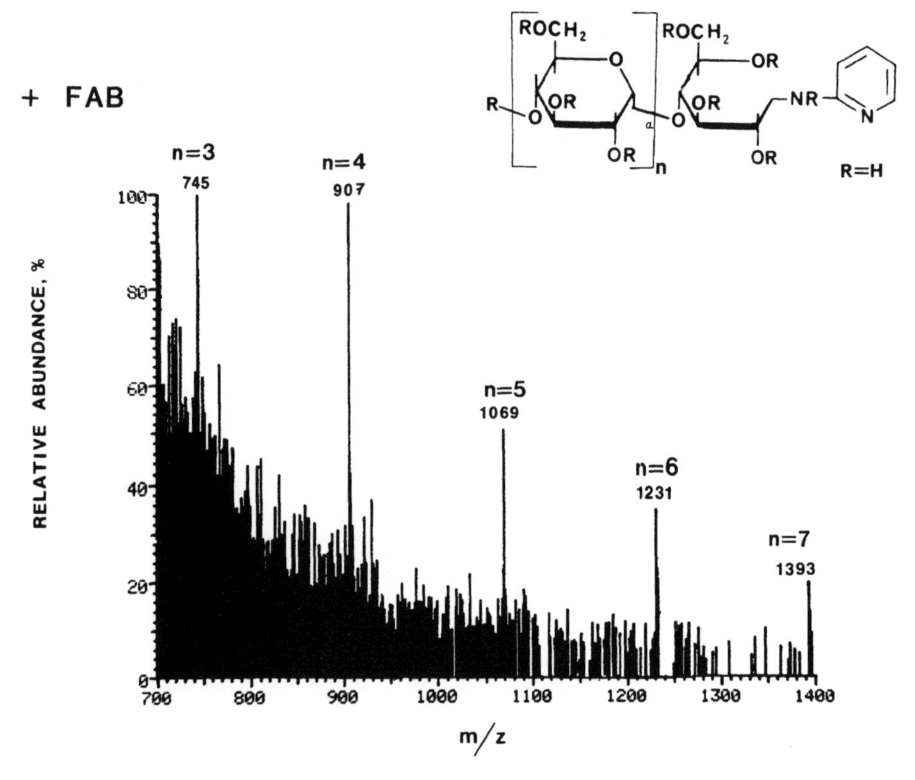

Fig. 14. Fast atom bombardment MS of corn syrup oligomers prepared as pyridinylamine derivatives. Thioglycerol used as matrix, (Carr, et al., 1985, with permission).

be obtained by CID. The fragments not consistant with this interpretation may well be accounted for as sample contaminants. An additional feature that MS/MS instrumentation brings to oligosaccharide characterization is the ability to select specific parent ions from complex mixtures and determine the structure of the isolated component. This capability is demonstrated in Figure 14 and 15. A mixture of oligosaccharides were labeled at the reducing end with a fluorescent group. The complex mixture was directly analyzed by FAB-MS using thioglycerol as the matrix (Fig. 14). To determine the structure of one of these components, (m/z 1069), the protonated molecular ion was selected and collided with helium. The products were scanned as described above with the EB analyzers (Fig.15). The spectrum shows a series of pyridinylamine sequence fragments differing by a hexose unit.

mw 1068

+FAB MS-CID-MS

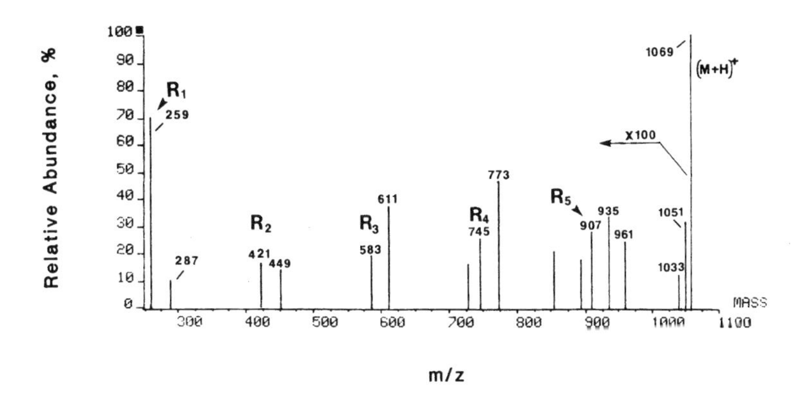

Fig. 15. Spectrum of linear hexasaccharide (m/z 1069) obtained with MS/MS instrument. Fast atom bombardment desorption, BE analyzer isolation, collision with helium gas and EB linked scan of daughter ion products, (Carr, et al., 1985, with permission).

SUMMARY

Many current problems in immunology depend on a detailed understanding of oligosaccharide and glycoconjugate structures. In contrast to other biopolymers, oligosaccharides do not have well established and sensitive analytical procedures for the determination of their structures. Additionally, oligosaccharides because of branching, linkage, and structural isomerism, pose specific complications unique to this biopolymer. We have discussed MS data utilizing DCI and FAB ionization and presented new information that indicates FAB-MS can provide a wealth of structural information including sequence, branching and molecular weight. In the final sections FAB ionization and collision induced dissociation, have been discussed, two adjunct techniques that amplify the analytical value of MS instrumentation in the pursuit of oligosaccharide structure.

ACKNOWLEDGEMENTS

The authors would like to express their appreciation to Dr. Hers for generous time spent discussing aspects of this chapter and reading the manuscript. We would further like to express our appreciation to Drs. Wu and Adams for organizing this symposium and the opportunity to present some of our work. Financial support has been provided by the National Science Foundation and the National Institutes of Health.

REFERENCES

1. Barber, M., Bordoli, R.S., Sedgwick, R.D. and Tyler, A.N. (1981). Fast Atom Bombardment of Solids as an Ion Source in Mass Spectrometry, Nature, (London) 293, 270-275.
2. Carr, S.A. and Reinhold, V.N. (1984). Structural Characterization of Glycosphingolipids by Direct Chemical Ionization Mass Spectrometry, Biomed. Mass Spectrom., 11, 633-642.
3. Carr, S.A., Reinhold, V.N., Green, B.N. and Hass, J.R. (1985). Enhancement of Structural Information in FAB Ionized Carbohydrate Samples by Neutral Gas Collision. Biomed. Mass Spectrom, 12, 288-295.
4. Chan, K.W.S. and Cook, R. D. (1983). Factors Affecting Mass Spectral Sensitivity for Ions Sampled by Field Evaporation from a Liquid Matrix, Anal. Chem. 55, 1306-1309
5. Cooks, R.G. (1978). *Collision Spectroscopy*, (ed. R.G. Cooks), Plenum Press, New York, p. 458.
6. Dell, A. and Ballou, C.E. (1983a). Fast Atom Bombardment, Negative-Ion, Mass Spectrometry of the Mycobacterial O-Methyl-D-Glucose Polysaccharide and Lipopolysaccharides, Carbohydr. Res., 120, 95-111.
7. Dell, A. and Ballou, C.E. (1983b). Fast Atom Bombardment Mass Spectrometry of 6-O-Methylglucose Polysaccharide, Biomed. Mass Spectrom., 10, 50-56.
8. Dell, A., Oates, J., Lugowski, C., Romanowska, E., Kenne, L. and Lindberg, B. (1984). The Enterobacterial Common-Antigen, A Cyclic Polysaccharide, Carbohydr. Res., 133, 95-104.
9. Dell, A. and Taylor, G.W. (1983). High-Field Magnet Mass Spectrometry of Biological Molecules, Mass Spectrom. Rev., 3, 357-394.
10. DeMark B.R. and Klein P.D. (1981). Ammonia gas: An Improved Reagent for Chemical Ionization Mass Spectrometry of Bile Acid Methyl Ester Acetates, J. Lipid Res., 22, 166-177.

11. Egge, H., Dell, A. and von Nicolai, H. (1983). Fucose Containing Oligosaccharides from Human Milk. I. Separation and Identification of New Constituents, Arch. Biochem. Biophys., 224,235-253.
12. Gower, J.L. (1985). Matrix Compounds for Fast Atom Bombardment Mass Spectrometry, Biomed. Mass Spectrom., 12,191-196.
13. Hanfland, P., Dabrowski, U., Peter-Katalinic, J., Kordowicz, M., Niermann, H., Egge, H., Dabrowski, J. (1983a). Isolation and Characterization of Three Different Blood Group B Active Complex Glycosphingolipids from Human Erythrocyte Membranes, *Proc. 7th Int. Symp. on Glycoconjugates, Lund, Sweden*, Chester, M.A., Heinegard, D., Lundblad, A., Svensson, S., Eds.: pp. 411-412.
14. Hanfland, P., Dabrowski, U., Kordowicz, M., Peter-Katalinic, J., Dabrowski, J., Egge, H. (1983b). Purification and Structural Elucidation of Three Different Blood Group B and I or i Glycosphingolipids from Rabbit Erythrocyte Membranes. *Proc. 7th Int. Symp. on Glycoconjugates, Lund, Sweden*, Chester, M.A., Heinegard, D., Lundblad, A., Svensson, S., Eds.; pp. 413-414.
15. Harada, K., Ito, S., Takeda, N., Suzuki, M. and Tatematsu, A. (1983). Sequence Determination of Permethylated Oligosaccharides by Chemical Ionization Mass Spectrometry, Biomed. Mass Spectrom., 10,5-12.
16. Harada, K., Ito, S., Suzuki, M. and Iwashita, T. (1983). Structural Investigation of the Antibiotic Sporaviridin. VII. Structural Studies on the Constituent Pentasaccharides, Viridopentoses, Chem. Pharm. Bull. (Japan), 31,3829-3843.
17. Harrison, A.G. (1983). *Chemical Ionization Mass Spectrometry* CRC Press, Inc., Boca Raton, FL.
18. Hunt, D.F. (1974). "Reagent Gases for Chemical Ionization Mass Spectrometry," in *Advances in Mass Spectrometry* (Ed. A.R. West), pp. 143-152. Applied Science Publishers, London.
19. Jensen, N.J., Tomer, K.B. and Gross, M.L. (1985). Gas-Phase Ion Decomposition Occurring Remote to a Charged Site, J. Am. Chem. Soc., 107,1863-1868.
20. Kamerling, J.P., Heerma, W., Vliegenthart, J.F.G., Green, B.N., Lewis, I.A.S., Strecker, G. and Spik, G. (1983). Fast Atom Bombardment Mass Spectrometry of Carbohydrate Chains Derived from Glycoproteins, Biomed. Mass Spectrom., 10,420-425.
21. Karlsson, K.-A., Pascher, I., Pimlott, W. and Samuelsson, B.E. (1974). Use of Mass Spectrometry for the Carbohydrate Composition and Sequence Analysis of 2-Hydroxy Fatty Acids, Biomed. Mass Spectrom., 1,49-55.
22. Keough, T. (1985). Cationization of Organic Molecules Using Fast Atom Bombardment Mass Spectrometry, Anal. Chem.,57,2027-2034.
23. Kondrat, R.W. and Cooks, R.G. (1978). Direct Analysis of Mixtures by Mass Spectrometry, Anal. Chem., 50,81A-92A.
24. Miller, J.K., Kennedy, E. and Reinhold, V.N. (1986). Osmotic Adaption by Gram-Negative Bacteria: Possible Role for Periplasmic Oligosaccharides, Science, 231,48-51.
25. McFadden, W.H., Schwartz, H.L. and Evans, S. (1976). Direct Analysis of Liquid Chromatographic Effluents, J. Chromatogr., 122,389-396.

26. McLafferty, F.W. (1983). *Tandem Mass Spectrometry*, Wiley-Interscience, New York.
27. McLafferty, F.W. and Bockhoff, F.M. (1978). Separation/Identification System for Complex Mixtures Using Mass Separation and Mass Characterization, Anal. Chem., 50,69-76.
28. Munson, B. (1977). Chemical Ionization Mass Spectrometry: Ten Years Later, Anal. Chem., 49,772A-776A.
29. Reinhold, V.N., Coles, E. and Carr, S.A. (1982). New Techniques for Oligosaccharide Sequencing, J. Carbohydr. Chem., 2,1-18.
30. Rudewicz, P. and Munson, B. (1985). Neutral Reactions in Gas Chromatography/Chemical Ionization Mass Spectrometry, Anal. Chem., 57,786-789.
31. Santikarn, S., Her, G.-R. and Reinhold, V.N. (1986). Fast Atom Bombardment Mass Spectrometry of Complex Carbohydrates Desorbed from a Moving Belt. (Manuscript in preparation).
32. Santikarn, S., Trimble, R.B. and Reinhold, V.N. (1985). Unpublished results.
33. Stroh, J.G., Cook, J.C., Milberg, R.M., Brayton, L., Kihara, T., Haung, Z. and Rinehart, K.L., Jr. (1985). On-Line Liquid Chromatography/Fast Atom Bombardment Mass Spectrometry, Anal. Chem., 57,985-991.
34. Tabet, J.C. and Fraisse, C. (1981). Reaction of Cyclohexanone with NH_4 Ion Under Chemical Ionization Conditions. 1. Formation of Protonated Unsubstituted Imines, Org. Mass Spectrom., 16,45-47.
35. Van Langenhove, A., and Reinhold, V.N. (1985a). Determination of Polysaccharide Linkage and Branching by Reductive Depolymerization: GC and GC-MS Reference Data, Carbohydr. Res., 140,1-20.
36. van der Peyl, G.J.Q., Isa, K., Haverkamp, J. and Kistemaker, P.G. (1981). Gas Phase Ion/Molecule Reactions in Laser Desorption Mass Spectrometry, Org. Mass Spectrom., 16,416-420.
37. van der Peyl, G.J.Q., van der Zande, W.J., Hoogerbrugge, R. and Kistemaker, P.G. (1985). Kinetic Energy Distributions of Organic Ions Sputtered from Solids and Liquids. Int. J. Mass Spectrom. Ion Proc., 67,147-159.
38. Yost, R.A. and Enke, C.G. (1979). Triple Quadrupole Mass Spectrometry for Direct Mixture Analysis and Structural Elucidation, Anal. Chem., 51,1251A-1264A.
39. Yost, R.A. and Fetterolf, D.D. (1983). Tandem Mass Spectrometry (MS/MS). *Mass Spectrom. Rev.*,2,1-45.

ANALYSIS OF COMPLEX CARBOHYDRATE PRIMARY AND SECONDARY STRUCTURE VIA TWO-DIMENSIONAL PROTON NUCLEAR MAGNETIC RESONANCE SPECTROSCOPY

Theodore A.W. Koerner*, Robert K. Yu, J. Neel Scarsdale, Peter C. Demou and James H. Prestegard

Departments of Neurology (School of Medicine) and Chemistry, Yale University, New Haven, Connecticut

The growing interest in the biological function of cell surface complex carbohydrates has stimulated the constant search for new methods for analyzing their primary and secondary structures. To determine the primary structure of a complex carbohydrate, it is necessary to establish the composition and configuration of its sugar residues, and the sequence and linkage sites of its oligosaccharide chain. This information has traditionally been obtained by application of a combination of such procedures as compositional analysis by gas-liquid chromatography, mass spectrometry, permethylation studies, Smith degradation, partial acid or enzyme hydrolysis, optical rotation measurements, etc. However, these techniques are time-consuming, require elaborate derivatization of the substance, and access to many different instruments. Furthermore, these procedures do not afford secondary structural information (conformation) which is important in understanding the biological activities of these compounds.

**The present address of TAWK is Dept. of Pathology, University of Iowa, College of Medicine, Iowa City, IA 52242*

Although x-ray crystallography has been used to provide accurate information about the conformation of a few complex carbohydrates by measurement of bond lengths, bond angles and interatomic distances, it is not certain whether the conformation that exists in a crystalline state is the one that is preferred in solution.

Proton nuclear magnetic resonance (NMR) spectroscopy is well-suited for providing the primary and secondary structure of a complex carbohydrate in solution because this method is rapid, quantitative, sensitive and nondestructive. However, previous utilization of proton NMR has yielded only fragmentary data concerning the primary structure of these substances. Its potential in providing complete structural information has not been fully realized until recently.

The marriage of sophisticated computers and high-field superconducting magnets approximately twelve years ago created the potential for a revolution in the analytical power of proton NMR spectroscopy. The full realization of this potential, however, had to await the imagination of R.R. Ernst and co-workers who, by developing pulsing programs that have allowed increasingly more complex and structurally-revealing spectrometer "interrogations" of sample nuclei, have transformed NMR spectroscopy into a method capable of the complete and independent structural analysis of biomolecules. The graphical products of these pulsing programs have become known as "two-dimensional" (2-D) NMR spectra, even though they contain three dimensions of information. Two-dimensional NMR methods and their physical basis have been reviewed (Jelinski, 1984; Benn and Günther, 1983; Bax, 1982; Bax and Lerner, 1986).

These 2-D methods have been extremely useful in the application of proton NMR to the primary and secondary structure of complex carbohydrates, specifically oligosaccharides and their conjugates (e.g. glycolipids, glycopeptides, steroid glycosides, etc.). This is because of two factors. First, the natural allocation of protons within oligosaccharides is well-suited to yield all or most of their primary and secondary structure. Second, through 2-D NMR the severe resolution problem peculiar to oligosaccharides and their conjugates (the "hidden resonance problem") has been generally overcome. This resolution problem results from the facts that almost all resonances of oligosaccharides occur between 3-5 ppm, regardless of solvent, and that within this same narrow region the substantial water-derived resonance occurs. By reducing spectra into subspectra, each attributable to only one oligosaccharide residue, 2-D NMR allows overlapping envelopes of resonances to be deciphered.

The reason that oligosaccharide structures lend themselves to proton NMR analysis is that protons are spaced consecutively around each oligosaccharide residue. Thus there is usually one proton at each optically-active ring carbon, which serves as a structural "reporter group" at that carbon. The C-2 of ketose residues is the only important exception. When more than one proton is present at a particular carbon it signifies the terminus of the residue or important constitutional data (e.g. the residue is a deoxysugar). Because of the consecutive allocation of protons around each residue ring, a vicinal *J*-coupled connectivity or "trail" exists for each residue such that each proton leads to the next proton, all around the ring, revealing the stereochemistry and constitution at each carbon. Conveniently, one terminus of each aldose residue *J*-connectivity, the H-1 or anomeric proton of each residue, resonates in a characteristic region (4.0-5.0 ppm) that contains few other signals. Thus the *J*-connectivity trail for such residues can easily be "picked up" in this region.

Just as important as these intra-residue *J*-connectivities is the potential proximity of anomeric protons and linkage site protons across the glycosidic bonds. Thus when through-space coupling is considered, oligosaccharides manifest a continuously coupled series of proton "reporter groups" from one end of the molecule to the other, making their structures readily accessible to proton NMR analysis. However, in order to undertake such an analysis the *J*-connectivities of each residue must be separated from each other and the through-space coupling revealed. Both of these problems have been solved via 2-D NMR.

TYPES OF 2-D NMR EXPERIMENTS

Two general classes of 2-D NMR experiments exist at this time. They are J-resolved and correlated spectroscopy. In the first type, spectra are characterized by one frequency axis (F_1) containing coupling (*J*) information and by another frequency axis (F_2) containing chemical shift information (ppm). The third dimension of resonance intensity is plotted as a contour. In the second class of 2-D experiments, both frequency axes (F_1 and F_2) contain chemical shift data and the third dimension is resonance intensity plotted as a contour. Two types of *J*-resolved and correlated 2-D spectra are possible, homonuclear and heteronuclear, that is those involving couplings between the same or differing nuclear types, e.g. ^1H and ^1H or ^1H and ^{13}C nuclei. Only proton homonuclear 2-D methods will be considered in this article, since ^1H-^{13}C heteronuclear 2-D methods require too much sample to be useful in studies of most oligosaccharide samples of biological origin. The pulsing sequences for the different types of homonuclear 2-D NMR experiments and their variants are shown in Table I. Though proton homonuclear *J*-resolved 2-D NMR is of some use in assigning the spectra of simple structures, its lack of sensitivity and inability to provide *J*-connectivity information render it a fairly useless procedure for the analysis of oligosaccharide structure. On the other hand, proton homonuclear correlated 2-D methods have been extremely useful for such studies.

Two types of correlated 2-D NMR experiments are possible, those involving scalar couplings through coherent transfer of transverse magnetization or those involving dipole couplings through incoherent transfer of magnetization. The significant difference between the two experiments is that scalar correlated spectra can be used to reveal through-bond connectivities and dipole corrrelated spectra can be used to reveal through-space connectivities. Two experimental variations of scalar correlated 2-D NMR (Table I) are correlated spectroscopy or COSY (Bax *et al*., 1981; Bax and Freeman, 1981) and spin echo correlated spectroscopy or SECSY

TABLE 1

PULSING SEQUENCES OF NMR EXPERIMENTS

Experiment	Number of Pulses	Pulsing Sequence[a]
1 - D	1	$90°-t_2 (t_1 = 0)$
2 - D J-resolved	2	$90°-0.5t_1-180°-0.5t_1-t_2$
Scalar Correlated		
2 - D SECSY[b]	2	$90°-0.5t_1-90°-0.5t_1-t_2$
2 - D COSY[c]	2	$90°-t_1-90°-t_2$
2 - D COSY,		
Multiple Quantum Filtered	3	$90°-t_1-90°-90°-t_2$
2 - D RECSY[d]	4	$90°-t_1-90°-\tau-180°-\tau-90°-t_2$
Dipolar Correlated		
2 - D NOESY,[e]		
SECSY matrix	3	$90°-0.5t_1-90°-\Delta t_1-90°-0.5t_1-t_2$
COSY matrix	3.	$90°-t_1-90°-\Delta t_1-90°-t_2$

[a] Symbols 90°, ninety degree pulse; 180° one-hundred-eighty degree pulse: t_1, evolution time; t_2, detection time during which the FID (free induction decay) signal is recorded; Δt_1, increment in t_1; and τ, a fixed delay time.

[b] SECSY, spin echo correlated spectroscopy.

[c] COSY, correlated spectroscopy.

[d] RECSY, relayed correlation (ie. relayed coherence transfer) spectroscopy.

[e] NOESY, nuclear Overhauser effect spectroscopy

(Aue *et al.*, 1976; Nagayama *et al.*, 1980). The dipole correlated experiment useful in oligosaccharide structure determination is 2-D nuclear Overhauser effect spectroscopy or 2-D NOESY (Jeener *et al.*, 1979; Kumar *et al.*, 1980). It should be pointed out that some authors have referred to scalar correlated spectra as "J-correlated" or "chemical shift correlated" spectra.

Recent improvements in scalar correlated experiments include multiple quantum filtered (MQF) spectroscopy (Piantini *et al.*, 1982; Rance *et al.*, 1983), also known as phase-sensitive or pure absorption spectroscopy, and relayed correlation spectroscopy (RECSY), also known as relayed coherence transfer spectroscopy (Eich *et al.*, 1982; Wagner, 1983). Multiple quantum filtering of 2-D scalar correlated experiments has the advantages of suppressing the aqueous (HOD) peak and simplifying spectra by eliminating singlets and suppressing dominant diagonal (or cross) peaks. RECSY allows a difficult J-connectivity to be followed by revealing scalar coupling to protons more distant than the next in the sequence. In effect we can "jump ahead," so that if the J-coupling of H-2→H-3 is obscure, we can see H-1→H-3, and know that the entire sequence is H-1→H-2→H-3.

APPLICATIONS OF 2-D NMR TO COMPLEX CARBOHYDRATES

In 1982 investigators in New Haven (Prestegard *et al.*) and Vancouver (Bernstein and Hall) reported almost simultaneously the first application of both scalar and dipole correlated 2-D NMR methods to complex carbohydrates. In these early studies the full power of 2-D methods was manifest as the complete analysis of oligosaccharide primary structure, including sequence and linkage sites, was possible. Since then, 2-D SECSY and NOESY have been applied to gangliosides and their asialo derivatives (Koerner *et al.*, 1983a and b), ganglioside G_{M1} oligosaccharide (Ong and Yu, 1986) and the neutral glycolipid globoside (Koerner et al., 1984; Yu *et al.*, 1984). Two-dimensional COSY, NOESY and RECSY experiments have been performed on peracetylated globo-series glycolipids (Gasa *et al.*, 1986). Two-dimensional NOESY (Homans *et al.*, 1983) and COSY studies (Geyer *et al.*, 1984) have also been carried out on *N*-linked oligosaccharides derived from glycoproteins. Early interest in *J*-resolved 2-D NMR studies of glycolipids (Yamada *et al.*, 1980; Dabrowski and Hanfland, 1982) and oligosaccharides (Hall *et al.*, 1980; Bruch and Bruch, 1982) has abated for the reasons mentioned above. Three recent review articles by Yu *et al.* (1984), Sweeley and Nunez (1985), and Koerner *et al.* (1987) on the application of 2-D NMR methods to glycoconjugates have been published and the majority of one issue of the new *Journal of Carbohydrate Chemistry* has been devoted to 2-D NMR spectroscopy of monosaccharides and oligosaccharides.(volume 3, number 4, 1984) In this article, the procedures for the analysis of oligosaccharide structure via correlated 2-D proton NMR spectroscopy will be presented.

DETERMINATION OF PRIMARY STRUCTURE

Preliminary studies and an Overview

The first step in any proton NMR study of an oligosaccharide or derivative should be obtaining an integrated, one-dimensional spectrum from 0 to 10 ppm. From this study, the purity and actual concentration of the sample, appropriateness of the solvent and temperature and the structural complexity of the oligosaccharide can be assessed. In the case of spectra of simple structures or of complex structures that are identical to ones already in a well-defined database, a one-dimensional study may suffice for structure determination. Examples of databases already in existence are oligosaccharides in deuterium oxide (Vliegenthart *et al.*, 1983; Carver and Brisson, 1984) and glycolipids in methylsulfoxide-*d*6-deuterium oxide (98:2 v/v) (Koerner *et al.*, 1983a; Dabrowski *et al.*, 1982; Gasa *et al.*, 1983; Clausen *et al.*,1985). Particularly useful in such one-dimensional analyses are the anomeric and methyl resonances located between 4 and 5 and between 1 and 2 ppm, respectively. In the case of an oligosaccharide whose one-dimensional spectrum does not resemble previously encountered structures but is technically good, then 2-D NMR analysis is indicated.

Such a 2-D analysis requires at least 0.5 mg of sample to be practical at the present time. For a trisaccharide, or more complex oligosaccharide, at least a 360 MHz spectrometer is necessary.

Analysis of primary structure via 2-D proton NMR involves two types of correlated experiments. First, one of the various scalar correlated spectra is obtained to reveal through-bond (*J*) couplings. From this first experiment, *J*-connectivities are established and the spectrum is thus divided into subspectra, each of which originates from and identifies one oligosaccharide residue. From the pattern of the chemical shifts and coupling constants found in each subspectrum, the identity of the residue of origin is established. The second type of 2-D NMR experiment carried out is 2-D NOESY, which reveals through-space couplings. By examination of the anomeric region of the NOESY spectrum of an oligosaccharide, linkage sites and sequence are determined for all glycosidic linkages except those involving ketose residues (Koerner *et al.*, 1987).

A key concept in the 2-D NMR analysis of an oligosaccharide is the comparison of scalar correlated spectral data with NOESY data through chemical shift identities of resonances. As an example, the comparison of 2D-SECSY and 2D-NOESY spectra of an idealized trisaccharide is shown in Fig. 1. From such a comparison, information on residue type, sequence and linkage sites can be combined to yield the complete primary structure of an oligosaccharide or derivative. It should be noted that in Fig. 1 *J*-multiplicities and H-6 resonances have been omitted for didactic clarity. Details of the use of 2-D NMR experiments to elucidate the primary structure of complex carbohydrates are presented below, using the examples of two glycolipids: asialo-G_{M2} (Fig. 2a) from human brain and the complex L-IV glycolipid from bivalve spermatozoa (Fig. 3a). We have recently described in great detail the application of 2-D methods to elucidate the primary structure of gangliosides (Yu *et al.*, 1984; Koerner *et al.*, 1987).

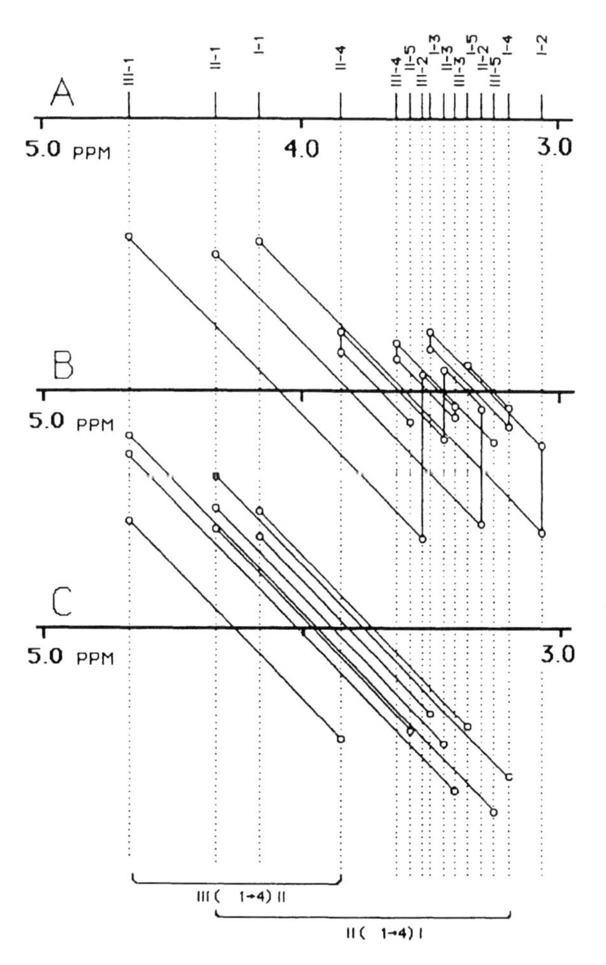

Fig. 1: Idealized 2D proton NMR analysis of a trisaccharide: A, 1D proton spectrum; B, 2D-SECSY spectrum; and C, 2D-NOE spectrum, involving anomeric protons.

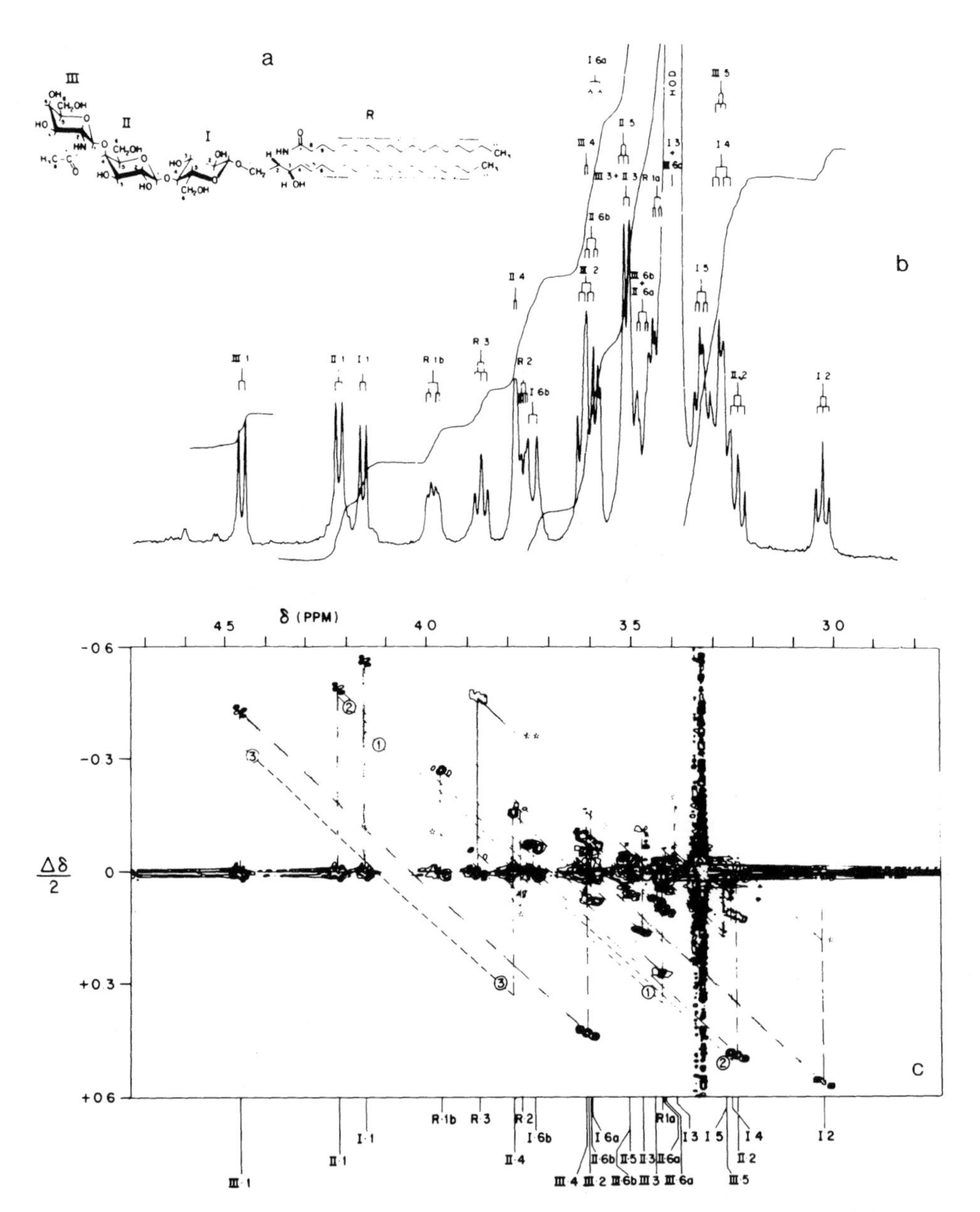

Fig. 2: (a) The primary structure of asialo-G_{M2}; (b) integrated, one dimensional proton NMR spectrum of asialo-G_{M2} between 3 and 5 ppm, obtained at 500 MHz; (c) 2-D SECSY spectrum of the same region as b. Graphical analyses revealing the four J-connectivities are labeled as follows: I, β-glucopyranosyl (-·-); II, β-galactopyranosyl (-··-); III, 2-acetamido-2-deoxy-β-galactopyranosyl (-···-); and R, ceramide (——). Also shown are the three (numbered) interresidue NOE couplings (---), obtained from the 2-D NOESY spectrum. Reprinted from Prestegard (1982) by permission of the American Chemical Society.

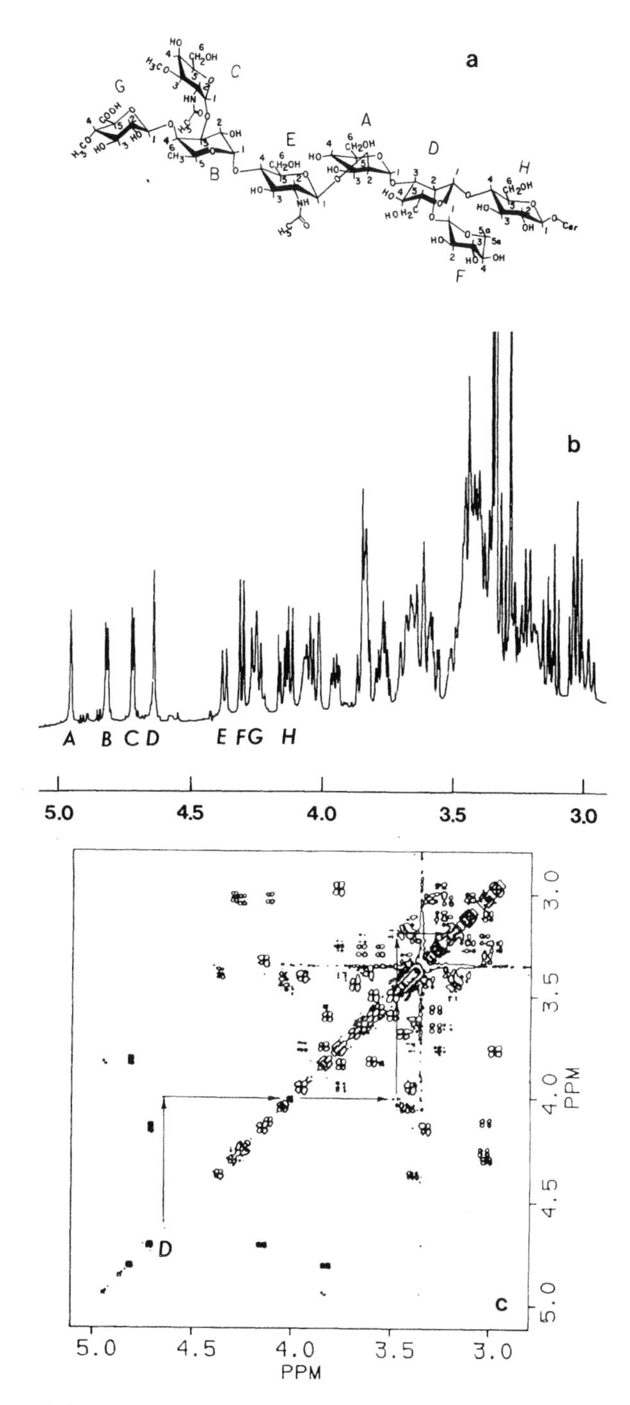

Fig. 3: (a) The primary structure of the L-IV glycolipid; (b) one-dimensional proton NMR spectrum of L-IV between 3 and 5 ppm, obtained at 490 MHZ. Anomeric protons are labeled A-H; (c)MQF-COSY spectrum of the same region as b. Graphical analysis revealing the J-connectivity of the D residue is shown. Reprinted from Scarsdale et al (1986a).

Generation of J-Connectivities via Scalar Correlated 2-D NMR

Both SECSY and COSY scalar correlated experiments have been performed with complex carbohydrates. Shown in Fig. 2c is the 2-D SECSY spectrum of asialo-GM2 and in Fig. 3c is the MQF-COSY spectrum of the L-IV glycolipid.

Having obtained a scalar correlated spectrum of an unknown oligosaccharide, a graphical analysis is undertaken to extract *J*-connectivities. In the case of the SECSY experiment, contour densities above and below the central horizontal axis identify *J*-coupled protons. Thus, sequential construction of vertical-upward, 135° diagonal, and vertical-upward lines of lengths *x*, 2.828 *x* and *x*, respectively, leads one through the *J*-connectivity (Fig. 2c). In the case of the COSY experiment, contour densities parallel to the diagonal central axis identify *J*-coupled protons. Thus, sequential construction of vertical and 90° horizontal lines of equal length leads one through the *J*-connectivity (Fig. 3c). It is easiest to begin in the well-resolved 4.0-5.0 ppm region in which the terminal anomeric protons of aldopyranose residues are found, then proceed upfield for each *J*-connectivity into the complex ring proton region (3.0-4.0 ppm). In this manner, *J*-connectivities are generated that are usually complete at high field (500 MHz). At lower resolution at least the first four protons of each oligosaccharide residue (H-1→H-4) can usually be followed. Once the *J*-connectivity is determined, the chemical shift and coupling constants are extracted for each subspectral proton from either the SECSY spectrum directly or by using also the accompanying 1-D spectrum. This data is then tabulated and each *J*-connectivity is analyzed for its proton number, pattern, vicinal coupling constant sequence and chemical shifts.

Identification of Residue Type via J-Connectivity Data

Probably the most important determinants of oligosaccharide residue constitution are the number and pattern of protons manifest in a residue's *J*-connectivity. These characteristic *J*-connectivity numbers and patterns and their use in the assignment of the constitution of complex carbohydrate residues have been thoroughly described (Koerner et al., 1987). Having established its constitution, the stereochemistry of the residue is determined by consideration of the values of the first four vicinal (^3J) coupling constants displayed in its *J*-connectivity. The correlation of all possible vicinal coupling-constant sequences with the stereochemistry of aldohexopyranose residues has been described (Koerner et al., 1987). The last types of data considered are the chemical shifts of protons within each *J*-connectivity. The use of this type data is more empirical than the other types of data derived from *J*-connectivities, but may be essential in establishing such structural features as methylation, acetylation, and sialylation sites (Koerner et al., 1987).

Oligosaccharide Sequence and Linkage Sites via 2-D NOESY

It is important that the 2-D NOESY experiment be carried out so that the spectrum produced can be easily compared to the scalar correlated spectrum already obtained. Graphical identification of dipole-coupled protons is as described above for the scalar correlated spectral analysis. Dipole-couplings are observed between anomeric (H-1) and nonanomeric resonances and between pairs of nonanomeric resonances; however, only the former will be considered, since only they are essential for establishment of oligosaccharide primary structure. Two types of dipole (or through-space) couplings are observed for anomeric protons. These are intraresidue or "across the ring" couplings, helpful in confirming resonance assignments, and interresidue couplings, which are across the glycosidic linkage and very important in identifying linkage sites and sequence.

The use of 2-D NMR spectroscopy to elucidate the primary structure of complex carbohydrates will be illustrated using the example of the L-IV glycolipid (Scarsdale et al., 1986a). Shown in Fig. 3b is the one dimensional spectrum of L-IV which contains most resonances from the oligosaccharide moiety. Higher and lower field regions contain information allowing characterization as a ceramide with a predominantly unsaturated acyl chain composition (Koerner _et al_., 1983a). From the one dimensional spectrum, it is possible to reach several preliminary conclusions concerning the saccharide composition of L-IV. From the anomeric region (4.0-5.0 ppm) it is possible to estimate the number of sugar residues. While it appears that there are doublets corresponding to nine anomeric protons, we shall show that one of these apparent doublets is not from an anomeric proton and there are actually eight sugars. It also appears there are two mannose residues on the basis of the small, poorly resolved $J_{1,2}$ couplings in resonances A and D at 4.95 and 4.64 ppm. Two additional α-linked sugars (B and C) and 4 β-linked sugars also exist. We have labeled the anomeric resonances using letters A-H working from low to high field. The letters will also have a one-to-one correspondence with residues in the final structure. In the region from 3.0-4.0, there are two singlets of an intensity corresponding to three protons each which can be assigned to _O_-methyl groups (3.28 and 3.35 ppm). In the region from 1.5-2.0 ppm (not shown), there are two resonances at 1.98 and 1.80 ppm which are typical of _N_-acetyl methyls, and finally, at 1.16 ppm a doublet appears which is characteristic of the methyl protons of a fucose residue. Extension beyond this preliminary analysis necessitates the use of two-dimensional methods.

Fig. 3c presents a contour plot of a multiple quantum filtered (pure absorption) COSY experiment. In this experiment, both cross and auto peaks occur as alternating pairs of positive and negative peaks. These phase properties are very useful in identifying and resolving overlapping cross peaks. Positive and negative contour levels are not distinguished in Fig. 3c but were color-coded in the plots used for analysis. Anomeric resonances and their cross peaks are well resolved in the low-field region of the F_2 dimension and provide a convenient starting point for spectral analyses. Beginning at the diagonal, successive vertical and horizontal lines, which correspond to scalar connectivities, are constructed to group resonances into subspectra belonging to a single residue. For example, beginning at 4.64 ppm with a D-1 resonance, and lowering the contour level somewhat, it is possible to find a weak cross peak showing a connectivity to the D-2 resonance at 4.01 ppm. That this connectivity is weak is a result of a small $J_{1,2}$ coupling constant. From the D-2 resonance, a connectivity to the D-3 resonance at 3.48 ppm can be found and, likewise, from D-3 to the D-4 resonance at 3.19 ppm. Thus, even in this rather difficult case, it is possible to assign resonances at 4.64, 4.01, 3.48 and 3.19 ppm, to the 1, 2, 3, and 4 protons of sugar residue D. This procedure is illustrated in Figure 3c. The shifts of the 1 and 2 resonances are consistent with those tabulated by Vliegenthart (1983) for a 3-substituted β-mannose residue. This, along with the appearance of a 1→3 intra-ring cross relaxation peak (see below) leads us to the conclusion that the D spin system arises from a β-mannose residue.

In some cases, assignment beyond the 2 proton of the J-connectivity is not possible due to spectral overlap. For example, the F anomeric resonance at 4.30 ppm can be connected to the F-2 proton at 3.01 ppm; however, the assignment is difficult to carry further due to overlap from the G-2 and H-2 protons which also have resonances near 3.01 ppm.

Figure 4a presents a contour plot of a homonuclear relayed coherence transfer spectrum (RECSY). The homonuclear relayed coherence transfer experiment is useful because it serves to establish remote as well as direct connectivities. This helps to resolve ambiguities due to spectral overlap. For instance, in the case of the F-1 resonance at 4.30 ppm, described above, it is possible to bypass the F-2 proton at 3.01 ppm and directly assign the F-3 proton at 3.09 ppm. This resolves the ambiguity due to overlap of the G-2 and H-2 proton resonances and allows connectivities F3 to F5 to be established. The unique occurrence of scalar connectivities of the 4 proton to two 5 protons and comparison to model compound data (Wise *et al*., 1984) assigns spin system F to a xylose residue.

The combination of MQF-COSY and RECSY experiments allows assignment of most of the ring protons of all carbohydrate

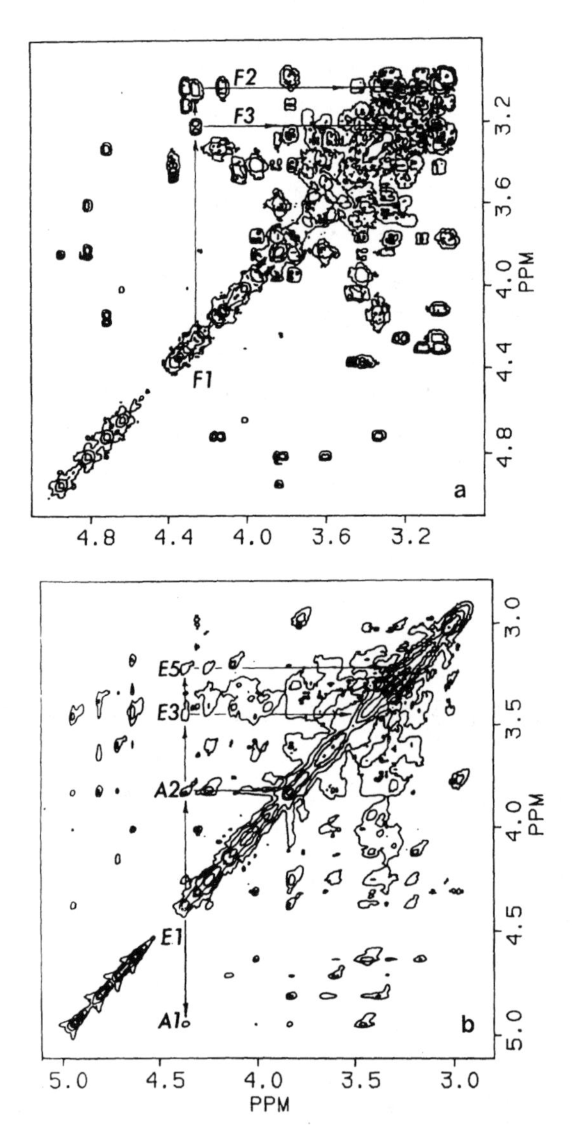

Fig. 4: (a) 2-D RECSY spectrum of the L-IV glycolipid between 3 and 5 ppm. Graphical analysis reveals the J-connectivity for the F residue. This illustrates the usefulness of this experiment in resolving ambiguities due to spectral overlap; (b) 2-D NOESY spectrum of the L-IV glycolipid between 3 and 5 ppm. Graphical analysis reveals the dipolar (through-space) connectivities of the E-1 resonance. Reprinted from Scarsdale *et al*. (1986a).

residues. Usually *J*-connectivities are followed from the anomeric end, as illustrated in the case of residues D and F. For galactose and fucose residues, connectivities beyond the 3 proton are difficult because of the small $J_{3,4}$ coupling and resultant low amplitude connectivity peak. In the case of the fucose residue, the B-6 resonance at 1.16 ppm provides a second useful starting point at the other end of the residue. Connectivities to the B-5 resonance at 4.25 ppm and B-4 resonance at 3.83 ppm are easily established. With the connection of B-1, B-2, and B-3 from the anomeric end and assignment to fucose on the basis of shift correlation, complete assignment of all fucose resonances is achieved. It is interesting that the B-5 resonance of the α-fucose falls in a region normally reserved for anomeric protons. The assignment of this resonance to a fucose 5 proton leaves only eight anomeric protons in the anomeric resonance region. Along with integration data to eliminate the possibility of sugars not having an anomeric proton, these data show the compound of interest to have an octasaccharide headgroup.

Assignments of resonances are summarized in Table II. As illustrated above, these result from consideration of a number of factors including chemical shift correlation, small or large scalar couplings as reflected in intensities of crosspeaks, and unique resonance positions or multiplicities as in the case of xylose, fucose, or the glucuronic acid 5 proton. Chemical shifts for analogous sites in previously assigned oligosaccharides were also used (Scarsdale *et al.*, 1986a). In most cases, agreement is within 0.1 ppm. This is acceptable given differences in solvent systems and the occurrence of some additional linkage sites in the compound of interest.

Having assigned the various resonances and sugar types, the major remaining structural problems involve linking residues together and locating sites of O-methyl attachment. Although some of this information can be deduced from the characteristic linkage shift seen in comparing model compound data to data for L-IV (Scarsdale *et al.*, 1986a), 2-D NOESY experiments which show connectivities based on cross relaxation are most useful in solving both problems. When

TABLE II

CHEMICAL SHIFTS FOR OLIGOSACCHARIDE PROTONS OF GLYCOLIPID L-IV[1]

Residue	Proton Number (1)	(2)	(3)	(4)	(5)	(6)	Structure
A.	4.95	3.83	3.55	3.26	3.65		(β1-2)Man(α1-)
B.	4.81	3.82	3.59	3.82	4.25	1.16	Fuc(α1-)
C.	4.72	4.15	3.32	4.07			GalNAc(α1-)
D.	4.64	4.01	3.48	3.19	3.42		(α1-3)Man(β1-)
E.	4.36	3.38	3.44	3.67	3.21		(β1-4)GlcNAc(β1-)
F.	4.30	3.01	3.09	3.27	2.96	3.76	Xyl(β1-)
G.	4.25	3.02	3.20	3.37	3.61		GlcA[2](β1-)
H.	4.11	3.01	3.33	3.44	3.18		(β1-4)Glc(β1-)

[1]Data taken from Scarsdale *et al.* (1986a). [2]GlcA, Glucuronic Acid

accumulated for short mixing times, peak intensities in 2-D NOE experiments are proportional to $1/\underline{r}^6$ where $\underline{r}$ is the distance between proton pairs. As practiced here, this means that one sees significant connectivities only for interproton distances less than 3 Å. Focusing on the anomeric resonances, one expects to see one or more intra-ring cross peaks to resonances from protons on the same residue. For a β-linked sugar residue, one sees axial 1,3 and 1,5 connectivities at 2.5 Å. In β-mannosides one expects to see an additional equatorial 1,2 connectivity at 2.5 Å. For an α-linked sugar residue, one sees only an equatorial 1,2 connectivity at 2.5 Å. In other words, except for β-mannose, one expects to see one or two intra-ring connectivities.

A 2-D NOESY spectrum is presented in Figure 4b. Inspection of data, however, shows three or more connectivities for most rings. Consideration of the energetics of glycosidic rotational conformation by Lemieux and co-workers (Lemieux *et al.*, 1980; Thorgerson et al., 1982; Bock *et al.*, 1982; Sabesan *et al.*, 1984) suggests that a small range of torsional angles are preferred and that in all of these cases the anomeric proton is expected to be with 3 Å of a proton on the linkage site. By eliminating NOE crosspeaks assignable to intra-ring interactions and choosing the largest remaining crosspeak as a route to the transglycosidic proton, it is possible to sequence and identify linkage sites for L-IV. For four of the eight anomeric protons resolution is adequate to give ambiguous linkages as follows: Man(α1-3)Man(β1-), GalNAc(α1-3)Fuc(α1-) Xyl(β1-2)Man(β1-), and Glc(β1-1')Cer. In the case of H-1 resonances for GlcNAc(β1-) and GlcA(β1-), three possible linkage sites are identified: 2Manα1, 2Fucα1 or 4Fucα1. In the case of GlcNAc(β1-), an additional weak cross peak gives an unambiguous connectivity to the anomeric resonance of Man(α1-) and suggests the GlcNAc(β1-2)Man(α1-) linkage to be correct. Of the two possible GlcA(β1-)Fuc(α1-) connectivities, GlcA(β1-4)Fuc(α1-) is consistent with the structure proposed by Hori *et al.* (1983). For the remaining two anomeric protons, Fuc(α1-) and Man(β1-), the strongest cross peaks that are not assignable to intra-ring NOEs, do not correlate with proposed linkages. It is possible that these cross peaks belong to unassigned intra (H-5) or inter (H-6) ring protons. In the case of Fuc(α1-), a weaker inter-ring cross peak correlates with the 4 protons on GlcNAc(β1-) establishing a linkage consistent with the proposed structure. In the case of Man(β1-), the cross peak for the proposed 4 linkage to Glc(β1-) would lie under an observed intra-ring cross peak. Thus, we would independently propose five of the linkages shown in the proposed structure in Fig. 3a, and our data are consistent with the remaining three linkages (Hori *et al.*, 1983).

Since methyl protons of O-methyl sugars are within 3 Å of the ring protons of the methylation site, some information on O-methylation sites can also be obtained from 2-D cross relaxation experiments. The methyl peaks at 3.28 and 3.35 ppm are, however, very intense singlets and as such produce a number of false NOE cross peaks which result from intersection of low amplitude tails on peaks in the F_1 dimension with tails of the methyl peak in the F_2 dimension, and vice versa. For the methyl resonance at 3.28 ppm, the strongest cross peak lies at 3.38, a position consistent with proposed methylation at the 4 position of glucuronic acid. The methyl resonance at 3.35 ppm lies very close to a resonance from residual HOD in the sample and a definitive assignment cannot be made from the

NOE data. However, the large deviation of chemical shift from the model compound at position 3 of residue C (Scarsdale *et al.*, 1986a) would be consistent with proposed methylation at the 3 position of GalNAc (Hori *et al.*, 1983).

DETERMINATION OF SECONDARY STRUCTURE

Since complex carbohydrates are involved in cell-cell recognition, enzyme substrate interaction, antigen-antibody interaction, and other molecular events that may involve recognition processes, it is reasonable to assume that these events are necessarily governed not only by the primary structure but also by the secondary structure of the oligosaccharide moieties. 2-D NMR has proven to be particularly useful in this regard. Using NOESY to measure the proton-proton dipolar cross-relaxation network in various glycolipids (Koerner *et al.*, 1983b and 1984; Yu *et al.*, 1984; and Scarsdale *et al.*, 1986a), we have found that the magnitude of the NOE between trans-glycosidic pairs of protons was equal to or greater than that between intra-residue 1,3 or 1,5-diaxial couplings. Since the cross-relaxation rate of protons (NOE) is inversely proportional to the distance to the sixth power, this suggests that the transglycosidic pairs of protons are close in space and conformers which bring these pairs close must be highly populated. In fact, semiquantitative estimation of the trans-glycosidic interprotonic distances (r) of the glycolipid globoside via 2-D NOESY reveals values of approximately 2.5 Å (Koerner *et al.*, 1984) in all cases. Such short distances for the r values of the glycolipid globoside have been confirmed using more rigorous one-dimensional NOE experiments (Yu *et al.*, 1984) and a combined 2D-NOESY NMR pseudo energy approach (Scarsdale *et al.*, 1986b).

Construction of Dreiding models of the glycosidic linkages of globoside with fixed interprotonic distances of 2.5 Å and systematic inspection of allowed dihedral angles Φ and Ψ (+60°, 0°, and -60°) reveal a likely conformation to be that shown in Fig. 5 (Yu *et al.*, 1984). While it is possible that such a conformation may be solvent dependent, several interesting features are apparent for this proposed structure:

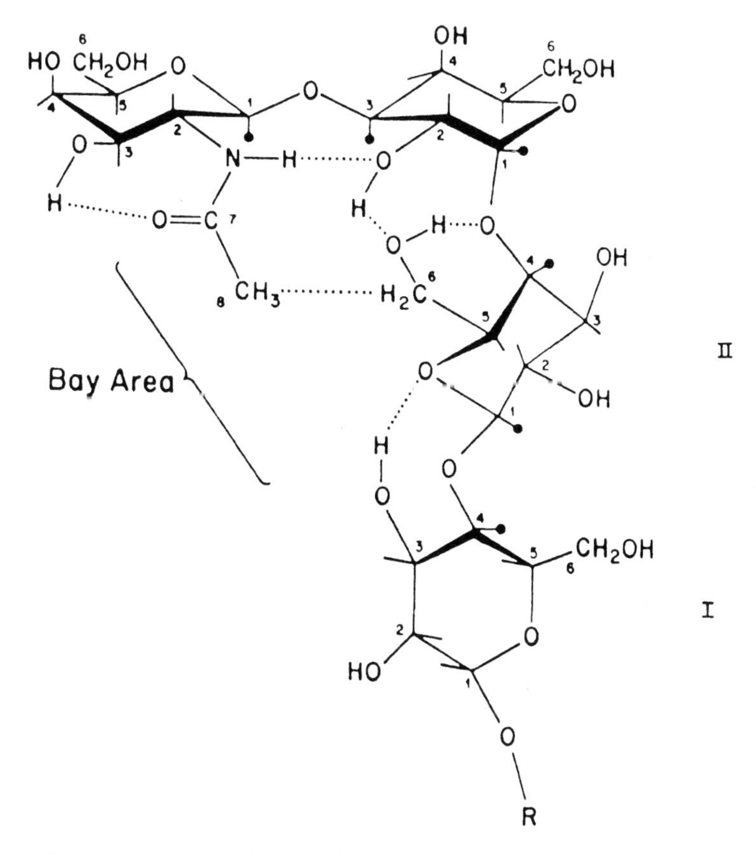

Fig. 5: Structure of the oligosaccharide moiety of globoside. The secondary structure shown is one that could be stabilized by an extensive network of H-bonds and a van der Waals interaction. It is one of two possible conformations that were determined by a pseudo energy approach (Scarsdale *et al.*; 1986b). Black dots represent transglycosidic protons. Reprinted from Yu *et al.* (1984).

1) an extended H-bond network involving ten atoms and four H-bonds is optimized. This begins with the IV-3 hydroxyl and ends with the II-4 linkage oxygen. The sequence of this extended H-bond network is O(II-4)...H-O(II-6)...H-O(III-2) ...H-N-C=O(IV-2)...H-O(IV-3), proceeding from H-bond acceptor terminus to the H-bond donor terminus. Such extended, vectoral H-bond networks could result in stabilized secondary structures for oligosaccharides, as has been noted by Jeffrey and Takagi (1978); 2) a van der Waals interaction between the IV-2 acetamido methyl and the II-6 methylene can exist; 3) an overall "L-shape" for the oligosaccharide with a hydrophobic inner-side or "bay area" (containing the van der Waals interaction and extended H-bond network) and a hydrophilic outer-side (presenting all non-H-bonded hydroxyl goups) is apparent. We are currently extending the above study to aqueous solvents and to other glycolipids with the hope of relating conformational properties to membrane surface interactions.

Recent differential scanning calorimety and lipid monolayer studies (Maggio *et al.*, 1985) have fully confirmed the existence of the L-shaped oligosaccharide conformation proposed above for globoside in solution (Fig. 5).

CONCLUSIONS

The introduction of multiple-pulse sequences into the nuclear magnetic resonance experiment has created "two-dimensional" proton NMR spectroscopy, a method capable of the independent and complete analysis of the primary and secondary structure of complex carbohydrates.

Primary structure is determined by carrying out two types of 2-D NMR experiments. First, a scalar correlated spectrum (SECSY, COSY, etc.) is obtained to reveal through-bond couplings. Subspectral (*J*-connectivity) patterns are obtained from this spectrum that determine monosaccharide composition, stereochemistry, and anomeric configuration. Second, a dipole correlated spectrum (NOESY) is obtained to reveal through-space connectivities. From analysis of transglycosidic (inter-residue) connectivities of the anomeric protons observed in this spectrum, the sequence and glycosidic linkage sites of all oligosaccharide residues are revealed.

The major limitation of this 2-D NMR method of primary structure analysis is its inability to determine the glycosidation linkage sites of ketoses (e.g., fructose and sialic acid). However, this limitation may be overcome through the use of empirically-derived glycosidation shifts.

Secondary structure (conformation) is determined by carrying out the dipole correlated experiment (2-D NOESY) and estimating the trans-glycosidic interprotonic distances (r) between residues via internal calibration. Since complex carbohydrate conformation is determined almost exclusively by the dihedral angles and about the glycosidic linkages, only a small number of additional nuclear Overhauser effect connectivities are necessary to define a unique oligosaccharide conformation.

ACKNOWLEDGEMENTS

The authors wish to thank Henry Weinfeld (Tulane University) for help in preparing computer graphics and Winifred Hunter for word processing. This work was supported by USPHS grant NS 11853.

REFERENCES

Aue, W.P., Bartholdi, E., and Ernst, R. R. (1976) Two-dimensional spectroscopy. Application to nuclear magnetic resonance, J. Chem. Phys. 64:2229-2246.

Bax, A. (1982) "Two-dimensional Nuclear Magnetic Resonance in Liquids", Riedel Publishing, Boston.

Bax, A., Freeman, R., and Morris, G. (1981) Correlation of proton chemical shifts by two-dimensional Fourier-transform NMR, J. Magn. Reson. 42:164-168.

Bax, A., and Freeman, R. (1981) Investigation of complex networks of spin-spin coupling by two-dimensional NMR, J. Magn. Reson. 44:542-561.

Bax, A., and Lerner, L. (1986) Two-dimensional nuclear magnetic resonance spectroscopy, Science 232:960-967.

Benn, R., and Günther, H. (1983) Modern pulse methods in high-resolution NMR spectroscopy, Agnew. Chem. Int. Ed. Engl. 22:350-380.

Bernstein, M.A., and Hall, L.D. (1982) De novo sequencing of oligosaccharide by proton NMR spectroscopy, J. Am. Chem. Soc. 104:5553-5555.

Bock, K., Arnurp, J., and Lönngren, J. (1982) The preferred conformation of oligosaccharides derived from the complex-type carbohydrate portions of glycoproteins, Eur. J. Biochem. 129:171-178.

Bruch, R.C., and Bruch, M.D. (1982) Two-dimensional *J*-resolved proton NMR-spectroscopy of oligomannosidic glycopeptides, J. Biol. Chem. 257:3409-3413.

Carver, J.P., and Brisson, J.R. (1984) The three-dimensional structure of N-linked oligosaccharides, in: "The Biology of Carbohydrates", V. Ginsberg and P.W. Robbins, eds., Vol. 2, pp 289-390, Wiley, New York.

Clausen, H., Levery, S.B., McKibbin, J.M., and Hakomori, S. (1985) Blood group A determinants with mono- and difucosyl type 1 chain in human erthrocyte membrane, J. Biol. Chem. 24:3578-3586.

Dabrowski, J. and Hanfland, P. (1982) Structure determination of a ceramide pentadecasaccharide by two-dimensional *J*-resolved and *J*-correlated NMR spectroscopy, FEBS Lett. 142:138-142.

Dabrowski, J., Hanfland, P., and Egge, H. (1982) Analysis of glycosphingolipids by high-resolution proton nuclear magnetic resonance spectroscopy, Meth. Enzymol. 83:69-86.

Eich, G., Bodenhauser, G., and Ernst, R.R. (1982) Exploring nuclear-spin systems by relayed magnetization transfer, J. Am. Chem. Soc. 104:3731-3732.

Gasa, S., Mitsuyama, T., and Makita, A. (1983) Proton nuclear magnetic resonance of neutral and acidic glycosphingolipids, J. Lipid Res. 24:174-182.

Gasa, S., Nakamura, M., Makita, A., Ikura, M., and Hikichi, K. (1986) Complete structural analysis of globoseries glycolipids by two-dimensional nuclear magnetic resonance, Eur. J. Biochem. 155:603-611.

Geyer, R., Geyer, H., Stirm, S., Hunsmann, G., Schneider, J., Dabrowski, U., and Dabrowski, J. (1984) Major oligosaccharides in the glycoprotein of friend murine leukemia-virus. Structure elucidation by one-dimensional and two-dimensional proton nuclear magnetic resonance and methylation analysis, Biochemistry 23:5628-5637.

Hall, L.D., Morris, G.A., and Sukumar, S. (1980) Resolution and assignment of the 270 MHz proton spectrum of cellobiose by homonuclear and heteronuclear two-dimensional NMR, J. Am. Chem. Soc. 102:1745-1747.

Homans, S.W., Dwek, R.A., Fernandez, D.L., and Rademacher, T.W. (1983) Solution conformation of biantennary complex type oligosaccharides. Determination of major conformers about the glycosidic linkages, FEBS Lett. 164:231-235.

Hori, T., Sugita, M., Ando, S., Tsukada, K., Shiota, K., Tsuzuki, M., and Iatsaka, O. (1983) Isolation and characterization of a 4-O-methylglucuronic acid-containing glycosphingolipid from spermatozoa of a fresh water bivalve, *Hyriopsis schlegelii*, J. Biol. Chem. 258:2239-2245.

Jeener, J., Meier, B.H., Bachman, P., and Ernst, R.R. (1979) Investigation of exchange process by two-dimensional NMR spectroscopy, J. Chem. Phys. 71:4546-4553.

Jeffery, G.A. and Takagi, S. (1978) Hydrogen-bond structure in carbohydrate crystals, Acc. Chem. Res. 11:264-270.

Jelinski, L.W. (1984) Modern NMR spectroscopy, Chem. Eng. News 62(Nov. 5):26-47.

Koerner, T.A.W., Prestegard, J.H., Demou, P.C., and Yu, R.K. (1983a) High resolution proton NMR studies of gangliosides. 1. Use of homonuclear two-dimensional spin-echo J-correlated spectroscopy for determination of residue composition and anomeric configuration, Biochemistry 22:2676-2687.

Koerner, T.A.W., Prestegard, J.H., Demou, P.C., and Yu, R.K. (1983b) High-resolution proton NMR studies of gangliosidesl 2. Use of two-dimensional nuclear Overhauser effect spectroscopy and sialylation shifts for determination of oligosaccharide sequence and linkage sites, Biochemistry 22:2687-2690.

Koerner, T.A.W., Scarsdale, J.M., Prestegard, J.H., and Yu, R.K. (1984) Structure determination of glycolipids using two-dimensional proton NMR spectroscopy: globoside, J. Carbohydr. Chem. 3:565-580.

Koerner, T.A.W., Prestegard, J.H., and Yu, R.K. (1987) Analysis of oligosaccharide structure via two-dimensional proton nuclear magnetic resonance spectroscopy, Meth. Enzymol., 138:38-59 (Complex Carbohydrates, Part E).

Kumar, A., Ernst, R.R., and Wüthrich, K. (1980) A two-dimensional nuclear Overhauser enhancement (2D NOE) experiment for the elucidation of complete proton-proton cross-relaxation networks in biological macromolecules, Biochem. Biophys. Res. Commun. 95:1-6.

Lemieux, R.U., Bock, K., Delbaere, L.T.J., Koto, S., and Rao, V.S. (1980) The conformation of oligosaccharides related to the ABH and Lewis human blood group determinants, Can. J. Chem. 58:631-653.

Maggio, B., Ariga, T., and Yu, R.K. (1985) Molecular parameters and conformation of globoside and asialo-G_{M1}, Arch. Biochem. Biophys. 241:14-21.

Nagayama, K., Kumar, A., Wüthrich, K., and Ernst, R.R. (1980) Two-dimensional spin-echo correlated spectroscopy (SECSY) for ^{1}H NMR studies of biological molecules, J. Magn. Reson. 40:321-334.

Ong, R.L. and Yu, R.K. (1986) Proton NMR assignments of G_{M1}-oligosaccharide in deuterated water at 500 MHz by two-dimensional spin-echo J-correlated spectroscopy, Arch. Biochem. Biophys. 245:157-166.

Prestegard, J.H., Koerner, T.A.W., Demou, P.C., and Yu, R.K. (1982) Complete analysis of oligsaccharide primary structure using two-dimensional high-field proton NMR, J. Am. Chem. Soc. 104:4993-4995.

Piantini, U., Sorensen, O.W., and Ernst, R.R. (1982) Multiple quantum filters for elucidating NMR coupling networks, J. Am. Chem. Soc. 104:6800-6801.

Rance, M., Sorenson, O.W., Bodenhausen, G., Wagner, G., Ernst, R.R., and Wüthrich, K. (1983) Improved spectral resolution in COSY H-1 NMR spectra of proteins via double quantum filtering, Biochem. Biophys. Res. Commun. 117:479-485.

Sabesan, S., Bock, K., and Lemieux, R.U. (1984) The conformational properties of the gangliosides G_{M2} and G_{M1} based 1H and 13C nuclear magnetic resonance studies, Can. J. Chem. 62:1034-1045.

Scarsdale, J.N., Ando, S., Hori, T., Yu, R.K., and Prestegard, J.H. (1986a) 1H NMR characterization of the structure of a novel octasaccharide glycolipid isolated from the spermatozoa of bivalves, Carbohydr. Res., in press.

Scarsdale, J.N., Yu, R.K., and Prestegard, J.H. (1986b) Structural analysis of a potentially flexible glycolipid headgroup using an NMR pseudo energy approach, J. Am. Chem. Soc., in press.

Sheatly, F., Scott, J.E., Jeanloz, R.W., Walkerna, E. (1982) Secondary structure in glycosaminoglycuronans: NMR spectra in dimethyl-sulfoxide of disaccharides related to hyaluronic acid and chrondroitin sulfate, Carbohydr. Res. 99:1-11.

Sweeley, C.C. and Nunez, H.A. (1985) Structural analysis of glycoconjugates by mass spectrometry and nuclear magnetic resonance spectroscopy, Ann. Rev. Biochem. 54:765-801.

Thorgersen, H., Lemieux, R.U., Bock, K., and Meyer, B. (1982) Further justification for the exo-anomeric effect. Conformational analysis based on nuclear magnetic resonance spectroscopy of oligosaccharides, Can. J. Chem. 60:44-57.

Vliegenthart, J.F.G., Dorland, L., and Van Halbeek, H. (1983) High-resolution, 1H-nuclear magnetic resonance spectroscopy as a tool in the structural analysis of carbohydrates related to glycoproteins, Adv. Carbohydr. Chem. Biochem. 41:209-374.

Wagner, G. (1983) Two-dimensional relayed coherence transfer spectroscopy of a protein, J. Magn. Reson. 55:151-156.

Wise, W.B., Pfeffer, P.E., and Kovac, P. (1984) A study of methyl β-xylobioside: An illustrative example of two-dimensional NMR methods, J. Carbohydr. Chem. 3:513-524.

Yamada, A., Dabrowski, J., Hanfland, P., and Egge, H. (1980) Preliminary results of J-resolved, two-dimensional 1H-NMR studies on glycosphingolipids, Biochem. Biophys. Acta 618:473-479.

Yu, R.K., Koerner, T.A.W., Demou, P.C., Scarsdale, J.N., and Prestegard, J.H. (1984) Recent advances in structural analysis of gangliosides: primary and secondary structures, Adv. Exp. Med. Biol. 174:87-102.

ABSTRACT

STRUCTURAL STUDIES ON THE CARBOHYDRATE CHAINS OF GLYCOPROTEINS

J.F.G. Vliegenthart

Department of Bio-Organic Chemistry, Utrecht University
P.O. Box 80.075, NL 3508 TB Utrecht, The Netherlands

In the past 10 years, the approach of the structure elucidation of carbohydrates and glycoconjugates has changed considerably. The availability of advanced physical techniques has now led to many applications in the carbohydrate field. For the carbohydrate chains of glycoproteins this development rendered possible determinations of numerous new primary structures. Furthermore, many details were disclosed of carbohydrate-microheterogeneity as a general feature. The introduction of high-resolution ^{1}H-NMR spectroscopy for these purposes has shown to be extremely useful. For the asparagine-linked carbohydrates it turns out to be possible to define the members of the oligomannose type of compounds occurring in higher organisms, by ^{1}H-NMR parameters merely. The *N*-acetyllactosamine type of chains can be easily recognized as to the type of branching of the core mannotriose on the basis of the ^{1}H-NMR spectra. In a vast number of cases, the details of the internal and peripheral parts also can be characterized in this way. However, the repeating elements like in polylactosamine type of structures are much more difficult to describe in terms of NMR spectroscopic data (1). For the carbohydrates which are *O*-linked to the polypeptide backbone through serine or threonine, a wide variety in structures exists. For these compounds a large collection of ^{1}H-NMR data could be obtained only recently (2). It turns out that in this case too the variability can adequately be defined in structural terms in cases when differences in sugar composition are involved. The location of branching points and the identity of carbohydrate chains can be elucidated. However, it may be clear that the occurrence of repeating units may complicate the picture. To gain insight into the functioning of carbohydrate chains it is essential to get information on the spatial structure of such chains in solution. It is particularly relevant to know for glycoproteins which part of the carbohydrate chain is exposed to the solution and thereby avaiable for interaction and recognition processes. The influence of the protein part on the presentation of the carbohydrate chains to the solution still needs further study.

REFERENCES: 1) Vliegenthart, J.F.G., Dorland, L., and Van Halbeek, H., Adv. Carbohydrate Chain and Biochemistry 41:209-374 (1983). 2) Lamblein, G., Boersma, A., Lhermitte, M., Roussel, Ph., Mutsaers, J.H.G., Van Halbeek, H., and Vliegnenthart, J.F.G., Eur. J. Biochem. 143:227-236 (1984). 3) Mutsaers, J.H.G., Van Halbeek, H., Vliegenthart, J.F.G., Wu, A.M., and Kabat, E.A., Eur. J. Biochem. 157:134-146 (1986). 4) Nasir-Ud-Din, Jeanloz, R.W., Lablein, G., Roussel, Ph., Van Halbeek, H., Musaers, J.H.G., and Vliegenthart, J.F.G., J. Biol. Chem. 261:1992-1997 (1986).

THE USE OF ENZYMES FOR STRUCTURAL DETERMINATION OF COMPLEX CARBOHYDRATES

Y.-T. Li and S.-C. Li

Department of Biochemistry,
Tulane University School of Medicine
New Orleans, Louisiana 70112 USA

It has been shown that sugar units in glycoconjugates carry specific biological messages. They have been found to serve as blood group determinants, cell surface antigens, and receptors for hormones, toxins and viruses. In order to understand the biological activity of a glycoconjugate, it is very important to accurately determine the structure of its complex carbohydrate chains. A complete structural analysis of a complex carbohydrate chain includes the determination of the anomeric configuration of each sugar unit, the sequential arrangement of sugar moieties and the linkage-points between pairs of monosaccharides within the sugar chain. Among the chemical methods, periodate oxidation and permethylation analysis are very useful, however, these two methods cannot provide information about the anomeric configuration and sequential arrangement of sugar units. We have shown that the stepwise release of monosaccharide units from the non-reducing terminus of a sugar chain by specific exo-glycosidases can provide us with this information. The use of endo-glycosidases to cleave the endoglycosidic linkages of a sugar chain has also been useful. The use of glycosidases for the structural analysis of complex carbohydrates is analogous to that of using nucleases and proteases for the structural analyses of nucleic acids and proteins, respectively. Since no single method can furnish complete structural information of a sugar chain, it is advantageous to use glycosidases together with other methods such as permethylation analysis and NMR spectroscopy to study the primary structure of complex carbohydrates.

This chapter gives specific examples to show the advantage of using glycosidases for the structural analysis of complex carbohydrates. Methods for the preparation of glycosidases can be found in Methods in Enzymology, Vol. 28, Part B, 1972; Vol 50, Part C, 1978; Vol. 83, Part D, 1982. Many glycosidases are now commercially available, however, the purity of these glycosidases often requires careful consideration.

DETECTION OF β-LINKED MANNOSE AT THE CORE REGION OF ASN-LINKED SUGAR CHAIN

While studying the biosynthesis of plasma glycoprotein, we isolated two glycopeptides from rat liver microsomes (Li *et al*. 1968). One of the asparagine-linked glycopeptides was found to contain NeuAc, Gal, GlcNAc, Man, in the proportion of 1:1:4:5. The sequential arrangement and the anomeric configuration of these saccharide units was studied by using various glycosidases. As shown in Table 1, treatment of the glycopeptide with only neuraminidase or α-mannosidase resulted in the liberation of 1 mole of NeuAc or 2 moles of Man per mole of glycopeptide. This result indicates that these two saccharides occupy the nonreducing ends of a branched heterosaccharide chain, on which NeuAc and Man are the two terminal sugars. None of the Gal in this glycopeptide was cleaved off by β-galactosidase until the NeuAc was removed. Therefore, Gal is located next to the NeuAc residue. After the treatment of the glycopeptide either with α-mannosidase or with neuraminidase plus β-galactosidase, one mole each of GlcNAc was liberated by β-*N*-acetylhexosaminidase. This result suggests that one residue each of GlcNAc is located internally to Gal and to Man, respectively. An exhaustive digestion of the glycopeptide with neuraminidase, α-mannosidase, β-galactosidase, and β-*N*-acetylhexosaminidase left 2 moles of GlcNAc and 1 mole of Man intact. Assuming that this glycopeptide contains a branched sugar chain, the sequential arrangement and the anomeric configuration of the monosaccharide units in this glycopeptide could be expressed as:

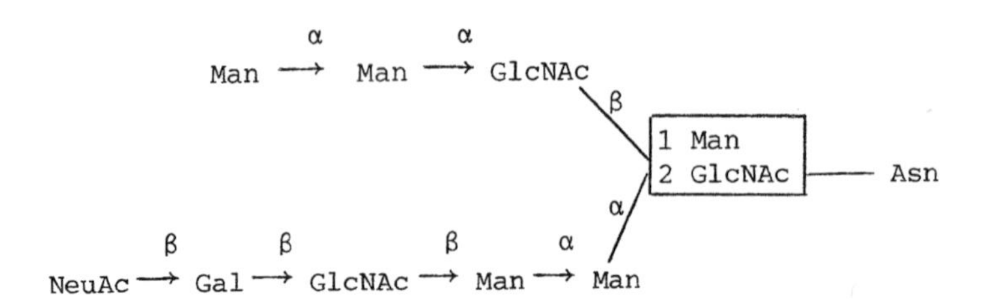

TABLE 1. Enzymatic Liberation of Monosaccharide Units from the Glycopeptide Isolated from Rat Liver Microsomes[a].

Description	NeuAc	Gal	GlcNAc	Man
		mole/mole	glycopeptide	
Neuraminidase only	1.0	---	---	---
α-Mannosidase only	---	---	---	2.1
β-Galactosidase only	---	0	---	---
Neuraminidase, then β-galactosidase	1.0	1.0	---	---
β-*N*-Acetylhexosaminidase only	---	---	0	---
α-Mannosidase, then β-*N*-acetylhexosaminidase	---	---	1.0	2.1
Neuraminidase, β-galactosidase, then β-*N*-acetylhexosaminidase	1.0	1.0	0.9	---
Exhaustive digestion with neuraminidase, α-mannosidase, β-galactosidase, and β-*N*-acetylhexosaminidase	1.0	1.0	2.0	4.2

[a]Neuraminidase purified from *Clostridum perfrigens* (0.1 units/5 μl) was obtained from Sigma. α-Mannosidase (5 units/5 μl), β-*N*-acetylhexosaminidase (4.5 units/5 μl) and β-D-galactosidase (1.2 units/5 μl) were isolated from jack bean meal (Li and Li, 1972a). Between 300 and 500 μg of glycopeptide were dissolved in 100 μl of sodium citrate buffer, pH 4.0 in a microcentrifuge tube and incubated with 5 μl of enzyme at 37° for 16 hours. For the stepwise degradation of monosaccharide from glycopeptide with the use of two or more enzymes, the enzyme in the incubation mixture was inactivated by immersing the tube in boiling water bath for 3 minutes before the next enzyme was added. After the incubation, free NeuAc and GlcNAc were determined by the thiobarbituric acid method (Warren, 1959) and the Morgan-Elson reaction (Reissig *et al*, 1955), respectively. Man and Gal were determined by using an automatic sugar analyzer (Lee, 1972).

This was the first indication of the possible presence of a β-linked Man and two GlcNAc at the core region of the asparagine linked oligosaccharide chain. It is now well established that Man(GlcNAc)$_2$ is the common core structure of the asparagine-linked sugar chains. In order to establish the anomeric linkage of the Man residue in the core region, we isolated a β-mannosidase from pineapple bromelain. As shown in Fig. 1, the bromelain β-mannosidase (Li and Lee, 1972) completely liberated the Man residue from Man → GlcNAc → GlcNAc → Asn isolated from ovalbumin. α-Mannosidases isolated from pineapple bromelain and jack bean were inactive toward this core glycopeptide.

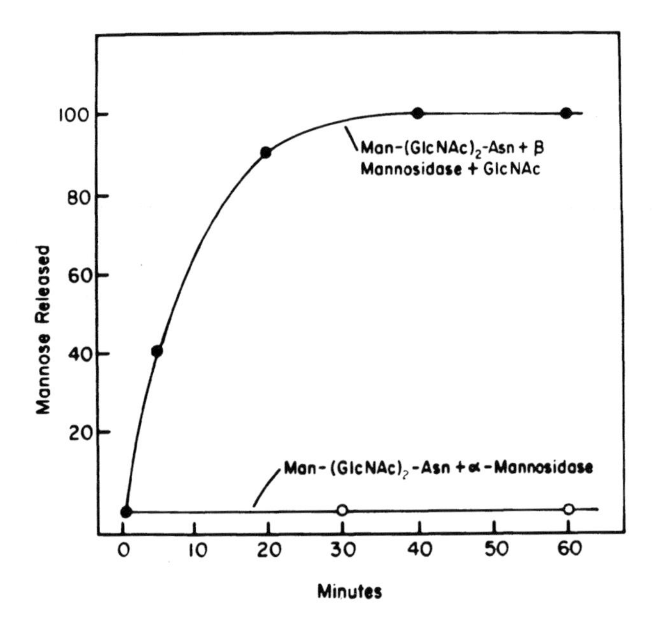

Fig. 1. Kinetics of the liberation of Man from ovalbumin core glycopeptide. Incubation mixtures contained the following components in 120 μl: Man-(GlcNAc)$_2$ -Asn, 0.08 μmole; 0.05 M glycine-HCl buffer, pH 3.5 for β-mannosidase (Li and Lee, 1972), or 0.05 M sodium citrate buffer, pH 4.5 for α-mannosidase (Li and Li, 1972a); 0.88 unit of β-mannosidase or 0.7 unit of α-mannosidase. After incubation at 37°C for various periods, 100 μl of the incubation mixture were analyzed for the liberated Man (Lee, 1972).

This result indicates the presence of β-linked Man in Man(GlcNAc)$_2$-Asn at the core region of the glycopeptide isolated from ovalbumin. The same result was obtained for the core glycopeptide isolated from Taka amylase and rat liver microsomes.

$GbOse_3Cer$ is a neutral glycosphingolipid found in visceral organs (Hakamori, 1983). In 1963 Sweeley and Klionsky isolated this glycolipid from the kidney of a patient with Fabry's disease. Although the structure of the carbohydrate moiety in this glycolipid was known to be Gal-Gal-Glc-Cer, the anomeric configuration of the two Gal in this glycolipid was a mystery until early 1970. Using glycosidases, $GbOse_3Cer$ was found to be converted into LacCer by fig α-galactosidase (Li and Li, 1972b), but not by jack bean β-galactosidase (Fig. 2). Lactosylceramide derived from $GbOse_3Cer$, in turn, could be converted into GlcCer by jack bean β-galactosidase (Li and Li, 1971). These results show that $GbOse_3$-Cer contains a terminal α-galactosyl and a penulimate β-galactosyl linkages (Galα → Galβ → Glc → Cer).

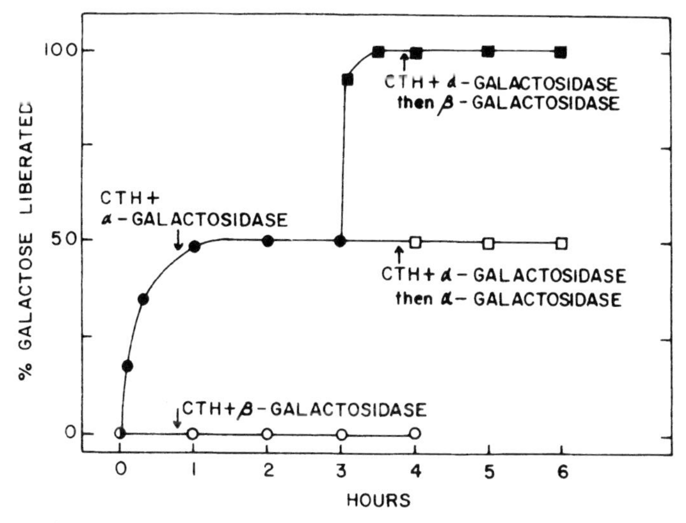

Fig. 2. Kinetics of the liberation of galactose from $GbOse_3Cer$ (Li and Li, 1971). ●, incubation of $GbOse_3Cer$ with fig α-galactosidase; ○, incubation of $GbOse_3Cer$ with jack bean β-galactosidase; ■, incubation of $GbOse_3Cer$ with fig α-galactosidase then again with jack bean β-galactosidase; □, incubation of $GbOse_3Cer$ with fig α-galactosidase and again with additional fig α-galactosidase. CTH = $GbOse_3Cer$.

Among various α-galactosidases examined, only the α-galactosidase isolated from ficin could efficiently cleave the terminal galactosyl residue from $GbOse_3Cer$. Crystalline α-galactosidase isolated from

Mortierella vinacea (Suzuki and Li, 1970), for example, was not able to hydrolyze this α-galactosyl residue. Therefore, it is very important to carefully consider the specificity of glycosidases to be used for the structural analysis of glycoconjugates.

A SEQUENCEGRAM SHOWING THE STEPWISE DEGRADATION OF MONOSACCHARIDE UNITS FROM A SIALOSYLHEXAGLYCOSYLCERAMIDE ISOLATED FROM BOVINE ERYTHROCYTES

Glycosphingolipids with different chain length can be separated by thin-layer chromatography. I is therefore very convenient to use thin-layer chromatography to analyze the sequential degradation of sugar units from a glycosphingolipid. Fig. 3 shows an example of the sequential

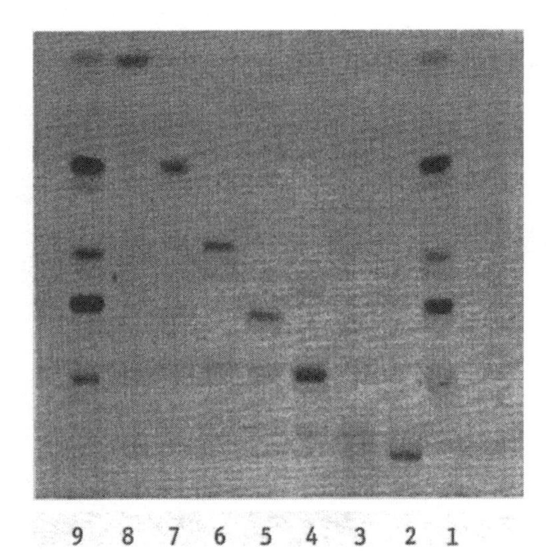

Fig. 3. Enzymatic hydrolysis of bovine stroma sialosylhexaglycosylceramide. Lanes 1 and 9 are the standards: from top, GlcCer, LacCer, $GbOse_3Cer$, $GbOse_4Cer$ and pentaglycosylceramide. 2, sialosylhexaglycosylceramide; 3, 2 + neuraminidase; 4, 3 + β-galactosidase; 5, 4 + β-hexosaminidase; 6, 5 + β-galactosidase; 7, 6 + β-hexosaminidase; 8, 5 + β-galactosidase + β-hexosaminidase.

enzymatic hydrolysis of a NeuGc-hexaglycosylceramide isolated from bovine erythrocyte stroma (Chien *et al.*, 1978). The intact ganglioside was resistant to α- and β-galactosidase and β-*N*-acetylhexosaminidase. Incubation of this ganglioside with neuraminidase produced a neutral glycosphingolipid with mobility slower than the pentaglycosylceramide isolated from bovine erythrocytes. β-Galactosidase, subsequently, converted this glycolipid into a compound with mobility slightly ahead of the pentaglycosylceramide. The alternative treatment of the newly formed glycolipid with β-*N*-acetylhexosaminidase and β-galactosidase produced a tetra-, tri-, di-, and glucosylceramide. From these results, one can deduce that the parent glycolipid has a structure of NeuGcα→Galβ→GlcNAc-β→Galβ→GlcNAcβ→Glc→Cer. By using this method, not only the sequential arrangement, but also the anomeric configulation of saccharide units was determined. It should be pointed out that the enzymatic hydrolysis of sugar units from glycosphingolipids requires the addition of a detergent such as sodium taurodeoxycholate in the reaction mixture (Li and Li, 1977 .

Table 2. Specificity of Jack Bean and *Escherichia coli* β-Galactosidase.

Substrate	Incubation time	% Hydrolysis	
		Jack bean β-galactosidase	*E. coli* β-galactosidase
	min	%	%
Galβ1→6GlcNAc	10	100	100
Galβ1→4GlcNAc	10	75	55
Galβ1→3GlcNAc	10	1	92
Galβ1→4Glc	10	42	100
Galβ1→6Man	10	19	N.D.[a]
Galβ1→6ManNAc	10	10	N.D.
Asialo α_1-acid glycoprotein	120	79	0
Lactosylceramide	120	100	0
GM1	120	9	0
Asialo-GM1	120	82	0

[a] Not determined.

For the specificity of *N*-acetyllactosamine isomers and various disaccharides, incubation mixture contained the following components in 100 µl: substrate, 0.125 µmol; sodium phosphate buffer, pH 7.0, 4 µmol (for *E. coli* β-galactosidase); enzyme, 0.3 units. For the liberation of Gal from asialo-α_1-acid glycoprotein, the reaction mixture contained following components in 150 µl:asialo-α_1-acid glycoprotein, 1 mg: sodium citrate buffer, pH 4.0 or sodium phosphate buffer, pH 7.0, 4 µmol; enzyme, 0.3 units. For the hydrolysis of sphingoglycolipids, incubation mixture contained the following components in 400 µl: substrate, 0.02 µmol; sodium citrate buffer, pH 4.0 or sodium phosphate buffer, pH 7.0, 15 µmol; sodium taurodeoxycholate, 300 µg; enzyme, 0.75 units.

LINKAGE AND AGLYCON SPECIFICITY OF GLYCOSIDASES

Glycosidases display glycon specificity as well as aglycon specificity. Table 2 summarizes the specificity of jack bean β-galactosidase and *E. coli* β-galactosidase on various saccharides, glycoproteins, and glycosphingolipids. Among three *N*-acetyllactosamines examined, the 1→3 linked isomer was very resistant to jack bean β-galactosidase, while this disaccharide could be cleaved by *E. coli* β-galactosidase. The effect of aglycon moiety on the rate of hydrolysis was revealed by the comparison of the hydrolysis of Galβ1→4Glc, Galβ1→6Man and Galβ1→6ManNAc. The results indicate that the change of aglycon can cause the change of the rate of hydrolysis. Again, the linkage and aglycon specificities of these two β-galactosidases can be seen in the hydrolysis of asialo α_1-acid glycoprotein, lactosylceramide, GM1, and asialo GM1. Jack bean β-galactosidase, but not *E. coli* β-galactosidase, was able to cleave these glycoconjugates.

MULTIPLE SPECIFICITY OF GLYCOSIDASES

In addition to linkage specificity, some glycosidases do not display a strict glycon specificity. For example, β-*N*-acetyl-hexosaminidase can hydrolyze β-linked GlcNAc as well as GalNAc residues (Fig. 4). Apparently

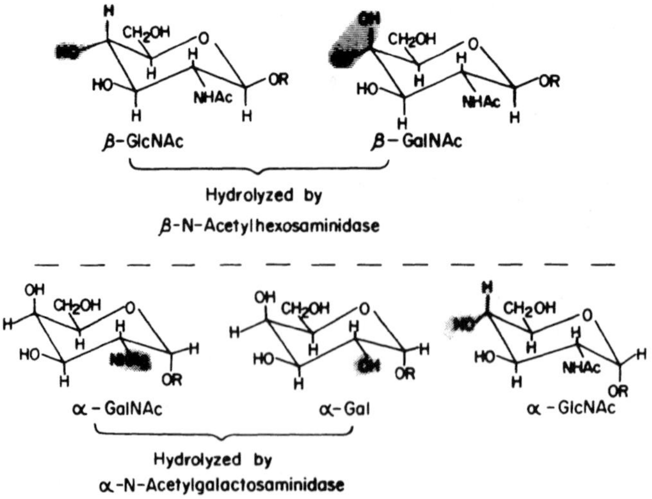

Fig. 4. Specificity of β-*N*-acetylhexosaminidase and α-*N*-acetylgalactosaminidase.

the enzyme does not have the absolute specificity toward the configuration of the hydroxyl group at C-4 of the pyranose ring. Another interesting example is the dual specificity of α-N-acetylgalactosaminidase. This enzyme can hydrolyze α-linked GalNAc as well as Gal (Fig. 4). This may indicate that the acetamido group at C-2 is not involved in forming the enzyme substrate complex. I should be pointed out that although α-N-acetylgalactosaminidase can hydrolyze α-linked Gal, α-galactosidase, has not been reported to hydrolyze α-linked GalNAc, and α-linked GlcNA is not hydrolyzed by α-N-acetylgalactosaminidase. Fig. 5 shows the sequential hydrolysis of sugar units from Forssman hapten (GalNAcα1→3-GalNAcβ1→3Galα1→4Galβ→4 Glc Cer). As expected, α-N-acetylgalactosaminidase converted this glycolipid into $GbOse_4$ Cer. From the structure of this glycolipid one should obtain $GbOse_3$ Cer when this glycolipid was incubated with both α-N-acetyl-galactosaminidase and β-N-acetylhexosaminidase. However, these two enzymes converted this glycolipid into LacCer instead of $GbOse_3$Cer. This is due to the ability of α-N-acetylgalactosaminidase to further cleave the terminal α-Gal residue exposed by β-N-acetylhexosaminidase.

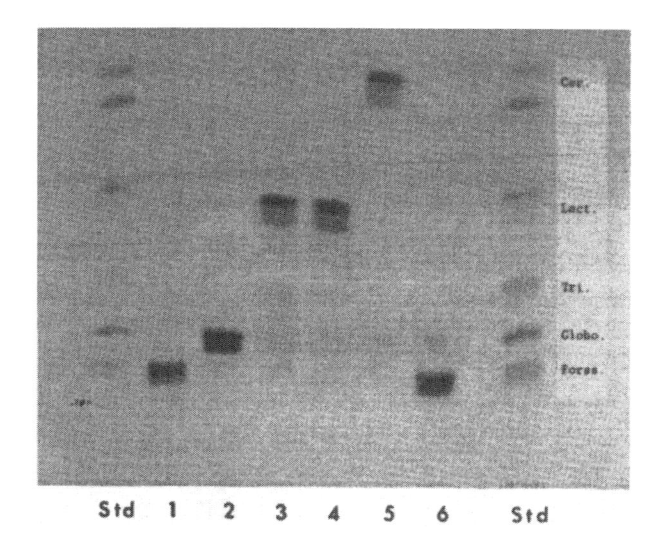

Fig. 5. Sequential hydrolysis of Forssman Hapten by α-N-acetylgalactosaminidase and β-N-acetylhexosaminidase Std, standards, from top, GlcCer, LacCer, $GbOse_3$Cer, $GbOse_4$-Cer, $GbOse_5$ Cer (Forssman's hapten); 1, Forssman's hapten; 2, 1 + α-N-acetylgalactossaminidase; 3, 1 + α-N-acetylgalactosaminidase and β-N-acetylhexosaminidase; 4, 1 + α-N-acetylgalactosaminidase, β-hexosaminidase and α-galactosidase. 5,1 + α-N-acetylgalactosamininidase, β-N-acetylhexosaminidase, α-galactosidase, and β-galactosidase.

REQUIREMENT OF A PROTEIN COFACTOR FOR ENZYMATIC HYDROLYSIS OF GLYCOSPHINGOLIPIDS

Using glycosidases to hydrolyze sugar units from glycosphingolipids often requires the assistance of a protein cofactor called activator protein (Li and Li, 1982). For example, the terminal Gal of a hexa-glycosylceramide (Galβ1→3GalNAcα1→4GalNAcβ1→4GlcNAcβ1→3Manβ1→4Glcβ1→1'Cer) isolated from *Calliphora vicina* (Dennis *et al.*, 1985) could be readily hydrolyzed by human hepatic β-galactosidase in the presence of GM1-activator isolated from human liver (Li and Li, 1976). However, this Gal was resistant to the same enzyme in the absence of the activator protein. This Gal was also found to be resistant to the β-galactosidases isolated from *Canavalia ensiformis* and jack bean.

Another example is the structural determination of GalNAc-GM1b (Itoh *et al.*, 1981). By using chemical methods, one cannot distinguish the position of NeuAc in GalNAc-GM1b from GalNAc-GM1 (Fig. 6). However, using the enzymatic method, we were able to show that this ganglioside could be converted into GM1b by human hepatic β-hexosaminidase A in the presence of GM2-activator (Li *et al.*, 1981). Thus, this ganglioside was identified by enzymatic method as GalNAcGM1b, not GalNAc-GM1 (Itoh *et al.*, 1981).

```
GalNAc-GM1:   GalNAcβ1→4Galβ1→3GalNAcβ1→4Galβ1→4Glcβ→Cer
                                           3
                                           ↑α
                                           2NeuAc

GalNAc-GM1b:  GalNAcβ1→4Galβ1→3GalNAcβ1→4Galβ1→4Glcβ→Cer
                         3
                         ↑α
                         2NeuAc
```

Fig. 6. Structures of GalNAc-GM1 and GalNAc-GM1b

STRUCTURAL ANALYSIS OF GLYCOCONJUGATES USING ENDO-β-GALACTOSIDASE

Until recently, keratan sulfate was the only glycoconjugate found to contain *N*-acetyllactosamine repeating units. Through the action of endo-β-galactosidase, this type of repeating unit was confirmed to exist in a variety of glycoconjugates. By using endo-β-galactosidase of *E. freundii* (Nakagawa *et al.*, 1980), we found that glycoproteins in human erythrocytes (Jarnefelt *et al.*, 1978; Mueller *et al.*, 1979) and the sugar chain in polyglycosylceramide (Nakagawa *et al.*, 1980) contained this repeating unit.

We have used this enzyme to study the structure of a hexasaccharide isolated from I-active ovalian cyst glycoprotein (Dua *et al.*, 1985). After the hydrolysis, a trisaccharide (Galβ1→3GlcNAcβ1→3Gal) was released from the hexasaccharide (Fig. 7).

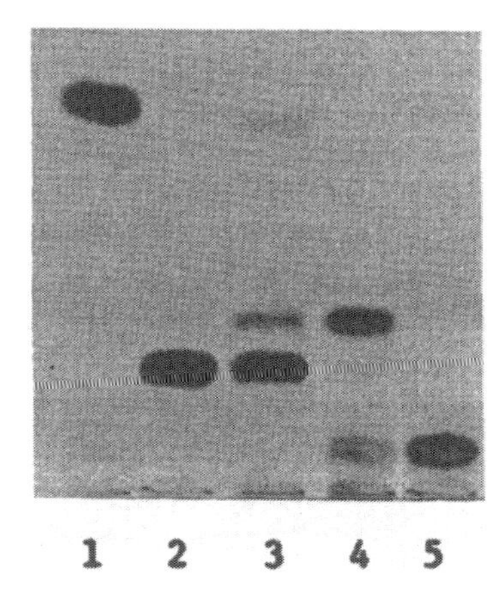

Fig. 7. Thin-layer chromatogram of lacto-*N*-tetraose and hexasaccharide and their endo-β-galactosidase digestion products. 1, Glc; 2, lacto-*N*-tetraose; 3, 2 + endo-β-galactosidase; 4, hexasaccharide + endo-β-galactosidase; 5, hexasaccharide only.

This trisaccharide could also be generated from lacto-*N*-tetraose, Galβ1→3GlcNAcβ1→3Galβ1→4Glc. The following scheme and Fig. 7 show the action of endo-β-galactosidase on the hexasaccharide:

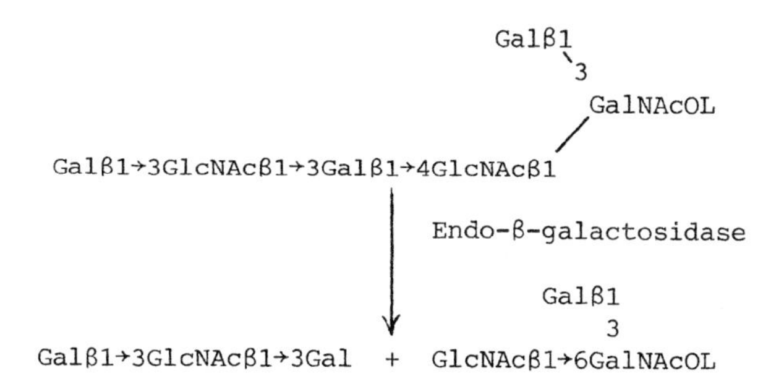

ANALYSIS OF ENDO-β-N-ACETYLGLUCOSAMINIDASE ACTIVITY BY HIGH-PRESSURE LIQUID CHROMATOGRAPHY AND THIN-LAYER CHROMATOGRAPHY

Endo-β-N-acetylglucosaminidase is one of the best studied endo-glycosidases that are useful for the structural analysis of glycoconjugates (Kobata, 1979). The conventional method for the assay of endo-β-N-acetylglucosaminidase involved the use of ^{14}C-labeled or dansylated asparaginyl oligosaccharides, such as $(Man)_5(GlcNAc)_2Asn-[^{14}C]Ac$ or $(Man)_5(GlcNAc)_2$-Asn-DNS, as substrate and analyzing the production of GlcNAc-Asn$[^{14}C]$Ac or GlcNAc-Asn-DNS by high voltage paper electrophoresis or paper chromatography (Tarentino and Maley, 1974; Koide and Muramatsu, 1974). We have examined the separation of dansyl asparaginyl oligosaccharides by high-pressure liquid chromatography (hplc) and developed a new method for the assay of endo-β-N-acetylglucosaminidase activity. Fig. 8 shows the hydrolysis of $(Man)_5(GlcNAc)_2$Asn-DNS by endo-β-N-acetylglucosaminidase C_{II} isolated from Clostridium perfringens (Koide and

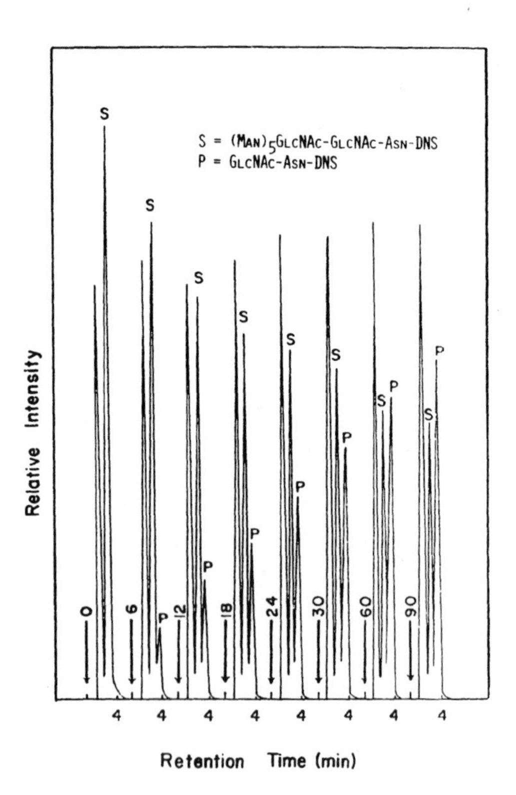

Fig. 8. Analysis of endo-β-N-acetylglucosaminidase reaction by hplc (Iwase et al., 1981). S = substrate, $(Man)_5(GlcNAc)_2$-Asn-DNS; P = product, GlcNAc-Ans-DNS. The peak eluted before S is due to the enzyme protein. Arrows indicate the injection points and the numbers corresponding to each arrow indicate the incubation time.

Muramatsu, 1974 . Using hplc one can detect the appearance of the product, GlcNAc-Asn-DNS, with concomitant disappearance of the substrate, $(Man)_5(GlcNAc)_2$Asn-DNS. (Iwase et al., 1981).

The oligosaccharide, $(Man)_6$GlcNAc, released from $(Man)_6(GlcNAc)_2$Asn can be also analyzed by thin-layer chromatography as shown in Fig. 9.

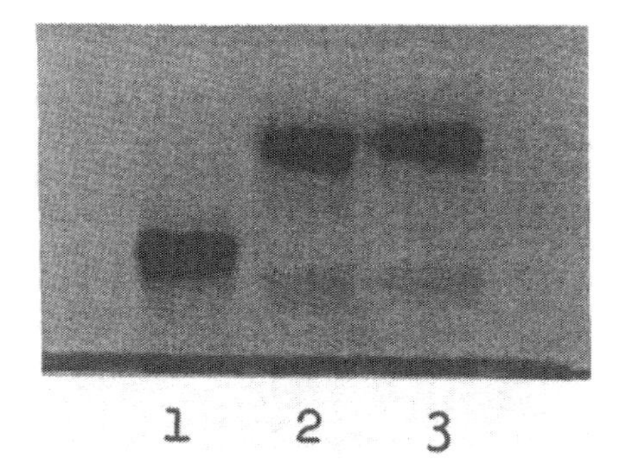

Fig. 9. Thin layer chromatography showing the production of $(Man)_6$GlcNAc from $(Man)_6(GlcNAc)_2$Asn. 1, $(Man)_6(GlcNAc)_2$-Asn; 2, 1 + Endo B (Morinaga *et al.*, 1983) 3, 1 + Endo C_{II}. The plate was developed with n-butanol: acetic acid H_2O (1:1:1).

CONCLUSION

Structural determination of complex carbohydrates is an integral part of glycoconjugate research. During the past twenty years, we have seen a great advance in using NMR spectroscopy and mass spectroscopy for the structural elucidation of glycoconjugates. The enzymatic method, a simple and economical method, will continue to be valuable for the structural analysis of glycoconjugates.

ACKNOWLEDGEMENTS

This investigation was supported by Grant PCM82-19489 from National Science Foundation and Grant NS09626 from National Institutes of Health. Y.-T. Li is a recipient of the Javitz Neuroscience Investigator Award.

REFERENCES

Chien, J.-L., Li, S.-C., Laine, R.A., and Li, Y.-T. (1978) Characterization of gangliosides from bovine erythrocyte membranes, J. Biol. Chem. 253:4031-4035.

Chien, S.-F., Yevich, S.J., Li, S.-C., and Li, Y.-T. (1975) Presence of endo-β-N-acetylglucosaminidase and protease activities in the commercial neuraminidase preparations isolated from Clostridium perfringens, Biochim. Biophys. Res. Commun. 65:683-691.

Dennis, R.D., Geyer, R., Egge, H., Peter-Katalinic, J., Li, S.-C., Stirm, S., and Wiegandt, H. (1985) Glycosphingolipids in insects - Chemical structures of ceramide tetra-, penta-, hexa-, and heptasaccharide from Calliphora vicina pupae (Insecta:Diptera), J. Biol. Chem. 250:5370-5375.

Dua, V.K., Duke, V.E., Li, Y.-T., and Bush, C.A. (1985) Reverse phase HPLC fractionation of the oligosaccharide alditols isolated from an I-active ovarian cyst mucin glycoprotein, Glycoconj. J. 2:17-30.

Hakomori, S. (1983) Chemistry of glycosphingolipids, in: Handbook of Lipid Research, Plenum Press, New York.

Itoh, T., Li, Y.-T., Li, S.-C., and Yu, R.K. (1981) Isolation and characterization of a novel monosialosylpentahexosyl ceramide from Tay-Sach's brain, J. Biol. Chem. 256:163-169.

Iwase, H., Morinaga, T., Li, Y.-T., and Li, S.-C. (1981) Analysis of endo-β-N-acetylglucosaminidase activity by high-pressure liquid chromatography on a silica-based chemically bonded octadecyl column, Anal. Biochem. 113:93-95.

Jarnefelt, J., Rush, J., Li, Y.-T., and Laine, R.A. (1978) Erythroglycan, a high molecular weight glycopeptide with the repeating structure [galactosyl(1→4 -2-deoxy-2-acetamidoglucosyl (1→3)] comprising more than one-third of the protein-bound carbohydrate of human erythrocyte stroma, J. Biol. Chem. 253:8006-8009.

Kobata, A. (1979) Use of endo- and exo-glycosidases for structural studies of glycoconjugates, Anal. Biochem. 100:1-14.

Koide, N., and Muramatsu, T. (1974) Endo-β-N-acetyl-glucosaminidase acting on carbohydrate moieties of glycoproteins - purification and properties of the enzyme from Diplococus pneumoniae, J. Biol. Chem. 249:4897-4904.

Lee, Y.C. (1972) Analysis of sugars by automated lipid chromatography, Methods Enzymol., 28 (Part B):63-73.

Li, S.-C., Hirabayashi, Y., and Li, Y.-T. (1981) A protein activator for the enzymic hydrolysis of GM2 ganglioside, J. Biol. Chem. 256:6234-6240.

Li, S.-C., and Li, Y.-T. (1976) An activator stimulating the enzymic hydrolysis of sphingoglycolipids, J. Biol. Chem. 251:1159-1163.

Li, Y.-T., and Lee, Y.C. (1972) Pineapple α-and β-D-manno-pyranosidase and their action on core glycopeptides, J. Biol. Chem. 247:3677-3683.

Li, Y.-T., and Li, S.-C. (1972a) α-Mannosidase, β-N-acetylhexosaminidase, and β-galactosidase from jack bean meal, Methods Enzymol., 28 (Part B):702-713.

Li, Y.-T., and Li, S.-C. (1972b) α-Galactosidase from Figs. Methods Enzymol. 28:714-720.

Li, Y.-T., and Li, S.-C. (1971) Anomeric configuration of galactose residue in ceramide trihexoside, J. Biol. Chem. 246:3769-3771.

Li, Y.-T. and Li, S.-C. (1977) Use of enzymes in elucidation of structure, "Mammalian Glycoproteins and Glycolipids," in "The Glycoconjugates" volume 1, W. Pigman and M.I. Horowitz, ed., A ademic Press, New York.

Li, Y.-T., and Li, S.-C. (1982) Biosynthesis and catabolism of glycosphingolipids, Adv. Carbohyr. Chem. Biochem. 40:235-286.

Li, Y.-T., Li, S.-C., and Shetlar, M.R. (1968) Isolation of glycopeptides from rat liver microsomes involved in the biosynthesis of plasma glycoprotein, J. Biol. Chem. 243:656-665.

Morinaga, T., Kitamikado, K., Iwase, H., Li, S.-C., and Li, Y.-T. (1983) The use of mannose-sepharose 4B affinity chromatography for the purification of endo-β-N-acetylglucosaminidase from *Bacillus alvei*, Biochim. Biophys. Acta 749:211-213.

Mueller, T.J., Li, Y.-T., and Morrison, M. (1979) Effect of endo-β-galactosidase on intact human erythrocytes, J. Biol. Chem. 254:8103-8106.

Nakagawa, H., Yamada, T., Chien, J.-L., Gardas, A., Kitamikado, M., Li, S.-C., and Li, Y.-T. (1980) Isolation and characterization of an endo-βgalactosidase from a new strain of *Escherichia freundii*, J. Biol. Chem. 255:5955-5959.

Reissig, J.L., Strominger, J.L., and Leloir, L.F. (1955) A modified colorimetric method for the estimation of N-acetyl amino sugars, J. Biol. Chem. 217:959-966.

Suzuki, H., Li, S.-C., and Li, Y.-T. (1970) α-Galactosidase from *Mortierella vinacea*, J. Biol. Chem. 245:781-786.

Sweeley, C.C., and Klionsky, B. (1963) Fabray's disease: Classification as a sphingolipidosis and partial characterization of a novel glycolipid, J. Biol. Chem. 238:3148-3150.

Tarentino, A.L., and Maley, F. (1974) Purification and properties of an endo-β-N-acetylglucosaminidase from *Streptomyces griseus*, J. Biol. Chem. 249:811-817.

Warren, L. (1959) The thiobarbituric acid assay of sialic acids, J. Biol. Chem. 234:1971-1975.

MASS SPECTROMETRY OF HEXOSAMINE CONTAINING OLIGOSACCHARIDES AS PERMETHYLATED N-TRIFLUOROACETYL DERIVATIVES

Bo Nilsson

BioCarb AB
Lund, Sweden

Structural analysis of carbohydrates often includes mass spectrometry of derivatized monosaccharides or oligo-saccharides. Mass spectrometry gives, combined with chemical analyses and NMR spectrometry, the complete structure. The major advantage with mass spectrometry is that structural information can be obtained on micro grams of material.

Gas chromatography-mass spectrometry (GLC-MS) offers separation as well as characterization. Complex mixtures of oligosaccharides containing six or less monosaccharide residues can be resolved by the gas chromatograph and subsequently characterized by the mass spectrometer.

The information that can be extracted from mass spectra of permethylated oligosaccharides is the monosaccharide sequence and in some cases the position of substitution of the monosaccharide residues. The identity of the monosaccharide residues and the anomeric configuration can, however, not be deduced.

N-Acetylhexosamine containing oligosaccharides are difficult to analyse by GLC-MS as permethylated derivatives due to their low volatility, resulting in long retention times on GLC. By converting the *N*-acetyl function to an *N*-trifluoroacetyl derivative the permethylated compound has a shorter retention time. Electron impact (EI) mass spectra of these permethylated *N*-trifluoroacetyl derivatives give more structural information than the corresponding *N*-acetyl derivatives. Fragmentation pathways worked out for EI-spectra can also be applied to fast atom bombardment (FAB) spectra. In this chapter will be discussed how sequence and linkages can be determined by EI or FAB mass spectrometry of hexosamine containing oligosaccharides as permethylated *N*-trifluoroacetyl derivatives.

TRIFLUOROACETOLYSIS

The trifluoroacetolysis reactions is carried out in a mixture of trifluoroacetic acid and trifluoroacetic anhydride.[1] In this reaction hydroxyl groups are rapidly *O*-trifluoroacetylated. The *O*-trifluoroacetyl groups will stabilize glycosidic linkages by the inductive effect of fluorine atoms.[2] Acetamido groups are slowly transformed to trifluoroacetamido functions. It has previously been shown that the *O*- and *N*-glycosidically linked oligosaccharides[3,4] are released from glycoproteins and that glycolipids are degraded by specific cleavage of the carbohydrate-ceramide linkage.[5] Hexosamine containing oligosaccharides are recovered as *N*-trifluoroacetyl derivatives. The protecting *O*-trifluoroacetyl groups are easily removed by aqueous methanol. The *N*-trifluoroacetyl groups are stable in water solutions and can be methylated using the Hakomori procedure.[6]

GLC-MS OF PERMETHYLATED N-TRIFLUOROACETYL HEXOSAMINE CONTAINING OLIGOSACCHARIDES

Introduction of an *N*-trifluoroacetyl function on the hexosamine has an influence on the chemical and physical properties of the oligosaccharide. The retention times on GLC of permethylated *N*-trifluoroacetyl derivatives are shorter than for the corresponding *N*-acetyl derivatives.[7] This means that these derivatives are eluted from the GLC-column at a lower temperature that the *N*-acetyl derivatives. Oligosaccharides containing up to six monosaccharide residues can be analysed by GLC-MS. Mass spectra of permethylated and *N*-trifluoroacetylated oligosaccharide alditols are characterized by strong fragments where the positive charge is on the trifluoroacetamido sugar. A systematic examination of fragments containing sequence and linkage information is presented below for EI-mass spectra.

A. *The Sequence: hexNTF-*

Oligosaccharides containing non-reducing terminal hexNTF are easily recognized by the primary fragment of m/z 314 formed by a homolytic cleavage of the hexosaminidic linkage. Secondary fragments are formed by eliminations of methanol to give m/z 282 and m/z 250. The relative abundance of these ions is m/z 314 > m/z 282 and m/z 250. The corresponding set of ions for *N*-acetyl derivatives are m/z 260, m/z 228 and m/z 196. The latter ion is of low abundance due to further elimination of ketene, which is not possible for the *N*-trifluoroacetyl derivatives.

B. *The Sequence: hexNTF-hex-*

The primary fragments m/z 314 and 518 determine a sequence of hexNTF-hex. The secondary fragments formed from m/z 314 have been discussed under A. The ion at m/z 518 gives rise to the secondary fragments m/z 486 by elimination of methanol and m/z 187 by elimination by elimination of hexNTF-OH. The abundance of m/z 187 can be used to determine the hexNTF-hex-linkage. For a hexNTF1-3hex-linkage, the abundance of m/z 187 is about 20% and for a 1-2 or 1-4 linkage the abundance is 2% or less.[5]

C. *The Sequence: hex-hexNTF-*

The primary and secondary fragments m/z 155, m/z 187 and m/z 219 represent a non-reducing terminal hex- and combined with m/z 518 the sequence hex-hexNTF is determined. Secondary fragments are formed from m/z 518 by elimination of hex-OH. The ion of m/z 282 can further eliminate methanol to m/z 250. The secondary fragments m/z 250 and m/z 282 can be used to determine the position of substitution of the hexNTF residue. For a hex1-3 hexNTF- linkage m/z 250 is smaller than m/z 282, while for a hex1-4 hexNTF- linkage m/z 250 is greater than m/z 282.[5]

D. The Sequence: deoxyhex-hex-hexNTF-

The sequence ions m/z 189, m/z 393 and m/z 692 determine a linear sequence of deoxyhex-hex-hexNTF. The secondary fragments formed from m/z 692 by eliminations are m/z 660 (elimination of methanol), m/z 282 (elimination of deoxyhex-hex-OH), and m/z 250 (elimination of deoxyhex-hex-OH and methanol). By comparing the relative abundances for m/z 282 and m/z 250 the position of substitution of the hexNTF-residue can be determined as discussed under C.

E. The Sequence: deoxy hex-hexNTF-

```
                          deoxy hex-hexNTF-
                                  |
                                 hex
```

The ions for non-reducing terminal deoxyhex m/z 189 and non-reducing terminal hex m/z 219, in combination with m/z 692 show a di-*O*-substituted hexNTF. Diagnostic secondary fragments are formed from m/z 692 by elimination of hex-OH to give m/z 456, elimination of deoxyhex-OH to give m/z 486 and elimination of both residues to give m/z 250. The secondary fragments m/z 456 and m/z 486 can be used to determine the substitution pattern of a 3,4 di-O-substituted hexNTF.

For a sequence of

```
hex1-3hexNTF-
         4
         |
  deoxyhex1
```

m/z 456 is greater than m/z 486 and for a sequence of

```
 hex1-4hexNTF-
        3
        |
deoxyhex1
```

m/z 456 is smaller than m/z 486. This is because elimination from position 3 of the hexNTF residue predominates over elimination from 4-position.[5] The substitution pattern of 3,4-di-O-substituted hexNTF residue is supported by ions containing the reduced terminal as demonstrated below:

The positive charge is always on the 3-position of the hexNTF residue. If X is a disaccharide alditol-1-d the ions are m/z 913 in I and m/z 943 in II.

F. The Sequence: hexNTF-OL

Oligosaccharides alditols containing reduced hexNTF are recognized by the fragment m/z 330.[9,10] Ions formed by cleavage within the alditol give the position of substitution. The fragment m/z 184 formed by cleavage between carbon 2 and carbon 3 shows a 3-O-substituted hexNTF-OL. Unlike the *N*-acetyl derivatives, no elimination of ketene is seen for the *N*-trifluoroacetyl derivatives. For a 4-O-substituted hexNTF-OL the series of primary fragments m/z 184 and m/z 228 are observed. A secondary fragment m/z 196 is formed from m/z 228 by elimination of methanol. Fragments formed from a 6-substituted hexNTF-OL are m/z 184, m/z 228, m/z 196 and m/z 272. The latter ion is formed by cleavage between carbons 4 and 5.

DIRECT-PROBE Ms OF PERMETHYLATED N-TRIFLUOROACETYL HEXOSAMINE CONTAINING OLIGOSACCHARIDES

Oligosaccharides which can not be analysed by glc-ms can sometimes be analysed by direct-probe ms.[8] The usefulness of permethylated *N*-trifluoroacetyl derivatives is demonstrated for some oligosaccharides containing terminal as well as internal hexNTF.

```
The sequence of hexNTF-hex-hexNTF-
                           |
                        deoxyhex
```

is determined by the primary fragments m/z 189, m/z 314, m/z 692 and m/z 991. Secondary fragments are formed from m/z 991 by elimination of

```
hexNTF-hex-OH
        |
     deoxyhex
```

to give m/z 282, followed by elimination of methanol to give m/z 250. The ions m/z 250 and m/z 282 are also formed from m/z 314 by elimination of methanol. The relative abundance for these ions is m/z 314 > m/z 282 > m/z 250 (see paragraph A). For a 3-*O*-substituted hexNTF m/z 282 is greater than m/z 250 (see paragraph C). If the internal hexNTF in the sequence above is 3-*O*-substituted, a contribution will be given to m/z 282 so that m/z 282 will be greater than m/z 314.[11] For a 4-*O*-substituted internal hexNTF the contribution would be given to m/z 250 resulting in that m/z 250 would be greater than m/z 314.

FAB MASS SPECTROMETRY OF PERMETHYLATED N-TRIFLUORO-ACETYL HEXOSAMINE CONTAINING OLIGOSACCHARIDES

Studies of hexosamine containing oligosaccharides with known structures have revealed that fragmentation rules worked out on EI-spectra can be used on FAB-spectra.[12] This opens possibilities to do structural analysis of large hexosamine containing oligosaccharides as permethylated *N*-trifluoroacetyl derivatives. FAB-spectra of these derivatives also contain the molecular ion plus 19. This is not observed for the *N*-acetyl derivatives which show the molecular ion plus 1. Fig 1-6 show FAB-spectra of some common human milk oligosaccharides as permethylated and *N*-trifluoroacetylated alditols. Primary and diagnostic secondary fragments are indicated.

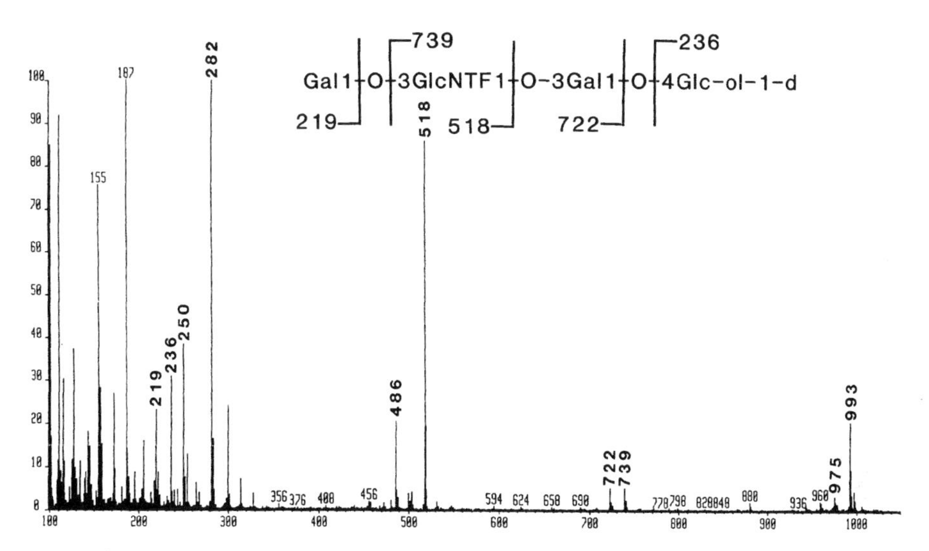

Fig 1. FAB mass spectrum of lacto-N-tetraose as its permethylated *N*-trifluoroacetylated alditol

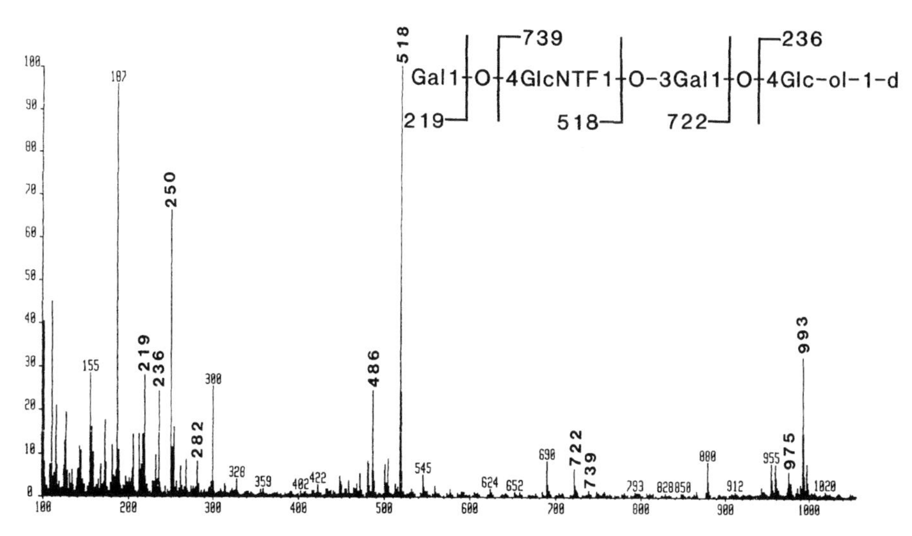

Fig 2. FAB mass spectrum of lacto-N-neotetraose as its permethylated *N*-trifluoroacetylated alditol

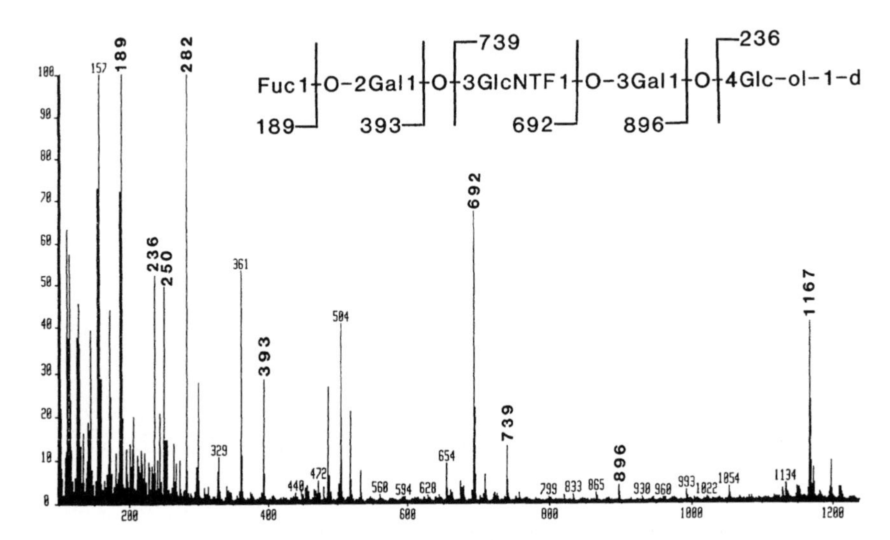

Fig 3. FAB mass spectrum of lacto-N-fucopentaose I as its permethylated *N*-trifluoroacetylated alditol

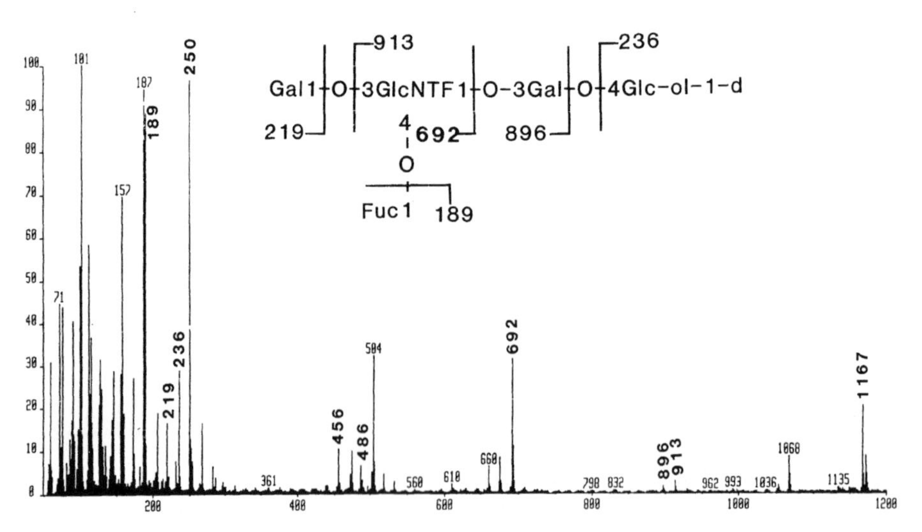

Fig 4. FAB mass spectrum of lacto-N-fucopentaose II as its permethylated *N*-trifluoroacetylated alditol

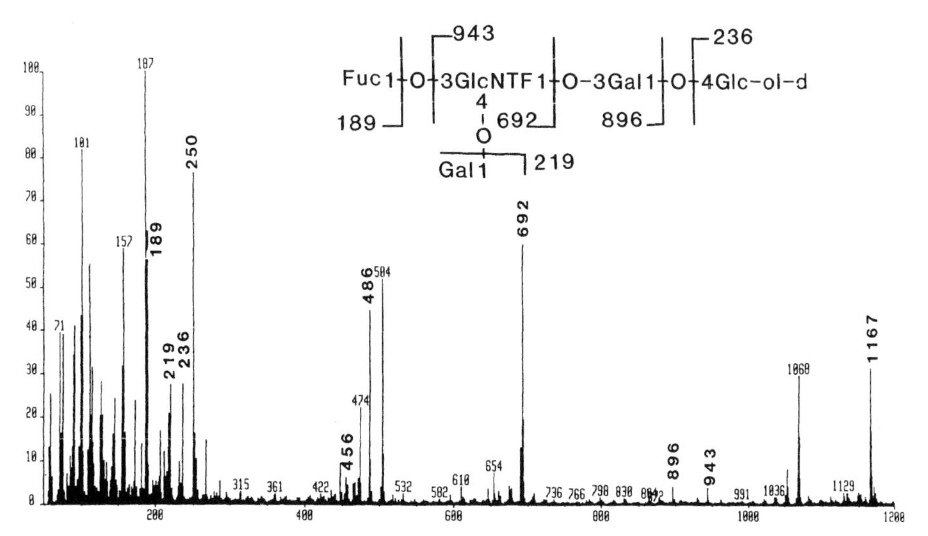

Fig 5. FAB mass spectrum of lacto-N-fucopentaose III as its permethylated N-trifluoroacetylated alditol

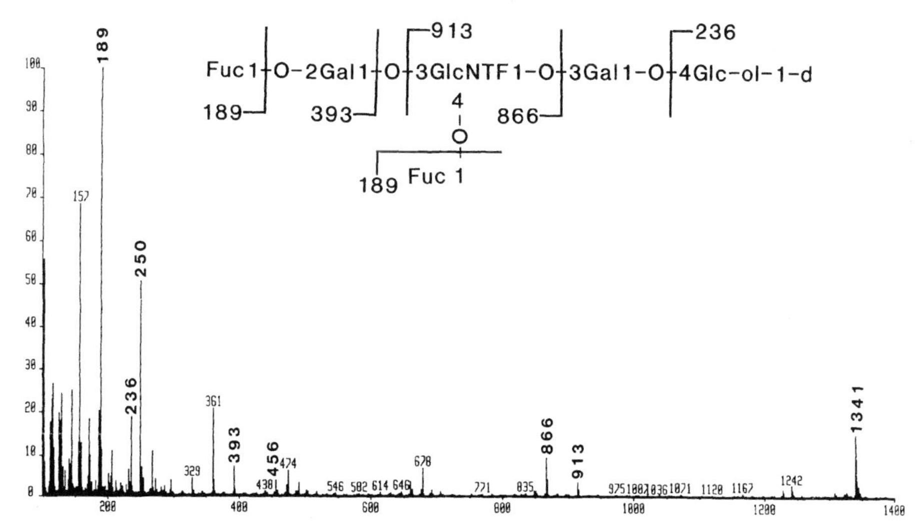

Fig 6. FAB mass spectrum of lacto-N-difucohexaose I as its permethylated N-trifluoroacetylated alditol

SUMMARY

Permethylated *N*-trifluoroacetyl hexosamine containing oligosaccharide alditols are favorable for analysis by mass spectrometry. Electron impact and fast atom bombardment mass spectra of these derivatives are characterized by strong primary and secondary fragments in which a positive charge is localized on the *N*-trifluoroacetyl hexosamine residue. From the mass spectra the monosaccharide sequence and the position(s) of substitution of a *N*-trifluoroacetyl hexosamine can be determined.

REFERENCES

1. Nilsson, B. and Svensson, S. (1978). A New Method for N-deacetylation of 2-acetamido-2-deoxy Sugars. Carbohydr. Res. 62: 377-380.
2. Nilsson, B. and Svensson, S. (1979). Studies on the Reactivity of Methyl Glycosides, Oligosaccharides and Polysaccharides towards Trifluoroacetolysis. Carbohydr. Res. 69: 292-296.
3. Lindberg, B., Nilsson, B., Norberg, T. and Svensson, S. (1979). Specific Cleavage of O-glycosidic Bonds to L-serine and L-threonine by Trifluoroacetolysis. Acta Chem. Scand. B33: 230-231.
4. Nilsson, B. and Svensson, S. (1979). A New Method for Degradation of the Protein Part of Glycoproteins; Isolation of the Carbohydrate Chains of Asialo Fetuin. Carbohydr. Res. 72: 183-190.
5. Nilsson, B. and Zopf, D. (1983). Oligosaccharides Released from Glycolipids by Trifluoroacetolysis can be Analyzed by Gas Chromatography/Mass Spectrometry. Arch. Biochem. Biophys. 222: 628-648.
6. Hakomori, S. (1964). Rapid Permethylation of Glycolipids and Polysaccharides, Catalyzed by Methylsulfinyl Carbanion in Dimethylsulfoxide. J. Biochem (Tokyo) 55: 205-208.
7. Nilsson, B. and Zopf, D. (1982). Gas Chromatography and Mass Spectrometry of Hexosamine Containing Oligosaccharide Alditols as their Permethylated N-trifluoroacetyl Derivatives. Methods in Enzymology. V. Ginsburg ed., Academic Press, New York, Vol. 83: 46-58.
8. Nilsson, B., Horne, M. and Gralnick, H. (1983). The Carbohydrate of Human Thrombin: Structural Analysis of Glycoprotein Oligosaccharides by Mass Spectrometry. Arch. Biochem. Biophys. 224: 127-133.
9. Nilsson, B., De Luca, S., Lohmander, S. and Hascall, V. (1982). Structures of N-linked and O-linked Oligosaccharides on Proteoglycan Monomer Isolated from the Swarm Rat Chondrosarcoma. J. Biol. Chem. 257: 10920-10927.

10. Wu, A., Kabat, E.,Nilsson, B. and Zopf, D.,Gruezo, F. and Liao, J. (1984). Immunochemical Studies on Blood Groups. J. Biol. Chem. 259: 7178-7186.
11. Sabharwal, H., Nilsson, B., Chester, A., Sjöblad, S. and Lundblad, A. (1984). Blood Group Specific Oligosaccharides from Faeces of a Blood Group A Breast-fed Infant. Mol. Immunology 21: 1105-1112.
12. Strömberg, S., Lindh, F. and Nilsson, B. (1985). Fast Atom Bombardment Mass Spectrometry of Oligosaccharides as Permethylated N-trifluoroacetyl Derivatives. Proceeding of the 8th International Symposium on Glycoconjugates, Houston, Texas, September 8-13, 1985.

ABSTRACT

IMMUNOCHEMICAL ANALYSIS OF LIPOPOLYSACCHARIDES WITH 2-D GEL ELECTROPHORESIS AND MONOCLONAL ANTIBODIES

Blair A. Sowa[1], Richard P. Crawford[2], Fred C. Heck[3], John D. Williams[3], Albert M. Wu[1], Katherine A. Kelly[1], and L. Garry Adams[1]

Departments of [1]Veterinary Pathology, [2]Veterinary Public Health, [3]Veterinary Microbiology and Parasitology, Texas Agricultural Experiment Station, Texas A&M University, College Station, Texas 77843

Phenol extracted, alkali treated lipopolysaccharide from vaccine strain (S19) *Brucella abortus* was demonstrated by two-dimensional gel electrophoresis to have at least ten silver staining analogues. When tested on nitrocellulose immune blots all ten were antigenically reactive with bovine anti-*B. abortus* polyclonal sera, but only six reacted with murine monoclonal anti-O-antigen antibody. Analogues of LPS focusing at distinct pIs were polydisperse and in several cases polyionic, suggesting the incorporation of groups in the O-antigen side chain which modified the pI. Analogues focusing at different pIs were concluded to arise from differences in either core or O-antigen side chain structure. While no qualitative differences were observed, LPS from a pathogenic *B. abortus* strain (S2308) had lesser amounts of analogues 1,2,5,6, and 8 than S19 when examined by 2-D Gel.
The 2-D Gel method has proven to be valuable in the study of LPS from *E. coli* and *B. abortus* and should be useful for study of LPS from a wide range of sources. Given molecules with discrete isoelectric points and affinities for sodium dodecylsulfate or other ionic detergents, the possibility exists that other large biopolymers may be studied with the 2-D method as well. Used in conjunction with electroblotting and monoclonal antibody techniques, complex mixtures or fragments of purified materials may be readily separated and studied. It should also be possible to generate monoclonal or polyclonal antibodies to selected LPS subfractions separated by 2-D gel by the use of methods such as those reported by Boulard, Lecroisey, and Tracey, *et al.* and discussed by Pearson and Anderson using LPS fractions as well as protein immunogens contained in excised acrylamide gel patterns.

References:

1) Baker, P. and Wilson, J.B., J. Bacteriol. 90:903-910 (1965). 2) Boulard, C.L. and Lecroisey, A. J., Immunol. Meth. 50:221-226 (1982). 3)Cousland, G., Poxton, I.R., FEMS Microbiology Letters 20:461-465 (1983). 4) Dubray, G. and Charriaut, C., Ann. Rech. Vet. 14:311-318 (1983). 5) Goldman, R.C., and Leive, L., Eur. J. Biochem. 107:145-153 (1980). 6) Holman, P.J., Adams, L.G., Hunter, D.M., Heck, F.C., Nielson, K.N., and Wagner, G.G., Vet. Immun. Immunopath. 4:603-614 (1983). 7) Moreno, E., Pit, M.W., Jones, L.M., Schurig, G.G. and Berman, D.T., J. Bacteriol. 138:361-369 (1979). 8) O'Farrell, P.H., J. Biol. Chem. 250:4007-4021 (1975). 9) Pearson, T. and Anderson, L., Anal. Biochem. 101:377-386 (1980). 10) Sowa, B.A., Crawford, R.P., Heck, F.C., Williams, J.D., Wu,. A.M., Kelly, K.A., and Adams, L.G., Electrophoresis 7:283-288 (1986). 11) Stiller, J.M. and Nielson, K.H., J. Clin. Microbiology 27: 323-326 (1983). 12) Tracey, R.P., Katzman, J.A., Kimlinger, T.K., Hunt, G.A. and Young, D.S. ,J. Immunol. Meth. 65:97-107 (1983). 13) Towbin, H., Staehelin, T., Gordon, J., Proc. Natl. Acad. Sci. USA 76(9):4350-4354 (1979). 14) Tsai, C. and Frash, C.E., Anal. Biochem. 119:115-119 (1982). 15) Wu, A.M., Heck, F.C., Adams, L.G., and Jones, K., Mol. Immunol. 21:1123-1129 (1984).

Appendix I

A Guide for Carbohydrate Specificities of Lectins

Albert M. Wu, Shunji Sugii, and Anthony Herp

A GUIDE FOR CARBOHYDRATE SPECIFICITIES OF LECTINS [2]

Albert M. Wu[1], Shunji Sugii+, and Anthony Herp++

1Department of Veterinary Pathology, College of Veterinary Medicine, Texas A & M University, College Station, Texas , 77843, USA
+Department of Serology and Immunology, School of Medical Technology, Kitasato University 1-15-1 Kitasato, Sagamihara, Kanagawa 228, Japan, and
++Department of Biochemistry, New York Medical College, Valhalla, New York, 10595, USA

Lectins are carbohydrate-binding proteins of nonimmune origin, which have been widely used in the fields of cell biology, biochemistry and histochemistry to isolate and/or to characterize cell surface carbohydrates(1-3). Lectins require configurational and structural complementarity of sugars for interaction to occur. All lectin molecules have two or more carbohydrate binding sites, a property essential to their ability to agglutinate cells or to precipitate complex carbohydrates(1,2,4). Until the early seventies, the carbohydrate specificities of lectins were mainly determined by the abilities of monosaccharides or their glycosides to inhibit lectin-induced haemagglutination(1,4). Mäkelä(5), in the late fifties, divided lectin-reactive monosaccharides into four classes, based on their configuration at C-3 and C-4 of the

[1] Corresponding author
[2] Abbreviations: Gal,D-galactopyranose; Glc, D-glucopyranose; Man,D-mannopyranose; LFuc or Fuc,L-fucopyranose;DFuc, D-fucopyranose; GalNAc, 2-acetamido-2-deoxy-D-galacto-pyranose; GlcNAc,2-acetamido-2-deoxy-D-glucopyranose; chitin oligosaccharide, (GlcNAcβ1→4)n, repeat unit of GlcNAc; NeuAc,N-acetylneuraminic acid; NeuGly,N-glycolylneuraminic acid; R, carbohydrate residue; Melibiose, Galα1→6Glc; Raffinose, Galα1→6Glcβ1→2Fruf; Stachyose, Galα1→6 Galα1→6Glcβ1→2Fruf;Cellobiose, Glcβ1→4Glc,and Laminaribiose, Glcβ1→3Glc.
[3] Coded for the source or properties of oligosaccharides tested.

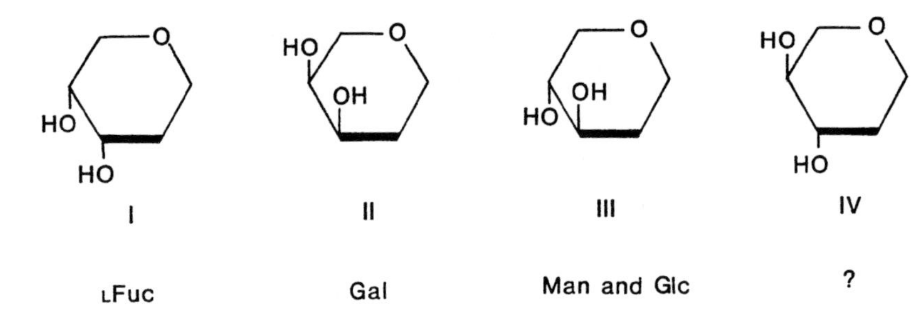

Fig.1. Classification of pyranoses of lectin reactive monosaccharides. Based on the configuration of the 3- and 4-hydroxyl groups suggested by Makela (5), four possible patterns are illustrated.

pyranose form, as shown in Fig. 1. Lectins that bind to Mäkelä's group II sugars are Gal-specific whereas those reacting with sugars of group III are Man and/or Glc-specific lectins. Fucose-binding lectins are specific for group I sugars. Lectins binding sugars belonging to group IV have not yet been reported. Based on this classification, group specificity also applies to GalNAc-, GlcNAc-, sialic acid-binding lectins and to the "complex"-binding lectins. Lectins reactive with ABO-blood group determinants show a strong preference for sugars in the α-anomeric configuration (4). Some show different degrees of reactivity for the configuration at C-2. ConA exhibits ten times less specificity for glucose than for mannose(6). An acetamido group at C-2 is important for the binding sites of SBA(soy bean) (17) and WGA(wheat germ) (1,130) but it blocks the binding of PNA (peanut) (44) and RCA-I(*Ricinus communis* agglutinin) (63). UEL-I(*Ulex europaeus*-I) and eel serum agglutinins with strong affinity for *L*-sugars are not well inhibited by most unsubstituted *D*-sugars(4,7,146,152).

During the past decade, the combining sites of many lectins have been well characterized based on their affinity for monosaccharides, glycosides and oligosaccharides. When oligosaccharides of defined structures from glycoconjugates became available, different lectins were found to recognize different saccharide sequences as their own receptor sites, although they are considered to be identical in terms of monosaccharide specificity. The combining sizes of some lectins may accommodate up to five or more sugar residues, and many multi-branched oligosaccharides exhibit stronger lectin-binding reactivities than linear ones, which is due to cooperative binding effects of lectin and carbohydrate complexes. For example, TAL shows over 480-fold higher specificity for a biantennary penta-2,6-saccharide, which contains two branches of Galβ1→4GlcNAcβ1→6Man and Galβ1→4GlcNAcβ1→2Man at nonreducing ends, than one of its branched trisaccharides. Some of the lectins possess dual or multiple affinities for different disaccharides, thus *Maclura pomifera* is specific for both Galβ1→3GalNAc and GalNAcα1→*O* to the Ser(Thr) of the protein core, *Bauhinia purpurea* for Galβ1→3GalNAc, GalNAcα1→Ser(Thr), and Galβ1→4(3)GlcNAcβ1→ linked residues, and *Datura stramonium*, for *N*-acetyllactosamine and GlcNAcβ1→4GlcNAcβ1→4GlcNAc. However, investigations are still hampered by the paucity of suitable reagents (oligosaccharides). The purpose of this guide is to give an updated list of the lectin specificities, which are useful to determine the binding properties of cell surface carbohydrates and to provide structural information about complex carbohydrates. In this guide, the carbohydrate specificities of various lectins are classified into six groups according to their specificities to monosaccharides. The subgroups are based on lectin affinities to GalNAcα1→*O* to Ser(Thr) of the peptide chain, disaccharides, the number and the location of LFucα1→ linked to oligosaccharides or those requiring further characterization. A scheme of the classification is shown as follows.

I. GalNAc-specific lectins

1. GalNAcα1→3GalNAc (Forssman, **F**) specific lectins
   Dolichos biflorus (DBL), *Helix pomatia* (HPL), Hog-peanut (ABL), and *Wistaria floribunda* (WFL) lectins.
2. GalNAcα1→3Gal (Blood Group A determinant disaccharide, **A**) specific lectins - *Griffonia(Bandeiraea) simplicifolia* A4 (GSI-A4), Lima bean (LBL), Soy bean (SBL), *Vicia villosa* (VVL), *Wistaria floribunda* (WFL), *Dolichos biflorus* (DBL) and *Helix pomatia* (HPL) lectins.
3. GalNAcα1→Ser or Thr, (**Tn** antigen) specific lectins
   Vicia villosa B4 (VVL-B4) *Salvia sclarea* (SSL), *Maclura pomifera* (MPL) and *Bauhinia purpurea alba* (BPL) lectins.
4. Other GalNAc-specific lectins

II. Gal-specific lectins

1. Galβ1→3GalNAc (**T** antigen, the mucin type sugar sequence on human erythrocyte membrane or Galβ1→3GalNAcβ1→ at the terminal

nonreducing end of the ganglioside) specific lectins - Peanut (PNA), *Bauhinia purpurea alba* (BPL), *Maclura pomifera* (MPL), and *Sophora japonica* (SJL).

2. Galβ1→3(4)GlcNAc (Lacto-*N*-biose/*N*-acetyllactosamine) specific lectins (Human blood group type I (Galβ1→3GlcNAc) and type II (Galβ1→4GlcNAc) carbohydrate sequences, **I** and **II** (7A), the disaccharide residues at nonreducing end of the carbohydrate chains derived from either *N*-glycosidic or *O*-glycosidic linkages.) - *Ricinus communis* agglutinin (RCA), *Datura stramonium* (Thorn apple, TAL), *Erythrina cristagalli* (Coral tree, ECL), *Sumbucus nigra* (SNL), and *Geodia cydonium* (GCL).
3. Galα1→3Gal(Human blood group B disaccharide, **B**) - *Griffonia (Bandeiraea) simplicifolia* B4 (GS1-B4).
4. Other Gal-specific lectins

III. Man and/or Glc-specific lectins

1. Man-linked oligosaccharide-specific lectins
   Concanavalin ensiformis (Jack bean, ConA); *Lens culinaris* (LCL); *Pisum sativum* (PSL, pea); *Vicia faba* (VFL) and *Phytolacca americana* (PAL).
2. Other Man and/or Glc-specific lectins

IV. GlcNAc-specific lectins

1. Chitin oligosaccharide specific lectins
   Wheat germ (WGA), *Griffonia (Bandeiraea) simplicifolia* II (GSL-II), Tomato (LEL), Potato (STL), Thorn apple (TAL).
2. Other GlcNAc-specific lectins

V. Fuc-specific lectins

1. Monofucosyl-specific lectins - *Ulex europaeus* I (UEL-I), *Ulex europaeus* II (UEL-II) and *Lotus tetragonolobus* (LTL).
2. Difucosyl-specific lectins - *Griffonia (Bandeiraea) simplicifolia* IV (GSL-IV).
3. Other LFuc-specific lectins

VI. Sialic acid specific lectins

Limulus polyphemus agglutinin (LPL).

I. GalNAc-specific lectins

I-1. GalNAcα1→3GalNAc(Forssman,F)specific lectins

DBL, *Dolichos biflorus* lectin (Horse Gram)(8,9)
GalNAcα1→3GalNAcβ1→3Galα1→4Galβ1→4Glc
(Forssman specific pentasaccharide) >>
GalNAcα1→3GalNAc >>
GalNAcα1→3[LFucα1→2]Galβ1→4GlcNAcβ1→6R >
(Blood group A active pentasaccharide, R_L, 0.52[3])
GalNAcα1→3Galβ1→3GlcNAc (A_5II[3]) and GalNAcα1→3Gal
(R_L,1.85[3]) > GalNAc.

HPL, *Helix pomatia* lectin (Edible snail)(9-10)
GalNAcα1→3GalNAcβ1→3Galα1→4Galβ1→4Glc

(Forssman specific pentasaccharide) >
GalNAcα1→3GalNAc >
GalNAcα1→3Galβ1→3GlcNAc (A_5II3) > GalNAc.

ABrL, *Amphicarpaea bracteata* lectin (Hog-peanut)(1,10A)
GalNAcα1→3GalNAc > GalNAcα1→3Gal > GalNAc.

WFL, *Wistaria floribunda* lectin (9,11)
GalNAcα1→6Gal > GalNAcα1→3Gal (R_L,1.85[3]),
GalNAcα1→3GalNAcβ1→3Galα1→4Galβ1→4Glc
(Forssman specific pentasaccharide) and
GalNAcα1→3Galβ1→3GlcNAc (A_5II^3) >
GalNAcβ1→3Galα1→4Galβ1→4Glc (Globoside) >
GalNAcα1→3GalNAc > GalNAc and
GalNAcα1→3[LFucα1→2]Galβ1→4GlcNAcβ1→6R(R_L,0.52[3])>
Galβ1→4GlcNAc(*N*-acetyllactosamine) > Lactose.

I-2. GalNAcα1→3Gal(Human blood group A) specific lectins

GSI-A_4, *Griffonia (Bandeiraea) simplicifolia* A_4 (12,13)
GalNAcα1→3Galβ1→3GlcNAc (A_5II^3), GalNAcα1→3Gal
(R_L,1.85[3]), and GalNAcα1→6Gal, GalNAcα1→R >
Galα1→3Gal, Galα1→6Glc and Raffinose > Gal.

LBL, *Phaseolus lunatus* lectin (Lima bean)(14-16)
GalNAcα1→3[LFucα1→2]Galβ1→3GlcNAcβ1→3Galβ1→4Glc >
GalNAcα1→3[LFucα1→2]Galβ1→4Glc >
GalNAcα1→3[LFucα1→2]Galβ1→4GlcNAcβ1→6R(AR,0.56[3]) >
GalNAcα1→3[LFucα1→2]Gal >> GalNAc >> Gal
LFucα1→2 to subterminal Gal is an important factor for binding.

SBL, *Glycine max* lectin (Soy bean)(17,18)
GalNAcα or β1-linked,GalNAcα1→3Galβ1→3GlcNAc(A_5II^3)
and GalNAcα1→3Gal (R_L 1.85[3])>>
GalNAcα1→3[LFucα1→2]Galβ1→4GlcNAcβ1→6R(R_L,0.52[3]),
Galα1→6Glc (Melibiose), Galβ1→6Glc,Raffinose,
Stachyose, Galβ1→4Glc (Lactose) and
LFucα1→2Galβ1→4GlcNAcβ1→6R(Human blood H active oligosaccharide, JS, R_L,0.75[3]) > Gal.
LFucα1→2 to subterminal Gal blocks Gal for binding.

VVL, *Vicia villosa* lectin (Hairy Vetch, it is assumed that the lectin used to study the following information is a mixture of three isolectins; A_4-minor, B_4-major, and A_2B_2-minor) (19,20)

GalNAcα1→3Gal (RL,1.85[3]) > GalNAcα1→6Gal >
GalNAcα1→3Galβ1→3GlcNAc (A5II[3]) and
GalNAcα1→3GalNAc >>
GalNAcα1→3[LFucα1→2]Galβ1→4GlcNAcβ1→6R (RL, 0.52[3]) >
Galα1→3Gal.

WFL, *Wistaria floribunda* lectin (see Forssman specific lectins, Section **I-1**).

DBL, *Dolichos biflorus* lectin (see Forssman-specific lectins, Section **I-1**).

HPL, *Helix pomatia* lectin(see Forssman-specific lectins).

I-3. GalNAcα1→Ser or Thr (Tn) specific lectins

VVL-B4, *Vicia villosa* B4 lectin (Hairy Vetch) (20-21)
A fetuin glycopeptide containing two GalNAcα1→*O* to Ser(Thr) of the peptide chains (two Tn structures)>> single Tn structure >> a fetuin glycopeptide containing one or two Galβ1→3GalNAcα1→*O* to Ser(Thr) of the peptide chain(s) (T structure) > NeuNAcα2→3Galβ1→3GalNAcα1→Ser(Thr) of the peptide chain.

SSL, *Salvia sclarea* lectin (22)
Glycopeptides containing two Tn structures > Glycopeptides containing one Tn structure or Glycopeptides containing three sequential Tn structures.

MPL, *Maclura pomifera* lectin
(see Galβ1→3GalNAc-specific lectins, Section **II-1**).

BPL, *Bauhinia purpurea alba* lectin
(see Galβ1→3GalNAc specific lectins, Section **II-1**).

I-4. Other GalNAc specific lectins requiring further characterization

APL, *Aegopodium podagraria* lectin
GalNAc > Lactose > Gal > Melibiose > Raffinose > Ara (23).

BDL, *Bryonia dioica* lectin (24)
GalNAc > Lactose > Melibiose > Raffinose > Stachyose.

CAL, *Caragana arborescens* lectin (Pea tree) (25)
GalNAc > Gal > Lactose and Raffinose > Stachyose > Ara > Rhamnose.

CTL, *Clerodendron trichotomum* lectin (26)
GalNAc > Lactose > Melibiose > Gal.

CNL, *Clitocyle nebularis* lectin (27)
GalNAc and Gal.

DDL-I, *Dictyostelium discoideum* I lectin (Discoidin I) (28,29)
GalNAc > DFuc > Melibiose > Lactose > Gal >LFuc.

EHy,L, *Eranthis hyemalis* lectin (Winter aconite root tubers) (30)
GalNAc > Lactose > Gal > Melibiose > LFuc > Ara > Raffinose.

EHe,L, *Euphorbia heterophylla* lectin (31)
GalNAc > Lactose > Gal > Melibiose > LFuc.

FFL, *Fomes fomentarius* lectin (27)
GalNAc > Raffinose > Gal.

HCL, *Hura crepitans* seed lectin (32)
GalNAc > Gal > Melibiose and DFuc > Raffinose.

MAL, *Macrotyloma axillare* anti-A lectin (33)
GalNAc.

MCL, *Momordica charantia* lectin (Bitter pear melon) (34,35)
GalNAc > DFuc > Gal > Raffinose and Stachyose > Melibiose and LAra.

OHL, *Ononis hircina Jacq* lectin (36)
GalNAc > Lactose > Gal > GlcNAc > Ara.

PTL, *Psophocarpus tetragonolobus* lectin (Winged bean) (37-39)
GalNAc > Gal and Melibiose > Raffinose > DFuc > Lactose.

RPL, *Robinia pseudacacia* lectin (40)
GalNAc.

TML, *Tridacna maxima* lectin (Röding clam) (41)
GalNAc > Galβ1→6Gal, Lactose and Gal > Melibiose, Raffinose and Stachyose.

VCL, *Vicia cracca* anti-A lectin (42)
GalNAc.

VAL-II & III, *Viscum album*-II & III lectins (Mistletoe)
GalNAc > Gal (43).

II. Gal-specific lectins

II-1. Galβ1→3GalNAc(T)specific lectins

PNA, *Arachis hypogaea* lectin (Peanut) (44,45)
Galβ1→3GalNAc >> Galβ1→4GlcNAc(*N*-acetyllactosamine)> Gal and Galβ1→3GlcNAc.

BPL, *Bauhinia purpurea alba* lectin (46,47)
Galβ1→3GalNAc > GalNAcα1→6Gal > GalNAc
Galβ1→3GlcNAcβ1→3Galβ1→4Glc (Lacto-*N*-tetraose) >
Galβ1→3GlcNAc, Galβ1→6GlcNAc and
GalNAcα1→3Galβ1→3GlcNAc (A_5II^3) >
GalNAcα1→3Gal > Galβ1→4GlcNAc(*N*-acetyllactosamine)
and Galβ1→3[LFucα1→4]GlcNAcβ1→3Galβ1→4Glc
(Lacto-*N*-fucopentaose II) >
Galβ1→4GlcNAcβ1→6Gal and Lactose > Gal.

J, *Maclura pomifera* lectin (Osage orange tree, Hedge apple tree) (48,49)
Galβ1→3GalNAc >> GalNAcα1→6Gal > Galα1→6Glc (Melibiose), Stachyose > GalNAc > Gal.

SJL, *Sophora japonica* lectin (Japanese pagoda tree) (50,51)
Galβ1→3GalNAcβ1→*N*-tosyl-L-serine >
Galβ1→3GlcNAc >
Galβ1→3GlcNAcβ1→3Galβ1→4Glc (Lacto-*N*-tetraose) >
Galβ1→3GlcNAcβ1→6Gal > Galβ1→4GlcNAcβ1→6Gal
(Blood group I Ma-active trisaccharide) and
Galβ1→4GlcNAc > GalNAc >> Lactose >
Gal and Melibiose.

Some unknown mild acid sensitive compound near the terminal group at the nonreducing end plays an important role in binding or precipitating this lectin.

RCA_2, Ricin (18,52,53)
From Baenziger *et al* (52)
[Galβ1→3GalNAcα1→Ser or Thr]$_4$ glycopeptide >>
Triantennary oligosaccharides containing Galβ1→4 GlcNAc at nonreducing end >
[Galβ1→3GalNAcα1→Ser or Thr]$_2$ glycopeptide with space between carbohydrate side chains,
Galβ1→3GalNAcα1→Ser or Thr >> GalNAcα1→Ser or Thr.

The ability of RCA_2 to bind to Galβ1→3GalNAc disaccharides appears to be markedly influenced by the relative locations of the disaccharides along the peptide backbone.

From Debray *et al.* (18) and Baenziger *et al.* (52)
Biantennary oligosaccharides containing Galβ1→4 GlcNAc at nonreducing end > Lac > GalNAc >
Melibiose and Gal ($\geq$ GalNAc, see ref. 53).

VGL, *Vicia graminea* lectin (Blood group N specific) (54-57)
Galβ1→3GalNAcα1-linked in clusters and
desialized blood group N glycopeptide >
untreated N glycopeptide and desialized M glycopeptide.

ABiL, *Agaricus bisporus* lectin (Mushroom) (58,59)
Galβ1→3GalNAcα-*N*-tosyl-L-serine.

ARL, *Agropyrum repens* lectin (Couch grass) (60)
Galβ1→3GalNAc > GalNAc and chitin oligosaccharide (with type A cells).

AIL, *Artocarpus integrifolia* lectin (Jacalin,61)
Galβ1→3GalNAc >>> Galβ1→3GlcNAc; Galβ1→4GlcNAc and Galβ1→4Glc.

ALL, *Artocarpus lakoocha* lectin (61A)
Galβ1→3GalNAc > GalNAcα1→*O* Ser/Thr > Melibiose > Raffinose > Gal.

II-2. Galβ1→3(4)GlcNAc (Human blood group type I and type II carbohydrate sequences, I & II) specific lectins (7A)

RCA_1, *Ricinus communis* agglutinin (Castor bean)(18,52,62,63)
Biantennary oligosaccharides containing Galβ1→4GlcNAc-linked units at nonreducing end > Galβ1→4GlcNAcβ1→6Gal (Blood group I Ma specific) > Galβ1→4GlcNAc > Galβ1→3GlcNAc > Lactose > Galα1→3Gal > Raffinose
RCA_1 is also specific for Galβ1→3GalNAcα1→ Ser (Thr). For further information, see reference 52.

TAL, *Datura stramonium* lectin (DSL,Thorn apple) (18,64-67)
Penta-2,6[4] >> Penta-2,4[4] > [GlcNAcβ1→4]$_4$ > [GlcNAcβ1→4]$_3$ > Galβ1→4GlcNAc(*N*-acetyllactosamine), GlcNAcβ1→4GlcNAc and Galβ1→4GlcNAcβ1→6Man > Manα1→4Manβ1→2GlcNAc > GlcNAcβ1→4Glc
More information is available in reference 65.

ECL, *Erythrina cristagalli* lectin (Coral tree)(68)
Tetra and tri-antennary oligosaccharides containing Galβ1→4GlcNAc linked units at nonreducing end > Galβ1→4GlcNAcβ1→6Gal (Blood group I Ma specific) > Galβ1→4GlcNAc > Galβ1→3GlcNAc >> Lactose >>> Galβ1→6Gal and, Galα1→3Gal > Raffinose.

GCL, *Geodia cydonium* lectin (69)
Galβ1→4GlcNAc, Galβ1→3GlcNAc and Lactose > GalNAc.

Other *Erythrina* lectins (70)
E. caffra[+] (ECaf,L), *E. corallodendron* (ECor,L), *E. flabelliformis* (EFL), *E. humeana* (EHL), *E. latissima* (ELat,L), *E. lysistemon* (ELys,L), *E. perrieri*[+] (EPL), *E. stricta* (ESL), *E. zeyheri* (EZL)

---

[4] Galβ1→4GlcNAcβ1,2↓
Man
Galβ1→4GlcNAcβ1,6↑

Penta-2,6

GlcNAcβ1,2↓
Man
GlcNAcβ1,6↑

Tri-2,6

Galβ1→4GlcNAcβ1,2↓
Man
Galβ1→4GlcNAcβ1,4↑

Penta-2,4

Triantennary or biantennary oligosaccharides containing Galβ1→4GlcNAc at non-reducing end > *N*-acetyllactosamine > Lactose ≥ GalNAc > Gal (GalNAc ≥ Lactose ≥ LGal)$^{+}$.

PHA-E, *Phaseolus vulgaris* (isolectin E_4) erythroagglutinin (Red kidney bean) (71-73).
Minimum structural unit:

```
        Galβ1→4GlcNAcβ1→2Manα1,6↓
                          GlcNAcβ1→4Manβ1→4GlcNAcβ1→4R3
             R1→GlcNAcβ1→2Manα1,3↑
                               ↑
                               R2
```

in which **R1** and **R2** represent either hydrogen atoms or sugars and **R3** represents either GlcNAc or (Fucα1→6) GlcNAc.

PAL, *Phytolacca americana* lectin (pokeweed mitogen)
Three different carbohydrate specificities have been reported. Their relationship has not been established. Branched poly (Galβ1→4GlcNAc)-complex(74); (GlcNAcβ1→4)$_{n>5}$→ units(75) or Manα1→2Manα1→2Manα1→3Manβ1→4GlcNAcβ1→4GlcNAc-AsN(76).

PHA-L, *Phaseolus vulgaris* (isolectin L_4) leukoagglutinin (71,77)
Tri- and Tetra-antennary oligosaccharides containing Galβ1→4GlcNAc-linked units at nonreducing end >> Penta-2,6^{4} > Tri-2,6^{4} >

```
Galβ1→4GlcNAcβ1→2Manα1,3↓
                        Man (hepta-) >
Galβ1→4GlcNAcβ1→2Manα1,6↑
```

Galβ1→4GlcNAcβ1→2Man > GlcNAcβ1→2Man.

EHL, *Eranthis hyemalis* agglutinin (from Winter-aconite root tubers)(78)
Galβ1→4GlcNAc > Galβ1→4Glc > Gal, Melibiose > Raffinose.

PNL, *Arachis hypogaea* lectin (see Galβ1→3GalNAc specific lectins, Section **II-1**).

BPL, *Bauhinia purpurea alba* lectin (see Galβ1→3GalNAc specific lectins, Section **II-1**).

Ricin, (see Galβ1→3GalNAc specific lectins, Section **II-1**).

UDL, *Urtica dioica* agglutinin (Stinging nettle)(129)
(see GlcNAcβ1→4GlcNAc specific lectin, Section **IV**).

SNL, *Sambucus nigra* lectin (79,80)(see sialic acid specific lectins, Section **VI**).

II-3. Galα1→3Gal(B)specific lectin

GSL-B4, *Griffonia (Bandeiraea) simplicifolia* I-B4 lectin (12,13)
Galα1→3Gal > Melibiose > Raffinose > Gal > GalNAc (inactive).

II-4. Other Gal-specific lectins and those with more information required for classification

AXP-I and II, *Axinella polypoides* I and II lectins(81)
Galβ1→6R > Galβ1→4R > Melibiose and Raffinose > Gal.
Myxococcus xanthus lectin (82)
NeuNAc-Galβ1→3GalNAc-Ser(Thr) (possibly).

APL, *Abrus precatorius* lectin (Jequirity bean)(83,84)
Lactose and Gal > Melibiose > LFuc. Abrin (83,84)
Gal > Lactose > Melibiose > LFuc and Ara > Glc.
ADL, *Aplysia depilans* lectin (85)
Galacturonic acid > Gal.

AML, *Ascidia malaca* lectin (86)
Melibiose > Raffinose > Gal > Lactose > Ara.

ABL, *Abramis brama* lectin (87)
Rhamnose > Gal.

BCL, *Bauhinia carronii* lectin (88)
Gal and Lactose.

BFL, *Butea frondosa* lectin (35)
Lactose > DFuc > Gal > GalNAc > Ara.

CJL, *Crotalaria juncea* lectin (Sunn Hemp) (89)
Lactose > Melibiose > Raffinose > GalNAc > Gal.

CML-I, *Cytisus multiflorus*-I lectin (90)
Gal, Lactose and Melibiose > Stachyose > Raffinose.

CSL, *Cytisus scoparius* lectin (91)
Gal and GalNAc.

DCL, *Didemnum candidum* lectin (92)
Lactose > Gal > Melibiose > Stachyose > GalNAc.

DDL-II, *Dictyostelium discoideum*-II lectin (Discoidin II) (29)
Lactose and DFuc > Gal > Melibiose > GalNAc > LFuc.

EAL, *Erythrina arborescens* lectin (93)
Lactose and GalNAc > Melibiose and Gal > DFuc.

EIL, *Erythrina indica* lectin (35,93)
Lactose > GalNAc and Melibiose > Gal and Raffinose > DFuc.

ELL, *Erythrina lithosperma* lectin (93)
Lactose and Melibiose > GalNAc and Raffinose > Gal > DFuc.

ESL, *Erythrina suberosa* lectin (93)
Lactose > GalNAc > Melibiose and Gal > Raffinose > DFuc.

ECL, *Euphorbia characias* L. lectin (Mediterranean spurge) (94) Lactose > Melibiose > Gal > DFuc > Ara.

HRL, *Halocynthia roretzi* lectin (95)
Melibiose > Lactose, Gal and DFuc>LFuc and Stachyose.

HCL, *Hardenbergia comptoniana* lectin (88)
Raffinose.

HCL, *Hura crepitans* latex lectin (Sand-box tree)(94)
Lactose > Gal and Melibiose > Ara.

LAL, *Laccaria amethystina* lectin (LAL)(Mushroom)(96)
Lactose > GalNAc.

OVL, *Octopus vulgaris* lectin (97)
Lactose and *N*-acetyllactosamine.

PAL, *Phaseolus aureas* or *Vigma radiata* lectin (Mung bean)(98) Gal > Xyl > Inositol.

PPL, *Polysphondylium pallidum* lectin (Pallidin)(28,99)
Lactose > Gal > GalNAc and DFuc > LFuc.

RRL, *Rutilus rutilus* lectin (87)
Rhamnose > Gal and Ara > Lactose.

SSL, *Sarothamnus scoparius* lectin (27)
Raffinose > GalNAc > Gal > Lactose.

SEL, *Scardinus erythrophtalmus* lectin (87)
Rhamnose > Arabinose > Galactose.

VVL, *Vimba vimba* lectin (87)
Rhamnose > Raffinose > Gal and Ara > Lactose.

TKL, *Trichosanthes kirilowii* lectin (100)
Lactose > Gal > Melibiose > Raffinose.

VAL-I, *Viscum album* I lectin (Mistletoe)(43,101)
Lactose > Melibiose, Raffinose and Gal.

VUL, *Vigna unguiculate* lectin (Cowpeas)(102)
Gal.

III. Man and/or Glc specific lectins

III-1. Manα1→linked oligosaccharide-specific lectins

ConA, *Concanavalia ensiformis* agglutinin (Jack Bean)(18, 103-106).

Manα1→linked > Glcα1→linked > GlcNAcα1→linked. The binding sugar requires the C-3, C-4 and C-5 hydroxyl groups of the Man or Glc ring (103). One of the essential core structures in glycopeptides reactive with this lectin is shown below (18,104).

$$\begin{array}{l} \mathbf{R_1}\text{-Man}\alpha1,3\downarrow \\ \qquad\qquad \mathbf{R_2}\text{-Man}\beta1\rightarrow4\text{GlcNAc}\beta1\rightarrow4\text{GlcNAc-Asn} \\ \mathbf{R_3}\text{-Man}\alpha1,6\uparrow \end{array}$$

ConA has a great affinity for the trimannosidic core structure substituted by two GlcNAc residues (**R1**) at the non-reducing end; the affinity is not affected when this structure is modified by addition of a β1→4 linked GlcNAc on the β-linked man:(**R2**)(18), but its affinity is reduced when these GlcNAc residues are substituted by Gal residues. **LFuc** is not as important for the binding of ConA as for LCL, PSL, and VFL (18).

LCL, *Lens culinaris* lectin (Lentil) (18,107-109)

Manα1→linked > Glcα1→linked >> GlcNAc. For comparison, see PSL, and VFA. LFucα1→6 linked to GlcNAc at the **reducing end** in the following structure is an important binding determinant (as shown in glycopeptide, GP-hLTF-1 in Fig 2, reference 18), but not LFuc α1→3 linked to GlcNAc near the nonreducing end (Glycopeptide, GP-hLTF-2 in Fig. 2, reference 18).

```
NeuAcα2→6Galβ1→4GlcNAcα1→2Manα1,3↓
                                   Manβ1→4GlcNAcβ1→4GlcNAcβ1→6AsN
NeuAcα2→6Galβ1→4GlcNAcα1→2Manα1,6↑          LFucα1,6↑
         LFucα1,3↑                  (Important factor for binding)
```

Removal of LFucα1→6 linked from the above glycopeptide with an αL-fucosidase yields a structure that is about 10-fold less active as compared to the untreated one.

PSL, *Pisum sativum* lectin (Pea) (18,107-109)

Manα1→ linked > Glcα1→ linked >> GlcNAc. For comparison, see LCL and VFA. LFucα1→6 linked to GlcNAc at the **reducing end** is also an important binding determinant as described for Lentil. The removal of the α1→6 linked fucose residue produces an aLFuc glycopeptide 17-fold less inhibitory. The biantennary structure is also an essential feature 110).

VFL, *Vicia faba* lectin (Fava bean) (18,107,109,110)

Manα1→ linked > Glcα1→ linked >> GlcNAc. For comparison see LCL and PSL. Same as Lentil and PSL, i.e. the binding properties of Lentil, *Vicia* and *Pisum* are similar in many aspects, but quite different from ConA. Specifically, LFuc α1→6 linked to GlcNAc at the reducing end is not required for ConA binding (18). The fine affinity of these three lectins can be differentiated by some of the mannose containing oligosaccharides such as sialoside 9 and asialo-oligosaccharides derived from sialoside 9, as shown in Table 2, reference 18.

PAL, *Phytolacca americana* Pa-2 lectin (pokeweed) (74-76)
See Galβ1→3 or 4GlcNAc specific lectin.

III-2. Other Man and/or Glc-specific lectins requiring further characterization

DGL, *Dioclea grandiflora* lectin (111)
Man > Fru > Glc.

LOL, *Lathyrus odoratus* lectin (Sweet pea) (112,113)
Man > Glc > GlcNAc and Sucrose.

LSL, *Lathyrus sativum* lectin (Chickling vetch) (112,114)
Man > Glc > GlcNAc and Sucrose.

LTL, *Lathyrus tingitanus* lectin (Tangier pea) (115)

Man > Glc > GlcNAc and Sucrose > Fru > Glucosamine.

OVL, *Onobrychis vicifolia* lectin (Sainfoin) (116)
Glc, Man, Maltose and Sucrose.

PFL, *Perca fluriatilis* lectin (87)
Man > LFuc > Gal and GalNAc > Rhamnose.

POL, *Ptilotis obovatus* lectin (88)
Maltose.

RCL, *Rhadogia crassifolia* lectin (88)
Maltose.

VCL, *Vicia cracca* lectin (Common vetch) (117)
Man > Maltose > Glc and Sucrose > GlcNAc > Fru.

VSL, *Vicia sativa* lectin (118)
Man > Sucrose and Maltose > GlcNAc.

IV. *GlcNAc specific lectins*

IV-1. Chitin oligosaccharide-specific lectins

APL-I, *Aaptos papillata* I lectin (119)
$(GlcNAc\beta 1\rightarrow 4)_4$ > $(GlcNAc\beta 1\rightarrow 4)_3$ >
GlcNAcβ1→4GlcNAc.

APL-II and III, *Aaptos papillata* II and III lectins (119)
$(GlcNAc\beta 1\rightarrow 4)_3$ > GlcNAcβ1→4GlcNAc.

CML, *Cucurbita maxima* lectin (Pumpkin) (120)
$(GlcNAc\beta 1\rightarrow 4)_4$ > $(GlcNAc\beta 1\rightarrow 4)_3$ > GlcNAcβ1→4GlcNAc.

CPL, *Cucurbita pepo* lectin (121)
$(GlcNAc\beta 1\rightarrow 4)_4$, $(GlcNAc\beta 1\rightarrow 4)_3$ and GlcNAcβ1→4GlcNAc >
GlcNAc-NeuNAc > GlcNAc.

CML-II, *Cytisus multiflorus*-II lectin (90)
$(GlcNAc\beta 1\rightarrow 4)_3$ and GlcNAcβ1→4GlcNAc >
Cellobiose and Salicin (*O*-hydroxymethylphenyl-β-D-glucopyranoside) > Lactose.

DSL, *Datura stramonium* lectin (Thorn-apple)
see DGalβ1→3(4)GlcNAc specific lectin, Section **II-2**.

GSL-II, *Griffonia (Bandeiraea) simplicifolia* II lectin
(122-124)
$(GlcNAc\beta 1\rightarrow 4)_4$, $(GlcNAc\beta 1\rightarrow 4)_3$, and GlcNAcβ1→4Glc >
GlcNAc.

LEL, *Lycopersicon esculentum* lectin (Tomato) (125,126)
$(GlcNAc\beta 1\rightarrow 4)_4$ > $(GlcNAc\beta 1\rightarrow 4)_3$ > GlcNAcβ1→4GlcNAc.

LAL, *Luffa acutangula* lectin (Ridge Gourd) (127)
(GlcNAcβ1→4)$_5$ > (GlcNAcβ1→4)$_3$ > GlcNAcβ1→4GlcNAc > GlcNAc.

PAL, *Phytolacca americana* Pa-2 lectin (Pokeweed) (74-76) See Galβ1→4(3)GlcNAc specific lectin.(**II-2**).

STL, *Solanum tubersum* lectin (Potato) (18,128)
(GlcNAcβ1→4)$_4$ > (GlcNAcβ1→4)$_3$ > GlcNAcβ1→4GlcNAc and (GlcNAc-MurNAc)$_2$ >>> GlcNAc.

UDL, *Urtica dioica* lectin (Stinging nettle) (129)
(GlcNAcβ1→4)$_4$ > (GlcNAcβ1→4)$_3$ >>
GlcNAcβ1→4GlcNAc and Galβ1→4GlcNAcβ1→6Man ≥
Manα1→3Manβ1→4GlcNAc >
Galβ1→4GlcNAc(*N*-acetyllactosamine)>>GlcNAc >> GalNAc.

WGA, *Triticum vulgaris* agglutinin (Wheat germ)(18,130-131)
(GlcNAcβ1→4)$_5$ > (GlcNAcβ1→4)$_4$ and (GlcNAcβ1→4)$_3$ >
GlcNAcβ1→4GlcNAc and sialic acid.

UEL-II, *Ulex europaeus* II lectin (see LFuc specific lectins, **V-I**).

IV-2. Other GlcNAc-specific lectins requiring further characterization

CSL, *Cytisus sessilifolius* lectin (132,133)
See LFuc specific lectins (Section **V-1**)
GlcNAcβ1→4GlcNAc > Salicin, Cellobiose and
Laminaribiose > Lactose and Galβ1→3GlcNAc.

LAL, *Laburnum alpinum* lectin (132)
GlcNAcβ1→4GlcNAc > Cellobiose and Laminaribiose >
Lactose > Galβ1→3GlcNAc.

BSL, *Brachypodium sylvaticum* lectin (False brome grass)(134)
Chitin oligosaccharide > GlcNAc.

DLL, *Dolichos lablab* lectin (135) GlcNAc.

HVL, *Hordeum vulgare* lectin (Barley) (136)
Chitin oligosaccharide > GlcNAc.

Rice lectin (137,138)
Chitin oligosaccharide > GlcNAc.

SCL, *Secale cereale* lectin (Rye) (136,139))
Chitin oligosaccharide > GlcNAc.

SPL, *Saxidomus purpuratus* lectin (140,141)
GlcNAc > GalNAc > ManNAc.

V. LFuc specific lectins

V-1. Monofucosyl specific lectins

AAurL, *Aleuria aurantia* lectin (Orange peel fungus) (142,143)
LFuc > LFucα1→2Galβ1→4Glc(2'-Fucosyllactose) >
Galβ1→3[LFucα1→4]GlcNAcβ1→3Galβ1→4Glc
(Lacto-*N*-fucopentaose II, L*N*FP-II)
Galβ1→4[LFucα1→3]Glc(3-fucosyllactose) >
LFucα1→2Galβ1→3GlcNAcβ1→3Galβ1→4Glc
(Lacto-*N*-fucopentaose I, L*N*FP-I).

CSL, *Cytisus sessilifolius* lectin (132,133)
LFucα1→2Galβ1→4Glc (2'-fucosyllactose) >
LFucα1→2Galβ1→3GlcNAcβ1→3Galβ1→4Glc (L*N*FP-I) >
Galβ1→3GlcNAcβ1→3Galβ1→4Glc(Lacto-*N*-tetraose) and
Galβ1→4GlcNAcβ1→3Galβ1→4Glc(Lacto-*N*-neotetraose).
Also see GlcNAcβ1→4 specific lectins (Section **IV-2**).

EEL, *Euonymus europeus* lectin (Spindle tree) (144,145)
Galα1→3[LFucα1→2]Galβ1→4GlcNAcβ1→6[Galβ1→3]GalNAc-ol
(Tij phenol insoluble R_{IM5},0.9[3]) and
Galα1→3[LFucα1→2]Galβ1→4GlcNAcβ1→6R
(Beach phenol insoluble R_L,0.44[3]) >
Galα1→3[LFucα1→2]Galβ1→3GlcNAcβ1→3(4)Gal-ol
(Beach P1 R_L,0.26[3]) >
LFucα1→2Galβ1→3GlcNAcβ1→3Galβ1→4Glc (L*N*FP-I) >
Galα1→3[LFucα1→2]Galβ1→4[LFucα1→3]*D*Glc
Urine B pentasaccharide >
Galα1→3Gal > Galβ1→4GlcNAc.

The extent of specificity can be summarized in the following order: (145)

1. best, tetrasaccharides with blood group B activity. Galα1→3[LFucα1→2]Galβ1→4(3)GlcNAc.
2. good, trisaccharides derived from blood group B tetrasaccharides. LFucα1→2Galβ1→4(3)GlcNAc, Galα1→3Galβ1→4GlcNAc or Galα1→3[LFucα1→2]Gal.
3. poor, disaccharides derived from blood group B tetrasaccharides. Fucα1→2Gal, Galβ1→4GlcNAc, or Galα1→3Gal.

LAL, *Laburnum alpinum* lectin (133)
LFucα1→2Galβ1→4Glc (2'-fucosyllactose) >
LFucα1→2Galβ1→3GlcNAcβ1→3Galβ1→4Glc (L*N*FP-I)
Galβ1→3GlcNAcβ1→3Galβ1→4Glc(Lacto-*N*-tetraose) and
Galβ1→4GlcNAcβ1→3Galβ1→4Glc(Lacto-*N*-neotetraose).

LTL, *Lotus tetragonolobus* lectin[5] (Asparagus pea) (18,146)
LFucα1→2Galβ1→4[LFucα1→3]GlcNAcβ1→6R >
LFucα1→2Galβ1→4[LFucα1→3]GlcNAc >
LFucα1→2Galβ1→4GlcNAc.

UEL-I, *Ulex europaeus*-I lectin[5] (Gorse or Furz seeds) (18,147,148)
LFucα1→2Galβ1→4GlcNAcβ1→6R >
LFucα1→2Galβ1→4[LFucα1→3]GlcNAcβ1→6R >
LFucα1→2Galβ1→4GlcNAc (2'-fucosyllactosamine) >
LFucα1→2Galβ1→4[LFucα1→3]Glc (Lacto-difucotetraose).

UEL-II, *Ulex europaeus*-II lectin (18,149,150)
LFucα1→2Galβ1→4GlcNAcβ1→6R (HJS, R_{IM}2.5) >
LFucα1→2Galβ1→4Glc (2'-fucosyllactose) >
$(GlcNAcβ1→4)_4$ > $(GlcNAcβ1→4)_3$ > GlcNAcβ1→4GlcNAc >
GalNAcα1→3Galβ1→3GlcNAc (A_5II) > GlcNAcβ1→6Gal >
Galβ1→3GlcNAcβ1→3Galβ1→4Glc (Lacto-*N*-tetraose) >
LFucα1→2Galβ1→3GlcNAcβ1→3Galβ1→4Glc (L*N*FP-I) >
GalNAcα1→3Gal (R_L,1.85[3]).

V-2. Difucosyl specific lectins

GSL-IV, *Griffonia (Bandeiraea) simplicifolia* IV lectin (151)
LFucα1→2Galβ1→3[LFucα1→4]GlcNAcβ1→3Galβ1→4Glc (Lacto-*N*-difucohexaose I) >
Galβ1→3[LFucα1→4]GlcNAcβ1→3Galβ1→4[LFucα1→3]Glc (Lacto-*N*-difucohexaose II) >
LFucα1→2Galβ1→4[LFucα1→3]Glc (Lacto-difucotetraose) and LFucα1→2Galβ1→4[LFucα1→3]GlcNAc >>
(Lacto-*N*-difucotetraose)
Galβ1→3[LFucα1→4]GlcNAcβ1→3Galβ1→4Glc (L*N*FP-II).

V-3. Other LFuc-specific lectins requiring further characterization

AAngL, *Anguilla anguilla* lectin (Eel serum) (152)
LFucα1→3Gal > LFuc > LFucα1→3Galβ1→4GlcNAc.

LAF, *Laccaria amethystina* lectin(LAF) (Mushroom) (96)
LFuc.

---

[5] For differential affinities among these lectins, see reference 145.

VI. Sialic acid specific lectins

CRCA, *Carcinoscorpius rotunda cauda* agglutinin (Indian horseshoe crab) (153)
NeuAcα2→6Gal > NeuAcα2→3 or α2→6Galβ1→4Gal > Glucuronic acid > 3-deoxy-2-oxo-octonate (KDO) > GalNAc and GlcNAc.

AAusL, *Androctonus australis* lectin (Saharan scorpion) (154)
NeuAc.

Avocado lectin (155) NeuAc > Melibiose > Stachyose.

BGL, *Biomphalaria glabrata* lectin (156)
NeuAc > Fru > Gal > Man > LFuc > Glc.

BLL, *Birgus latro* lectin (Coconut crab) (157)
NeuAc and NeuGly.

HAL, *Homarus americanas* lectin (Lobster) (158)
NeuAc, NeuGly and ManNAc.

LFL, *Limax flavus* lectin (Slug) (159) NeuNAc and NeuGly.

LPL, *Limulus polyphemus* lectin (Horseshoe crab) (160,161)
Sialic acid, GlcNAc, Glucuronic acid, and Teichoic acid.

SNL, *Sambucus nigra* lectin (79,80)
Bovine fetuin triantennary carbohydrate chains[6] or porcine thyroglobulin biantennary carbohydrate chains[7] > Neu5Acα2→6Galβ1→4GlcNAcβ1→3Galβ1→4Glc > Neu5Acα2→6Galβ1→4Glc > Neu5Acα2→6Galβ1→4Glc-ol**>>** Neu5Acα2→3Galβ1→4Glc > Neu5Acα2→3Galβ1→3GlcNAcβ 1→3Galβ1→4Glc >Galβ1→4DGlcNAc and GalNAcβ1→6Gal > Galβ1→4Glc and GalNAcα1→6Gal > GalNAc and GalNAcα1→3Gal ≥ Gal > Melibiose, Raffinose and Stachyose.

WGA, *Triticum vulgaris* agglutinin, (see GlcNAc specific lectins, Section **IV-I**).

---

[6]Neu5Acα2→3 (60%)
or }Galβ1→4GlcNAcβ1→2Manα1,6↓
Neu5Acα2→6 (40%) Manβ1→4GlcNAcβ1→4GlcNAc
Neu5Acα2→6Galβ1→4GlcNAcβ1→2Manα1,3↑
Neu5Acα2→3Galβ1→4GlcNAcβ1,4↑

[7][Galα1→3] ± Galβ1→4GlcNAcβ1→2Manα1,6↓
Manβ1→4GlcNAcβ1→4GlcNAc
Neu5Acα2→6Galβ1→4GlcNAcβ1→2Manα1,3↑

Acknowledgement

1. This work was aided by grants from the Texas Agricultural Experiment Station (TAES H6194), USDA/SEA/ARS (Cooperative Agreement No. 58-6125-5-4), USDA/SEA Formula Animal Health Funds (Project 6648), and The Robert J. Kleberg and Helen C. Kleberg Foundation.

2. The authors thank Ms. Josie Arevala and Ms. Debbie Brantley for their secretarial assistance in preparing this manuscript.

REFERENCES

1. Goldstein, I.J. and Poretz, R.D. (1986) Isolation, physicochemical characterization, and carbohydrate-binding specificity of lectins. *In* "The Lectins. Properties, Functions, and Applications in Biology and Medicine." (Liener, I.E., Sharon N. and Goldstein, I.J. eds.). pp,. 33-247. Academic Press.
2. Lis, H. and Sharon, N. (1986) Lectins as Molecules and as Tools. Ann. Rev. Biochem., 55:35-67.
3. Bog-Hansen, T.C., ed. 1981-1985. Lectins: Biology, Biochemistry, Clinical Biochemistry, Vol. 1 (1981); Vol. 2 (1981); Vol. 3 (1983); Vol. 4 (1985); Vol. 5 (1987). Berlin/New York: de Gruyter.
4. Gallagher, J.T. (1984) Carbohydrate-binding properties of lectins: a possible approach to lectin nomenclature and classification. Biosci. Rep., 4:621-632.
5. Mäkelä, O. (1957) Studies *In* Hemagglutinins of Leguminosae Seeds. Ann. Med. Exp. Biol. Fenn. Suppl. 11,35:1-156.
6. Loontiens, F.G., Van Wauwe, J.P. and De Bruyne, C.K. (1975) Concanavalin A: Relation between hapten inhibition indexes and association constants for different glycosides. Carbohydr. Res. 44:150-153.
7. Springer, G.F., Desai, P.R. and Kolecki, B. (1964) Synthesis and immunochemistry of fucose methyl ethers and their methylglycosides. Biochemistry, 3:1076-1085.

7A Watkins, W.M. (1972) Blood-Group Specific Substances in Glycoproteins (Gottschalk, A. ed.) 2nd Ed. pp. 830-891. Elsevier, Amsterdam.

8. Etzler, M.E. and Kabat, E.A.. (1970) Purification and characterization of a lectin (plant hemagglutinin) with blood group A specificity from *Dolichos biflorus*. Biochemistry, 9:869-877.
9. Baker, D.A., Sugii, S., Kabat, E.A., Ratcliffe, R. M., Hermentin, P. and Lemieux, R.U. (1983) Immunochemical studies on the combining sites of Forssman hapten reactive hemagglutinins from *Dolichos biflorus*, *Helix pomatia* and *Wistaria floribunda*. Biochemistry, 22:2741-2750.
10. Hammarström, S. and Kabat, E.A. (1969) Purification and characterization of a blood-group A reactive hemagglutinin from the snail *Helix pomatia* and a study of its combining site. Biochemistry, 8:2696-2705.

10A Maliarik, M.J., Roberts, D.D. and Goldstein, I.J. (1987) Properties of the lectin from the Hog Peanut (*Amphicarpaea bracteata*). Arch. Biochem. Biophy., 255:194-200.

11. Sugii, S. and Kabat, E.A. (1980) Immunochemical specificity of the combining site of *Wistaria floribunda* hemagglutinin. Biochemistry, 19:1192-1199.
12. Murphy, L.A. and Goldstein, I.J. (1977) Five α-D-galactopyranosyl-binding isolectins from *Bandeiraea simplicifolia* seeds. J. Biol. Chem., 252:4739-4742.
13. Wood, C., Kabat, E.A., Murphy, L.A. and Goldstein, I.J. (1979) Immunochemical studies on the combining sites of the two isolectins, A_4 and B_4, isolated from *Bandeiraea simplicifolia*. Arch. Biochem. Biophys., 198:1-11.
14. Hammarström, S., Murphy, L.A., Goldstein, I.J. and Etzler, M.E. (1977) Carbohydrate binding specificity of of *N*-acetyl-D-galactosamine-"specific" lectins: *Helix pomatia* A hemagglutinin, soy bean agglutinin, lima bean lectin and *Dolichos biflorus* lectin. Biochemistry, 16:2750-2755.
15. Roberts, D.D. and Goldstein, I.J. (1984) Effect of carbohydrate and metal ion binding on the reactivity of the essential thiol groups of lima bean lectin. J. Biol. Chem., 259:903-908.
16. Sikder, S.K., Kabat, E.A., Roberts, D.D. and Goldstein, I.J. (1986) Immunochemical studies on the combining site of the blood group A-specific lima bean lectin. Carbohydr. Res. 151:247-260.
17. Pereira, M.E.A., Kabat, E.A. and Sharon, N. (1974) Immunochemical studies on the specificity of soybean agglutinin. Carbohydr. Res., 37:89-102.
18. Debray, H., Decout, D., Strecker, G., Spik, G. and Montreuil, J. (1981) Specificity of twelve lectins towards oligosaccharides and glycopeptides related to *N*-glycosylproteins. Eur. J. Biochem., 117:41-55.
19. Kaladas, P.M., Kabat, E.A., Kimura, A. and Ersson, B. (1981) The specificity of the combining site of the lectin from *Vicia villosa* seeds which reacts with cytotoxic T-lymphocytes. Mol. Immunol., 18:969-977.
20. Tollefsen, S.E. and Kornfeld, R. (1983) Isolation and characterization of lectins from *Vicia villosa*. Two distinct carbohydrate binding activities are present in seed extracts. J. Biol. Chem., 258:5166-5171.
21. Tollefsen, S.E. and Kornfeld, R. (1983) The B_4 lectin from *Vicia villosa* seeds interacts with *N*-acetyl-galactosamine residues α-linked to serine or threonine residues in cell surface glycoproteins. J. Biol. Chem., 258:5172-5176.
22 Piller, V., Piller, F. and Cartron, J.P. (1986) Isolation and characterization of an *N*-acetylgalactosamine specific lectin from *Salvia sclarea* seeds. J. Biol. Chem. 261:14069-14075.
23. Peumans, W.J., Nsimba-Lubaki, M., Peeters, B., and Broeckaert, W.F. (1985) Isolation and partial characterization of a lectin from ground elder (*Aegopodium podagraria*) rhizomes. Planta, 164:75-85.
24 Peumans, W.J., Nsimba-Lubaki, M., Carlier, A.R. and Van Driessche, E. (1984) A lectin from *Bryonia dioica* root stocks. Planta, 160:222-228.
25. Bloch, R., Jenkins, J., Roth, J., and Burger, M.M. (1976) Purification and characterization of two lectins from *Caragana arborescens* seeds. J. Biol. Chem., 251: 5929-5935.
26. Kitagaki, H., Seno, H., Yamaguchi, H. and Matsumoto, I. (1985) Isolation and characterization of a lectin from

the fruit of *Clerodendron trichotomum*. J. Biochem. (Japan), 97:791-799.

27. Horejsi, V. and Kocourek, J. (1978) Studies on lectins. XXXVI. Properties of some lectins prepared by affinity chromatography on *O*-glycosyl polyacrylamide gels. Biochim. Biophys. Acta., 538:299-315.
28. Rosen, S.D., Simpson, D.L., Rose, J.E. and Barondes, S.H. (1974) Carbohydrate-binding protein from *Polysphondylium pallidum* implicated in intercellular adhesion. Nature, 252:128, 149-151.
29. Frazier, W.A., Rosen, S.D., Reitherman, R.W. and Barondes, S.H. (1975) Purification and comparison of two developmentally regulated lectins from *Dictyostelium discoideum*. Discoidin I and II. J. Biol. Chem., 250:7714-7721.
30. Cammue, B.R., Peeters, B. and Peumans, W.J. (1985) Isolation and partial characterization of an *N*-acetylgalactosamine-specific lectin from winter-aconite (*Eranthis hyemalis*) root tubers. Biochem. J. 227:949-955.
31. Nsimba-Lubaki, M., Peumans, W.J. and Carlier, A.R. (1983) Isolation and partial characterization of a lectin from *Euphorbia heterophylla* seeds. Biochem. J., 215:141-145.
32. Falasca, A., Franceschi, C., Rossi, C.A. and Stripe, F. (1980) Mitogenic and haemagglutinating properties of a lectin purified from *Hura crepitans* seeds. Biochim. Biophys. Acta., 632:95-105.
33. Haylett, T. and Swart, L.S. (1982) Isolation and characterization of an anti-A1 lectin from *Macrotyloma axillare*. S. Afr. J. Chem., 35:33-36.
34. Mazumder, T., Gaur, N. and Surolia, A. (1981) The physicochemical properties of the galactose-specific lectin from *Momordica charantia*. Eur. J. Biochem. 113:463-470.
35. Horejsi, V., Tichá, M., Novotny, J. and Kocourek, J. (1980) Studies on lectins XLVII. Some properties of D-galactose binding lectins isolated from the seeds of *Butea frondosa*, *Erythrina indica* and *Momordica charantia*. Biochim. Biophys. Acta., 623:439-448.
36. Horejsi, V., Chaloupecka, O. and Kocourek, J. (1979) Studies on lectins. XLII. Isolation and characterization of the lectin from restharrow roots (*Ononis hircina* Jacq.). Biochim. Biophys. Acta., 538:287-293.
37. Appukutan, P.S. and Basu, D. (1981) Isolation of an *N*-acetyl-D-galactosamine-binding protein from winged bean (*Psophocarpus tetragonolobus*). Anal. Biochem. 113:253-255.
38. Kortt, A.A. (1984) Purification and properties of the basic lectins from winged bean seed [*Psophocarpus tetragonolobus* (L.) DC]. Eur. J. Biochem., 138:519-525.
39. Kortt, A.A. (1985) Characterization of the acidic lectins from winged bean seed [*Psophocarpus tetragonolobus* (L.) DC]. Arch. Biochem. Biophys. 236:544-554.
40. Horejsi, V., Haskovec, C. and Kocourek, J. (1978) Studies on lectins. XXXVIII. Isolation and characterization of the lectin from black locust bark (*Robinia pseudacacia*, L). Biochim. Biophys. Acta., 532:98-104.
41. Baldo, B.A., Sawyer, W.H., Stick, R.V. and Uhlenbruck, G. (1978) Purification and characterization of a galactan-reactive agglutinin from the clam *Tridacna*

maxima (Röding) and a study of its combining site. Biochem. J., 175:467-477.
42. Rüdiger, H. (1977) Purification and properties of blood-group-specific lectins from *Vicia cracca*. Eur. J. Biochem., 72:317-322.
43. Franz, H., Ziska, P. and Kindt, A. (1981) Isolation and properties of three lectins from mistletoe (*Viscum album* L) Biochem. J., 195:481-484.
44. Pereira, M.E.A., Kabat, E.A., Lotan, R. and Sharon, N. (1976) Immunochemical studies on the specificity of the peanut (*Arachis hypogaea*) agglutinin. Carbohydr. Res., 51:107-118.
45. Lotan, R. and Sharon, N. (1978) Peanut (*Arachis hypogaea*) agglutinin. Method. Enzymol., 50:361-367.
46. Osawa, T., Irimura, T. and Kawaguchi, T. (1978) *Bauhinia purpurea* agglutinin. Method. Enzymol., 50:367-372.
47. Wu, A.M., Kabat, E.A., Gruezo, F.G. and Allen, H.J. (1980) Immunochemical studies on the combining site of the D-galactopyranose and 2-acetamido-2-deoxy-D-galactopyranose specific lectin isolated from *Bauhinia purpurea alba* seeds. Arch. Biochem. Biophys. 204:622-639.
48. Bausch, J.N. and Poretz, R.D. (1977) Purification and properties of the hemagglutinin from *Maclura pomifera* seeds. Biochemistry, 16:5790-5794.
49. Sarkar, M., Wu, A.M. and Kabat, E.A. (1981) Immunochemical studies on the carbohydrate specificity of *Maclura pomifera* lectin. Arch. Biochem. Biophys., 209:204-218.
50. Poretz, R.D., Riss, H., Timberlake, J.W. and Chien, S-M. (1974) Purification and properties of the hemagglutinin from *Sophora japonica* seeds. Biochemistry 13:250-256.
51. Wu, A.M., Kabat, E.A., Gruezo, F.G. and Poretz, R.D. (1981) Immunochemical studies on the reactivities and combining sites of the D-galactopyranose- and 2-acetamido-2-deoxy-D-galactopyranose specific lectin purified from *Sophora japonica* seeds. Arch. Biochem. Biophys., 209:191-203.
52. Baenziger, J.U. and Fiete, D. (1979) Structural determinants of *Ricinus communis* agglutinin and toxin specificity for oligosaccharides. J. Biol. Chem., 254: 9795-9799.
53. Nicolson, G.L., Blaustein, J. and Etzler, M.E. (1974) Characterization of two plant lectins from *Ricinus communis* and their quantitative interaction with a murine lymphoma. Biochemistry, 13:196-204.
54. Uhlenbruck, G. and Dahr, W. (1971) Studies on lectins with a broad agglutination spectrum. XII. *N*-acetyl-D-galactosamine specific lectins from the seeds of *Soja hispida*, *Bauhinia purpurea*, *Iberis amara*, *Moluccella laevis* and *Vicia graminea*. Vox Sang., 21:338-351.
55. Glöckner, W.M., Newman, R.A. and Uhlenbruck, G. (1975) Carbohydrate structure and serological behavior of "antifreeze" glycoproteins from an antarctic fish. Biochem. Biophys. Res. Comm., 66:701-705.
56. Duk, M. and Lisowska, E. (1981) *Vicia graminea* anti-N lectin: partial characterization of the purified and its binding to erythrocytes. Eur. J. Biochem., 118: 131-136.
57. Duk, M., Lisowska, E., Kordowicz, M. and Warniowska, K. (1982) Studies on the specificity of the binding site

of *Vicia graminea* anti-N lectin. Eur. J. Biochem., 123:105-112.

58. Presant, C.A. and Kornfeld, S. (1972) Characterization of the cell surface receptor for the *Agaricus bisporus* hemagglutinin. J. Biol. Chem., 247:6937-6945.
59. Sueyoshi, S., Tsuji, T. and Osawa, T. (1985) Purification and characterization of four isolectins of mushroom (*Agaricus bisporus*). Biol. Chem. Hoppe-Seyler, 366:213-221.
60. Cammue, B., Stinissen, H.M. and Peumans, W.J. (1985) A new type of cereal lectin from leaves of couch grass (*Agropyrum repens*). Eur. J. Biochem., 148:315-322.
61. Sastry, M.V.K., Banarjee, P., Patanjali, S.R., Swamy, M.J., Swarnalatha, G.V. and Surolia, A. (1986) Analysis of saccharide binding to *Artocarpus integrifolia* lectin reveals specific recognition of T-antigen (β-D-Gal(1→3)-D-GalNAc). J. Biol. Chem., 261: 11726-11733.

61A Chatterjee, B.P., Chowdhury, S., and Ahmed, H. Further Characterization of *Artocarpus lakoocha* Lectin Purified by Rivanol. *In:* Abstract (B-8) 4th European Carbohydrate Symposium July 12-17, 1987, Darmstadt. FRG.

62. Kaifu, R. and Osawa, T. (1976) Synthesis of *O*-β-D-galactopyranosyl-(1→4)-*O*-(2-acetamido-2-deoxy-β-D-glucopyranosyl)-(1→2)-D-mannose and its interaction with various lectins. Carbohydr. Res., 52:179-185.
63. Wu, A.M., Sugii, S., Gruezo, F.G and Kabat, E.A. (1987) Immunochemical studies on the Galβ1→4 or 3GlcNAc specificity of *Ricinus communis* agglutinin. Carbohydr. Res., in press.
64. Desai, N.N., Allen, A.K. and Neuberger, A. (1981) Some properties of the lectin from *Datura stramonium* (thorn-apple) and the nature of its glycoprotein linkage. Biochem. J., 197:345-353.
65. Crowley, J.F., Goldstein, I.J., Arnarp, J. and Lonngren, J. (1984) Carbohydrate binding studies on the lectin from *Datura stramonium* seeds. Arch. Biochem. Biophys., 231:524-533.
66. Yamashita, K., Totani, K., Ohkura, T., Takasaki, S., Goldstein, I.J. and Kobata, A. (1987) Carbohydrate binding properties of complex-type oligosaccharides on immobilized *Datura stramonium* lectin. J. Biol. Chem. 262:1602-1607.
67. Cummings, R.D., and Kornfeld, S. (1984). The distribution of repeating [Galβ1,4GlcNAcβ1,3] sequences in asparagine-linked oligosaccharides of the mouse lymphoma cell lines BW5147 and PHAR2.1. Finding of oligosaccharides containing these sequences to immobilized *Datura stramonium* agglutinin. J. Biol. Chem. 259:6253-6260.
68. Kaladas, P.M., Kabat, E.A., Iglesias, J.L., Lis, H. and Sharon, N. (1982) Immunochemical studies on the combining site of the D-galactose/*N*-acetyl-D-galactosamine specific lectin from *Erythrina cristagalli* seeds. Arch. Biochem. Biophys., 217:624-637.
69. Bretting, H., Phillips, S.G., Klumpart, H.J. and Kabat, E.A. (1981) A mitogenic lactose-binding lectin from the sponge *Geodia cydonium*. J. Immunol., 127:1652-1658.
70. Lis, H., Joubert, F.J. and Sharon, N. (1985) Isolation and properties of *N*-acetyllactosamine-specific lectins

from nine *Erythrina* species. Phytochemistry. 24:2803-2809.

71. Cummings, R.D. and Kornfeld, S. (1982) Characterization of the structural determinants required for the high affinity interaction of asparagine-linked oligosaccharides with immobilized *Phaseolus vulgaris* leukoagglutinating and erythroagglutinating lectins. J. Biol. Chem., 257:11230-11234.
72. Irimura, T., Tsuji, T., Tagami, S., Yamamoto, K. and Osawa, T. (1981) Structure of a complex-type sugar chain of Human Glycophorin A. Biochemistry, 20:560-566.
73. Yamashita, K., Hitoi, A. and Kobata, A. (1983) Structural determinants of *Phaseolus vulgaris* erythroagglutinating lectin for oligosaccharides. J. Biol. Chem., 258:14753-14755.
74. Irimura, T. and Nicolson, G.L. (1983) Interaction of Pokeweed mitogen with poly(*N*-acetyllactosamine)-type carbohydrate chains. Carbohydr. Res., 120:187-195.
75. Yokoyama, K., Terao, T. and Osawa, T. (1978) Carbohydrate-binding specificity of pokeweed mitogens. Biochim. Biophys. Acta., 538:384-396.
76. Katagiri, Y., Yamamoto, K., Tsuji, T. and Osawa, T. (1983) Structural requirements for the binding of high-mannose-type glycopeptide to immobilized pokeweed Pa-2 lectin. Carbohydr. Res., 120:283-292.
77. Hammarström, S., Hammarström, M.L., Sundblad, G., Arnarp, J. and Lönngren, J. (1982) Mitogenic leukoagglutinin from *Phaseolus vulgaris* binds to a pentasaccharide unit in *N*-acetyllactosamine-type glycoprotein glycans. Proc. Natl. Acad. Sci., 79:1611-1615.
78. Cammue, B.P., Peeters, B. and Peumans, W.J. (1985) Isolation and partial characterization of an *N*-acetylgalactosamine-specific lectin from winter-aconite (*Eranthis hyemalis*) root tubers. Biochem. J. 227:949-955.
79. Broekaert, W.F., Nsimba-Lubaki, M., Peeters, B. and Peumans, W.J. (1984) A lectin from elder (*Sambucus nigra* L.) bark. Biochem. J., 221:163-169.
80. Shibuya, N., Goldstein, I.J., Broekaert, W.F., Nsimba-Lubaki, M., Peeters, B. and Peumans, W.J. (1987) The elderberry (*Sambucus nigra* L.) bark lectin recognizes the Neu5Ac(α2→6)Gal/GalNAc sequence. J.Biol. Chem., 262:1596-1601.
81. Bretting, H. and Kabat, E.A. (1976) Purification and characterization of the agglutinins from the sponge *Axinella polypoides* and a study of their combining sites. Biochemistry, 15:3228-3236.
82. Cumsky, M. and Zusman, D.R. (1979) Myxobacterial hemagglutinin: a development-specific lectin of *Myxococcus xanthus*. Proc. Natl. Acad. Sci., 76:5505-5509.
83. Olsnes, S., Saltvedt, E. and Pihl, A. (1974) Isolation and comparison of galactose-binding lectins from *Abrus precatorius* and *Ricinus communis*. J. Biol. Chem., 249: 803-810.
84. Olsnes, S. (1978) Toxic and nontoxic lectins from *Abrus precatorius*. Method. Enzymol., 50:323-330.
85. Gilboa-Garber, N., Susswein, A.J., Mizrahi, L. and Avichezer, D. (1985) Purification and characterization of the gonad lectin of *Aplysia depilans*. FEBS Lett., 181:267-270.

86. Parrinello, N. and Canicatti, C. (1982) Carbohydrate binding specificity and purification by biospecific affinity chromatography of *Ascidia malaca* traust hemagglutinins. Dev. Comp. Immunol., 6:53-64.
87. Krajhanzl, A., Horejsi, V. and Kocourek, J. (1978) Studies on lectins. XLII. Isolation, partial characterization and comparison of lectins from the roe of five fish species.Biochim. Biophys. Acta.532:215-224.
88. Flower, R.L.P., Wilcox, G.E., Chugg, V. and Neal, J.R. (1984) Lectins from indigenous Australian wildflowers - detection of lectins from *Bauhinia carronii*, *Hardenbergia comptoniana*, *Ptilotis obovatus* and *Rhadogia crassifolia*. Aust. J. Exp. Med. Sci., 62:763-769.
89. Ersson, B., Aspberg, K. and Porath, J. (1973) The phytohemagglutinin from sunn hemp seeds (*Crotalaria juncea*). Purification by biospecific affinity chromatography. Biochim. Biophys. Acta., 310:446-452.
90. Konami, Y., Yamamoto, K., Tsuji, T., Matsumoto, I. and Osawa,T. (1983) Purification and characterization of two types of *Cytisus multiflorus* hemagglutinin by affinity chromatography. J. Pharmacobiodyn., 6:737-747.
91. Young, N.M., Watson, D.C. and Williams, R.E. (1984) Structural differences between two lectins from *Cytisus scoparius*, both specific for D-galactose and *N*-acetyl-D-galactosamine. Biochem. J., 222:41-48.
92. Vasta, G.R. and Marchalonis, J.J. (1986) Galactosyl-binding lectins from the tunicate *Didemnum candidum*. Carbohydrate specificity and characterization of the combining site. J. Biol. Chem., 261:9182-9186.
93. Bhattacharyya, L., Das, P.K. and Sen, A. (1981) Purification and properties of D-galactose-binding lectins from some Erythrina species: comparison of properties of lectins from *E. indica*, *E. arborescens*, *E. suberosa* and *E. lithosperma*. Arch. Biochem. Biophys., 211:459-470.
94. Barbieri, L., Falasca, A., Franceschi, C., Licastro, F., Rossi, C.A. and Stirpe, F. (1983) Purification and properties of two lectins of the latex of the euphorbiaceous plants *Hura crepitans* L. (sand-box tree) and *Euphorbia characias* L. (Mediterranean spurge) Biochem. J., 215:433-439.
95. Yokosawa, H., Sawada, H., Abe, Y., Numakunai, T. and Ishii, S-I. (1982) Galactose-specific lectin in the hemolymph of solitary ascidian, *Halocynthia roretzi*: isolation and characterization. Biochem. Biophys. Res. Comm., 107:451-457.
96. Guillot, J., Genaud, L., Gueugnot, J. and Damez, M. (1983) Purification and properties of two hemagglutinins of the mushroom *Laccaria amethystina*. Biochemistry, 22:5365-5369.
97. Rögener, W., Renwrantz, L. and Uhlenbruck, G. (1985) Isolation and characterization of a lectin from the hemolymph of the cephalopod *Octopus vulgaris* (Lam.) inhibited by alpha-D-lactose and *N*-acetyl-lactosamine. Dev. Comp. Immunol., 9:605-616.
98. Hankins, C.N. and Shannon, L.M. (1978) The physical and enzymatic properties of a phytohemagglutinin from mung beans. J. Biol. Chem., 253:7791-7797.
99. Rosen, S.D., Kaur, J., Clark, D.L., Pardos, B.T. and Frazier, W.A. (1979) Purification and characterization of multiple species (isolectins) of a slime mold lectin

implicated in intercellular adhesion. J. Biol. Chem., 254:9408-9415.
100. Yeung, H.W., Ng, T.B., Wong, D.M. and Li, W.W. (1986) Chemical and biological characterization of the galactose binding lectins from *Trichosanthes kirilowii* root tubers. Int. J. Pep. Protein. Res., 27:208-220.
101. Ziska, P. and Franz, H. (1981) Studies on the interaction of the mistletoe lectin I with carbohydrates. Experientia., 37:219.
102. Roberson, B.J. and Strength, D.R. (1983) Characterization of a lectin from cowpeas. Prep. Biochem., 13:45-56.
103. Goldstein, I.J., Hollerman, C.E. and Smith, E.K. (1965) Protein-carbohydrate interaction. II. Inhibition studies on the interaction of concanavalin A with polysaccharides. Biochemistry, 4:876-883.
104. Baenziger, J.U. and Fiete, D. (1979) Structural determinants of concanavalin A specificity for oligosaccharides. J. Biol. Chem., 254:2400-2407.
105. Allen, A.K., Desai, N.N. and Neuberger, A. (1976) The purification of the glycoprotein lectin from the broad bean (*Vicia faba*) and a comparison of its properties with lectins of similar specificity. Biochem. J., 155:127-135.
106. Hardman, K.D., and Goldstein, I.J. (1977) *in* The Structure and Activity of Concanavalin A. Immunochemistry of proteins. (Atassi, M.Z., ed.). 2. pp. 373-416. Plenum Press, New York.
107. Kornfeld, K., Reitman, M.L. and Kornfeld, R. (1981) The carbohydrate-binding specificity of pea and lentil lectins. Fucose is an important determinant. J. Biol. Chem., 256:6633-6640.
108. Yamamoto, K., Tsuji, T. and Osawa, T. (1982) Requirement of the core structure of a complex-type glycopeptide for the binding to immobilized lentil- and pea-lectins. Carbohydr. Res., 110:283-289.
109. Matsumoto, I., Uehara, Y., Jimbo, A. and Seno, N. (1983) Immunochemical and spectral studies on *Vicia faba* agglutinin. J. Biochem., 93:763-769.
110. Katagiri, Y., Yamamoto, K., Tsuji, T. and Osawa, T.. (1984) Structural requirements for the binding of glycopeptides to immobilized *Vicia faba* (fava) lectin. Carbohydr. Res., 129:257-265.
111. Moreira, R.A., Barros, A.C.H., Stewart, J.C. and Pusztai, A. (1983) Isolation and characterization of a lectin from the seeds of *Dioclea grandiflora* (Mart). Planta, 158:63-69.
112. Kolberg, J. and Sletten, K. (1982) Purification and properties of a mitogenic lectin from *Lathyrus sativus* seeds. Biochim. Biophys. Acta, 704:26-30.
113. Tichá, M., Zeineddine, I. and Kocourek, J. (1980) Studies on lectins XLXII. Isolation and characterization of lectins from the seeds of *Lathyrus odoratus* L. and *Lathyrus silvestris* L. Acta. Biol. Med. Ger., 39:649-655.
114. Gupta, B.K.D., Chatterjee-Ghose, R. and Sen, A. (1980) Purification and properties of mitogenic lectins from seeds of *Lathyrus sativus linn* (chickling vetch). Arch. Biochem. Biophys., 201:137-146.

115. Rougé, P. and Chabert, P. (1983) Purification and properties of a lectin from *Lathyrus tingitanus* seeds. FEBS Lett., 157:257-260.
116. Hapner, K.D. and Robbins, J.E. (1979) Isolation and properties of a lectin from sainfoin (*Onobrychis vicifolia,* scop). Biochim. Biophys. Acta., 580:186-197.
117. Baumann, C., Strosberg, A.D. and Rüdiger, H. (1982) Purification and characterization of a mannose/glucose-specific lectin from *Vicia cracca.* Eur. J. Biochem., 122:105-110.
118. Gebauer, G., Schiltz, E., Schimpl, A. and Rüdiger, H. (1979) Purification and characterization of a mitogenic lectin and a lectin-binding protein from *Vicia sativa.* Hoppe-Seylers Z. Physiol. Chem., 360:1727-1735.
119. Bretting, H., Kabat, E.A., Liao, J. and Pereira, M.E.A.. (1976) Purification and characterization of the agglutinins from the sponge *Aaptos papillata* and a study of their combining sites. Biochemistry, 15:5029-5038.
120. Read, S.M. and Northcote, D.H.. (1983) Subunit structure and interactions of the phloem proteins of *Cucurbita maxima* (pumpkin). Eur. J. Biochem., 134:561-569.
121. Allen, A.K. (1979) A lectin from the exudate of the fruit of the vegetable marrow (*Cucurbita pepo*) that has a specificity for β-1,4-linked *N*-acetylglucosamine oligosaccharides. Biochem. J., 183:133-137.
122. Iyer, P.N.S., Wilkinson, K.D. and Goldstein, I.J. (1976) An *N*-Acetyl-D-Glucosamine binding lectin from *Bandeiraea simplicifolia* seeds. Arch. Biochem. Biophys. 177:330-333.
123. Ebisu, S., Iyer, P.N.S. and Goldstein, I.J. (1978) Equilibrium dialysis and carbohydrate-binding studies on the 2-acetamido-2-deoxy-D-glucopyranosyl-binding lectin from *Bandeiraea simplicifolia* seeds. Carbohydr. Res. 61:129-138.
124. Wood, C., Kabat, E.A., Ebisu, S. and Goldstein, I.J. (1978) An immunochemical study on the combining sites of the second lectin isolated from *Bandeiraea simplicifolia* (BS-II). Ann. Immunol., 129C:143-158.
125. Nachbar, M.S., Oppenheim, J.D. and Thomas, J.O. (1980) Lectins in the U.S. diet. Isolation and characterization of a lectin from the tomato (*Lycopersicon esculentum*). J. Biol. Chem., 255:2056-2061.
126. Kilpatrick, D.C. (1980) Purification and some properties of a lectin from the fruit juice of the tomato (*Lycopersicon esculentum*). Biochem J., 185:269-272.
127. Anantharam, V., Patanjali, S.R., Swamy, M.J., Sanadi, A.R, Goldstein, I.J. and Surolia, A. (1986) Isolation, macromolecular properties and combining site of a chitooligosaccharide-specific lectin from the exudate of ridge gourd (*Luffa acutangula*). J. Biol. Chem., 261:14621-14627.
128. Allen, A.K. and Neuberger, A. (1973) The purification and properties of the lectin from potato tubers, a hydroxyproline-containing glycoprotein. Biochem. J. 135:307-314.
129. Shibuya, N., Goldstein, I.J., Shafer, J.A., Peumans, W.J. and Broekaert, W.F. (1986) Carbohydrate binding properties of the stinging nettle (*Urtica dioica*) rhizome lectin. Arch. Biochem. Biophys., 249:215-224.
130. Goldstein, I.J., Hammarström, S., and Sundblad, G. (1975) Precipitation and carbohydrate binding spec-

ificity studies on wheat germ agglutinin. Biochim. Biophys. Acta. 404:63-67.
130A Allen, A.K., Neuberger, A. and Sharon, N. (1973) The purification, composition and specificity of wheat-germ agglutinin. Biochem. J., 131:155-162.
131. Gallagher, J.T., Morris, A. and Dexter, T.M.. (1985) Identification of two binding sites for wheat-germ agglutinin on polylactosamine-type oligosaccharides. Biochem. J., 231:115-122.
132. Osawa, T. (1966) Inhibition tests of plant agglutinins from *Laburnum alpinum* and *Cytisus sessifolius* with various sugars and sugar derivatives. Indication of the type of linkage in certain disaccharides. Biochim. Biophys. Acta., 115:507-510.
133. Matsumoto, I. and Osawa, T. (1971) On the specificity of various heterologous anti-H hemagglutinins. Vox Sang., 21:548-557.
134. Peumans, W.J., Spaepen, C., Stinissen, H.M. and Carlier, A.R. (1982) Isolation and partial characterization of a lectin from a false brome grass (*Brachypodium sylvaticum*). Biochem. J., 205:635-638.
135. Güran, A., Tichá, M.,. Filka, K. and Kocourek, J. (1983) Isolation and properties of a lectin from the seeds of the Indian bean or lablab (*Dolichos lablab* L.). Biochem. J., 209:653-657.
136. Peumans, W.J., Stinissen, H.M. and Carlier, A.R. (1982) Isolation and partial characterization of wheat-germ-agglutinin-like lectins from rye (*Secale cereale*) and barley (*Hordeum vulgare*) embryo. Biochem. J., 203:239-243.
137. Tsuda, M. (1979) Purification and characterization of a lectin from rice bran. J. Biochem., 86:1451-1461.
138. Peumans, W.J. and Stinissen, H.M. (1982) Rice lectin: purification ,properties, molecular structure and relationship to cereal lectins. Arch. Int. Physiol. Biochem., 90:B210-B211.
139. Kubanek, J., Entlicher, G. and Kocourek, J. (1982) Studies on lectins. LII. Isolation and characterization of the lectins from rye germ (*Secale cerale* L). Acta. Biol. Med. Ger., 41:771-780.
140. Tatsumi, M., Hosokawa, N., Arai, Y. and Itoh, T. (1982) Purification of lectin from some shellfish and ascidiacea. Carbohydr. Res., 108:148-152.
141. Tatsumi, M., Arai, Y. and Itoh, T. (1982) Purification and characterization of a lectin from the shell-fish *Saxidomus purpuratus*. J. Biochem., 91:1139-1146.
142. Kochibe, N. and Furukawa, K. (1980) Purification and properties of a novel fucose-specific hemagglutinin of *Aleuria aurantia*. Biochemistry, 19:2841-2846.
143. Yamashita, K., Kochibe, N., Ohkura, T., Ueda, I. and Kobata, A. (1985) Fractionation of L-fucose-containing oligosaccharides on immobilized *Aleuria aurantia* lectin. J. Biol. Chem., 260:4688-4693.
144. Petryniak, J., Pereira, M.E.A. and Kabat, E.A. (1977) The lectin of *Euonymus europeus*: purification, characterization and an immunochemical study of its combining site. Arch. Biochem. Biophys., 178:118-134.
145. Petryniak, J. and Goldstein, I.J. (1986) Immunochemical studies on the interaction between synthetic glycoconjugates and α-L-fucosyl binding lectins. Biochemistry, 25:2829-2838.

146. Pereira, M.E.A. and Kabat, E.A. (1974) Specificity of purified hemagglutinin (lectin) from *Lotus tetragonolobus*. Biochemistry, 13:3184-3192.
147. Matsumoto, I. and Osawa, T. (1969) Purification and characterization of an anti-H(O) phytohemagglutinin of *Ulex europeus*. Biochim. Biophys. Acta., 194:180-189.
148. Pereira, M.E.A., Kisailus, E.C., Gruezo, F. and Kabat, E.A. (1978) Immunochemical studies on the combining site of the blood group H-specific lectin I from *Ulex europeus* seeds. Arch. Biochem. Biophys., 185:108-115.
149. Matsumoto, I. and Osawa, T. (1970) Purification and characterization of a cytisus-type anti-H(O) phytohemagglutinin from *Ulex europeus* seeds. Arch. Biochem. Biophys., 140:484-491.
150. Pereira, M.E.A., Gruezo, G.F. and Kabat, E.A. (1979) Purification and characterization of lectin II from *Ulex europeus* seeds and an immunochemical study of its combining site. Arch. Biochem. Biophys., 194:511-525.
151. Kaladas, P.M., Kabat, E.A., Shibata, S. and Goldstein, I.J. (1983) Immunochemical studies on the binding specificity of the blood group Le^b specific lectin *Griffonia simplicifolia* IV. Arch. Biochem. Biophys., 223:309-318.
152. Rege, Y.P., Painter, T.J., Watkins, W.M. and Morgan, W.T.J. (1964) Isolation of serologically active fucose-containing oligosaccharides from human blood-group H substance. Nature, 203:360-363.
153. Mohan, S., Dorai, D.T., Srimal, S. and Bachhawat, B.K. (1982) Binding studies of a sialic acid-specific lectin from horseshoe crab *Carcinoscorpius rotunda cauda* with various sialoglycoproteins. Biochem. J., 203:253-261.
154. Vasta, G.R., Ilodi, G.H.U., Cohen, E. and Brahmi, Z. (1982) A comparative study on the specificity of *Androctonus australis* (Saharan scorpion) and *Limulus polyphemus* (horseshoe crab) agglutinins. Dev. Comp. Immunol., 6:625-634.
155. Yaakobovich, Y. and Neeman, I. (1983) Partial isolation and characterization of a hemagglutinating factor from avocado seed. Arch. Toxicol., Suppl. 6:52-57.
156. Bretting, H., Stanislawski, E., Jacob, G. and Becker, W. (1983) Isolation and characterization of a lectin from the snail. *Biomphalaria glabrata* and a study of its combining site. Biochim. Biophys. Acta., 749:143-152.
157. Vasta, G.R. and Cohen, E. (1984) Carbohydrate specificities of *Birgus latro* (coconut crab) serum lectins. Dev. Comp. Immunol., 8:197-202.
158. Hall, J.L. and Rowlands, D.T., Jr. (1974) Heterogeneity of lobster agglutinins. II. Specificity of agglutinin-erythrocyte binding. Biochemistry, 13:828-832.
159. Miller, R.L., Collawn, J.F., Jr. and Fish, W.W. (1982) Purification and macromolecular properties of a sialic acid-specific lectin from the slug *Limax flavus*. J. Biol. Chem., 257:7574-7580.
160. Nowak, T.P. and Barondes, S.H. (1975) Agglutinin from *Limulus polyphemus*. Purification with formalinized horse erythrocytes as the affinity adsorbent. Biochim. Biophys. Acta., 393:115-123.
161. Brandin, E.R. and Pistole, T.G. (1983) Polyphemin: a teichoic acid-binding lectin from the horseshoe crab, *Limulus polyphemus*. Biochem. Biophys. Res. Comm., 113:611-617.

CONTRIBUTORS

Domenico Acquotti (II-7), Study Center for the Functional Biochemistry of Brain Lipids, Department of Biological Chemistry, The Medical School, University of Milan, Milan, Italy.

L. Garry Adams (II-10, and IV-6 Abstract), Department of Veterinary Pathology, Texas A&M University, College Station, Texas, USA.

Byron Anderson (III-1), Department of Molecular Biology, Northwestern University Medical School, Chicago, Illinois, USA.

Hägen Bretting (II-15 Abstract), Zoologisches Institut and Zoologisches Museum, Hamburg, Federal Republic of Germany.

Klaus Bock (I-7), Department of Organic Chemistry, The Technical University of Denmark, Lyngby, Denmark.

Carol Borelli (II-6), Department of Biochemistry, New York Medical College, Vahalla, New York, USA.

Reinhard Brossmer (I-6), Department of Biochemistry 2, University of Heidelberg, Heidelberg, West Germany.

M. Alan Chester (I-3), Department of Clinical Chemistry, University of Lund, Lund, Sweden.

Karen F. Cleary (III-3), Department of Pathology, The University of Texas M.D. Anderson Hospital and Tumor Institute, Houston, Texas, USA.

Richard P. Crawford (IV-6), Veterinary Public Health, Texas A&M University, college Station, Texas, USA.

Lyman E. Davis (III-1), Department of Molecular Biology, Northwestern University Medical School, Chicago, Illinois, USA.

Peter C. Demou (IV-2), Departments of Neurology (School of Medicine) and Chemistry, Yale University, New Haven, Connecticut, USA.

Ten Feizi (II-2), Applied Immunochemistry Research Group, Clinical Research Centre, Harrow, Middlesex, United Kingdom.

Giuliano Gazzotti (II-7), Study Center for the Functional Biochemistry of Brain Lipids, Department of Biological Chemistry, The Medical School, University of Milan, Milan, Italy.

Riccardo Ghidoni (II-7), Study Center for the Functional Biochemistry of Brain Lipids, Department of Biological Chemistry, The Medical School, University of Milan, Milan, Italy.

Tian-Jue Gu (III-4), Department of Biochemistry, Shangai First Medical College, Shangai, People's Republic of China.

Gunnar C. Hansson (II-8, III-2), Department of Medical Biochemistry, University of Göthenburg, Göthenburg, Sweden.

Fred C. Heck (IV-6), Veterinary Microbiology and Parasitology Texas A&M University, College Station, Texas, USA.

Anthony Herp (II-6), Department of Biochemistry, New York Medical College, Vahalla, New York, USA.

Yoshio Hirabayashi (III-6), Department of Biochemistry, Shizuoka College of Pharmacy, Shizuoka-shi, Japan.

Tatsuro Irimura (III-3), Department of Tumor Biology, Surgery, The University of Texas M.D. Anderson Hospital and Tumor Institute, Houston, Texas, USA.

Gunter Jacobs (II-15), Zoologisches Institut and Zoologisches Museum, Hamburg, Federal Republic of Germany.

H.J. Jennings(II-9), Division of Biological Sciences, National Research Council of Canada, Ottawa, Ontario, Canada.

Elvin A. Kabat (I-1), Departments of Microbiology, Human Genetics and Development, College of Physicians and Surgeons, Columbia University, New York, New York, USA.

Karl-Anders Karlsson (I-7), Department of Medical Biochemistry, University of Göteborg, Göteborg, Sweden.

Shiro Kato (III-6), Department of Pathology, Research Institute for Microbial Disease, Osaka University, Osaka, Japan.

Ulrike Knels (II-15), Zoologisches Institut and Zoologisches Museum, Hamburg, Federal Republic of Germany.

Theodore A. W. Koerner (IV-2), Departments of Pathology College of Medicine, Iowa city, University of Iowa, Iowa, USA.

Roger A. Laine (II-3), Department of Biochemistry, Louisiana State University and the Louisiana State University Agricultural Center, Baton Rouge, Louisiana, USA.

Y.C. Lee (I-5), Biology Department, The Johns Hopkins University, Baltimore, Maryland, USA.

Y.-T. Li (IV-4), Department of Biochemistry, Tulane University School of Medicine, New Orleans, Louisiana, USA.

S.-C. Li (IV-4),Department of Biochemistry, Tulane University School of Medicine, New Orleans, Louisiana, USA.

Mats Lindahl (I-6), Department of Veterinary Microbiology, Swedish University of Agricultural Sciences, Uppsala, Sweden.

Elwira Lisowska (II-1), Department of Immunochemistry, Institute of Immunolgy and Experimental Therapy, Polish Academy of Sciences, Wroclaw, Poland.

Arne Lundblad (I-3), Department of Clinical Chemistry, University of Lund, Lund, Sweden.

Neil E. Mackenzie (II-10), Department of Pharmaceutic Science, College of Pharmacy, University of Arizona, Tucson, Arizona, USA.

John L. Magnani (III-5), Laboratory of Structural Biology, NIDDK, National Institutes of Health, Bethesda, Maryland, USA.

Makoto Matsumoto(III-6), Department of Biochemistry, Shizuoka College of Pharmacy, Shizuoka-shi, Japan.

Hubert Mayer (II-11), Max-Planck-Institut für Immunbiologie, Freiburg, Federal Republic of Germany.

M. Moreau (II-13), Division of Biological Sciences, National Research Council of Canada, Ottawa, Canada.

Motowo Nakajima (III-3), The University of Texas M.D. Anderson Hospital and Tumor Institute, Houston, Texas, USA.

Garth L. Nicolson (III-3), Department of Tumor Biology, The University of Texas M.D. Anderson Hospital and Tumor Institute, Houston, Texas, USA.

Bo Nilsson (IV-5), BioCarb AB, S-223 70 Lund, Sweden.

Toshiaki Osawa (I-4), Division of Chemical Toxicology and Immunochemistry, Faculty of Pharmaceutical Sciences, University of Toyko, Toyko, Japan.

David M. Ota, (III-3), Departments of Tumor Biology and Surgery, The University of Texas M.D. Anderson Hospital and Tumor Institute, Houston, Texas, USA.

Malcolm B. Perry (II-12, II-13, II-14), Division of Biological Sciences, National Research Council of Canada, Ottawa, Canada.

James H. Prestegard (IV-2), Departments of Neurology (School of Medicine) and Chemistry, Yale University, New Haven, Connecticut, USA.

Roberta Pugh(II-10), Department of Veterinary Microbiology and Parasitology, Texas A&M University, College Station, Texas, USA.

Joanna Radziejewska-Lebrecht (II-11), Institute of Microbiology, University of Lódz, Poland.

James C. Richards (II-12, II-13, II-14), Division of Biological Sciences, National Research Council of Canada, Ottawa, Canada.

Vernon N. Reinhold (IV-1), Department of Nutrition, Harvard School of Public Health, Boston, Massachusetts, USA.

Jeffrey S. Rush(II-3), Department of Biochemistry, University of Kentucky-Medical, Lexington, Kentucky, USA.

Sitthivet Santikarn (IV-1), Department of Nutrition, Harvard School of Public Health, Boston, Massachusetts, USA.

J. Neel Scarsdale (IV-2), Departments of Neurology (School of Medicine) and Chemistry, Yale University, New Haven, Connecticut, USA.

Roland Schauer (I-2), Biochemisches Institut, Christian-Albrechts-Universität, Kiel, Federal Republic of Germany.

Stefan Schramek (II-11), Institute of Virology, Slovak Academy of Sciences, Bratislava, CSSR.

Zhao-Wen Shen (I-8), Shanghai Institute of Biochemistry, Chinese Academy of Sciences, Shanghai, People's Republic of China.

Sandro Sonnino (II-7), Study Center for the Functional Biochemistry of Brain Lipids, Department of Biological Chemistry, The Medical School, University of Milan, Milan, Italy.

Blair A. Sowa (IV-6), Department of Veterinary Pathology, Texas A&M University, College Station, Texas, USA.

Nicklas Strömberg (I-7), Department of Organic Chemistry, The Technical University of Denmark, Denmark; Department of Medical Biochemistry, University of Göteborg, Göteborg, Sweden.

Shunji Sugii (I-9), Department of Serology and Immunolgy, School of Medical Technology, Kitasato University, Kitasato, Kanagawa, Japan.

SUSA.nn Teneberg (I-7), Department of Organic Chemistry, The Technical University of Denmark, Denmark; Department of Medical Biochemistry, University of Göteborg, Göteborg, Sweden.

Guido Tettamanti (II-7), Study Center for the Functional Biochemistry of Brain Lipids, Department of Biological Chemistry, The Medical School, University of Milan, Milan, Italy.

Jochen Theim (II-5), Organisch-Chemisches Institut, Munster, Federal Republic of Germany.

Mario Venegas (III-1), Department of Molecular Biology, Northwestern University Medical School, Chicago, Illinois, USA.

J.F.G. Vliegenthart (IV-3), Department of Bio-Organic Chemistry, Utrecht University, P.O.Box 80.075 NL-3508 TB Utrecht, The Netherlands.

Torkel Wadström (I-6), Department of Medical Microbiology, University of Lund, Sölvegatan 23 S223 62 Lund, Sweden.

W.M. Watkins (II-4), Division of Immunochemical Genetics, MRC Clinical Research Centre, Harrow, Middlesex, United Kingdom.

Albert M. Wu (I-9, II-5, II-6, II-10, and Appendix I), Department of Veterinary Pathology, Texas A&M University College Station, Texas, USA.

Takao Yamori (III-3), The University of Texas M.D. Anderson Hospital and Tumor Institute, Houston, Texas, USA.

Robert K. Yu (IV-2), Departments of Neurology (School of Medicine) and Chemistry, Yale University, New Haven, Connecticut, USA.

David Zopf (III-2), Laboratory of Pathology, National Cancer Institute, National Institutes of Health, Bethesda, Maryland, USA.

SUBJECT INDEX

Printed in the United States
By Bookmasters